Handbook of Sugar Refining

Handbook of Sugar Refining

A Manual for the Design and Operation of Sugar Refining Facilities

edited by

Chung Chi Chou

Sugar Processing Research Institute, Inc.
New Orleans, Louisiana

JOHN WILEY & SONS, INC.

New York • Chichester • Weinheim • Brisbane • Singapore • Toronto

This book is printed on acid-free paper. ♾

Published simultaneously in Canada.

Library of Congress Cataloging-in-Publication Data:

Chung, Chi Chou
Handbook of sugar refining: a manual for design and operation of sugar refining facilities / edited by Chung Chi Chou.
p. cm.
Includes index.
ISBN 0-471-18357-1 (alk. paper)
1. Sugar—Manufacture and refining—Handbooks, manuals, etc. I. Chou, Chung-Chi, 1936–
TP377 .H27 2000
664'.115dc21

99-042174

Printed in the United States of America

10 9 8 7 6 5 4 3 2 1

Contents

Foreword

Until now the cane sugar industry has not had a handbook devoted entirely to the subject of refining. More often than not, the technology of refining is included in a few chapters in handbooks on more general subjects. The *Handbook of Sugar Refining* is the first publication dedicated to cane sugar refining. The detail and breadth of coverage required on this subject is possible only in a handbook such as this.

Cane sugar refining has undergone many changes in this century, with changes accelerating in the last two decades. By 1930, bone char was the "standard" decolorizer in all large refineries, even though its mode of action was not understood. Large quantities of char were used requiring large use of energy. The Bone Char Research Project (1939) was the industry's first cooperative research effort, which successfully defined what bone char did and how it did it. This resulted in satisfactory decolorization win the use of less char and opened the way for other technologies, such as ion-exchange resins, functionalized granular activated carbons, and new clarification procedures. Even these technologies have been proven to be energy intensive, and with the competitive pressures of the latter half of the twentieth century, ever more innovative processes, such as membrane technologies, waste treatment and recycling, and higher-quality raw sugars, have allowed entire portions of the traditional refining process to be eliminated. Today, the refining industry is in a very early stage of reinventing its processes, and timing for a handbook on sugar refining could not be better.

A wide range of experts in various fields of sugar refining have contributed to this handbook under the able editorship of Dr. Chung Chi Chou. It is hoped that this handbook will find wide use throughout the world, and follow in the footsteps of the *Cane Sugar Handbook* to become the Sugar Bible II.

Frank Carpenter
Formerly Director
Cane Sugar Refining Research Projects
New Orleans, Louisiana
August 1999

Preface

As we enter the twenty-first century, the demand for high-quality white sugar around the world has continued to increase. Consumers demand a high-quality product every bit as much as does the industrial user, for beverage, bakery, confectionery, and other applications. At the same time, the competitive pressures on the sugar industry have also increased, leading the industry to seek innovative ways to produce refined sugar. With all of these changes ongoing, it was felt that a handbook devoted entirely to the subject of sugar refining was needed.

In this first handbook of sugar refining, I have attempted to give both a general overview of the many aspects involved in refining, as well as to go into great detail in the critical areas. With the help of my very able co-authors, experts in many areas, I believe that we have come close to achieving our objective, which is to provide, in a single book, the most comprehensive treatment to date on the refining of cane sugar.

Since the refining of cane sugar begins with raw sugar, there are chapters devoted to raw sugar storage and handling, automation in a raw sugar factory, and integration of raw and refined sugar operations. The traditional unit processes in a refinery are covered in detailed chapters in Part II. In this section are also included some of the newer processes being introduced to refining, such as membrane technology. Refinery design is very important to achieve efficiencies, and an entire chapter is devoted to this subject. In Part III we cover subjects dealing with instrumentation and control, computers, automation, and the always difficult subject of loss control.

I believe that value-added products in the form of specialty sugar products provide an excellent avenue for increasing profit margins and utilizing a refinery's capacities to its maximum. Thus, Part IV is devoted to that topic, with four chapters detailing various types of specialty products.

It is hoped that this book will be useful to sugar technologists around the world.

Chung Chi Chou, Ph.D.
Managing Director
Sugar Processing Research Institute, Inc.
New Orleans, Louisiana
July 2000

Acknowledgments

It takes many people to make possible a book of this magnitude. As editor, I wish to express my sincere appreciation and thanks to the 29 contributors who, by giving freely of their expertise, have made this book possible. The collaboration was truly international, with experts from many countries contributing. I would also like to thank my editor as well as the publishers for seeing the need for such a handbook and agreeing to publish it, and for their patience and forbearance with us as we finalized the book.

I would like especially to thank Dr. James C. P. Chen, my co-author and chief editor of the 12th edition of the *Cane Sugar Handbook,* the Sugar Bible I, for his support and encouragement during my long association with the world sugar industry.

Chung Chi Chou

Contributors

Walter Altenburg
New York Sugar Trade Laboratory
300 Terminal Avenue W
Clark, NJ 07066

Leon Anhaiser
C&H Sugar Company
P.O. Box 308
Aiea, HI 96701

Francisco Carlos Bŕatolo
Refinarias de Açúcar Reunides, SA
RUA Manuel Pinto De Azevedo
272/4100 Porto
Portugal

Paul Benecke
CSR Refined Sugars Group
Level 5, 11 Help Street
Chatwood NSW 2067
Australia

Luis San Miguel Bento
Refinarias de Açúcar Reunides, SA
RUA Manuel Pinto De Azevedo
272/4100 Porto
Portugal

Denis Bouree
Marseilles Refinery
336 Rue de Lyon
13343 Marseille Cedex 15
France

Ray T. Burke
Savannah Sugar Refinery
P.O. Box 710
Savannah, GA 31402

Hsin Sui Chang
7417 N. Peters St.
Arabi, LA 70032

Chung Chi Chou
Sugar Processing Research Institute, Inc.
1100 Robert E. Lee Boulevard
New Orleans, LA 70124

Stephen J. Clarke
Flo-Sun Sugar Co.
Okeelanta Corporation
P.O. Box 86
Hamilton, OH 45102

Donal F. Day
Audubon Sugar Institute
Louisiana State University
Sugar Station Building
Baton Rouge, LA 70803

Joseph F. Dowling
Refined Sugars, Inc.
One Federal Street
Yonkers, NY 10702

G. Fawcett
487 Beaulieu Ave
Route 3
Savannah, GA 31406

Peter J. Field
Field Technology Consulting
Westleigh NSW 2120
Australia

G. Clive Grimwood
Broadbent Thomas & Sons Ltd.
Queens Street South
Huddersfield, Yorkshire
England HD1 3EA

Willem H. Kampen
Audubon Sugar Institute
Louisiana State University
Sugar Station Building
Baton Rouge, LA 70803

Robert J. Kwok
Hawaiian Commercial & Sugar Company
1 Hanson Road
Puunene, HI 96784

Eddy H. Lam
Process Control Engineer
formerly of
Hawaiian Commercial & Sugar Company
Puunene, HI 96784

Ju-Hwa Liang
Taiwan Sugar Corporation
No. 266 Chien-Kuo S. Rd; Sec 1
Taipei 106
Taiwan
Republic of China

G. R. E. (Raoul) Lionnet
University of Natal
Private Bag X10
4014 Dalbridge

Dave Meadows
Tongaat-Hulett Sugar Ltd.
Private Bag 3
Glenashley 4022
Kwa Zulu-Natal
South Africa

Naotsugu Mera
1-8-1-1106 Futamata
Ichikawa City Chiba Prefecture
Japan 272-0001

Jean-Paul Merle
C&H Sugar Company
830 Loring Avenue
Crockett, CA 94525

Thomas N. Pearson
14 Epping Forest Way
Sugar Land, TX 77479

Peter Rein
Audubon Sugar Institute
Louisiana State University
South Stadium Road
Baton Rouge, LA 70803

Richard Riffer
1401 Walnut St., #1B
Berkeley, CA 94702-1402

Michael Saska
Audubon Sugar Institute
Lousiana State University
Sugar Station Building
Baton Rouge, LA 70803-7305

Walter Simoneaux, Sr.
Glenwood Cooperative, Inc.
5065 Highway 1006
Napoleonville, LA 70372

John C. Thompson
United States Sugar Corporation
111 Ponce de Leon Avenue
Clewiston, FL 33440

Joseph C. Tillman, Jr.
Savannah Sugar Refinery
P.O. Box 710
Savannah, GA 31498-4710

PART I

Introduction to Sugar Refining

CHAPTER 1

Glossary of Terms and Definitions*

Affination Treatment of raw sugar crystals with a concentrated syrup to remove the film of adhering molasses. This is achieved by mixing sugar with syrup and then centrifuging the magma with or without water washing.

Affined sugar Sugar purified by affination.

Agglomeration Sticking together of two or more crystals during the purging and drying operations. It is usually a minor problem.

Ash content Solid residue determined gravimetrically after incineration in the presence of oxygen. In analysis of sugar products, sulfuric acid is added to the sample, and this residue as *sulfated ash* heated to 800°C is taken to be a measure of the inorganic constituents. Sometimes determined indirectly by measurement of electrical conductivity of the product in solution.

Bagacillo Fine fraction of bagasse obtained by sieving.

Bagasse Cane fiber leaving extraction apparatus after extraction of juice (e.g., first mill bagasse).

Batch From opening of the feed valve to discharge of the massecuite.

Blowup Holding tank or tank where clarification is taking place.

Boiling point of sugar solution The boiling temperature of a sugar solution depends on (1) vacuum and (2) sugar concentration [either % RDS (refractometric dry substance) or supersaturation]. Since the purpose of the pan operation is to grow sugar crystals, we are more interested in supersaturation than % RDS. Also, all sugar boiling is done under a vacuum so that the operating temperatures are kept as low as possible.

Boiling point of water The boiling temperature of any pure liquid, such as water, depends on the pressure at which it is boiled. If the pressure is greater than normal atmospheric pressure, its boiling temperature will be raised. If the pressure is less than atmospheric (a

*By Chung Chi Chou.

vacuum), its boiling temperature will be lowered. All sugar boiling is done under reduced pressure.

***Boiling point rise* (*BPR*)** Difference between the boiling temperature of a sugar solution and the boiling temperature of pure water, both measured at the same pressure. The BPR can be used to measure the supersaturation of the sugar solution and is the basis for some pan supersaturation instruments.

Brix Measurement of total solids in a sugar liquor or syrup using a Brix hydrometer. For solutions containing only sugar and water, Brix = % sugar.

***Bulk density* (*volume weight*)** Weight of a specific volume of a solid. The volume is usually 1 cubic foot (a container 1 foot long, 1 foot wide, and 1 foot high). Granulated sugar weighs 50 to 52 lb/ft^3. Coarse sugars weigh 48 to 50 lb/ft^3 because the small crystals pack very closely.

Calandria Tubular heating element situated in and constituting most of the bottom third of a vacuum crystallizer known as a *calandria pan.*

Candy crystal Large sugar crystals produced by a special crystallization process.

Carbonation Introduction of carbon dioxide gas into limed juice or syrup (e.g., first and second carbonations) to remove nonsugar solid. The alternative term *carbonatation* is sometimes used.

Carbonation gas Gas rich in carbon dioxide for use in carbonation.

Carbonation sludge Carbonation slurry after concentration with vacuum filters to 45 to 50% dry solids.

Carbonation slurry Turbid liquid consisting of juice and carbonation precipitate (e.g., first carbonation slurry, second carbonation slurry).

Carbonation slurry concentrate Carbonation slurry after concentration with thickening filters or decanters.

Clarifier Apparatus for the elimination by sedimentation or flotation of suspended solids from a turbid liquid.

Color Attenuation index, determined under defined conditions.

Color type Result of the visual assessment of white sugar against standards.

Condenser water Mixture of condensate and cooling water produced by a direct-contact condenser.

Conglomerate Two or more crystals grown together during pan boiling. Conglomerates are undesirable because they are difficult to wash in the centrifugal and difficult to dry in the granulators. Conglomeration is a major problem in sugar boiling.

Conglomeration Intergrowth cluster of several crystals.

Cooling crystallization Crystallization by cooling of the magma.

Cooling crystallization effect Difference in the purity of the mother liquor of the magma at the beginning and end of the crystallization by cooling.

Cooling crystallizer Vessel for crystallization by seeding.

Cossette fines Portion of cossettes less than 1 cm long.

Crystal content Proportion by weight of crystals in magma.

Crystal growth rate Crystal weight produced per time and surface unit or crystal length per unit time.

Crystallizate Crystal fraction obtained by crystallization.

Crystallization Nucleation and growth of crystals.

Crystallization scheme Defines the number of crystallization stages involved in producing sugar.

Crystallizer Cylindrical vessel equipped with cooling and agitating elements used for completing the crystallization of a massecuite discharged from a pan.

Curing Crystallization process that takes place in a crystallizer.

Cut a pan Discharge a portion of the massecuite from a pan, retaining a footing upon which to feed more syrup for crystallization.

Cycle time (a) *In batch evaporating crystallization*: time from start of drawing in a charge to the beginning of the next; (b) *in centrifuging*: time from start of loading the centrifugal to the beginning of the next charge.

Deteriorated cane Cane of reduced suitability for processing due to external causes (e.g., frost).

Drop a pan Discharge all of the massecuite from a pan.

Dropping *See* Drop a pan.

Entrainment separator Apparatus for removing syrup or magma from vapor.

Evaporating crystallization Crystallization by evaporation of the solvent (e.g., water).

Evaporating crystallization effect Difference between the purity of the magma fillmass and the purity of the mother liquor of the magma on discharge.

Evaporating crystallization process Process of evaporating crystallization.

Evaporating crystallizer Apparatus for crystallization by evaporation. *See also* Vacuum pan vapor.

Evaporator effect Evaporators operating in series at a given steam pressure (e.g., first effect, second effect). Condensates and vapors are labeled correspondingly (e.g., first condensate or vapor: condensate or vapor from the first effect).

False grain Undesirable small crystals, formed when the concentration in the pan is carried too high.

Feed syrup Supply syrup for crystallization.

Fillmass *See* Massecuite.

Filter cake Material retained on the screens or cloths of the filters.

Filtrate Liquid passed through the screens or cloths of the filters.

Final or final pan Last remelt strike boiled, which is dropped to a crystallizer and then centrifuged to yield final remelt sugar and blackstrap molasses.

Footing Base massecuite retained in or transferred to a pan to which a syrup is fed (generally of the same or lower purity than that of the mother liquor of the footing) until massecuite fills the pan to constitute a full strike.

Green run off First syrup produced on centrifuging a magma.

Head Depth of massecuite between the top tube sheet of the calandria and the top level of the boiling massecuite.

High or high pan First remelt strike boiled from affination syrup and/or concentrated low-purity sweetwater and the lowest-purity white sugar syrup.

Inboiling Practice of finishing the boiling of a strike by feeding a syrup to the pan of lower purity than initially fed to the footing.

Injection water Water for jet condenser.

Intermediate or intermediate pan Second remelt strike boiled, usually coming from a high-pan footing fed on mother liquor centrifuged from a high.

Invert sugar Mixture of approximately equal parts of glucose and fructose resulting from the hydrolysis of sucrose, or inversion. Unlike sucrose, glucose and fructose are reducing sugars.

Jet Commonly used rather than *syrup* in Europe (e.g., first jet, second jet, instead of first syrup, second syrup).

Liming Process step in juice purification in which quicklime or milk of lime is introduced into the sugar liquor.

Liquid density Weight of a unit volume of liquid. Density is often expressed as pounds per gallon. The density of any pure liquid depends on its temperature. As temperature increases, density is reduced. The density of a sugar solution depends on its sugar content [Brix or RDS (refractometric dry substance)] in addition to its temperature.

Liquid sugar Refined sugar products in liquid sugar (e.g., liquid sucrose, liquid invert).

Liquor Sugar solution obtained by dissolving (melting) crystals. Raw liquor is produced by melting the washed raw sugar from the melt house centrifugal.

Loose pan Massecuite with a high ratio of syrup to crystals, one wherein the ratio of crystal weight to syrup solid weight would be less than about 47:53.

Low or low pan *Same as* Intermediate or intermediate pan.

Magma Mixture of crystals and syrup. It is occasionally also used to mean massecuite.

Magma mixer Mingler.

Massecuite The mixture of crystals and syrup (mother liquor) resulting from boiling a seeded syrup or a footing, fed with more syrup, sufficient to maintain a fluid mixture within the vacuum pan.

Massecuite mixer Apparatus to distribute magma to the centrifugal.

Melt Quantity and/or capacity of raw sugar processed (e.g., daily melt).

Molasses Sugar-bearing product of the sugar end, whose purity has been reduced to the point that further crystallization of sugar is not possible.

Molassed dried pulp Commercial term under feedstuff regulations for a mixture of dried pressed pulp and molasses.

Mother liquor Liquid phase remaining after a crystallization; often refers to syrup between the crystals of a massecuite.

Nonsucrose Substances contained in raw material and its products except sucrose and water.

Nonsugar Common overall term for substances contained in the raw materials and products of the sugar industry, except sugar and water.

Nonsugar content Difference between dry substance and sugar content.

Nucleation Generation and development of small crystals (protocrystal aggregates) capable of growth.

Pan Vacuum evaporator used in the sugar industry to boil and crystallize sugar from liquor or syrup.

Phosphatation Clarification using phosphoric acid and lime, in which certain nonsugar content is removed by flotation.

Polarization Term customarily used in sugar analysis for the optical rotation of a sugar industry product, measured under conditions defined by the International Commission for Uniform Methods of Sugar Analysis (ICUMSA), as a percentage of the rotation of pure sucrose measured under the same conditions.

Precipitated calcium carbonate (PCC) Carbonation slurry after concentration with filter press to about 70% dry solids.

Pressure Roughly defined as the force bearing down on something (such as the weight of a pallet of sugar sitting on the floor) or the force required to hold something inside a container (such as air inside a tire). Pressure is measured with a pressure gauge. The reading is in pounds per square inch (lb/in^2 gauge, or psig). Normal air pressure is defined as 0 psig.

Propinquity Relative proximity of crystals in a massecuite.

Purity Sugar content as percent of dry substance content. The solids consist of sugar plus impurities, such as invert, ash, and colorants. The measurement does not include water. Since sugar can be expressed as polarization on sucrose and dry solid as Brix, refractometer solid, and so on, the purity can be expressed as apparent purity, refractometer purity, and so on.

Raw juice Juice obtained from cane after extraction, pressing, or milling.

Reducing sugars Generally referred to and/or interpreted as invert sugar.

Refining Purification of sugar through recrystallizing and chemical and physical methods.

Refractometric dry substance (RDS) Measurement of total solids in a sugar liquor or syrup using a refractometer. For solutions containing only sugar and water, % RDS = Brix = % sugar. The temperature is usually controlled to 20°C (68°F).

Remelt Massecuite or centrifuged sugar boiled from syrups (both filtered and unfiltered) which are too low in purity and/or dark in color to yield the white granulated sugar of commerce. All remelt referred to in the book is derived from affination syrup or last white sugar strike syrup and concentrated low-purity sweetwater.

Remelt syrup Solution of washed or affined sugar.

Retail package Package up to 10 kg in size.

Runoff General term for syrups produced on centrifuging a magma.

Safety factor Number to indicate keeping quality of raw sugar.

Sand separator Apparatus that separates sand and earth from a liquid stream, such as raw juice, press water, or flume water.

Saturation Sugar solution that will not dissolve any more crystals at the temperature of the solution. Saturation depends on temperature. If the temperature is reduced, crystals will grow. Alternatively, saturation can be defined as "solubility at the limit of thermodynamic stability."

Seeding (a) Introducing crystal fragment to induce nucleation, as a means of initiating the crystallization process; (b) introduction of a finite number of crystal fragments growth (similar to full seeding).

Seed magma Suspension of crystals as the basis for controlled crystal growth.

Skip or skipping Terms used for *strike* in pan boilings in a refinery.

Solids extraction Ratio of crystal weight to syrup solids weight in a massecuite.

Stiff Massecuite characterized by excessive propinquity.

Strike Massecuite composing a completed boiling, all of which is discharged from the pan.

Sucrose Common term for the disaccharide α-D-glucopyranosyl-β–D-fructofuranoside.

Sugar Term for the disaccharide sucrose and products of the sugar industry, essentially composed of sucrose.

Sugar content (a) *Sucrose content:* as this term is generally used, it refers to the fact that in addition to sucrose, other sugars are present (usually determined by polarimetry); (b) *total sugar content:* for technical products with a substantial amount of other sugars (e.g., liquid sugars), the term *total sugar content* should be used; (c) *recoverable sugar content:* part of the sugar in cane, which can be obtained as white sugar.

Sulfitation Introduction of sulfur dioxide to juice or liquor.

Supersaturated Sugar solution containing more sugar than its natural solubility limit (i.e., contains more sugar than would be expected). *Supersaturation* is defined as the ratio of the sugar concentration of the liquor to that of a saturated liquor at the same temperature. All pan operations, from seeding to dropping, must be performed to maintain a supersaturated liquor so that crystals will always continue to grow. However, the ratio must not be allowed to get too high since this would force new crystals to form (i.e., false or smear grain).

Supersaturation coefficient Quotient formed by dividing the sugar/water ratio of a supersaturated solution by the sugar/water ratio of a saturated solution under the same conditions (temperature and purity or nonsugar/water ratio).

Supersaturation, critical Supersaturation at which nucleation begins spontaneously.

Sweetwater Wash water or water containing sufficient sugar to warrant recovery.

Syrup Mother liquor obtained from a crystallization process. Syrup is removed from the crystals in the centrifugal.

Target purity Reference purity of final molasses, taking into account the effect of nonsucrose on its exhaustion.

Tighten the pan The final step in boiling a strike, wherein syrup fed to the pan is discontinued and boiling continued for a short period of time, to increase the massecuite propinquity to a point that will make its consistency and flowability optimum for centrifuging.

Tight pan *Opposite of* Loose pan.

Treacle English term for molasses.

Undersaturated Sugar solution that is below its saturation point (i.e., more crystals will dissolve).

Vacuum Any pressure that is less than normal air pressure. There are two scales: the absolute pressure scale and the vacuum pressure scale, both measured in inches of mercury and both used on pans. Absolute pressure is shown by a mercury manometer and vacuum pressure is shown by a dial gauge. Their total will always add to about 30.

Vacuum pan vapor Vapor created in evaporating crystallization.

Washing Washing of sugar crystals during centrifuging with syrup, water, or steam.

Wash syrup Syrup produced in washing sugar during centrifuging.

CHAPTER 2

Sugar Refining Processes and Equipment*

INTRODUCTION

In this chapter we define briefly the technical terms in each process unit and the equipment used for two standard sugar refineries, one with carbonation as the primary decolorization process, the other with phosphatation. Both refineries are rated for 1200 tonnes of raw sugar per day. Attempts have been made to indicate the number of units for each type of apparatus and the relative size of equipment for each process operation. The criteria for the design of a new sugar refinery are also presented.

2.1 REFINING WITH CARBONATION

We describe next the equipment and operating condition requirements for a typical carbonation refinery handling 1200 tonnes of sugar per day. In Section 2.2 we do the same for a phosphatation refinery.

2.1.1 Raw Sugar Handling

When raw sugar is received, it is weighed, sampled, and stored in a warehouse with careful placement based on its origins, to minimize the need for blending. A raw sugar hopper is usually provided with a smaller surge hopper before the sugar is fed into the balance or scale. The hopper should be equipped with high and low indicators. Belt conveyors should be wide enough to avoid spillage (e.g., 0.61 m wide). The weighing scale generally has a maximum capacity of 700 kg per dump. A 600-kg dump with a 45-s cycling time will produce a melt rate of about 1200 tonnes/day.

2.1.2 Affination

Figure 2.1 shows a typical refinery process flow diagram. Raw sugar is weighed first and discharged through a dry crush roller. The weighed raw sugar is then fed into a magma

*By Chung Chi Chou.

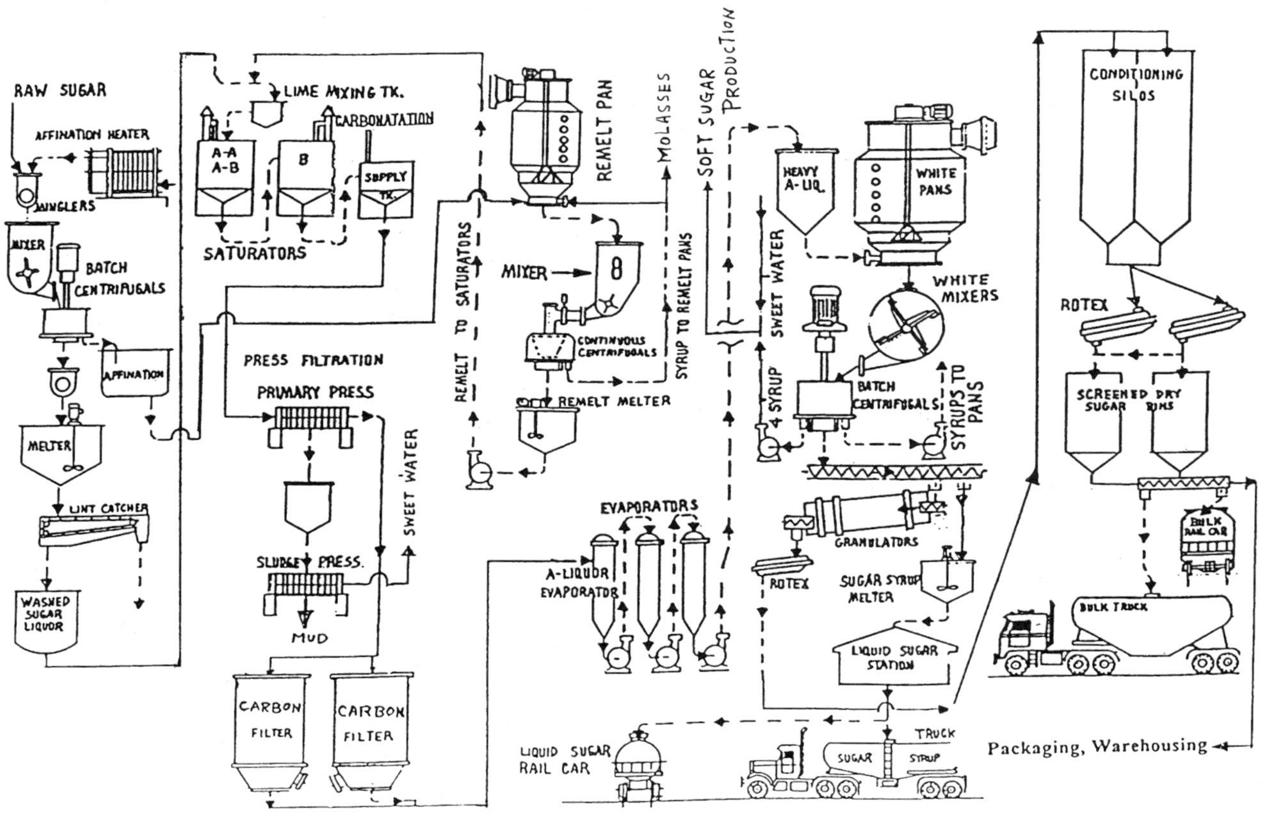

FIGURE 2.1 Typical flow process flow diagram.

mingler, where it is mixed with affination syrup to a desired consistency. The magma temperature is maintained at about 40°C and mingling with the affination syrup is held at 72 Brix. The water temperature for the affination heater should not exceed 85°C. The affination syrup can be added automatically each time the raw sugar discharge gates open. Alternatively, it can be controlled through monitoring the motor load on the mingler drive motors. The motor load is correlated with the fluidity of the magma in the mingler.

The magma from the minglers overflows to the mixers, which feed the seven Western States batch automatic centrifugals, with a basket size of 48 by 36 by 7 in. Each basket has a capacity of 0.5 m^3 with a loading speed of 250 rpm. The Allis-Chalmers single-winding motors have two speeds and 250 A on acceleration with full load. The wash water temperature and pressure are 82°C and 85 psi, respectively. The amount of wash water is controlled to give a washed raw sugar normally with 0.1% of invert and 0.1% of ash content.

After centrifugation, the washed raw sugar is melted in high-purity sweetwater from a carbonate cake sludge press filter to 72 Brix at 75°C. Low-pressure steam and/or vapor from the evaporators can be used to melt the washed raw sugar. The affination syrup collected from centrifugals is adjusted with lime slurry to give a pH of 7.2. Part of the syrup is reheated to 60°C for mingling raw sugar, and the excess portion is directed to the recovery side of the refinery for sucrose recovery. The recovered remelted sugar is then blended with washed raw sugar liquor from the affination operation. The resulting liquor, commonly called melt liquor, is then pumped to the next stage of carbonation.

2.1.3 Carbonation

Typically, a two-stage carbonation process is used. The melt liquor is blended with 20 Brix milk of lime at a ratio of 0.7 to 1% lime solid to total soluble sugar solids, depending on the quality of the raw sugar being processed. A typical lime slaker might be a Wallace & Tiernan model A-758 with a capacity of 900 kg/h. A single-deck No. 10 mesh screen is used for grit handling. The first stage consists of two parallel tanks, AA and AB (Fig. 2.1), which are evenly split for the melt liquor flow. The second stage consists of a single tank, B, which receives the combined flow from AA and AB. Each tank has a volume of 40 m^3. The retention time in the AA/AB saturation is about 45 min individually, and 22.5 min in tank B. Scrubbed flue gas, recovered from natural gas–fired boilers, is pumped or dispersed through all three tanks as a source of carbon dioxide, to react with the lime in the carbonation liquor to form calcium carbonate in the tanks. About 75 to 90% carbon dioxide gasing is carried out in the first stage (AA and AB). The resulting pH is about 9.6 (when measuring cold). The B saturator is gassed to a pH of 8.2. The carbonated liquor leaving the first stage is reheated to 85°C. The calcium carbonate formed in the melt liquor inside the saturators entraps wax, gum, polysaccharides, colorants, and ash, mostly sulfate. Most inverts in the melt liquor are also destroyed by carbonation.

2.1.4 Mud Filtration

Carbonate cake (mud) in the carbonated liquor is removed, as shown in Fig. 2.1, by filtration through eight Sweetland presses using magnetic flowmeters to monitor and control the liquor flow. Each press has 72 leaves, with a filtration area of 93m^2. The polyester cloth carries 502/yd^2 multifilament. Occasional in situ acid and caustic treatments are required to keep the cloth clean. Recirculation of liquor is needed at the beginning of each press cycle to minimize excessive turbulence while precoating. Sequencing of the press cycles is handled automatically. A typical productive cycle length is about 45 to 50 min. The color of the pressed liquor ranges from 600 to 800 ICUMSA. The press filter liquor is pumped into a supply tank and the carbonate cake (mud) is sliced out of the primary filter for desweetening in two automated presses. The filter employs compressed air for final drying of the mud.

2.1.5 Bone Char Decolorization and De-ashing

Bone char filtration can be used to further reduce the processed liquor color to 100 to 150 ICUMSA with low ash content for the production of white sugar and soft sugar. Fifteen char cisterns (filters) are needed, of which 10 are in the productive cycle. Nine are for the filtered pressed liquor and one on jet 2. All are fed at a flow rate of about 8 m^3/h by gravity at a maximum Brix value of 67.5 and a temperature of 80°C. The total liquor cycling time through char cisterns ranges from 80 to 120 h. All cisterns are settled with filtered liquor at 60 Brix. After the capacity of the char cisterns to remove color and ash is exhausted, sweetening-off is carried out with hot water at a flow rate of 3 m^3/h followed by hot-water washing at 80°C for about 25 h. Following the washing, spent char is revivified in a Nichols–Herreshoff kiln or conventional pipe kiln, fired on natural gas or oil. A Hereshoff kiln has eight hearths with a static louver dryer and three forced-air coolers. Char burning at hearth 8 proceeds at about 4.5 tonnes/h with a temperature of 540°C.

2.1.6 Evaporation

The char-filtered A liquor at 65 Brix is evaporated to 76 Brix using a triple-effect falling film evaporator. The Brix value of the liquor being discharged from the evaporator is monitored and controlled by a density instrument that regulates the steam flow to the evaporator. Sweetwater evaporation is accomplished by using the vapors from the last effect of the A liquor evaporator. The vapor from the first body can be used to melt the raw sugar.

2.1.7 White Sugar Boiling and Crystallization

Thick A liquor at 76 Brix is pumped into the vacuum pan supply tanks. Vacuum pans are of stainless steel construction equipped with (1) an agitator, using amperage output for mobility control; (2) a barometric condenser with water ring vacuum pumps; (3) a crystalloscope and LSC refractometer for density control; and (4) steam at 26 psi with an Anubar flowmeter for flow control and the liquor feed controlled by a TigerMag flowmeter.

A typical automated white pan boiling cycle includes an initial charge of 18 m;^{3} seeded with 700 mL of fondant slurry at 1.15 supersaturation or 78.5 Brix at 157 mmHg absolute, and a pan to drop at a load of 50 A using mobility control with a strike volume of approximately 40 m^3. The total cycling time of one pan is about 85 min.

A straight three-boiling scheme is used to produce white sugar. The first strike is boiled from thick A liquor (fine liquor) with a small amount of jet 1 back-boiling, the second strike is boiled strictly from jet 1 syrup, and the third strike is boiled from jet 2 syrup. The jet 3 and 4 syrup, together with concentrated low-purity sweetwater originated in the char decolorization process, is used for soft sugar production. The resulting soft syrup is sent to the recovery house for further sucrose recovery. Three vacuum pans with a 40-m^3 strike volume are needed for white sugar production.

Two pans with an average strike capacity of 35 m^3 are used on the recovery (remelt) side to boil three strikes for sucrose recovery: highs, mediums, and finals. Melted highs and mediums constitute the remelt liquor, which is combined with the washed sugar liquor for the carbonation. The melted finals are used to boil high or seeds strikes. The first two (first and second crops) boilings are machined immediately, while the finals (third crop) are cured for approximately 50 h before being machined.

For first-crop boiling (highs), the initial charge is 20 m^3 of syrup. The pan is seeded with 150 mL of fondant. The graining purity is controlled at 78 Brix. The target purities for the first and second masses are 88 and 78, with a strike volume of 40 m^3, respectively. The third crop (finals) is seeded with 500 mL of fondant slurry. The graining (seeding) charge and third masses purity are controlled at 76 and 64 Brix, respectively. All initial charged syrup should consist of about 80% affination syrup, to facilitate the initial stage of crystal growth.

2.1.8 Centrifugal Operation

For white sugar centrifugation, eight machines, identical in size to the affination machines, are used. Those machines have a nine-nozzle wash tree with a high-speed spin cycle of 90 s. In addition, two BMA G1500 batch centrifugals are used. These machines have a basket volume of 0.87 m^3 and 0.18-m caps and a wash tree with five fully retractable nozzles. The machines are also equipped with a variable-speed induction drive with a high-speed spin of 70 s. The moisture content of white sugar discharged from the centrifugal should be about 1% or less.

Two BMA K1100 continuous centrifugals with a maximum of 2000 rpm and a 30° basket angle is used for the first- and second-crop boiled masses. For the third-crop (final) masses, two Western States C5 continuous centrifugals with a maximum of 2200 rpm and 30° basket angles are used.

2.1.9 Sugar Drying and Conditioning

Major drying of sugar discharged from white centrifugals is achieved in three rotary granulators. Wet sugar with about 1% moisture enters a granulator countercurrent to airflow and exits at about 50°C with a moisture content of about 0.05%. Air enters and exits the granulators at about 85 and 42°C, respectively.

Sugar discharged from the granulators is first coarse screened in Rotex screeners, which allow the sugar to cool down further and dissipate the additional moisture. The sugar is then stored in bins for about 2 h to "repose" for further conditioning in two silos. In the silos, conditioned air is percolated through the sugar from the bottom. After about 24 h of resident time, the sugar leaves the silos with a moisture content of about 0.02%. For railcar bulk shipment, further cooling of the sugar in fluidized bends immediately prior to loading into the car is helpful. Following conditioning in the silos, the granulated sugar is screened again to produce sugar meeting the size requirements of various customers.

2.1.10 Packaging

Typically, granulated sugar is packaged in 1-, 2- and 4-kg bags with one ply of natural kraft and one ply of printed bleached kraft, and closed with a roll-top seal and in 10-, 20-, and 40-kg in multiwalled bags. Icing sugar, produced by a micropulverizer hammer mill, is usually packed in 0.5- and 1-kg bags and 20- and 40-kg packs. In conjunction with the production of white sugar, natural brown or soft sugar is also produced in two grades, light and dark. Soft sugar is packed in 1- and 2-kg bags and in 20- and 40-kg open-mounted multiwalled bags with a vapor barrier.

2.1.11 Liquid Sugar Production

In North America, liquid sugar consists primarily of either 67.5 Brix sucrose syrup or invert syrup with 50% sucrose at 76 Brix. Sucrose syrup is produced by melting the granulated sugar collected at the point of discharge from centrifugation and before entering the granulators. The melted sugar is treated with activated carbon and a filter aide and then press filtered. The finished product is shipped to customers via railcar or truck. Invert syrup is produced by acid hydrolysis of sucrose syrup using 31% hydrochloric acid. The inverted syrup is then neutralized by 20% sodium hydroxide and flash evaporated to 76 Brix for shipment.

2.2 REFINING WITH PHOSPHATION

The equipment needed for a phosphatation refinery with a 1200-tonne/day capacity is described in this section. The information is based on material presented at the Sugar Industry Technology, Inc. 56th Annual Meeting in Montreal, Canada by Lantic Sugar Ltd.

2.2.1 Sugar Receiving and Storage

The receiving wharf is 183 m long. There are two cranes, each rated at 300 tonnes/h. The belt conveyors are 1.2 m wide and have a maximum rate of 800 tonnes/h. Two weighing bins have a capacity of 2 tonnes each per weighing. The raw sugar shed has a total storage capacity of 44,000 tonnes.

2.2.2 Affination

The raw sugar elevator has a maximum rate of 55 tonnes/h. The mingler is 0.6 m in diameter and 9 m long, the magma header is 0.9 m wide and 10 m long with a capacity of 28.5 m^3. The affination centrifugals have a total rate of 50 tonnes/h with four batch machines 1219 mm in diameter and 762 mm in height and two batch machines 1219 mm in diameter and 914 mm in height. The washed raw sugar setscrew conveyor is 508 mm in diameters and 8839 mm long, and the washed raw sugar melter setscrew conveyor is 508 mm in diameter and 4724 mm long. The washed raw sugar melter has a capacity of 13.6 m^3 with a diameter of 2667 mm and a height of 2515 mm. Three units of 1524-mm-diameter screens with a total capacity of 80 tonnes/h are required to screen raw sugar.

2.2.3 Clarification

The coagulant tank has a 200-L capacity; the flocculant solution tank is 1.8 m in diameter and 1.9 m high with a capacity of 5.1 m^3. There are two lime sucrate tanks, each 1.7 m in diameter and 2.1 m high with a volume of 4 m^3 each. The phosphoric acid tank has a capacity of 0.2 m^3 (610 mm in diameter and 838 mm high). The raw sugar tank has a capacity of 35 m^3 and is 4.3 m in diameter and 2.7 m in height. The main aeration and reaction tanks have a capacity of 0.3 and 3.8 m^3, respectively. The main clarifier has a volume of 24.6 m^3 with a rate of 50 tonnes/h at 65 Brix, and the remelt liquor tank has a capacity of 17.8 m^3 and is 3048 mm in diameter by 2667 mm high. The remelt liquor aeration and reaction tanks have a capacity of 0.3 and 0.5 m^3, respectively. The remelt liquor clarifier has a volume of 4.6 m^3 at a rate of 10 tonnes/h at 65 Brix. The scum feed tank has a capacity of 9.3 m^3 and is 3048 mm in diameter and 1422 mm in height. The three-units three-stage scum clarifier has a total volume of 7.5 m^3 at a rate of 60 tonnes/h. The three scum mixing tanks, one preceding each scum clarifier, each has a total capacity of 0.7 m^3. The final tank for the scum from the last stage of the scum clarifier has a capacity of 8.6 m^3 and is 2591 mm in diameter and 1829 mm in height. The scum centrifugal decanter can produce 2 tonnes/h of heavy scum from 10.8 tonnes/h of diluted scum at 3000 rpm. In parallel with the centrifugal decanter, a Parkson twin-belt press with various speeds would also produce 2 tonnes/h of heavy scum from 10.8 tonnes/h of diluted scum.

2.2.4 Filtration

The clarified liquor tank has a capacity of 32.8 m^3 with a diameter of 3658 mm and a height of 3023 mm. The precoat liquor tank, 1219 mm in diameter and 1118 mm high, has a capacity of 1.4 m^3. Four Sparkler filter presses for clarified liquor each have a filtration surface of 92.9 m^3. The filter liquor transfer tank, 3048 mm in diameter and 2438 in height, has a capacity of 16.1 m^3.

2.2.5 Decolorization and Regeneration

Two filtered liquor tanks, each 3.6 by 3.0 by 1.8 m have a total capacity of 42.5 m^3. Thirty bone char filters, each with a volume of 29.5 m^3 have a total rate of 55 tonnes/h. Each refiltering B liquor transfer tank and B liquor feed tank has a capacity of 42.5 m^3 and is 3.6 by 3.0 by 1.8 m. Five bone char blow filters, each with a volume of 38.6 m^3 have a total wet char rate of 8.75 tonnes/h. One Nichols–Herreshoff kiln furnace with a predryer regenerates 8.75 tonnes of char per hour. There are 950 tonnes of stock char in the decolorization char system.

2.2.6 White Sugar Boiling and Centrifugal Operation

There is one No. 1 liquor tank with a volume of 15 m^3 (3 m in diameter and 2.1 m in height) and two No. 1 liquor polished presses, each with a capacity of 6.2 m^2 of filtration surface. The fine liquor preheater has a maximum rate 68.5 tonnes/h with 83 m^2 of heat surface. Two single-effect A liquor evaporators 1.8 m in diameter each has a heating surface area of 145.1 m^2. The concentrated A liquor is discharged into two concentrated No. 1 liquor tanks each, with a capacity of 73.1 m^3 (3.8 m in diameter and 6.4 m in height). The two white sugar pans, equipped with an agitator and tube calendria each has heating surface of 287.1 m^2 and a capacity of 48.2 m^3. There are two magma mixers, each with a capacity of 51 m^3 (3.2 m in diameter and 6.3 m in height). The header drum for the eight batch centrifugals is 0.9 m in diameter and 1.6 m in height with a capacity of 10.7 m^3. The eight white sugar centrifugals have a total rate of 68.6 tonnes/h; each is 1.2 m in diameter by 0.76 m in height. The distributing screw conveyor for granulators is 0.5 m in diameter and 27.7 m long. The five granulators for sugar drying, each 1.8 m in diameter and 7.8 m in length, have a total rate of 60 tonnes/h. The 0.6-m-wide belt conveyor used to convey sugar to the silo, has a capacity of 17.2 m^2.

2.2.7 Conditioning Silo

All screw conveyors to and from the conditioning silos are 0.4 m in diameter. The lower and upper scales for weighing sugar going into and out of silo, respectively, have a capacity of 500 tonnes/h, with 544.3 kg per weighing. The two silos (each 15 m in diameter and 33 m in height) have a total capacity of 9400 tonnes. The belt conveyor used to carry sugar beyond the upper scale is 0.5 m wide.

2.2.8 Screening, Storage, and Shipping

The distributing screw conveyor for screening has a diameter of 0.5 m. There are 10 screens, each with four levels, 1 m by 1.8 m, with 7.4 m^2 of screening surface for each screen. The maximum total rate for screening is 50 tonnes/h. There are 10 storage bins each with a capacity of 5.5 tonnes. In addition, there are 10 storage silos 3.6 m in diameter and 9.4 m in height, each with a total capacity of 639 tonnes. The railcar bulk rate is 83 tonnes/h and the truck bulk rate is 30 tonnes/h.

2.2.9 Soft (Brown) Sugar Boiling and Centrifugal Operation

In this particular refinery, raw sugar/affination syrup is treated, pressed, and char filtered before boiling into soft/brown products. These are high-profit items that any refinery should try to develop and market. The following equipment is used in this operation.

The raw syrup treating tank, 3.8 m in diameter and 1.9 m in height, has a capacity of 22.4 m^3. The raw syrup filter press has a filtration surface of 51.7 m^3; the filtered raw syrup tank (1.5 m in diameter and 2.2 m high) has a volume of 3.8 m^3. Seven bone char filters, each 29.5 m^3, are needed, with a total rate of 11 tonnes/h. The char-filtered liquor, termed *soft liquor,* is subjected to further treatment. The soft liquor treating tank, 3.8 m in diameter by 1.6 m in height, has a volume of 18.2 in^2. The soft liquor filter press has a filtration area of 33.7 m^2; the filtered soft liquor tank (1.5 m in diameter and 2.2 m in height) has a capacity of 3.8 m^3. The single-effect soft liquor evaporator (1.8 m in diameter) has a heating surface of 8.5 m^2.

The concentrated soft liquor tank (3.8 m in diameter and 6.4 m in height) has a volume of 73.1 m^3. The soft sugar is boiled in a vacuum pan with 158 m^2 of heating surface and 37.7 m^3 of maximum capacity. The soft sugar magma mixer, which is 2.9 m in diameter and 6.4 m in height, has a capacity of 41.8 m^3. The soft sugar is boiled from a pan with a capacity of 38 m^3 and a heating surface of 158 m^2. The two soft sugar centrifugals, each 1200 mm in diameter and 760 mm in height, have a total rate of 17 tonnes/h. The soft sugar is carried to a packaging machine using a oscillating conveyor at a rate of 7 tonnes/h.

2.2.10 Recovery (Remelt Operation)

Both high- and low-purity sweetwater are pressed filtered through two filter presses with a filtration area of 81 m^2 per press. The combined press-filtered sweetwater is concentrated using a triple-effect evaporator; each effect has a heating surface of 85 m^2. Two pans are needed for remelt boiling; one has a capacity of 35.4 m^3 and 163 m^2 of heating surface and the other has a capacity of 21 m^3 and a heating surface of 126 m^2. Two batch centrifugals with a total capacity of 17 tonnes/h are used for remelt sugar recovery, and two continuous centrifugals with a total rate of 13 tonnes/h are used for final strike sugar recovery.

2.2.11 Liquid Sugar Production

Both liquid invert and liquid sucrose are produced. The equipment requirement depends on the volume of the market.

2.2.12 Steam and Power Plant

Three natural gas– or oil-fired boilers, have a total output of 82 tonnes/h. Each boiler has a heating surface of 710 m^2. Steam is produced at 115 psig and 172°C. The turbo-alternator has a 375-kVA output with exhaust steam at 15 psig. The deaerator (1980 mm in diameter and 3660 mm in height) is rated at 112 m^3/h. The blowoff tank is rated at 4.4 m^3/h. Five compressors, each with a separate desiccant dryer, produce a total of 1840 ft^3/min compressed air at 100 psig for plant use.

CHAPTER 3

Automation in a Raw Cane Sugar Factory*

INTRODUCTION

Factory automation in the raw cane factory ranges from fully automated facilities to facilities without any automation. It is evident that for countries where labor cost is expensive, sugar factories are highly automated. On the other hand, in some countries where the need is to employ as many people as possible, the level of automation is sometimes nonexistent.

Today, factory managers and owners are gearing up to implement automation in their factories. The push toward automation is the result of increasing complexity in equipment and processes. In some cases the rising cost of labor and lack of qualified workers are forcing managers to move into automation and computerization. In others, the benefits of automation to optimize process and sugar quality are the driving force in moving into automation.

3.1 BRIEF HISTORY OF PROCESS AUTOMATION

Automation in raw sugar factories started shortly after World War II with what was state of the art at that time, the pneumatic systems. Pneumatic systems were primarily single-loop controllers, and in some instances, cascade loops were implemented. Later, the advent of some electrical functionality in controllers enabled signal characterization such as consistency controls. The advent of powerful electronic controllers in the 1960s and 1970s helped revolutionize control systems in the factory. With the new electronic controllers, complex control strategies are possible. Also, electrically carrying a control signal allowed for extended distances between control elements. Island-type control strategies were the primary method of control until the emergence of the data highway and distributed control systems (DCSs). DCSs took advantage of advanced communication protocols to stretch the control system beyond the geographic limit imposed in earlier systems. Signals could now be transmitted thousands of feet away on a wire. Devices were equipped with smart features to report health and diagnostic information about itself. Processes could now be monitored from

* By Robert J. Kwok and Eddy H. Lam.

several stations. Sharing of information among the various parts of the factory became an integral part of the overall operation. Further advances in communication yielded increasingly sophisticated devices and controllers. The move from analog systems to digital systems marks the arrival of a new era in the control industry.

3.2 CONTROL SYSTEM IN A RAW SUGAR FACTORY

The general layout of the control system consists of regulatory controllers, programmable logic controllers (PLCs), operator workstations, control input/output (I/O), and configuration workstations. In addition, several stations on office personal computers (PCs) may be part of the general network. The system is distributed throughout the plant and the operator consoles and man–machine interface (MMI) are divided among several control rooms. The controllers are centrally located in different areas with their I/O distributed in the field. The configuration station is located at a remote location easily accessible by the engineering staff and technical staff. The PLCs are located in the motor control centers (MCCs) and interface with the rest of the control system. PLCs use man–machine interface for operator accessibility. The main purpose of using PLCs is to handle most of the discrete logic in the system. The cane yard and milling area presents a wide array of process and equipment. It is important to protect and properly control these equipment. An intricate combination of interlock and regulatory control scheme is therefore required.

Control system in a raw cane sugar factory features a wide variety of control strategies. The system ranges from material handling to crystallization, evaporation, drying, and boiler controls. The basic system in a sugar factory may be divided into four distinct unit operations, the cane yard/milling or diffusion area, the clarification, filtration, evaporation, and vacuum pan crystallization area, the crystallizer and centrifugal area, and the boiler and turbogenerator area.

Each area mentioned previously can be considered in its own entity. However, in a modern automated sugar factory where efficiency is important, sharing information between the various unit operations is necessary. The control system makes use of state-of-the-art communication capabilities to transmit a huge amount of data to all the devices linked in the control system. The distributed control system (DCS), in a raw sugar factory is usually supplemented by one or several PLCs. The control system and the PLC system are usually linked together through serial communication ports. The combination of the DCS and PLC functionality makes for powerful control strategies. This combination enables the use of complex interlock systems to work in perfect conjunction with a regulatory control scheme. The result is an overall control scheme that combines safety and efficiency.

3.2.1 Operator Interface

The control system features a graphic operator interface that makes extensive use of custom displays and dynamic representation of unit operations. The operator interface is through an X-Window environment. The functionality of the operator consoles are very graphics driven and intuitive. Point-and-click features on graphic representation of a process offer a real feeling for the process flow and operation. X-Window environment allows for use of X-Window-compatible applications, making interfacing with the laboratory computer and plant local area network (LAN) possible. Up-to-the-minute reporting capabilities and trending may be mapped out in detailed and accurate manner. Historical trending and data gathering offer powerful tools to management staff in predicting and correcting any problems that have developed through time. The system has the capabilities to support multiple operator consoles scattered throughout the plant. Operating from a single database, the entire plant is therefore linked on a real-time basis. Total plantwide awareness of operation enables operators to make decisions based on overall situations.

The consoles may operate in a simplex or redundant configuration. Since all the devices are linked together in the network, an operator may log on to any operator workstation from any console device regardless of area. Given the proper privilege, one can view or operate an area of the plant from any remote station. This sort of flexibility demands tight security features which the system offers. The system divides the plant into several areas as determined by the engineering and management staff. Each type of operator is then assigned area of responsibility as per their function. The operators may operate only from the equipment assigned to them. In addition, tuning privilege may be removed from the functions of an operator if needed. Remote log on such as an office manager may have only viewing privilege from his/her personal computer.

Console configuration is performed from the configuration workstation. All configuration work is done off line and does not disrupt operation. A download is saved on hard disk and configuration update at the console is done at everybody's convenience so as to minimize disruption. A configuration update usually takes up to 5 min.

3.2.2 Controllers

The controllers themselves are powerful electronic machines at high levels of efficiency and are easily programmable and modifiable. The controller hardware consists of several multiloop, simplex, or redundant controllers. These controllers have the ability to run several hundred loops simultaneously and offer the capability to perform very complex control algorithms. The controllers make use of distributed control I/O. Through the configuration interface, the controllers offer an extensive choice of control algorithms and allow for custom programming of loops and processes. Continuous and batch capabilities may be programmed as well as interlocking and discrete logic functions. The I/O files are linked through a serial interface and may be scattered up to 5000 ft apart. This feature enables the controller hardware to be housed in a central controlled environment, and the I/O could then be dispersed in the field. Several controllers may be linked through the data highway. Information may be extended to the control I/O and communication levels.

Programming of controllers is performed off-line. A program check is available in the configuration software at the time of programming for syntax errors or program errors. In addition, the system features a trace/tune option where programs may be simulated off-line before download. This is an extremely powerful tool for engineers to verify and simulate a program before download to the controller.

Controller downloads are possible on-line without disruption of the operation. A standard feature known as a *partial matching download* allows engineers to make changes and download only the changes to the controller. Although the controller may operate on several hundreds of loops at the time of download, only the modified controllers are affected. In the case of redundant controllers, the controls switch briefly to the backup controller during download.

3.2.3 Diagnostics

The system diagnostic is accessed from the configuration workstation. Since the configuration workstation can act only as a host device, it is possible to log on from an operator station. The engineering staff and technician make extensive use of the diagnostic feature to correct system problems. This tool vastly facilitates the work of a technician troubleshooting the system. The diagnostic program offers overall system integrity as well as detailed integrity of a particular device.

The general LAN status may be looked at in determining if there is any problem in the entire system. This display shows brief information about all the devices making up the control system. For further information on a particular device or area, there are options available to navigate down to the level of interests.

The controllers offer extensive diagnostic capabilities. At the workstations or operator consoles, diagnostic screens may be displayed. These screens display general information on the health of the controller and its backup. Several options offer the possibility to view down to the I/O level of the controller. Controller loading information and communication status, along with slippage data, may be accessed through the diagnostic program.

Similar to the controller, the console also offers a wealth of information about itself through the diagnostic program. At a glance, the overall health of the device may be viewed. If there are problems, the option to dig deeper through the levels of diagnostic helps pinpoint them.

3.2.4 Communication

All communication in the system is redundant using either coax or fiber-optic cables. Communication in the system takes place at three levels. The I/O bus links all the I/O files to the controllers. The data highway links all the devices (controllers, consoles, and workstation) in the system. The normal data highway communicates at a speed of 5 or 10 MHz. Finally, devices such as consoles and configuration workstations feature a LAN using Ethernet. The LAN offers an avenue to the outside world. Important information may be passed on to the management information services (MIS) department and agricultural department of the performance of the plant. The LAN uses TCP/TP protocol and is protected from unsolicited data. For security interests, a bridge between itself and the outside world protects the control system network. The capability to provide a window into the process to a manager on a regular personal computer as an application adds awareness throughout the ranks of management on a real-time basis.

3.2.5 Configuration

The system is configured from a configuration workstation. The workstation is a powerful multitasking computer capable of compiling a massive amount of data. The configuration station uses a relative database to compile programs. The database system speeds up configuration time and allows data to be retrieved and massaged in various different ways. Engineers have all the tools to configure and test their program off-line before downloading the controller. Taking advantage of a window environment, such luxuries as cut-and-paste makes it easy to modify a program. Finally, the system offers the capability to store programs on tapes or on the hard disk itself. The backup and restore feature makes it easy to restore the system in case of disastrous breakdown. A typical plant layout for a fully automated raw sugar factory is shown in Figure 3.1.

3.3 CONTROL LOGIC USED IN RAW SUGAR FACTORIES

3.3.1 Cane Yard and Cane Preparation

For steady feed of cane to a milling or diffusion plant, it is very important that the cane handling system be controlled automatically. In the raw sugar factory, cane is received in

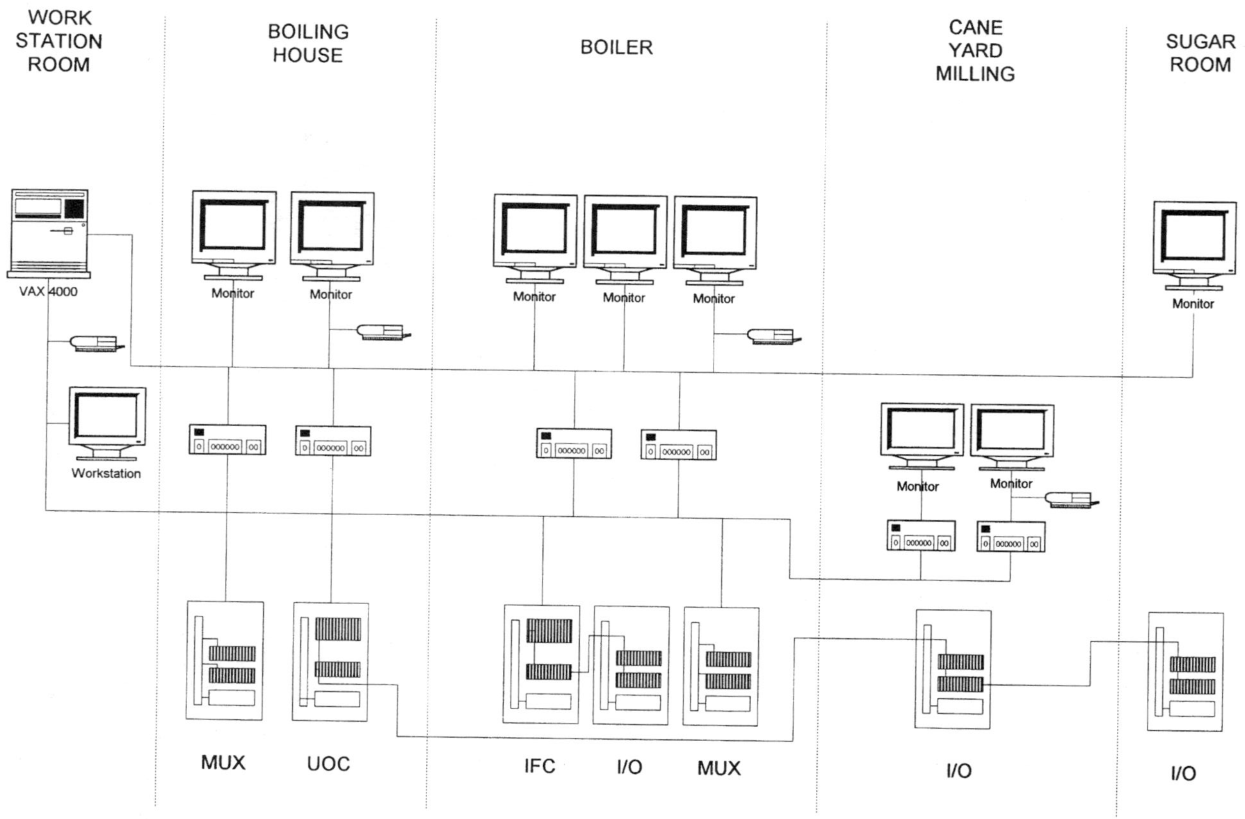

FIGURE 3.1 Typical plant control layout.

the mill yard in various ways and forms depending on the harvesting methods used. In some factories, such as those in Australia, Florida, and Texas, only billeted canes are processed. Whole-stalk cane is received in most other sugar factories. This is delivered in loose form, small bundles tied with cane leaf (typical in developing countries), or in large bundles wrapped with chains. At the mill yard, the cane is unloaded into a feeder table that feeds the main cane carrier to the mill. In Hawaii, a complex cane washing plant is incorporated between the mill yard and the milling train, to remove rocks, soil, and other extraneous matters. The control system therefore becomes more complicated. A typical mill yard control system is shown in Figure 3.2. The control strategy used in the example consists of measuring cane levels with ultrasonic level devices at various points on conveyors. A preset cane height is used on the main cane carrier to determine the amount of cane to be processed. Based on the cane-level measurement, the main conveyor will speed up or slow down to match the desired set point. Override protection is incorporated at the knife or shredder load to prevent overfeeding and choking of the knife set or shredder. This could be a current transformer when a motor is used, or ring pressure when a turbine is used as a prime mover. A high load detected at the knife or shredder will automatically slow down the main conveyor. Interlock for high loads will automatically stop the feed conveyor until the load clears out. The feeder table is controlled automatically by the level of cane at the drop chute. A level detector is also installed at the head shaft of the feeder table to move the cane forward when gaps are detected. In factories where a belt conveyor is used after the shredder, a weigh belt is used as the primary cane rate control to the mill. The desired cane rate in tons per hour is then set to control the speed of the main carrier.

3.3.2 Milling Train and Diffusion

Most modern milling trains are equipped with a four-roller mill and Donelly chutes. The control philosophy in the mill varies from factory to factory. A typical control strategy consists of measuring the bagasse level in the Donelly chute using a capacitance device or conductivity buttons. The level signal is used to speed up or slow down the mill turbine to keep a constant bagasse level in the chute. Override protection from the turbine nozzle pressure is normally used to prevent chokes and to reduce mill torque. This measurement is also used to control the Donelly chute opening. Besides the level control loop for the individual mill, interlocks are incorporated to monitor the turbine for low oil pressure, high oil temperature, and low steam pressure. Other interlocks are gearbox high temperature, gearbox low oil pressure, mill zero speed, and a high chute level switch. In addition, all intermediate conveyors and juice pumps are also interlocked for automatic startup and shutdown. Mill maceration water is measured and controlled based on grinding rate and dilution required. A typical mill control is shown in Figure 3.3.

For a diffusion plant the basic controls used are diffuser speed, cane-level profile, juice heater temperature, maceration flow, juice level, juice flow out of the diffuser, juice draft, press juice pH and temperature, and press juice Brix value. The stage pumps are normally interlocked for automatic startup and shutdown.

3.3.3 Juice Clarification and Filtration

For optimum clarification results, final mixed juice temperature control, limed juice pH control, mixed juice level and flow control, flocculant flow control, and lime density control are used. Mixed juice tank level is also interlocked with the mill to prevent overflow of juice in case of a high level. For factories using a trayless rapid settling clarifier, it is important to ramp the juice flow rate to avoid upsets inside the clarifier, especially on mill startup.

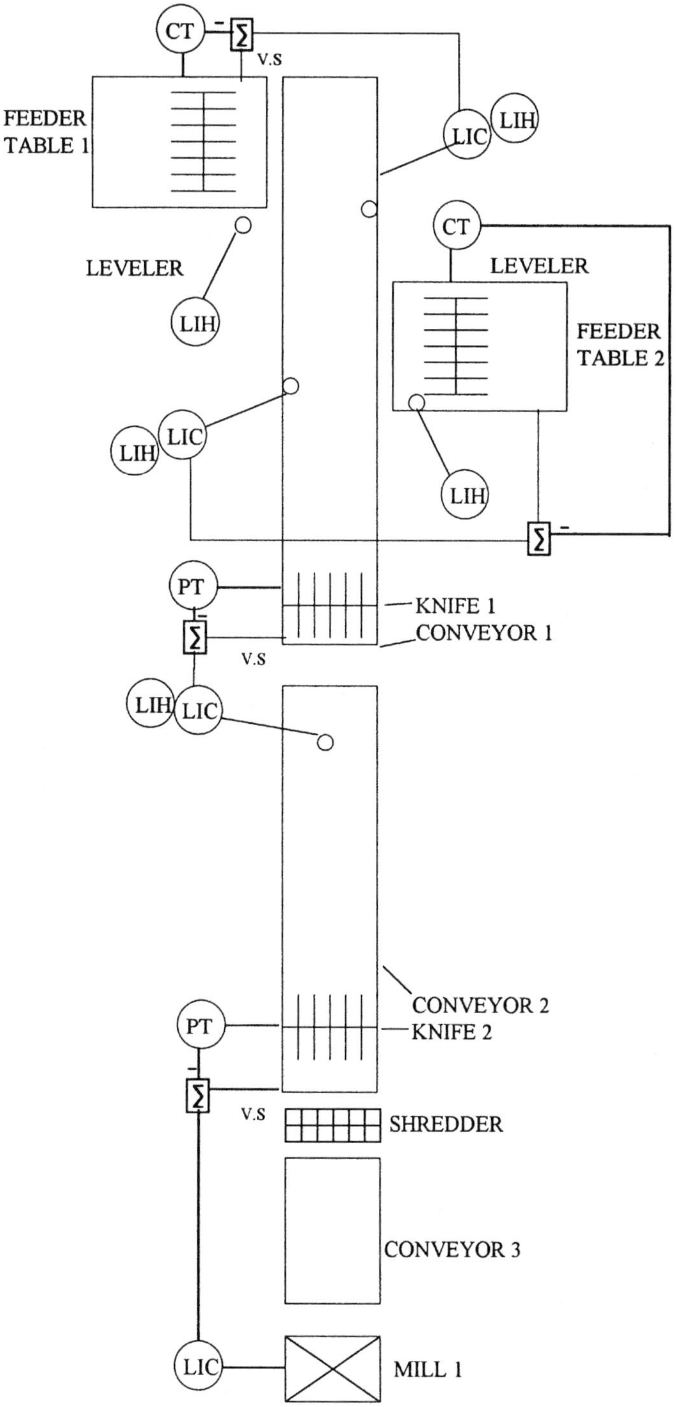

FIGURE 3.2 Mill yard and cane preparation control system.

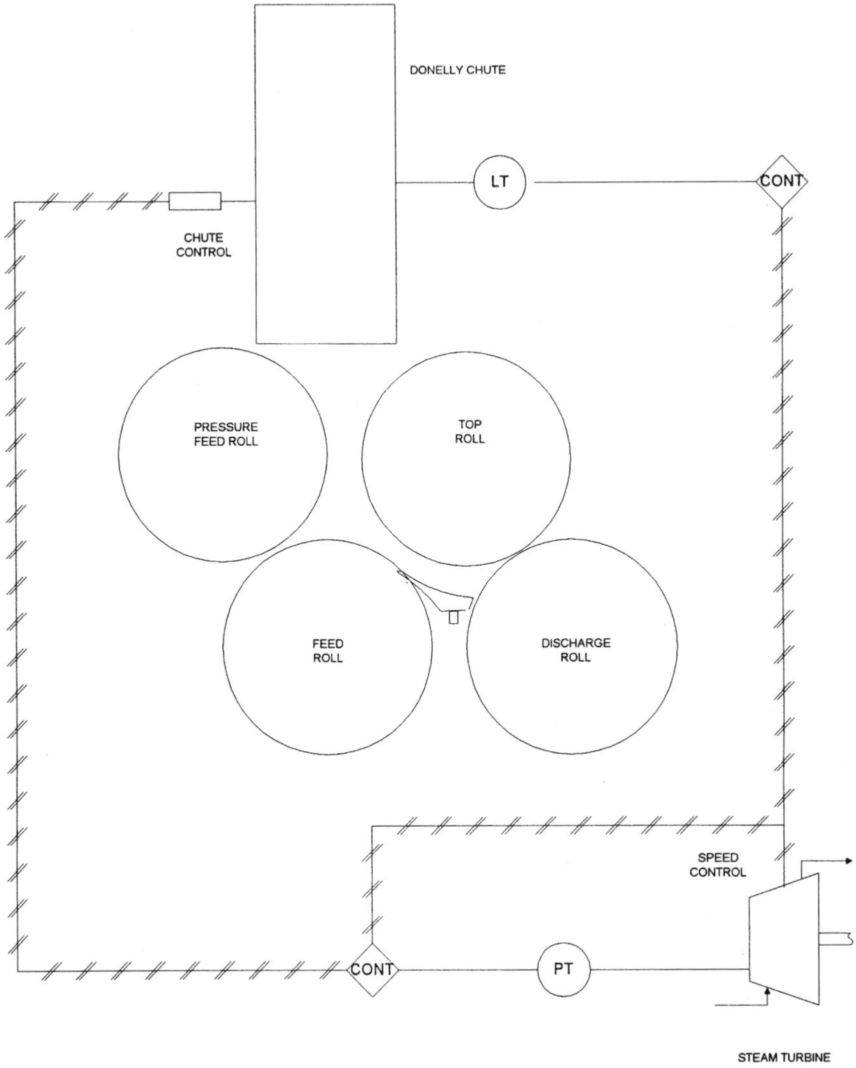

FIGURE 3.3 Mill control system.

Clarified juice pH is also used as a feedback control to regulate lime addition. A PLC is used to prepare the lime saccharate so that the right ratio of lime and syrup is obtained consistently. Continuous turbidity of clarified juice is also used as a measurement to help optimize juice clarification.

Capacitance-level sensors are used in the vacuum filter to regulate mud flow from the mud mixer to the mud filter so that no mud is overflowed and recirculated. A typical control diagram for clarification and filtration is shown in Figure 3.4.

3.3.4 Evaporator Station

For sugar factories that cogenerate power for a local utility company, the evaporator station is the most important unit operation for energy optimization. The control logic starts with

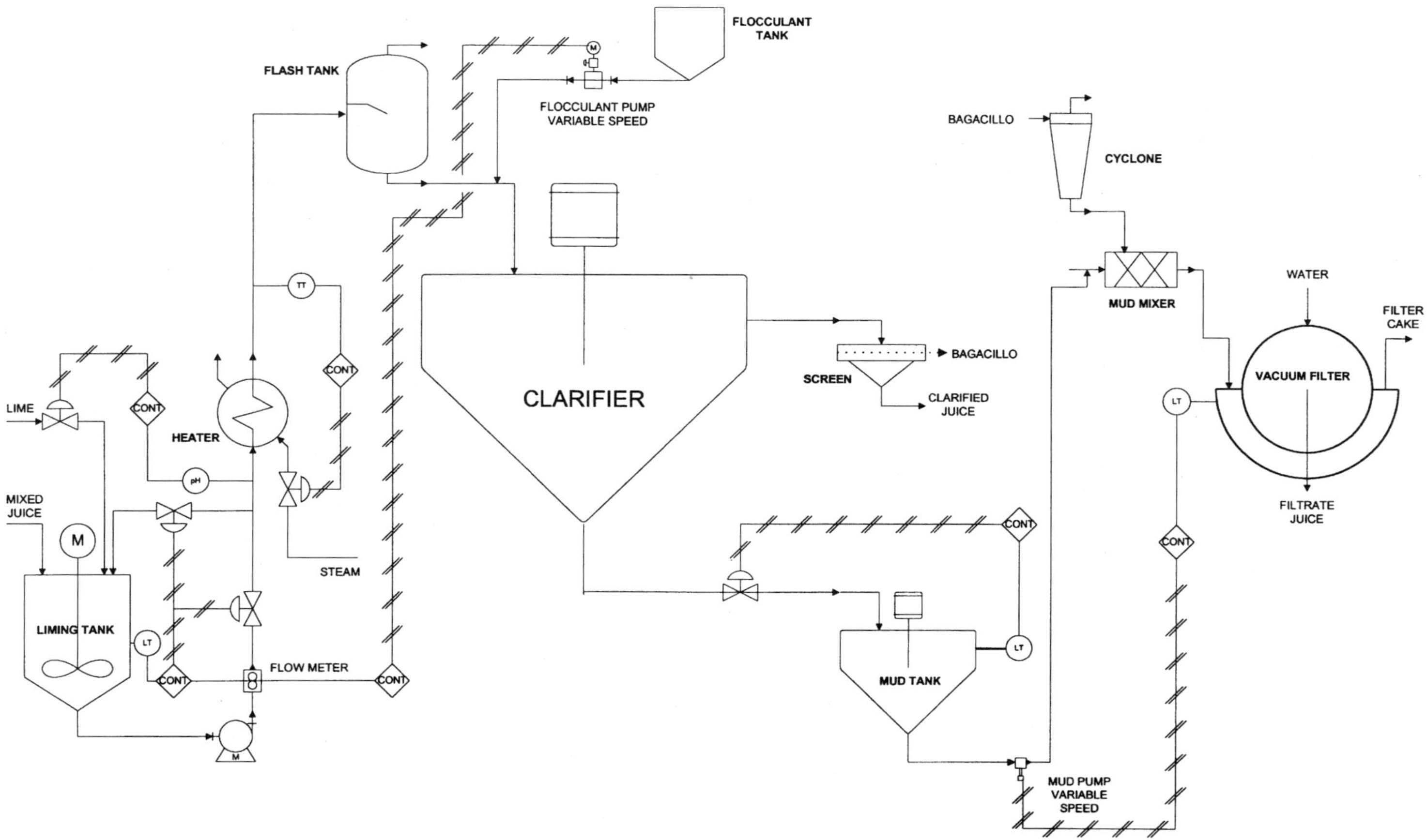

FIGURE 3.4 Clarification / filtration control system.

the evaporator supply tank level. This level measurement is used to optimize steam consumption at the evaporator station and also to maximize electrical energy production at the powerhouse. A rate control is used to slow down the third, fourth, and fifth vessels when the supply tank level is low and falls within a deadband. Automatic control for juice level at each vessel is necessary for optimum performance and to minimize juice entrainment or carryover. Syrup density control and last vessel vacuum control are other controls that must be incorporated at the evaporator station. A first vapor makeup control must be provided to supply sufficient steam to the vacuum pans when the factory is down. A typical control scheme for an evaporator train is shown in Figure 3.5.

3.3.5 Vacuum Pan Control

Vacuum pan control is perhaps the most difficult unit operation to understand and control. A fully automatic pan control system utilizes a combination of PLC and DCS where complete batching can be programmed. Each individual batch pan control will include level, consistency or conductivity, temperature, and vacuum controls. For the seed pan, an automatic blend of syrup and molasses to achieve a desired footing purity is used to produce very consistent and uniform seed. Radio-frequency (RF) or refractometer Brix is used as the primary measurement for seeding. The control algorithms has been fully described by Lawler and Lam [1]. Automatic pan startup, cut over, and discharge are initiated from the computer console. Water ratio control is also used to hold the pan at the end of the boiling cycle. Continuous vacuum pan will normally be equipped with RF probes at each pan compartment. The RF signal is used to activate the intermittent injection of syrup, molasses, or water. A PLC is normally used to open and close the feed valves at each compartment. A fully automated batch pan flow diagram is shown in Figure 3.6.

3.3.6 Crystallizer and Centrifugal Station

Basic automation used at a crystallizer station are massecuite temperature and motor load. Hot- and cold-water temperature controls are also used to maximize massecuite exhaustibility. At the centrifugal station the high-grade centrifugals are normally of batch type and are controlled by PLCs. Molasses density and temperature controls are used to melt small sugar crystals. The low-grade centrifugals are in most cases continuous type. Individual automatic control is based on motor load. Hot-water temperature control for the reheater is used to reduce massecuite viscosity. Magma Brix is also controlled automatically so that consistent magma seed is sent back to the pan floor.

3.3.7 Boiler and Turbogenerator

The boiler and turbogenerator are perhaps the most critical equipment in a raw sugar factory. The performance of the entire factory depends on the efficiency of steam and power production from this equipment. The primary control objective at the boiler is to maintain a constant output steam pressure. This is normally done with a boiler master, which measures the output heat flow and generates a feedforward signal to the bagasse rate controller. The feedforward signal is dynamically compensated when bagasse is fired because of the dead time and lag between a bagasse rate change and the associated heat rate change. Under bagasse firing conditions, the airflow control is a simple proportional with bias and a zero-set-point device driven by the bagasse feeder signal. The proportional gain provides an air/feed ratio and the bias is used to modify the air under the low load and high moisture of bagasse conditions. The output of this controller provides set points for the overfire and

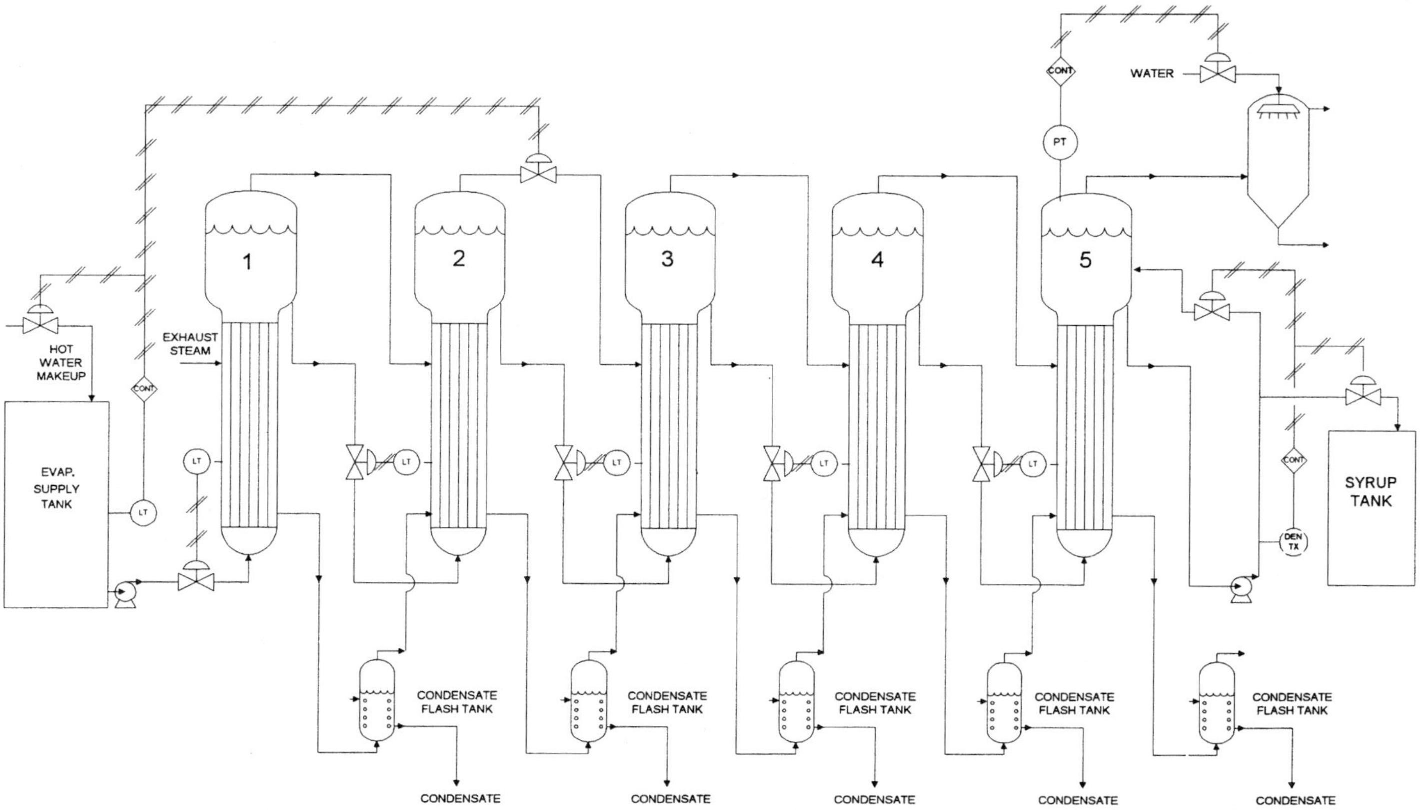

FIGURE 3.5 Evaporator control system.

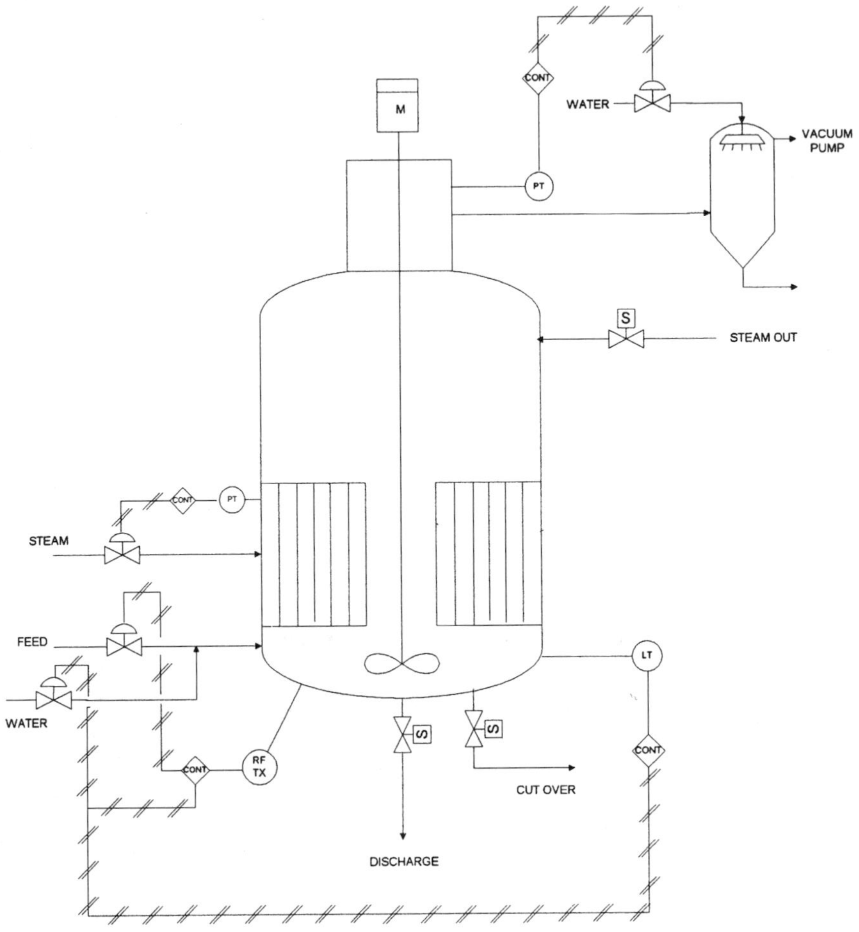

FIGURE 3.6 Pan control system.

undergrate air controllers. Excess oxygen controller is used during oil firing. The remainder of the boiler controls are normal textbook controllers for drum level, feedwater, forced-draft pressure, induced-draft pressure, steam temperature, desuperheating, furnace draft, and grate temperature.

The turbogenerator control is normally done by a microprocessor-based unit or electronic governor in which kilowatt or extraction pressure is controlled automatically based on exhaust steam demand from the boiling house. The microprocessor-based unit includes autostart control, overspeed safety, kilowatt limiter, automatic synchronization, and frequency control in case of a tie-line breaker trip. Safety devices on the turbine generator include vibration monitoring of all main bearings, bearing oil temperature, and low-oil-pressure, low-vacuum, and high-temperature safety trips.

The rest of the boiler area control consists of individual PLC-based units for controlling bagasse house management; demineralizers; boiler feedwater conditions, such as level, temperature, conductivity, and pH; chemical mixing and dosing; and oil heating and firing system.

3.4 ADVANTAGES OF AUTOMATION AND COMPUTERIZATION IN A RAW SUGAR FACTORY

Factory automation and computerization in a raw sugar factory have shown to provide a number of benefits:

- Increased factory throughout
- Increased sugar recovery
- Increased grinding time efficiency
- Improved sugar quality
- Increased safety
- Reduced plant maintenance
- Increased plant life
- Increased power production
- Reduced steam use
- Reduced manning and operating costs

3.5 REFERENCE

1. R. P. Lawler and E. Lam, *Proc. 48th Annu. Hawaii. Sugar Technol. Conf.*, 1989.

CHAPTER 4

Raw Sugar Storage and Handling*

INTRODUCTION

Distances between raw sugar mills and refineries vary greatly. In some cases it is necessary for a raw sugar producer to store large quantities of raw sugar during the grinding season and then deliver to the refinery over a several-month period; or months of shipment may be involved. Upon receipt of the raw sugar, the refinery may further store the product for several months as situations dictate. Thus, from the time of production of raw sugar until the time of use at the refinery, several months to 1 or 2 years may have passed. Therefore, to prevent the deterioration of raw sugar quality, it is very important to avoid adverse conditions during raw sugar storage and shipment. Since the raw sugar quality is affected during storage, an understanding of the impact of the storage condition on the raw sugar is critical to minimize the loss and quality deterioration during storage. In this chapter we focus on raw sugar storage conditions and handling process, both of which have detrimental effects on raw sugar quality.

4.1 CHARACTERISTICS OF RAW SUGAR

4.1.1 Molasses Film

Raw sugar delivered to the refinery is a mixture of various grain sizes, surrounded by the original film of molasses from the final crystallization step at the raw sugar mill. This film of syrup offers an incubator for microbial growth, causing sucrose deterioration, and therefore is directly related to the polarization loss during storage. Other characteristics of raw sugar adopted to indicate the quality change during storage, such as color and moisture, are also governed by this molasses film.

4.1.2 Desirable Characteristics

Raw sugar of high invert content can become sticky, of low flowability, and be hard to handle in bucket elevators. Particles of bagacillo or other insoluble matter hold moisture

* By Hsin-Sui Chang and Chung Chi Chou.

and serve as breeding grounds for microorganisms, and poor filterability has been found due to excessive insoluble material. Also, colloidal matter and dispersoids of smaller particle size can affect crystal formation in the vacuum pan [1].

General considerations regarding the quality of raw sugar desired to minimize the loss on storage include that it be (1) comparatively free of insoluble matter, such as bagacillo and soil; (2) of uniform grain size and free of conglomerates; (3) surrounded by the original film of molasses (i.e., unwashed); (4) with a safety factor of less than 0.25 to avoid deterioration of all sugars; and (5) manufactured under sanitary conditions to minimize contamination by microorganisms.

4.1.3 Effects of the Raw Sugar Purging Process

Removal of molasses from the massecuite discharged from a vacuum pan is carried out in a centrifugal machine, and the term *purging* is used to refer to this process. *Purging efficiency* (PE) is an indication of the extent of the molasses removal that has been accomplished during the purging process:

$$\text{PE} = \frac{(\text{purity of sugar} - \text{purity of massecuite})(100 - \text{purity of molasses}) \times 100}{(\text{purity of sugar} - \text{purity of molasses})(100 - \text{purity of massecuite})}$$

It is an important control figure used to determine the operating efficiency at the low-grade section of a factory.

Determination of purging efficiency is an attempt to achieve the maximum removal of impurities while minimizing the transfer of sugar to molasses during the purging operation. The formula shows the components that play a role in removing a high level of impurities. Other formulas used to evaluate the centrifugal efficiency are

$$\text{massecuite rate} = (\text{molasses rate})\,\frac{(\text{DS}_{\text{mol.}})(\text{purity of sugar} - \text{purity of mass.}) \times 100}{(\text{DS}_{\text{mass.}})(\text{purity of sugar} - \text{purity of molasses})}$$

$$\text{crystal content (\% on massecuite)} = \frac{(\text{purity of mass.} - \text{purity of mother liquor})(\text{DS}_{\text{mass.}})}{(100 - \text{purity of mother liquor})}$$

where DS denotes dry solids.

4.1.4 Effect of Massecuite Properties and Conditioning

A high coefficient of variation (CV) will affect dissolution during the purging process. With a batch centrifugal, the primary cause of this situation is the physical blocking of drainage channels by movement of smaller grains within the crystal mass. It has also been observed that blockage occurs when a layer of molasses forms on the inner vertical sugar wall within the basket. The purging efficiency is greatly reduced under these conditions.

The influence of the viscosity of the mother liquor on the separating performance of batch and continuous centrifugals has been mentioned by Wells and James [2]. The presence of dextran will increase viscosity and make removal of molasses film more difficult. A study by Mitchell and Bechne [3] on the effects of viscosity change during separation demonstrated an important connection with the timing of wash applications during the separation cycle of batch centrifugal operation. That is, once the free-flowing molasses between the crystals has been removed during the early stages of purging, the crystal bed becomes porous and air moves freely through it. The viscosity of the molasses film surrounding the crystal will rise rapidly when the air removes moisture from it. This action will then become a dominant factor in determining the purging efficiency. Therefore, the wash application should be timed properly to follow the last of the free-flow molasses from the crystal bed. As for a continuous centrifugal, optimization of the wash-water effect in continuous separation is a matter of wash application rather than the timing of wash application as in the batch operation.

The crystal content of the massecuite affects purging in various ways. A low-crystal-content material normally will have a reduced purging rate because of a higher percentage of molasses to be removed. However, this factor should not be referred to as a prime indicator of the separating ability of the material. Low-grade massecuite is usually "conditioned" to improve its separating properties (e.g., a reduction of viscosity by heating). The heating device used in recent years includes the electrical resistance heater [4],7 and finned tube economizer tube heater [5,6]. A most recent development is a fineless heater, which has been designed to operate with a low head of massecuite. An alternative to heating to reduce the viscosity is the dilution of low-grade massecuite by molasses or water. However, the main problem with the dilution method is the difficulty in achieving sufficient control to avoid periods of undersaturation.

4.2 DRYING, COOLING, AND PACKING OF RAW SUGAR

4.2.1 Drying and Cooling of Raw Sugar

Raw sugar can be dried with a fluidized bed or spouted bed. Figure 4.1 shows typical fluidized and spouted beds. Weilant et al. [7] reported that for drying a batch raw sugar, the drying time of 1 min for a fluidized bed and 2 min for a spouted bed are required to reach the equilibrium moisture content of wet sugar. The advantage of the spouted bed drier is its greater stability, its resistance to shock loads, and its lower air requirement than that of the fluidized-bed drier. However, the spouted drier seems to produce finer crystals than the fluidized drier.

A fluidized-bed dryer system has been carried out successfully for drying of raw sugar in a South Africa factory with a grinding capacity of 500 t/h (12,000 tonnes/day). This drying system is shown in Figure 4.2. The air velocities vary between 1.0 and 1.4 m/s, the

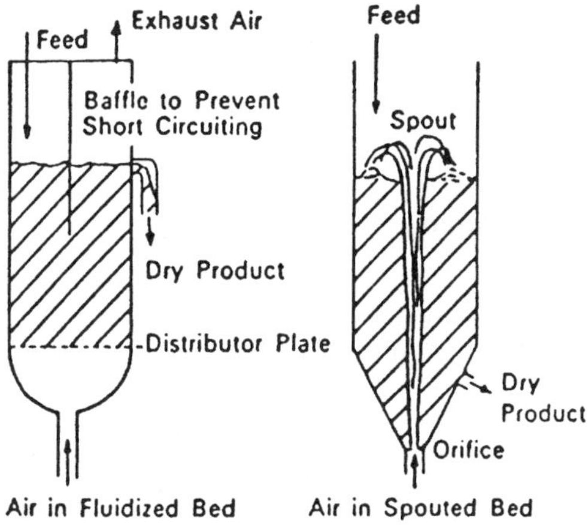

FIGURE 4.1 Typical fluidized and spouted beds.

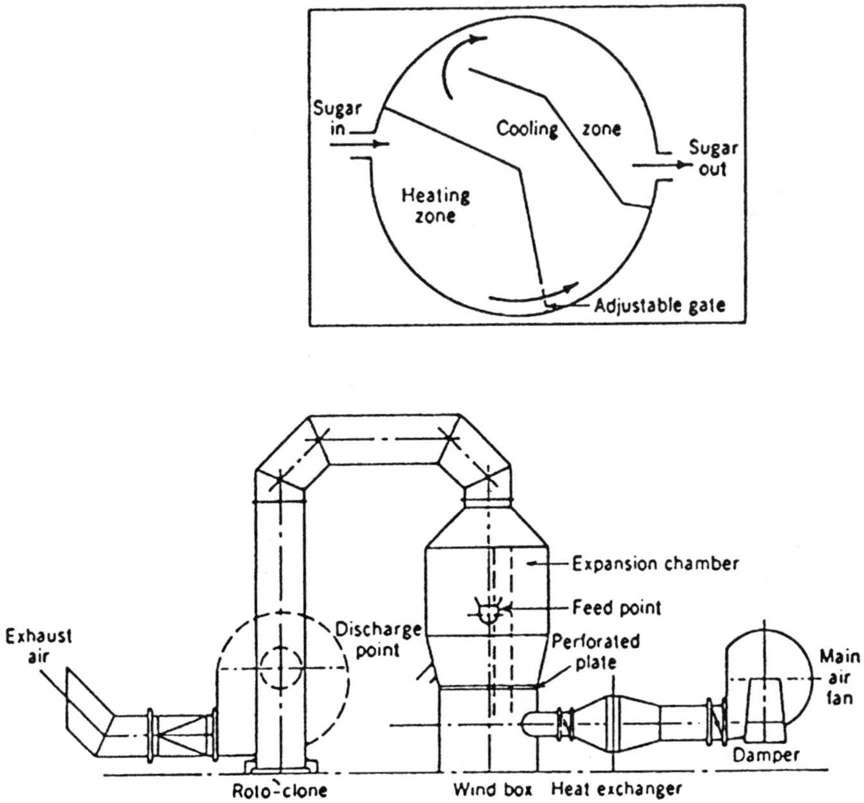

FIGURE 4.2 Raw sugar drying with fluidized-bed system (insert: baffle arrangement). (From Ref. 8.)

temperatures of the drying zone between 40 and 50°C, and the bed pressure differentials between 2.0 and 2.5 kPa. The average sugar output is 35 tonnes/h, the inlet sugar moisture 0.38%, and the outlet sugar moisture 0.11 to 0.16%, depending on the process variables.

The same principles can be applied to cooling. Cooling raw sugar is important to prevent high temperatures during storage, which tend to increase raw sugar color both in the film and inside the crystals.

4.2.2 Cooler–Drier System

A simple, low-cost, efficient cooler–drier system was developed successfully for drying sugar at Illovo mill in South Africa during mid-1970. Over the years the production rate of sugar at Illovo mill almost doubled, yet the overall moisture level remained within the limits specified. The plant at Illovo consists of two vertical cooler–drier units working in parallel and fed at the top by a dual-screw conveyor system, as illustrated in Figure 4.3 and 4.4. The fixed cones prevent the sugar from falling straight down and thereby increase the contact time between the sugar and air. The air flows countercurrent to the cascading sugar [9]. Results of the performance of the cooler–drier at Illovo are shown in Table 4.1. Despite the relatively high ambient temperature and relative humidity, the sugar moisture was reduced from 0.33 to 0.1%, which is within specifications.

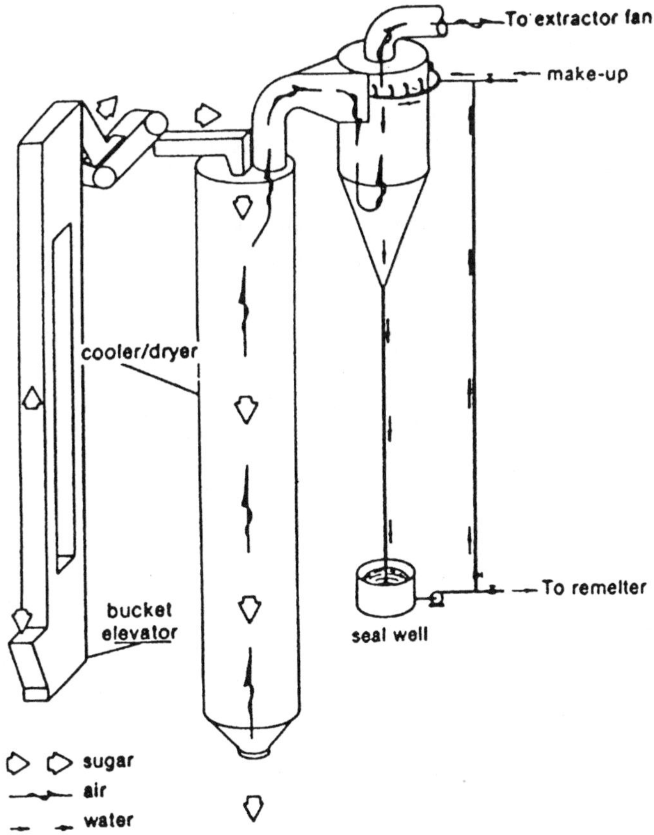

FIGURE 4.3 Cooler–drier plant.

4.2.3 Entrainment Arresting System

Sugar dust is removed from the exhaust air by means of a wet cyclone, where water is being recycled after spraying. A newly designed entrainment arresting system, using induced draught rather than forced draught, resulted in a higher airflow without excessive sugar spillage [10]. Table 4.2 shows the performance of this drier system. To avoid the growth of *Leuconostoc* bacteria in the entrainment arrester area, it is necessary to heat to 85°C and treat the water system with chlorine once a day (concentrated calcium hypochloride).

4.2.4 Raw Sugar Packing

Various types of conveyors—for example, ribbon or screw conveyors or grasshopper conveyors—are used in raw sugar houses to transport sugar from centrifugals or driers to packing bins. The ribbon or screw conveyor consists of a spiral ribbon or screw that revolves in a trough, carrying sugar with it. The cross section of the trough is usually parabolic rather than U-shaped. The grasshopper conveyor, preferred in eastern hemisphere factories, is a wide, flat trough supported on flexible strips that vibrate by means of an eccentric such that sugar is thrown forward through a short arc with each vibration of the trough. The advantage of the grasshopper is that there is no breakage of sugar crystals in handling, whereas

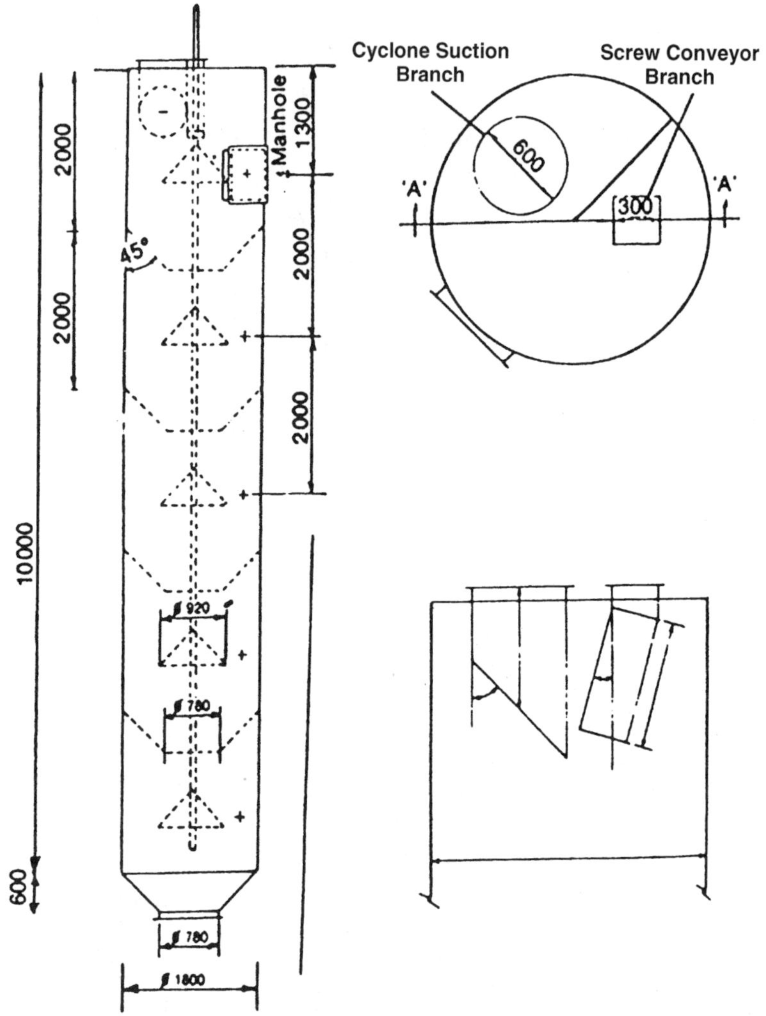

FIGURE 4.4 Cooler–drier design.

TABLE 4.1 Average Performance of Cooler–Drier at Illovo (1983)

Air				Sugar			
Temperature (°C)		% Relative Humidity	Velocity of Air inside Cooler / Drier (m / s)	Temperature (°C)		% Moisture	
Dry Bulb	Wet Bulb			Inlet	Outlet	Inlet	Outlet
27.15	24.35	78	0.6	55.7	32.9	0.327	0.099

TABLE 4.2 Sugar Drier Performance (1984)

Sugar Temperature (°C)		Air Temperature (°C)		Sugar Dilution Indicator (%)		Flow Rate		
In	Out	In	Out	In	Out	Sugar (metric tonnes / h)	Air (m^3 / s)	Air / Sugar Ratio
58.6	31	22	47	49	31	51.3	10.35	0.87
66.6	30	23	44	80	30	50.1	12.75	1.10
61.7	32	27	44	51	30	49.9	12.75	1.10
64.4	28	28	40	93	28	49.5	12.75	1.11
66.1	28	28	43	135	30	51.5	11.55	0.97
63.0	32	28	46	183	27	57.2	6.75	0.51
63.0	32	28	44	223	31	39.8	12.75	1.38
63.7	31	29	42	112	30	57.9	12.75	0.95
67.4	30	29	47	122	34	46.6	12.75	1.18
70.4	35	29	48	95	28	48.8	6.75	0.60
63.6	32	29	45	173	21	46.1	12.75	1.20
57.7	32	30	46	62	29	54.8	8.00	0.63

little crystal breakage occurs with screw conveyors even if the trough is filled only about one-third with sugar. It is a common practice to deliver the raw sugar to the bin from the elevator, where the sugar strikes a spreader or fan that throws it out into the air and causes some cooling and evaporation. This concept is used widely in bagged sugar and bulk sugar handling and storage (Section 4.3.4).

4.3 IMPACT OF HANDLING AND STORAGE CONDITIONS ON RAW SUGAR QUALITY

Handling problems can occur if there are lumps of damaged sugar. Such lumps can often be missed when sampling, thereby giving the raw sugar a false quality. Lumps are usually due to poor centrifugal operation, water leaking into the sugar, or unstable storage conditions of temperature and humidity.

4.3.1 Moisture, Safety Factor, and Dilution Indicator

Moisture levels in the environment affect the raw sugar quality directly during storage. The moisture level of the air surrounding the sugar crystals alters the moisture content of the raw sugar instantly. The *safety factor* (SF) and *dilution indicator* (DI), which are related to the moisture content of raw sugar, serve as quality criteria for raw sugar storage. The safety factor found by Whalley [11] is defined as

$$SF = \frac{\text{moisture (\%)}}{100 - \text{polarization}}$$

The Colonial Sugar Company of Australia suggested that the safety factor, which compares the relationship of the moisture content to the nonpolarization components of the sugar, should not exceed 0.333% of the raw sugar. That is, to avoid deterioration of the sucrose, the moisture content must not exceed 33% of the total nonpolarization components of the raw sugar.

The dilution indicator, another way of expressing the relationship between moisture and nonpolarization, is

$$\mathrm{DI} = \frac{\text{moisture (\%)}}{100 - \text{polarization moisture}}$$

The dilution indicator, expressed as a whole number, is related to the safety factor. For example, a DI of 33 equals an SF of 0.25, and a DI of 50 equals an SF of 0.33 (Fig. 4.5). It is generally accepted that when the safety factor is below 0.25 (i.e., DI below 33), spoilage of raw sugar due to microbial growth can be effectively stopped.

4.3.2 Equilibrium Relative Humidity

The moisture in sugar will quickly reach an equilibrium with the moisture surrounding the crystals in the air. Moisture will migrate in or out of sugar according to the equilibrium relative humidity (ERH) of the air surrounding the sugar. The ERH of raw sugar is defined as the point at which the sugar is in equilibrium with the relative humidity of the atmosphere and does not gain or lose moisture to the surrounding air at a given temperature.

The suitable relative humidity for sugar storage is about 65 to 70%. If the humidity of the air is greater than this, sugar will absorb moisture and the safety factor will exceed the safety zone, allowing microbial infestation to occur, resulting in rapid deterioration of the sugar. If the humidity of the air is lower than this, moisture in the sugar will evaporate and the syrup film will reach supersaturation, causing crystallization within the syrup film. As the sugar crystals grow, they tend to cement (bind) themselves together, producing caking or hardening. Thus, it is essential that the relatively humidity be kept near the equilibrium point so that the sugar will neither absorb moisture nor dry out during storage.

The storage condition of the warehouse determines the quality of raw sugar to a great extent, especially when long-term storage is involved. To prevent sugar from deterioration, the warehouse should be kept sealed to prevent air leaks and closed whenever possible to ensure that as little moisture as possible comes into contact with the sugar.

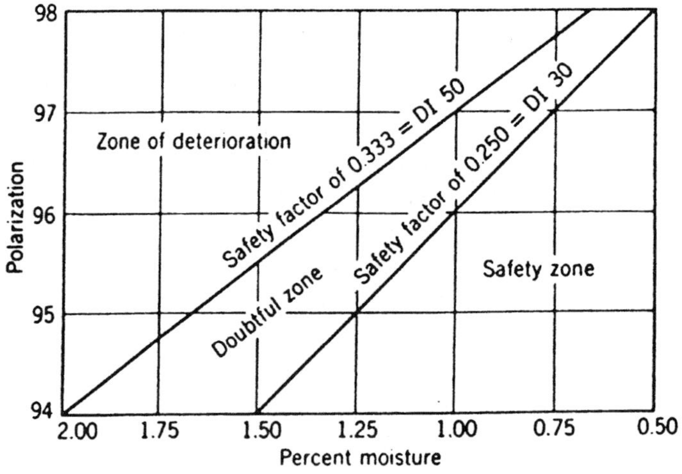

FIGURE 4.5 Safety factor (or dilution indicator) for raw sugars.

4.3.3 Temperature

The significant effect of temperature on quality is the development in color of stored raw sugar. It has been well documented that raw cane sugar must be cooled sufficiently to prevent a drop in polarization and an increase in color [12], a marked decrease in polarization, drop in pH, and an increase in color for raw sugar stored at high temperature [13]. It has been observed that color increase deep inside a large pile of raw sugar was twice as high as on the surface. Temperatures inside the pile were shown to be twice that of the average temperature. The effect of high temperature is the destruction of sucrose to form color. It should be a common practice to keep the temperature as low as possible during storage. There have been suggested optimum temperatures for different storage periods (Table 4.3). A constant temperature control should be maintained inside the warehouse. Moisture disturbance in raw sugar can occur if the ambient temperature varies constantly, even to the degree of causing the safety factor to be exceeded, and initiating microbiological deterioration.

4.3.4 Bulk Raw Sugar

Raw sugar quality is susceptible not only to the surrounding environment but also to the manner in which raw sugar piles is stored in the warehouse, particularly in the case of bulk storage (Fig. 4.6). For example, the ideal angle of repose for bulk raw sugar pile is about 40. The angle of slide is about 35° and the angle of draining is about 55° (Fig. 4.7). Sugar for bulk handling should also be noncaking and free running. That is, a sugar with a uniform grain and low moisture content. Usually, a sugar with uniform grain size above 0.8 mm, with less than 10% fines below 0.4 mm, and with moisture content below 0.4%, will unload well even under adverse environmental conditions (e.g., bad weather). However, the compressibility/pressure of raw sugar piles, which creates high density or develops heat inside the piles, will change the temperature profile of the raw sugar body, resulting in the deterioration of quality (e.g., caking).

Compacting of sugar in a pile of 2000 tons 30 ft (9 m) high showed that 20 ft (6 m) down the sugar had a bulk density of 62.3 lb/ft^3 (998 kg/m^3). The bulk density varies with crystal size, volume of measuring vessel, vessel shape, angle of charging funnel, and height between funnel and vessel [15]. The density obtained from large and tall vessels varied between 49.12 and 53.10 lb/ft^3 (786.9 and 850.5 kg/m^3), respectively. For designing of a warehouse, 55 lb/ft^3 is generally used.

To minimize the change inside a raw sugar pile during storage, the practice in which the raw sugar is stowed into a warehouse is examined carefully. A sugar *slinger* is a thrower for bulk sugar handling. This device consists of a high-speed short belt (2000 rpm) rolling over two large pulleys, in which the forward motion can throw the sugar crystals out in a stream (Fig. 4.8). When slinging the sugar through the air from the end of belt in a trajectory, it will reduce the heat development inside the piles, thus ensuing a less serious caking problem.

TABLE 4.3 Sugar Storage Temperature

Storage Period	Optimum Sugar Temperature (°F)
Weeks	<113
Months	<100
Years	<86

Source: Data from Ref. 14.

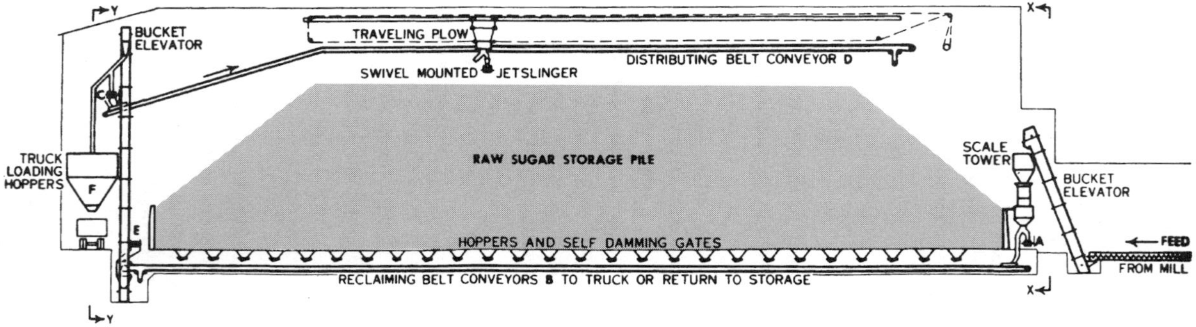

FIGURE 4.6 Bulk raw sugar storage. (Courtesy of Link Belt Company.)

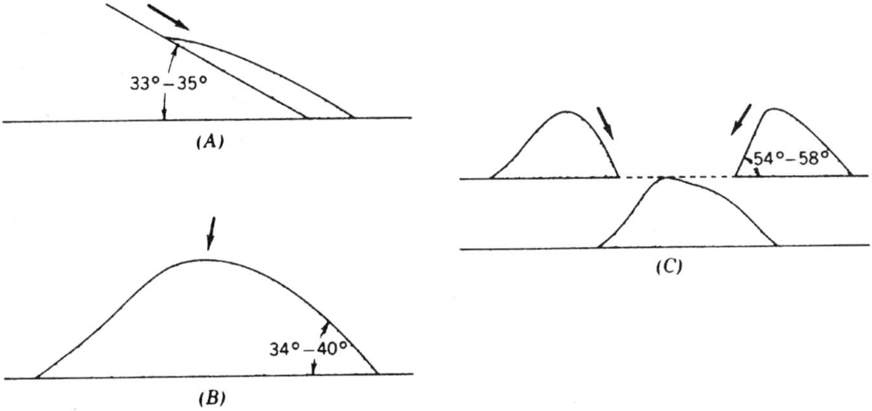

FIGURE 4.7 Bulk sugar storage angles: (*A*) angle of slide; (*B*) angle of repose; (*C*) angle of draining.

4.4 EXPLOSION HAZARDS IN STORAGE

4.4.1 Dust Explosion

It is generally accepted that the combustion of individual dust particles (e.g., sugar) causes an increase in the local volume. This phenomenon, in turn, leads to the development of a pressure wave that will pass through the vessel containing the dust clouds, and subsequent damage follows. The pressure wave can cause an increase in flame velocity and dust concentration because it can bring into suspension dust that had settled onto the floors previously, for example. In other words, it can cause an explosion to continue into areas where only settled dust was present prior to the explosion.

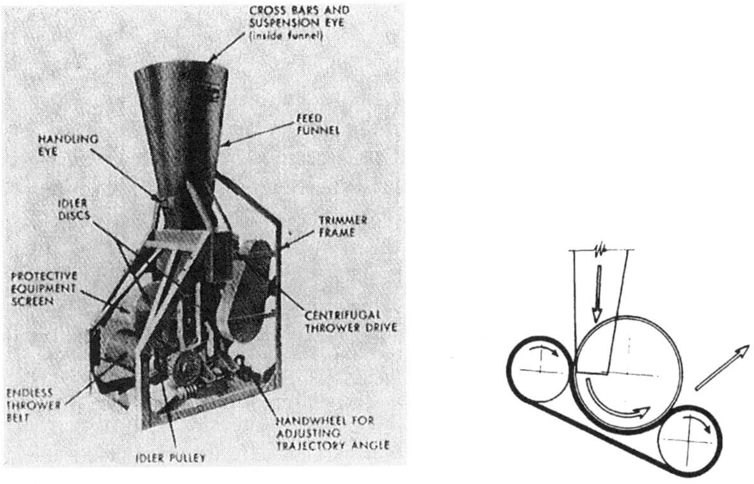

(A) (B)

FIGURE 4.8 Thrower for bulk sugar handling: (*A*) Rhe thrower; (*B*) diagram of thrower. (From Ref. 1.)

Generally speaking, for a sugar dust explosion to occur, four conditions must be fulfilled simultaneously: (1) sufficient oxygen must be present to support combustion; (2) concentration of sugar dust must be within the explosive limits; (3) the ignition source must be hot enough to ignite the sugar dust cloud; and (4) the ignition source must release sufficient energy to ignite the sugar dust cloud. Other factors influencing the explosion include average shape and size of the dust particle and the degree of agglomeration of the particle.

The minimum oxygen content of an atmosphere that will support flame propagation in sugar dust clouds has been found to be 9% by volume [16]. The values of ignition source temperature may vary between 350 and 805°C, depending on the purity, concentration, and particle size of the dust, and the rate of heating [17].

4.4.2 Dust Concentration

A dust concentration of 20 g/m^3 or below for particles smaller than 100 μm is considered safe from explosion. A dust concentration of 20 g/m^3 is rather like a very hard rain with a visibility of approximately 8 to 10 ft. Since there is a decrease in minimum explosive concentration with a decrease in particle size [18] and considering that dust explosions are transient phenomena, it is better to employ a much safer, lower level in a factory.

The flame velocity in a sugar dust cloud of concentration 60 g/m^3 averaged 530 ft/s, but was also found to increase with the distance of travel to a maximum of 810 ft/s [19]. Theoretically, the flame velocities increase with decreasing particle size and increasing dust concentration, to about 300 g/m^3 [20].

4.4.3 Source of Sugar Dust

Source and formation of sugar dust were observed to come from rotary driers, bucket elevators, cylindrical sieves [21], chipping from crystal surfaces, and broken fractions of agglomerates [22]. An increase in conglomerates caused by bad vacuum pan operation and an increase in sugar temperature will enhance dust generation.

4.4.4 Effect of Humidity

High humidity normally will tend to promote the agglomeration of the dust in the air, which will help to reduce the risk of explosion. The water content of sugar dust depends on the atmospheric relative humidity, and the explosion hazard can be reduced if it causes the airborne particles to become sticky and agglomerate. The National Fire Protection Association (NFPA) statistics on explosions show that more sugar dust explosions occur in winter than in summer: The higher the humidity, the less the risk of explosion.

4.4.5 Fire Hazard

Although fires in bulk sugar are rare, the hazards of fire cannot be ignored. Two distinct types of fire are quite recognizable: primary fire and secondary fire. *Primary fire* will spread in all directions or widen gradually. All such fires are quite unstable. Primary fires are very easily extinguished (e.g., a puff of wind, a slight fall of sugar, or a very light spray of water). *Secondary fires* apparently are the most damaging types. The fire spreads over the surface and into the stack by a combination of flaming of the pyrolysis gases and glowing of the residual char. Both sources of the heat are evident over the entire area involved.

4.5 QUALITY CHANGE DURING STORAGE

4.5.1 Polarization

Polarization in storage is very significant. It is generally recognized that there is a high correlation between polarization loss and the safety factor of raw sugar. As the ratio of moisture to nonpolarization or the safety factor increases, the surrounding molasses film of raw sugar becomes a perfect breeding ground for microorganisms. The propagation of microorganisms during storage will result in a significant deterioration of sucrose, which is a major economical concern.

4.5.2 Color

Another primary change in raw sugar quality during storage is the increase in color. The increase in color during storage can easily place the sugar in the penalty category. Development of color, not only in the molasses film but also inside the crystal [13], is determined by the storage temperature (Section 4.3.3) and the original color level of the raw sugar, reducing the sugar and amino acid content. In addition, good-quality raw sugar with a lower starting color increases less color than with a higher starting color.

The coloring matter formed during sugar manufacture can be attributed to (1) reaction of reducing sugars with amino acids and proteins, usually termed a *browning* reaction; (2) caramelization caused by thermal degradation of sucrose or other carbohydrates; (3) the presence of phenolic compound due to enzymatical reactions between flavonoid and cinnamic acid precursors; and (4) invert degradation products [e.g., 5-(hydroxymethyl)-2-furaldehyde], which are very susceptible to the darkening process to produce brown polymers.

The principal color substances of raw sugar are melanoidins and caramels. The primary cause of the color increase is the Mallard reaction, which tends to become more significant at higher temperatures.

4.5.3 Flowability

The flowability property is particularly important in the handling and shipping of raw sugar. Loss of flowability or gain in strength of raw sugar during storage has been reported by Bagster [23]. An experimental study by Bagster found that over about a month there was an enormous increase in strength which subsequently fell again, and this behavior became less significant as the sugar aged. Bagster also observed that when raw sugar was stored after drying, it had a tendency to cake, but as time progressed, caking became less severe. Caking, which occurs as moisture migrates or evaporates from the regions of contact between crystals [24], will reduce the flowability of raw sugar. Generally speaking, an initial gain in strength was due to the formation of a crystalline bridge, but this decreased again due to a depletion of sucrose from the molasses film surrounding the crystals.

REFERENCES

1. J. C. P. Chen and C. C. Chou, *Cane Sugar Handbook,* 12th ed., Wiley, New York, 1993.
2. W. D. Wells and G. P. James, *Proc. Queensl. Soc. Sugar Cane Technol.,* 1976, pp. 287–293.
3. Mitchell and Bechne, *Bur. Sugar Exp. Stn. Tech. Commun.,* No. 12, 1937.

4. D. J. Wright, *Proc. Queensl. Soc. Sugar Cane Technol.*, 1964, pp. 281–288.
5. L. K. Kirby et al., *Proc. Austral. Soc. Sugar Cane Technol.* (*Queensl.*), 1976, pp. 255–262.
6. D. J. Wright, *Proc. Austral. Soc. Sugar Cane Technol.* (*Queensl.*), 1987, pp. 201–205.
7. R. H. Weiland et al., *Proc. Int. Soc. Sugar Cane Technol.*, 1974, pp. 1561–1566.
8. G. F. Mann, *Proc. S. Afr. Sugar Technol. Assoc.*, 1983, pp. 56–59.
9. J. P. M. De Robillard, *Proc. S. Afr. Sugar Technol. Assoc.*, 1984, pp. 61–63.
10. R. G. Attard and J. H. King, *Proc. Austral. Soc. Sugar Cane Technol.* (*Queensl.*), 1985, pp. 189–193.
11. T. G. Whalley, Paper 54-426, *Proc. Queensl. Soc. Sugar Cane Technol.*, 1954, pp. 27–39.
12. C. C. Chou, *Proc. Sugar Process. Res. Inst. Workshops*, 1990, p. 64.
13. J. C. P. Chen, *Sugar J.*, Sept. 1969, p. 13.
14. Cane Sugar Refining Research Project (CSRRP) No. 48, Aug. 1979.
15. O. Prozem, Paper 81-1236, *Spozyw.*, 35(2):58–60, 1981.
16. P. Beyersdorfer, *Int'l Sugar Jour.*, 24(753), 1922.
17. M. Jacobsen et al., *Report of Investigations*, NQ 5753, U.S. Bureau of Mines, Washington, DC, 1961.
18. R. L. Meek and J. M. Dallavalle II, *Eng. Chem.*, 46(363), 1954.
19. I. Hartmann, *Chem. Eng. Prog.*, 53(3):107-M, 1957.
20. F. Schneider and D. Schliephake, *Zucker*, 14(569), 1961.
21. W. H. Geck, Zucker, 4(31), 1951.
22. G. Andersson, B. Hoglund, O. Wiklumo, and H. Socker, *Handl. II*, 18(3):56, 1963.
23. D. F. Bagster, *Proc. Austral. Soc. Sugar Cane Technol.*, 1981, pp. 61–65.
24. D. F. Bagster, *Sugar J.*, 72:263–267, 298–302, 1970.

PART II

Refining Process and Operations

CHAPTER 5

Affination*

INTRODUCTION

The purpose of the affination station is to remove the molasses film surrounding the raw sugar crystal and to minimize any dissolving of the crystal. The raw sugar received today will polarize from 98 to 99.5° compared to 97 to 98° twenty years ago, which means that there are fewer nonsucrose solids to handle in the remelt recovery system. As a result, melt rates have been increased without increasing the remelt end in most refineries.

The molasses film removed from the raw sugar crystal along with the syrup produced from hot-water spraying of these crystals in the centrifugal machine is called *affination syrup.* The syrup should be 74 to 75 Brix; however, due to the preflush water used to clean the basket screen and milk of lime for pH control, the Brix value can be several degrees lower. The purity will also vary depending on the molasses film purity of the raw sugar and the amount of wash water used. It can be as low as 75 but will normally be in the range 80 to 85°. Affination syrup is used for mingling with the raw sugar, the excess being press filtered and charred for soft sugar production or sent to the remelt recovery process.

The amount of work that the refinery has to do to produce a refined sugar is to a certain extent determined by how well the raw sugar mill did its job. If the raw sugar crystals are the correct size, fairly uniform in distribution, free of conglomerates, and do not have impurities boiled in the crystal, the affination process will produce an affined sugar at a purity of 99.5 or more. The amount of centrifugal wash water used will be kept to a minimum, resulting in less sugar being dissolved from the crystal and a lower affination syrup purity. If the crystals are mixed or small, the task of removing the molasses film will be more difficult.

5.1 MINGLING

The raw sugar coming into the melt house is weighed over a melt scale, then dropped into a hopper that feeds the mingler. The mingler is a U-shaped trough containing shaft-mounted

* By Thomas N. Pearson.

paddles or a ribbon scroll. The paddles are mounted at angles to convey and mix the raw sugar with the affination syrup. For a refinery producing 4 million pounds per day, a mingler about 30 to 40 ft long, 3 ft wide, and 4 ft deep is required to mix the raw sugar and affination syrup properly. At the end of a mingler is a dam that the mixture, called *magma*, must flow over. This provides more retention time and a better mingled magma. The dam is made so that it can be lowered to empty the mingler for shutdown. When the mingler is emptied and washed for shutdown, it should be inspected for bent or improperly positioned paddles. The strings that came in with the raw sugar and wrapped around the mingler shaft are also removed at this time.

The control system for mixing the raw sugar and affination syrup varies with refineries. One method is to weigh the raw sugar and meter the affination syrup in. Another way is to add the affination syrup based on the load on the motor used to drive the mingler. The magma should be about 92 Brix after mingling. A well-mingled magma will split in half when it hits the loading cone of the centrifugal machine. If it wraps around the spindle, it is too loose, and if it just hits the loading cone, it is too tight. A properly mingled sugar is critical to good centrifugal performance. There is usually a coarse grating that the magma flows through to trap the foreign matter that was not removed from the raw sugar. This also traps hard lumps, which are removed and broken up.

The removal of the molasses film can be facilitated by reducing the viscosity of the film, which is done by raising its temperature. The affination syrup is usually heated for this purpose, but the magma can also be heated in the mixer using a coil [1]. Heating the affination syrup in a plate heat exchanger to about 145 to 155°F using superheated water instead of steam will minimize burning in the heat exchanger and reduce sucrose losses. The hot syrup should not be circulated to a large tank because it will invite microbiological growth and sucrose losses. The affination syrup should be free of crystals from accidental overcharging of the centrifugals and other foreign matter that would plug the heat exchanger. It takes about 2 gal 75 Brix affination syrup to 100 lb of raw sugar (99° polarization) to give a magma of about 93 Brix.

The magma from the mingler flows into a mixer. The mixer size varies with the size of the refinery, but a typical size for six centrifugals would be 3 ft 4 in. wide by 37.5 ft long and 8.5 ft high, with a rounded bottom to which the centrifugal chutes are attached. A steel pipe of 3 or 4 in. diameter is coiled about a center hollow shaft through which hot water is circulated to heat the magma. A baffle plate is located above the coil to form a heating compartment. The hot water is heated to about 170°F and circulated through the coil, then flows back through the center to the supply tank. The coil rotates at about 45 rpm. A temperature sensor in the mixer starts and stops the hot water pump to maintain the desired magma temperature. The temperature of the magma may be varied according to the raw sugar impurities and will range from 110 to 140°F.

A level sensor measures the mixer level that controls the flow of raw sugar into the mingler, thus keeping the mixer level fairly constant. A constant mixer level makes it easier to control the magma flow into the centrifugal and to maintain a fully loaded basket.

5.2 CENTRIFUGAL MACHINES

A centrifugal machine used in the sugar industry is a mechanical device for separating a solid from a liquid. It does this by rotating the basket of sugar and syrup at a high rate of speed, causing the syrup to be expelled from the mass and through the retaining screen. The residual syrup film is removed by hot-water washing. The centrifugal force is directly proportional to the mass of the rotating body and to the square of the peripheral speed, and inversely proportional to the radius of rotation [2].

The performance of the centrifugal machine is affected greatly by other factors, such as crystal size, molasses film viscosity, and depth of the sugar wall. A large uniform crystal will purge the molasses faster and more efficiently because of larger pathways between the crystals than will a mixed- or small-grain crystal. Mixed- and small-grain sugar will pack in the centrifugal basket, which reduces the openings between the crystals and slows or prevents the molasses from leaving the sugar.

In most of the cane sugar refineries in the United States, the 48 in. by 30 in. or 48 in. by 36 in. batch centrifugal has been the standard machine in use for the past 40 or more years. This centrifugal has proven to be a workhorse for the industry. Now a new generation of centrifugals are being manufactured that feature capacities of more than double that of older centrifugals, energy-efficient motors, and advanced electronic controls. Greater reliability and less maintenance are other advantages.

The continuous centrifugal, which has been used for years in remelt recovery operations, promises additional savings when concerns such as capacity, foreign matter in the magma feed, color control, and affination purity are all solved. The typical operating cycle of the batch centrifugal is from 150 to 180 s or more and will have the approximate function times listed in Table 5.1. Most affination stations are now controlled by computers, which have replaced mechanical timers and relays, making the stations more reliable and efficient.

The 48 in. by 36 in. basket uses a three-nozzle spray system to wash the sugar. The nozzles will deliver about 1.2 qt/s at 60 psig. Water at 180°F or more is used to wash the crystals. Some refineries use superheated water, which atomizes into small droplets for washing. Selection of a centrifugal wash water pump is important to getting good results. The pump curve should be almost flat from the point of rated flow to the shutoff point in order to get constant pressure and flow to the nozzles. A proper line size and strainer are also necessary.

The washed crystals are plowed from the centrifugal into a large scroll that conveys the sugar to the melter. Sometimes the scroll is used as a premelter by dosing sweetwater into it. This starts the wetting process and aids in dissolving the sugar in the melter.

The melter is a round tank about 12 ft in diameter and 6 ft high. There may be a plate perforated with $\frac{1}{2}$-in. holes located about halfway down in the tank. The grating retains foreign matter that comes in with the raw sugar, as well as lumps of sugar, giving them more time to dissolve. An agitator with paddles above and below the grating provides the necessary agitation to pull the sugar down into the liquor and keep it suspended until it dissolves. High-purity sweetwater is used to dissolve the affined sugar and is added automatically as needed by a Brix controller to maintain the desired Brix value, usually about 66. A large refinery will usually have two such melters. Exhaust steam should be used for heating and is usually through an open steam coil, which aids in agitation. Char sweetwater should not be used for melting because of its color and other impurities; however, some

TABLE 5.1 Times for Centrifugal Functions

Function	Time (s)
Feed gate closed to first wash on	15–22
First wash time	2–4
Time to second wash	8–10
Second wash time	2–10
Time to regenerative breaking	50–60
Mechanical break on centrifugal stops	10–12
Discharging	20–30
Centrifugal starts until the feed gate opens	3–5
Feed gate closes	8–10

refineries use it to wash the sugar in the centrifugals. Foreign matter such as cane fibers, small rocks, and strings make it necessary to fine screen the liquor before pumping it to the clarification process. A vibrating screen of about 80 mesh is used for this purpose.

5.3 OPERATING THE AFFINATION STATION

Almost all affination stations are now automated to the point where only one attendant is required to monitor the operations, and this attendant may have others duties as well. The automation does not lessen the need for someone well trained in the technical operations of this station. A major part of the purification process occurs here, and if the job is not done properly, the rest of the refinery will have a much more difficult task.

Although the mingling process appears to be very simple, it is a very important part of the process of preparing the sugar to be separated from the molasses in the centrifugal machine. If the magma is at the correct consistency, the mass will hit the loading cone, split in half, and fill the centrifugal basket uniformly. If it is too fluid, it will splatter, possibly going into the discharge screw, and fill the basket before the excess syrup can spin off. As a result, the basket will not have a full charge, causing lost production. With a fluid mass the centrifugal will usually have an overcharge, causing the magma to go over the basket cap and into the syrup compartment. This causes the pump strainer to clog as well as excessive wear on the pump impeller and excess affination syrup at a higher purity.

On occasion a raw sugar cargo will be received that contains numerous strings and fibers caused by dumping raw sugar from bags into the cargo hole of the ship. These strings will wrap around the mingler shaft, forming a ball that interferes with the mingling process. When this happens, the mingler will have to be emptied and the strings cut from the shaft. The mixer level should be kept fairly constant and at a sufficient height that the magma will flow into the centrifugal quickly but not cause the basket to sway. Loading should be completed in about 7 s. It goes without saying that the refinery melt rate cannot be anymore than the throughput of the centrifugals. It is up to the centrifugal operator to maintain the quantity and quality of the washed raw sugar produced at this station.

The color of the affined sugar should be compared to that obtain from laboratory washed raw sugar. The laboratory washed sugar will be better than that obtainable in the refinery. The ratio of the laboratory washed sugar to the affined sugar from the refinery should be 85 to 90%. If it is less than this, the centrifugals must be inspected to find the cause. Some of the causes for poor centrifugal performance are:

- Clogged spray nozzles
- Improper spray nozzle angle
- Bent spray pipe
- Preflush not cleaning the screen
- Plow leaving too much sugar on the screen
- Low water pressure
- Insufficient washing time
- Not enough drying time
- Wash water coming on before the syrup purges

To check for the proper washing, look at the sugar wall before the centrifugal plows out. If a straightedge is held against the wall of the sugar, it should make contact all the way up and down. If there are ripples in the sugar wall, the nozzles need to be repositioned. The rippling will become more pronounced if an excess of wash water is applied and is a better way to check for proper washing. If a fine- or mixed-grain raw sugar is being processed, special adjustments to the normal washing times must be made to produce a well-washed sugar. When the grain is fine (more than 52% through a U.S. 30 mesh screen) or mixed, the sugar packs together, reducing the pores for the affination syrup to flow through to the screen and into the syrup compartment. It will take several seconds longer (4 to 12 s) for the syrup to leave the sugar wall, and if the wash is applied too soon, the water and syrup will mix, forming a shiny surface, producing an effect called *mirroring*. The affined sugar in this case will not be properly washed and will have a higher color.

To make the maximum production, there are several checks to make on the centrifugal operation. The centrifugal should load up to the lip of the cap with a properly mingled sugar. A short charge not filling the basket will lose many pounds of melt over a day's time. If all of the centrifugal functions, especially the plowing, are tuned properly, the cycle time can be set at a minimum (150 s), and maximum throughput can be obtained. A stopwatch should be used by the centrifugal operator to check the major points in the cycle, especially the discharging operation. The discharge sequence is as follows:

1. Centrifugal stopped until the reverse motor starts
2. Plow starts in
3. Plow in
4. Plow starts down
5. Plow down
6. Plow starts up
7. Discharge motor stops
8. Plow up

The up-and-down movement of the plow is usually where the time is lost. The shaft of the discharger must be kept free of sugar and lubricated for smooth operations. Proper air pressure and clean air are essential for good centrifugal work. If the plow is shimmed properly, it will remove the sugar from the entire screen but not scrape the screen. A short preflush of water should finish the cleaning of the screen. The discharging sequence should not require more than 20 to 30 s.

The use of loading cone mounted on the spindle has eliminated the mechanical problems of the bottom discharge valve and its lifter, but it has created some new problems. Windage created by the open bottom basket interferes with the spray pattern. It blocks the view of the bottom spray nozzle and pipe. For this reason the centrifugals should be inspected visually from the bottom daily to see if the bottom spray is functioning properly. Magma that is too heavy can fall through the bottom opening during charging and end up in the melter. If the centrifugal trips out during loading or before it picks up speed, the majority of the charge will end up going to the melter, resulting in a very high liquor color. There should be a convenient place to view the washed sugar crystals as they are being conveyed to the melter. This makes it easy to detect improper washing of the sugar and to pinpoint which centrifugal it came from.

The quantity of the wash water used should be the least to do the job. The timer settings for the wash water should be adjusted each shift to ensure that the minimum water is being used. The affination syrup purity will sometimes be the controlling factor for the amount of wash-water use. If the remelt station is congested, either the water use must be adjusted or the melt rate reduced to accommodate the remelt station.

Obtaining maximum throughput at the centrifugal station involves paying attention to details. Some useful information showing how small changes affect the capacity of the batch centrifugal are:

1. Two seconds of cycle time changes production by 1%.
2. A $\frac{1}{4}$-in. smaller sugar wall changes production by 3%.
3. One second of wash-water changes production by 1%.
4. Six pounds of residual sugar changes production by 1%.

Maintenance of the centrifugals is an important part of how well they perform their job. Routine checks and repairs of the feed gate operation, wash-water valves, plow tips and springs, brakes, and plows will keep the centrifugal producing a high-quality product.

REFERENCES

1. R. A. McGinnis, *Beet-Sugar Technology,* Beet Sugar Development Foundation, Ft. Collins, CO, 1982, pp. 406–467.
2. P. Honig, *Principles of Sugar Technology,* Vol. III, Elsevier, New York, NY, 1963, pp. 223–230.

CHAPTER 6

Phosphatation for Turbidity and Color Removal*

INTRODUCTION

The clarification of raw juice/liquor using heat and lime is a long-established practice. The process was studied intensively in the late nineteenth and early twentieth centuries, to try to understand its theoretical basis and to modify the procedure for optimization. Evidence accumulated that the acidity and phosphate content of the juice were among the most important factors in efficient clarification. Consequently, the addition of phosphate to refractory juices became common practice throughout the sugar world.

Some insoluble and suspended matter—such as bagacillo, soil, starch, and gums—was known to escape clarification in the raw house and persist in washed raw liquor. Additional treatment by defecation or pressure filtration was found to be necessary subsequent to affination. Today we know that the vague designation "gums" refers to polysaccharides of highly specific and regular molecular structure that, unlike starch, are soluble in cold juice. We also know that some of these substances are of bacterial origin and can be found as a result of microbiological infection in the field, mill, or refinery. The most important of these bacterial slimes by far is of course dextran; the other polysaccharides are indigenous to sugarcane.

In the refinery it had become common practice in the 1880s to defecate washed raw liquor using lime and phosphoric acid. However, the calcium phosphate floc produced using this technique could be filtered only with considerable difficulty. The use of cotton bag filters was labor intensive and unsanitary, and after 1915, when leaf-type pressure filters using inert filter aids were introduced, the Sweetland, Kelly, and Vallez systems largely superseded defecation. The first Sweetland use in the western hemisphere was at Cardenas, Cuba, in 1916.

There had been English patents for phosphate–lime clarification of raw liquor as early as 1850, but the traditional use of ox blood and lime continued for about another 30 years. Upon heating, the blood albumin coagulated, entrapping fine suspended matter that could

* By Richard Riffer.

then be removed by bag filtration. It has been postulated by Bennett [1] that the suspended particles (isoelectric point about pH 3.5) became coated with an adsorbed layer of the protein (isoelectric point of pH 5.0 to 5.5), bringing the particles closer to neutrality at operating pH values and reducing repulsive forces that inhibit flocculation. When large-scale refining developing around 1880, the use of blood was almost entirely eclipsed by phosphate–lime treatment. However, chemical defecation was usually avoided for affination syrups, which tended to clarify poorly and form a large volume of precipitate.

In 1918, George Williamson at the Gramercy refinery in Louisiana patented a continuous clarification system based on phosphate and lime. He had found that by shaking a mixture of sugar liquor and phosphate in a test tube, thus trapping air bubbles in the liquor, he could float the scum to the top of the tube. His invention consisted of impregnating the defecated raw liquor with air and then heating to about 99°C in flat vessels so that the air bubbles would rise and carry with them the calcium phosphate precipitate and occluded material. This scum, containing entrapped impurities, could then be drawn off by a suitable means, originally a slowly rotating wooden roller. Scums adhered to the roller and were removed by a scraper, while clear liquor was withdrawn below. The original Williamson system was 2 ft deep, 6 ft wide, and 12 ft long, heated by transverse steam pipes. It was able to treat 1000 gal or more per hour. Essentially all present-day phosphatation systems, also called *frothing clarifiers* or *dissolved air flotation* (DAF) *units*, are based on the Williamson patents. DAF units are also commonly encountered as elements of wastewater treatment systems.

The Williamson clarifier provided excellent color removal and clarity, greatly reducing the load on char systems. Chemical costs were much lower than for filter aids. On the other hand, the system required more floor space than pressure filters, and careful control of pH, temperature, and density were essential—what we today would call steady-state conditions. There was also concern that the high operating temperature would increase sugar losses through thermal degradation and inversion. In addition, the frothing clarifier appeared to be more sensitive to changes in raw sugar quality than was pressure filtration. This was further evidence that the process required uniform operating conditions to perform well. Some early detractors of the Williamson system claimed that the color removal achieved was ephemeral and hence of no consequence. This was, of course, later demonstrated to be entirely false.

For about 12 years following the Williamson invention, the trend away from defecation and toward pressure filtration continued. The new leaf-type pressure filters seemed to be the wave of the future. However, the superior performance of continuous clarification gradually became known, especially its removal of colloidal material and substantial color, which could not be accomplished by mechanical filtration. Clarification produced liquors of obvious brilliance that could not be obtained by filtration. Although we have no information to support this, it is possible that the introduction of frothing clarification on a broad scale was delayed until Williamson's patent protection ran out, to avoid payment of license fees. We do know that during the period from 1915 to 1930, the average polarization of raw sugars improved from about 95.3 to 96.6. Thus the color and colloid removal that had been lost when phosphate defecation was being discontinued might have gone unnoticed.

Although the trend in polarization of raw sugars had, of course, been upward throughout the twentieth century, there was a retrograde interlude beginning about 1930. Raw sugar quality became markedly poorer as variety canes replaced the older native types. The Williamson system now attracted new attention as refiners were faced with a more difficult challenge.

It was formerly the practice to generate phosphoric acid for defecation by treating discard bone char with hydrochloric acid, generating a black paste containing about 11%

P_2O_5. However, this acid preparation was high in chloride and increased the ash level in the product. To a lesser degree, sulfuric acid was also used on discard bone char to generate phosphate, but the high-sulfate liquors performed poorly on bone char because (as we know today) this species competes with anionic color bodies for cationic (i.e., calcium) sites on the char surface. Improvements in manufacturing methods for phosphoric acid gradually led to the commercial availability of the very high quality acid with which we are familiar today.

The level of phosphoric acid used in defecation varies from 0.012 to 0.05% on melt solids, as P_2O_5. Higher levels are reserved for more refractory raws, that is, those containing more suspended matter—generally, lower polarization. However, poor-quality raw sugars may perform poorly even at high phosphate levels and result in a high level of mud that interferes with continuous operations. When performance is poor, it is often helpful to increase retention time. The ratio of phosphate to suspended impurities, *suspensoids,* also affects the *rate* of scum formation, a topic we return to below. Reaction rate considerations would clearly affect retention time and hence throughput.

Although the relationship is too complex to permit generalization, it can be said that for a given raw quality, higher phosphate addition would tend to provide better clarity and decolorization. Furthermore, a small percentage improvement in decolorization at the clarifier can translate into a much larger percentage decrease in bone char utilization. Thus the optimal level of phosphate addition properly requires a cost-effectiveness computation, not simply a raw sugar quality evaluation. This would necessitate not merely an appraisal of the quality of the clarified liquor but also of the subsequent bone char effluent.

In phosphatation, no one actually adds phosphorus pentoxide, which reacts explosively with aqueous systems. But basing the level of addition on this acid anhydride of phosphoric acid eliminates the ambiguity of concentration, since it is commonly added in aqueous solution of various strengths. In essence, the P_2O_5 value serves to normalize the phosphoric acid level on solids. To convert from kg P_2O_5 to kg A% phosphoric acid, multiply by the factor 138/A. Thus 100 kg of phosphorus pentoxide is equivalent to 162 kg of 85% phosphoric acid.

As the Williamson system came into broad application, several improvements were developed. One worthy of comment was the prescreening of raw liquor before defecation. Coarse screening had long been standard practice to remove macrodebris, but it was now found that fine sieving on vibrating screens reduced P_2O_5 requirements and at the same time improved clarification. The basis for this is no doubt that more phosphate remains available for removal of colloidal material, because it is not being expended on microscopic, but relatively larger particles, such as bagacillo, which are removable without defecation.

An improvement of highest significance was the introduction of flocculating chemicals to improve phosphatation performance. This topic is addressed in a separate section. It was also found that Williamson-type systems could be operated at lower temperatures, close to 82°C. This was of considerable importance, of course, since it reduced energy costs and lessened the very considerable concerns about sugar losses associated with treatment. In addition, the scum roller gave way to a metal strip which on each rotation pushed scums over a weir.

Under optimal performance, phosphatation is equivalent to a 0.1-μm filtration, which can never be achieved by ordinary Sweetland or similar leaf-type systems. About one-third of the suspended solids in raw sugar consists of particles smaller than 0.5 μm, which is close to the lower limit of removal for pressure filtration. In other words, pressure filtration removes only about two-thirds of what can be eliminated using phosphatation. It is for this reason that clarified liquors display a brilliance that can never be attained using pressure filtration.

Phosphatation systems are also used to clarify lower-purity refinery syrups, such as components of boiled soft sugars. Such streams are typically very difficult to filter, requiring reduced feed densities and short press cycles. Replacing conventional kieselguhr operations by clarification affords significant savings in sweetwater evaporation and, of course, reduced filter air consumption. A description of the C & H refinery system was published by Dorn and Harris [2].

6.1 CALCIUM PHOSPHATE CRYSTAL MORPHOLOGY

The reaction of lime (i.e., calcium oxide) with phosphoric acid in the frothing clarifier produces what we may call simply calcium phosphate. It is the case, however, that there are a number of *different* calcium phosphates, each with characteristic solubilities, crystal structures, and other properties that would distinguish a set of distinct but related species. It was recognized early in phosphatation studies that to understand the process on a molecular level, with an eye toward maximizing cost-effectiveness, one had to know the details of what was going on when lime and acid were mixed in the clarifier. For this reason, considerable attention was directed toward identifying the exact nature of the calcium phosphate formed. At the same time, it was recognized that simple aqueous systems were inappropriate models for study since they did not reproduce the conditions in sugar liquors, in which the sucrose presence has a marked effect not only on the water activity but on the availability of calcium.

Clarke and Carpenter at the Cane Sugar Refining Research Project (CSRRP) in New Orleans [3] studied calcium phosphate precipitation in sugar solutions under varying conditions approximating those of refinery clarifier systems, that is, near 60 Brix and with a 15- to 20-min holding time for precipitation. A plot of critical ion activities (i.e., calcium activities in the mother liquor against phosphate concentration) provided an inverse slope giving the Ca/P ratio at the initiation of precipitation. (*Activity* in this context refers to effective concentration.) However, the ratio found, 0.55, did not correspond to any of the precipitates usually ascribed to the calcium phosphate precipitation as shown in Table 6.1. It was acknowledged that this could have been a consequence of the fact that the study had been conducted at room temperature, since the calcium-specific ion electrode used did not permit high-temperature measurements. Nevertheless, this study represented an important milestone in efforts to characterize the calcium phosphate precipitate.

TABLE 6.1 Calcium Phosphates

Compound	Ca/P Ratio[a]
Tetracalcium phosphate, Ca_2PO_4OH	2.0
Hydroxyapatite, $Ca_{10}(OH)_2(PO_4)_6$	1.67
Tricalcium phosphate, $Ca_3(PO_4)_2$	1.5
Octacalcium phosphate, $Ca_8H_2(PO_4)_6$	1.33
Amorphous calcium phosphate	1.3–1.7
Dicalcium phosphate, $CaHPO_4$	1.0
Monocalcium phosphate, $Ca(H_2PO_4)_2$	0.5

[a]The ratio is indicative of the degree of basicity, with the highest number representing the most alkaline form. On this basis, phosphoric acid itself would have a ratio of zero, while that of lime would be undefined ("division by zero"). A high ratio is also indicative of low solubility. Degree of hydration in the forms shown has been omitted for simplicity.

Researchers at the CSRRP also examined clarifier scum components using several analytical techniques, including x-ray diffraction, refractive index measurement, infrared spectroscopy, and microscopy. By studying calcium phosphates from a functioning real installation, any question about reproducing clarifier conditions in a laboratory setting could, of course, be sidestepped. They concluded that the predominant basic calcium phosphate present was octacalcium phosphate.

Knowing the principal crystal system present in clarifier mud is of considerable practical significance, not a mere detail or footnote of academic interest. The size, structure, rate of formation, and stability of the calcium phosphate precipitate are of paramount importance because these features are on a fundamental level responsible for performance in a phosphatation clarifier. The crystals were found to be lamellar in structure, that is, composed of very thin overlapping platelets on the order of 1 μm that could provide a large surface for entrapment of impurities. By contrast, hydroxyapatite crystallizes in needles, tiny hexagonal rods up to 100 μm in length. This substance functions in bone and teeth to fill spaces surrounding collagen fibers, a purpose not necessarily requiring a high surface area.

As noted above, the more acidic forms of phosphate are more soluble, which is why the principal inorganic phosphates in blood are HPO_4^{2-} and $H_2PO_4^{-}$, present in a ratio of approximately 4:1 to help keep plasma buffered near pH 7.4. More pertinent to our purposes here, however, is the relationship between the pH of clarifier operation and performance. Bennett [1] determined the amount of carryover, measured as effluent attenuation index at a near-infrared wavelength (900 nm), as a function of operating pH (Fig. 6.1). He found that carryover increases gradually with decreasing pH over the pH range 9 to about 7. However, below about pH 6.5, the amount of carryover increases sharply, with a point of inflection near pH 5.8. Near pH 5, the carryover increase levels off once again, because clarification has become so ineffectual at this point that little damage is incurred by increasing the acidity further still.

The curve of carryover versus pH bears a strong resemblance to that which would be obtained if we were to titrate HPO_4^{2-} with strong alkali, except that we are not accustomed to seeing pH on the abscissa on titration plots. The inflection point in this case would be slightly below pH 5, but we should not be surprised at an offset when going from a dilute

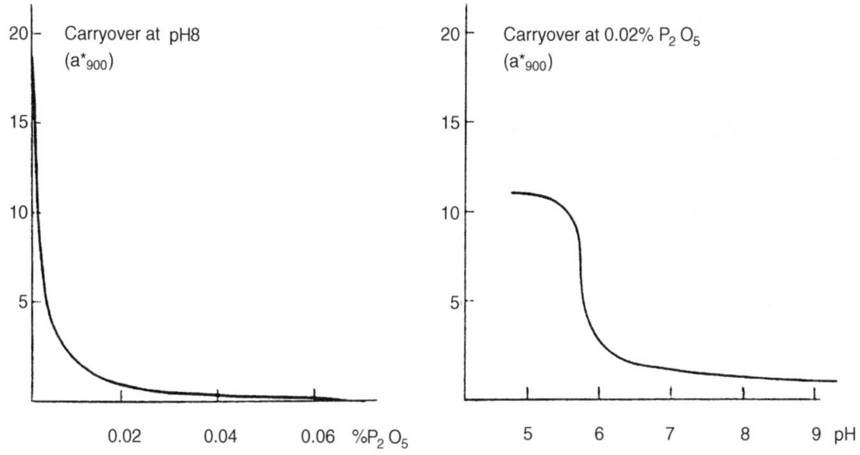

FIGURE 6.1 Carryover in phosphatation flocculation as a function of % P_2O_5 and pH. (From Ref. 4.)

aqueous solution at ambient temperature to a hot environment of relatively low water activity in a medium containing lime and sucrose. The two systems are sufficiently dissimilar that predicting the direction or magnitude of the offset would be a daunting task.

Although refractive index, x-ray diffraction, and microscopic examination have been used to study clarifier scum, the identification of octacalcium phosphate as the principal phase was based primarily on infrared spectra. The scum spectrum, however, shows very little fine structure and could very well represent an intracrystalline mixture. There is some evidence that several forms can be present within the same crystal, with continuous transitions between crystal types. The pH behavior described above suggests that to the extent that octacalcium phosphate can be considered the equivalent of a cocrystallized 50:50 mixture of di- and tricalcium phosphates,

$$2CaHPO_4 + 2Ca_3(PO_4)_2 = Ca_8H_2(PO_4)_6$$

it may be the dicalcium phosphate component that is active in clarification. This should not be surprising, because in a pure aqueous system, PO_4^{3-} does not exist at appreciable levels below pH about 8.5.

To carry this analysis one step further, one might conclude that any tricalcium phosphate presence is a consequence of slow kinetics, since this substance would not be thermodynamically stable under ordinary refining conditions, that is, in the pH region of about 6.8 to 8.0. In fact, Bhangoo and Carpenter [5] have pointed out that the basic calcium phosphates are notoriously slow at approaching equilibrium, requiring at least days and even month or years. Thus the equilibrium phases of the precipitates are normally not present and hence of no immediate significance to sugar refining problems.

The range of retention times in operational clarifiers is on the order of minutes to tens of minutes, probably not sufficiently broad to observe such thermodynamic effects (i.e., tricalcium phosphate being supplanted by dicalcium phosphate), but one might expect performance outcome to vary with extended holding times. Such effects no doubt contribute to the commonly encountered problem of *afterfloc,* further calcium phosphate precipitation in clarified liquor. Such material is typically carried over to the absorbent, such as bone char, in a subsequent unit operation, forming a blanket of mud at the top of the filter that impedes liquor flow.

Bennett [1] also showed (Fig. 6.1) that at a fixed operating pH within the favorable zone, in this case pH 8, the carryover, again measured as effluent attenuation index in the near infrared, decreases approximately linearly from 0.02% P_2O_5 to 0.06% and beyond. This portion of the curve shows, not surprisingly, that although performance improves with increased dosage, beyond a certain point one is overdosing to little effect. The clarifier feed quality in such a test must necessarily be uniform, but in practical terms, to achieve a uniform level of performance, one must accommodate raw liquors containing higher levels of suspended matter by increasing the P_2O_5 dosage. As noted above, depending on raw sugar quality, refiners typically add phosphate at levels in the range 0.012 to 0.05% on melt solids. At uniform feed quality, the dosage is, of course, proportional to flow. Not surprisingly, poor-quality raws result in increased costs for process chemicals.

For installations in which phosphatation is followed by pressure filtration using kieselguhr, even lower addition levels can be used, below 0.005%. Calcium phosphate scums are not easily filtered, so in such applications a reduced level of phosphate addition may be unavoidable. Or, stated somewhat differently, phosphatation in these cases might be considered an adjunct to pressure filtration, comparable at least conceptually to the use of powdered activated carbon for added decolorization capacity. This is clearly different from using filtration as an alternative to clarification, or from filtration as an alternative to flotation in a phosphatation at higher phosphate addition levels. Phosphatation followed by pressure filtration is sufficiently different quantitatively from the flotation procedure that it might be considered *qualitatively* different as well.

Phosphatation used with pressure filtration provides much improved clarity over filtration alone, which, as noted above, cannot remove colloidal matter. Depending on the phosphate level used, color removal as great as 30% can be achieved, compared to decolorization levels as high as 40% commonly obtained with flotation systems. These figures might appear to make the filtration option more attractive than is warranted. In fact, such an installation requires a much large filter press station than that used for pressure filtration alone and requires higher-porosity kieselguhr. However, pressure filtration with phosphate should not be confused with polish pressure filtration following standard clarification, practiced by some refineries.

6.2 FLOCCULATION REACTION

Based on the premise that octacalcium phosphate is the principal species formed in frothing clarification, the flocculation reaction may be written as

$$8Ca(OH)_2 + 6H_3PO_4 = Ca_8H_2(PO_4)_6 + 16H_2O$$

Note that lime and phosphoric acid are *not* added on a 1:1 basis, whatever the addition level. Water is generated in the process, as it would be in any neutralization reaction of the classical Arrhenius acid–base type. Clarification must therefore be accompanied by a small drop in Brix value in addition to that resulting from the dilute additives themselves. How significant is this? It can be calculated that phosphatation using 75% phosphoric acid and lime sucrate at typical usage levels would reduce the Brix value by several tenths. In practice, however, this is offset by evaporative losses of water resulting from holding hot liquor in open clarification vessels with large surface exposure. Refining costs are highly sensitive to operating densities because of fuel requirements for evaporation.

It has been shown that phosphoric acid and lime may be added in either order, although it can be argued that even transient low pH exposure is a less than perfect choice for high-temperature sugar liquors. In a typical installation, food-grade phosphoric acid is added first, and the pH is autoadjusted to about 7.3 using lime sucrate. Initial vigorous mixing to disperse the additives is followed by gentle agitation, to allow the calcium phosphate floc to grow. Any mixing in the clarifier body itself should be minimal (Fig. 6.2).

Air enters the system via injectors or pumps, and bubbles attach to the floc. If the bubbles are too large, flotation may be adversely affected, and unwanted mixing can occur. The temperature in the clarifier body is raised somewhat to reduce viscosity, providing faster and more efficient flotation. The mud then rises to the top of the clarifier and can be skimmed off. As noted above, uniform operations are essential to good performance, which can easily be disrupted by non-steady-state conditions.

As indicated above, much suspended particulate matter in the raw liquor is entrapped in the growing coagulum and removed. Some impurities are weak acids and precipitate as their insoluble calcium salts. Since most colorants in raw liquor are weakly acidic, the question might be asked why decolorization across the clarifier is partial rather than total.

To address that issue, we can begin by examining the acidities of these substances. The p*K* values of caramels, melanoidins, and alkaline degradation products have been estimated to be 4.7, 6.9, and 5.1, respectively [7], while phenolic pigments would be expected to have a value near that of catechol, 9.4. Using these figures and the Henderson–Hasselbalch equa-

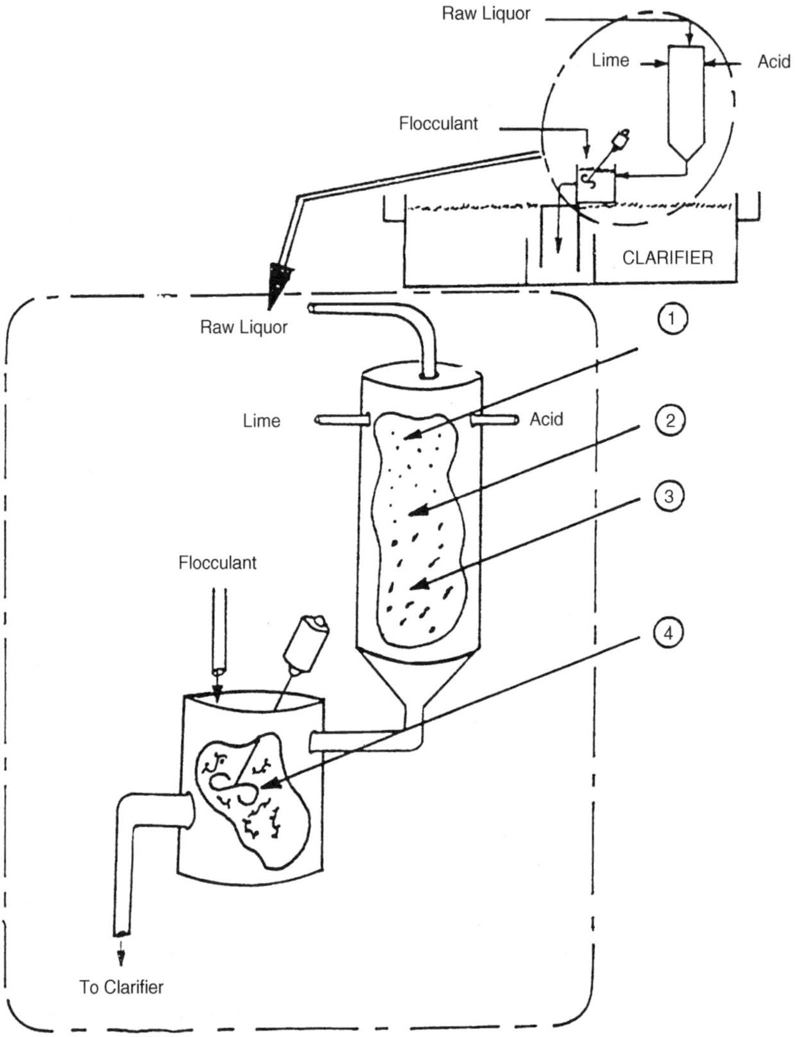

FIGURE 6.2 Decolorization sequence in raw liquor clarification. (1) Lime and phosphoric acid are added to the raw liquor stream. (2) The two react to form insoluble calcium phosphate. (3) Calcium phosphate and suspended particles form a coagulum. (4) System pressure is released to effect flotation. Added polyacrylamide flocculant attaches to the coagulum, improving its stability and the performance of the process. (From Ref. 6.)

tion, one can calculate to what degree these classes of compounds are ionized at clarification pH values of say 7.2 (Table 6.2). The values in the table suggest that virtually all carboxylic acid functions in the caramel and alkaline degradation product components of raw liquor colorant are present in their salt forms in the clarifier environment. In addition, most of the melanoidin fraction is in this category. Phenolics, on the other hand, would require a much higher pH value to be converted to their respective salts.

However, there is an obvious second criterion for these substances to be precipitated in the clarifier mud: Their calcium salts must be insoluble under the conditions of clarifi-

TABLE 6.2 Degrees of Ionization at Clarification pH

Class of Compound	% Ionized
Caramels	99.7
Melanoidins	66.7
Alkaline degradation products	99.2
Flavonoids and cinnamic acid derivatives	0.6

cation pH and temperature. Here we run into a roadblock, because of the complexity of the colorant material. Many components are polymeric and have been characterized only as classes, not as distinct pure compounds of determinate composition. The colorant fraction is known to be highly heterogeneous; because of differences encountered in raw juice composition, from genetic and environmental factors, and almost limitless possibilities for variations in color-forming reactions during processing, it is likely that no two colorant fractions are exactly alike.

We know, of course, that many calcium salts of organic acids are at least sparingly soluble, such as acetate and gluconate. Colorant species would necessarily be of higher molecular weight than these substances mentioned, but this would not necessarily result in lower solubility, because of the presence of multiple carboxyl functions and other polar groups in these substances. Note also that virtually all of the colorants are present at extremely low concentrations, so that even a low solubility would suffice to keep their salts in solution, despite the low water activity. Kinetics could also play a role here, if the solubility product of the salt is exceeded but the retention time does not suffice for precipitation.

It should not be inferred from this discussion, however, that little is known about the nature of materials separated during clarification. Roberts et al. [8] analyzed the turbidity-causing fractions in raw juice and clarified juice (Table 6.3). The latter material should approximate the particulate fraction in raw sugar, since we know that although there is clearly some partition into the molasses, crystallization is not particularly effective at removing such substances. If it were, clarification would be unknown.

It is noteworthy that the carbohydrate contribution to turbidity is only about one-fifth; thus this study suggests that it is incorrect to think of the suspended matter in raw sugar as being composed primarily of polysaccharide material. Furthermore, plant lipids could be understated in Table 6.2, because the turbidity preparation to be analyzed was extracted five times with chloroform, which removed about 20% by weight of the turbidity-causing material. The inorganic portion is also substantial, with finely divided silica, iron, and aluminum, perhaps in part from soil contamination. The phosphorus component is very likely inorganic orthophosphate, although some DNA could be present; the sulfur presence is probably sulfate.

Bennett [1] has suggested that during liming, calcium ions bind to the adsorbed proteinaceous and/or polysaccharide layer on the surface of particulates, providing an essential link in the bridging mechanism that is the basis for flocculation.

Published data describing removal of specific categories of nonsugars in quantitative terms are not necessarily reliable predictors of the performance one may expect in a specific phosphatation installation. James et al. [1] monitored two such systems and found broad ranges in the reduction of phenolics, aminonitrogen, and polysaccharides, as shown in Table 6.4.

To some degree these wide ranges no doubt reflect variability in raw liquor quality rather than characteristics of individual clarifier systems. The relationship between sugar quality and removal of impurities during phosphatation was studied by Murray et al. at the Sugar Milling Institute in South Africa [10]. With poor-quality, high-starch raw sugars,

TABLE 6.3 Turbidity-Causing Materials in Raw and Clarified Juice

Material	% in Raw Juice	% in Clear Juice
Protein	1.01	0.35
Total nitrogen	4.05	2.38
Phosphorus	0.71	0.57
Calcium	0.51	1.78
Magnesium	0.21	0.20
Aluminum	0.67	1.20
Iron	2.7	1.04
Silica	3.59	8.12
Sulfur	0.03	0.02
Sulfated ash	15.38	29.22
Lipids	20.0	12.6
Polysaccharides	17.96	21.7
Arabinose	2.07	2.6
Rhamnose	0.16	0.2
Xylose	0.99	1.0
Aconitic acid	0.09	0.1
Galactose	0.99	1.2
Glucose	10.0	12.2
Uronic acid	0.58	0.7
Unknown	3.08	3.7

Source: Data from Ref. 8.

turbidity removal was about 95%, compared to essentially quantitative (99 to 100%) removal for other raw sugars. Removal of color (about 25%), gums (15%), and starch (30%) were fairly insensitive to raw sugar quality. On the other hand, clarified liquor from poor-quality raw sugars was much more difficult to filter, which indicated that the particles in clear liquor responsible for filter blockage are not necessarily important contributors to clarified liquor turbidity.

Murray et al. [10] also studied removal of the components of starch, amylose, and amylopectin. Sugarcane amylose is an uncharged linear molecule of molecular weight 10^5 to 10^6, whereas amylopectin is a highly branched, much larger molecule, molecular weight 10^7 to 10^8, containing charged phosphate ester groups. Amylopectin is far more detrimental to filterability than is amylose. Murray and his co-workers suggested that amylopectin could wrap itself around calcium phosphate particles and bind chemically to the crystallite surface, hindering coagulation and requiring increased phosphate levels. Amylose, on the other hand,

TABLE 6.4 Percent Removal of Nonsugars in Two Phosphatation Systems

Class of Nonsugar	Refinery A	Refinery B
Phenolics	9.32	71.0
Aminonitrogen	88.2	24.0
Polysaccharides	78.0	—

Source: Data from Ref. 9.

would bind only weakly and thus would have less of a detrimental effect. This suggests that in assessing the quality of a raw sugar, one should measure amylopectin rather than "starch." This is in fact what is done at the C & H refinery in the United States.

6.3 ZETA-POTENTIAL CONSIDERATIONS

Bennett [1] studied significant factors affecting stability and flocculation in cane juice at the Imperial College of Tropical Agriculture in Trinidad. His findings are also of considerable relevance to raw liquor phosphatation.

Stable suspensions in a liquid/solid interface system are characterized by high zeta potentials, which inhibit aggregation that ultimately results in settling. Although settling is slow in the case of intermediate potentials, the particles are free to move into positions of close packing in the sediment. If flocculation in a stable system is initiated by bringing about a decrease in the zeta potential, both the settling rate and the sediment volume increase as the potential approaches zero.

When settling is rapid, as in a flocculated system, packing is random and the sedimentation volume is large. In a clarifier, particulates are made to floc by the precipitation in situ of calcium phosphate. In this case, destabilization apparently occurs without an appreciable change in zeta potential.

Bennett studied the effect on sedimentation behavior of the addition of two detergents, cetyl trimethylammonium bromide and sodium lauryl sulfate. The hydrophobic organic portions of these substances are cationic and anionic, respectively, and thus can be used to adjust the zeta potential in either direction. Aqueous systems of naturally occurring materials, such as cells of microorganisms or cane juice components, typically exhibit a negative surface charge. As cationic detergent is added, the charge decreases, passes through neutrality, and then becomes increasingly positively charged. It should be evident that it is possible to overdose with detergent, reversing the charge and once again imparting resistance to settling.

Bennett found that at the point of reversal of charge the system displays a maximum settling rate accompanied by a minimum sedimentation volume. This has obvious implications for throughput rate and floc stability.

According to Bennett [1], the suspended particles all carry a layer of adsorbed strongly hydrated proteinaceous and/or polysaccharide material which cushions the particles from close approach, inhibiting flocculation. The degree of hydration can be reduced by the addition of cationic surfactants or simply by heating the liquor to boiling.

Freeland et al. [6] investigated the role of zeta potential in raw liquor clarification. They first prepared sugar-free colorant fractions using dialysis and adsorption on polystyrene beads and found that their isoelectric points—at which the net charge is zero—were close to pH 3.0 in all cases. Thus at refining pH values near neutrality, they would be anionic. The zeta potentials were demonstrated to be pH dependent, as increasing proton levels reversibly neutralized a portion of the anionic surface charge. In other words, the absolute value of the zeta potential diminished as the isoelectric point was approached, as one would expect.

It was further shown that colorant viscosity increased with pH at constant ionic strength. Freeland et al. [6] suggested that this was a consequence of unfolding of the molecules together with increased solvation and the drag effect of counterions. These findings have important implications not only for phosphatation but also for other charge-dependent decolorization mechanisms, such as the use of color precipitants or adsorption on bone char and ion-exchange resin. (The portion of resin decolorization that proceeds by matrix adsorption rather than by electrostatic attraction at the ionic sites is relatively insensitive to charge density, except insofar as this affects diffusion rates.)

It was found that raw sugars containing relatively low zeta-potential color at neutrality decolorized better during clarification than did those of higher potential. Thus a secondary destabilization mechanism sensitive to surface charge appeared to enhance removal of particulates. Specifically, a high potential seemed to inhibit the coagulation step. However, more refractory raw sugars not only contained higher zeta-potential color but were also more likely to contain higher levels of those polysaccharides that inhibit effective flocculation, such as amylopectin. Cell-wall arabinogalactan, which exhibits a strong negative charge down to pH 3 as a consequence of its uronic acid residues, would also be expected to have a pH-sensitive zeta potential affecting its removal.

Freeland et al. [6] also studied modification of the zeta potential using added detergents. The goal was not to bring the potential to zero but rather, to maintain it at slightly negative values, since this status was a requisite for the flocculant used. Conversely, one might select a flocculant with properties appropriate to the zeta-potential status of the particulates to be removed. In fact, the optimal flocculant for a given phosphatation system is likely to depend on particular raw liquor characteristics (i.e., the properties of nonsugar impurities).

6.4 USE OF FLOCCULANTS AND COLOR PRECIPITANTS

In the 1960s, Dow introduced synthetic water-soluble polyacrylamide flocculants as sugar processing aids. These were initially used in raw juice clarification, where they provided important benefits: reduced inversion (and subsequent invert destruction), reduced new color development, and reduced losses to molasses. Their later application to refinery phosphatation systems allowed processing temperatures to be reduced by at least 5°C and processing densities to be raised to 65 Brix or even higher. At the same time, flocculants improved clarity and dramatically shortened the time required to separate the liquor from the mud, from about 45 min to 20 min, with accompanying benefits of reduced losses and reduced caramelization (i.e., development of new thermal color). More rapid separation would obviously also allow greater throughput through a single clarifier unit.

Polyacrylamide flocculants are high-molecular-weight copolymers of sodium acrylate and acrylamide (Fig. 6.3). They are very long liner molecular chains containing a high concentration of mobile anionic charges. Fully extended, they can be as much as 0.05 mm in length. The primary calcium phosphate floc has positive surface charge sites, which form bridges to negatively charged colorants and other nonsugars. These clusters can be cross-linked into a secondary floc, a three-dimensional matrix held together by electrostatic at-

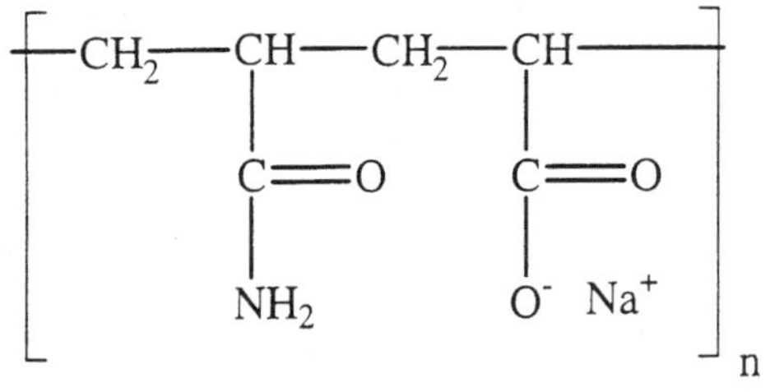

FIGURE 6.3 Polyacrylamide.

traction between negative polymer sites and positive calcium sites. Such cross-linkage would be expected to greatly improve the stability of the floc, such as to shear forces that are inevitable in scum removal and treatment. It might be less self-evident why this would at the same time improve the kinetics (i.e., increase the *rate* of flocculation); this has to do with the *size* of the floc elements. According to Bennett [1], the secondary flocs can be as much as 1000 times larger than the primary.

In addition to the copolymerization process described above, polyacrylamide flocculants may also be manufactured by a less costly technique of partial hydrolysis, which, however, does not allow the reaction to be controlled as well. *Polyacrylamide* is not an entirely satisfactory descriptive term because it implies that all acrylic acid functions in the chain are present as their amides, which for sugar processing aids is not at all the case. On the other hand, calling the substances *partially hydrolyzed polyacrylamides* better describes their structure but makes unwarranted assumptions about their manufacture.

Increasing the molecular weight of the chain during manufacture increases the settling rate and provides improved clarity, but increases the risk of chain cleavage by mechanical forces, such as in a high-speed mixer. Higher-molecular-weight material is also more difficult to manufacture, because of the increased tendency to cross-link during production, which renders the affected sites inaccessible to the primary floc. Improved manufacturing technique in recent years has made possible the production of material as high as 26 million molecular weight. However higher-molecular-weight polymer is more difficult to handle: is less soluble, is more difficult to prepare and dose, and its solutions are more viscous and more difficult to mix into the sugar liquor. Waste is apt to be higher.

On the other hand, a high concentration of negative charge (i.e., a high sodium acrylate content) results in a high settling efficiency. Elvin [12] has reported that for low turbidity and fast settling rate, the optimal polyacrylamide properties are about 23 million molecular weight and 30% charge density, but this applies to juice clarification, not raw liquor. As the purity of the liquor to be treated increases, a lower charge density is required. The choice of polymer for best performance at a given phosphatation installation is likely to depend on characteristics of the liquor treated. Within a given refinery, the best polymer choice for raw liquor treatment (purity 99%) is apt to be different from that for syrup clarification (purity perhaps 80%).

6.5 COLOR PRECIPITANTS

Color precipitants may be used as adjuncts to phosphatation, providing markedly enhanced decolorization with essentially no capital investment, but with increased operating costs for the relatively expensive chemical itself. These substances are high-molecular-weight materials bearing a positively charged strong base site, which can form an adduct with colorants, most of which are negatively charged at processing pH values. The polymer itself is borderline soluble, so that forming a complex with a colorant molecule, also of high molecular weight, produces a hydrophobic insoluble precipitate that can be removed with the clarifier scum. These additives are also used in production of Blanco Directo sugar and in conjunction with carbonatation processes.

The first such chemical to become available was dimethyl ditallow ammonium chloride, Talofloc, a waxy material developed by Bennett at Tate & Lyle Ltd. (Fig. 6.4). *Tallow* is not a term of organic chemical nomenclature such as *ethyl* or *propyl*, but instead, refers to the fact that this material is manufactured from a by-product of the meatpacking industry. Talofloc is a semisolid at ambient temperatures and a solid under cold conditions. Hence

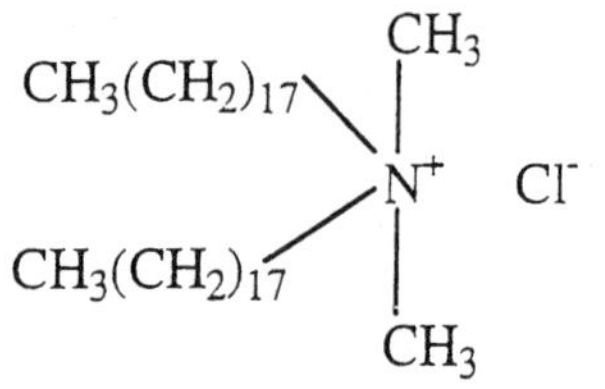

FIGURE 6.4 Talofloc.

it must be heated to be liquefied before it can be pumped, and spills are difficult to clean up. It is also corrosive to some stainless steels. The U.S. Food and Drug Administration (FDA) permits this substance to be used at levels up to 700 ppm on solids.

An alternative color precipitant developed by American Cyanamid is an epichlorohydrin–methylamine copolymer (Fig. 6.5). This substance is prepared as a 50% aqueous solution and hence is more easily handled than Talofloc. However, the monomer epichlorohydrin is a carcinogen, and chronic exposure can cause kidney injury. Although the polymer does *not* exhibit these toxicological characteristics, the FDA allows use at only 300 ppm on solids. At comparable dose levels, Talofloc and the epichlorohydrin–methylamine copolymer perform similarly, but J. R. Elvin has pointed out that if the raw liquor is high in ash, the former outperforms the latter. A third type of precipitant, the cyclic polymer diallyldimethyl ammonium chloride, Talomel, is used mainly in the production of Blanco Director sugars.

6.6 SCUM HANDLING

Typically, about 3 to 6% of incoming feed liquor is removed with the scum and must be recovered by secondary and tertiary clarification. The scum is diluted and reseparated several times to recover the sugar as high-purity sweetwater. During primary clarification, scums may also be diluted with spray water, used to keep the scums moving, which further increases sweetwater generation.

In a typical recovery system the primary scum is diluted with 1 to 3 Brix hot clear sweetwater to about 20 Brix and treated with a 3 Brix calcium hydroxide suspension and polyacrylamide. Scums from this system are diluted with water to 1 to 3 Brix and treated a second time, with polymer, as required. The scum from the tertiary system may be sweetened-off on a rotary vacuum filter or discharged to a settling tank. At this point the sugar content is well below 1%, and the scum can be handled as waste. This often means discharge

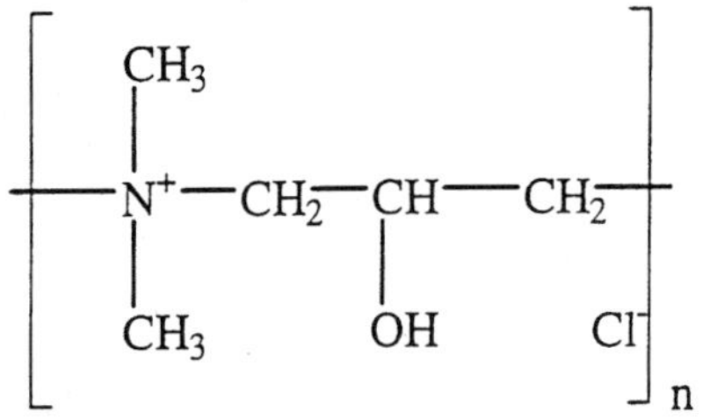

FIGURE 6.5 Epichlorohydrin–methylamine copolymer.

to a secondary waste treatment facility, where the residual sugar is consumed microbiologically. The desweetened sum typically is about 0.2 to 0.4% of raw sugar milled on a dry basis.

Sugar losses associated with scum handling can be estimated on the basis of retention times, operating temperatures, and pH. Although the fraction of the sugar that is subject to this recovery processing is small, the retention time in the several recovery systems is likely to be much longer than that of primary clarification itself. The use of lime can, of course, maintain pH values in the zone in which inversion is minimal, but under alkaline conditions, especially above pH 8, noninversion losses can be a concern. Among the degradation products formed during scum treatment are lactic and formic acids, which are common sweetwater components. If this evaporated material is later passed over bone char, there is likely to be a large circulating load of organic acids that is alternatively adsorbed on char and then desorbed into sweetwater. With proper technical and process control, inversion losses associated with phosphatation are typically about 0.025%.

6.7 CARBONATION VERSUS PHOSPHATATION WITH COLOR PRECIPITANT

Chen and Chou [13] describe the important differences between the two processes as follows:

1. The carbonation procedure precipitates about 30 times more calcium phosphatation. Consequently, the former does a better job of removing impurities that form sparingly soluble calcium salts, such as sulfate, phosphate, and acidic polysaccharides.
2. For polysaccharides containing few anionic functions, such as starch, carbonation appears to do a better job [14], (Tables 6.5 and 6.6), although phosphatation may be

TABLE 6.5 Polysaccharides in a Phosphatation Process Refinery

Sample	Total Polysaccharides (ppm starch on solids)
Raw sugar	1410
Washed raw sugar	671
Affination syrup	7434
Melt sweetwater	2669
Melt liquor	614
Clarified liquor	588
Liquor off char	
Beginning of cycle	350
End of cycle	608
Liquor to resin	494
Liquor of resin, new column	552
No. 1 liquor to pans	437
First strike sugar (wet)	150
First strike syrup	1041
Second strike syrup (wet)	284
Second strike syrup	1971
Third strike sugar	473
Third strike syrup	3712

TABLE 6.6 Polysaccharides in a Carbonation Refinery

Sample	Total Polysaccharides (ppm starch on solids)
Raw sugar	930
Washed raw sugar	450
Melt sweetwater	562
Melted washed sugar liquor	519
Remelt liquor	2040
Melt liquor (mixture)	675
Carbonated liquor	
Before filtration	394
After filtration	378
Char-filtered liquor to pans (No. 1)	244
First strike sugar (wet)	50
First strike syrup	603

more effective at removing the amylose component. Neither process is particularly effective at removing dextran.

3. The two processes may remove a similar range in color, 25 to 50%, and both produce excellent clarity. However, the clarity from carbonation is slightly better because the final liquor is pressure filtered.
4. Carbonation produces considerably more solid waste than does phosphation. The former technique requires a higher capital investment but uses cheaper process materials. There is also a considerable maintenance cost associated with the gas compressors used in carbonation.
5. Carbonation uses more extreme pH values and temperatures; hence sugar losses by degradation are *potentially* higher. On the other hand, phosphatation results in a somewhat higher invert gain, about 0.025%, compared to 0.010 to 0.015% after carbonation.

REFERENCES

1. M. C. Bennett, *Rev. Agric. Surc. Ile Maurice,* 55(1–2):275–284, 1976.
2. E. L. Dorn and R. A. Harris, Paper 640, *Proc. Sugar Ind. Technol.,* Toronto, 1993.
3. M. A. Clarke and F. G. Carpenter, *Proc. Tech. Sess. Cane Sugar Refining Research,* 76–81, 1972.
4. M. C. Bennett, Paper 640, *Proc. Sugar Ind. Technol.,* Vancouver, 1990.
5. M. S. Bhangoo and F. G. Carpenter, *Proc. Tech. Session Cane Sugar Refin. Res.,* New Orleans, LA, 1966.
6. D. V. Freeland, R. Riffer, and J. G. Penniman, *Int. Sugar J.,* 81:196–200, 1979.
7. G. A. Chikin, V. I. Sigova, and T. M. Makeeva, *Teor. Prak. Sorbtsionnykh Protsessov,* 11: 101–102, 1976.
8. E. J. Roberts, M. A. Clarke, and L. A. Edye, *Proc. Sugar Process. Res. Conf.,* Helsinki, Finland, 1994.
9. A. H. James, M. A. Clarke, and R. S. Blanco, Paper 543, *Sugar Ind. Technol.,* Baltimore, 1986.
10. J. P. Murray, F. M. Runggas, and G. S. Shephard, *Proc. S. Afr. Sugar Technol. Assoc.,* June 1976.

11. M. C. Bennett, *Nature,* No. 4647, Nov. 22, 1958, pp. 1439–1440.
12. J. R. Elvin, *Proc. Workshop Sep. Process. Sugar Ind.,* Sugar Processing Research Institute, New Orleans, LA, 1996.
13. J. C. P. Chen and C. C. Chou, *Cane Sugar Handbook,* 12th ed., Wiley, New York, 1993.
14. Roberts et al. *Sugar J.,* Feb. 1978, pp. 21–23.

CHAPTER 7

Carbonatation for Turbidity and Color Removal*

INTRODUCTION

Carbonatation is a process used in sugar refining to purify and clarify sugar liquor. It involves the precipitation of calcium carbonate through the addition of lime and gasing with a gas containing carbon dioxide. The crystalline mass so formed removes impurities by incorporation in the crystals and constitutes a filter aid for the pressure filtration process. Carbonatation is generally applied to melt liquor in a refinery, ahead of any decolorizing process. A recent survey [1] has shown that more than half the refineries surveyed utilized carbonatation. It is therefore used more widely as a clarification process than are phosphatation and its variations. In general terms, carbonatation is a more capital-intensive process, requiring a large filtration area compared with the cheaper plant required for phosphatation. However, the operating costs for the carbonatation process are significantly lower than those for phosphatation for a comparable amount of decolorization.

Oliver Lyle [2] has said: "Carbonatation is something of a mystery. It really has a miraculous effect on sugar liquors." Apart from the low cost of operation, carbonatation has a number of other advantages. As well as being a filtration process, it is also a purification process, enabling a color reduction of about 45% to 55%, together with a reduction in ash content of about 20%. Carbonatation has been applied to juices and syrups from beet and cane for the removal of ash, color and suspended matter. It is now applied extensively in the purification of beet juices but is no longer often seen in raw cane sugar factories. In this chapter we consider carbonatation only in the context of cane sugar refining.

The carbonatation process incorporates a number of physical and chemical processes. First, there is a chemical reaction between lime and carbon dioxide to form a calcium carbonate precipitate. The vessels in which this occurs are generally called *saturators*, but would more correctly be referred to as *reactors*. The operating conditions and flow characteristics of the saturators need to be considered in the light of getting a high conversion efficiency. Second, a process of crystallization takes place and the requirements are to pro-

* By Peter Rein.

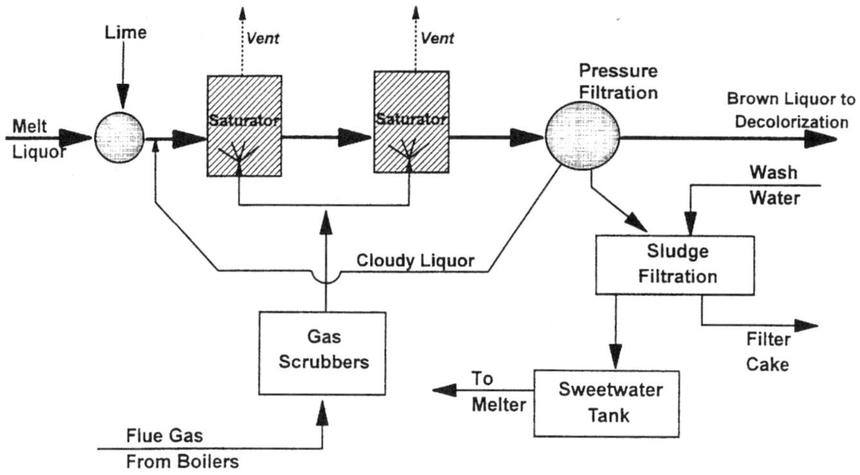

FIGURE 7.1 Schematic diagram of a carbonation station.

duce a mass of calcium carbonate crystals, which are of the right size, size distribution, and degree of conglomeration to facilitate the subsequent filtration step. Third, the conditions of pH and temperature dictate the extent to which inorganic species are precipitated, thus effecting ash removal, while destruction of sucrose and/or monosaccharides and filterability of the carbonate cake are also affected by pH and temperature conditions. Finally, the filtration process itself needs to be properly designed and operated to ensure that the required throughput is obtained while minimizing the loss of sugar in filter cake.

A block diagram of the process is shown in Figure 7.1. Lime is added to melt liquor immediately ahead of the saturators. Generally, boiler flue gas is used, after scrubbing out particulate matter and sulfur dioxide, to gas the liquor down to target pH values. Saturators are generally arranged in two or three stages, with progressively lower pH values in each stage. The liquor then passes to a pressure filtration stage, and the clear liquor filtrate, generally called *brown liquor* or pressed filtered melt liquor, passes on to the next stage of decolorization. Some cloudy liquor is generated at the start of each filtration cycle, which is usually recycled to the feed to the first saturator but could be fed into either of the saturator stages or, in fact, be returned to the melter. In some operations, the cloudy liquor is returned to a tank to be refiltered. The cake that is sluiced off the pressure filters goes to a second stage of filtration, where it is desweetened by the addition of wash water. Filter cake is generally carted off as a by-product for soil conditioning or as a waste product and sweetwater generated in the desweetening process is utilized elsewhere in the refinery, generally for melting sugar.

7.1 DESIGN OF SATURATORS

7.1.1 Configuration

There are some conflicting requirements with regard to the ideal arrangement of saturators. To maximize the absorption of CO_2, the pH value should be kept as high as possible. However, to improve the filterability, the carbonate crystals should be grown at a pH as low as possible, definitely below 10. The final exit liquor from the final saturator stage should be at a pH of less than 8.5 if maximum ash reduction is to be achieved. This has led to a compromise arrangement of saturators incorporating either two or three stages of gasing. In the first stage, the pH is kept at around 9.5 and most of the CO_2 absorption occurs in this stage. Gasing down to a lower pH in subsequent stages satisfies the requirements for minimum ash in brown liquor. The pH of the liquor is normally measured at room temperature. However, from a reactor design viewpoint and with the requirement to get a uniform mass of carbonate crystals, a plug flow system is required, which is more closely approached the greater the number of saturator stages. In this respect, three saturator stages will always be better than two.

7.1.2 Sizing and Design of Saturators

The liquid volume contained in the saturators is generally sized to give an overall residence time of liquor in the system of about 1 h. Residence times reported to give satisfactory performance seem to vary between 40 and 80 min. Longer residence times generally give higher color removal and better filterability, but above 1 h the benefit is much reduced, and losses of reducing sugars are increased. The saturator vessels themselves are generally made to be the same size, although in some cases the volume in the first stage is larger, to achieve a lower pH value at which carbonate crystals are grown. In addition, the last vessel may be slightly smaller, since the primary objective of the last vessel is to get the pH value down to 8.5 and most of the crystal growth occurs in the preceding saturators. The practice in North America is to have two saturators in parallel in the first stage with a retention time ranging from 40 to 50 min. The partially carbonated liquor is then fed to the second stage with only one saturator of the same size.

Various designs of saturator have been used in beet juice carbonatation (e.g., Ref. 3), in an attempt to improve gasing efficiency. For instance, some of these incorporate draft tubes, with gasing inside the tube. This achieves strong circulation of liquid within the vessel. It should be noted that more CO_2 gas is needed in beet operation because of the high lime dosage used. However, if the gas supply is not a restriction, simpler vessel designs with lower gasing efficiencies are preferred.

In some cases, some or all of the saturators are fitted with steam calandrias to maintain the required temperatures in the vessels. In general, however, the control of temperature in the saturators is not critical and it is more common to use a higher temperature of melt liquor going into the first saturator and live with the reduction in temperature through the system which occurs due to the cooling and evaporation caused by the sparging of gas. In general, the goal is a final temperature for liquor leaving the last saturator of about 85°C.

Liquid height in the saturator vessel is normally limited by the head the gas pumps can accommodate. This could be as high as 7 m, but is generally limited to about 2.5 m. Gas piping is arranged to feed gas from the top of the saturator down to a distribution system at the bottom of the vessel so that there is no danger of liquor finding its way back into the gas reticulation and pumping system. The outlet pipe from the last saturator is generally positioned at the correct height to give a constant level in all the saturators.

7.1.3 Gas Distribution

The requirements of the gas distribution system are (1) to distribute the gas uniformly over the cross-sectional area of the saturator, (2) to produce a large number of small bubbles, since this provides a larger area for mass transfer to take place, and (3) to ensure sufficient agitation in the vessel so that uniform conditions occur in the vessel and the boundary layers between bubble and liquid are kept as thin as possible. A good description of the physicochemical processes involved is given by Napp et al. [3]. They suggest that the right degree of mixing is obtained with superficial gas velocities in the range 10 to 15 cm/s, which should be considered in assessing the geometry of the saturator vessels. The degree of mixing can be improved in a large-diameter saturator by recirculating liquor by an external centrifugal pump. However, it has been suggested that the pumping of the liquor damages the floc [2].

Gas distribution in the vessels is via sawtooth launders or a system of gas nozzles designed to distribute gas as evenly as possible across the cross section of the vessel. In some cases, making the change from launders to nozzles has been reported to improve gas absorption, with the result that less gas is required. However, scaling in saturators can be severe, and the nozzle system is particularly prone to scaling, which can lead to a maldistribution of gas unless the nozzles are inspected and cleaned on a regular basis. It is anticipated that the use of nozzles could lead to smaller bubble sizes with beneficial effects on utilization of the gas. For efficient operation, CO_2 gas entering the saturators should not exceed 85°C.

Some beet sugar factories use Richter tubes for gas distribution. This system uses a set of parallel tubes with downward-pointing slots. The slots are kept free of encrustation by mechanical means and are therefore more complicated. On the order of 75 to 90% of the carbonatation is achieved at the higher pH value in the first stage of saturation. Gas piping should be sized accordingly.

7.1.4 Entrainment and Frothing

It is important to have a system for entrainment separation on each saturator to ensure that droplets of liquid are not carried out with the gas leaving the vessel. This must be sized for the highest gas velocity that will be expected, ensuring that the velocity through the entrainment separator is not sufficiently high to entrain droplets from the surfaces of the separator (i.e., <6 m/s). It is also important to ensure that the separator is of an open type, since the scaling characteristics of the liquor easily lead to blockages either in the entrainment separator or in the liquid return pipe. Otherwise, conventional types of entrainment separator can be used.

Excessive frothing has also been observed to occur in saturators under certain conditions. These may be associated with high temperatures [4] or other factors. Nonetheless, there should be sufficient vapor space above the liquid level to cater to frothing. Most entrainment separators do not handle froth well, and excessive frothing can lead to very high entrainment losses. Chapman [5] suggests that 1.5 m is sufficient freeboard above the liquid level.

7.2 SATURATOR OPERATION, PERFORMANCE, AND CONTROL

7.2.1 Color and Ash Removal

Values of color removal have been reported in the range 20 to over 60%, but are generally in the range 30 to 50%. Various color bodies are trapped in the calcium carbonate floc, and very high molecular weight colorants may be removed in filtration. Bennett [6] has shown that the mechanism whereby impurities are removed from the liquor involves chemical inclusion of impurity within the growing calcium carbonate crystals, and simple adsorption of impurities is unlikely. A higher lime addition rate could lead to improved color removal in some cases, but the lime addition rate is generally set taking into account the need to have sufficient calcium carbonate floc to achieve adequate filtration rates.

Roughly 10 to 20% of the ash in feed liquor can be removed, due largely to the precipitation of insoluble calcium salts, $CaSO_4$ in particular. To minimize ash content, it is important that the final pH of the material leaving the last saturator be below 8.7. Murray and Runggas [4] have shown very clearly that above 8.7, the conductivity ash of the carbonated liquor rises very steeply. In practice, it would be advisable to aim for an exit pH value of 8.5. Residence time in the saturators does not appear to have a significant effect on ash removal.

Although it is important to control the final pH of the saturators, it is equally important to run at the correct pH conditions in the first saturator. Reaction conditions within the first saturator generally determine the filterability of the final carbonated liquor. Maximum filterability is achieved if gasing is done at a pH of about 8.5. This is not feasible in practice because of the very low CO_2 absorption that will occur at this low pH, and so in general, the first-stage pH is run at about 9.5. If higher pH values are used in the first stage, the filterability is impaired significantly. The observed increase in filterability with drop in reaction pH is attributable to the propensity of calcium carbonate to form better and larger conglomerates at lower pH values [4].

It is possible to increase the amount of color removed by the use of color precipitation chemicals. Cationic polyamines (e.g., Talocarb, sold by Tate & Lyle) have been shown to be able to reduce brown liquor color by 30% at a dosage rate of 200 ppm on Brix. However, it has been shown that same degree of decolorization is not achieved in the final crystal color and since the cost of the chemicals is fairly high, this is a measure that in general is used only when a boost is required to color removal, due, for example, to an unusually high color of input sugars.

7.2.2 Destruction of Invert Sugars and Ash Gain

A disadvantage of carbonatation is the fact that high pH levels lead to a certain degree of destruction of monosaccharides. The breakdown products are generally organic acids, which require more lime in neutralization and increase the ash in the final molasses, thereby increasing the loss of sugar in molasses.

A comprehensive study was undertaken at the Hulett refinery when an investigation into the ash balance showed increases in ash of up to 40%. It was found that this was due to the formation of lactic acid in the lime mixing tank ahead of the first saturator as a by-product of the alkaline degradation of reducing sugars [7]. This tank was eliminated and a mixer installed to mix the lime into the liquor. The ash gain dropped to below 10%. It was found that the ash gain was positively correlated with the lactic acid content in the A saturator. The monitoring of lactic acid levels across carbonatation by gas chromatography is now a routine quality control procedure. Perhaps more important, though, is for a refinery to calculate regularly the ash gain across the entire refinery and monitor its value. In general, the ash gain across the refinery should be less than 15%, and a high level could point to such problems in carbonatation. Improper control of pH in the last saturator also leads to ash gain across the carbonatation.

7.2.3 Carbon Dioxide Absorption

A number of attempts have been made over the years to measure the efficiency of absorption of carbon dioxide. In general, these figures are quite low, in the range 20 to 50%. Higher absorption efficiencies can easily be obtained at higher pH, but the requirement to run the saturators at as low a pH as possible reduces the absorption efficiency.

In general, lower efficiencies of absorption are obtained with high gasing rates; the efficiency of absorption may be lower, but the quantity of CO_2 transferred could conceivably be higher. However, attempts are made to keep gasing rates as low as possible to reduce pumping power costs and minimize the effects of frothing and entrainment. A higher absorption efficiency is obtained in vessels with a high syrup level because of the longer bubble residence time within the liquid.

The reaction between dissolved carbonic acid and dissolved calcium hydroxide is very fast and takes place in a narrow reaction zone in the proximity of the bubble surface. If the mixing in the liquid phase is inadequate, the transport of calcium hydroxide to the reaction zone is the rate-determining step and the CO_2 concentration and pressure of the gas will have no significant influence on the mass transfer. A low pH also results in the supply of calcium hydroxide to the reaction zone being somewhat reduced. Only when there is good mixing in the saturator and a high rate of dissolution of the calcium hydroxide into the liquid phase will the transfer of carbon dioxide through the phase boundary layer be rate determining. In this case the CO_2 concentration and the pressure of the gas should influence the carbon dioxide absorption efficiency.

7.3 LIME PREPARATION AND USE

7.3.1 Quantity of Lime

In general, the amount of lime used is on the order of 0.5 to 0.8% CaO on melt. With very poor sugars this can be considerably higher, and values of over 1% have occasionally been reported. Since larger quantities of lime require more gas for neutralization, the gas pump capacity may sometimes determine the highest permissible lime addition rate. Alternatively, insufficient residence time in the vessels could place a limit on the amount of lime added if the right pH conditions in the first and last saturators are to be achieved.

In practice, the intention is always to add as little lime as possible consistent with an adequate filtration rate on the filters. This is done to reduce the cost of lime and also to minimize the amount of cake, since a larger quantity of cake leads to increased sugar loss in the cake, more sweetwater, and a higher cost of disposal. Generally, however, the dosage rate should be sufficiently in excess of the minimum requirement to ensure relative insensitivity to minor variations in the quality of sugar being processed. For consistency in operation, the quality of the lime in terms of total lime, available lime, and temperature rise in slaking should also be monitored.

7.3.2 Preparation and Conditioning of Lime

In general, calcium hydroxide is produced by slaking quick lime (CaO). This is usually brought on site in bulk and stored in silos and used as required. Some refineries have lime kilns, which have the advantage of producing a carbon dioxide–rich gas for carbonatation as well as producing quick lime. In some cases, recycling of calcium carbonate cake to a lime kiln has been adopted, but the accumulation of impurities limits return of lime to about 50% of consumption [5].

Lime is generally prepared at a concentration of about 10 to 15° Baumé, at a specific gravity of about 1.1. In some quarters it is believed that slaked lime should be conditioned by storing it for up to 15 h before use. However, laboratory trials [4] have shown that an aging time of 2 h is sufficient and that no improvement in filterability is achieved for longer lime storage periods.

7.3.3 Addition of Lime to Liquor

It is important to ensure good mixing between the lime and liquor before it enters the saturators. It is not advisable to have a mixing tank ahead of the first saturator, since holding liquor at pH values of around 11 is detrimental (see Section 7.2.2). The ideal is to have a high efficiency in-line mixer, just upstream of the first saturator.

7.4 FLUE GAS PRODUCTION AND HANDLING

7.4.1 Flue Gas Supply

In general, flue gas from boilers is used in the saturators. The CO_2 content depends on the fuel being fired in the boilers; with coal firing, a CO_2 content of around 10 to 12% is expected and a slightly lower value would be expected from an oil-fired boiler. With gas-fired boilers a CO_2 content of 7 to 9% is expected. It is often the case that the ducts from the boiler develop leaks; in this respect it is an advantage to have the pressure in the ducts above rather than below atmospheric pressure. If duct pressures are subatmospheric, the ingress of air reduces the CO_2 content of the flue gas.

In some cases, lime kiln gases are used, which have a CO_2 content of about 30%. It is to be expected that this should be an advantage in terms of a higher concentration of CO_2, promoting mass transfer, reduced quantities of gas to be pumped, and the absence of SO_2 in the gas, eliminating the need for soda ash scrubbing. However, quantities of gas sparged into the saturators should not be so low as to reduce the degree of mixing within the saturator to unacceptable levels (see Section 7.2.3).

The quantity of gas required depends on the amount of lime added that has to be neutralized. The gas quantity can easily be calculated if some assumption is made about the efficiency of gas absorption. In the normal range of lime additions, the quantity of gas required is on the order of 60 to 120 m^3 of gas per tonne of melt. In practice, a design figure of about 150 m^3 at STP per tonne of melt should cope with lime/melt ratios up to about 1%.

7.4.2 Flue Gas Handling

It is generally the practice to scrub flue gases to remove particulate matter and traces of sulfur dioxide. However, in the case of gas-fired boilers, scrubbers can be dispensed with entirely. In most cases a two-stage scrubbing system is required. In the first stage, the flue gas is washed with water to remove particulate matter, and in the second stage, the flue gas is washed with a soda ash solution to remove the SO_2 down to a level of about 5 ppm. In practice, most of the SO_2 is removed in the first scrubber, but the soda ash wash is necessary to protect the pipes and gas pumps from excessive corrosion.

The scrubbers are themselves simple scrubbing columns packed with ceramic or other packings, or simply fitted with baffles and liquid sprays. Peabody tray–type scrubbers are used in some refineries. Materials of construction need to be chosen to withstand corrosive conditions if the flue gas contains SO_2. Stainless steel, ceramics, and plastics can be considered.

It is not clear why scrubbers are necessary other than to neutralize the gas and ensure that gas pipes and gas pumps do not corrode. Particulates in the form of carbon particles should not cause any problems in the process and would, in any event, get filtered out, and small traces of SO_2 could be of benefit in introducing a mild sulfitation, which could, if anything, improve color removal.

In some cases, Rootes blowers are used to pump the gas into the saturators, but more often, liquid ring pumps are used. The liquid ring pump seal generally consists of a mild soda ash solution to minimize corrosion. Even with soda ash washing, corrosion in piping and pumps can be significant. Corrosion-resistant steels are advisable and glass-lined pipes and gas pumps have been used with success at the Hulett refinery.

7.5 LIQUOR FILTERABILITY

7.5.1 Factors Affecting Filterability

Filterability of the liquor is a quality of prime importance in the performance of the carbonatation station. The ease of separation of impurities from carbonatated liquors is determined by the filterability of the precipitated calcium carbonate. Apart from the quantity of material to be removed, which is determined largely by the lime addition rate, the filterability depends on the conditions under which the calcium carbonate crystals have been grown. In general, the lower the pH at which the carbonatation occurs, the better will be the filterability. Murray and Runggas [4] found in the laboratory that maximum filterability is achieved at a pH of 8.2. It is essential that the pH in the first saturator be kept below 10 and preferably not more than 9.5.

According to Bennett [6], higher temperatures increase the degree of conglomeration of crystals up to 82°C, which leads to high filterabilities. According to Murray and Runggas [4], maximum filterability is achieved when operating at 86°C. In addition, the Brix value is reported to have an effect, with a reduction in Brix giving an increase in filterability over and above the effect of liquor viscosity. For every liquor, there is a particular lime addition rate at which the filterability will be greatest. This optimum lime dose varies from 0.4% CaO on solids in some liquors to over 1.2% CaO in others. Many workers have investigated the factors affecting liquor filterability. It has been found that both soluble and insoluble solids affect filterability. Starch, dextran, and soluble phosphate have at various times been shown to be factors implicated in reducing filterability.

The two main factors affecting the filtering quality of raw sugar are insoluble suspended matter and starch [8]. Suspended solid material has a mechanical effect on the filterability of liquor and needs to be filtered out either with the addition of filter aid or in the carbonatation process with the addition of precipitated calcium carbonate. In either case, more filter aid is required with high levels of suspended solids; in the case of carbonatation a higher lime dosage rate is required. Lee and Donovan [9] have shown that to achieve satisfactory filtration performance in a refinery, a sugar with a filterability of at least 50% is desired, which corresponds to a level of suspended solids of 300 ppm. They showed silica to be a major component of suspended solids, but that part of the suspended solids consists of organic components.

Because carbonatation is a crystallization process, it is to be expected that impurities in the liquor can have a profound effect on the nucleation, growth, and flocculation processes involved in formation of the final calcium carbonate mass. In the case of starch, it has been found that the amylose fraction of starch acts as a protective colloid, coating the surface of the growing crystals. This prevents the formation of agglomerates and reduces the interparticle repulsion charge. Both factors lead to a tightly packed cake with low filterability [10].

In practice it has been found that starch enzymes added to carbonatation sweetwater recycled to the melter can remove the starch to a low enough level for it not to affect filterability. This has been carried out at the Hulett refinery since 1983 [11]. A dosage rate of about 10 ppm on sweetwater, or 3 ppm on white sugar output, is used. Temperature-tolerant enzymes are used, which eliminate all starch in sweetwater and appear to achieve additional starch removal even at the high Brix values obtained in the melter. In practice, it has been found that starch does not significantly reduce the cake resistance at concentrations in sugar of less than 150 ppm [9], and starch values below 150 ppm do not affect filterability.

7.5.2 Equations for Filtration

In the case of an incompressible cake, the resistance to flow of a given cake volume is not affected appreciably either by the pressure difference across the cake or by the rate of deposition of material. On the other hand, with a compressible cake an increase in the pressure difference or of the rate of flow causes the formation of a denser cake with a higher resistance.

For incompressible cakes [12],

$$\frac{1}{A}\frac{dV}{dt} = \frac{\Delta P}{r\mu z} \tag{7.1}$$

where A is the total cross-sectional area of the filter cake (m^2), V the volume of filtrate (m^3) that has passed time t (s), ΔP the applied pressure difference (Pa), μ the viscosity of the filtrate (Pa·s), and z the cake thickness (m). r is termed the specific resistance and has units of m^{-2}. In some cases, r values are reported in units of cm/g (i.e., divided by the solids bulk density in the cake). For incompressible cakes, r is taken as constant, but will depend on the rate of deposition, the nature of the particles, and on forces between the particles. If v is the volume of cake deposited by a unit volume of filtrate, then

$$z = \frac{vV}{A} \tag{7.2}$$

Substituting for z in equation (7.1), we have

$$\frac{dV}{dt} = \frac{A^2\,\Delta P}{r\mu vV} \tag{7.3}$$

This equation can be regarded as the basic relation among ΔP, V, and t.

Two important types of operation need to be considered: (1) where the rate of filtration is maintained constant and (2) where the pressure difference is maintained constant. For filtration at constant rate,

$$\frac{dV}{dt} = \frac{V}{t} = \text{constant}$$

so that

$$\frac{V}{t} = \frac{A^2 \, \Delta P}{r\mu V v} \tag{7.4}$$

that is,

$$\frac{t}{V} = \frac{r\mu v}{A^2 \, \Delta P} \tag{7.5}$$

and ΔP is directly proportional to V. For a filtration at constant pressure difference,

$$\frac{V^2}{2} = \frac{A^2 \, \Delta P}{r\mu v} \tag{7.6}$$

that is,

$$\frac{t}{V} = \frac{r\mu v}{2A^2 \, \Delta P} V \tag{7.7}$$

Thus for a constant-pressure filtration, there is a linear relation between V^2 and t and between t/V and V.

In this analysis, the resistance of the cloth has been ignored. The resistance of the cloth plus initial layers of particles deposited is important since the latter not only form the true medium but also tend to block the pores of the cloth, thus increasing its resistance. Cloths may have to be discarded because of high resistance well before they are worn mechanically. No accurate analysis of the buildup of resistance is possible because the resistance will depend on the way in which the pressure is developed, and small variations in support geometry can have a big effect. It is therefore usual to combine the resistance of the cloth with that of the first few layers of particles and assume that this corresponds to a thickness of cake as deposited at a later stage. Alternatively, the resistance of the cloth is ignored, as being small in relation to the resistance of the cake.

The specific resistance of the cake, r, is a property of the solids forming the cake and represents the resistance to flow of liquid per unit volume of cake. Since v is the volume of solid cake deposited per unit volume of filtrate, the product rv represents the total resistance to flow of filtrate. Bennett [13] has proposed that the filterability, F (in m^2), be defined as

$$F = \frac{1}{rv} \tag{7.8}$$

Substituting this in the equation for constant-pressure filtration [equation 7.7)] leads to

$$\frac{t}{V} = \frac{\mu V}{2A^2 F \, \Delta P} \tag{7.9}$$

A value for F can therefore be determined by plotting t/V against V under constant-pressure conditions, and using the slope of the line to calculate F given the area, pressure difference, and viscosity of the liquid.

7.5.3 Laboratory Filtration Measurements

Filterability of sugar has been studied over many years, and attempts have been made to measure it in the laboratory through the use of standard laboratory filtration tests. These have generally been based on the filtration test method and apparatus of Nicholson and Horsley [14], in which a kieselguhr filter aid is used to compare the filtration characteristics of pure sucrose and of the unknown raw sugar, expressing the ratio of filtration rate of the raw sugar as a percentage of that of pure sugar. The filterability of most raw sugars so measured lies in the range 40 to 70%. Details of a slightly modified procedure are given by Lee and Donovan [9]. A simpler filtration measure used by Domino Sugar Corporation merely measures the volume of filtrate collected in a given time from a laboratory filter run under standard conditions.

Standard filterability tests can be used to predict filtration performance only if the starch content in the sugar is less than 200 ppm [9]. Other methods of laboratory testing of filterability have been developed which involve a batch carbonatation process before measuring the filtration rate of the liquor. This is more representative from the point of view of a carbonatation refinery since this method assesses not only the mechanical effects on the filter cake, but the chemical effects on the crystallization of calcium carbonate. Since the entire test is carried out at constant conditions of pH (8.5), temperature (80°C), and Brix (60) for a fixed time (90 min), the conditions are different from those of industrial continuous carbonatation. However, these conditions produce a slurry similar in structure and behavior to refinery slurry [10]. Lee and Donovan [9] use a similar approach but run the test at 70°C, with the pH dropping to 7.4 over the period of the test. From these laboratory tests, values of r and F can be calculated using equation (7.9). In some cases, the value of $\sqrt{F}$ rather than F is reported, since flow rate is proportional to $\sqrt{F}$ [13]. Measured values of $\sqrt{F}$ lie in the range 2.5 to 7.5×10^{-7} m. Bennett reports that refinery filter stations experience problems when $\sqrt{F}$ drops below 4×10^{-7} m.

7.6 LIQUOR FILTRATION

7.6.1 Types of Filters

It is common practice to filter the carbonatated liquor in a set of pressure filters and then to sweeten off the cake in a second set of sludge filters. This is done in two stages, so that each set of filters can be designed specifically for a single duty. By so doing, the overall cost of the filter station, which is a large proportion of the refining capital cost, can be kept to a minimum. In addition, because syrup and water have widely differing viscosities, the efficiency of sweetening off in the liquor filters is low, and less sweetwater is generated if the mud is repulped with water and sweetened off in a separate filtration stage. The de-sweetening filters are discussed in Section 7.7.

The pressure filter cycle ends when the filtration rate has dropped to a low level or when the maximum cake thickness has developed on the filter elements. Then the mud is sluiced off and goes to the sludge filtration step. A major feature of the pressure filters is that the proportion of time off for mud sluicing and refilling is relatively low, giving them a much higher on-line factor than plate and frame filters. The total time out of service on a pressure leaf filter for draining, mud sluicing, and preparing for the next cycle can be as little as 15 min.

The features of the most commonly used filters are summarized in Table 7.1. The Sweetland pressure leaf filter, introduced in the early 1920s, has been the best known and

TABLE 7.1 Comparison of Liquor Pressure Filters

Parameter	Sweetland	Suchar Auto Filter	U.S. Filter	Gaudfrin
Type	Fixed vertical circular disks	Rotating vertical disks	Rotating vertical circular disks	Drained cloth
Filter body	Horizontal	Horizontal	Horizontal	Vertical
Filter medium	Cloth-covered screens	Cloth on steel frame	Cloth-covered screens	Vertically suspended, seamed cloth elements
Maximum operating pressure (kPa)	420	420	490	400
Replacement of elements	Individual	Individual	Remove shaft	Individual
Mud sluicing	Oscillating jets between plates	Moving water jets	Water jets	Countercurrent steam / water flow
Filtrate sight glasses	One per disk	One per two disks	None	One per filter element
Access to elements	Bottom half drops down	Upper half of shell opens	—	Hinged top cover
Maximum area / filter (m^2)	100	200	200	400

most widely used filter. The Suchar filter or Auto Filter, introduced later, has the advantage that the rotating elements lead to a more even buildup and more positive sluicing off of cake. A sketch of the Rota Filter, a later model of the Auto Filter is shown in Figure 7.2.

The filtering elements are made of cloth. Initially, cotton was used, but cotton elements were replaced by nylon and then by polypropylene, which gives much better service. Monofilament polypropylene woven fabric cloths now appear to be used universally. Periodic washing of the filter cloths is required to restore permeability. At periodic intervals, approximately weekly, they are washed in situ with sulfamic acid or inhibited hydrochloric acid. At less frequent intervals, they need to be removed from the steel support frames for laundering.

The horizontal filters are expensive because of the heavy filter frame required. They are now sometimes being replaced by vertical filters, using supported cloth filter elements. The Gaudfrin filter has been installed more recently in some refineries (see, e.g., Ref. 15). It is simple, generally cheaper, and can be made with large surface areas. This type of filter is preferable to candle filters for this duty because of the large quantity of cake to be handled and the smaller filter volume per unit surface area.

7.6.2 Filter Area Required

The installed filter capacity needs to be sized for the worst quality of sugar that it is anticipated can be accepted for processing without leading to a reduction in melt rate. In this respect, a minimum sugar filterability of the raw sugar should be defined. In global terms, an installed filter area requirement of 15 m^2 of filtering area per tonne of melt per hour is

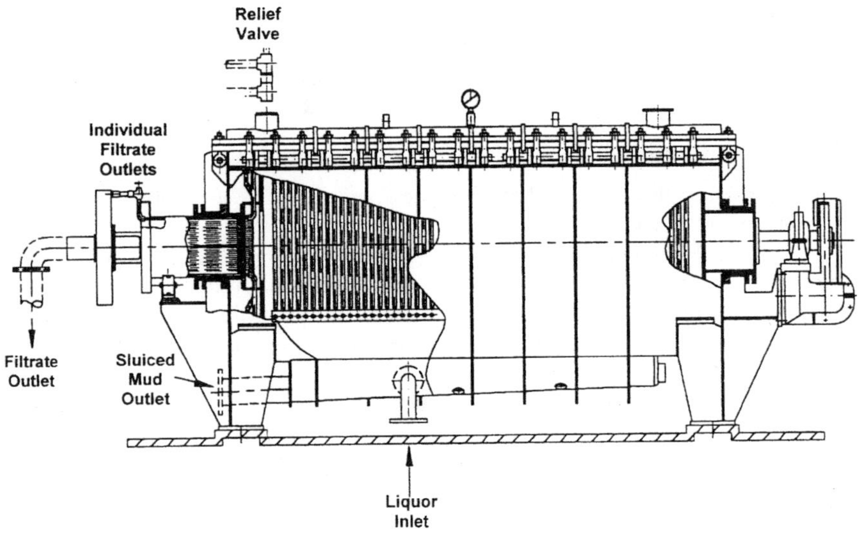

FIGURE 7.2 Sectional elevation of the Rota filter. (Courtesy of Fletcher Smith.)

required. Expressed as a volumetric rate, a filtration rate of 0.1 m^3/m^2 per hour will generally be acceptable. This should be adjusted for the amount of downtime that will be expected on each filter, due to mud sluicing and refilling, and cloth cleaning.

An indication of the filter area required can be deduced from measured filterability figures using the equation for constant rate filtration [equation (7.5)], introducing the filterability F:

$$\frac{t}{V} = \frac{\mu V}{A^2 F \, \Delta P} \tag{7.10}$$

Rearranging, and setting $Q = V/t$ leads to

$$Q = \left(\frac{F \, \Delta P}{\mu t}\right)^{0.5} A \tag{7.11}$$

If we require the area of a filter, which for a fraction α of the time produces clear liquor, rearranging equation (7.11) gives

$$A = \frac{1}{\alpha}\left(\frac{\mu t}{F \, \Delta P}\right)^{0.5} Q \tag{7.12}$$

where A is the filter area (m^2), α the fraction of total cycle time filter producing clear liquor, μ, the viscosity (Pa·s), t the time (s), F filterability (m^2), ΔP the pressure difference (Pa), and Q liquor flow rate (m^3/s) ($= V/t$).

As an example, take a refinery handling 100 tonnes/h of liquor at 66 Brix and 80°C (0.0219 m^3/s), filtering on a 2-h cycle, which produces clear filtrate for 90% of the time, with a pressure difference of 350 kPa. This equation predicts an area of 868 m^2 for a filterability F of 16×10^{-14} m^2 ($\sqrt{F} = 4 \times 10^{-5}$ cm). This represents a liquor filtration rate of 0.09 $m^3/m^2 \cdot h$, and for $\alpha = 1$, would be exactly 0.1 $m^3/m^2 \cdot h$. Note that for this relationship to be useful, the value of F should have been determined at about the same ΔP value as is used in the plant, since the cake is not incompressible.

7.6.3 Filter Operation

The effect of syrup Brix value and temperature is shown in the effect of viscosity on the area required. Either a lower Brix or a higher temperature will reduce the area requirement. Linear regression of filtration data at the Hulett refinery showed that the area required is reduced by 2.5% for each 1-unit reduction in Brix. Calculation using equation (7.10) suggests that Brix reductions leading to viscosity reductions will reduce the area required by about 5% for each 1-unit Brix reduction. However, lower Brix values lead to higher steam use, so are avoided where possible.

In general, the temperature of the carbonatated liquor is chosen to suit the filtration rate as well as carbonatation itself. Again using equation (7.10), a reduction of 5°C can be calculated to increase the area required by about 8%. In addition, the requirements of the downstream decolorization process could dictate limits to the temperature range for filtration.

If poor-filtering raw sugars encountered are due to a high starch content, the problem can be largely overcome by the use of enzymes [11]. If it is due to other factors, the refinery generally has to reduce the melt Brix or slow down the refinery. Blending of bad and good raw sugars has limited usefulness, since the filterability of a blend is not the arithmetic mean of the individual filterabilities [16].

The thickness of the cake can be calculated from the area of the filter and the amount of cake filtered out, as shown in equation (7.2). Since 100 kg of cake is produced from 56 kg of CaO (in the ratio of molecular weights), cake thickness can be calculated as

$$z = \frac{vV}{A} \tag{7.2}$$

$$= \frac{56\ Q\rho Bt}{10^6 A\rho_b} \times \text{lime \% solids} \tag{7.13}$$

where z cake thickness (m), v the volume of cake deposited by unit volume of filtrate, V the filter area (m^2), A the filter area (m^2), Q the liquor flow rate (m^3/s), ρ the liquor density (kg/m^3), B the liquor Brix (°), t the filtration time (s), and ρ_b the bulk density of dry cake (kg/m^3).

At the start of the filter operation, the filtrate is cloudy until a cake has been established on the cloth which catches all the particles. This liquor is recycled, generally back to the saturators, until the liquor is clear. The establishment of the initial cake at the start of a cycle (sometimes called *smearing*) is very important. The most important requirement is for this to be done at reduced pressure, generally achieved by a gravity feed to the filter. This results in a less compact layer close to the cloth, which gives better filtering characteristics. Once the cake is established, a change to pumped feed is made.

As outlined in Section 7.5.2, there are two possible modes of operation, constant pressure and constant flow. The latter is preferred [17], particularly since the cake is subjected to full pressure only at the end of the cycle, and the average cake resistance is lower. It also enables the optimum operation to be established, in terms of reducing the overall number of filter cycles. In fact, a filter running at constant rate can itself be used to monitor filterability. The length of time a filter runs at constant rate can be used to compute filterability from equation (7.11).

As filtration continues, the thickness of cake on the filter leaves increases. The filter should not be run long enough for the cakes on each leaf to touch, or the cake becomes very difficult to remove. In general, however, filtration is not usually as good, and the cycle is terminated because the filtrate flow has dropped too much. Once the liquor is shut off, a valve at the bottom of the filter is opened. The filter is drained, and then the mud is sluiced off the filter elements, into a mud tank ahead of desweetening.

7.7 CAKE HANDLING AND DESWEETENING

7.7.1 Filter Equipment

The conventional plate and frame press has most often been used for mud filtration and desweetening. It does an excellent filtration job but has some serious disadvantages: Its capacity is low, it is out of action for roughly half the time for cleaning, it requires much labor, and it is prone to leak. However, large filter areas are possible. Newer models are automated but are then far more costly. Automation is facilitated through the use of programmable logic controllers, and new installations are generally always completely automated.

Cloth-covered rotary vacuum filters were used for many years in some refineries. However, sweetwater quality was poor, the cake moisture was high, and the cake used to be sluiced to the sewer. This option is no longer allowed in most countries, and these filters have been replaced with filters that yield low-moisture-content cakes, which are transported to suitable solid disposal sites or to farms as soil conditioners.

The introduction of polypropylene plates in the 1970s was a major step forward, making the presses far easier to handle and operate. Since that time, some other changes in design have occurred. The first change was the development of the recessed plate or chamber press, with plates recessed on both sides so that unlike plate and frame presses, all the press leaves are identical. A further development is the membrane filter press, which incorporates a membrane on every alternate plate, which can be pressurized to squeeze the cake, improve desweetening, and reduce the moisture content of the carbonate cake to a figure as low as 30%. A value of 35% seems easily achievable. The sequence of operations and flow arrangements in the leaves of a membrane filter press are shown in Figure 7.3.

Recent installations of automatic filter presses have been supplied by Putsch, Faure, and Hoesch. Carbonatation cakes are well suited to these filters, as the degree of compression of the cake is limited, so that membrane fatigue due to excessive flexing does not occur [18]. It is important in sizing membrane filters to ensure that the chamber capacity as well as the filtration rate are considered. This is best determined on a pilot unit [18]. A filtration rate of 0.6 $m^3/m^2 \cdot h$ can be assumed on carbonatation mud duty. Newer installations by Larox and Hoesch of the automatic horizontal frame press type have been reported. They can be fully automated, and the endless filter cloth acts as a belt conveyor to facilitate cake discharge. A report on the use of a Larox filter [19] indicates cloth and membrane lives somewhat shorter than expected. However, a high filtration rate of 0.95 $m^3/m^2 \cdot h$ was achieved, which compensates partially for the low areas currently available on these filters (about 32 m^2).

7.7.2 Desweetening

It is important that a high recovery of sugar from the cake be achieved, to maximize refinery yield. The sweetwater so produced generally finds use in the refinery as a source of melter water. The moisture of the cake should also be as low as possible, to minimize disposal

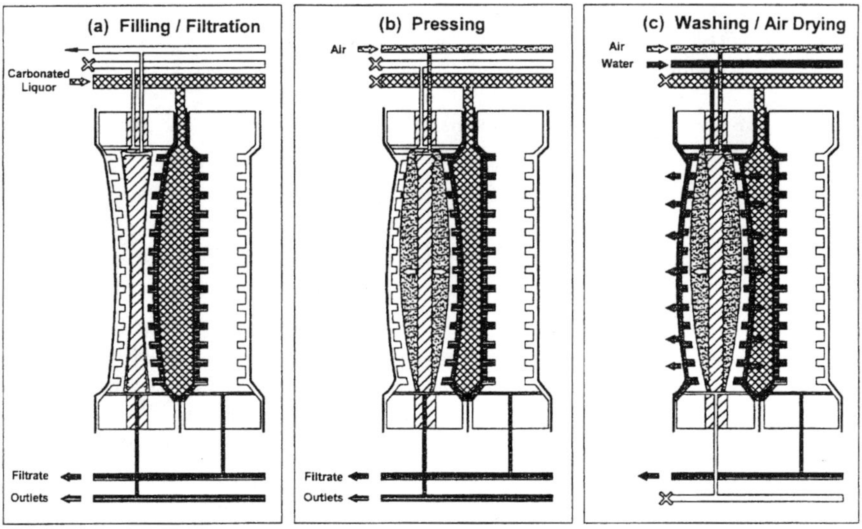

FIGURE 7.3 Stages in membrane filter press operation. (Courtesy of Putsch.)

costs. The calcining of the cake to regenerate lime is generally not regarded as a feasible operation. As with pressure leaf filters, each cycle should be started at reduced pump pressure, to avoid overcompacting the initial layers of cake. Ideally, each filter should be fed by a single variable-speed pump, with the feed pressure gradually increasing through the course of the cycle. In some cases, wash water is limed to ensure that the impurities in the cake are not redissolved in the washing process and recycled to the melter. It is important to attempt to minimize recycling of impurities.

7.8 PROCESS CONTROL

7.8.1 Control of Lime Addition

In the ideal arrangement, the amount of lime added is controlled in direct proportion to the flow of liquor to the first saturator to get the required lime percent solids ratio for the sugar being processed. This involves measuring the liquor flow to the saturator and adjusting the lime flow rate automatically to suit. In practice, handling of lime causes a number of problems because of its propensity to scale and block pipes. The flow of lime cannot easily be measured and controlled successfully, and a different system has been used widely by refineries engineered by Tate & Lyle. This consists of lime and liquor wheels. In this arrangement the changes in liquor flow change the speed at which the liquor wheel rotates, and the lime wheel speed is adjusted mechanically to give the same proportion of lime and liquor to the first saturator. A number of lime wheels are still in use in refineries. However, this system requires that lime and liquor be fed together into a mixing tank upstream of the first saturator. It has been shown that the time that the liquor is kept at high pH in this mixing tank can be detrimental in terms of invert destruction and additional ash production (see Section 7.2.2).

It is believed that the ideal system for controlling lime addition is via a variable-speed peristaltic pump. The scheme, recommended shown in Figure 7.4, involves measuring the

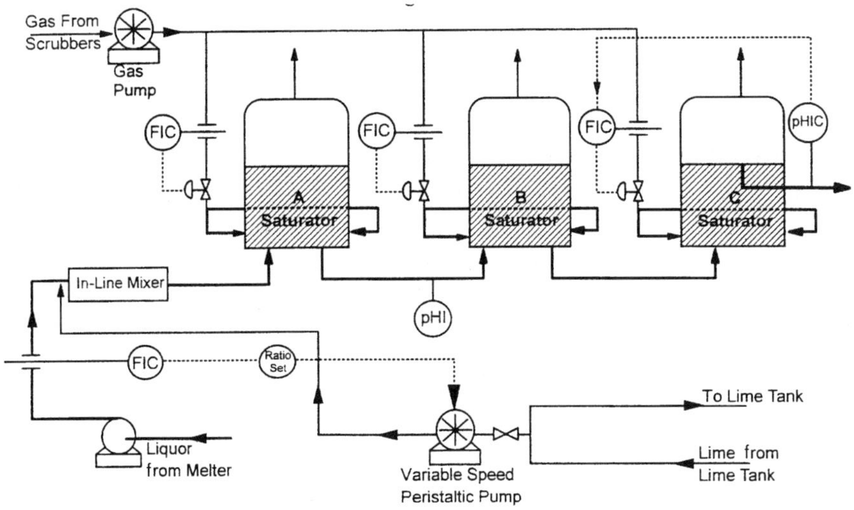

FIGURE 7.4 Process control systems recommended for carbonatation.

liquor flow, probably via a magnetic flow meter, and controlling the lime flow in a given ratio to an in-line mixer just upstream of the first saturator by adjusting the pump motor speed. Various types of in-line mixers are available. The simplest consists of a series of segmental baffles in an enlarged pipe section with a pressure drop of greater than 10 kPa to get the right degree of mixing of liquor and lime ahead of the first saturator.

7.8.2 pH Control

Because of the important effect that pH has on both filterability and the final ash content of carbonatated liquor, it is vital that the pH be controlled. This is done once the right lime/melt ratio is established by controlling gas flow to the saturators. In practice, however, it is extremely difficult to measure the pH in carbonated liquor because of the severe scaling problems that are encountered. In some refineries, conductivity is used, instead, as an indirect measurement of pH in the first-stage saturator. Scraped surface pH electrodes have been found to be successful, but even so, sample lines to the pH sampling point will often choke, due to scaling up of the sample lines. Batch sampling to a pH probe, with periodic air purging of the sample line to keep it clear, was reported by Alberino [20]. For this reason, sometimes only the pH in the last saturator is controlled, and the gas flow to the first saturator is set at maximum gas flow to ensure the lowest possible pH in the first saturator at all times. Details are shown in Figure 7.4. It should be noted in this figure that the gas flow to each saturator needs to be independently controlled. If not, throttling down of gas to one saturator will inevitably cause the gas flow to the other saturator to increase correspondingly.

7.8.3 Filter Feed Control

Since constant-flow operation of pressure filters is preferred to constant-pressure operation, it is necessary to measure and control the flow to each pressure filter. The set points on the flow controllers to all filters should be able to be set remotely, so that the optimum flow for the particular melt rate can be run on the filters at all times to minimise the number of filter cycles. Nevertheless, constant-pressure operation is a common practice in North American refineries.

7.9 POWER AND ENERGY IMPLICATIONS

The addition of lime to liquor leads to a reduction in Brix of brown liquor. This is counteracted to some extent by the evaporation of water by the carbonating gas. A reduction in Brix of about 1 unit or more from melt to brown liquor is generally experienced. This means that additional water has to be evaporated. Chapman [5] reports that heat loss increases directly with gas volume and rises steeply with increasing liquor temperature. Gasing out at low pH and high temperature can be expensive in both gas pumping and heat loss. This heat loss can be equivalent to a value as high as 2% steam on melt. Recovery of heat from outgoing gas could be considered as an energy-saving option and has been practiced at some refineries.

REFERENCES

1. M. A. Clarke, *Proc. Sugar Ind. Technol. Symp. Colour Removal,* 1997.
2. O. Lyle, *Technology for Sugar Refinery Workers,* 3rd ed., Chapman & Hall, London, 1957.
3. W. Napp, T. Cronewitz, and G. Witte, *Zuckerindustrie* 112:859–867, 1987.
4. J. P. Murray and F. M. Runggas, *Proc. S. Afr. Sugar Technol. Assoc.,* 55:90–93, 1975.
5. F. M. Chapman, *Int. Sugar J.,* 69:231–236, 1967.
6. M. C. Bennett, *Int. Sugar J.,* 76:40–44, 68–73, 1974.
7. A. A. Bervoets, and M. G. S. Cox, *Proc. Sugar Ind. Technol.,* 51:21–35, 1992.
8. J. P. Murray, *Proc. S. Afr. Sugar Technol. Assoc.,* 52:116–132, 1972.
9. F. T. Lee and M. Donovan, *Proc. Conf. Sugar Process Res. Inst.,* 1996, pp. 103–120.
10. J. P. Murray, F. M. Runggas, and M. Vanis, *Proc. Int. Soc. Sugar Cane Technol.,* 15:1296–1306, 1974.
11. J. W. Besijn, *Proc. Sugar Ind. Technol.,* 54:197–209, 1995.
12. J. M. Coulson and J. F. Richardson, *Chemical Engineering,* Vol. 2, 3rd ed., Pergamon Press, Elmsford, NY, 1978.
13. M. C. Bennett, *Int. Sugar J.,* 69:101–103, 1967.
14. R. I., Nicholson and M. Horsley, *Proc. Int. Soc. Sugar Cane Technol.,* 2(10):271–287, 1956.
15. M. Mabillot, *Proc. Sugar Ind. Technol.,* 52:59–72, 1993.
16. F. T. Lee and M. Donovan, *Proc. Int. Soc. Sugar Cane Technol.,* 22:195–204, 1995.
17. J. E. Morton, *Proc. Sugar Ind. Technol.,* 37:286–309, 1978.
18. C. C. Chou and S. J. Clarke, *Proc. Sugar Ind. Technol.,* 53:255–268, 1994.
19. B. Nichols and R. Smith, *Proc. Sugar Ind. Technol.,* 48:306–319, 1989.
20. J. W. Alberino, *Proc. Sugar Ind. Technol.,* 20:55–66, 1961.

CHAPTER 8

Granular Carbon Decolorization System*

INTRODUCTION

For conventional sugar refineries processing raw sugars of more than 2000 ICUMSA color and producing white refined sugar of less than 30 color, adsorbent-based decolorization is an essential process stage. While refining processes such as affination and clarification remove significant amounts of color from the raw sugar being processed, an additional color removal step is needed ahead of the final crystallization stage to ensure that the white sugar product always meets the product color specification. This additional color removal process is almost always adsorbent based, using bone charcoal, powdered carbon, granular carbon, or ion-exchange resins. The adsorbents may be used singly or sequentially, depending on the color loading and effectiveness of other process stages.

Carbon-based adsorbents have been the major decolorizing agent for sugar refining since the early nineteenth century, with bone charcoal as the primary adsorbent. Powdered activated carbon has also been used, with the application of granular carbon first reported in the early 1950s [1]. The development of granular carbon grew out of the search for a replacement for bone charcoal, which despite its valuable process attributes was becoming more difficult to obtain and more costly to use. Carbon, with about 10 times the decolorizing power of bone charcoal, and lower operating costs, is an attractive alternative to the traditional adsorbent and is gradually replacing it in new or upgraded factories.

A recent indication of the comparative numbers of bone charcoal, granular carbon, and ion-exchange resin decolorization plants in current use was shown in a survey of 47 refineries reported by Clarke at the 1997 Sugar Industry Technologists' Symposium on Sugar Colour as follows [2]:

Total number of system:

Bone char	18
Ion-exchange resin	22
Granular carbon	17

*By Peter J. Field and H. Paul Benecke.

Number of single-adsorbent systems:

Bone char	7
Ion-exchange resin	8
Granular carbon	8

Granular carbon decolorization is also widely used in the corn syrup industry, where many installations use pulsed rather than fixed beds.

General principles for the design and operation of granular carbon decolorization plants, together with information about the properties and characteristics of carbon, are set out below. There is an emphasis on a practical approach based on experience in operating refineries. The design principles in particular have been presented so that readers may better understand the influence that design parameters have on plant performance. Where the detailed design of an actual plant is required, it is recommended that further advice be sought from carbon suppliers and specialist plant manufacturers. A typical granular activated carbon plant is shown in Fig. 8.1.

8.1 GENERAL PROPERTIES OF GRANULAR CARBON

Granular carbon adsorbents are used for a variety of purposes in the food and process industries, including gas and water purification as well as sugar decolorization. The granular activated carbons used for sugar refining may contain up to 90% carbon and are generally made from a low-ash bituminous coal base. The manufacturing process includes an initial carbonization stage in a reducing atmosphere at 600 to 700°C, followed by activation with steam at 900 to 1000°C. The activation process both develops and opens up the pore structure, which gives the carbon particle its high surface area and consequent high adsorption capacity. Careful control of the activation process allows some tailoring of pore size distribution to produce a range of granular carbons optimized for particular applications.

New carbon (sometimes termed *virgin carbon*) intended for sugar decolorization is typically about 1.0 mm in particle size and 0.45 in bulk density. Mitsubishi Kasei quoted a surface area of 1135 m^2/g for Cane Diahope carbon, with most pores being in the range 10 to 30 Å in diameter [3]. The activation process also influences the hardness of the carbon particles. In general, high activation produces a softer carbon.

Unlike bone charcoal, granular carbon has no buffering capacity, and sugar liquors treated with carbon show a measurable pH drop in the liquor off carbon. Consequently, it is the usual practice to include about 5% powdered magnesite (magnesium carbonate) in carbons intended for use in sugar refining. The magnesite dissolves slowly over time, simulating the buffer capacity of bone charcoal and minimizing the pH drop in liquors being decolorized. Magnesia (magnesium oxide) may be added instead of magnesite, and there is a view that the kilning process reduces the magnesite to the oxide in any case. Because the carbon surface is nonpolar, granular carbon has no ash removal capacity. This is another difference between carbon and bone charcoal, and one that may sometimes present a problem where high-ash raw sugars are being processed.

8.2 COLOR REMOVAL WITH GRANULAR CARBON

8.2.1 Sugar Colorants

The sugar liquors and syrups processed in a refinery are colored by the impurities they contain. Sucrose in solution is colorless. These colored impurities, generally termed *colorants,* can be classified into several groups, depending on their origin.

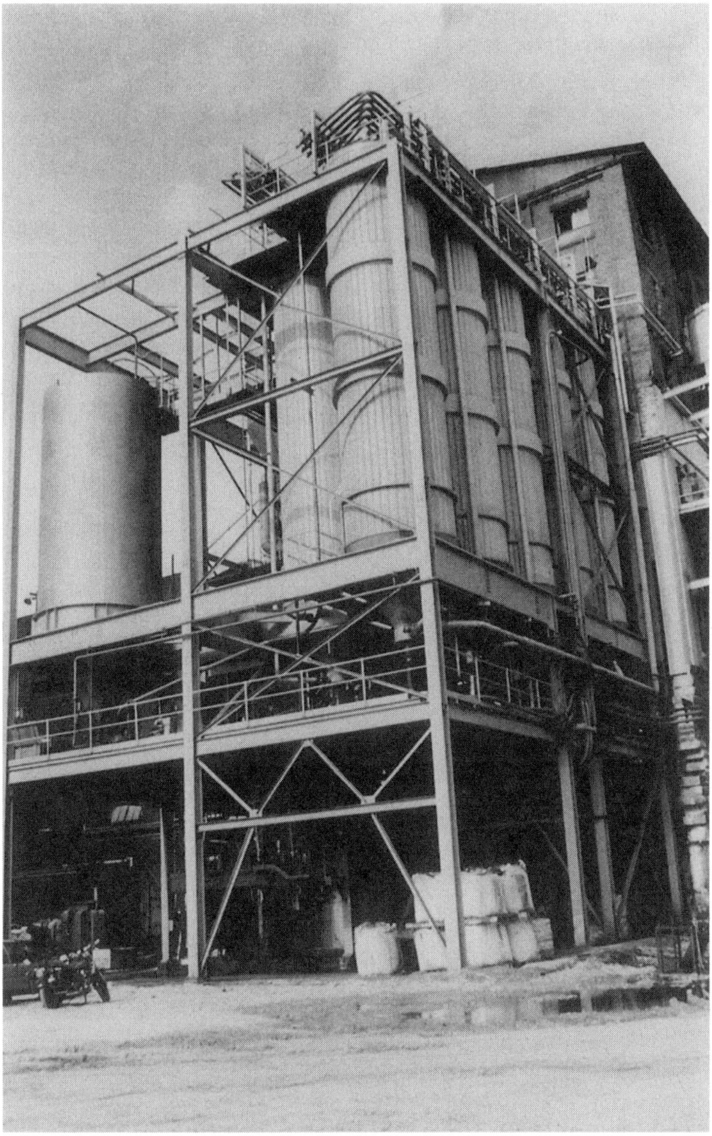

FIGURE 8.1 Granular activated carbon plant at CSR Ltd. Pyrmont refinery. Liquor adsorbers (5 + 3) at right, regenerated carbon storage tank at left. Plant capacity on eight columns of 61 tonnes of melt per hour.

a. Sugarcane Plant Origin. Phenolic, polyphenolic, and flavonoid colorants were originally attached to the plant cell walls. They are generally of low molecular weight and feature aromatic ring structures. These colorants are pH sensitive, with indicator values (the ratio of color measured at pH 9 to color measured at pH 4) greater than 5. Color precursors, also called colorless color, generally impurities that react during processing to form colored compounds, includes simple phenolics and aminonitrogen compounds. Iron compounds, if present or accessible from process plant construction materials, may catalyze color-forming reactions in the refining process.

b. Factory-Formed Colorants. Melanoidins are formed from the reaction of glucose and fructose with amino acids in Maillard reactions. These intensely colored compounds have a low indicator value (about 1) and polymerize to form relatively high molecular weight colorants which are generally difficult to remove. Caramels are formed by the thermal degradation of sugar and other carbohydrates that occurs during processing at high temperatures and long residence times, such as during evaporation or crystallization at low vacuum. Indicator values are typically 1 to 1.5. Alkaline degradation products of fructose and to a lesser extent, glucose, particularly at higher pH levels (>7.5 pH) are additional colorants. A more detailed description of the various classes of sugar colorants and their properties was set out by Clarke [4].

8.2.2 Carbon Decolorization Mechanisms

The term *adsorption* refers to the existence of a higher concentration of any particular component at the surface of a liquid or solid phase than is present in the bulk. The key attributes of granular carbon are the very high surface area available for adsorption, and the porosity of the particles. Because the carbon surfaces are nonpolar, adsorption is probably due to the effects of van der Waals forces. These are relatively weak intermolecular forces (estimated at 5 kcal/g). The importance of high surface area is amplified because colorant molecules are likely to be in the form of a monolayer at the carbon surface.

Colorants that have an aromatic ring structure are likely to be adsorbed through a hydrophobic bonding mechanism with carbon, similar to the mechanism that occurs with a polystyrenic ion-exchange resin matrix. The wide range of colorants present in sugar liquors makes a quantitative theoretical treatment of adsorption virtually impossible for sugar decolorization. Each color component present has its own adsorption potential, so that any theoretical treatment requires summation of the effects of a large number of colorants. No practical solution to this problem has been developed to date.

8.2.3 Colorant Adsorption Specificity

Carbon-based adsorbents such as granular carbon are strong general adsorbents showing little specificity in the sense that one colorant is adsorbed to the exclusion of another. However, it has been shown [5] that there are some preferences in terms of colorant removal between carbon, bone char, and ion-exchange resins. In particular, carbon removes almost all of the flavonoid colorants in the feed liquor, whereas some flavonoid materials are not removed by resins. Carbon is also effective in removing most phenolic compounds.

8.3 DECOLORIZATION OPERATIONS WITH GRANULAR CARBON

Granular carbon is used in both fixed- and moving-bed installations with the carbon held inside an adsorption column. Fixed-bed installations have some advantages in simplicity in small to medium sized installations up to six or eight columns. However, moving-(pulsed) bed installations are likely to be appropriate for very small installations (where only one adsorber column is required) and also for large-capacity installations where more than eight fixed columns would be needed. The benefit from having a single pulsed-bed column in a small installation is that a more consistent product liquor quality is obtained.

In fixed-bed installations, liquor flow is down through the carbon bed held in the column. In pulsed beds, fresh carbon is introduced into the top of the column and spent carbon removed periodically from the bottom. Liquor flow is countercurrent to carbon movement and is thus from the bottom to the top of the column. Spent carbon is desweetened and reactivated before returning to the adsorber columns for further use. The carbon is transferred to and from the columns hydraulically as a water–carbon slurry. Whereas granular carbons may be reactivated thermally, chemically, or by use of superheated steam, sugar colorants are so tightly adsorbed that thermal reactivation at 900 to 1000°C is almost universal. However, successful operation of an alkali regenerable granular carbon has been reported from Japan [6]. Fixed- and pulsed-bed systems have been described in the literature on a number of occasions [7–10].

8.3.1 General Design Principles

Typical design targets for granular carbon decolorization plants are shown in Table 8.1. As with most process plant, the design of a granular activated carbon plant requires the solution of a problem that may have multiple answers. The best design can only be found through careful examination of the local factors affecting the general design. However, adherence to several basic principles is necessary. Below we describe briefly methods for determination of the critical design parameters. These parameters are independent of the selection of a fixed- or pulsed-bed system.

Laboratory trials are the most effective way of determining the critical design parameters. Generally, however, colorants in cane sugar liquors worldwide fall into common categories as described in Section 8.2.1, and for preliminary design purposes, estimates of the design parameters can be made without carrying out laboratory decolorization tests. The expected ranges of these design parameters are indicated in the following sections. We include an example based on the data in Table 8.2.

a. Carbon Type Selection. The design begins with the selection of the most suitable carbon for the duty proposed. Established carbon suppliers have a number of potentially suitable carbons available, with and without added magnesite. Laboratory testing for adsorption isotherm, carbon dosage, and contact time using samples of the material to be decolorized provides the data for correct selection of the most suitable carbon. The final

TABLE 8.1 Typical GAC Plant Design Targets

Performance Indicator	Design
Decolorization (%) (Overall)	85–90
Fixed-bed lead column	80
Fixed-bed trail column	50
Product fine liquor (pH)	<7.0
Liquor pH drop over decolorization (pH units)	<1.0
Invert gain over decolorization (%)	<0.05
Carbon use (burn) (% feed liquor solids)	0.8
Carbon loss (% carbon reactivated)	<5.0

TABLE 8.2 Data for Example

Parameter	Value
Capacity (m^3 / h)	43.8
Decolorization (% of feed color)	90
Carbon dosage (% on dry tonnes of liquor)	0.8
Density (Brix)	65
Contact time (h)	6
Superficial velocity [(m^3 / h) / m^2 bed]	3.7

decision on the type and brand of carbon to be used is an economic optimization balancing the capital cost of plant and the operating costs associated with the design selected.

b. Contact Time **(τ).** A number of experiments are conducted in which identical aliquots of the material to be decolorized are contacted for varying amounts of time with identical amounts of the activated carbon selected. A plot of contact time versus percent adsorbate remaining in the liquid phase is produced (Fig. 8.2). To achieve full utilization of the carbon, the designer is looking for the system to have achieved an equilibrium between adsorbate in the liquid phase and adsorbate in the adsorbed phase. Typical Australian raw sugar liquors require contact times of between 4.5 and 6 h with available commercial carbons.

c. Carbon Dosage **(r_{AC}).** Once the contact time is established, the rate of carbon dosage can be determined by generating what is known as an *adsorption isotherm.* Aliquots of the material to be decolorized are mixed with varying quantities of carbon. All the samples are contacted for the predetermined contact time and then the solutions are analyzed for the color remaining in the liquid phase. A plot (normally in log-log format) of concentration of adsorbate left in the liquid phase versus adsorbate adsorbed per unit weight of carbon is produced. A straight line should be produced if all components of "color" are adsorbed similarly. If nonlinear isotherms are produced, this is an indication that there are color components within the system that are adsorbed at significantly different rates by the carbon; perhaps some components may not be adsorbed at all. For further detail in this area, con-

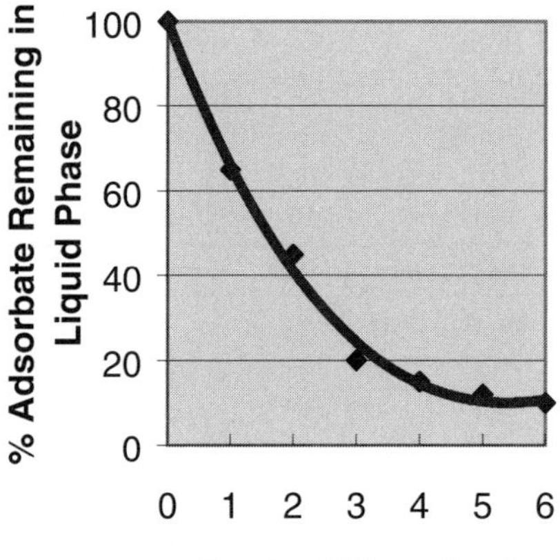

FIGURE 8.2 Typical contact time plot.

sultation with a carbon supplier is recommended. The plot will indicate the point at which the adsorbed phase on the carbon is in equilibrium with the influent concentration. This point represents the ultimate capacity of the carbon under the test conditions and so represents the *minimum* carbon dosage required. In practice, the actual carbon dosage is higher than this value because in the plant, it is unlikely that the carbon removed from the system will be saturated with respect to the color in the liquor. There are also other dynamic considerations to consider (e.g., mass transfer mechanisms).

The adsorption isotherm technique is useful to (1) establish whether the decolorization performance required is feasible, (2) compare the performance of two or more carbons, and (3) compare the performance of a carbon at differing temperatures and pH values. Generally, for decolorization of sugar liquors to a level of 90% from approximately 1200 ICUMSA feed color to 120 ICUMSA color requires a carbon dosage rate of about 0.008 tonne of carbon per dry tonne of liquor. Each tonne (dry basis) of liquor processed requires 8 kg of carbon.

d. Dynamic Considerations. The adsorption of color onto the carbon is governed by the general diffusion equation

$$m = DA \frac{\Delta c}{\Delta t} \tag{8.1}$$

The concentration term c, the driving force for the mass transfer, is governed by the characteristics derived in the adsorption isotherm test. The area A is the surface area of the carbon through which mass transfer takes place. The mass transfer coefficient D is concerned with the dynamics of the system and accounts for superficial velocities, viscosity, and density. A pilot plant test setup is the usual method of determining the influence of these factors.

***e. Superficial Velocity* (v).** Superficial velocity is generally defined in units of $(m^3/h)/m^2$ of bed cross-sectional area. A higher superficial velocity will increase the pressure drop through the bed but may increase decolorization efficiency through thinning of the boundary layer surrounding each carbon particle through which the mass transfer of colorant in the desorbed phase to the adsorbed phase occurs. For granular carbons of approximately 1 mm mean aperture, superficial velocities (v) are typically in the range 3.5 to 4.5 $(m^3/h)/m^2$ of bed. One consequence of superficial velocities in this range is that columns are relatively long and of small diameter. In some instances it may be necessary to design for superficial velocities in the range 2 to 3 $(m^3/h)/m^2$ of bed area to reduce the overall plant height to fit building constraints. This is likely to reduce decolorization performance slightly as a result.

f. Carbon Use and Kiln Rate. The carbon dosage rate and plant flow rate are used to calculate the carbon use that is equivalent to the *average kiln production rate:*

$$m_{AC} = \frac{r_{AC} F \rho \mathrm{Bx}}{100} \tag{8.2}$$

$$\left[m_{AC} = \frac{0.008 \times 43.8 \times 1.27 \times 65}{100} = 0.29 \right]$$

where m_{AC} is the mass rate of activated carbon exiting the kiln (tonnes/h), r_{AC} the carbon dose (tonnes of carbon per dry tonne of liquor), F the plant flow rate (m^3/h), ρ the specific gravity of the feed liquor, and Bx the percent solids in the feed liquor (°Brix).

Average feed rate to the kiln is calculated from the average production rate and then adjusted for losses within the kiln system, losses within the carbon transport system, and differences in bulk densities between spent carbon [ca. 0.6 tonnes/m^3 (dry basis)] and regenerated carbon [ca. 0.5 tonnes/m^3 (dry basis)]. Carbon losses through burning vary, but good manufacturing practice gives values of about 3 to 5% loss (carbon kiln production rate basis).

$$m_{SC} = m_{AC}(1 + l_k)(1 + l_T)\,\frac{BD_{SC}}{BD_{AC}} \tag{8.3}$$

$$\left[m_{SC} = 0.29(1 + 0.03)(1 + 0.02)\left(\frac{0.6}{0.5}\right) = 0.37\right]$$

where m_{SC} is the mass rate of spent carbon to the kiln (tonnes/h), m_{AC} the mass rate of activated carbon exiting the kiln (tonnes/h), l_K is the level of carbon loss within the kiln (%), l_T the level of carbon loss within the carbon transport system (%), BD_{SC} the bulk density of spent carbon (tonnes/m^3), and BD_{AC} the bulk density of activated carbon (tonnes/m^3).

g. Bed Cross-Sectional Area (A). Given the plant flow rate and the required superficial velocity, the total cross-sectional area of the bed is calculated.

$$A = \frac{F}{v} \tag{8.4}$$

$$\left[A = \frac{43.8}{3.7} = 11.8\right]$$

where A is the cross-sectional area of the bed (m^2), F the plant flow rate (m^3/h), and v the superficial velocity [($m^3/h)/m^2$ of bed]. Note that this is the total cross-sectional area and not actual column cross-sectional area. If two sets of columns are running in parallel, the individual column cross-sectional area would be halved.

h. Total Liquor Volume On-Line. Contact time required and plant flow rate required allow calculation of the total volume of liquor required.

$$V_{TOT} = \tau F \tag{8.5}$$

$$[V_{TOT} = 6 \times 43.8 = 262.8]$$

where V_{TOT} is the total volume of liquor in the columns (m^3), τ the contact time (h), and F the plant flow rate (m^3/h).

i. Total Column Volume On-Line. The column contents are made up of the carbon bed and liquor filling the pores and interstitial space. The characteristics of the carbon bed can be obtained from the carbon manufacturer. For the example, assume that a Calgon Cane-Cal carbon is chosen with a typical voidage of 38% and a pore volume of 66%. This means that in a packed bed, 38% of the volume of the bed is interstitial space, and of the remaining 62%, 66% is carbon pore space (i.e., a further 41%). It is the sum of these spaces that is filled with liquor (i.e., 79%). We have already calculated that the on-line columns will hold 262.8 m^3 of liquor. This liquor is contained in 79% of the total volume of the column (i.e., a total column volume of 332.7 m^3 is required).

8.3.2 Fixed-Bed Systems

a. Column Features. Figure 8.3 shows a schematic for a typical fixed-bed column. Significant features are as follows:

1. *Inverted conical bottom.* This ensures that there is no holdup of carbon in the column during a spent carbon transfer.

2. *Internal screens in base of column.* These screens may take a number of forms. In Australia, the predominant type is a wedge wire screen formed into a cylinder. Up to six screens are mounted radially in the base of the column. See Figure 8.4 for a partial cross section of the base of a typical column.

3. *Positioning the carbon slurry inlets and outlets.* Carbon slurries are transported into and out of the column via nozzles mounted on the vertical axis of the column. The central

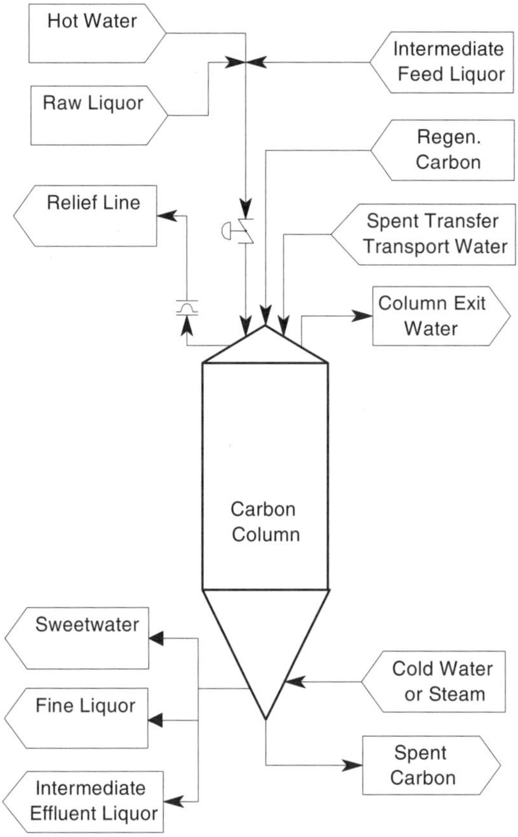

FIGURE 8.3 Fixed-bed column schematic.

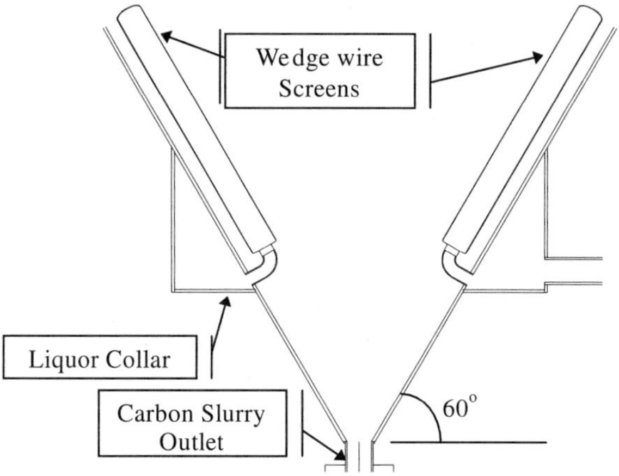

FIGURE 8.4 Fixed-bed column cone detail.

outlet together with the conical bottom ensures that no carbon remains in the column after a spent transfer. The central inlet allows even filling of the column.

4. *Overpressure protection.* The columns are pressure vessels, and pressure vessel standards apply to their design. The design pressure varies according to many conditions, including flow rate, bed pressure drop (which depend on the type of carbon used), and number of columns in series with no intermediate pumping. The column must also have systems in place or be sufficiently robust in design to accommodate full vacuum. (Sterilization steam condensing in the column creates a vacuum.)

5. *Insulation.* The liquor should be kept at approximately 80°C. Lower temperatures cause lower efficiency of the adsorption process. Also, lower temperatures lead to increases in the viscosity of the liquor. This in turn raises the resistance to flow through the bed and will lower flow rates for the same inlet pressure.

b. Number of Column Sets. The number of column sets running in parallel is primarily an economic decision by the process designer but with some important principles to consider:

1. The greater the number of parallel sets, the better utilized the carbon, as the situation more closely approaches the theoretical countercurrent optimum.

2. The greater the number of parallel sets, the greater the frequency of column changeovers.

3. The greater the number of parallel sets, the greater the capital cost. As the number of sets increases, the cross-sectional area of each column decreases (total cross-sectional area is set by the superficial velocity). It is significantly less expensive to construct five large columns than 10 small columns where the total cross-sectional area in each case is identical. Ten columns also requires replication of individual column controls 10 times.

Generally, fixed-bed plants should not be designed with only one set of columns in series. Utilization of carbon is poor. In plants where there is more than one set, the newer sets produce liquor with color lower than the color standard required. Their effluent is offset by older sets producing liquor above the color standard. Where only one set exists, the effluent color will be below standard at the start of the cycle (although blending/bypassing of the column can alleviate this problem), but when the effluent color reaches the maximum allowable color, the cycle must end. There is no offsetting benefit from other columns.

For the example, two column sets are chosen. This, in conjunction with the bed cross-sectional area (Section 8.3.1), allows calculation of an individual bed cross section (i.e., 5.9 m^2).

c. Column Sizing. With the diameter of the column set, it is now appropriate to determine the number of stages required. The more stages there are, the closer the system is to describing a true countercurrent system, however, there is significant cost penalties with a large number of stages. The majority of plants around the world are constructed with two passes and occasionally, three passes. This is regarded as the most cost-effective selection.

For the example, two stages are selected. The total column volume on line was 332.7 m^3, there were two column sets, and the individual bed cross section is 5.9 m^2. We have already determined that there are two column sets, each with two columns in series. The total volume of a column is therefore

$$\frac{332.7}{4} = 83.2 \text{ m}^3$$

Columns have a 60° conical bottom and a torospherical top. Usually, the column is filled only to the top of the cylindrical section, so the volume is calculated on this section only. For an 83.2-m^3 column of 5.9-m^2 cross-sectional area, the diameter is 2.74 m, the height of the cylindrical section is 13.32 m, and the total height of the column is 15.69 m.

d. Flow Patterns. The ideal flow pattern for contacting the sugar liquor to be decolorized with the activated carbon is countercurrent. This arrangement is not practical for fixed beds, so a compromise is used. Fixed-bed carbon plants often run multiple sets of columns in series. Figure 8.5 shows a typical arrangement for a five-column plant. Two columns run in the leading position and two in the trailing position. The columns have staggered starts.

e. Column Changeovers. A schedule for column changeovers from one liquor duty to the next or for sweetening off one column and bringing a new column on line can be calculated. This sets the expected liquor running period (i.e., the actual period when a column is decolorizing liquor). The actual time for changeover may vary depending upon, for example, base color loading, kiln performance, and factory rate. In the design stage, however, the calculation of "expected" column changeovers is useful in verifying the appropriateness of the column and kiln designs. Small columns mean more frequent changeovers. Some of the processes in the off-line cycle (sweeten off, carbon transfer, column sterilization, column fresh carbon fill, and sweeten on) have periods that are only partially dependent on the column size, so halving column size does not necessarily mean that the cycle time can be halved. Where the off-line cycle is under pressure from the design, a second, spare column is sometimes installed to provide the extra buffer within the system.

For the example, the total column volume on-line is 332.7 m^3 in four columns (two by two). If a standard carbon bulk density of 0.5 tonne/m^3 is assumed, the tonnes of carbon on-line is approximately (0.5 × 332.7, i.e., 166.4). The carbon dose is assumed to be 0.8%; hence this carbon will treat (166.4/0.008) 20,800 tonnes of liquor. At the design flow rate

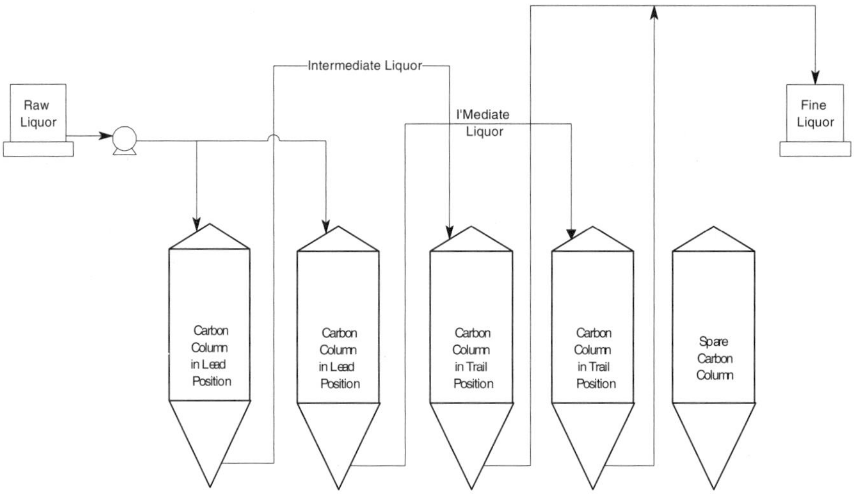

FIGURE 8.5 Typical column arrangement for a five-column fixed-bed system.

of 43.8 m^3/h [36.2 tonnes (dry) per hour at 65 Brix], the columns will be exhausted after (20,794/36.2) 574.4 h. There are four columns and the changeovers will be equally spaced; therefore, the changeover will occur every (574.4/4) 144 h.

f. Column Cycle Steps. Each column in the plant follows a cycle. Table 8.3 shows typical steps in one such cycle. The starting point assumes that the column is full of carbon and water.

8.3.3 Pulsed-Bed Systems

a. Column Features. A typical arrangement for a pulsed-bed system is shown in Figure 8.6. The main features of the layout are:

1. *A 60° conical bottom* assists in the mass flow exit of carbon from the column. The nature of the pulsed-bed system makes it imperative that carbon flow through the column during pulses is mass flow. "Rat-holing" is not permissible.

2. *The desweetening column* is placed low down in the plant so that carbon transfer from the column will occur under gravity.

3. *The reactivated carbon feed tank* is sited directly above the main decolorizing column. Carbon transport into the column is again by gravity.

4. *Screen filters are placed on the liquor outlets* from the column. These can be either internal or external and are used to screen carbon granules from the liquor. They will not screen out fine carbon particles. The bed structure should accomplish this task except for the period just after a slug or pulse.

b. Pulse Cycle. The reactivated carbon feed tank above the column should preferably be full or at least contain sufficient carbon for a complete pulse prior to beginning the pulse cycle. When the pulse is scheduled, liquor flow to the column is stopped. The spent carbon

TABLE 8.3 Column Cycle Steps

Step	Source of Column Feed	Destination of Column Effluent	Comments
Sweetening on			
Stage 1	Intermediate liquor	Sweetwater	This step finishes when the effluent Brix is too high to send to sweetwater (typically, 40 Brix).
Stage 2	Intermediate liquor	Intermediate liquor	Carbon fines sometimes contaminate the initial effluent from the column. This column is recirculated until clear.
Trailing position	Intermediate liquor	Fine liquor	The column goes into the trail position. The fresh carbon in the column contacts with the liquor of lowest color. As the color loading on the carbon increases (on both the trail and lead column), the feed color will increase and the effluent color will also increase. Gradual loading of the carbon with color occurs.
Leading position	Raw liquor	Intermediate liquor	When the effluent color is no longer acceptable for addition to the fine liquor stream, the column is moved to the leading position. The column should stay in the leading position for the same time as the trailing position.
Sweetening-off			
Stage 1	Hot water	Intermediate liquor	Once the effluent color is no longer acceptable for continuing to intermediate liquor, hot water is introduced to the top of the column and the sugar solution is washed out.
Stage 2	Hot water	Sweetwater	Once the effluent Brix is too low for addition to intermediate liquor (approximately 40 Brix), the effluent is redirected to sweetwater. Desweetening the column uses approximately three bed volumes of hot water.
Washing	Hot water	Drain	Some ash components left on the carbon leach out during washing.
Spent carbon transfer	Contaminated transport water	Spent carbon tank	The column pressurizes with transport water, and the carbon transports hydraulically to the spent carbon tank. The overflow of water from the spent carbon tank returns to the contaminated transport water tank.
Column drain	—	Drain	After transfer, the column is full of contaminated transport water that contains a significant proportion of carbon fines. This water is dumped to waste.
Column clean	Steam	—	The empty column is sterilized, usually with stream. Biocides may be used as an alternative.
Fresh water fill	Fresh water	—	Filling with carbon uses a wet fill technique. The column fills with fresh water in preparation for this step.
Carbon fill	Carbon slurry	Regenerated carbon tank	As carbon slurry feeds to the top of the column, water overflows to the regenerated carbon tank. Overflow from this tank goes to the clean transport water tank.

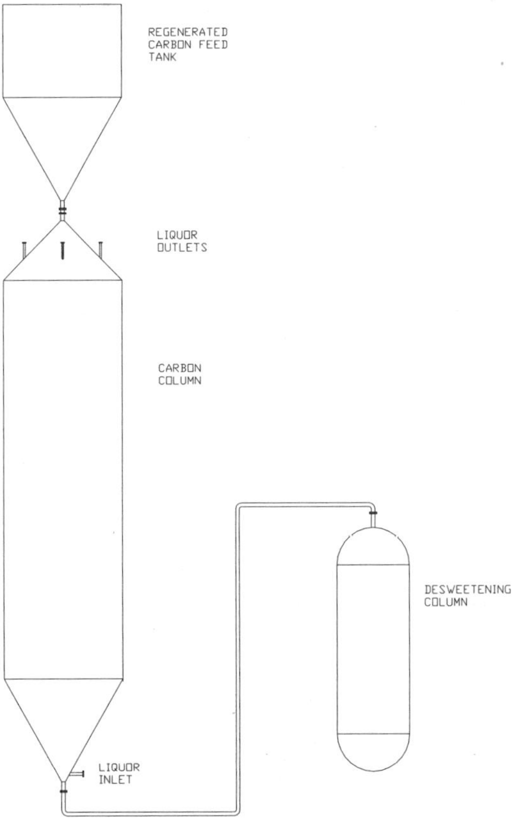

FIGURE 8.6 Typical pulsed-bed layout.

valve at the base of the column and the reactivated carbon fill valve at the top of the column are opened and spent carbon transfers to the desweetening column under gravity. The reactivated carbon enters the column simultaneously. When sufficient carbon has transferred, the spent carbon valve is closed, followed by the reactivated carbon valve. Liquor flow is restored.

The first runnings of the new bed will contain fines that have been disturbed from the bed and the solution will be dilute (the carbon slurry in the regenerated feed tank is carbon–water). This effluent can be recirculated to the feed tank if the amount is considered to be insignificant, sent to the sweetwater tank, sent to product (if the low density is considered insignificant and the trap filters are considered to be adequate), or a combination of the three.

The carbon–liquor slurry in the desweetening column is desweetened using hot water. The effluent is usually sent to sweetwater for recovery of the sugar with final runnings (below 1 Brix) sent to drain if the water is not required within the factory. Approximately three bed volumes of water are used to desweeten the carbon completely. The rate of desweetening is determined by trials. There is an optimum rate for minimization of water usage that is affected by bed density, column diameter, and hot-water temperature.

Carbon must be fully desweetened before sending to the kiln feed tank. Failure to desweeten will increase the loading on the kiln, as oxidation of the organics must occur to fully reactivate the carbon. In some cases, the extra load may increase the loading to such an extent that insufficient oxygen is available within the kiln to complete the oxidation, and odors may be produced. Once the carbon is desweetened, it is transported as a water slurry to the carbon feed system. There it is dewatered and fed though the kiln for reactivation. On leaving the kiln, the carbon is quenched by dropping into cold water and then pumped to the reactivated carbon feed tanks above the columns ready for reuse.

c. Flow Arrangements. There are several ways of designing the flow through the pulsed-bed plant. In general, however, the liquor and carbon are both single pass (i.e., regenerated carbon enters the top of the column and is spent by the time it reaches the bottom). Filtered liquor (from the clarification plant) enters the bottom of the column and is fully decolorized by the time it reaches the top. In theory, the columns could be placed in a multiple-pass system for carbon and/or liquor. Although this may be desirable from a process point of view, it is usually too expensive and too complicated to consider in practice.

d. Column Sizing. Column sizing within a pulsed-bed system uses the same basic requirements as those for a fixed-bed plant (i.e., for the example, the contact time is 6 h and the superficial velocity is 3.7 m^3/m^2 of bed area. The difference in design occurs because each pulsed-bed column performs the entire decolorization duty in one column. Section 8.3.1i established the total column volume required on line (332.7 m^3). The cross-sectional area of the plant bed has also been determined at 11.8 m^2. If one column were selected, the required total column height would be 30.4 m. For two columns, where the column cross section is half the plant bed cross section, the column height required is 29.8 m (diameter of 2.74 m). For three columns, the total column height is 29.5 m with a diameter 2.24 m. It can be seen that the column height varies only slightly, irrespective of the diameter. This is because the superficial velocity requirement is the main determinant of column height. In all cases, the superficial velocity is identical, so the column height should also be identical. The slight differences in total height are due to the differing heights attributable to the 60° conical bottom.

As noted earlier, while the use of high superficial velocities may enhance decolorization performance, it also means that the decolorization columns are relatively long and of small diameter (almost 30 m high and 2.74 m in diameter in the example above). Practical considerations may require shortening the columns by increasing diameter and reducing the superficial velocity into the range 2 to 3 $(m^3/h)/m^2$ of bed area. As a result, a slight increase in carbon use may be required to achieve design decolorization.

e. Pulses and Number of Columns. The selection of the number of columns and the frequency of pulsing is a multivariable system with multiple answers. To illustrate the issues to be taken into account in determining the optimum solution, the previous design example data will be used. The average carbon rate required was 0.29 tonne/h (Section 8.3.1c). For a single column, this might be one pulse of 290 kg each hour or a 6.96-tonne pulse each 24 h. For two columns each treating half the flow, the pulse might be 145 kg each hour for each column or a 3.48-tonne pulse for each column each 24 h. For three columns, the pulse might be 96.7 kg per column per hour or a 2.32-tonne pulse each 24 h.

Each pulse within the pulsed-bed system disturbs the carbon bed. This disturbance causes fine carbon particles to be removed from the carbon and leave the column with the product liquor. Small frequent pulses result in numerous disturbances to the bed and are undesirable, as the plant spends a high proportion of time recirculating and not producing. Large, less frequent pulses allow the bed to settle in between disturbances, but carbon loading is not as consistent as it would be with smaller pulses. Common operating practice is to use a pulse volume of 5 to 9% of the column volume at a pulse rate of one per day.

The disturbance to production is lessened by having more than one column, as only that proportion of production being treated by the pulsing column will be affected. Consideration of the size of the buffer tanks both upstream and downstream of the plant can also affect selection of the frequency of pulse and number of columns. Between each pulse, the spent carbon must be desweetened and transported to the kiln feed supply tank. The period between pulses must allow sufficient time for this to occur. Increasing the number of desweetening columns effectively increases the available time to desweeten but at extra capital cost.

8.3.4 Pulsed-Bed Versus Fixed-Bed Systems

Pulsed-bed fixed-bed systems are compared in Table 8.4.

8.4 REACTIVATION OF GRANULAR CARBON

Several processes are available for reactivation of carbon: chemical reactivation, steam reactivation, and thermal reactivation. Almost exclusively, however, large-scale applications within the sugar industry use thermal activation because of its relatively high reactivation efficiency. In this section we examine first the process of thermal reactivation of the granular activated carbon. We then go on to compare the more common equipment used for reactivation.

8.4.1 Reactivation Reactions

Thermal reactivation of carbon consists of four basic steps: drying, desorption/distillation of volatile compounds, carbonization/calcination/pyrolysis of nonvolatile compounds, and finally, gasification of the carbonaceous residue left after the carbonization/pyrolysis step (Table 8.5).

8.4.2 Residence Time

Residence time in the kiln is important. Too long a residence time and the carbon is overactivated and loses its hardness. The crystalline structure of the carbon base is slowly broken down at the elevated temperatures in the kiln. The crystalline structure is what gives the carbon its strength. Higher attrition rates occur with subsequent handling of this overacti-

TABLE 8.4 Comparison of Pulsed-Bed and Fixed-Bed Systems

Parameter	Pulsed Bed	Fixed Bed
Capital cost	Lower capital cost	Higher capital cost
Working capital	Lower working capital cost	Higher working capital cost
Plant simplicity	Simple	Complex
Operational simplicity	Complex	Simple
Kiln	High reliance on kiln availability	Low reliance on kiln availability
Hygiene	Microbial infections in column difficult to eradicate	Columns are sterilized after each cycle
Technology base	Fewer plants and lower knowledge base	Known technology and high knowledge base

TABLE 8.5 Thermal Reactivation of Carbon

Stage	Description	Temperature (°C)	Bulk Density (tonnes / m^3)
Initial conditions	The spent carbon tank stores carbon as a slurry. The carbon doses to a kiln feed mechanism (often, a slow-speed helical screw conveyor) that dewaters it to approximately 50% water (w / w). At this water content there is no free water. The spent carbon consists of a carbon base structure, included water, organic adsorbate bound to the activation sites, and some inorganic adsorbate.	Ambient	1.1 –1.3
Drying	Water is removed early in the process. This includes both the water in the interstitial voids and water in the pores. Some volatile components of the adsorbate may evaporate.	~100	0.55–0.65
Carbonization / calcination	As the temperature of the carbon rises, organic adsorbates decompose and / or polymerize to form a carbonaceous residue. This residue remains in the pores. Inorganic adsorbate is calcined.	100–800	0.52–0.55
Activation	The process of activation oxidizes the carbonaceous residue to CO. The carbonaceous residue fouls the potential active sites on the base carbon structure and must be removed from the pores of the carbon. This is done through gasification of the residue through reaction with H_2O and / or CO_2. The simplified reactions are $C + H_2O \leftrightarrow CO + H_2$ $C + CO_2 \leftrightarrow 2\,CO$ Both reactions are endothermic. Steam is usually preferred, as it is readily available and much more readily controlled. Industry sources give steam addition rates ranging from 0.5 to 1.5 kg / kg of carbon. The rate of reaction below 700°C is negligible but increases exponentially above that temperature. The temperature reached in the reactivation kiln is generally about 900 to 1000°C. The CO and H_2 generated are oxidized to CO_2 and H_2O using excess air from the burners. The oxidization reactions provide supplemental heat to that of the burners. As the gasification reaction essentially removes carbon from the system, care must be taken to ensure that only carbonaceous residue is removed and not the activated carbon. Some activated carbon will always be lost and this comprises most of the kilning loss referred to earlier. Physical chemical considerations indicate that if the spent carbon is heated rapidly to around 800°C (where the carbonization reactions are complete), the rate of carbonization will preclude the formation of any crystalline structure within the residue; an amorphous state prevails. This amorphous residue exhibits a greater gasification activity than does a crystalline structured carbon (the structure of the carbon base).	800–1000	0.49–0.55

vated carbon. A short residence time will not permit the gasification reaction to be completed, and the carbon will not be fully reactivated.

The appropriate residence time is governed by the rate of heat transfer to the carbon and, at times, by the rate of mass transfer of the impurities (water and organics) from the granule to the surrounding air and, of course, the rate of reaction between the impurities and the air. In practice, the residence time is determined by experimentation after the kiln is on-line, but as a general guideline, multiple-hearth kiln residence times are around 45 min, whereas fluidized-bed kilns are around 30 min. The lower residence time in the fluidized-bed kiln is due to the much better heat transfer characteristics of this type of kiln.

8.4.3 Reactivation Kilns

The multiple-hearth kiln is the most widely used type for reactivation of carbons used in sugar refining. Other types, such as pipe kilns and fluidized-bed kilns, may be used. Most of the discussion in this section focuses on multiple-hearth kilns. Fluidized-bed kilns are examined and their characteristics compared to those of the multiple-hearth kiln.

a. Multiple-Hearth Kiln

Description. The multiple-hearth kiln (Fig. 8.7), also commonly known as a Herreshoff-type kiln, consists of a number of hearths (usually six to eight in the case of reactivation of carbon from sugar liquor decolorization duty) situated above one another. A central shaft fitted with rabble arms on each hearth slowly rotates and rakes carbon across the surface of the hearths. The shell is constructed of refractory-lined mild steel, and each hearth is also constructed of refractory steel. The central shaft and rabble arms are constructed of high-temperature cast steel and are air cooled. The exhaust air from the shaft cooling system is fed to the afterburner as preheated air.

After passing through a dewatering screw conveyor, the carbon is fed to the top hearth. The action of the central shaft rakes the carbon toward the center of the kiln, where it falls through holes to the next hearth. The scraper blades fitted on the arms of this hearth rake the carbon to the perimeter of the kiln, where again the carbon falls through holes to the next hearth. This procedure continues down through the kiln. Carbon leaving the bottom hearth is quenched and pumped away to the regenerated carbon storage vessel.

Burners control the temperature on individual hearths and are generally located on the bottom four or five hearths. Steam addition occurs in the bottom one or two hearths. Product gases rise through the kiln countercurrent to the carbon flow and leave the kiln to pass through a simple cyclone and then an afterburner. The kiln operates under a slight vacuum. Pressure is controlled by an induced-draft fan sited in the exhaust gas line from the kiln (between the cyclone and the afterburner).

Kiln Design. Correctly sizing and designing the kiln is an important part of the carbon plant design. A kiln that is too large spends a large proportion of its time either idling with no feed available (a waste of energy) or shut down. Shutting down and restarting kilns is generally considered to be undesirable because of the large thermal stresses induced in the hearths, ultimately leading to mechanical failure. However, a kiln that is too small cannot maintain the regeneration rate to supply the decolorization columns. Kiln burner turndown is usually poor; hence the option of running a large kiln slowly is not valid. One potential solution is to run the large kiln with some burners turned off. This can be done only on kilns designed with this capability in mind.

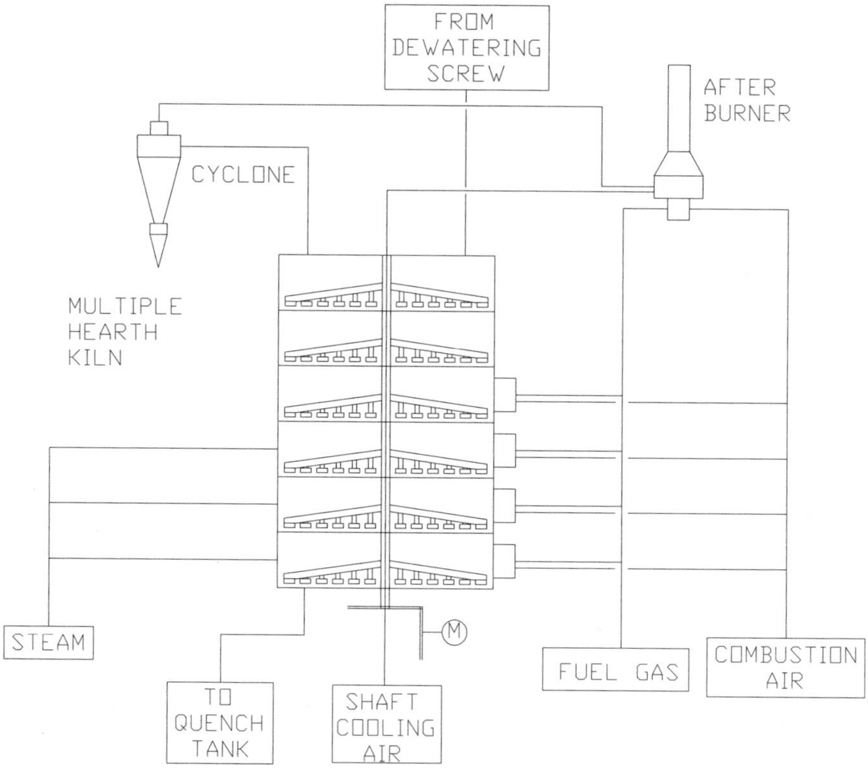

FIGURE 8.7 Schematic of a rotary multiple-hearth kiln.

Kiln Rate. From the example, as calculated in Section 8.3.1f, the calculated kiln rate is 0.29 tonne/h carbon exiting the kiln or 0.37 tonne/h entering into the kiln. If the kiln were designed at 0.29 tonne/h (the exiting kiln rate), there is no allowance for any delay or breakdown to the system. At higher rates than this, if it is assumed that the kiln has zero turndown, there will be idle time. The best solution is to determine the maximum turndown that the kiln can offer, and that capacity should be the minimum carbon rate required by the process. Hence there is a need to understand the factory rate as well as the kiln rate to design the kiln rate correctly.

Assume a factory production turndown of 30% (i.e., the factory may run stably at 70% design capacity). For the example, then, the minimum carbon rate is $(100 - 30)/100 \times 0.29 = 0.20$ tonne/h. Assume that the kiln burner turndown is 60% (i.e., stable operations can be maintained at 40% of the kiln's rated capacity). The minimum carbon rate is equated to the kiln turndown rate (i.e., 40% of kiln rated capacity $= 0.20$ tonne/h). The kiln design rate is therefore 0.5 tonne/h.

Diameter and Number of Hearths. The process of regeneration (Section 8.4.1) has three distinct areas: drying, carbonization/calcining, and reactivation. The drying process should be slow, as too short a drying period may result in some water remaining within the carbon pore structure. At higher temperatures this may vaporize and "explode" the carbon granule, resulting in a high attrition rate through the kiln. The calcining reaction can occur relatively

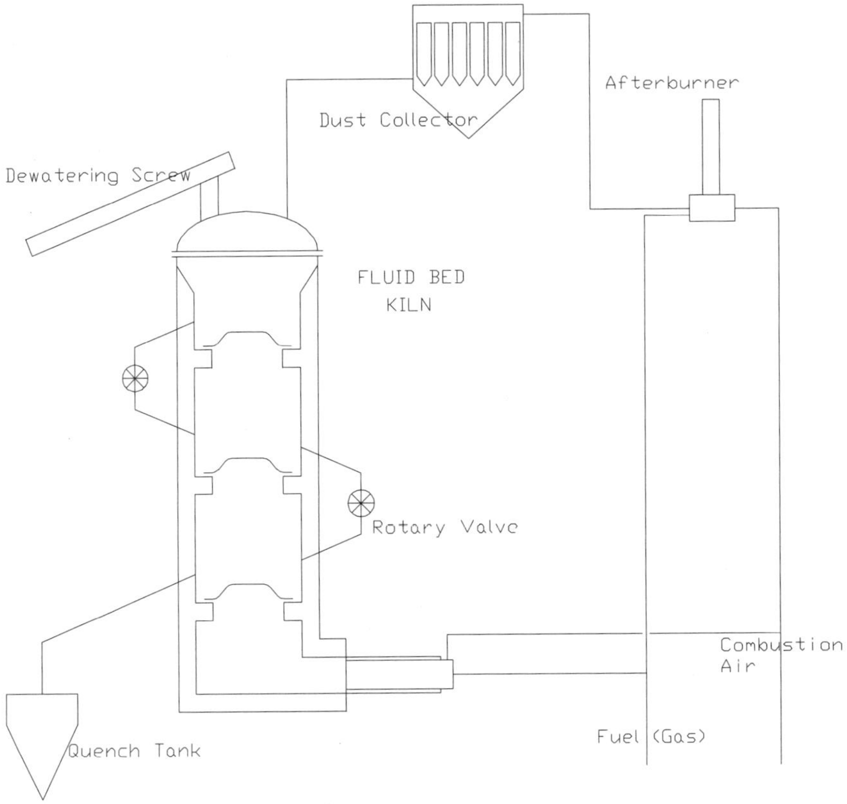

FIGURE 8.8 Schematic of a typical fluidized-bed kiln.

quickly on one hearth, but the reactivation section requires around 20 min at the required carbon temperature of 870°C.

The recommended minimum number of hearths is five and the maximum number is eight. The maximum number is generally arrived at from an economic viewpoint. More hearths mean more burners and higher capital and operating cost, but reduce the temperature gradients between hearths. The general design rule for physically sizing the kiln is to allow 23.7 kg/h carbon throughput per square meter of hearth area. For the design example of 0.5 tonne/h, the total hearth area required is 21.1 m^2. Five hearths would give an individual hearth area of 4.22 m^2 (a diameter of 2.3 m); six hearths, an area of 3.52 m^2, (a diameter of 2.1 m); and eight hearths, an area of 6.72 m^2 (a diameter of 1.8 m).

Shaft Speed. This is set between 0.5 and 1.5 rpm. For the same feed rate, slowing the speed will increase the kiln residence time but will also increase the carbon bed depth within the kiln. The bed should not be too deep, as some of the carbon at the bottom of the bed may not receive enough contact time with the kiln gases. In this case, the effect of increasing the residence time within the kiln is offset by the decrease in available time for the mass transfer to occur. In practice, the shaft speed should be set so that the hearth is just covered entirely with a bed of carbon, which shows a uniform sawtooth pattern made by the rabble arm rakes.

Operating Regimes. Due to the nature of the work they perform, rotary multiple-hearth kilns can be subject to substantial operating stresses. It is well known that the reducing atmosphere that exists within the kiln can also weaken the structure of the hearths. Thermal shocks induced by burner outages, by heating or cooling of the hearths through critical temperature ranges, or by sudden feed changes can also damage the integrity of the hearth structure. Consideration of these points alone would lead the designer to recommend a kiln that is at the operating temperature continuously.

Balanced against this is the relatively high cost of operating the kiln at the elevated temperature. Burner control turndown is usually quite poor and the lack of control when the kiln is at low firing rates can lead to instability of the flame and consequent shutdown, causing the shutdown one was trying to avoid. A further argument against continuous running may be dictated by the local environmental regulations. In some countries, the pollution control system on the exhaust of the kiln must be run if the kiln is on-line. If this pollution control system is an afterburner, it can contribute up to 40% of the fuel bill.

Pollution Control Systems. Environmental pollution control systems on kilns vary widely with the requirements of local statutory authorities. The information presented here summarizes the most popular components of these systems. It is recommended that selection of the appropriate design be done in conjunction with the local authorities.

Cyclones are the lowest-cost but least effective of the pollution control devices. They can only separate solid pollutants and then only based on the relative ease of gravity separation of these particles. The design of the discharge of the material collected from the cyclone should be considered carefully. The gas stream passing through the cyclone may be saturated with water, and condensation in the cyclone can bond with separated particles and block the outlet. Cyclones are rarely used alone but are used in conjunction with other pollution control devices.

The afterburner is another low-capital solution for kiln emission control. However, the low capital cost is balanced by a high operating cost. The afterburner must run at a high excess oxygen content to ensure complete oxidation of the kiln emissions. As mentioned earlier, the fuel cost could be up to 40% of the total fuel cost for the kiln.

Water scrubbers, using water or soda solutions, can be used to reduce pollutant emissions. The disadvantages of water scrubbers include the operating costs due to pumping and chemicals supply and the steam plume that exists over the exhaust from the scrubber. This can be counteracted by reheating the effluent from the kiln through contact (in a heat exchanger) with the raw emissions from the kiln but at an additional capital cost.

Fabric filters, commonly used in boiler station baghouses, can also be used to reduce particulate emissions but in general do not alone control gaseous pollutants. They are generally used in conjunction with a scrubber or afterburner. Figure 8.9 is another carbon installation of CSR Limited New Farm Refinery.

b. Fluidized-Bed Kiln. A typical fluidized-bed kiln (Fig. 8.8) for reactivation of carbon used for sugar liquor decolorization would consist of three hearths. The base of each hearth is constructed of a perforated plate through which the hot fluidizing gases pass. The material of construction of the hearths is generally high-temperature steel (301SS or similar). Walls consist of a refractory-lined mild steel shell. Spent carbon is fed from the spent carbon storage tank to a dewatering screw, which reduces the water content of the feed. The feed is introduced to the top hearth, gradually makes its way across the hearth, and passes out through a rotary valve and thence onto the next hearth. After exiting the bottom hearth, the carbon is quenched and pumped to the regenerated carbon storage vessel.

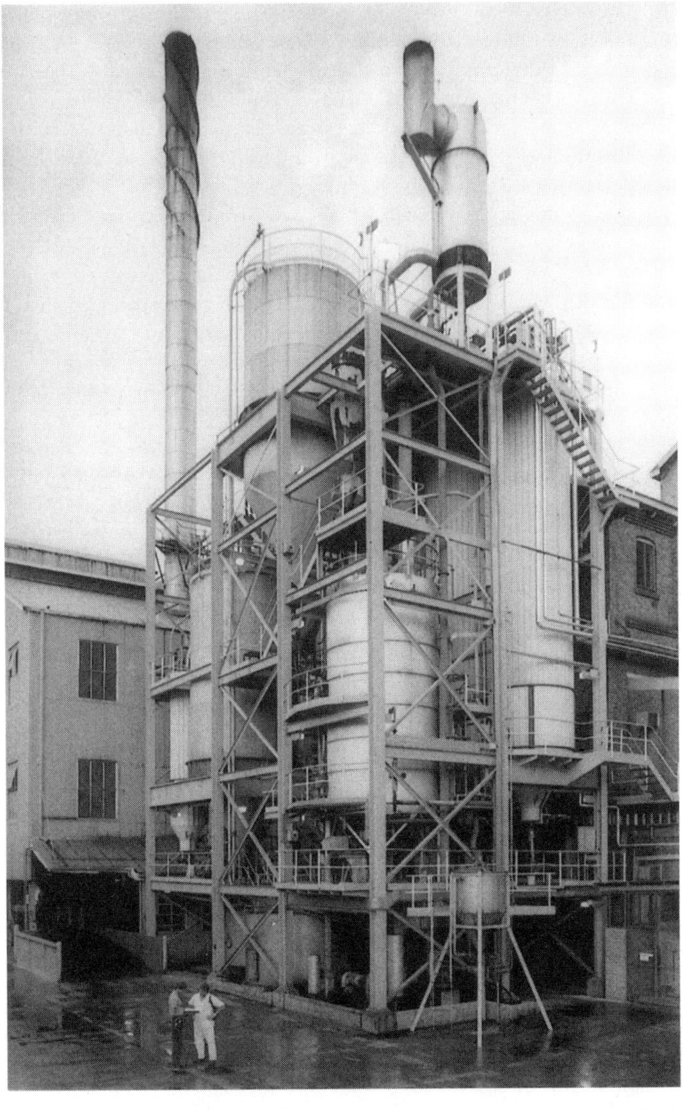

FIGURE 8.9 Granular carbon plant at CSR Ltd. New Farm refinery. Kiln and afterburner in foreground, liquor adsorbers (5 off) at right, spent carbon tank high in structure, regenerated carbon tank lower in structure toward back, small golden syrup adsorber at back left. (Courtesy of CSR Limited.)

Only one burner is associated with the kiln. The exhaust gases from this burner provide the fluidizing medium as well as the heating energy to reactivate the carbon. Steam and air are added on each hearth to control the temperature and aid the reactions. Exhaust from the kiln is passed though some dust-arresting device (cyclone or multiclone) and thence to an afterburner. The afterburner ensures complete oxidation of the gaseous products leaving the kiln and also incinerates any entrained particles that have managed to pass through the dust arrestor. The kiln runs at a slight positive pressure and care must be taken to ensure that there are no leaks of hot combustion gases from the kiln.

8.4.4 Equipment Comparison: Multiple-Hearth Versus Fluidized-Bed*

a. Process Comparison. The fluidized-bed and multiple-hearth kiln processes are compared in Table 8.6. Generally, the fluidized-bed kiln is more energy efficient than a multiple-hearth kiln, due to the much more intimate contact between the carbon and the heating medium. The fluidized-bed kiln does run at a slight positive pressure and care must be taken to minimize the risk of leakage of hot combustion gases. Conversely, by running at a slightly positive pressure, the risk of ingress of air that can upset the stoichiometric atmosphere is lessened. Carbon losses are reported as being similar [11], but the fluidized-bed kiln is thought to be more likely to increase carbon losses due to the fluidizing action required to move the carbon through the kiln. Multiple-hearth kilns are thought to provide less opportunity for loss by attrition, due to a less vigorous action. That perception has meant almost universal acceptance of multiple-hearth kilns in the refining industry, despite their relatively high maintenance requirements.

b. Physical Comparison. The physical characteristics of the fluidized-bed and multiple-hearth kils are compared in Table 8.7. The construction of a fluidized-bed kiln allows it to tolerate much higher rates of temperature change than a multiple-hearth kiln. This is important when it is intended to shut down the kiln regularly. It also affects how long it takes to restart after a shutdown. In the case of multiple-hearth kilns, fast temperature changes

TABLE 8.6 Comparison of Fluidized-Bed and Multiple-Hearth Kiln Processes

Parameter	Fluidized-Bed Kiln	Multiple-Hearth Kiln
Feed moisture content (%)	50	50
Carbon loss (% of feed)	5	5
Steam rate (kg steam / kg carbon)	0.8–1.0	0.8–1.0
Energy use (MJ / kg carbon)	5.8	8.0 (not including afterburner; including the afterburner, energy use is typically 15.0 MJ / kg carbon)
Temperature, solid (gas) (°C)	Top: 200–300 [200–300] Middle: 600–700 [600–700] Bottom: 700–800 [700–800]	Top two hearths: 95 [300–450] Middle hearths: 750 [450–900] Bottom hearths: 870 [900–950]
Pressure	Top: slight overpressure Bottom: 100 mbar	Middle: −20 mm / Hg
Burner conditions	Normal: $\lambda = 0.7–0.9$ Startup: $\lambda = 1.0$	1.2 1.6
Dust loading (g / m^3)	2.0	1.6

*Much of the data in this section originate from a report by Lurgi (Australia) Pty Ltd. Commissioned by CSR Ltd. and are reprinted here with their kind permission.

TABLE 8.7 Comparison of Fluidized-Bed and Multiple-Hearth Kiln Physical Characteristics

Physical dimensions (m; data for a 700 kg/h kiln)	Height: 8.5 Diameter: 2.8	Height: 9.4 Diameter: 3.3
Refractory thickness (mm)	400	340
Number of hearths	3	8
Stage construction	High-temperature grate	Brick or cast arch, self-supporting
Stage spacing (mm)	2100	740
Mechanical equipment	No moving parts inside the reactor; two high-temperature rotary values	Air-cooled central shaft driving rabble arms (two to four per hearth), drive beneath kiln
Feed discharge	In: chute from inclined dewatering screw; out: chute to quench tank	In: chute from inclined dewatering screw; out: chute to quench tank
Burner configuration	One special burner attacked to lower side of kiln	Up to nine burners
Fans	One low- to moderate-pressure combustion air fan	Three low-pressure fans; combustion air, induced-draft, and shaft cooling
Dust control equipment	Multiclone	Simple cyclone
Afterburner	Required to burn CO and H_2 and to destroy all odorous components	Required to burn CO and H_2 and to destroy all odorous components
Estimated capital cost index[a] (multiple-hearth = 1)	1.15	1.00

[a]Costs do not include import duty, structural steel, civil work, instrumentation (apart from local burner control), and electrics (apart from electric motors).

within the kiln (>50°C/h) will cause weakening and eventual collapse of the self-supporting hearths. The fluidized-bed kiln has much less ancillary equipment and fewer burners. There are no moving parts within the kiln itself.

8.5 PROCESS CONTROL AND MANAGEMENT

Typical performance from a soundly operated and designed carbon plant (fixed or moving bed) compared to design would be as listed in Table 8.8. Achievement of operating performance similar to that above requires close attention to day-to-day operating practice, using the results of regular laboratory testing for process management. The important practices and test results are set out below.

8.5.1 Process Management: Liquor

a. Liquor Running Period Management. For fixed-bed decolorization systems, using two columns in series, typical liquor running periods are of the order of 200 h in the trail position, followed by 200 h in the lead position. At design flow rates, these times correspond to a carbon use of about 0.8% on liquor solids treated. In practice, depending on feed liquor color, carbon use at this level may produce fine liquor of color significantly above or below the required pan feed color. As a result, the liquor running period may be reduced if the

TABLE 8.8 Carbon Plant Performance

Performance Indicator	Design	Actual[a]
Decolorization (%) Overall	85–90	86–88
Fixed-bed lead column	80	76–78
Fixed-bed trail column	50	45–50
Product fine liquor pH	>7.0	7.7–7.9
Liquor pH drop over decolorization (pH units)	<1.0	0.3–0.5
Invert gain over decolorization (%)	Nil	0.02–0.03
Carbon use (burn) (% feed liquor solids)	0.8	0.7–0.8
Carbon loss (% carbon reactivated)	<5.0	3–5

[a]Actual performance data drawn from CSR operating data for carbonation refineries in Australia.

color is too high or extended if the color is low. The extent of the change is generally based on experience and it is usual to make small incremental changes. For conventional raw sugars it is unlikely that the operating range would exceed 0.6 to 1.0% carbon use. For pulsed-bed systems, the equivalent adjustment is to alter the pulse frequency and/or the pulse volume so as to vary the carbon use in a similar range.

b. Carbon Stock Level. It is important to maintain an adequate stock of service carbon to ensure that all adsorber columns are full and there is sufficient buffer stock to allow continuous kiln operation. In practice, for fixed-bed installations this approximates a carbon tonnage needed to fill $(N + 2)$ columns, where N is the number of full columns on decolorizing duty. Because pulsed-bed systems provide small spent carbon tonnages at shorter intervals than fixed-bed systems, the carbon stock required is less. With a carbon loss of about 5% over reactivation, it is important to have regular makeup of new carbon into the system so as to maintain the total stock at a safe level.

c. Column Filling. When adsorber columns are filled with a water–carbon slurry, it is important that the filling process be continuous and that the column be completely full of carbon. Any stoppages in filling or free headspace at the top of a column may allow deposition of a layer of carbon fines that can, in turn, cause an increase in pressure drop over the column. Normal pressure drops for fixed beds operating at the design levels outlined above with 1.0-mm-diameter carbon should be on the order of 160 to 200 kPa over each column.

d. Fines Removal. Granular carbon systems are generally designed with several facilities for removal of carbon fines. These include launders in the spent and reactivated carbon tanks and in the quench tank. Transport water overflows at each of these points, carrying carbon fines with it. In addition, carbon fines that escape the kiln cyclone are incinerated in the afterburner. Each of these systems is important for continuous removal of fines from the service carbon and minimizing adsorber column pressure drops. Failure to maintain these systems can allow a buildup of carbon fines with resultant high column pressure drops which may become rate limiting.

e. Magnesite Addition. As noted earlier, granular carbon has no pH buffering capacity, and magnesite is added during manufacture to provide pH support for liquor being decolorized. As this process means that the magnesite originally added dissolves over time, it is necessary to maintain at least 2% MgO in service carbon by adding magnesite or magnesia. This may be added either directly to the columns or to the spent carbon in the dewatering screw. Automated addition systems are used at the dewatering screw, allowing more precise addition rates.

f. Recirculation. Where adsorber columns are required to be shut down for extended periods (more than 12 h) while still full of liquor, it is normal practice to recirculate fine liquor over the columns at a reduced rate to maintain continuous flow through each column. Experience shows that allowing flow to stop for long periods results in invert formation in the liquor held in the column, with fine liquor subsequently analyzing at up to 1.0% reducing sugars. Recirculation ensures little or no sucrose loss during stoppages.

8.5.2 Cross-Contamination of Liquor and Transport Water

Granular activated carbon plants have two main circuits: the liquor circuit and the transport water circuits. It is important that these two circuits remain separate. Contamination of the transport water circuit with liquor will lead to microbial degradation of the sugar introduced into the transport system. The waste products of microbial degradation stay in the system and in subsequent carbon transfers contaminate the columns themselves. Transport water in the liquor system leads to dilution of the product liquor from the plant with consequent higher energy use in later evaporation stages.

Several methods are used to isolate the transport systems from the liquor systems. These include a double block and bleed system which ensures that there is no leakage from one system to the other. Other systems incorporate hose connections between the column and the carbon transport water systems. When the column is on liquor, the hoses are physically disconnected to ensure complete isolation of the liquor circuit from the transport water circuit. The double block and bleed system is expensive, as it requires three valves on each line into the column. However, there are proprietary valves available that utilize the one actuator to drive all three valves giving a reduced cost.

Other systems utilize temperature-activated flow switches to detect leaks through valves. In any case, it is not possible to ensure that leaks or operating errors never occur. Consequently, the design of the column must always consider the best way to minimize any damage caused by a leak. Physical disconnection is one solution but presents problems with regard to the connection of heavy hoses and, in the case of automated systems, determining if the hoses are correctly connected. The beverage industry uses key panels quite extensively, but care must be taken that the flow path through the return bends is suitable to carbon transport flow conditions (where appropriate). Another system consists of separate bleed systems that direct flow to a "safe" place when a failure occurs.

8.5.3 Process Management: Carbon

Generally, changes in carbon quality likely to affect performance take place slowly. However, routine testing of the critical carbon quality criteria is essential to monitor unfavorable trends and to ensure that day-to-day plant performance is satisfactory. Accepted test methods for carbon, similar to those used for bone charcoal, are in the *Cane Sugar Handbook* [12] or provided by carbon manufacturers.

a. Carbon Sampling. Before measurements of carbon properties can be made, a reliable sample of the carbon used to fill an adsorber column is required. Common practice is to take a sample of carbon from the kiln discharge pipe leading to the quench tank using a short sample pipe branch fitted with inlet and outlet valves to allow capture and partial cooling of the hot carbon. As the carbon sample may still be above its ignition point when removed from the sample pipe and exposed to air, systems such as these present some safety hazards in recovering the carbon sample for laboratory analysis. An alternative system samples carbon as a water slurry pumped from the kiln quench tank. While the carbon requires drying before laboratory analysis, and this may result in an 8- to 12-h delay in providing a result, sampling wet carbon is inherently safer than sampling hot carbon from the kiln at perhaps 800°C.

b. Carbon Apparent Density. Carbon apparent density (CAD), the most important carbon quality indicator for process management, should be measured at least once a day. The test measures the weight of a standard volume of dried carbon to express the bulk density as grams per milliliter of carbon. Virgin carbon typically has a CAD value of about 0.45. In use, some organics are held on the carbon, and reactivated working carbon from the kiln is typically controlled to 0.50 CAD by adjusting the bottom hearth temperature, steam addition rate, and the colorant loading. Colorant loading is a function of feed liquor color and quantity of liquor decolorized each cycle.

In general, adjusting kiln temperatures to reduce the carbon CAD below 0.50 is likely to increase carbon loss during reactivation without any improvement in color removal. CSR experience in Australia is that a single pass through a multiple-hearth kiln will reduce the carbon CAD by about 0.05 to 0.06 unit without having to use excessive kiln temperatures and risk carbon losses above the 5.0% level quoted above.

On this basis it is useful to monitor spent carbon CAD levels to ensure that the actual colorant loading used does not result in spent carbon exceeding 0.56 to 0.57 CAD. If necessary, column liquor running periods would be reduced to achieve a spent carbon CAD at this level. Where unusual operating conditions may have produced higher loadings, it is better to avoid significant increases in kiln bottom hearth temperatures and accept that it may take more than one reactivation cycle to restore the carbon to its normal 0.50 CAD level.

c. Particle Size. Carbon particle size is important because fine carbon will increase the pressure drop over the adsorber columns, while increasing particle size may reduce the decolorization achieved. Carbon used in sugar refining is typically about 1.0 mm in diameter, representing the best compromise between conflicting process requirements. Particle size is monitored by carrying out routine screen analyses of carbon used to fill adsorber columns. It is possible that the mean particle size of a service carbon stock may slowly fall below the usual 0.9 to 1.0 mm working range if the carbon hardness is below specification or fresh carbon is not added to the service stock to maintain a satisfactory stock level. Weekly analyses will provide sufficient information for process management.

d. Abrasion Number. Abrasion number measurements provide information about the relative hardness of carbon particles and thus potential for loss of carbon by attrition in use. The test is based on subjecting a weighed sample of carbon to abrasion by standard steel balls in a special hardness testing pan using a Ro-tap machine. The abraded carbon is then screened using a U.S. No. 50 or BSS 52 screen (nearest equivalent to a 0.300-mm opening). The hardness number is then calculated as

$$\text{hardness number} = \frac{(\text{weight of material retained on screen}) \times 100}{\text{weight of original sample}}$$

A hardness number of 100 would mean no abrasion loss of carbon or change in carbon particle size. New carbon should have a hardness number of at least 75 units (Calgon or

Mitsubishi tests) and maintain this level in use. A fall in abrasion number below this level may mean that kiln hearth temperatures are too high.

e. Iodine and Molasses Numbers. Measurement of iodine and molasses numbers provides information about possible changes in the carbon pore size distribution over time. The molasses number test is based on measuring the ratio of the optical densities of a molasses solution treated with a reference carbon and with the service carbon. It is empirical in the sense that it relies on having sufficient reference carbon and molasses material available to cover testing over a reasonable period. When the original reference materials have been used, the test needs to be restandardized using new materials. The iodine number test measures the quantity of iodine (as the number of milligrams of iodine) adsorbed by 1 g of carbon from 50 mL of 0.1 *N* iodine solution. Both of these tests attempt to provide an indication of the available adsorption sites for molecules of the size of colorants found in molasses and of molecules similar in size to those of iodine. Similar tests may be carried out using other materials, such as methylene blue.

New carbon would typically show test results of at least 240 for molasses number and 800 to 1000 for iodine number. Typical results for service carbons in CSR Australian factories are 500 to 700 for iodine numbers and 280 for molasses numbers. An increase in molasses number, with a fall in iodine number, is an indication that the reactivation process is increasing the average carbon pore size. CSR experience is that this change does not appear to affect decolorization or relate to bulk density changes.

f. Other Tests. Other tests include measurement of magnesia content, the pH of a water extract of carbon, and decolorization of process liquor. Periodic testing for magnesia content is important to ensure that the service carbon still retains some capacity to "buffer" the pH levels of the liquor being treated. Magnesia levels in service carbon should not be allowed to fall below 1.5 to 2% MgO.

8.6 CONCLUSIONS

In summary, granulated activated carbon is a strong adsorbent for sugar colorants and is certainly capable of acting as a single-adsorbent system without the need for supplementary decolorization capacity such as bone charcoal or ion-exchange resin in series. The carbon decolorization process is robust and simple, with no undesirable environmental impact, such as the need to treat spent regenerant from ion-exchange resins. In practical terms the process is forgiving of minor process upsets, and changes to carbon properties due to poor process management take place slowly, giving time for plant managers to detect and correct unsatisfactory trends.

Acknowledgments

The authors would like to acknowledge assistance from the following people and organizations in compiling this paper: P. Blundell, Jord Engineers Australia Pty. Ltd., for information on multiple-hearth kiln design; CSR Limited, Refined Sugars Group, for access to refinery technical data and permission to publish selected data; and Lurgi Australia Ltd. for information on fluidized-bed reactivation kilns.

REFERENCES

1. E. D. Gillette, Paper 132, *Proc. Sugar Ind. Technol.,* 1956.
2. M. A. Clarke, *Proc. Sugar Ind. Technol. Symp. Colour Removal,* 1997.
3. *Cane Diahope Handbook,* Mitsubishi Kasei, Tokyo, 1989.
4. M. A. Clarke and M. A. Godshall, eds., *Chemistry and Processing of Sugarbeet and Sugarcane,* Elsevier, Amsterdam, 1988, pp. 186–207.
5. N. Paton and P. Smith, *Proc. Conf. Sugar Process. Res.,* 1982, pp. 1–23.
6. H. Kato and S. Fukui, *Int. Sugar Jour.,* 95:441, 1993.
7. F. M. Williams, *Proc. Cane Sugar Refin. Res. Project,* 1968, pp. 19–23.
8. A. M. Moult, *Proc. Sugar Ind. Technol.,* 1969, pp. 174–193.
9. P. J. Field, *Proc. Sugar Ind. Technol.,* 1987, pp. 85–104.
10. M. Mabillot, *Proc. Sugar Ind. Technol.,* 1996, pp. 9–22.
11. H. R. Johnson and B. H. Kornegay, *Proc. Cane Sugar Refin. Res. Project,* 1978, pp. 17–32.
12. G. P. Meade and J. C. P. Chen, *Cane Sugar Handbook,* 11th ed., Wiley, New York, 1985, pp. 602–609.

CHAPTER 9

Pulsed-Bed (Moving-Bed) Granular Activated Carbon System*

INTRODUCTION

In a granular activated carbon adsorber, the movement of liquor is countercurrent with the spent granular activated carbon. In general, the liquor is pumped into the bottom of the adsorber, passes up through the granular activated carbon (GAC) bed, and the decolorized liquor leaves at the top (Fig. 9.1). Periodically, depending on the decolorization requirement, a fixed amount of spent carbon is removed from the bottom of the bed during each pulse. New carbon is fed to the adsorber simultaneously from the top. The sequence of operation is called *pulse operation.* Except during the pulsing period, the liquor passes through the adsorber continuously and the carbon is removed, regenerated, and replaced continuously (Fig. 9.2). This entire system is called pulsed-bed (moving-bed) GAC. Advantages of the pulsed-bed adsorber are as follows:

- Lower capital investment
- Small space requirement
- Easy maintenance
- Fewer personnel required per shift
- Lower fuel requirement for the regeneration
- Improved production quality
- Adaptability to automation
- Lower operating cost

9.1 SPECIFICATION OF GRANULAR ACTIVATED CARBON

Granular activated carbon has innumerable numbers of fine inner pores, so that it has an extremely large surface area and therefore exhibits an excellent adsorption performance on

* By Ju-Hwa Liang.

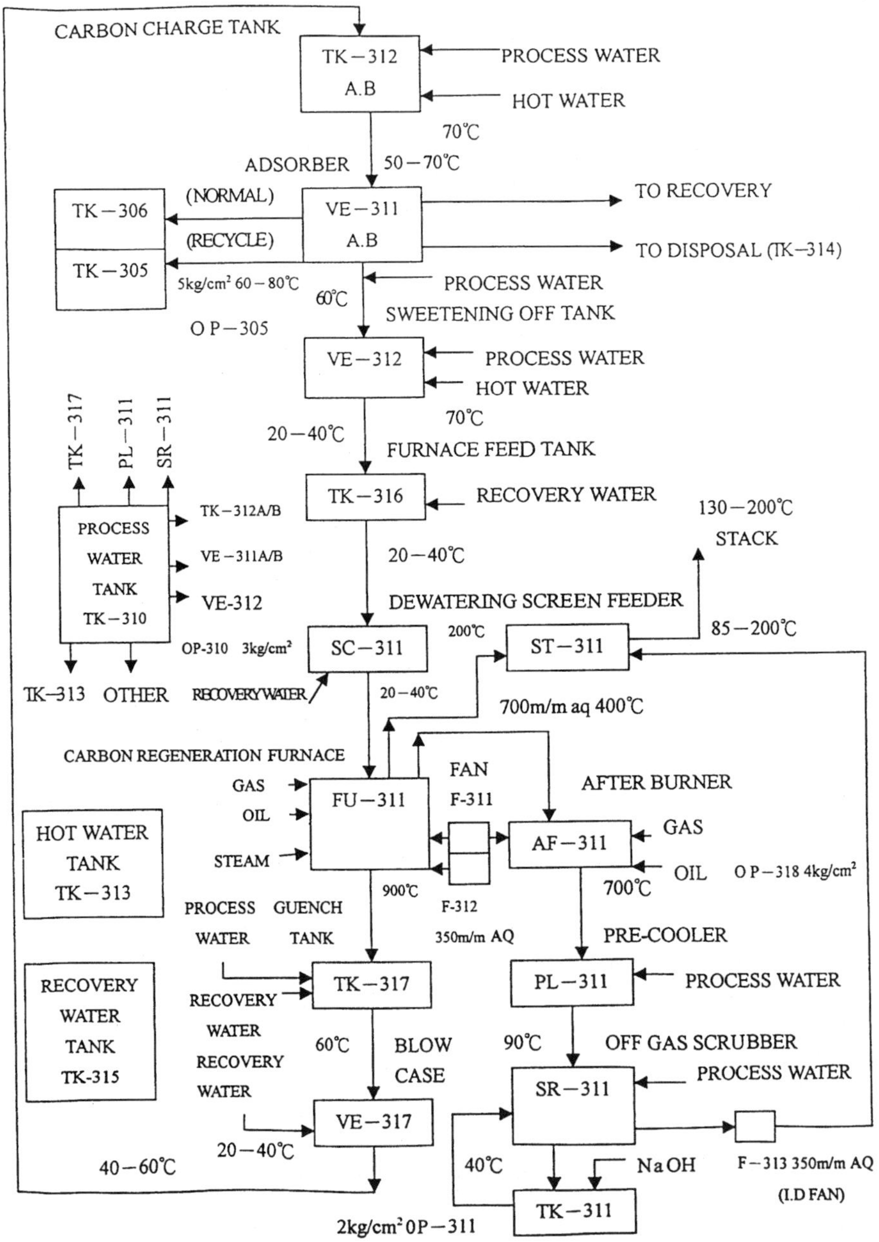

FIGURE 9.1 Liquor decolorization and carbon regeneration process.

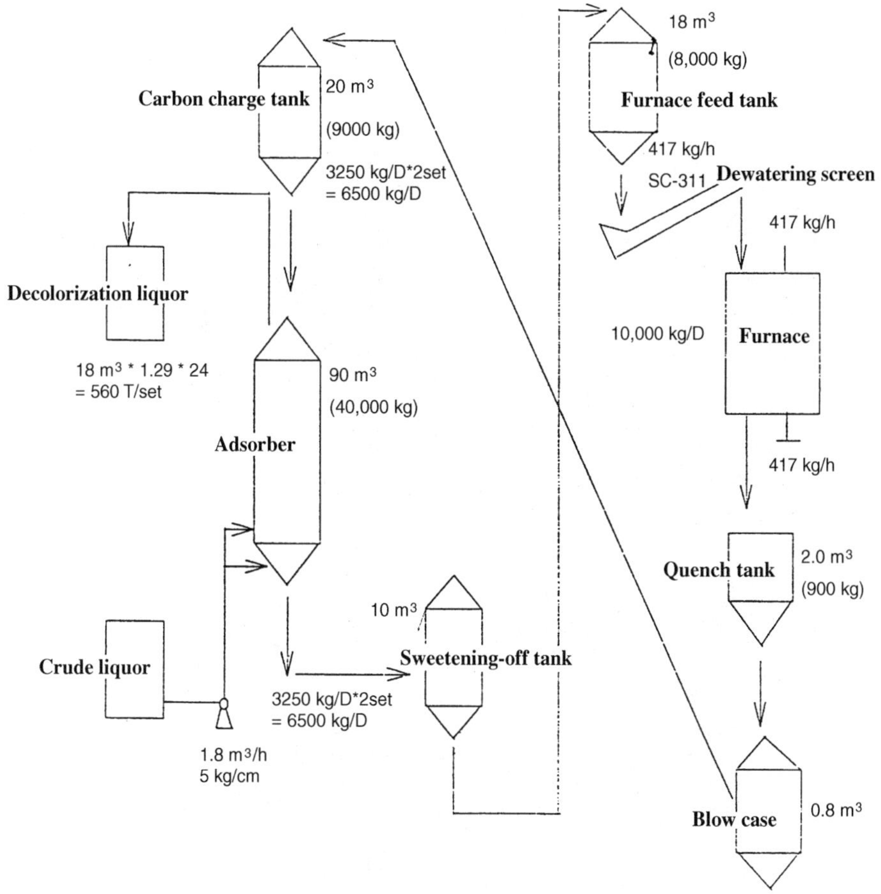

FIGURE 9.2 Flowchart of GAC decolorization process.

organic material (Fig. 9.3). The coal is pulverized and kilned in a low-oxygen atmosphere. GAC is manufactured from bituminous coal in a process that involves heating to about 1000°C and steam activation, which means that volatile organics are flashed off and what is left behind is a porous carbon structure with a large surface area. Because it lacks a component for buffering capacity, to prevent pH drop in decolorizing liquor, about 5% magnesia (MgO) is mixed with carbon or manufactured with carbon as the magnesia compound in the granules. Because the homogeneous mixing of MgO and carbon is difficult, in the carbonation process we control the liquor pH endpoint at 8.5. In the phosphatation process we adjust the pH from 7.5 to 8.5 with calcium saccharate before feeding it into the pulsed-bed adsorber. GAC physical properties are listed in Table 9.1.

9.2 GAC PACKING IN THE ADSORBER

Adsorber must be packed completely with GAC (except in pulse operation) to maintain the quality of the decolorizing liquor and the function of the septums. If the adsorber is not completely GAC packed, the result is as follows:

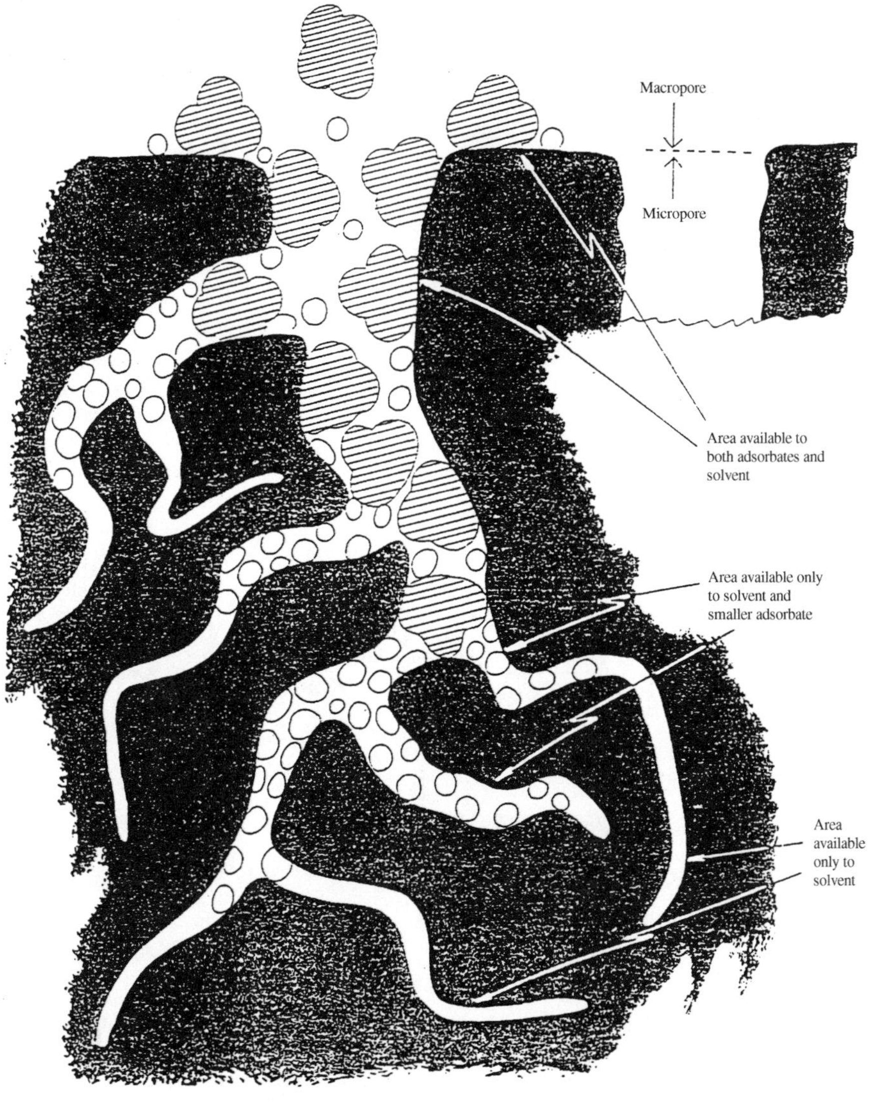

FIGURE 9.3 Pore model of activated carbon.

1. After pulse operation, the septums will break.
2. A liquor and GAC shift will occur and the GAC will move from the bottom to the top, resulting in losses of GAC and producing GAC powder.
3. The GAC powder will block the screens of the septums; the pressure increase results in difficulty in hardly sweetening-on.

The density of complete GAC packing is 1.25 to 1.6 kg/L.

TABLE 9.1 Physical Properties of Granulated Activated Carbon

Parameter	Standard
Iodine number	≥1000
Molasses number[a]	≥210
% Caramel decoloration	≥86
% Total MgO	2–7
% Ash[a]	≤13
% Moisture as packed	≤2
Abrasion number[a]	≥68
Mesh size: U.S. standard sieve series 10–40	
% Larger than 10 mesh	≤4.5
% Smaller than 40 mesh	≤4.5
Apparent density	0.45–0.5
pH	≥8.5

[a]Important property.

9.3 DESCRIPTION OF OPERATION

9.3.1 Liquor Flow in Adsorber

Feed liquor flows into the bottom of the adsorber from the seven points of the ring headers adhering to the adsorber. The flow of liquor must be homogeneous and not contain bubbles. The cone at the bottom has an auxiliary feeder to maintain temperature and prevent the growth of microorganisms. The inlet pressure must be controlled under 4 kg/cm^2.

9.3.2 Decolorizing Liquor Outlet from the Adsorber

The decolorizing liquor passes through the septums and outlet from the distributor at eight points at the top of the adsorber. The septums are filled with granular carbon, which is used to filter suspended solids and remove colloidal materials. The outlet pressure is 0.2 to 0.4 kg/cm^2. During pulse operation, some of the activated carbon fines adhere to the outside of the septums, so the decolorizing liquor passes through an automatic valve which drains the liquors back to the feed liquor tank. When the decolorizing liquor is clarified to the point where it has no fines at all, which takes 5 to 20 min, the automatic valve is closed and the liquor moves to the fine liquor tank.

9.3.3 Pulse Operation

As the carbon layers are plugged progressively, the line pressure increases progressively. When the pressure reaches its maximum allowable limit or when a running timer is reached, in about 24 h, the decolorization step is stopped automatically and a pulse step is initiated. The new carbon is passed from the carbon charge tank to the adsorber from the top, and the exhausted carbon is drawn from the bottom to the sweetening-off tank. This flow, which is about 7 to 8% of the total volume in the adsorber, contains high-Brix-value syrup and impurity-saturated carbon. In the sweetening-off tank, sugar containing spent carbon is immersed with hot water for about 3 h and 1 to 30 Brix sweetwater is recovered to the sweetwater tank. Sweetwater above 30 Brix is sent back to the liquor tank. Spent carbon slurries in the sweetening-off tank are drained away, and then the water valve is opened to

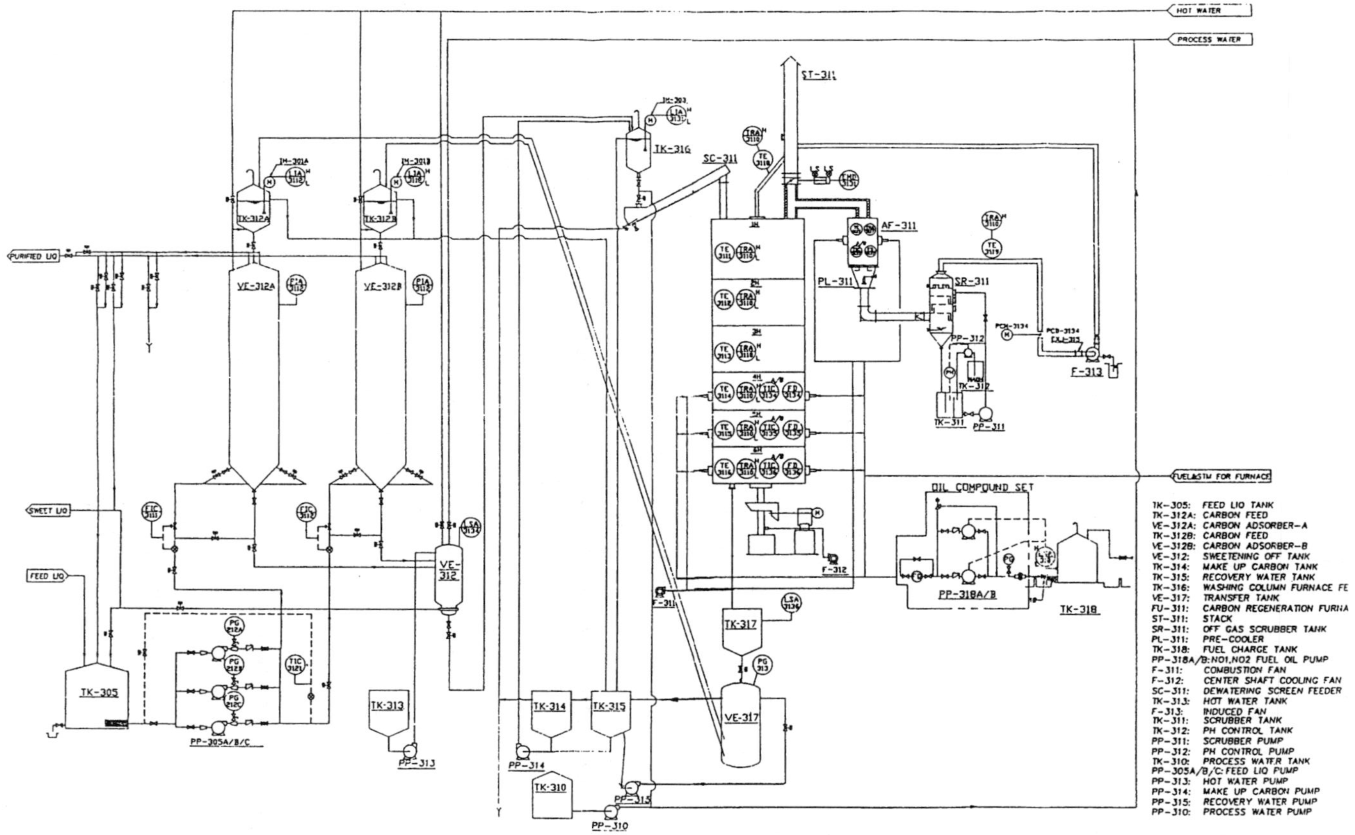

FIGURE 9.4 GAC flowchart.

carry spent carbon to the furnace feed tank. The desweetened carbon is conveyed hydraulically. The time of pulse operation is nearly 30 min.

9.3.4 Regeneration of GAC (Figs. 9.4–9.9)

Spent carbon from the furnace feed tank passes to the furnace through a dewatering screen. The dewatering screen serves two purposes. The first is to remove water from the spent carbon. This protects the furnace from thermal shock by reducing the mass of cold material entering the furnace and reduces the energy necessary to evaporate water prior to carbon activation. The second purpose is to create a water seal on the top of the furnace. When carbon leaving the screen has a moisture content below 50%, the dewatering screen is operating properly. The carbon regeneration furnace is a vertical multihearth furnace. The carbon in the furnace is moved from the top to the bottom by 1-rpm rotary arms. Temperature in the furnace is exceedingly high (800 to 1000°C). In the dry zone, moisture of wet carbon is driven off by heating. In the decomposition zone, the carbon is heated further to pyrolyze adsorbed organic material and drive off volatiles. Finally, in the gasification

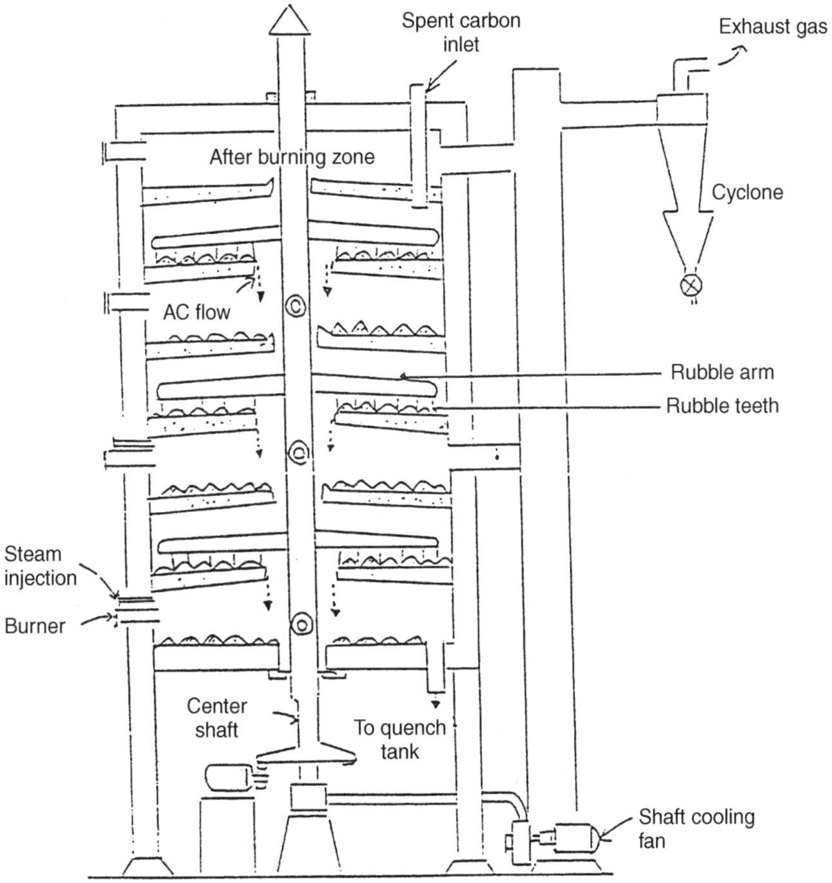

FIGURE 9.5 Multiple-hearth furnace type of regenerator.

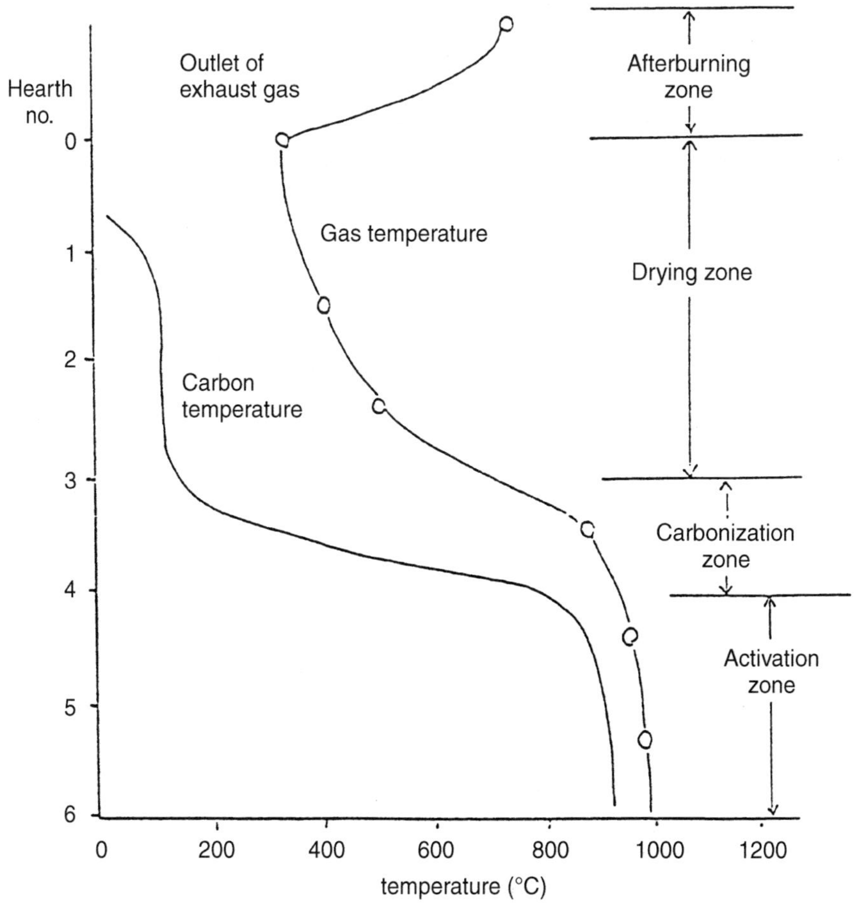

FIGURE 9.6 Temperature distribution in multiple-hearth furnace type of regenerator.

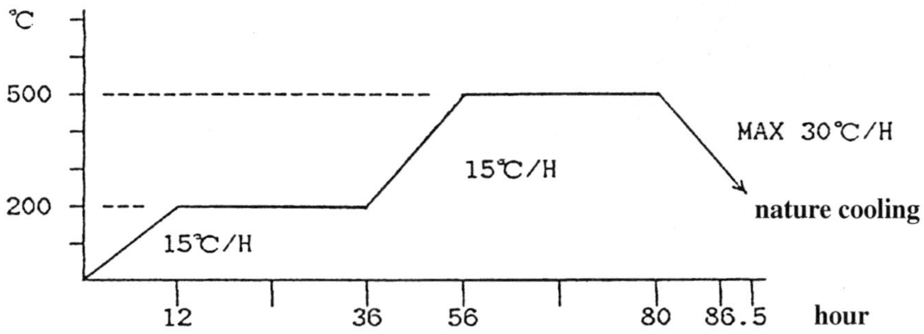

FIGURE 9.7 Furnace warm-up.

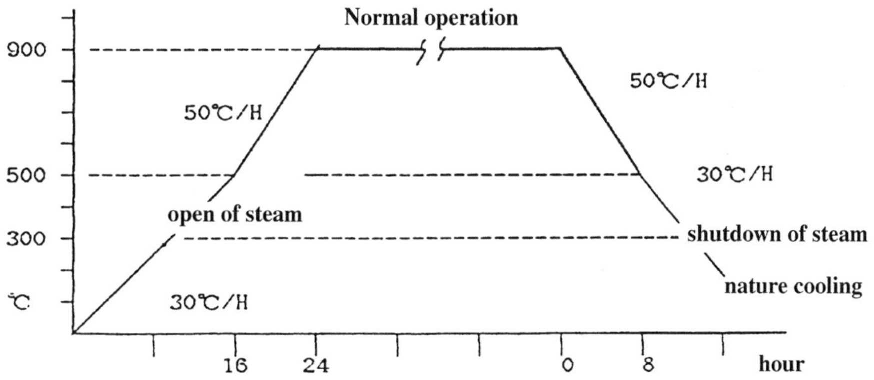

FIGURE 9.8 Temperature increase and decrease in furnace.

zone, the residual organic matter is pyrolyzed to minimize damage to the pore structure of carbon. Steam reacts with impurities adhering to GAC to produce H_2 and CO. The quantity of steam is 1.0 to 1.5 kg steam per kilogram of carbon. The loss of carbon is 3 to 5% of spent carbon. Excessive loss is an indication of overburn by a high level of O_2 in the furnace or excessive temperature in the furnace. The oxygen concentration must be controlled at about 1%.

9.3.5 Transportation of Regeneration GAC to Adsorber

Regenerated carbon is discharged from the bottom of the furnace and dropped by gravity into the quench tank, where it is cooled to 40°C by direct contact with fresh cold water. Excess water overflows from the quench tank and is pumped out, which serves two purposes. Heat from the carbon is carried away with the water, and carbon fines are entered in the overflow water and removed from the system. If the fines are not removed, they will create pressure drops across the adsorber. Periodically, the wetted carbon in the quench tank is discharged by gravity into the blow case beneath. When the blow case reaches a preset high level, the vent and top fill valve are closed, and the bottom discharge and water inlet valves are opened, the regenerated carbon is pushed to the carbon charge tank above the adsorber.

9.4 TREATMENT PROCEDURES FOR SHUTDOWN

For short-term shutdown (less than 1 week), the adsorber can be kept flowing with hot water or syrup to prevent anaerobic bacteria growth. For long-term shutdown, the adsorber must be washed with hot water to an effluent solid content of about 1 Brix and then the carbon must be emptied out.

9.5 OPERATION RESULTS

The loss of activated carbon is 0.4 kg/tonne Sugar (2.8% spent carbon) and the Oil consumption is 3.3 L/tonne sugar. The color values are listed in Table 9.2.

1. Spent carbon

   Water content; 40 ~ 45%

   Adsorbed material on AC:

$$\frac{BD_{(\text{spent carbon, dry base})}}{BD_{(\text{AC before adsorption})}} - 1 = 0.1 \sim 0.3 \text{ kg adsorbate / kg AC}$$

Activated carbon	1.0
Water	About 1.0
Organic adsorbate	0.1 0.3
Inorganic adsorbate	A little

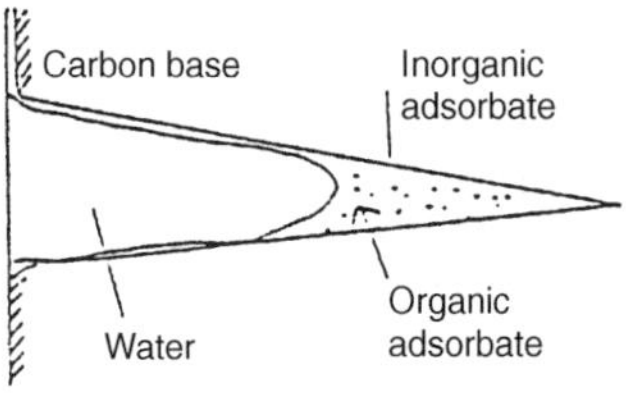

2. Drying

Activated carbon	1.0
Water	0.0
Organic adsorbate, some fraction vaporized	0.1~0.3
Inorganic adsorbate	No change
Temperature	~100°C

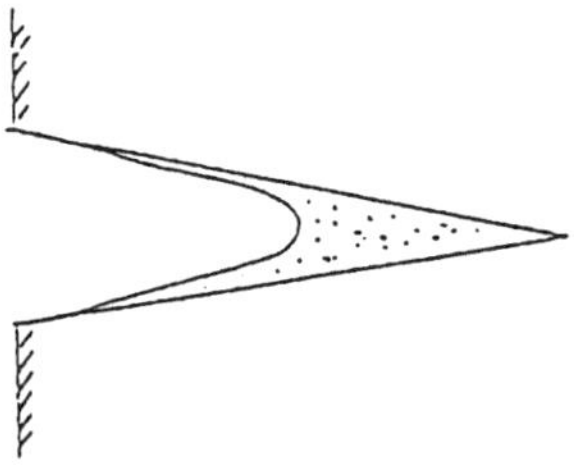

3. Carbonization

Activated carbon	1.0
Organic adsorbate	0.1~0.3
decomposition polymerization	↓
Carbonacious residue	0.03~0.09
Inorganic adsorbate calcined	
Temperature	100~800°C

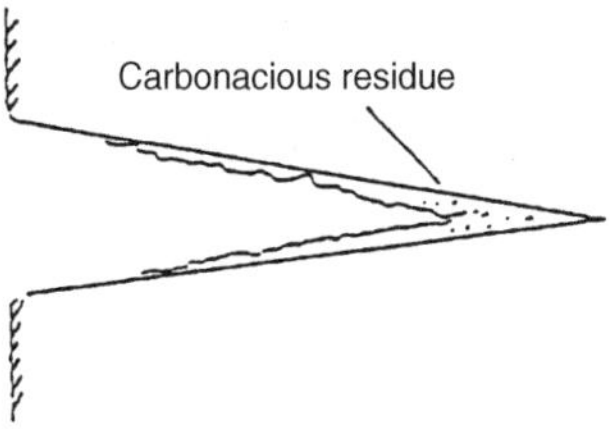

Carbonacious residue on AC:

$$\frac{BD_{(\text{carbonized AC})}}{BD_{(\text{AC before adsorption})}} - 1 = 0.03 \sim 0.09$$

4. Activation

Activated carbon	1.0
C ÷ H_2O → O + H_2	↓
	0.95 ~ 0.98
Carbonacious residue	0.03 ~ 0.09
C + H_2O → CO ÷ H_2	↓
	0.0
Inorganic adsorbate calcined	

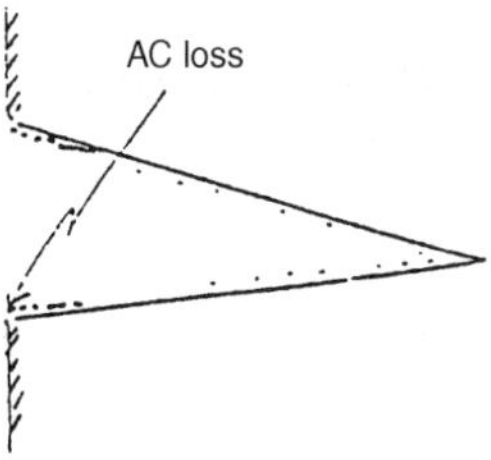

Total 0.95 ~ 0.98

Temperature 900 ~ 1000°C

FIGURE 9.9 Mass balance in regeneration process.

TABLE 9.2 Color Value of Powder- and Granular-Activated Carbon

	Color Value	
Item	PAC[a]	GAC[a]
Crude syrup[c]	1962	1744
Fine liquor[c]	633	314
White sugar[d]	4.0	2.8
% Color removal	68	82

[a]Average value of August 1, 1993 to June 30, 1994.
[b]Average value of December 1, 1993 to October 31, 1995.
[c]RBU color unit.
[d]CTU color unit.

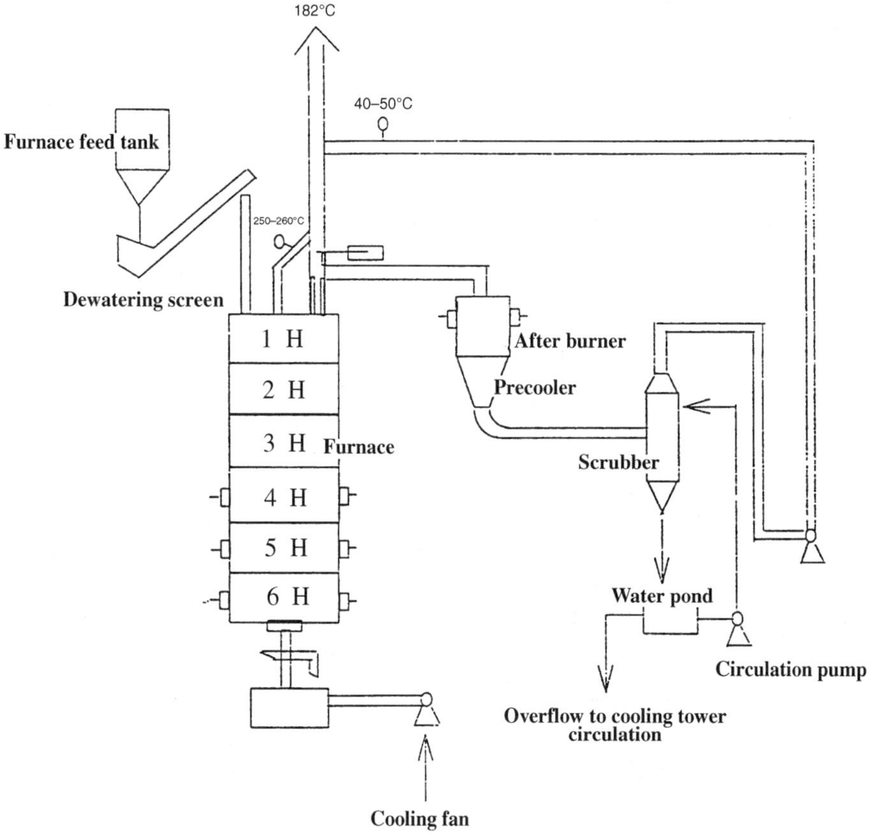

FIGURE 9.10 Flowchart of furnace waste gas treatment.

TABLE 9.3 Qualities of the Waste Gas

Parameter	Average Result	Government Standard
Suspension solid (mg / N·m^3)	35.5	448.3
SO_x (ppm)	39	500
NO_x (ppm)	15	400
Moisture (%)	10.0	
Temperature (°C)	152	
Flow speed rate (m / s)	3.73	
Quantity waste gas (N·m^3 / min)		
Dry base	48	
Wet base	43	
Composition of gas (%)		
CO_2	5.3	
O_2	13.5	
CO	0.1	

9.6 WASTE GAS TREATMENT (Fig. 9.10)

Waste gas treatment has three phases:

1. *Afterburner.* Waste gas from the furnace passes through the afterburner chamber for combustion of CO and residual organic compounds to CO_2. The temperature of the afterburner is 600 to 850°C.
2. *Precooler.* The precooler cools waste gas from 700°C to 90°C.
3. *Scrubber.* The function of the scrubber is to decrease the temperature of, and remove the dust from, the waste gas.

See Tables 9.3 and 9.4 for the quality and composition of the waste gases.

9.7 KEY POINTS OF OPERATION AND DESIGN

To avoid GAC clogging, the liquor from the filter station must pass through the 50-μm check filter to assure that the liquor has no suspended substances. The suspension solid should not exceed 12 ppm.

Since the filtered liquor is fed in at the bottom of the adsorber and the regenerated carbon is added to the top of the adsorber, the liquor with the lowest color will always be

TABLE 9.4 Composition of Furnace Gas

Temperature (°C)	O_2 (%)	CO (ppm)	CO_2 (%)	NO_x (ppm)	SO_x (ppm)
360	0.4	1947	14.8	89.5	140
445	0.25	1998	14.8	86.5	145.5
530	0.2	1998	14.9	79.0	154.5
875	0.23	1998	14.9	13.7	185
900	0.27	1998	14.8	15.0	193
900	0.53	1714	14.7	22.0	50

in contact with the newest carbon. This provides the greatest decolorizing efficiency. Because the newest regenerated carbon contains fines generated during the regeneration of carbon, fines are likely to escape with the liquor. The decolorizing liquor needs to be filtered by the septums. Generally, a 50-μm check filter is needed to assure that the fine liquor has no fines.

As the liquor is fed to the bottom of the adsorber, the liquor will lift the carbon up and expands the GAC bed slightly, which results in reducing the pressure drop across the adsorber. This is why we design the liquor flow with an upward direction.

REFERENCES

1. Carlos H. Tupas, Jr. and A. Garcia, *Sugar Azucar,* May 1980.
2. *Calgon Granular Activated Carbon Decolorization Process for Taiwan Sugar Corporation,* Jan, 1989.
3. *Operators Training Manual for Taiwan Sugar Corporation,* Refined Sugars, Inc., Yonkers, NY, Jan. 17, 1997.
4. Y.-C. Hsiao, W.-D. Hsieh, J.-C. Liu, and Y.-H. Lin, *Taiwan Sugar,* Nov.–Dec. 1996.

CHAPTER 10

Ion-Exchange Resin Process for Color and Ash Removal*

INTRODUCTION

In this chapter we review some of the main features and developments of ion-exchange technology used for the decolorization and de-ashing of the sugar melt in cane sugar refineries. Ion-exchange technology, introduced in the 1970s, is one of the more recent decolorization processes to be used. It has been installed either in combination with another decolorization process, such as activated carbon or bone char, or as the only decolorization process used in refineries.

A survey carried out by Margaret A. Clarke for the Sugar Industry Technologists 1997 Congress in Montreal [1] shows that in 47 sugar refineries studied, 22 were using ion exchange as a single decolorizer or in combination with other decolorization processes (regardless of the clarification system used). The number of existing ion-exchange decolorization plants in sugar refineries is increasing constantly.

Ion-exchange technology has been improved significantly over the past two decades, thanks to better plant design and improved performance obtained from resins [18]. The principal improvements were focused on (1) full process automation of various production and regeneration sequences, (2) up- or down-flow distribution, (3) the regenerant chemicals used, and (4) the variety of resins available, especially macroporous resins.

The decolorization rate and final color of the decolorized syrups are quite comparable with those of other decolorization technologies, with the advantage of significantly cheaper operating and capital costs. The more recent developments provide a significant reduction in liquid waste emission, especially by recycling the sodium chloride and sodium hydroxide used in regeneration of the resin. This can be undertaken thanks to a new recovery process using nanofiltration membranes.

10.1 PRINCIPLE OF COLOR REMOVAL BY ION EXCHANGE

The ion-exchange resins are usually considered to be similar to electrolytes. Most are manufactured from styrenic or acrylic polymers and are chemically activated with various func-

* By Denis Bourée, St. Louis Sucre François Rousset, and Applexion.

tional groups. Cation-exchange resins are activated by negative functional groups such as sulfonates, and anion-exchange resins are activated by positive functional groups such as quaternary ammonium.

When in contact with a chemical solution, the ion-exchange resin reacts according to a simple acid–base chemical reaction. For a strong cationic-exchange resin,

$$\text{NaCl} + \text{H—R} \leftrightarrow \text{Na—R} + \text{HCl}$$

For a strong anionic-exchange resin,

$$\text{NaCl} + \text{OH—R} \leftrightarrow \text{Cl—R} + \text{NaOH}$$

According to the relative concentration of the various ions in the solution and to the affinity of the resin for these ions, the ion-exchange resin will be saturated (from left to right above) or regenerated (from right to left above).

The ion-exchange resins are characterized by the following parameters:

- The structure of their copolymer (styrenic, acrylic, gel type, macroporous type)
- The nature of their functional group (sulfonate, quaternary ammonium)
- Their maximum theoretical and practical exchange capacity
- Their pK value as an electrolyte
- The average size and size distribution of the resin beads
- Their moisture content and density

When an ion-exchange resin comes in contact with a complex organic solution such as the melt liquor in cane sugar refining, many reactions take place at the same time and in competition. It is usually considered that the removal of sugar colorants by ion-exchange resins is the result of various types of reactions [2,3]:

- An ion-exchange reaction between the colorants mostly charged negatively with alkaline pH and the resin functional groups
- A hydrophobic interaction between the nonpolar part of the colorants and the styrenic resin matrix
- Adsorption inside the pores of the resin beads of the coloring materials in relation to their molecular weight and the size of the resin pores

The first-generation ion-exchange resins were limited to polishing after char or activated carbon decolorization, because the gel-type structure of their copolymer was not well adapted to high color loads. Neither were they also not adapted to a large variation in osmotic pressure when alternatively in contact with water or with dense sugar solution such as refinery syrups.

Today, two principal types of strong macroporous anionic resins are used for color removal in cane sugar refining: styrenic and acrylic (Table 10.1). These resins, manufactured from a strongly reticulated polymer, contain some artificially opened pores. A soluble porogenous agent is used during the polymerization process, giving channels with a diameter of approximately 100 nm. The solution processed, especially larger organic molecules can therefore percolate inside the matrix of the resin beads more easily. These macroporous ion exchangers are better adapted to decolorization of sugar solutions, as they have better resistance to high temperatures, variations in osmotic pressure, and contact with large organic molecules such as sugar colorants. They can now fully replace the bone char or

TABLE 10.1 Characteristics of Styrenic and Acrylic Resins

Characteristic	Styrenic Resins	Acrylic Resins
Matrix	Polystyrene, macroporous Reticulated with divinylbenzene	Polyacrylic, macroporous Reticulated with divinylbenzene
Functional groups	$-N^+(CH_3)_3$, strong base type 1	Quaternary ammonium, strong base
Mobile ion	Chloride	Chloride
Exchange capacity	Minimum 1.0 Eq/L (total capacity Cl^- form)	Min. 0.8 Eq/L (total capacity Cl^- form)
Moisture	58–64% (Cl^- form)	66–72% (Cl^- form)
Density	1.05–1.08 (Cl^- form in water)	1.05–1.08 (Cl^- form in water)
Average size	0.6–0.8 mm	0.65–0.85 mm

activated carbon decolorization systems used previously for decolorization of the melt liquor or work in combination with preexisting systems.

Acrylic resins, having less hydrophobic interaction with nonpolar parts of colorants, are more resistant than styrenic resins to high color feed levels. On the other hand, styrenic resins have a higher decolorization rate and better capacity with lower color feed levels. This is why in some existing ion-exchange decolorization plants, one or several beds of acrylic resins are used as front decolorizers, and one or several beds of styrenic resins are used as polishers.

10.2 DESCRIPTION OF AN ION-EXCHANGE DECOLORIZATION PLANT

The main components of an industrial plant with in situ regeneration of resin beds in ion-exchange columns is shown in the process flow sheet of Figure 10.1. The sugar solution to be decolorized is pumped from a feed tank into the first ion-exchange column. Usually, the sugar solution has a concentration of 60 to 65 Brix, color of 800 to 1500 ICUMSA (420

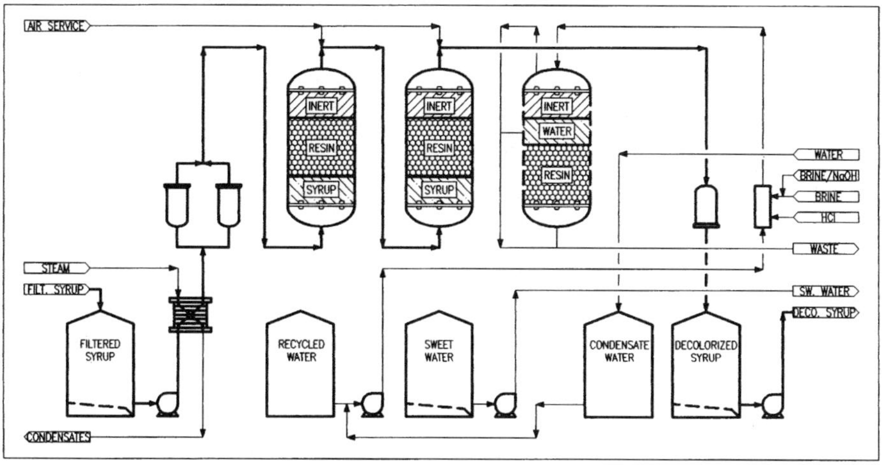

FIGURE 10.1 Typical process flowchart for an ion-exchange decolorization plant.

nm), and a temperature of 80 to 85°C. If this sugar solution has not been clarified perfectly, it is highly recommended that a safety filter be used, to avoid plugging of resin beds with suspended solids.

A safety filtration system must consist of a precoat leaf filter using diatomite or cellulose, a self-cleaning filter, or a simple bag filter (10 to 50 μm). Safety filtration must be selected carefully according to the characteristics of the sugar syrup to be processed: content in suspended solids, density, and temperature. Decolorization is achieved by percolation through one or several resin beds with syrup flow in the range 1.5 to 4 BV per hour (the volume of one active resin bed). The number of passes and the syrup flow depend on the required decolorization rate and initial color. Most often a single-pass process can be used for an average decolorization rate of about 65%, a double-pass process will be necessary to reach 85 to 90%.

The first industrial ion-exchange decolorization vessels were designed with the syrup percolating top to bottom in the resin beds. As the density of the sugar syrup at 60 to 65 Brix is about 1.3 and the density of the resin is 1.1 to 1.2 when hydrated with syrup, these decolorization vessels are difficult to operate, because resins have a natural inclination to float in the syrup. Considering the relative viscosity of the sugar syrup (about 5 cP at 65 Brix and 85°C), the possible flow through the resin beds is also soon limited by the pressure drop, and it is therefore difficult to maintain a well-compacted resin bed with this vessel's design. Good compacting of the resin bed is necessary for uniform flow distribution in the column without channeling. It is essential to obtain the best performance from the ion-exchange resin during decolorization and regeneration.

For this reason, most of the latest ion-exchange decolorization vessels have been installed with the syrup percolating from bottom to top into the resin beds. The resin is compacted against an upper plate or an inert material bed during up-flow decolorization of the syrup, and against a bottom plate or a sand material bed during down-flow regeneration (Fig. 10.2). To prevent accidental resin leakage likely to pollute the sugar circuits, a resin trap can be installed at the outlet pipe of the decolorization plant. It is recommended that an on-line 100 to 200 μm screening filter be used rather than a decanter.

In modern sugar refineries, all pipes in contact with sugar solutions in a decolorization station are usually manufactured in stainless steel. The decolorized syrup, also called *fine liquor,* is collected in a buffer tank and pumped to the refined sugar crystallization feed tank. Sometimes a fine liquor evaporator is installed to increase syrup concentration to 70 to 75 Brix before crystallization, to improve the thermal efficiency of the refinery.

Between decolorization and regeneration, the ion-exchange vessels must be sweetened off and on. The dilute sugar solutions produced during these intermediate sequences are collected separately and called *sweetwater.* Their average concentration is about 25 to 30 Brix, so they must be recycled quickly in the refinery process, usually at the sugar melt station, and represent 2 to 3% of the sugar dry matter processed.

The sweetening-on, sweetening-off, and regeneration steps use process water, usually condensates or softened water. Sometimes an additional water recovery tank can be installed, as described in Figure 10.1, to reduce the process water requirement by internal water recycling in the regeneration sequences. Up to 40% of the process water can be recycled by optimizing regeneration.

Figure 10.3 describes the brine and regeneration effluents station installed for the regeneration of ion-exchange vessels with acid or caustic brine solutions. First a saturated brine solution is obtained by dissolving NaCl crystals in cold water. At ambient temperature the saturated brine concentration is 330 g of NaCl per liter. As the efficiency of a neutral brine regeneration is usually not sufficient for complete removal of all coloring material from the resins, acid or caustic brine saturated solutions must be prepared by using HCl

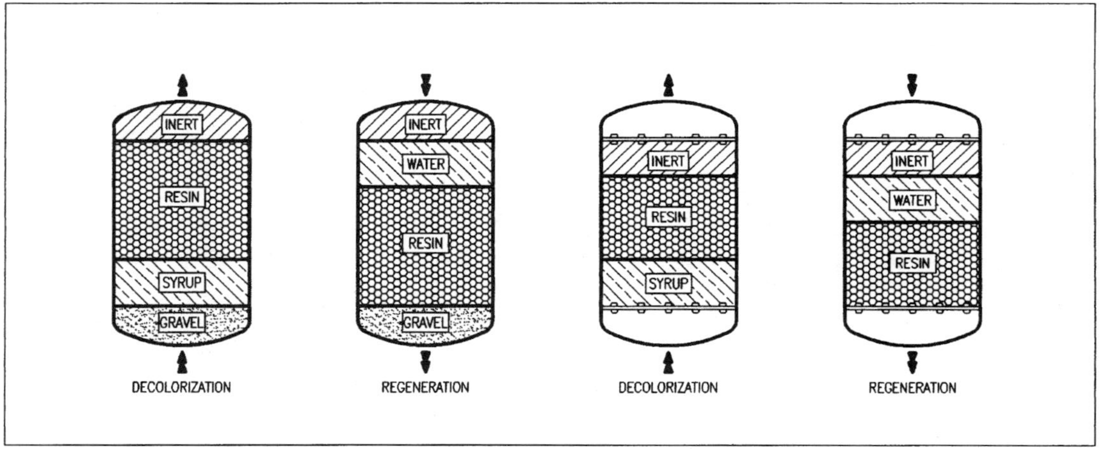

FIGURE 10.2 Typical design of upflow decolorization ion-exchange vessels.

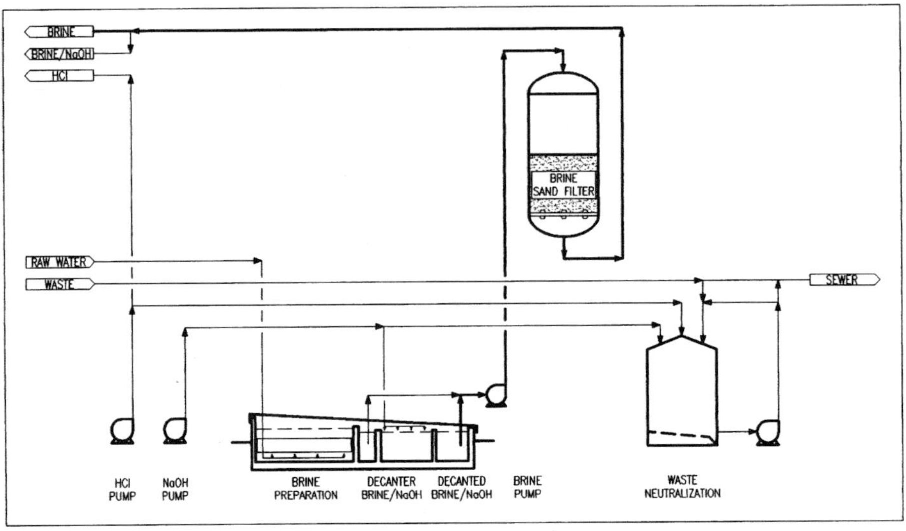

FIGURE 10.3 Caustic and acid brine preparation system.

and NaOH dosing pumps. Although the preparation of acid brine is very simple, the addition of a caustic solution 35 or 45% into the neutral saturated brine must be managed carefully. The impurities contained in the commercial NaCl crystals, especially Mg, floculate in the presence of NaOH into a hazy Mg hydroxide precipitate.

Before using the caustic brine solution for the regeneration of the resins, it is necessary to incorporate decantation step, usually followed by sand filtration. The design of the brine preparation system depends on the impurities content of the NaCl crystals commercially available on-site (Table 10.2). In Figure 10.3 the brine saturation, basification, and decantation is carried out in a multiple-compartment brine preparation tank, which can be manufactured in concrete. All internal concrete walls should be built with an antiacid finish. Vertical plastic dissolution tanks are installed when high-purity NaCl crystals are available. Usually, industrial NaCl originating from mining water or seawater is used in crystal form because of the expense of refined caustic solutions or refined commercial crystals.

The regeneration effluents, containing NaCl, sugar colorants, and water, can be divided in two fractions. The first, composed primarily of rinse water and backwash effluents, can be sent directly to waste. The second has either a very high pH, about 10 to 12 if caustic brine regeneration has been performed, or a low pH, 2 to 5 for acid brine regeneration. This effluent must be neutralized before it can be sent to waste, depending on the pH sensitivity of the waste treatment plant.

All regeneration effluents are collected at 80 to 85°C and must be cooled before being sent to waste. If softened water is used as process water for the regeneration of the decol-

TABLE 10.2 NaCl Specifications for an Ion-Exchange Decolorization Plant

NaCl	Minimum 99.5% on dry matter
Mg + Ca	Maximum 0.3% on dry matter
SO_4	Maximum 0.5% on dry matter
Fe	Less than 1 ppm on dry matter
Cu	Less than 1 ppm on dry matter
Moisture	Less than 6 wt%

orization station, the regeneration effluents can be used to preheat the soft water before regeneration at about 70°C. All heat-exchanger surfaces in contact with brine and regeneration effluents must be made of titanium, as they handle solutions at 80°C and very high Cl content. The pumps used on brine and effluent circuits are usually centrifugal plastic pumps with magnetic drive or bronze centrifugal pumps. The pipes should be made of polypropylene or fiberglass.

The diluted brine circuits and the internal parts of the ion-exchange vessels are protected against corrosion by a coating material: soft or hard rubber lining material manufactured by vulcanization, or fiberglass manufactured by vitrification. The membrane valves traditionally used in the past on the piping manifolds are now replaced by butterfly valves with air actuators, much simpler and cheaper to maintain. The butterflies in contact with the regenerant solutions must be made of coated steel to avoid corrosion.

Some existing sugar ion-exchange decolorization plants, especially in Russia and England, have been designed with an external centralized regeneration system. Once saturated with sugar colorants, the resin bed is transferred into a separate vessel for regeneration and rinsing. It is then transferred back into the decolorization vessel for the next cycle. One regeneration vessel can be used in combination with several decolorization ion-exchange vessels, especially if the decolorization step is very long.

All modern ion-exchange decolorization plants are fully automated and controlled by distributed control systems or programmable logic controllers. The automatic valves installed on the pipe manifolds are opened and closed by means of the sequences programmed in the process operation cycle. The flows in the various sequences are controlled by flowmeters and automatic control valves installed on inlet pipes of the plant. Once the sequence times or volumes and the flow set points have been adjusted on the automation system, the plant can run fully automatically without any special action required from the operator.

The efficiency of the ion-exchange decolorization of the syrup is controlled by inlet/outlet sampling and laboratory analysis, typically per shift, or more frequently in case of trouble. The following analyses are made at the inlet of the plant and at the outlet of each decolorization line:

- Brix
- pH
- Color of the filtered solution (ICUMSA 420 method)
- Color of the solution as is (ICUMSA 420 method)

10.3 OPERATING CYCLE

The sequential operation cycle of an ion-exchange decolorization plant can typically include the following sequences. The temperature of all fluids must be adjusted to 80 to 85°C in all sequences.

1. Sweetening-on
2. Production I: polishing or second pass
3. Production II: primary decolorization or first pass
4. Sweetening-off
5. Backwashing
6. Regeneration
7. Slow rinse
8. Fast rinse
9. Standby

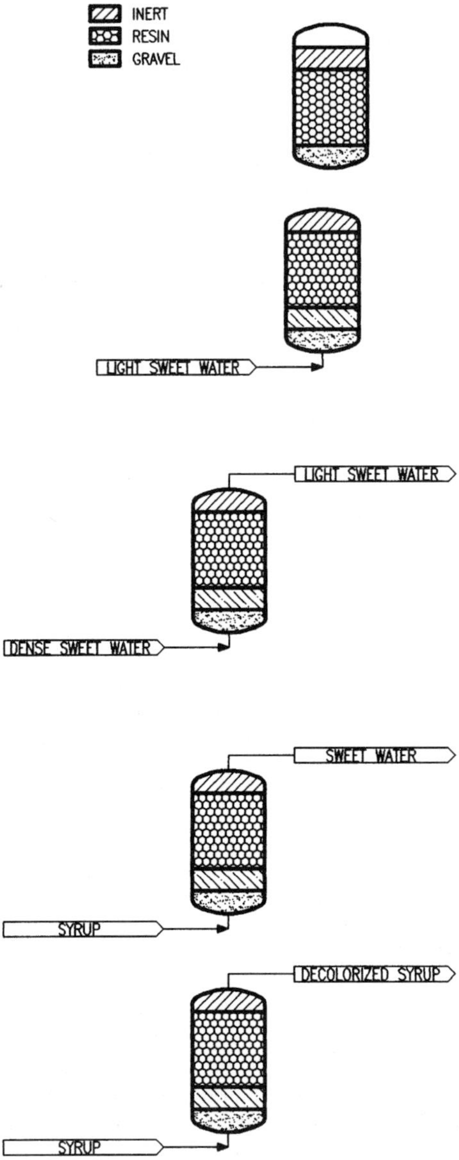

FIGURE 10.4

10.3.1 Sweetening-on

After the previous cycle, the column is full of hot water. This water must be replaced by syrup to be decolorized, in several successive steps.

a. Filling. Water is pumped from the light sweetwater tank to the primary column, then to the polishing column, until both columns are full. During this sequence the polymer loads rise to the top of the columns and the resins float.

b. Sweetening-on 1. Sweetwater is pumped from the dense sweet water tank, and water contained in the columns is sent to the light sweet water tank.

- *Duration:* 30 min
- *Quantity of water:* 1 BV*
- *Brix:* 0 to 20

c. Sweetening-on 2. Syrup is pumped from the feed tank and sweetwater is sent to the sugar melt station.

- *Duration:* 15 min
- *Quantity of syrup:* 1 BV
- *Brix:* 20 to 35

d. Sweetening-on 3. Using the same pump and the same circuit, sweetwater is sent to the decolorized syrup tank.

- *Duration:* 15 min
- *Quantity of syrup:* 1 BV
- *Brix:* 60 to 65

*1 BV = volume of one active resin bed.

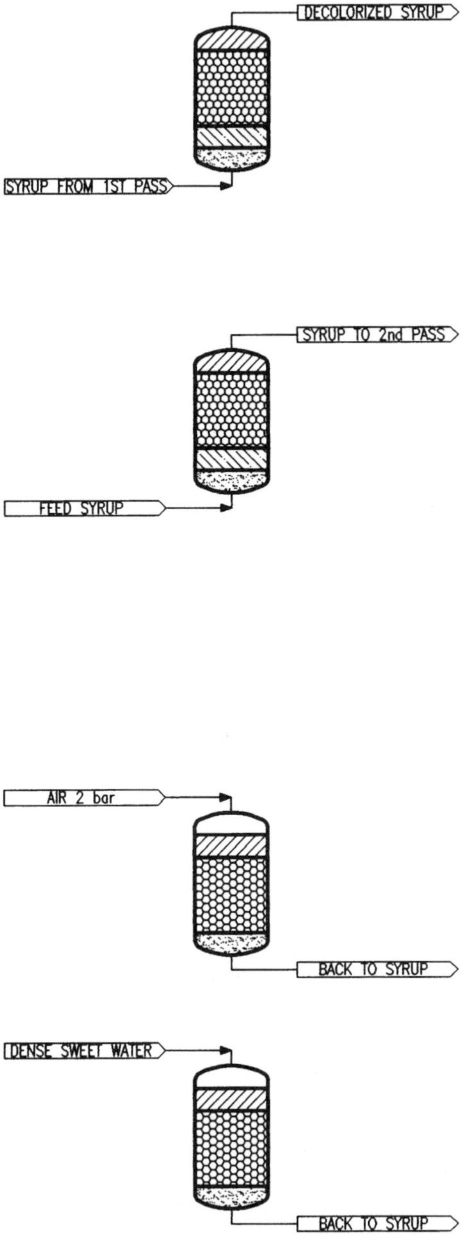

FIGURE 10.5

10.3.2 Production I: Polishing or Second Pass

The column is processing a syrup collected at the outlet of the primary decolorization vessel. The decolorized syrup, after the second pass, is collected in the decolorized syrup tank.

- *Duration:* 16 h
- *Quantity of syrup:* 30 to 40 BV
- *Brix:* 65

10.3.3 Production II: Primary Decolorization or First Pass

The column is processing the feed syrup. The decolorized syrup, after the first pass, is feeding the polishing column, if it exists.

- *Duration:* 16 h
- *Quantity of syrup:* 30 to 40 BV
- *Brix:* 65

10.3.4 Sweetening-off

The aim is to recover the syrup contained in the column with as little water dilution and sugar loss as possible.

a. Partial Draining. At the upper part of the column, air at 2 bar is injected to speed up the partial draining and lower the level of syrup and resins. Drained syrup is recycled in the syrup feed tank.

- *Duration:* 10 min
- *Quantity of air:* 0.5 BV at 2 bar

b. Sweetening-off 1. Dense sweetwater with a Brix value of about 40 is sent countercurrently to push the syrup back to the syrup feed tank.

- *Duration:* 30 min
- *Quantity of water:* 1 BV
- *Brix:* 65 to 55

c. Sweetening-off 2. Same pump with light sweetwater of Brix about 10; outcoming sweet-water is sent to the dense sweetwater tank.

- *Duration:* 30 min
- *Quantity of water:* 1 BV
- *Brix:* 55 to 35

d. Sweetening-off 3. Using the same pump with the same sweetwater, Brix about 10, outcoming sweetwater is sent to the sugar melt.

- *Duration:* 30 min
- *Quantity of water:* 1 BV
- *Brix:* 35 to 20

e. Sweetening-off 4. Process water is pumped to push the sweetwater to the light sweetwater tank.

- *Duration:* 30–60 min
- *Quantity of water:* 1–2 BV
- *Brix:* 20 to 0.5–1.5

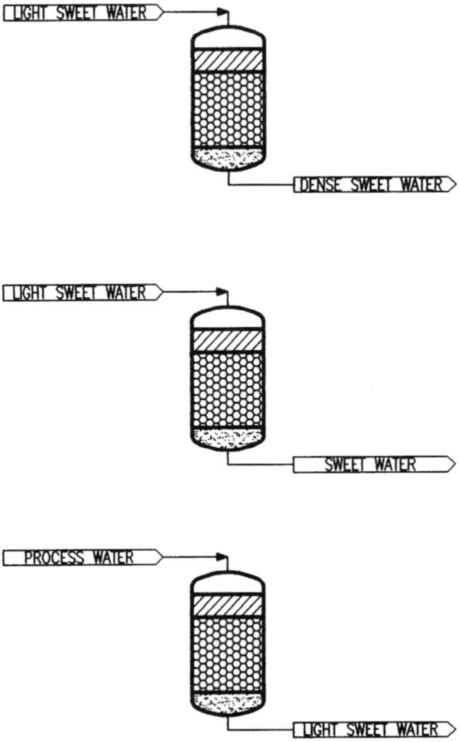

FIGURE 10.5

10.3.5 Regeneration

The quality of regeneration depends primarily on the compactness and homogeneity of the resin bed.

a. Air Scouring. At the lower part of the column, air at 1.1 bar is injected to ensure homogeneity of the resin bed.

- *Duration:* 5 min
- *Quantity of air:* 0.2 BV at 1.1 bar

b. Column Filling. The column is filled with water completely to push the inert polymer to the upper part of the vessel.

- *Duration:* 10 min
- *Quantity of water:* 0.5 BV

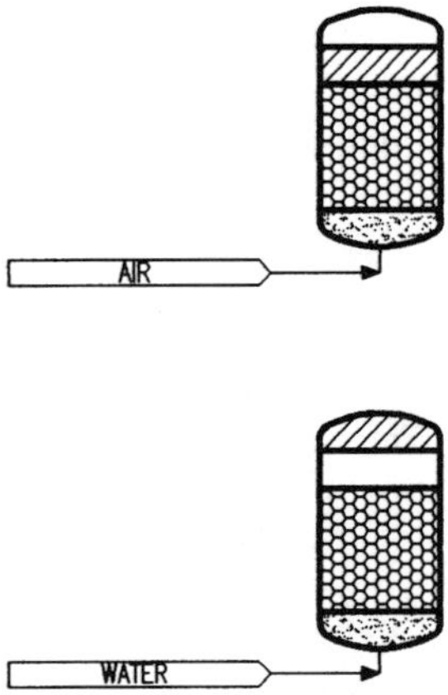

FIGURE 10.6

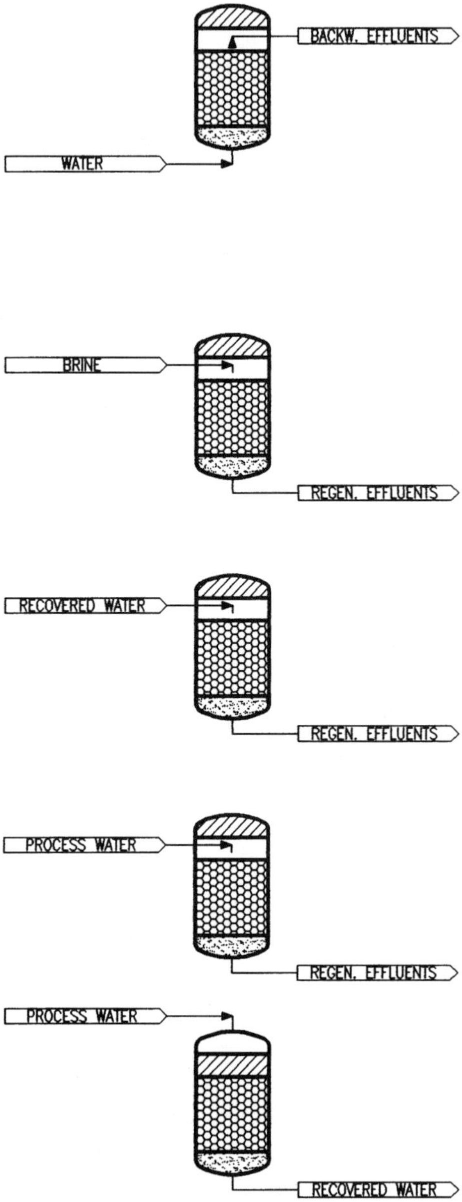

FIGURE 10.6

c. Backwashing Process water is used to backwash the resin bed and remove small broken particles of resins, called *fines,* and any suspended solids plugged in the resin bed during syrup decolorization. After passing through a resin trap, this backwash water is sent to waste.

- *Duration:* 15 min
- *Quantity of water:* 0.5 BV
- *Linear velocity:* 35 m/h

d. Regeneration with Caustic Brine. Regeneration is carried out countercurrently to minimize the regenerant consumption and improve the decolorization rate. The regeneration effluent is sent to the waste neutralization tank.

- *Duration:* 60 min
- *Dilution water:* 0.85 BV
- *Neutral brine 317 g/L:* 0.45 BV
- *Caustic soda 30.5%:* 0.04 BV

10.3.6 Slow Rinse

Recovered salted water is used to push the regeneration effluent into the waste neutral tank.

- *Duration:* 60 min
- *Quantity of water:* 1.6 BV

10.3.7 Fast Rinse

a. Fast Rinse 1. Process water is pumped to the column, outcoming effluent is sent to the waste neutral tank.

- *Duration:* 30 min
- *Quantity of water:* 1.6 BV

b. Fast Rinse 2. Process water is pumped to the column, outcoming water is sent to the saltwater recovery tank.

- *Duration:* 30 min
- *Quantity of water:* 1.6 BV

10.4 PERFORMANCE

With a feed syrup color of 800 to 1000 ICUMSA 420, two industrial double-pass ion-exchange decolorization plants filled with acrylic or styrenic resins give decolorization rates of 70% and 90%, respectively. A brief outline of their decolorization performance is summarized below in Table 10.3. These plants have, respectively, 20 and 18 years of operation as front decolorizers of melt liquor into fine liquor, without additional decolorizing agent used in the refinery, and still give consistent results.

TABLE 10.3 Decolorization Plant Performance

Parameter	Hulett Refinery, Durban, South Africa	Saint Louis Sucre, Marseille, France
Installation of ion-exchange decolorization	1978 / 1988	1980
Sugar melt capacity (tonnes / h)	90	50
Decolorization resins used	Acrylic	Styrenic
Single- or double-pass	Double-pass	Double-pass
Ion-exchange resin inventory (m^3 / tonne melt h)	1.8	1.7
Decolorization flow (BV / h)	1.8	2.5
Decolorization capacity [BV (first pass)]	43	40
Feed color (ICUMSA)	750–950	1000–1100
Average outlet color (ICUMSA)	220–230	100–150
Average decolorization rate (%)	65–70	85–90
Resin lifetime		
First (cycles polishing)	220	250
Then (cycles primary)	220	250
Resin consumption (L / tonne melt)	0.15	0.10

Source: Data from Refs. 4 and 5.

When the resins become exhausted after 440 or 500 cycles of decolorization / regeneration, the consumption in regenerants starts to rise and the decolorization efficiency can barely be maintained. Usually, one resin bed is replaced at a time in the multiple-column system, as it is not recommended to mix new resins with old resins in the same resin bed. Most ion-exchange decolorization plants installed in the 1990s have further optimized resin use by reducing the cycle time and resins inventory, extending the resin life to 500 cycles in polishing, then 500 cycles in primary decolorization, and reducing resin consumption to about 0.05 L / tonne of sugar melt.

These industrial results can be compared with laboratory experiments realized on various resin types with the same cane sugar syrup having the following characteristics:

- *Brix:* 65
- *pH:* 7
- *Color:* 2000 ICUMSA method 420
- *Syrup flow:* 1.5–3 BV / h single-pass
- *Temperature:* 80°C

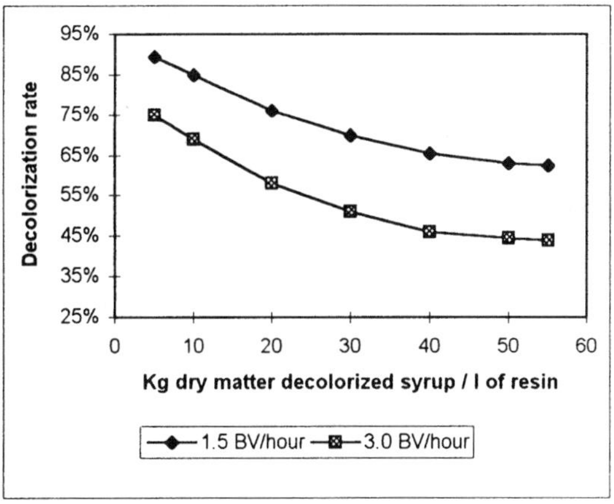

FIGURE 10.7 Acrylic strong anionic resin. (Courtesy of Applexion SA.)

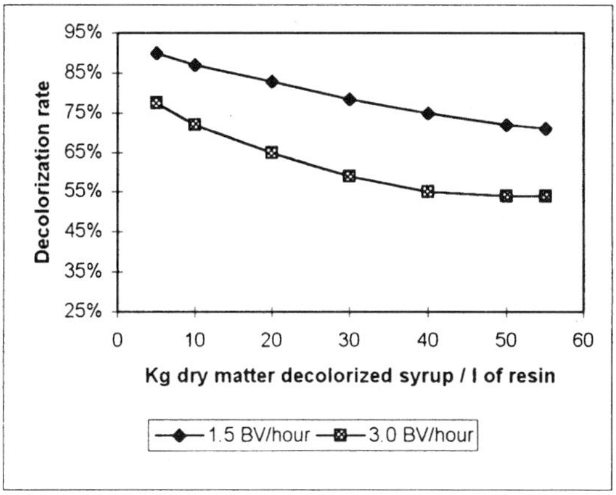

FIGURE 10.8 Styrenic strong anionic resin. (Courtesy of Applexion SA.)

10.5 CAPITAL COST AND OPERATING COST

One of the most interesting features of the ion-exchange decolorization process is the relative cheap operating cost and capital cost of the industrial plant, especially if compact up-flow ion exchangers are installed with limited resin inventory. The capital cost for a cane sugar refinery having a capacity of 1000 tonnes of sugar per day can be estimated as shown in Table 10.4. The operating costs for the same plant is estimated as shown in Table 10.5.

Usually less than one operator per shift is required to control the ion-exchange decolorization plant, as the operation is fully automatic. The same operator can monitor the decolorization with the front part of the refinery: sugar melt station, clarification and filtration, and decolorization. The maintenance cost is also typically cheap, because the only rotary machines used are all the various pumps around the ion-exchange columns.

TABLE 10.4 Capital Cost Estimation (U.S. dollars)

Ion-exchange decolorization plant including ion-exchange pressure vessels, front valves and piping manifold, automation hardware and software, instrumentation, connecting pipes and cables, safety bag and sand filters, pumps, and accessories	\$1,100,000
Resins, silex, and inert beds first loads	250,000
Tanks and erection works	1,150,000
Building and civil work	Not estimated
Total estimated capital cost	\$2,500,000

TABLE 10.5 Operating Cost Estimation

Item	Cost (U.S. dollars / tonne melt)
Regenerants	
NaCl 100%: 4 kg / tonne melt at \$100 / tonne	0.4
NaOH 100%: 0.4 kg / tonne melt at \$300 / tonne	0.1
Process water: 0.3 m^3 / tonne melt at \$1 / m^3	0.3
Resins: 0.08 L / tonne melt at \$4 / L	0.3
Effluents waste treatment: 0.25 m^3 / tonne at \$1 / m^3	0.25
Personnel: \$25.000 / y	0.15
Maintenance: \$50,000 / y	0.3
Total estimated operating cost	1.8

10.6 ION-EXCHANGE DE-ASHING AND PRODUCTION OF LIQUID SUGAR

The fine liquor usually obtained after clarification and ion-exchange decolorization has an ash content of 0.10 to 0.15% of dry matter (DM) and a color of 100 to 150 ICUMSA 420. This clarified and decolorized syrup has a sugar purity of about 99.5% DM and is perfectly suitable to production by crystallization: refined sugar <25 ICUMSA in the first refinery strike, and refined sugar <45 ICUMSA in the second strike. Sometimes the sugars obtained from the first, second, and third refinery strikes are mixed together, to produce an average quality of refined sugar of <45 ICUMSA. However, further de-ashing of this fine liquor may sometimes be necessary:

- If a high-quality refined sugar is to be produced by one or two crystallization strikes with a substantial recycling of the mother liquor, ash and color must be purged from the system to maintain the sugar quality.

- If fine liquor is used for the direct production of commercial liquid sugar without crystallization, the fine liquor must be polished by a final removal of both ash and color, according to the desired specifications of the commercial liquid sugar.

Both processes are in use in various sugar refineries to reduce energy cost and crystallization capital cost [6].

The conventional H—OH demineralization process described in Section 10.2 is difficult to use in this case, because the fine liquor is a dense syrup at 65 to 68 Brix. Sugar would be partially inverted at low pH on the strong cationic resin in H^+ form unless the temperature is lowered below 10 to 15°C, which seems hardly acceptable because of energy conservation and high viscosities. Mixed-bed ion-exchange technology has been used successfully for this application.

As shown in Figure 10.9, a cationic resin in H^+ form and an anionic resin in OH^- form are mixed in the same ion-exchange vessel. During the processing of fine liquor, the cationic and anionic resin must be mixed together perfectly, so an exchange liquor/H^+ resin is followed immediately by an exchange liquor/OH^- resin. In this way the pH of the fine liquor is maintained between 7 and 8 during the de-ashing process, and sucrose inversion can be limited to less than 0.1 to 0.5%. This technology also provides very efficient stabilization of the pH of the fine liquor or the liquid sugar, the pH adjustment being optimized by the ratio of cation resin volume to anion resin volume. If possible, the sugar solution should be processed at 65 to 67 Brix and a temperature not exceeding 45 to 60°C, to extend the lifetime of the anionic resin, which is not very stable in OH^- form at high temperature.

During the de-ashing process, competition between color and ash removal has been observed, the demineralization capacity being far better with fine liquor low in color. From a standard decolorized liquor, it is usually possible using a single-pass technique to remove about 90% of the ash content and 65% of the color. A double-pass process can be used to improve the demineralization rate to 95% and the decolorization rate to 80%.

When saturated with ash and colorants, the resins are sweetened-off and backwashed with water or a brine solution. The anionic resin has a lower density than the cationic resin and is flushed to the top of the resin bed. Selection of the cationic and anionic resin used is critical and includes their ability to be separated in water or brine solution. When the design of the mixed-bed system and the backwash procedure are done properly, the resins are perfectly separated, as shown in Figure 10.9. An intermediate distribution/collection network is installed at the interface between the two resin beds inside the ion-exchange vessel. The cationic resin is regenerated with HCl or H_2SO_4, while the anionic resin is regenerated separately with NaOH. After a final common rinsing and pH stabilization, the resins are remixed by air scouring injected at the bottom of the ion exchanger.

The separated resin beds can be regenerated either sequentially, producing acid and basic effluents to be neutralized, or simultaneously, producing a mixed NaCl effluent at the outlet of the ion exchanger. Because of the high density and the viscosity of the fine liquor, the process flow is limited to 0.8 to 1.5 BV/h. Down-flow processing must be used to avoid separation of cationic and anionic resins during the production step.

The operating cost of such a de-ashing and polishing process can be estimated as shown in Table 10.6.

For the production of a superior grade of industrial liquid sugar, a final polishing of the liquid sugar with powdered activated carbon is generally used, with a low carbon dosage of 0.2 to 1 kg/tonne dry matter. The disks or plates and frame filters used to separate the spent carbon usually produce a high-clarity industrial liquid sugar. Pasteurization or deep sterile filtration on cellulose plates may be used to extend the life of the liquid sugar, usually considered to be about 1 month.

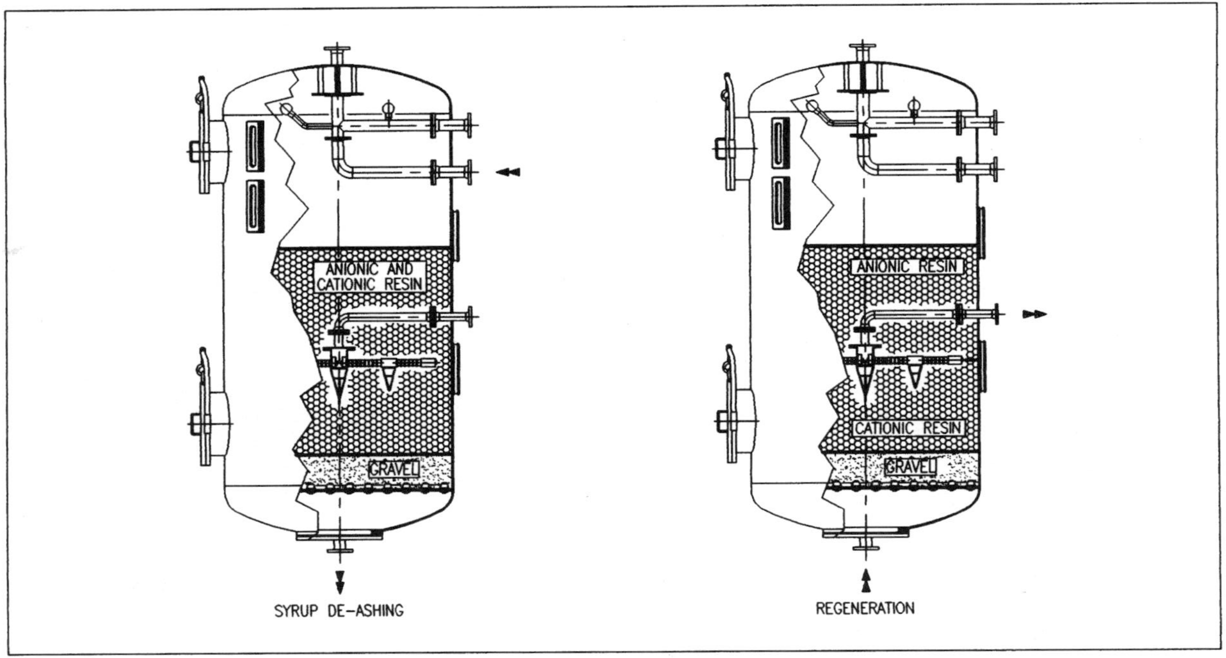

FIGURE 10.9 Typical design of mixed beds ion exchange vessels.

TABLE 10.6 Operating Cost Estimation: De-ashing and Polishing

Item	Cost (U.S. dollars / tonne sugar)
Regenerants	
HCl 100%: 2.2 kg / tonnes at $300 / tonnes	0.7
NaOH 100%: 4.4 kg / tonne sugar at $300 / tonne	1.3
Process water: 0.55 m^3 / tonne sugar at $1 / m^3	0.55
Resins: 0.20 L / tonne sugar at $4 / L	0.8
Effluents waste treatment: 0.45 m^3 / tonne sugar at 1 / m^3	0.45
Personnel: $15,000 / y	0.1
Maintenance: $25.000 / y	0.1
Total estimated operating cost	4

10.7 NEW DEVELOPMENTS

During the 1980s, extensive work was undertaken to improve the efficiency of decolorizing resin beds in ion-exchange decolorization plants. Up-flow decolorization vessels using countercurrent regeneration technology and full process automation have been designed and optimized to obtain the decolorization performance mentioned in Section 10.4 with a lower resin inventory, lower regenerant consumption, and lower resin consumption, as described in Section 10.5.

During the current decade, disposal of regeneration effluents has gradually become more crucial, as awareness of environmental protection is rapidly growing in most countries. In addition to providing a clean and hygienic process for raw sugar refining at an attractive capital and operating cost, an ion-exchange decolorization plant generates a daily volume of liquid effluents that is proportional to the volume of resins regenerated.

Typically, an ion-exchange decolorization plant handling 1000 tonnes of melt per day will generate a total of 250 to 300 m^3 of regeneration and rinsing effluents with the following approximate composition:

- *NaCl:* 20 to 25 g/L
- *Chemical Oxygen Demand:* 4500 mg O_2/L
- *Total Organic Carbon:* 2500 mg/L
- *pH:* 8 to 9 with caustic brine regeneration
- *Temperature:* 80°C

Although this composition is very close to the salinity of seawater, its high color and high chemical oxygen demand (COD) usually prevent straightforward disposal. The combination of very high salinity and high COD content, including barely degradable colorants such as polyphenols, does not simplify the disposal of this effluent.

10.7.1 Brine Recovery by Nanofiltration Membranes

The potential of cross-flow filtration to treat regeneration effluents has been investigated by several workers [7–11]. This recovery process is based on the ability of the membranes to concentrate the COD and coloring material in the retentate (rejected components), while smaller molecules such as sodium chloride, sodium hydroxide, and water can easily cross the membrane (permeate). According to Wilson and Percival, ultrafiltration induced a 45% reduction in organic matter in the portion recycled as regenerant, which represents 40% of the total effluent [7].

This low organics retention suggests that ultrafiltration membranes are not tight enough. Tongatt-Hulett reports on pilot experiments realized in its Durban central refinery, where tubular organic nanofiltration (NF) membranes were used to treat the salt-rich fraction of the regeneration effluent [8,9]. Such tubular configuration allows operation at high temperature (<70°C) and high pH (<12).

Using MPW SelRO MPT-30 and MPT-31 membranes at 45 and 60°C and an operating pressure of 30 bar, the experimental work focused on a portion including 86% of the sodium chloride and 37% of the organic compounds contained in the effluent, corresponding to a pH between 8 and 9. The NF process was recognized as being technically feasible on a pilot-plant scale and demonstrated a potential of 30% reduction in effluent volume and a 60% reduction in salt consumption at the Hulett refinery.

Applexion has developed a similar technology with Saint Louis Sucre at the Marseille refinery, using organic spiral-wound NF membranes and a higher concentration factor of effluent [10]. The spiral-wound membrane type was selected because its capital and operating costs are generally two to three times as low as those of the tubular configuration. After extensive pilot experiments, the first commercial plant was commissioned in Marseille in 1997, corresponding to the full capacity of the decolorization plant. Each effluent from a caustic brine regeneration is processed batchwise in a cross-flow filtration skid equipped with spiral nanofiltration membranes. Under a pressure of 20 bar, the spent brine is highly purified to give a new load of recovered brine, which is used for the next regeneration cycle.

The membranes are very easy to clean and the efficiency of resin regeneration with nanofiltered brine has been reported to be even better than with fresh caustic brine. A 70% reduction in salt consumption has been achieved on average, the plant being extendable to 90% salt recycling if a brine concentrator is installed for additional treatment of the salt-poor fractions. The COD and coloring materials are concentrated in a reduced volume of effluents, with much lower salinity, and can possibly be incorporated into the molasses. Diafiltration with water can be used in the NF plant to adjust the salinity of the retentate to an acceptable level.

Techsep has experimented at Tate & Lyle with a similar technology based on a new tubular mineral membrane. This membrane is much more expensive but would be able to treat the spent brine at its normal working conditions: a temperature of 80 to 85°C and a pH of 11 to 13 [11].

The capital cost of the brine recovery by nanofiltration can typically be balanced in two years by the savings in operating costs of the decolorization with a 70 to 90% savings on the NaCl consumption and a corresponding reduction of the tax costs associated with chloride disposal. Furthermore, the ability to send the NF concentrate to the molasses can be a key factor in the development of an effluent-free ion-exchange decolorization system operating under the most critical environmental conditions.

10.7.2 Further Reduction of the Resin Inventory

As an alternative to fixed-bed ion-exchange systems, AST has suggested using its ISEP contactor with 30 rotating columns as a possible method of reducing further the resins inventory used for decolorization and to implement a continuous ion-exchange decolorization process. Pilot testing has been reported at the Lantic Saint-John refinery [13] and the Hulett refinery

[12]. Decolorization rates of 50 to 60% have been obtained with acrylic resins and 600 to 900 ICUMSA 420 color feed while operating at 10 BV/h [13].

A 30% reduction in salt and chemicals consumption was reported compared with that in a fixed-bed ion-exchange decolorization plant without brine recycling. The ISEP pilot also offered more than 50% savings in water consumption, and in sweetwater and strip effluent to the sewer compared to conventional fixed-bed operation. Resin lifetime, mechanical stability, and feasibility of the short-sequence processing will be difficult to estimate in this multiple-port device until a commercial plant is in operation.

10.7.3 New Regeneration Process

Brine regeneration of anionic decolorizing resins is the simplest and easiest to practice, but it is far from being the most efficient, as demonstrated by several workers [14–17]. Only 5 to 10% of the NaCl used for regeneration is really exchanged on the resin, the remaining 90 to 95% being necessary as an excess of regenerant. This explains the high salt recovery obtained by nanofiltration of the spent brine, even when treating only the salt-rich fraction of effluents. Furthermore, the desorption of organic matter during brine regeneration is generally not 100% of the organic matter absorbed, and this progressive irreversible fouling is a probable cause of the limited lifetime of the resins.

An alternative regeneration process was proposed by Bento and used on a pilot scale at Refinarias Azucar Reunidas [14]. This process involves a separation of colorants in resin regeneration effluents based on their affinity to the resin. This separation is performed by regeneration using two solutions at different concentrations. In the first solution a salt concentration between 30 and 50 g/L is used. In this first regeneration salt, colorants with a low affinity to the resins and a low anionic charge are more efficiently removed from the resin. This effluent, with a low salt concentration, is discharged from the system.

The second regeneration process consists of passing a salt solution with a concentration between 70 and 110 g/L of NaCl. This regeneration provides more efficient removal of anionic-charged colorants that are subsequently precipitated with lime solution containing 100 g/L of $Ca(OH)_2$. Excess calcium in solution is precipitated with sodium carbonate, CO_2, or phosphoric acid. After filtration, recovered brine is recirculated for resin regeneration. In a test comprising 200 regeneration cycles, salt recovered was 80% using 22 kg of $Ca(OH)_2$ and 16 Kg of Na_2CO_3 per cubic meter of resin per regeneration. A salt makeup of 44 kg per cycle was necessary to maintain salt concentration.

A new and attractive proposal made by the same worker is to use salt at low concentrations with a mixture of calcium hydroxide in sucrose solution [15]. It was observed that this mixture enhances greatly the removal of colorants fixed to ion-exchange resins. The formation of a complex sucrose-calcium-colorant, soluble in alkaline sucrose solution, can explain the dislocation of the equilibrium reaction to the regeneration direction. In this way, chloride ions, even at low concentration, can efficiently dislocate colorants fixed ionically to resins. Also, at this low salt concentration, apolar colorants are easily removed.

Also tried was a regeneration process using 3 BV of a mixture containing $CaCl_2$ (0.4 *N*) and CaO (10 g/L) in sucrose solution (150 g/L) [16]. After 120 cycles in a 1-L resin column, an average decolorization of more than 90% was obtained. The results were better than with normal regeneration with alkalinized salt solution. This regeneration process uses a lower quantity of chemicals than that used in classical brine regeneration. However, it is necessary to use sucrose to maintain calcium hydroxide in solution. To recover sucrose from regeneration effluents, tangential filtration techniques were used. An ultrafiltration membrane was used to concentrate colorants in retentate, and permeate was recycled in the regeneration process.

The retentate, part comprising sucrose, calcium, and chloride ions, with the major part being sugar colorants, can be sent to the affination and/or recovery section of a refinery or to the low-grade section of a sugar factory. As this effluent has high alkalinity, it can be used to make pH corrections, rather than the lime that is normally used. Provided that the retentate can be recycled as above, this new regeneration process can be considered to be effluent-free. Other authors refer to a regeneration process using calcium chloride, the effluent being precipitated with calcium hydroxide and reused [17]. With this system a 47% reduction in chemical costs was obtained, compared with the cost of classical brine regeneration.

There is clearly a common trend in all these proposed new regeneration processes: (1) to increase the regeneration efficiency, for more complete desorption of colorants from the resins, and subsequent better decolorization effects, and (2) to design an effluent-free regeneration system in which most of the regenerant can be reused, and the concentrated colorants fraction can be returned to the sugar process. We are probably very close to seeing the first complete effluent-free ion-exchange decolorization processes used in sugar refineries.

REFERENCES

1. M. A. Clarke, *Proc. Sugar Ind. Technol. Conf., Symp. Sugar Color,* Sugar Processing Research Institute, New Orleans, LA, 1997, pp. 3–7.
2. L. S. M. Bento, *Proc. Sugar Ind. Technol. Conf.,* 1992, pp. 201–220.
3. L. Bento, *Proc. Sugar Ind. Technol. Conf., Symp. Sugar Color,* New Orleans, LA, 1997, pp. 49–69.
4. D. Tayfield, *Int. Sugar J.,* 98(1169), 1996
5. R. Celle and D. Hervé, *Ind. Aliment. Agric.,* July–Aug. 1980.
6. R. Tamaye, Paper 665, *Proc. Sugar Ind. Technol. Conf.,* 1994.
7. R. J. Wilson and R. W. Percival, *Proc. Tech. Sugar Refin. Res. Conf.,* pp. 116–125.
8. D. M. Meadows and S. Wadley, *Proc. S. A. Sugar Technol. Assoc.,* 1992, pp. 159–165.
9. S. Wadley, C. J. Brouckaert, L. A. D. Baddock, and C. A. Buckley, *J. Membr. Sci.,* 102: 163–175, 1995.
10. S. Cartier, M. A. Theoleyre, and M. Decloux, *Desalination,* 113:7–17, 1997.
11. G. Cueille, V. Thoraval, A. Byers, S. McGrath, D. Segal, and R. Kahn, *Filtr. Sep.,* Jan–Feb. 1997, pp. 25–27.
12. G. M. Hubbard and G. B. Dalgleish, *Proc. Sugar Ind. Technol. Conf.,* 1996, pp. 32–42.
13. S. E. M. McDonald and J. C. Thompson, *Proc. Sugar Process. Res. Inst. Conf.,* 1996, pp. 29–54.
14. L. S. M. Bento, *Proc. Sugar Ind. Technol. Conf.,* 1989, pp. 176–200.
15. L. S. M. Bento, *Proc. Sugar Process. Res. Inst. Conf.,* 1996, pp. 121–136.
16. L. S. M. Bento, *Proc. Sugar Ind. Technol. Conf.,* 1996, pp. 43–58.
17. R. Schick and G. Reiche, *Zuckerindustrie,* 120:1037–1042, 1995.
18. X. Lancrenon and D. Hervé, *Sugar Technol. Rev.,* 14:207–274, 1988.

CHAPTER 11

Filtration Processes*

INTRODUCTION

Filtration is a relatively old and well-established unit operation in which suspended particles are removed or concentrated from a particle–fluid mixture. This is achieved by moving the mixture toward a porous medium which stops the particles but allows the fluid to pass through. Filtration belongs to the family of mechanical separations characterized by having a mechanical rather than a thermodynamic interface. In sugar refining, filtration is used to remove calcium carbonate particles from carbonated liquors as well as sweetwater in a carbonation refinery, or to recover sucrose from the scum in a phosphatation refinery [1]. Filtration is also the final clarifying step at a liquid sugar station.

Filtration as a distinctive industrial process can be traced back as early as 1789, when Joseph Amy obtained the first French patent on a commercial filter [2]. The technique has since evolved into numerous designs and has helped shape modern chemical process industries. In a filtration operation there is always a filtering medium across which a pressure difference is applied to force the fluid to flow through the medium. The effluent flowing out of the medium is known as the *filtrate.* The quality of the filtrate is usually measured by the percentage removal of the suspended solids. In sugar refineries, turbidity, a photometric measurement of liquid clarity, is used to monitor filtrate quality. The efficiency of the filtration is usually measured by the flux of the filtrate, that is, the flow rate of the filtrate per unit filter area.

To design and operate a filtration process more effectively and efficiently, an engineer needs to understand the fundamental relationships between the flux, turbidity of the filtrate, and the operating parameters, such as viscosity, pressure gradient, and the particle characteristics. In this chapter we first provide essential information on the process analysis of filtration to help establish such a knowledge base. Then we discuss how to select a filter and how to operate it to achieve better filtration results.

* By Chung Chi Chou.

11.1 PROCESS ANALYSIS

11.1.1 Classifications of Filtration Processes

Based on the mechanism for the capture and collection of suspended particles, filtration is traditionally classified into two types: (1) deep-bed filtration and (2) cake filtration. In the literature, cross-flow filtration may also be added to the list. An actual filtration process may involve more than one mechanism.

In deep-bed filtration, the filtering medium is a packed bed of sands, diatomaceous earth, or fibers. "Pores" are formed in the interstitial spaces of the packing materials. The feed suspension travels into the filtering medium. In the course of migrating through the filtering medium, the suspended particles are trapped inside the pores by virtue of impingement (i.e., inertial retardation, combined with surface forces) [2–4]. Because of the limited capacity of the packed bed, deep-bed filtration is suited only for clarifying liquids or gases that contain very small amounts of suspended particles, typically less than 1% on the scale of total suspended solids. The deep-bed mechanism is observed in municipal drinking water systems and various cartridge filters used in liquid clarification applications. Because the content of the suspended particles is sufficiently low, the accumulation of particles inside the filter medium is not significant and deep-bed filtration is treated approximately as a steady state. Its design and equipment operation are relatively simple.

Cake filtration is the most commonly used industrial process for separating fine particles from a slurry material. Contrary to deep-bed filtrations, the septum in cake filtration has openings smaller than the particles, at least on average. The particles are caught by the septum at its first contact and then accumulated outside the septum to form a cake. This makes cake filtration capable of removing a relatively large amount of suspended particles. In many situations, cake filtration is the only choice for the engineer. Because the content of suspended particles is relatively high, the thickness of the cake increases noticeably as filtration proceeds. After an initial period of cake buildup, the cake becomes the primary flow resistance. Being a packed bed of particles, the cake also serves as the primary filtering medium. It has been found that the characteristics of the cake, rather than the septum itself, often dictates the filtrate flux and the filtrate turbidity. Since the thickness of the cake is a function of time, both the turbidity of the filtrate and the flow resistance are time dependent. For submicron particles, however, the flow resistance of the cake can be so great that the flux could approach zero. In this case, cake filtration is not effective.

Cross-flow filtration is the newest technological development in filtration. It is mostly used in ultrafiltration and reverse osmosis, where membranes are the filtering medium. In a cross-flow operation, a strong, tangential cross-flow is created by a circulating pump to keep the cake (retained colloidal species) from building up on the surface of the membrane. This is an effective way of minimizing the flow resistance. Cross-flow filtration is a special variation of cake filtration, different in the way the retained "particles" are distributed.

Although cake filtration is the oldest solid–liquid separation technique, it is widely used in sugar refining as well as in other manufacturing industries because of its unique solid capacity. In comparison with cross-flow membrane processes, cake filtration remains to be more cost-effective. A number of designs of cake filtration equipment have evolved, with differences in the pressure differential applied across the cake, such as pressure filters and vacuum filters, as well as in the way the cake is removed, either continuously or discontinuously. The mainstream literature regards such centrifugals as used in sugar manufacturing as special designs of cake filtration, simply because of the existence of a mechanical screen [3]. Others leave separations caused by a body force outside the filtration category [4].

11.1.2 Calculations of Filtration

For deep-bed filtration, flux is treated as in steady-state, but the quality of filtrate collapses at the end of the filter life. The required filter area is calculated from the average flux and the filtrate volume required. The schedule for the renewal or replacement of the filter is calculated according to the filter life, which is determined experimentally.

For cake filtration, the flux is time-dependent and the quality of filtrate improves as the cake builds up. We have to use integration to map a complete filtration cycle in which the cake thickness progresses from zero to the final value at the end of the cycle. A cycle is ended mathematically at the time the cake is discharged. Unlike steady-state processes, the filter area in cake filtration is calculated from the number of cycles and the flux as a function of time within a complete filtration cycle. A theoretical derivation is presented here to help illustrate the important relationships between operating parameters of a cake filtration process. The principles of filtration can be found in a typical chemical engineering text book such as, *Unit operation of chemical engineering,* by McCabe, Smith & Harriott (McGraw-Hill, Inc. 1993). A schematic profile is illustrated in Figure 11.1.

Cake filtration is not a steady-state process, regardless of continuous or discontinuous operation, automated or manual. We take a filtration cycle as the subject of our analysis.

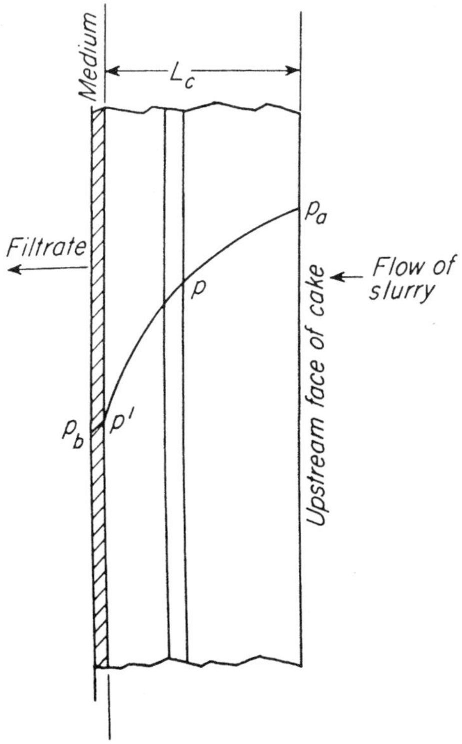

FIGURE 11.1 Schematic pressure profile across the flow axis in cake filtration. Note that p stands for fluid pressure at distance of L from the septum. $(p_a - p_b)$ is the overall driving force for the filtrate flow.

Within the cycle, the thickness of the cake increases with time. As a result, the flow resistance will also increase. In what is known as the constant-pressure mode, the overall pressure difference, the driving force of the filtration, is held constant and the filtrate flux decreases with time, as is the case in the carbonation filter presses. Theoretically, if the pressure differential is progressively increased to counter balance the increase in cake thickness, a cake filtration process can be operated in constant-flux mode. Nonetheless, most industrial applications are run in constant-pressure mode.

The cake is a packed bed of particles. For flow calculations, a packed bed is simulated by a bundle of straight tubes in each of which the flow is laminar and thus obeys the famous Poiseuille equation,

$$\langle v \rangle = \frac{d^{*2}}{32\mu}\left(\frac{dp}{dL}\right)_c \tag{11.1}$$

where $\langle v \rangle$ is the average flow velocity in the imaginary tube (which is the average linear velocity of the filtrate inside the cake), d^* the diameter of the imaginary tubes, μ the filtrate viscosity, and dp the pressure drop over the thickness of the cake dL (which is the length of the imaginary tubes). Assume that the stimulated tubes have the same total void volume (flow space) and the same total surface area (viscous drag) as those of the cake. This will lead to

$$d^* = \frac{2\Phi\langle d \rangle \varepsilon}{3(1 - \varepsilon)} \tag{11.2}$$

where Φ is a shape factor of the particles, $\langle d \rangle$ the mean diameter of the actual particles, and ε the void fraction of the actual cake, also known as the cake porosity.

The average liner velocity of the filtrate inside the cake is related to the actual filtrate flux as follows:

$$\langle v \rangle = \frac{1}{A}\frac{dV/dt}{\varepsilon} = \frac{\text{flux}}{\varepsilon} \tag{11.3}$$

where A is the cross-sectional area of the cake (which is the same as the septum area) and V is the total volume of the filtrate. Substituting equations (11.2) and (11.3) into equation (11.1), we get

$$\text{flux} = \frac{1}{A}\frac{dV}{dt} = \frac{1}{\dfrac{72\mu\varepsilon^3\Phi^2\langle d \rangle^2}{(1 - \varepsilon)^2}}\left(\frac{dp}{dL}\right)_c \tag{11.4}$$

Equation (11.4) has a number of very important predictions that will guide the engineer throughout the process analysis. According to this equation, the larger the particles are, the greater the cake porosity is. If the average particle diameter is doubled, the flux will be quadrupled. All this shows how important it is to grow bigger particles in the carbonation station. Since the flux is also inversely proportional to the filtrate viscosity, efforts to reduce viscosity either by raising operating temperature or by lowering liquor Brix will increase the filtration throughput. This equation shows that as the cake thickness increases, the flux will have to decrease if the pressure drop across the cake is constant.

If we assume that all parameters in equation (11.4) are independent of L, we can integrate it over the entire cake thickness L_c and the result is further rearranged:

$$\Delta p_c = \frac{\mu \alpha m_c}{A} \frac{1}{A} \frac{dV}{dt} \tag{11.5}$$

where m_c is the total mass of the cake not including the filtrate and α is the specific cake resistance factor, which combines all concerned cake characteristics:

$$\alpha = \frac{k(1 - \varepsilon)}{\varepsilon^3 \Phi^2 \langle d \rangle^2 \rho_p} \tag{11.6}$$

where ρ_p is the density of the particles. At any given time t, the overall pressure difference is equal to the sum of the pressure drop over the medium, Δp_m, and the pressure drop over the cake, Δp_c:

$$p_a - p_b = \Delta p_m + \Delta p_c \tag{11.7}$$

Using an analogy between cake and the septum, one obtains that

$$\Delta p = p_a - p_b = \frac{\mu \alpha m_c}{A + R_m} \frac{1}{A} \frac{dV}{dt} \tag{11.8}$$

where R_m stands for the flow resistance contributed by the medium itself. In cake filtration, the increase in the amount of cake is proportional to the increase in the filtrate volume. If c is defined as the mass of the cake (the particles) per unit volume of total filtrate, that is,

$$c = \frac{m_c}{V} \tag{11.9}$$

then equation (11.8) can be written as

$$\frac{dt}{dV} = \frac{\mu}{A\,\Delta p} \left(\frac{\alpha c V}{A} + R_m \right) \tag{11.10}$$

For constant-pressure cake filtration, we integrate equation (11.10) over time and obtain the following famous linear relationship between t/V and V:

$$\frac{t}{V} = \frac{K_c V}{2} + \frac{1}{q_0} \tag{11.11}$$

where q_0 is just a time-independent integration constant, and

$$K_c = \frac{\mu c \alpha}{A^2\,\Delta p} \tag{11.12}$$

Traditionally, at t/V versus V plot is prepared using actual process data to guide constant-pressure cake filtration design and operation. While equation (11.11) is used to calculate filtration cycle time and filter area, equation (11.4) is used to pinpoint opportunities to improve process efficiency.

11.1.3 Filtration-Impeding Impurities in Sugar Refining

To achieve optimal performance, the engineer needs to take into account not only the equipment but also the properties of the feed material that is to be filtered. In sugar refining applications, the filtration-impeding impurities in particular have been the subject of a number of in-depth studies [1]. The following impurities are generally responsible for causing difficulties in filtration in the sugar industry:

- Suspended particles with average size of less than 0.5 μm
- Phosphate particles, which are often ultrafine and fragile
- Starch, which is a viscosity-enhancing high-molecular-weight polymer
- Dextran, another species of polymer that is more soluble in water

In light of these important findings, the engineer needs to make every effort to reduce these impurities in order to improve filtrate flux.

As for carbonated liquor, the challenge is to operate the carbonation station to grow large calcium carbonate grains and to aid their aggregates to form. Recycling a small portion of carbonated liquor to the saturators is one example of a way to make larger crystals. In the carbonation station, impurities are entrapped in the calcium carbonate aggregates and finer grains sometimes are favorable for colorant removal because of larger surface areas. Therefore, the optimal process conditions are those that accommodate both high color removal and large particle size. To deal with very fine suspensions, a *filter aid* is often used to obtain economical flux. A filter aid is defined as an inert particulate material added in filtration feed to improve the cake porosity and protect the septum from being plugged.

11.2 EQUIPMENT AND MATERIALS

11.2.1 Filters

Although deep-bed filters such as cartridge filters are often suitable for water and wastewater clarification, the majority of filtration operations in sugar refining are of cake filtration type. There are two types of cake filtration equipment: pressure filters and vacuum filters. Pressure filters use above-atmospheric pressure to achieve high flux for viscous liquids such as sugar liquors. The positive pressure, however, makes it difficult to discharge the cake at the end of a filtration cycle. Therefore, pressure filters are usually operated discontinuously. A vacuum is a weaker pressure differential, according to our analysis in the preceding section, and the flux will be significantly lower for the same cake thickness. But the cake can be readily discharged under vacuum, and thus we can shorten the filtration cycle to allow the filter to operate at its thin cake region where the flux averages high. The choice between pressure filter and vacuum filter is made on overall cost, which may be affected by individual situations.

Currently, most carbonation refineries use pressure filters to filter carbonated liquor because the viscosity is high. Vacuum filters are found in some phosphatation refineries for scams filtration. For sweetwater, either filter type is adequate. Among the many designs of pressure filters, the filter press, with the configuration of plates and frames, and the leaf filter, characteristic of a pressure vessel containing a number of septum-wrapped leaves, are the primary choices.

Figure 11.2 shows a typical filter press. In a filter press, square plates alternate with open frames. The plates vary in materials and thickness, with grooved surfaces. Both sides of the plates are covered by the septum. The frames can be up to several inches thick. Plates and frames are mounted vertically on a horizontal metal rack that is on the two sides of the filter assembly. The alternating plates and frames are tightly clamped together by a hydraulic ram at the end of the metal rack. At the corner of each frame, there is a drilled

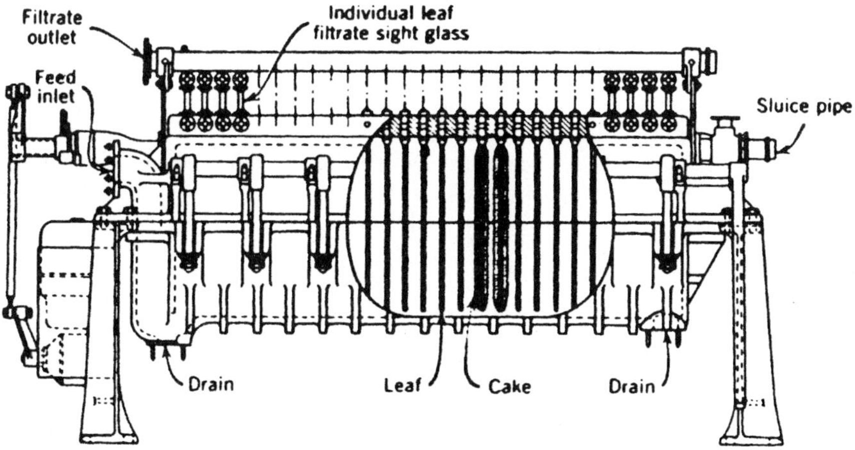

FIGURE 11.2 Filter press.

passage to allow the feed to enter the individual compartment formed between two adjacent plates. The cake is accumulated inside the compartment. The feed is pumped into the compartments, and the filtrate flows through the septum and flows out of the plate in the grooves between the cover cloth and the plate surface. The filtrate is collected at the corner of each plate where it joins with flows outgoing from other plates. When a cycle is completed, the cake in the frames is washed by water and then dried by air, then discharged.

In a leaf filter, a set of vertically oriented leaves made of corrugated metal screens are mounted on a retractable metal rack. The septum is wrapped on the outside of each leaf. The set of leaves is enclosed inside the pressure vessel during the filtration cycle. The feed is pumped into the pressure vessel. The filtrate is forced to flow through the septum into the leaves, and through the screened passages of the leaves, is collected at the nipples of the leaves, and is turned to a filtrate manifold. The cakes form on the outside of the leaves.

When we compare a filter press with a leaf filter, we find that it is much easier to wash the cake in the leaf filter than in the filter press. Also, it is easier to withstand a high pressure in a leaf filter than in a filter press, because a filter press has so many compartments that need to be sealed by a hydraulic ram.

According to the literature [1], the Sweetland leaf filter, developed in 1915, is by far the best known and most widely used filter in the sugar industry. The leaves in sugar refineries often are circular with $\frac{1}{2}$-in. spacings. The pressure vessel is normally a horizontal cylindrical tank. This lower half of the tank can be opened and closed for the sluice of cake. Leaf filters with a vertical tank design, such as Pronto leaf filters, are also used in sugar refineries, especially in liquid sugar stations. Most leaf filters are of stationary design; however, some have rotary leaves to allow the cake to form uniformly on the leaves [1].

Because pressure filters are discontinuous and there is up to 30 min of downtime between filtration cycles to perform cake sluice, they are considered to be labor intensive. One of the innovations brought to leaf filters over the last decade is automation of the leaf filter operation.

Vacuum filters are not commonly used in sugar refineries because the refineries deal with high-Brix syrup and the viscosity is inadequately high for vacuum filters. Vacuum filters are a better choice for sweetwater filtration. The most popular design of the vacuum filter is the continuous rotary-drum filter. In a rotary-drum filter, a horizontal drum with a slotted face is rotated slowly in an agitated feed trough. The drum is covered by a septum. The space inside the rotary drum is divided into annular compartments by radical metal partitions. Only the compartments that are exposed to the feed are doing filtration. The other compartments are doing either sluice or cake discharging. The only filter area is the drum surface immersed in the trough. Every revolution of the rotation is a complete filtration cycle. There is no downtime between cycles. This type of cake filtration is ideally suited for low-capacity and low-viscosity applications.

11.2.2 Septum

Among many criteria [3], the most important consideration in choosing a septum is that it must not plug or blind. As already discussed in Section 11.2.1, the cake is the primary filtering medium in a cake filtration. The septum resistance, R_m, is mostly contributed by the cake clog-up in the septum openings. The septum commonly used is made of either canvas cloth or synthetic fabrics such as polypropylene. Canvas cloth tends to be more ragged and more effective in removing very fine particles. Synthetic septum is, however, more chemically resistant and is often hydrophobic and less absorbing. The engineer needs to understand the chemical characteristics of the cake and determine what material is best in preventing the septum from plugging or blinding. As reported [1], synthetic fabrics such as polypropylene cloth and nylon cloth have been favored by sugar refineries due to their long life and superior performance, despite their high cost.

There is a detailed analysis in the literature on applicable criteria for choosing a septum, generally including the ease of cleaning, the durability, the cost, and the particle retaining capability. In a cake filtration cycle, the first 5 to 10% filtrate is often returned to the feed because the cake needs to be built up to obtain clear filtrate. Therefore, the turbidity of initial filtrate should not be a consideration in choosing a septum.

11.2.3 Filter-Aid Filtration*

Filtration is the process by which particles are separated from a fluid by passing the fluid through a permeable material. The filtration discussed below concerns the removal of suspended solids, including semicolloids, from a liquid. Ideally, the liquid passes through the filter and the solids remain, building a permeable cake on the filter screen or septum. This ideal can be approached only when the solids are large, incompressible particles. In practice, fine particles often pass through the filter cake, leaving only the large particles. If the solids are compressible or deformable, the flow is often reduced to an uneconomical level and the solids stick to the screen. To overcome the problem of solids sticking to the filter septum, and to aid in the removal of small as well as large suspended solids, filter aid can be used to coat the septum before forward filtration begins. Two types of filter aid are diatomaceous earth and perlite.

a. Nature of Diatomite. Diatomaceous earth or diatomite is the skeletal remains of single-celled, aquatic, waterless plants called *diatoms.* These microscopic alga have the unique capability of extracting silica from their water environment to produce their skeletal structure. When diatoms die, their skeletons settle to form a diatomite deposit (Fig. 11.3). Diatomite is a soft powdery mineral resembling chalk and distinguished by a variety of shapes. When processed, to become a filter aid, it is virtually inert and is predominantly pure silica.

*This section is contributed by World Minerals, Inc.

FIGURE 11.3 Diatomite.

Deposited on a filter septum, the diatomite forms a rigid but porous filter cake that sieves out the particulate matter in liquid as it passes through the filter. Properly processed, diatomite has virtually no effect on the odor and taste of any liquid filtered through it. Diatomite deposits may be of saltwater (marine) or freshwater origin. Marine diatomite deposits are characterized by having thousands of different-shaped species of diatoms. The intricate shapes form very rigid filter cakes well suited to the high demands of pressure filtration. Freshwater deposits typically contain far fewer diatom species and are often dominated by one or two species. Both sources are capable of producing high-quality filter aids for removing fine solids from a sugar syrup.

b. Expanded Perlite. Perlite filter aid is produced from perlite ore, a naturally occurring form of siliceous volcanic glass (Fig. 11.4). The distinguishing feature setting perlite apart from other volcanic glasses is that when heated to a suitable temperature in its softening range, perlite expands from 4 to 20 times its original volume. Expansion is due to the

FIGURE 11.4 Perlite.

presence of 2 to 6% combined water in the crude perlite rock. When heated to above 1600°F (871°C) the particles of crushed ore pop or expand in a manner similar to popcorn as the moisture vaporizes and creates countless bubbles of glass. This accounts for the dramatic decrease in density, from approximately 75 lb/ft^3 for the crushed ore to 3 to 5 lb/ft^3 for the expanded material. The expanded particles are then milled to fragment the bubble clusters so that a three-dimensional structure of fragments of controlled particle size is obtained to achieve the desired permeability.

c. How Filter Aid Works. Filtration using filter aid is a two-step operation. First, a thin layer of filter aid, called a *precoat,* is built up on the filter septum by recirculating a filter aid slurry through the filter. After precoating, small amounts of filter aid (body feed) are regularly added to the liquid to be filtered. As filtering progresses, the filter aid, mixed with the suspended solids from the unfiltered liquid, is deposited on the precoat. Thus a new filtering surface is formed continuously; the minute filter aid particles provide countless microscopic channels that entrap suspended impurities but allow clear liquid to pass through, without clogging.

An efficient, economical filter aid must (1) have rigid, intricately shaped, porous, individual particles; (2) form a highly permeable, stable, incompressible filter cake; (3) remove even the finest solids at high rates of flow; and (4) be chemically inert and essentially insoluble in the liquid being filtered. Diatomaceous earth meets these requirements, due to the wide variety of intricately shaped particles and inert composition, which makes it practically insoluble in all but a few liquids.

d. Selection of Filter-Aid Grade. Diatomite filter-aid grades are produced in a wide range of particle sizes to meet practically any industrial filtration requirement. The relative flow rates of these grades are determined by a standard filtration test. It is axiomatic in the use of filter aids that the ability of the filter aid to remove small particles of suspended matter decreases as the particle size and thus the flow rate increases. Conversely, as the filter-aid particle size, and therefore the flow rate, decreases, the ability of the filter aid to remove small particles increases. The extent to which this takes place will depend on the type and particle size distribution of the undissolved solids being removed.

Selection of the proper grade is a compromise between high clarity and low flow rate, and lower clarity and higher flow rate. The best filter aid is that grade which provides the fastest flow rate (or greatest throughput per dollar's worth of filter aid) while maintaining an acceptable degree of clarity, which must be determined and specified by the filter aid user. Desired clarity, or the amount of acceptable suspended solids in the filtrate, can be determined in a number of ways: (1) visual examination of a sample of filtrate; (2) comparing a sample of filtrate with a standard; (3) the use of electronic turbidity instruments; (4) filtering a sample of filtrate on a fine white or black filter paper, such as a membrane filter, and observing the impurities on the paper; (5) chemical or biological analysis; and (6) gravimetric analysis.

It is difficult to state the particle size of the solid that will be removed by any given grade of filter aid. This depends on the method used for measuring the particle size of the contaminent, the type of liquid involved, the shape of the particle, the compressibility of the particle, and filtration conditions. For example, a needlelike particle might be easily removed if it approaches the filter cake sideways, but could conceivably pass through the cake if it approaches on end. A soft, compressible particle might extrude its way through a filter cake, whereas a rigid particle of the same shape and size would not. Variations in pressure, flow rate, vibration, and air bubbles may also affect clarity.

e. Filtration System. In the operation of a filtration system, the filter is first precoated by circulating a mixture of filter aid and clear or filtered liquid from the precoat tank through the filter and back to the precoat tank. This is continued until all the filter aid is deposited on the filter septum. The body feed injection system is then started and the filter is changed over, with minimum fluctuations in pressure, from precoating to filtering.

The purpose of the precoat is threefold: (1) to prevent the filter septum from becoming clogged by impurities, thus prolonging septum life, (2) to give immediate clarity, and (3) to facilitate cleaning of the septum at the end of the cycle. Precoat quantity should be from 0.1 to 0.2 lb of filter aid per square foot of filtering surface, depending on the type of filter aid and the style of filter. For the sake of convenience, the quantity is typically rounded to the nearest full bag when dispensing from paper bags. Using a bulk or semibulk system, the quantity can be adjusted to a volumetric quantity, or weighed incrementally to a precise amount.

Precoat slurry concentration will depend on the particular style of filter and on the volume and shape of the precoat tank. To achieve the crowding effect necessary to precoat many woven septums, the concentration should never be less than 0.3%. The most important principle involved in precoating is to first push all the air out of the filter with clear liquid before adding filter aid to the slurry tank. It is also important not to generate air bubbles in the slurry tank from the agitator blades, and any liquid streams entering the precoat tank should enter below the liquid level once flow to the filter begins.

Precoat recirculation rate will also depend on the type or style of filter, and particularly to the location from which the liquid flow enters the filter. Internal baffling and the amount of space between the filtering surfaces also affects the flow rate. For a pressure leaf filter with good baffling and water viscosity precoat, the flow could be 1 to 2 gal/ft^2 of filtering surface per minute. Using a filter press, flow would be between 0.25 and 0.5 gal/ft^2 per minute with water viscosity and depending on the type of filter press. Precoating with sugar syrup, the precoat flow rate through a pressure leaf filter would be closer to 0.1 to 0.3 gal/ft^2 per minute because of the increase in viscosity.

f. Body Feed Addition. Body feed is added to the filter feed liquor to provide a continuously new filtering surface to the liquid stream. Body feed physically separates the suspended solids and provides a path for the liquid to pass through the filter. It extends the life of the precoat by preventing the precoat layer from being coated with solids that would restrict flow to an unacceptable level.

The amount of body feed added to a filter feed stream depends on the type and amount of suspended solids in the system. Coarse, rigid particles might not require any body feed. Fine rigid particles or deformable particles require body feed to maintain permeability of the filter cake. If deformable particles are also very fine, the pressure or the drag of the liquid flow can cause them to work their way through the filter cake. Additional body feed to provide a more tortuous path and reduced pressure, or reduced flow, or both, may be necessary when such conditions exist.

Typical body feed rates in various sugar syrups will vary depending on the purification process and the particular syrup stream. Typical rates will range between 0.1 and 0.3% by weight based on Brix. Any filter has a certain available filter cake space located between the filter plates. In a filter press the full available cake space can be utilized. At a filter cake density of 20 lb/ft^3 one can calculate how many pounds of filter aid can be added before the filter cycle must be terminated. The filter aid should be added at a rate to just fill the cake space as the maximum pressure across the filter is reached. The rules for a pressure leaf filter are different in that the filter cake on two adjacent filtering surfaces must never touch. Therefore, the total usable cake space must be calculated in order to know how much total precoat and body feed can safely be added before the cycle must be terminated. The maximum amount added should coincide with reaching the maximum pressure differential for maximum economy. This should provide the minimum amount of filter aid used per unit of sugar produced. Since the actual amount of suspended solids can vary from day to day, an average typical amount or slightly less than the theoretical optimum amount is a safe rate for day-to-day operation unless the amount of solids can be determined in advance.

Attempts to measure turbidity in the line ahead of the filter have been made to control the body feed rate with measured turbidity. However, there are several types of turbidimeters using beams of light transmitted at various wavelengths through the liquid flow at various angles to measure suspended solids content. Only recently has success been reported by some manufacturers of this equipment on some process streams. The customary approach has been to add a typical amount found effective for a specific process stream and allow operator adjustment based on experience and input from other known process variables.

g. Filter Cake Removal. At the end of the cycle the filter may be drained of unfiltered heel if appropriate for the filter in use. The filter might be sweetened off with water inside the filter, or the filter cake might be sluiced off the filter and sweetened off in another part of the process. Sluicing might be internal or from an external source, depending on the design of the filter. Spent filter cake might go to a lagoon or to sewer, or be collected, dewatered, and go to landfill or composting.

h. Summary. Filter aid is capable of removing finely divided suspended solids, including bacteria. Filter aid grades are available in a wide range of permeabilities to meet nearly liquid filtration requirement in the sugar industry. The advantage of filter aid is its inertness, its low cost per unit of production, and its versatility. If a particular grade no longer meets the needs of the refinery, another grade more coarse or more fine is available for use to meet the changing requirement without having to replace capital equipment.

11.3 TYPICAL OPERATING PROCEDURE OF A LEAF FILTER

A typical procedure for the washed sugar liquor filtration using a leaf filter is summarized as follows:

1. Precoating
   a. Use 40 lb of earth (regenerated and new makeup) to precoat each press.
   b. Make precoat slurry with clarified liquor (same liquor).
   c. Use precoat pressure of 10 psi.
   d. Control precoat operation with timer or turbidity meter. Approximate time is 4 min with a flow rate of about 200 gal/min.
2. Pressure filtration
   a. Use 100 lb of earth (regenerated and new makeup) as a body feed per cycle.
   b. Control internal pressure increase in body by pressure control device or orifice. Pressure should not exceed 60 psi. Line pressure should be 60 psi.
   c. Average flow rate is 0.50 gal/min. Typical pressure cycle time is 120 min at 65 Brix.
   d. Provide adequate leaf drainage to gutter or manifold with at least $\frac{3}{4}$-in. pipe diameter and with minimum length.
3. Body liquor blowdown
   a. Use 20-psi air to maintain cake on leaf and to express liquor from leaf and to drain press.
   b. Draining time is about 5 min with about 50% of the liquor going to the clear gutter and 50% drained to the cloudy liquor tank.
   c. Average flow rate is 150 gal/min.
4. Sweetening off
   a. Control high sweetwater by volume or time in order to have a uniform amount to melt washed sugar.
   b. Use thin mud slurry derived from sluicing washed sugar liquor presses only.
   c. Maintain line pressure at about 45 psi.
   d. Average flow rate is about 350 gal/min for a total time of about 5 min.
5. Drain body sweetwater
   a. Allow approximately 2 min with about 700 gal returned to the high thin mud slurry tank.
6. Sluicing
   a. Use fresh hot water plus condensate. Sluice between leaves, not across.

b. Maintain line pressure of 95 psi. Average available pressure in the sluicing manifold should be less than 80 psi.
c. Maintain flow rate of about 275 gal/min.
d. Divert first sluicings to thick mud tank (about 2 min) and the second portion to the high-purity thin mud tank.

REFERENCES

1. J. C. P. Chen and C.-C. Chou, *Cane Sugar Handbook,* 12th ed., Wiley, New York, 1993.
2. C. Orr, *Filtration Principles and Practices,* 2nd ed., Marcel Dekker, New York, 1977.
3. W. L. McCabe, J. C. Smith, and P. Harriott, *Unit Operations of Chemical Engineering,* 4th ed., McGraw-Hill, New York, 1993.
4. J. H. Perry, *Chemical Engineers' Handbook,* 6th ed., McGraw-Hill, New York, 1984.

CHAPTER 12

Evaporation Theory and Practices*

INTRODUCTION

Evaporation is the removal of a solvent from a solution by vaporization. In the sugar industry the solvent is water and the solute is largely sucrose, which at the temperature of evaporation has a very low vapor pressure. This distinguishes evaporation from distillation where all or most of the components in the liquid phase have an appreciable vapor pressure and appear in the overhead vapors. The overhead vapor in the sugar factory is largely water, contaminated only by small amounts of sucrose and impurities originally present in the liquid phase. This is called *entrainment* [1].

In a raw sugar or white sugar factory, the extracted and clarified juice is concentrated through evaporation and boiling. The principle involved is the same, but the evaporator station concentrates the juice only to a syrup of approximately 60 to 65% total solids, while boiling produces a total solids concentration (sucrose crystals plus some dissolved solids) of about 90%. Evaporation is to concentrate the thin juice into syrup as rapidly as possible at reduced boiling temperatures to minimize sucrose losses (inversion) and color formation at minimal steam consumption. During the boiling process the sucrose first appears in solid form, and in addition to the conditions required for evaporation, the maximum quantity of good-quality sucrose crystals have to be produced in as short a time as possible, while easy separation or sucrose crystals from the mother liquor is also of importance. The point at which clarified juice begins to crystallize is in the neighborhood of 78 to 80 Brix. Hence, it is possible to carry out the evaporation to 72 to 75 Brix. In practice, the sugar boiler requires a syrup still capable of dissolving further crystals, so that they may dissolve the false grain that may form at the beginning of a charge. This condition is indispensable for obtaining good results at the pan stage. For this reason a Brix of 70 is normally not exceeded. The evaporators are designed and operated in such a way that the Brix of the syrup lies between 60 and 70 Brix in the raw sugar factory and between 50 and 60 Brix for white sugar manufacture.

In the refinery it is mainly liquor concentration and sweetwater evaporation that have to be accomplished in evaporators. No matter how fast the evaporation takes place with

*By Willem H. Kampen.

present equipment, some sucrose decomposition or inversion, color formation and/or caramelization, will occur. If a syrup is diluted to the original feed concentration, its color will be darker than that of the feed. Most of the color is picked up in the first effect, where the temperature is highest. The increase in color formation is reduced with reducing temperatures [2]. Obviously, good circulation and mixing as well as short residence times are important.

The quantity of the water (W) to be evaporated can simply be calculated from the Brix of the feed (b_f), the Brix of the syrup (b_s, and the weight of the feed (F).

$$W = F\left(1 - \frac{b_f}{b_s}\right)$$

12.1 HISTORICAL DEVELOPMENT

To the author, the oldest known depiction of a complete cane sugar (factory) process is a copper engraving from 1570, made by Jan Stradanus, a Dutch artist working in Florence, Italy. The engraving is in the British Museum in London and shows the manually harvested whole stalks being brought to the small factory by donkey. Cane preparation consisted of manually chopping whole stalks into small pieces from which the juice is squeezed in a manually operated screw press (mill). The juice is concentrated through evaporation in two open copper kettles in series over open fires. In the first kettle, egg whites and some milk of lime are added as clarifying agents and the reasonably concentrated juice is ladled into the second kettle, where it is concentrated until saturation or supersaturation is achieved. To prevent splashing and bumping of the concentrated juice, some butter was added. The syrup obtained is then ladled into cone-shaped clay forms and the sucrose allowed to crystallize. The resulting loaves of sugar were a valuable commodity. Farmhands, at the time, had to work three long days to earn the money needed to purchase 1 pound of (crude) sugar [3]. Process improvements came slowly.

The clarified juice was later concentrated in the long pans or Chinese battery, several (often, three) open pans in series. An open fire was fed and maintained under the last pan, containing the most concentrated juice, while the clarified or thin juice entered the first pan, which is the farthest away from the fire. The hot gases flow past all pans and typically to a small chimney to provide sufficient draft. During concentration the floating impurities were removed, hence the mechanical purification worked well, but due to the (uncontrollable) high temperatures, color formation and sucrose losses were a problem, while energy consumption was high [2].

Evaporation in open pans occurred until the end of the eighteenth century, when steam heating at atmospheric pressure was introduced. This gave better temperature control and reduced energy consumption. The greatest increase in steam economy is achieved by reusing the vaporized solvent (water). This is done in a multiple-effect evaporator by using the vapor from one effect as the heating medium for another effect, in which boiling takes place at a lower pressure and thus temperature. This invention revolutionized the sugar, food, and chemical industries. The inventor, Norbert Rillieux, born in New Orleans, was a *quadroon libre* (a free person 75% white and 25% black) and the son of a French planter/inventor and a slave mother. He studied engineering in France. Rillieux developed the following idea: Since steam is being used for heating juice to evaporate the water it contains, why not use the vapor so furnished by the juice to heat a further portion of juice or finish the evaporation already begun by ordinary steam. The problem was that the vapors had a lower temperature than the steam. Rillieux solved this problem by putting the effects following the first one at a reduced pressure or under vacuum. This does require an installation to produce vacuum but presented two great advantages: it increases the total temperature difference between steam and juice by a quantity equal to the drop in boiling point of the juice between the pressure of the first and that of the last body, and it permits evaporation to be carried out at temperatures proportionately less dangerous from the point of view of inversion and color formation of the juice as it becomes more concentrated and more viscous. Because of these problems, the sugar industry uses only feedforward evaporation systems. After several trials, the Rillieux device was patented in 1846 as U.S. Patent 4,879. The evaporating pans enclosed a series of condensing coils in vacuum chambers. It took much of the manual labor out of making syrup, saved on fuel, and produced a much better product.

Rillieux's three principles developed for multiple-effect evaporators are: (1) 1 pound of steam will evaporate (theoretically) as many pounds of water as there are effects; (2) vapor bleeding of a given effect for use elsewhere will save steam amounting to (number of withdrawal effect)(pounds of vapor withdrawal) / (total number of effects); and (3) noncondensable gases must always be removed continuously wherever steam or vapors are condensed [4].

12.2 EVAPORATION PROCESS

The rate of evaporation of a technical sugar solution depends on the turbulence produced by boiling. At quite low rates, little turbulence results and the rate of heat transfer is little more than would be expected from natural convection. At reasonable rates, the boiling does produce turbulence and in effect promotes itself to fairly high steady-state boiling. This steady state, termed *nucleate boiling,* is quite sensitive to the temperature difference and such other factors as cleanliness of the heated surface and physical properties of the liquid [5]. As the temperature difference increases, the boiling rate reaches a maximum beyond which the vapor cannot escape and the surface becomes partially vapor bound, causing the evaporation rate to drop with increased temperature differences. This relatively unstable film boiling condition becomes stable film boiling at somewhat higher temperature differences wherein all the surface is completely vapor bound [5]. Nucleate boiling rates may be estimated by the following equation: $W = 110\rho_i^{2/3}\rho_v^{1/3}$ lb / h, where $\rho_{i,v}$ are the densities of liquid and vapor, respectively. Values computed by this equation represent a reasonable approximation [5]. Rates observed with different surfaces vary somewhat but are usually obtainable and frequently exceeded. Boiling coefficients are typically high, and conservative design frequently represents small cost differentials and therefore an entirely acceptable solution. Induced or forced circulation increases boiling coefficients. The *thermal syphon* or *airlift principle* is used to induce natural circulation. The difference between the densities of the liquid and those of the boiling mass provide a substantial head and improves the circulation. The smaller the Δt value, the higher the required liquid level in a Robert-type evaporator to maintain proper circulation. The juice enters the bottom of the tubes in the calandria at a temperature that corresponds to the prevailing boiling point at the top of the calandria. When the juice enters the tubes, it is preheated until evaporation begins. The vapor pressure above the calandria corresponds to the boiling-point temperature at the top of the calandria. Toward the bottom of the tubes, the boiling point increases progressively due to the increasing hydrostatic pressure. Hence, the temperature gradient decreases, thus decreasing the overall heat transfer coefficient.

When steam flows through a tube that is submerged in a pool of liquid, minute bubbles of vapor form at random points on the surface of the tube. The heat passing through the tube wall where no bubbles form enters the surrounding liquid by convection. Some of the heat in the liquid then flows toward the bubble, causing evaporation from its inner surface into itself. When sufficient buoyancy has been developed between the bubble and the liquid, the bubble breaks loose, from the forces holding it to the tube and rises to the surface of the liquid pool [6]. Kelvin postulated that for this behavior to prevail, the liquid must be hotter than the saturation temperature in the incipient bubble. This is possible since the spherical nature of the bubble establishes liquid surface forces on it so that the saturation pressure inside the bubble is less than that of the surrounding liquid. The saturation temperature of the bubble being lower than that of the liquid surrounding it, heat flows into the bubble. The surface tension of water is relatively high: about 75 dyn/cm^2 versus 20 to 30 dyn/cm^2 for organics at room temperature, and Figure 12.1*c* shows the effect of interfacial tension on bubble formation. Intermediate surface tensions, as shown in Figure 12.1*b*, may create a momentary balance between bubble and tube wall, and it is necessary to form larger bubbles for the buoyant force to free it from the surface. Liquids with very low surface tensions tend to wet the surface, so that the bubble in Figure 12.1*a* is readily occluded by the liquid and rises [6]. Consider the typical boiling coefficient curve of McAdams [7] based on the data of several investigators for water, as shown in Figure 12.2. From a $(\Delta t)_w$ value above 5°F, there is a relatively straight logarithmic relationship between the coefficient of vaporization and the temperature difference, where $(\Delta t)_w$ is the difference between the tube wall and the vapor temperatures. This relationship changes, however, at the critical temperature difference, which occurs at about 45°F for water evaporating from pools. The high rate of heat throughput (flux; Q/A in Btu/ft^2-h) produces too much vapor at the tube wall and blanketing or vapor binding occurs. This creates bubbles almost exclusively of the type in Figure 12.1*c*. When vaporization takes place directly at the heating surface, it is called *nucleate boiling*, and when it takes place through a blanketing film of gas, it is called *film boiling* [6]. With $Q = AU\Delta t_m$, where Q is heat flow in Btu/h, A is heat transfer surface in ft^2, U is overall heat transfer coefficient in Btu/h-ft^2-°F, and Δt_n is the (logarithmic) mean temperature difference, the maximum attainable heat flux is $(Q/A)_{\max}$ or $(U/\Delta t)_{\max}$. The heat flux versus $(\Delta t)_w$ is plotted for boiling water as the upper curve in Figure 12.2.

If the coefficient for vaporization from a pool has been reported for a fluid at atmospheric pressure, it can be converted to subatmospheric pressure by the equation of Jakob [8]:

$$h_v = h_{v,14.7\text{psia}}(\rho/\rho_{14.7\text{psia}})^{1/2}$$

The enthalpy of a liquid more generally depends on its concentration as well as its tem-

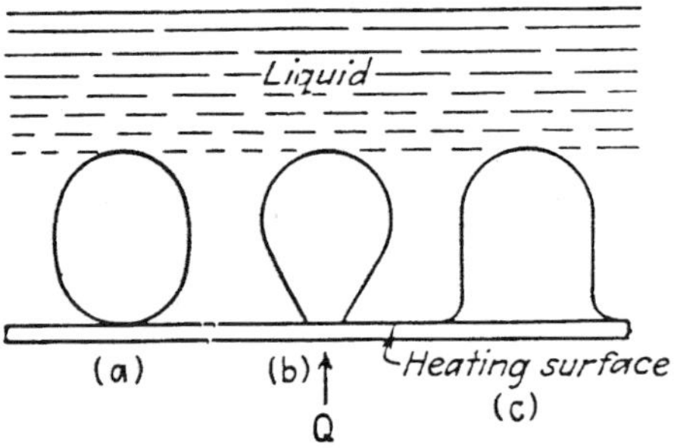

FIGURE 12.1 Effect of interfacial tension on bubble formation. (After Ref. 8.)

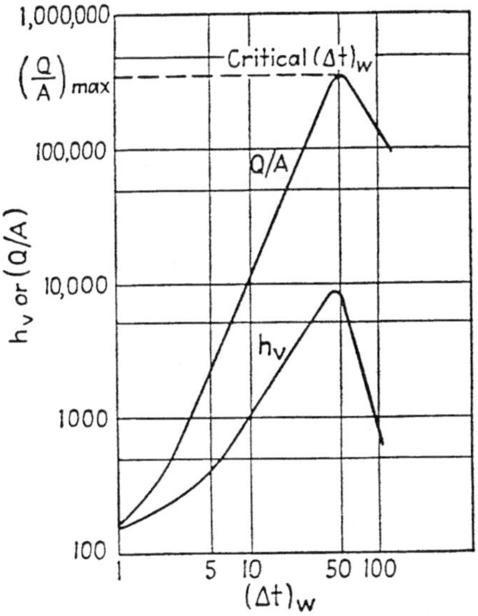

FIGURE 12.2 Boiling curve of water from pools. (After Ref. 7.)

perature. Whenever a solute is added to water, the partial pressure of the water is reduced. According to Raoult's law, $p_1 = x_1 p_1^0$, or the partial pressure of component 1 equals the mole fraction of component 1 times the vapor pressure of this pure component at a temperature T.

Adding any solute to water causes $x_w < 1$, so $p_w < p_w^0$. This implies that the solution must be heated to a higher temperature than the boiling point of the pure solvent before it will boil. This is the boiling point rise (BPR), or elevation due to addition of solute. A 30 wt% sucrose solution has a water mole fraction of $x_w = 0.9779$ at 212°F (100°C), which is the boiling point of pure water at 14.696 psi (1 atm) and $p_w = (0.9979)(1) = 0.9979$ atm (14.371 psi). For the solution to boil, the partial pressure must be raised to 1 atm. Thus for the same mole fraction of sucrose, the vapor pressure must be raised to 0.9779^{-1} or 1.0226 atm (15.028 psi). According to the steam tables, this equals 213.33°F and the BPR is 1.33°F. Since Raoult's law is asymptotically correct only as $x_w = >1$, the actual BPR is 1.8°F. Figure 12.3 shows the BPR, and Figure 12.4 shows the specific heat for sucrose solutions.

The driving force for heat transfer in evaporators is provided by making the pressure in the vapor space of the downstream effect lower than the vapor space upstream. This requires a vacuum pump or compressor: $p_2 < p_1$, $\Delta T_2 = T_1 - T_2 > 0$, and $q_2 = U_2 A_2\, \Delta T_2$. Steam is supplied to the first effect and $\Delta T_1 = T_s - T_1$ and $q_1 = U_1 A_1\, \Delta T_1$. Neglecting BPR and enthalpy changes (dilution/concentration), the same latent heat added to the steam in the first effect is recovered by condensing in the second effect and theoretically $q_1 = q_2$.

12.3 EVAPORATOR DESIGN

In evaporator design, heat transfer is the most important single factor, since the heating surface represents the largest part of the evaporator cost. Other things being equal, the type of evaporator selected is the one having the highest heat transfer coefficient under desired operating conditions in Btu/h-ft^2-°F per dollar of installed cost [9]. Evaporator performance

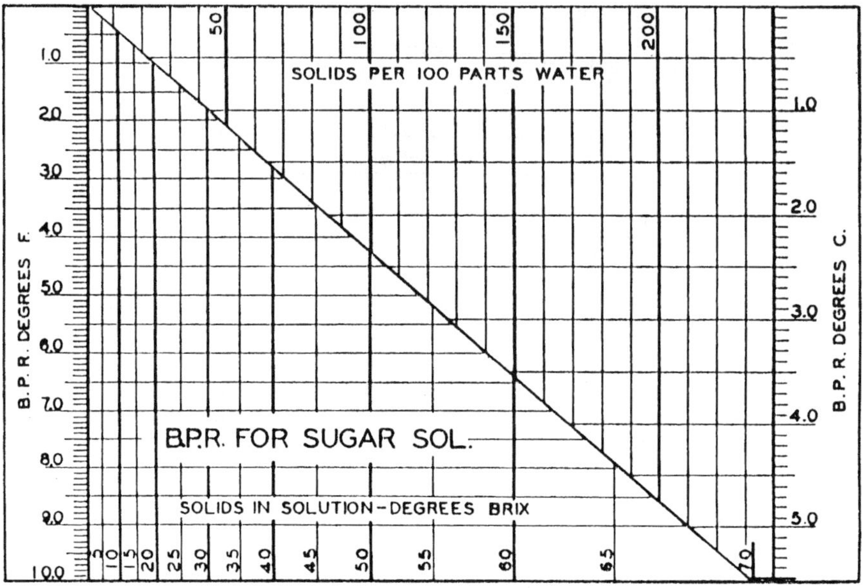

FIGURE 12.3 BPR for sugar solutions. (After Ref. 4.)

is rated on the basis of pounds of water evaporated per pound of steam consumed. Heat is required (1) to heat the feed from its initial temperature to its boiling point, (2) to provide the minimum thermodynamic energy to separate water from the feed, and (3) to vaporize the water. Thermocompression may be used to increase the steam economy. The vapor is compressed so that it will condense at a temperature high enough to permit its use as the heating medium in the same evaporator.

Low holdup time and low-temperature operation will avoid or minimize thermal degradation of sucrose in terms of inversion and color formation. The mean residence time of an evaporator is the evaporator liquid volume divided by the volumetric flow. The Nusselt equation allows the calculation of the mean thickness of the laminar liquid film: $Z = (3v^2/g)^{1/3}\mathrm{Re}^{1/3}$ for $\mathrm{Re} < 400$, where Z is the thickness of laminar fluid film in meters, v the fluid velocity in m/s, g the acceleration due to gravity in m/s^2, and Re the

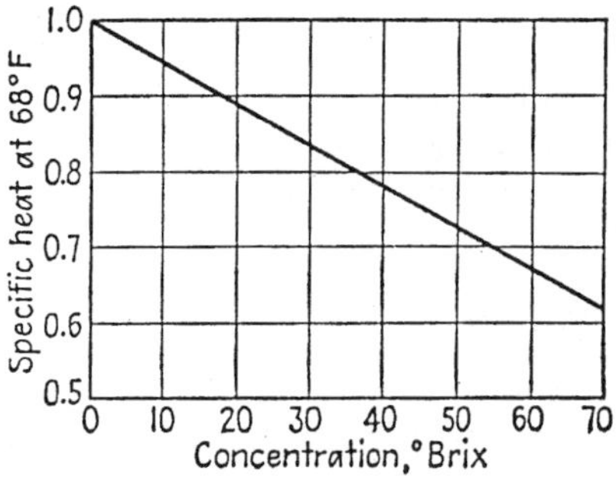

FIGURE 12.4 Specific heats of sugar solutions. (After Ref. 4.)

dimensionless Reynolds number. Scaling and fouling can be reduced by producing a clarified juice of very low turbidity and thus very low suspended solids, and treating this with electrocoagulation or possibly electromagnets. Present tests at the Audubon Sugar Institute and at a factory in Louisiana indicate that electromagnets installed before the juice heaters seem to have a positive effect on overall heat transfer coefficients for the heaters and on clarified juice turbidity; however, no significant decrease in cation concentration was observed. Electrocoagulation in batch tests on clarified juice at the Audubon Sugar Institute removed over 90% of the Si and over 95% of the Al, Fe, Ca, and Mg cations present in the clarified juice. This technology looks very promising for scale reduction in evaporators. Factory tests are being planned for the 1998 grinding season.

The raw sugar industry typically uses Robert or standard (calandria) evaporators, which are short-tube vertical evaporators, to evaporate clarified juice. Refiners may also use these for sweetwater evaporation but are using primarily long-tube vertical falling film (LTVFF) evaporators with bottom or external vapor–liquid disengagement. More total evaporation is accomplished in LTV evaporators than in all other types combined; it is normally the cheapest per unit of capacity. The equipment consists of a simple one-pass vertical shell-and-tube or plate heat exchanger discharging into a relatively small vapor head. The liquor residence time is measured in seconds. Tube diameters are normally from 0.75 to 1.5 in. and tube lengths from less than 20 to 35 ft.

Clarified juice and sweetwaters already contain 15 to 20% dissolved solids with sucrose purities of roughly 80 to 85 and about 98, respectively, while the highly concentrated syrups have properties far different from those of pure liquids in terms of thermal conductivity, specific heat, viscosity, boiling-point rise, and so on. In these forward-feed evaporation systems, heat recovery may be by use of feed preheaters heated by vapor bled from each of the evaporators. The heating surface of evaporators is defined as that in contact with the juice, whereas most other heat-exchange equipment is rated on the basis of outside area of the tubes.

The falling film evaporator is widely used for concentrating heat-sensitive materials, because the holdup time is very small, the juice is not overheated during passage through the tubes, and heat transfer coefficients are high even at low temperatures. The tube-side pressure drop is usually very low and there will be a corresponding loss in available temperature difference, due to some friction and acceleration of the vapors generated in the tubes. The juice is fed to the top of the tubes and flows down in the tubes as a film. Vapor–liquid separation typically takes place at the bottom. All tube surfaces must be wet continuously. This requires recirculation and excellent feed distribution. Falling film evaporators are usually not well suited for scaling or fouling liquids as concentrated clarified cane juice but can handle sweetwaters very well. Tubular falling film evaporators can be used as a factory's preevaporator or first and possibly second effects.

The sugar division of FCB has a patented juice distribution system, which consists of three successive devices: (1) a rough distribution of the entering juice through several inlet necks; (2) a juice distribution container perforated with 15-mm-diameter holes, which is large enough to avoid scaling problems; and (3) a Noryl plug in each tube assures proper juice film formation. The tube layout is on a triangular pitch and hexagonal sections of seven tubes, six surrounding one central tube, make up the calandria. Good wetting is obtained over a range of 500 to 3000 kg/h per meter of tube perimeter. Optimum wetting is already obtained at 600 kg/h · m or 400 lb/h-ft. This reduces power requirements and minimizes recirculation requirements. The basic calandria consists of 35-mm-diameter tubes with standard lengths of 8, 10, and 12 m. Tube length can be nonstandard. A steam distribution jacket with one or two inlets assures proper steam distribution into the tube bundle. Noncondensable gases are removed near the top and the bottom. Efficient removal is required for good heat transfer, but excess venting increases energy consumption (see Figure 12.5).

Vapor–liquid disengagement is accomplished at the bottom with a single- or double-pass centrifugal separator, a knitted wire-mesh separator with double water or juice-spraying unit, and a catch-all or separate vessel with a combination of the foregoing separators. Minor flashing typically does occur in falling film evaporator bottoms. The centrifugal cyclones may have either top or bottom outlets. The higher the velocity, the better the separation efficiency, but this is at the expense of some pressure drop and corresponding available temperature difference. The pressure drop is typically 10 to 16 times ($v^2/2g$), where v is the velocity in the inlet pipe [9]. Velocities of about 100 ft/s at atmospheric pressure are used. Increased sizes will reduce losses in available temperature differences.

Separators of knitted wire mesh have low pressure drops of about 0.4 to 1 in. of water column and collection efficiencies of over 99.8% in the range of vapor velocities from 8 to 20 ft/s. The vapor entrains droplets and the gravitational force must be larger than the vapor's entraining force in order for them to be separated from one another. The entraining force is equal to the square of the vapor velocity, and the velocity should normally not exceed 3.3 ft/s. For a given mass flow of vapor, this fixes the minimum diameter of the liquid–vapor disengagement chamber.

BMA falling film evaporators may be fitted with an upper and lower cone plus impingement baffle in the vapor–liquid disengagement chamber. This minimizes entrainment to such levels (<10 and >3 ppm) that the qualitative α-naphthol reaction for sucrose is negative. HPLC analysis should be used to detect levels of sucrose, glucose, and fructose below 3 ppm.

Vertical-flow mist eliminators such as the Munters T-271 and T-272 also are high-efficiency separators, designed for vapor velocities up to 60 ft/s. Each profile of a T-271 module is a multiturn, patented profile with opposing-angle chevron collection grooves on each surface (see Fig. 12.6). These grooves provide a low-velocity zone where droplets accumulate and drain to the edges of the profile subsections. The TM-272 uses the same primary profiles as the T-271; however, the spacing between the profiles is twice that of TM-271. This is better suited for high-liquid and particulate-laden gases. Higher allowable operating velocities result in smaller vapor head diameters. With droplet removal capabilities of less than 25 μm in diameter, a +99% removal of droplets can easily be achieved. Modules are fabricated from a wide range of nonmetallic and metallic materials. Polypropylene and glass-coupled polypropylene have maximum operating temperatures of 165 and 265°F, respectively, and weigh only 4.9 and 3.4 lb/ft^2, respectively. In stainless steel it would weigh almost three times as much. Limit drop sizes of 35 μm at 1000 ft/min are normal, but due to fractional efficiency, practical removal of smaller droplets is achievable. Factors that affect performance include vapor distribution, relative gas–liquid densities, temperature, pressure, viscosity, moisture content, and gas density.

Tube bundle falling film evaporators have high overall heat transfer coefficients and require relatively small calandria and floor space. Evaporation is from a thin film and the heating zone in the upper part of the tube bundle is rather short. There is no increase in BPR due to hydrostatic juice columns. This means that the temperature difference is essentially the same anywhere along the tube length. Color formation is controlled primarily by temperature and residence time. The ratio between the mean retention time of the juice in the tubes of a falling film evaporator and the mean retention time of the juice in the tubes of a Robert evaporator, each with the same heating surface area, is, according to BMA, $t = 4fL/d$, where t is the time ratio, f the film thickness in the falling film evaporator, L the apparent juice level in the Robert evaporator, and d the tube diameter. If $d = 31$ mm, $f = 0.5$ mm, and $L = 30\%$ in the first effect, and $f = 1$ mm and $L = 60\%$ in the fifth effect,

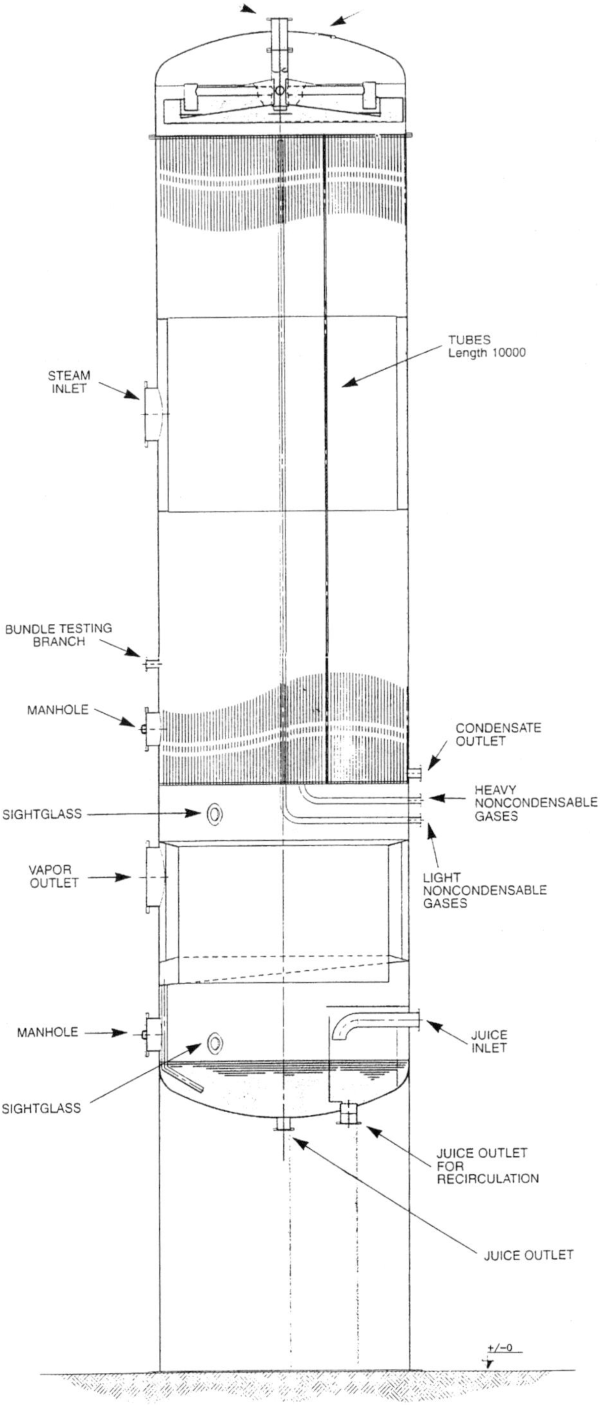

FIGURE 12.5 Tubular falling film evaporator. (Courtesy of FCB Sugar Division.)

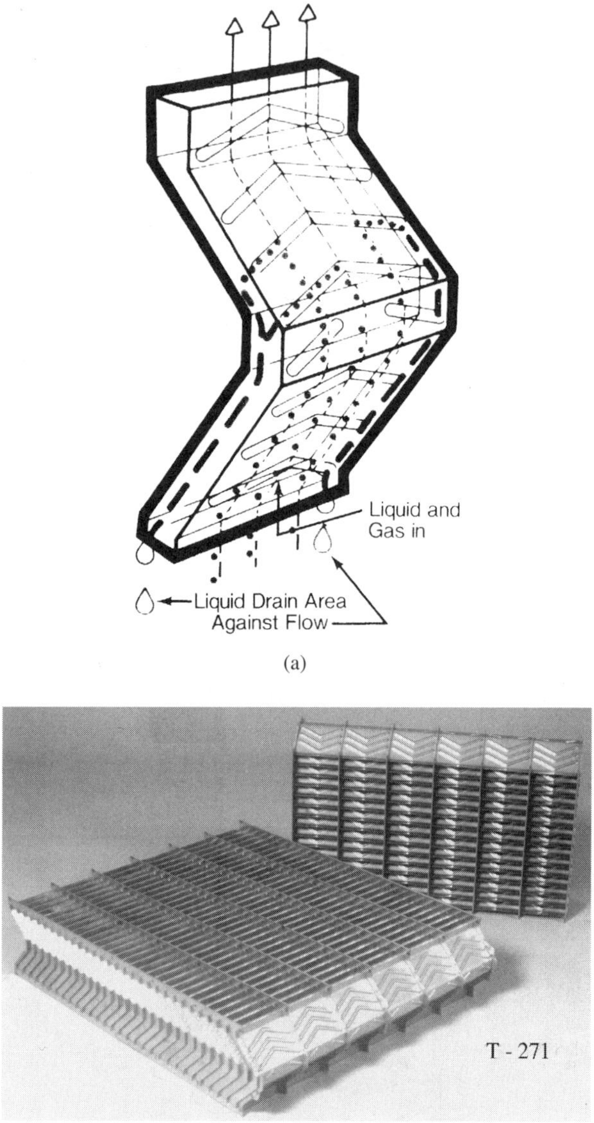

(a)

(b)

FIGURE 12.6 Vertical flow mist eliminators.

then in both effects of the falling film evaporator the juice retention time in the tubes is less than 25% of that in a Robert evaporator.

Research has shown that an increase in temperature of 26 K is almost equal to a tenfold increase in retention time. A 2°C temperature reduction in a first effect will equal roughly an increase in heating surface area corresponding to an increase in retention time of 20%. BMA segmental falling film evaporators (Fig. 12.7) show increased heat transfer coefficients, smaller heating surface areas, and even lower retention times. Each segment goes with two

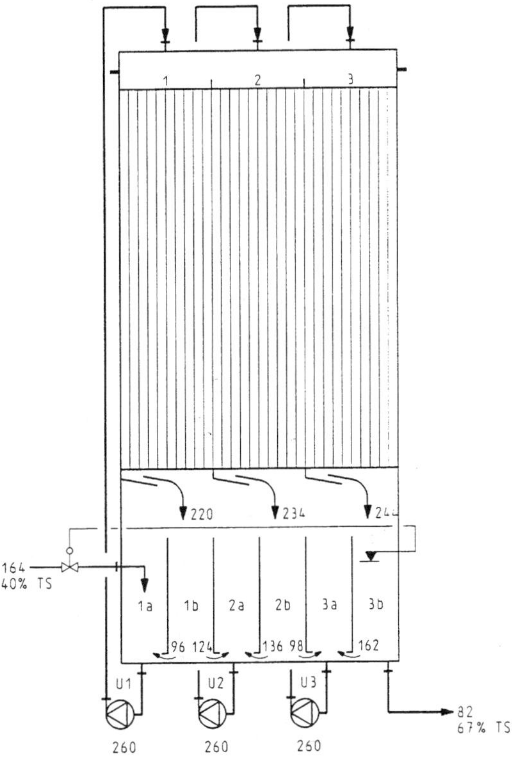

FIGURE 12.7 Segmental (tubular) falling film evaporator (BMA).

bottom chambers underneath the calandria. A unit having three segments would have six chambers, which are all connected like communicating vessels; however, the design minimizes remixing. The feed distributor at the top would also consist of three segments. The liquid level in the chambers of the last segment should be controlled. A segmental falling film evaporator typically requires 100% total pump capacity or three times one-third of total capacity (per segment), while a nonsegmental evaporator requires 200% (100% feed plus 100% standby).

Alfa-Laval developed a plate-type evaporator (EC 500) which can be installed in a vessel (open outlet model) or the design for a traditional evaporation station. In 1987, units were installed at Südzucker AG (Germany) and British Sugar plc. Plate-type falling film evaporators are the latest and most advanced type of evaporator. The falling film plate-type accomplishes excellent water evaporation at a small effective temperature difference. The (Balcke–Dürr) plates form almost circular tubules of diameter 6 or 9 mm on the product side and wavelike ducts at right angles to the tubes on the steam side. The plates are only 0.6 mm thick as they support one another and form a normal total height of 4.00 m (see Fig. 12.8). These stacks of plates may be installed in the shells of Robert evaporators and improve their performance. Plate-type falling film evaporators have thinner walls than tubular units and shorter tubular passages (Table 12.1), liquid films, and thus even higher overall heat transfer coefficients than the tubular falling film evaporator. Due to the short tube length in the plate-type falling film evaporator, the condensate film near the bottom

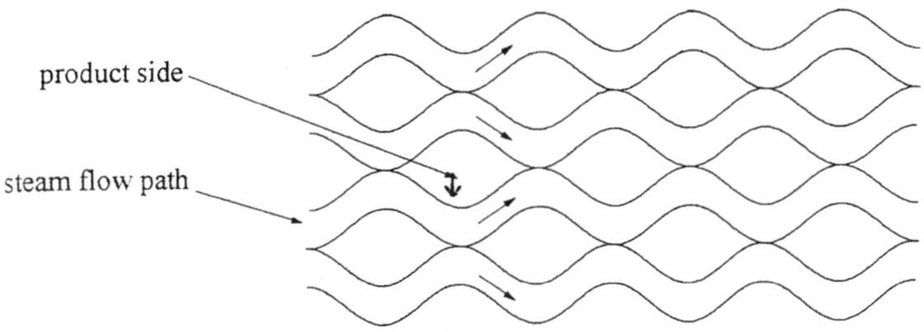

(a)

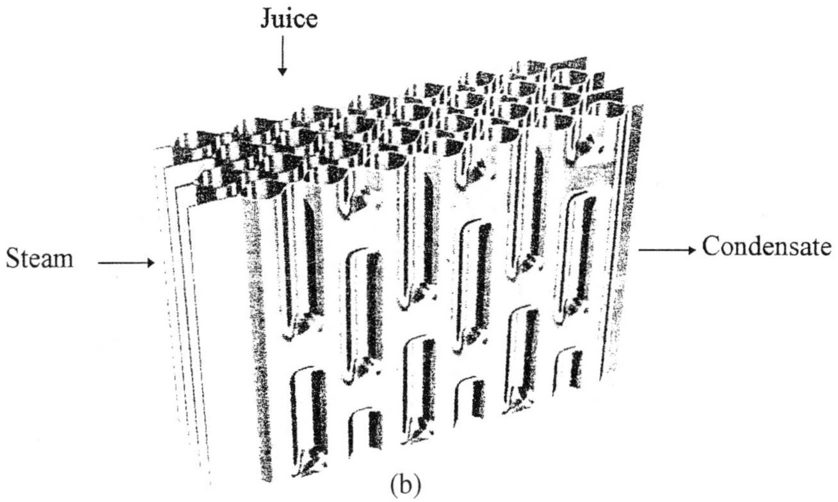

(b)

FIGURE 12.8 Balcke–Dürr plate packet flow pattern: (*a*) top view; (*b*) side view.

is much thinner than that in tubular falling film evaporators (ca. 0.05 versus 0.15 mm). Figure 12.9 shows the basics of a plate-type falling film evaporator.

In 1995 the largest plate-type falling film evaporator (Balcke–Dürr AG, ca. 66,000 ft^2) in the sugar industry was installed as a preevaporator at the beet sugar factory of Coöperatie Suiker Unie U.A. in Groningen, The Netherlands. It is a single-pass (no recirculation) pre-evaporator before the quintuple-effect evaporator station (see Fig. 12.10). It evaporates approximately 3.9 lb/h-ft^2 at an overall mean heat transfer coefficient of 3152 W/(m^2 · K) or 605 Btu/ft^2-°F and an effective temperature difference of 3.89 K. Temperature in/out was

TABLE 12.1 Tubular Passage Dimensions for Falling Film Evaporators

Parameter	Plate-Type FFE	Tubular FFE
Length (mm)	320	10,000
Outside diameter [mm (in.)]	7.2 (0.283)	35 (1.378)
Inside diameter [mm (in.)]	6.0 (0.236)	32 (1.260)

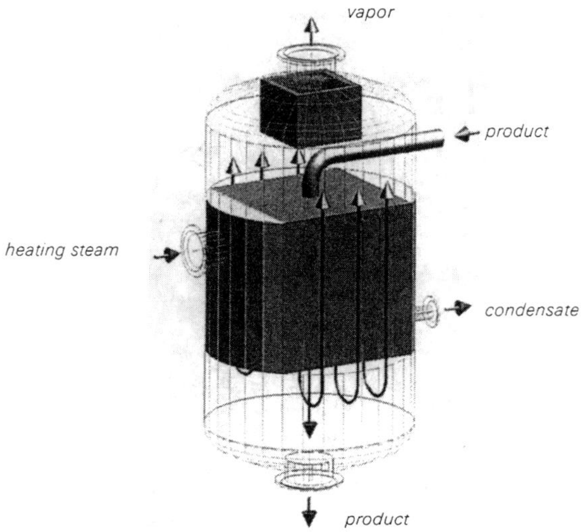

FIGURE 12.9 Balcke–Dürr plate package in BMA falling film evaporator.

220.5/222.4°F and Brix in/out 16.11/18.79. This corresponds to an overall coefficient that is approximately 30% higher than in a similar tubular falling film evaporator. Steam temperature was 229.5°F and vapor 221.8°F at 18.0 psia. Tube length/plate height is 0.32, height of heating chamber 0.32 m, juice flow number is 0.6 to 1.0 L/(h)(cm), heating surface density 240 m^2/m^3 versus 40 to 50 for Robert evaporators, inner tube diameter or gap between plates is 6 mm (available range 5 to 11 mm), and wall thickness is 0.6 mm (available range 0.5 to 0.8 mm). The unit operated the entire 1995 campaign and was treated with alkali at the end of the campaign only. No increase in pressure drop over the unit was observed, and thus minimal fouling occurred. The liquid residence time was about 75 s, and together with the aftereffects of thin juice sulfitation, this prevented any color formation. The thin juice in $IU_{420/560}$ units had 164/156 before and 156/1047 color after the preevaporator. After the first and fifth effects of the quintuple-effect evaporator, the color formation was 171/1122 and 247/1715, respectively. After the campaign a cyclone separator was installed to assure that sugar entrainment losses in vapor would be less than 20 mg/L. The preevaporator reduced energy requirements for evaporation by approximately 20% [10]. Figure 12.11 shows the overall heat transfer coefficients for a plate-type falling-film evaporator versus Robert and tube bundle falling-film evaporators.

The plate-type falling film evaporator is a cross of plate and tube heat exchanger technology [11]. It is made of preformed (pressed) welded stainless steel plates and has no seals. Steam and juice follow a cross-flow pattern. A single plate element of the plate pack is manufactured from two prepressed plates with a standard wall thickness of 0.6 mm and a width of 0.32 m, which are laterally inverted and laid on top of each other, then roll spot welded. The length of the prepressed elements is variable and may amount to 8 m or over 26 ft. The single shaped plate elements are stacked one above the other and welded together at the front ends by cross-seams. A plate module with variable dimensions is formed. Vertical, almost elliptical juice tubules are formed, which are at right angles with wavelike, undulating steam channels. Steam flow is smooth and no dead spaces occur. The flow is reversed once and the noncondensable gases accumulate at the end of the flow path, allowing for proper venting. Both the very short residence time and the low pressure/temperature

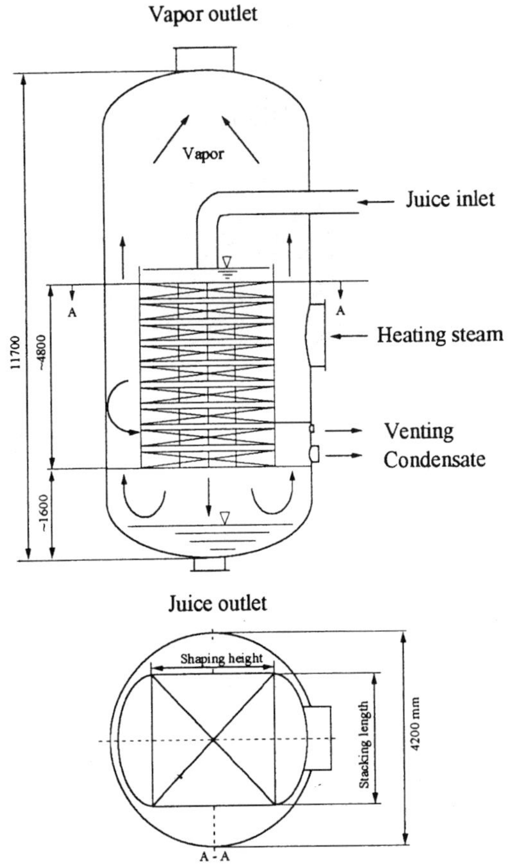

FIGURE 12.10 Details of a Robert-type evaporator converted to a plate-type falling film evaporator (66,000 ft^2) by Balcke–Dürr at the Suiker Unie beet sugar factory in Groningen, The Netherlands.

lead to minimal color formation and/or sucrose inversion. The plate-type falling film evaporator does not have a liquid level as does a rising film version or a Robert evaporator, hence there is no hydrostatic boiling-point increase. This means smaller temperature differences between steam and vapor. The seal-less plate modules with an overall height of 0.32 m are properly arranged based on steam and vapor pressure losses, the juice flow number, and the vessel dimensions. Vapor–liquid separation may also be accomplished external to the evaporator. Robert evaporators may have their calandria removed and replaced with a falling-film plate pack. Only minor nozzle modifications have to be made and the heating surface will be increased. Juice flow numbers are only 0.6 to 1.0 L/cm · h or 4.8 to 8.1 gal/ft-h.

It is critical that proper wetting of the heating surface occur at all times. Control systems allowing for a constant inlet flow work best. The use of antiscalants and optimum juice purification ensure uninterrupted operation of the entire evaporation station during the normal beet sugar campaign of up to 100 days. CIP methods with hot solutions containing

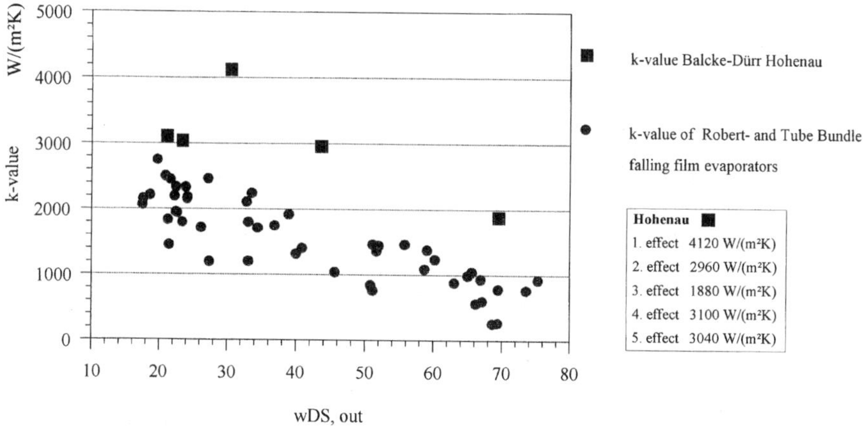

FIGURE 12.11 Initial overall heat transfer coefficients (k) of plate-type falling film evaporator at Hohenau, Germany, compared to typical values for Robert and tube bundle falling film evaporators. (From Ref. 10.)

caustic, soda ash, a wetting agent, and a dispersant may be used. This may be followed by a warm acid wash and a water or neutralization step if the unit is not immediately started up again. Figure 12.12 shows a simplified Piping & Instrument Diagram (P & ID) of a pilot plate evaporator at the Technical University of Berlin. These units are not yet in use in the cane sugar industry. Some work with South African clarified juice showed promising results. The Audubon Sugar Institute is in the process of installing a pilot unit as well.

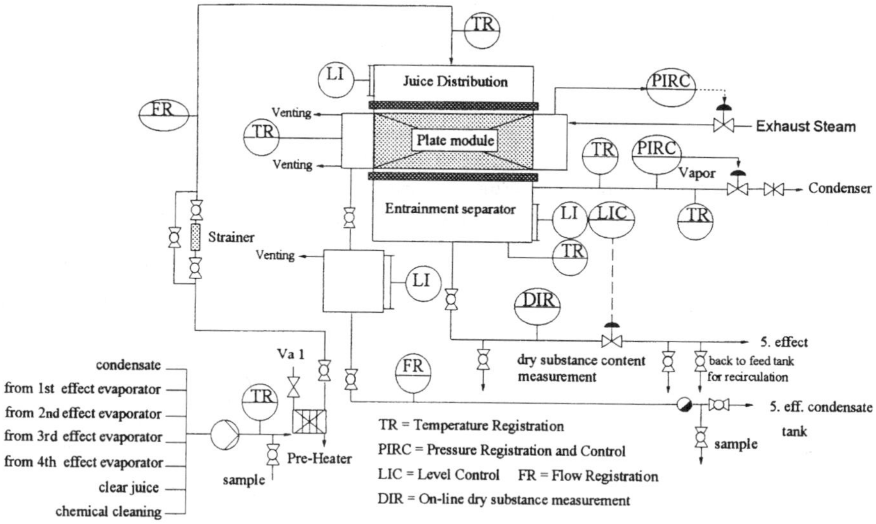

FIGURE 12.12 P & ID of Balcke–Dürr's plate-type pilot plant at the Technical University of Berlin. (From Ref. 10.)

Evaporator design is to enable production of the necessary amount of satisfactory product, without sucrose losses and/or color formation, at the lowest total cost. This requires economic balance calculations, and several factors have to be considered, such as:

- Initial steam pressure versus cost or availability
- Final vacuum versus water temperature, water cost, heat transfer performance, and product quality.
- Number of effects versus steam, water, and pump power cost
- Distribution of heating surface between effects versus evaporator cost
- Type of evaporator or combination of types versus cost
- Materials of construction versus product quality, tube or plate life, evaporator life, and cost
- Corrosion, erosion, and power consumption versus tube velocity
- Scaling and fouling and ease of cleaning
- Entrainment separation
- Bleeding of vapor for preheating services

As a general rule, the optimum number of effects increases with an increase in steam cost and/or plant size. However, it is not required to use the same heating surface in each effect nor that all effects in an evaporator have to be of the same type. Bonilla [12] has developed a simplified method for distributing the heating surface in a multiple-effect evaporator to achieve minimum cost. If the cost of the evaporator per square foot of heating surface is constant throughout, minimum cost and area will be achieved if the ratio of area to temperature difference $A/\Delta T$ is the same for all effects. If the cost per square foot z varies, as when different tube materials or evaporator types are used, $zA/\Delta T$ should be the same for all effects.

The vapor from the last effect of large evaporators is typically removed in a countercurrent barometric condenser, in which the vapor is condensed by rising against a deluge of cooling water. The condenser is set high enough (ca. 35 ft) so that water can discharge by gravity from the vacuum in the condenser. Although massive, these condensers are inexpensive and economical on water consumption. They generally maintain a vacuum corresponding to a saturated vapor temperature within 5°F of the water temperature leaving the condenser. The ratio of water consumption to vapor condensed is approximately lb water/lb vapor = $[H_v - (T_2 - 32)]/(T_2 - T_1)$, where H_v is the vapor enthalpy in Btu/lb and $T_{1,2}$ is the water temperature entering and leaving the condenser in °F. Jet condensers, also of the direct contact type, and surface condensers, where mixing of condensate with condenser cooling water is not desired, may also be used.

In evaporator work, the overall heat transfer coefficient is generally used and $Q = UA\ \Delta T$, where ΔT is the temperature difference between steam temperature and liquid temperature leaving the vapor head for natural circulation effects and log mean ΔT for forced circulation effects. The basic design equation for evaporators can be summarized as

$$(\text{lb vapor})(H_v) - (\text{lb feed})(H_f) + (\text{lb liquor})(H) = Q = (\text{lb steam})(H_s - H_c)$$

For preliminary designs, one can approximate T_m = approximately constant = $T_s - T$, where T is the boiling point of the liquor at the pressure in the vapor space. This is approximate since there typically is a BPR because T will vary along the length of the tubes as the concentration of the solute changes, and some subcooling of the condensed steam may occur. The cooling-water temperature in the condenser essentially fixes the vacuum of the vapor space in the last effect. In evaporator design one is to make an energy and mass balance around each effect of the evaporator. A well-insulated system will show radiation losses on the order of 2 to 3%.

The effective temperature difference in a plate-type falling film evaporator is

$$\Delta T_{eff} = \Delta T - T_{be} - T_{stat} - T_{hydr}$$

where ΔT_{eff} is the effective temperature difference, ΔT is temperature difference between saturated steam and vapor, ΔT_{be} is boiling-point elevation at the syrup outlet, T_{stat} is hydrostatic boiling-point rise, and T_{hydr} is hydrodynamic boiling-point rise.

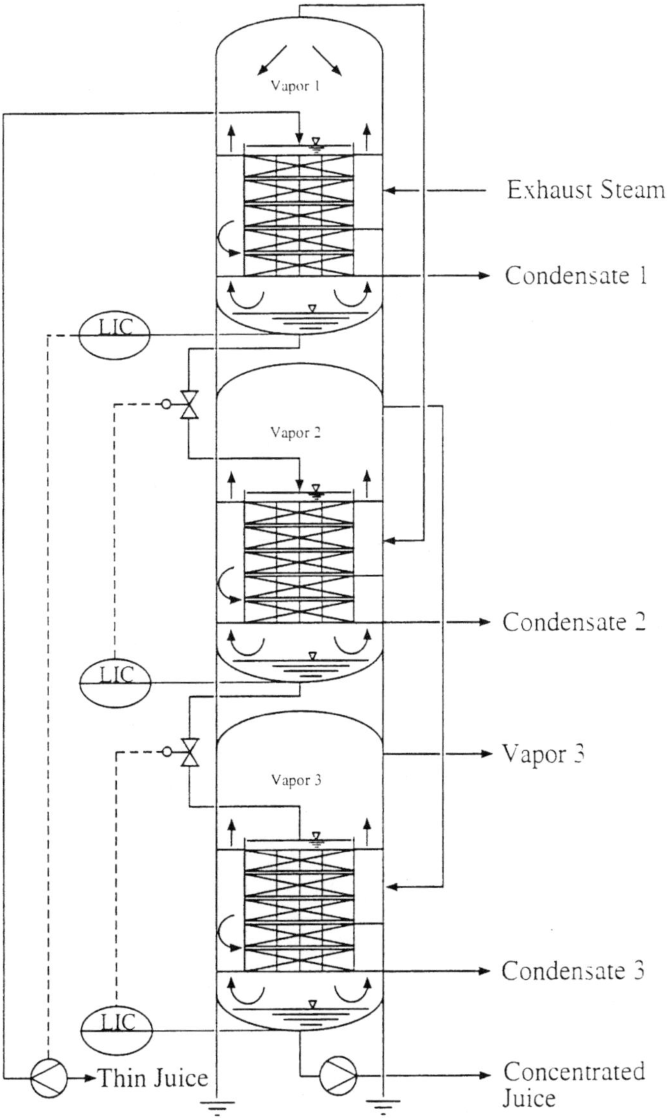

FIGURE 12.13 Vertical evaporator tower. (From Ref. 11.)

In the beet sugar industry, the customary Robert or standard evaporators have or are being replaced stagewise with conventional tube bundle falling film and now plate-type falling film evaporators. Several factories already have installed complete falling film evaporator stations combined with plate-type falling film units. This type of evaporator should work well in refineries. Morgenroth et al. [11] have designed a quintuple-effect evaporator system for a 12,000-tonne cane/day cane sugar factory. Steam consumption is only 20.7% cane. The plate-type evaporator station is linked to continuous crystallization pans and a vertical crystallization tower. Morgenroth et al. [11] propose a vertical triple-effect plate-type falling film evaporator with 5000-m^2 heating surfaces per effect, an overall height of 32 m (105 ft) and a diameter of 4 m (13.1 ft). Figure 12.13 shows a vertical evaporation tower. Expected advantages are:

- Reduced operational and investment costs.
- Very short residence time.
- Reduced pressure drop.
- Reduced power consumption, as there are no pumps between the effects and there is no juice recirculation. The specific juice flow number is low.
- Easy venting and entrainment control.
- Reduced energy consumption.

The Audubon Sugar Institute carried out tests at the CoraTexas Manufacturing Company in Louisiana during the 1998 and 1999 grinding seasons with an Alfa-Laval rising-film plate evaporator. The unit contained 256 ft^2 of heat exchange surface and was operated in parallel with the fourth effect of one of the factory's evaporator trains. A feed flash tank was installed and flow rates up to 24 gpm (on average 17 gpm) at an approximate Brix value of 32 were tested. Final syrup Brix was around 60. The residence time was around 1 min and the evaporation rate over a 2-week period decreased from 18 to 7 lb/(h)(ft^2) versus 6.5 to 4 lb/(h)(ft^2) for the factory's fourth effect. Typical feed conditions were : 32.0 Brix, color 12,000, and pH 6.46, and typical syrup conditions were: 58 to 62 Brix, color 11,400 to 12,200, and pH 6.48 (diluted to 32 Brix with DI water of pH 7.0). Cleaning was easy, color formation and sugar inversion minimal, energy requirements reduced, and the average evaporation load was approximately 2.25 times that of the Robert-type evaporator, while the unit was very easy to operate. Hence, both rising- and falling-film plate evaporators will make good substitutes for Robert evaporators.

REFERENCES

1. P. A. Schweitzer, Ed.-in-chief, *Handbook of Separation Techniques for Chemical Engineers,* McGraw-Hill, New York, 1979.
2. H. C. Prinsen-Geerligs, *Handboek ten dienst van de suikerriet-cultuur en de rietsuiker-fabricage op Java,* deel III, Vereenigde Proefstations voor suikerriet in West-en Oost Java, 1907.
3. H. H. Krützfeld, *Rum-Sonne der glücklichen Inseln,* H.H. Pott, Hamburg, Germany, 1969.
4. J. C. P. Chen and C. C. Chou, *Cane Sugar Handbook,* 12th ed, Wiley, New York, 1993.
5. L. Clarke and R. L. Davidson, *Manual for Process Engineering Calculations,* 2nd ed., McGraw-Hill, New York, 1962.
6. D. Q. Kern, *Process Heat Transfer,* McGraw-Hill, New York, 1950.
7. W. H. McAdams, *Heat Transmission,* McGraw-Hill, New York, 1942.
8. M. Jakob, *Tech. Bull. Armour Inst. Technol.,* 2(1), 1939.

9. J. H. Perry and C. H. Chilton, *Chemical Engineer's Handbook,* 5th ed., McGraw-Hill, New York, 1973.
10. B. Morgenroth, W. Jonker, and A. Lehnberger, *Zuckerindustrie,* 121(7), 1996.
11. B. Morgenroth, K. E. Austmeyer, and W. Mauch, *Zuckerindustrie,* 120, 1995.
12. J. Bonilla, *Trans. Am. Inst. Chem. Eng.,* 41:529, 1945.

CHAPTER 13

White Sugar Boiling and Crystallization*

INTRODUCTION

Both the theory and practices of sugar boiling/crystallization are thoroughly presented in the *Cane Sugar Handbook* [1]. This chapter is intended to provide and supplement additional and updated information on the subject.

13.1 WHITE SUGAR CRYSTALLIZATION OBJECTIVES

Some specific objectives are summarized as follows:

1. Produce a high-quality food product.
   a. Crystalline
   b. Purity: 100% sucrose
   c. Free of conglomerates <5%
   d. Low color: <35 I-420
   e. Low ash: <0.015%
   f. Specific desired crystal size distribution
2. Incorporate cost optimization.
   a. Low energy consumption
      (1) High feed syrup densities: 75.5 to 76 Brix
      (2) High operating vacuums: 27 to 26 in. Hg
          or
          Low operating temperature: 140 to 150°
      (3) No addition of water
      (4) No vacuum leaks in pans
      (5) Good steam balance with boiler house
   b. Low staffing requirements

*This chapter is updated by Chung Chi Chou on the lecture material presented by Peter Maria at the Cane Sugar Refiners' Institute, Nicholls State University, Thibodaux, Louisiana.

        (1) Well-trained operators
        (2) Pan automation
    c. Low maintenance costs
        (1) Good design features/materials of construction
        (2) Good preventive maintenance program
        (3) Use equipment to design parameters
        (4) Well-trained craftspeople
3. Establish a well-documented operating program.
    a. Reduction of color formation
        (1) Good circulation (pan mechanical agitators)
        (2) High operating vacuums: 26.0 to 27.0 in. Hg
            or
            Low operating temperature: 140 to 150°
    b. Control of feed materials
        (1) Good fundamental boiling scheme
        (2) Avoid mixing of syrups
    c. Good training program for operators
    d. Avoidance of conglomerates
    e. Enough tankage
4. Employ good instrumentation.
    a. Accurate, reliable, easy to interpret

The ultimate goals of sugar crystallization should be:

- Minimal chemical loss
- Minimal physical loss
- Minimal color formation
- Minimal conglomeration
- Maximum uniform grain size
- Maximum throughput
- Maximum yield
- Minimal energy consumption
- Minimal staffing
- Maximum repeatability

13.2 PAN INSTRUMENTATION

Figure 13.1 is a typical pan with condenser and locations of each inlet and outlet. Figure 13.2 is another typical pan highlighting the calandria section. Figure 13.3 shows pan instrumentation. Figure 13.4 gives a typical layout for a pan control system.

13.3 PAN OPERATING STEPS

The following 17 steps outline the operating steps of boiling a pan with a complete cycle. To design a pan control system and instrumentation, each step needs to be carefully studied and detailed.

1. Heat the pan.
    a. Exhaust steam enters pan.

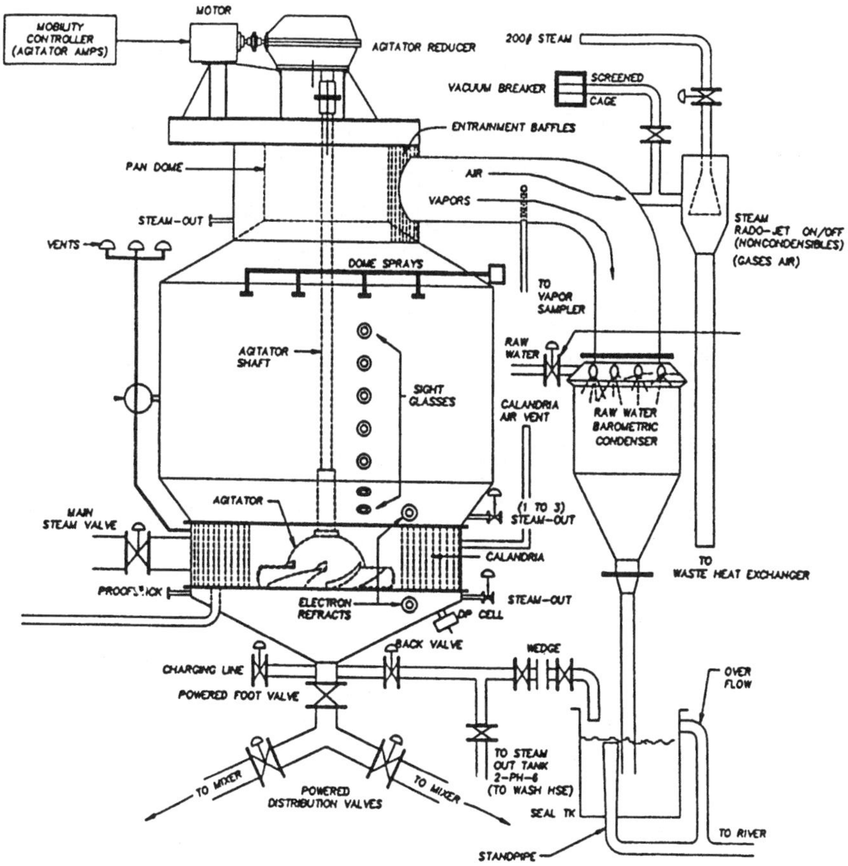

FIGURE 13.1 Typical pan with condenser.

b. Temperature raises to 212°F.
c. Time required is 15 to 20 min.
d. Initial raising of vacuum occurs.

2. Raise the Vacuum.
a. Condensibles (water vapors) are removed.
(1) Barometric condenser
b. Noncondensibles (air) are removed.
(1) Vacuum pump
(2) Venturi steam jet

3. Establish the initial vacuum set point.
a. Control vacuum by modulating quantity of water through barometric condenser.
b. Set point is 3.6 in. Hg absolute pressure/26.5 in. Hg vacuum.

4. Charge the pan.
a. Liquor (syrup) enters pan.
(1) Calandria pans 6 in. above tube sheet
(2) Normally 30 to 40% of pan's volume

5. Start the agitator.

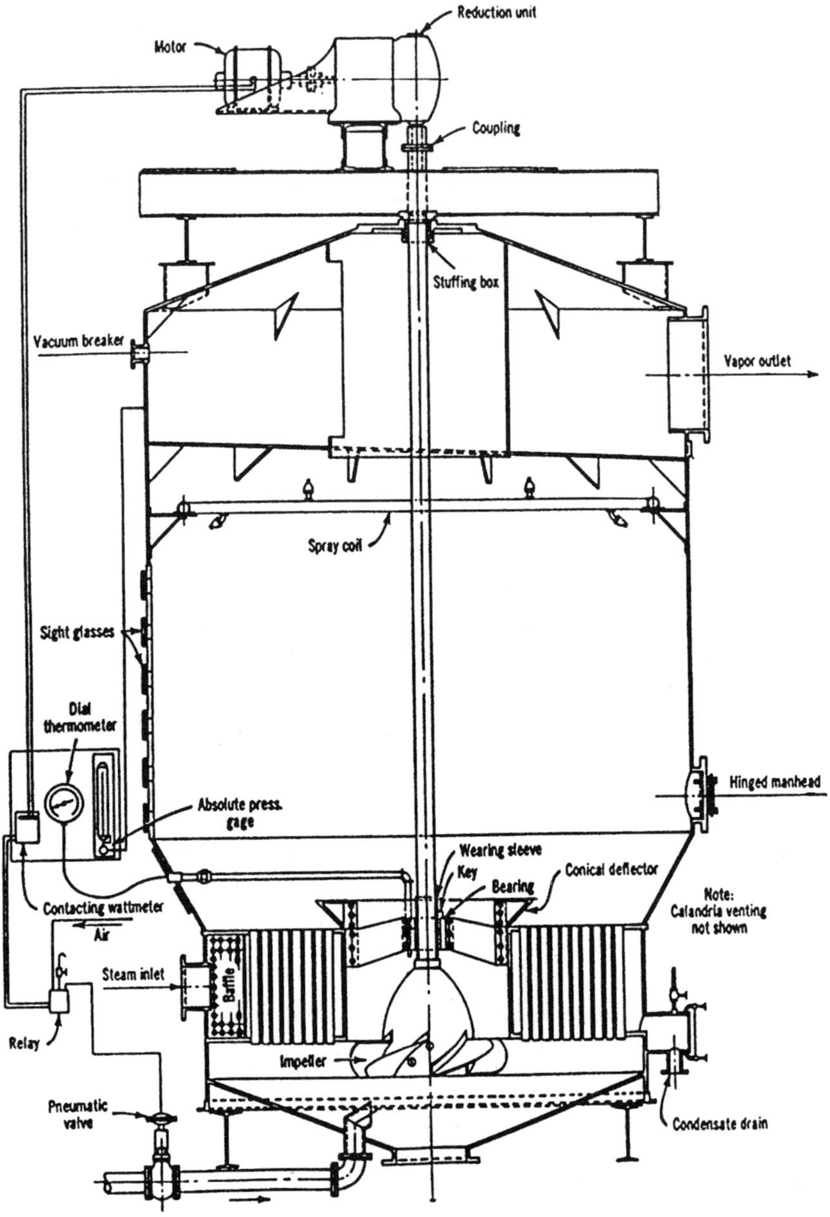

FIGURE 13.2 Typical pan.

a. Agitator starts off level contents of pan.
b. Liquor circulates through calandria tubes.

6. Turn the steam on.
   a. Main steam value opens, introducing steam into the calandria (8 to 10 psig).
7. Boil down to the concentration desired.

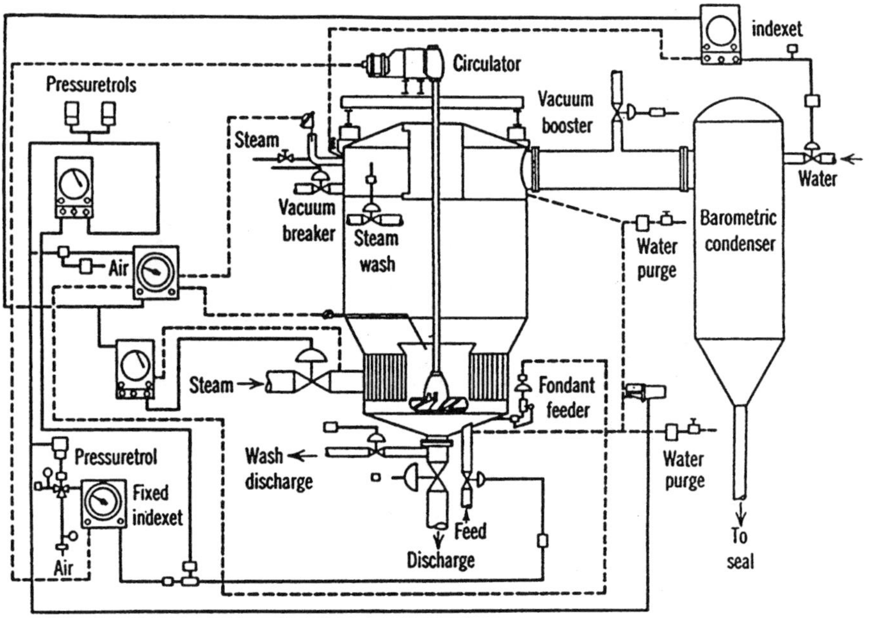

FIGURE 13.3 Pan instrumentation.

   a. Density of pan increases through evaporation.
   b. Liquor changes from a unsaturated/saturated solution to a supersaturated solution.
8. Seed the pan.
   a. At a specific solution density (78.5 Brix) and temperature (145°F) seed is introduced into the pan.
   b. One pound of fondant sugar per 2000 ft^3 of massecuite.
   c. Average seed size is 10 μm.
9. Apply minimum steam control.
   a. Calandria steam pressure is reduced from 8 to 10 psig to 3 to 5 psig.
10. Initiate the absolute press program.
    a. Absolute press/vacuum has programmed changes implemented.
11. Introduce minimum feed control.
    a. Specific programmed density of pan contents is controlled by the introduction of liquor/syrup to maintain 1.15 to 1.25 supersaturation.
12. Introduce mobility control.
    a. Off agitator motor load liquor/syrup is added to maintain specific massecuite consistency.
13. Tighten the pan.
    a. Tightening is done step by step over time, increasing the density of the massecuite.
14. Tighten the pan to drop.
    a. Increasing the density of contents optimizes centrifugal's operation.
    b. Stop adding liquor/syrup.
15. Break the vacuum.
    a. Turn the steam off.
    b. Turn the condenser water off.

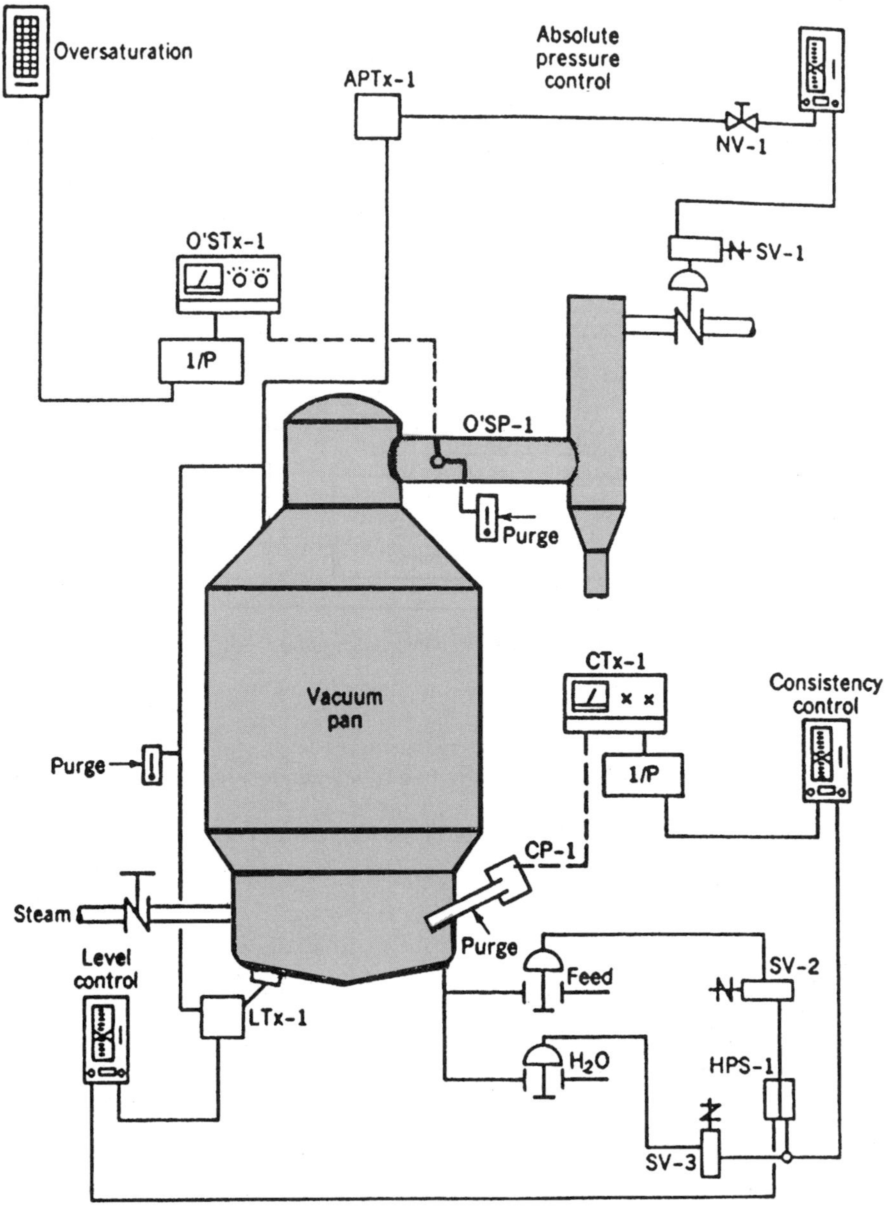

FIGURE 13.4 Typical layout of pan control system.

  c. Turn the noncondensible system off.
  d. Open the vacuum breaker.

16. Drop the pan.
  a. Empty contents of pan to mixer for centrifuging.

17. Steam out the pan.

a. Introduce exhaust steam to clean massecuite off the sides of the pan.

13.4 MATERIAL BALANCE

For a progress engineer to optimize the sugar boiling operation, the first step is to obtain material balance, energy balance, and to complete the study, a crystal population balance. Figure 13.5 shows the relationship between the weight of seed to that of the product. It can be seen that the crystal weight increases 60,000-fold. Figure 13.6 indicates the distribution of sucrose, as a percentage, in various strikes. It should be noted that the percentage of sugar in the fourth strike is only 6% of total sugar. The color and purity of each strike are also shown. Figure 13.7 is a typical material balance for the first-strike white sugar boiling. The crystal yield is shown to be 60%, which should be the target of any sugar refining operation.

13.5 PROCESS AUTOMATION OPTIMIZATION

The operating variables of sugar boiling include the following:

- Vacuum (temperature/viscosity)
- Degree of supersaturation
- Steam pressure

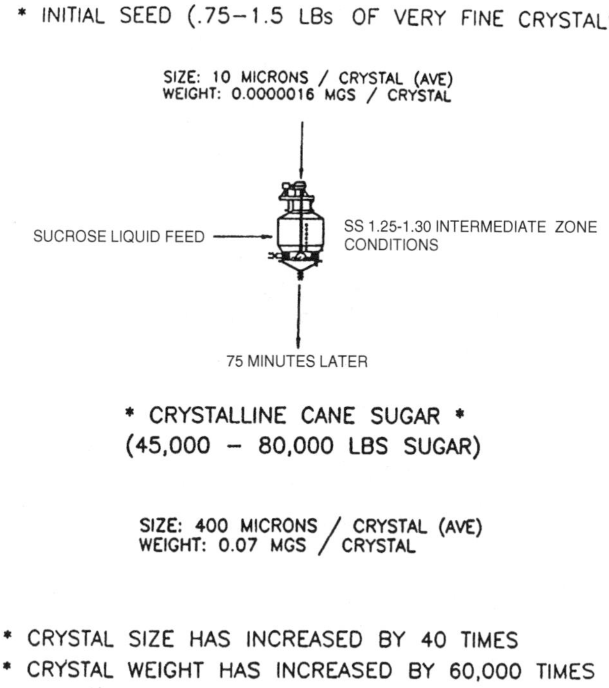

FIGURE 13.5 Seed and product balance.

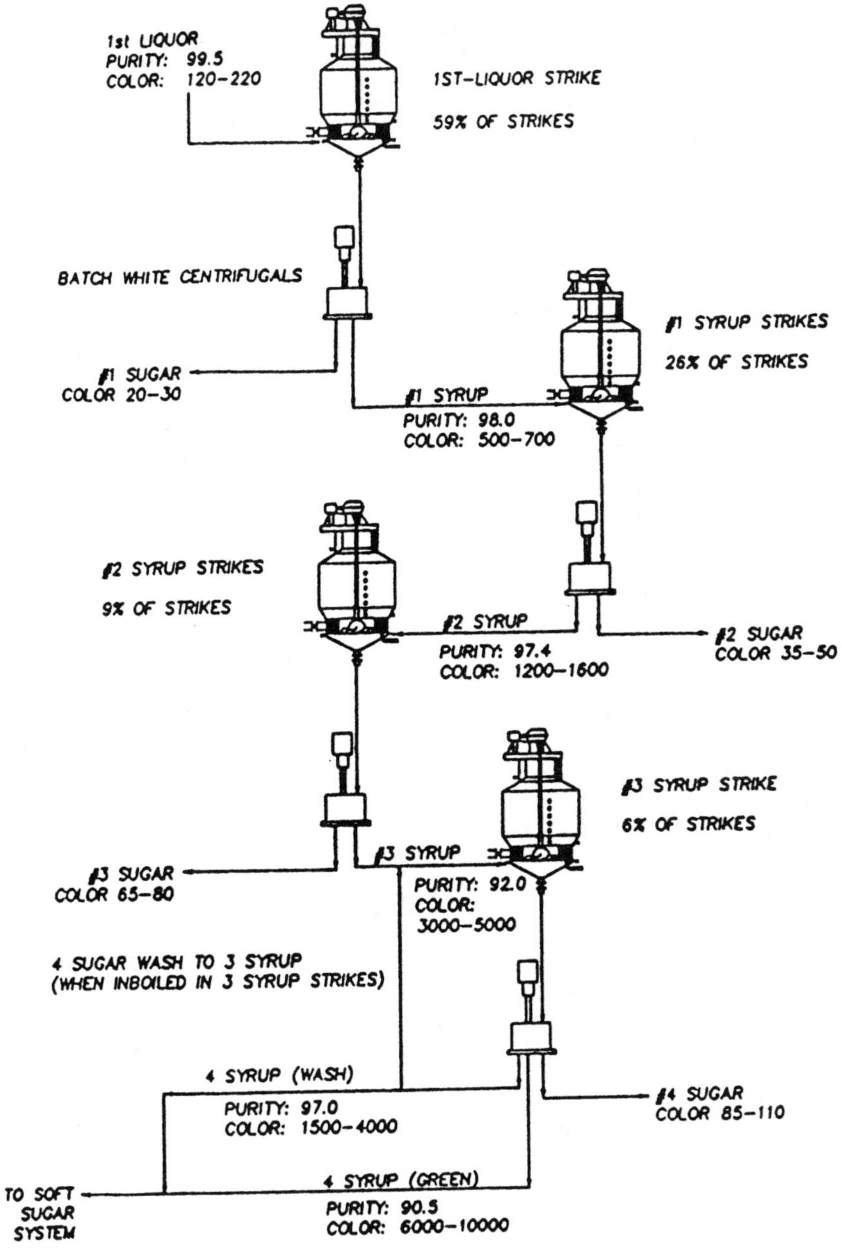

FIGURE 13.6 Distribution of sugar in various strikes.

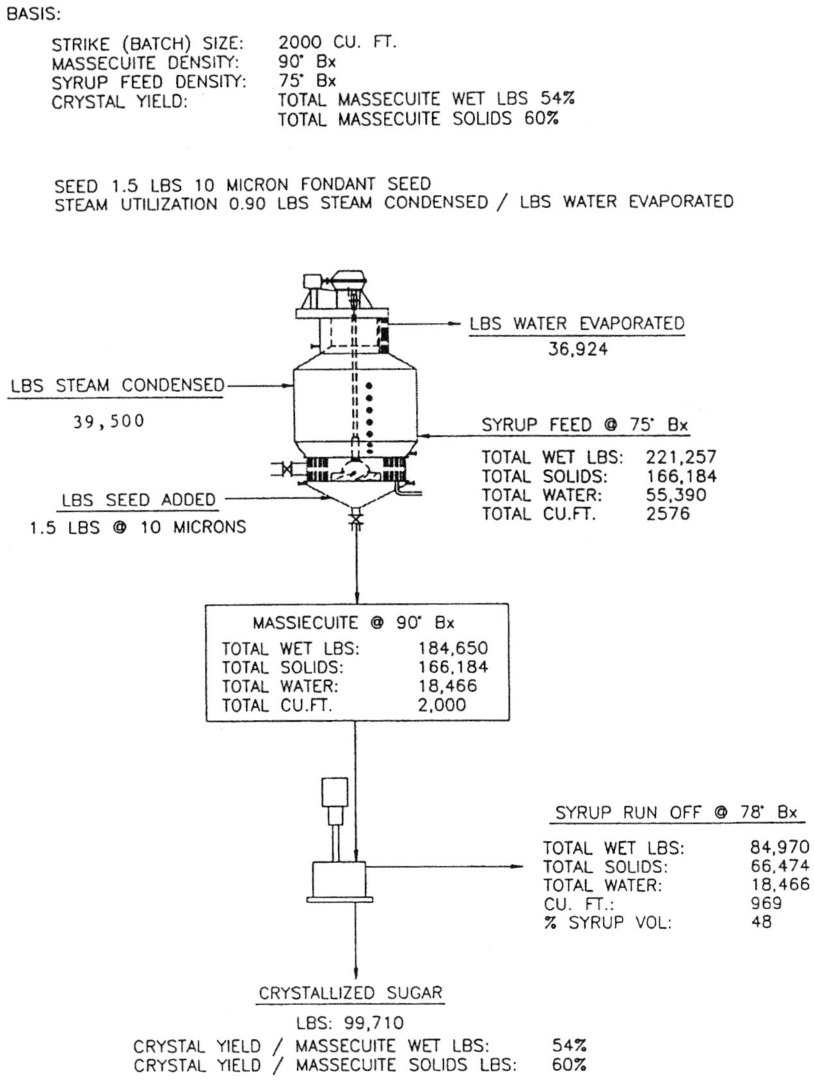

FIGURE 13.7 Typical material balance of white sugar boiling.

- Final massecuite level and Brix
- Seeding point and method
- Boiling cycle program
- Feed liquor Brix
- Feed liquor quality
- Heat transfer coefficient
- Pan maintenance
- Pan scheduling

The effect of each of the parameters needs to be fully understood. The *Cane Sugar Handbook* [1] has an excellent discussion on the subject.

The objectives of optimization are:

1. To produce sugar crystal with maximum uniformity
   a. Increase centrifugal capacity
   b. Reduce wash water (energy saving)
   c. Improve sugar quality (more premium and less penalty)
2. To accelerate sugar crystal growth
   a. Increase pan capacity
   b. Reduce sucrose loss/increase yield
   c. Reduce pan condenser water use
3. To optimize boiling scheme
   a. Increase capacity
   b. Improve yield/reduce molasses purity

and the benefits of automation are:

1. To reduce staffing (labor cost)
2. To achieve the highest possible efficiency by eliminating the human factor
3. To ensure consistency in sugar quality
4. To increase production capacity

Figure 13.8 shows a white sugar pan cycle for refinery C. The profile of absolute pressure as a function of pan cycle in minutes indicates that the boiling scheme is not optimized to any significant extent. The pan boiled about 1400 ft^3 in 78 min. In contrast,

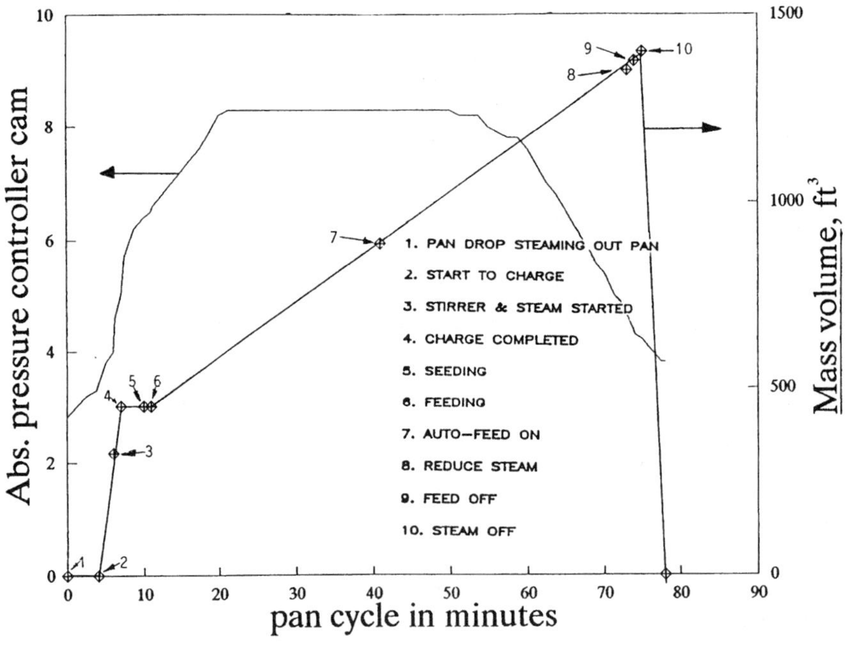

FIGURE 13.8 White sugar pan cycle: refinery C.

a pan in refinery A boiled about 3000 ft^3 in 90 min as shown in Figure 13.9. The absolute pressure profile in Figure 13.10 shows many distinct steps throughout the cycle. Each step has a specific purpose as to its absolute pressure and duration of time for each step. The productivity of refinery A is almost twice that of refinery C. With respect to pan capacity simply through process optimization via research and development, the optimization process obviously includes the design of the pan i.e., mass volume to heating surface ratio.

13.6 VACUUM PAN DESIGN CRITERIA

Pan design criteria as practice at Amstar Sugar Corporation (predecessor of Domino Sugar) was discussed by Tipper [2]. Generally, the following parameters needed to be considered for a modern vacuum pan:

1. *Design strike elevation* (above the top tube sheet). Too high an elevation will increase the boiling-point elevation at the heating surface, therefore reducing the heat transfer. Too low a strike elevation will reduce the pan house capacity: 7 to 8 ft should be appropriate.
2. *Vapor disengagement elevation* (between the design strike level and the center of the vapor outlet pipe). To minimize sucrose loss from carryover, the elevation should range from 8 to 11 ft, depending on the diameter of the pan.
3. *Calandria tube length.* The practice ranges from $2\frac{1}{2}$ to 6 ft. The long tube increases friction and reduces circulation and heat transfer coefficient. This writer prefers a short tube of about 3 ft.
4. *Calandria tube diameter and number of tube per square foot* (*tube density*). Tube diameter and tube density should be such that the cross-sectional area of all tube openings (up-

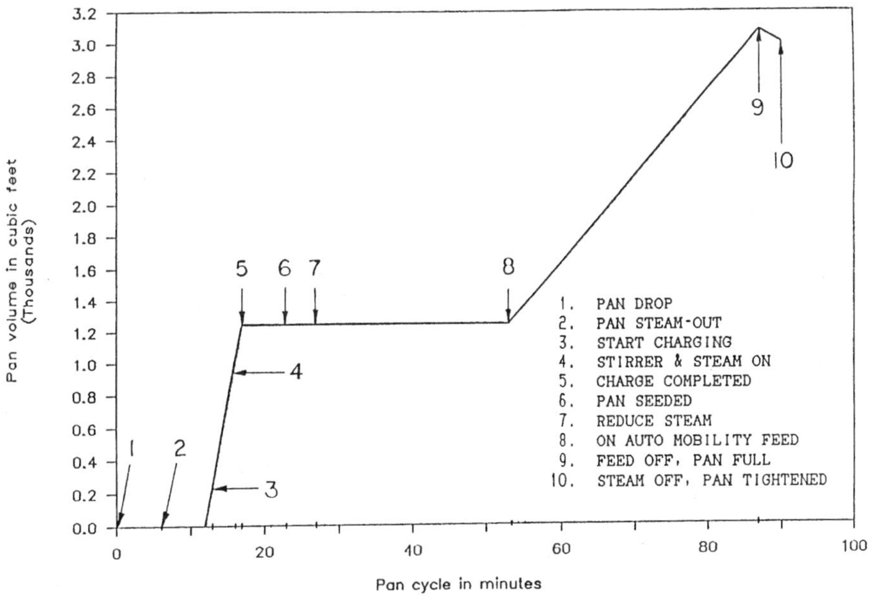

FIGURE 13.9 White sugar pan cycle: refinery A.

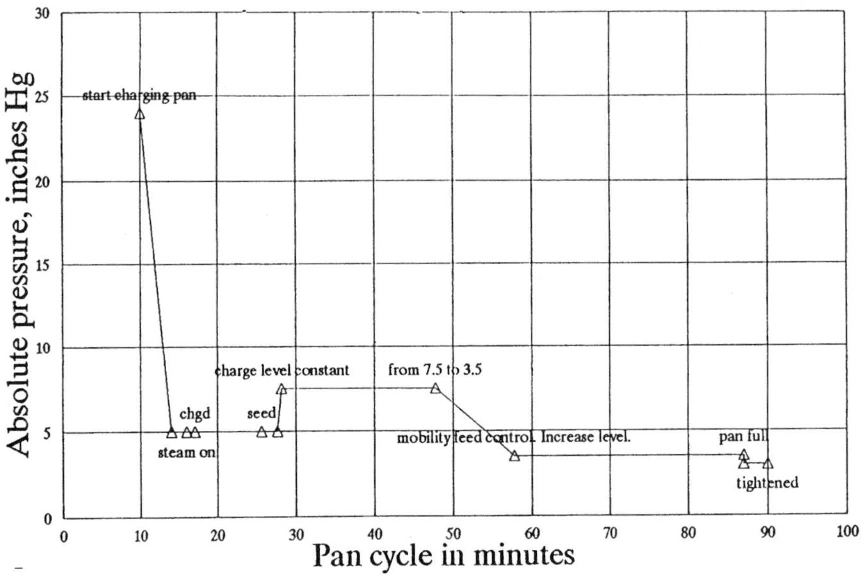

FIGURE 13.10 White sugar pan cycle: refinery A, 80-min boil time.

ward cross-sectional square footage) is at least three times the cross-sectional area of the center (center well downward cross-sectional square footage). This is to compensate for the friction (resistence) through the tube. Too small or too long a tube would decrease the circulation and therefore the heat transfer coefficient. Even with three times more opening, a mechanical circulator is still a necessity.

5. *Charge volume as a percentage of the design strike volume.* The charge volume is measured 6 in. about the top tube sheet and needed to cover the calandria. The percentage should not be more than 35. The ratio depends on the grade of sugar boiled, the footing (seed) volume requirement, and the final crystal size desired. This design ratio is particularly important when a pan is to be used for various strikes with different purities targets in the recovery (remelt) operation. A figure of 30 to 35% is generally adequate for white sugar boiling with a product crystal size of about 420 μm.
6. *Ratio of the tube heating surface area to design strike volume.* A ratio of 1.3 is adequate for white sugar boiling if the boiling cycle is to be within $1\frac{1}{2}$ h. For low-grade sugar boiling, a higher ratio would be needed because of the lower overall heat transfer coefficient.
7. *Flared body pan versus straight body pan.* A school of thought claimed that a "reduced" boiling head (or strike elevation) can be accomplished easily by incorporating a flared body design. However, critics insisted that a flared body will inhibit good circulation. A literature survey on discussions of "pro" and "con" proved inconclusive. Both types of pan are widely used.

13.7 ONE-STRIKE WHITE SUGAR BOILING

Some sugar refineries use a one-strike boiling scheme, particularly in Portugal, as compared to the conventional four-strike boiling schemes for white sugar. To achieve a one-strike system, runoff syrup is recycled and used for boiling the first strike until the combined purity falls below a certain level. Only then, some runoff syrup is "kicked out" of the system for remelt boiling. In some cases, the color of the first-strike sugar is used as criterion to "kick out" runoff syrup. In an in-boiling scheme the runoff syrup is combined with white fresh liquor before being fed to the pan, as shown in Figure 13.11. In a back-boiling scheme, runoff syrup is kept in a separate tank and used toward the end of a pan cycle, as shown in Figure 13.12. The back-boiling scheme is preferred, due to its flexibility in controlling the boiling process.

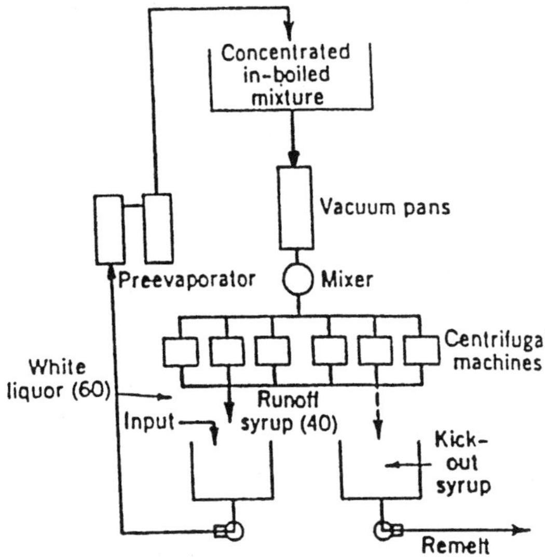

FIGURE 13.11 In-boiling scheme.

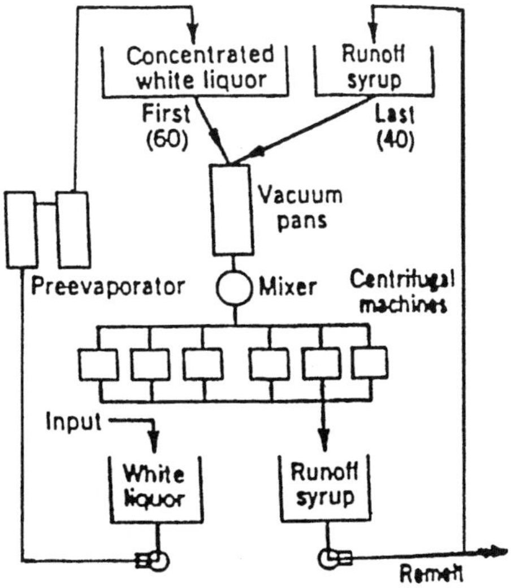

FIGURE 13.12 Back-boiling scheme.

TABLE 13.1 Sugar Production Data

	Mixture (%)	Color (ICU)	Rejection (%)
Fine liquor	55–60	300–400	—
Refined syrup	40–45	2000–6000	—
Sugar	—	45–60	—
Syrup to recovery	—	>6000	7–10

Luis Bento of the RAR refinery in Portugal discusses the in-boiling scheme as follows: "Commercial white sugars are crystallized from evaporated fine liquors and refined syrups using different boiling schemes. In the RAR in-boiling scheme, each strike is a mixture of high-grade liquor and runoff syrups. Usually, evaporated fine liquor has a Brix between 70 and 72 and purity higher than 99%. Color and other impurities in the runoff syrups of each strike increase and the cycle ends when the sugar color attains the quality limit."

13.7.1 White Sugar

White sugar production is done in cycles of 14 strikes. Table 13.1 summarizes the basic aspects. After the last strike, the syrup is boiled in another strike before being removed to the recovery station. Sugar color is higher than 60 ICU and must be blended. The big disadvantage of this scheme is the high purity content of the syrup transferred to recovery.

13.7.2 White Sugar (UE No. 2)

The number of strikes for each cycle is reduced to 10. On average, the percentage of fine liquor is near 65% and the syrup is sent to recovery when color reaches values greater than 5000 ICU. Sugar color is placed in the range 35 to 45 ICU.

13.7.3 Extra White Sugar

Usually, four strikes of 100% of fine liquor are done, giving a sugar color in the range 10 to 20. The resulting syrups are incorporated in the cycle of white sugar.

REFERENCES

1. J. C. P. Chen and C. C. Chou, *Cane Sugar Handbook,* 12th ed., Wiley, New York, 1993.
2. D. Tipper, Paper 341, *Proc. Sugar Ind. Technol. Conf.,* 1972, p. 81.

CHAPTER 14

Centrifugation*

INTRODUCTION

The need to separate solids from a liquid is widespread throughout the process industries of the world, and an enormous range of separators have been developed to satisfy this need. In the sugar industry the separation of the crystals from massecuite is a necessary step in the affination, refined product, and recovery stages in the production of refined sugar. All modern refineries producing crystalline sugar use centrifugal filtration to achieve these separations. Prior to the early nineteenth century the sugar crystals were separated from the syrup in molds under normal gravity. Batch centrifuges were first used in the mid-1850s and many developments have taken place to improve and extend their performance. A number of these improvements found application in other process areas during the industrial revolution in the late nineteenth century, such as in the chemical industry. It is fair to say that the batch centrifugal was developed initially for the textile and sugar industries.

Honig [1] describes some of the different types of centrifugals that have been used over the last century to separate sugar from massecuite. The sugar industry is now dominated by two types of filtering centrifugals: the batch centrifugal and the continuous conical centrifugal. Early centrifugals in the sugar industry were all batch machines. Continuous machines were introduced toward the end of the nineteenth century but became commonplace for low-grade massecuites only over the last few decades. In this chapter we consider only these two centrifugal types, their recent derivatives, and their application to the separation of sugar from massecuite. In the sections that follow we consider the theoretical basis of centrifugal filtration, the process performance, the design of centrifugals, and the basic safety and maintenance procedures.

14.1 FACTORS AFFECTING CENTRIFUGAL FILTRATION

Filtration is the separation of a liquid from a solid by means of a filter screen that allows the liquid to pass through but retains the solids. Centrifugal filtration enhances the process

* By G. Clive Grimwood.

by increasing the forces on the solids and liquids by spinning them at high rotational speed in a perforated cylinder or basket that supports the filter screen. At the start of the filtration the sugar crystals are immersed in the mother liquor to form a suspension. As filtration proceeds the crystals are pressed together by the centrifugal forces. The mother liquor or syrup flows through the gaps between the crystals and then through the holes in the filter screen, thereby effecting the separation.

Filtration proceeds by several phases or mechanisms. The first phase occurs when all the crystals are fully immersed in the mother liquor and the liquor is draining through the gaps between sugar crystals. This continues until the liquor level drops to the level of the sugar crystal whereupon the second phases commences. As the second phase proceeds the mother liquor level recedes through the sugar cake and gaps between the crystals are no longer all filled with mother liquor. The exposed sugar crystals are covered in a thin layer of liquor which drains through the sugar cake. This drainage of the thin film is the third phase of filtration. Phase I proceeds rapidly and is replaced by phase II. Phase II and III start at virtually the same time; however phase II predominates until close to the end of filtration. Finally, when phase II is complete all that remains is phase III, plus non filtration effects such as air drying. Ignoring non filtration effects phase III fixes the final crystal dryness.

Clearly, filtration is a complex physical process with many variables. Section 14.1.1 attempts to give an overview of some of the theoretical considerations governing the filtration behaviour of phases I and III. This overview is necessarily superficial and is presented to provide an understanding of the main physical mechanisms involved.

14.1.1 Physical Basis of Filtration

A comprehensive treatment of centrifugal filtration can be found in Refs. 2 and 3. The approach adopted here is to gain a physical understanding of the processes and the main factors which govern centrifugal filtration in both batch and continuous centrifugals.

It is possible to idealize the flow of mother liquor through a sugar cake by considering the flow down a thin pipe. The flow rate Q down a capillary was derived by Poiseuille [4] in 1840 when investigating the flow of blood in the arteries of horses:

$$Q = \frac{\pi r^4 P}{8\mu l}$$

where Q is the flow rate, r the radius of the capillary, P the pressure difference across the capillary, μ the dynamic viscosity of the fluid, and l the length of the capillary. The flow rate Q can also be expressed as

$$Q = \frac{V}{t}$$

where V is the volume that flows in time t.

The pressure P at the outlet of a vertical capillary of length l filled with liquid of density ρ resulting from the acceleration due to gravity g is

$$P = \rho g l$$

The volume V flowing out of the capillary in time t is obtained by combining the three equations above to give

$$V = \frac{\pi r^4 \rho g t}{8\mu}$$

The volume of the capillary is $\pi r^2 l$ and the volume of liquor expressed as a proportion of the capillary volume that flows out in time t is

$$\frac{V}{\pi r^2 l} \propto \frac{r^4 \rho g t}{8\mu \pi r^2 l}$$

The liquor density can be taken as constant, giving a flow over time t proportional to

$$\frac{r^2 g t}{\mu l} \tag{14.1}$$

This expression can now be used to compare the amount of liquor flowing from capillaries of different lengths and radii filled with liquor of different viscosities in different times, and so on. The use of this expression for centrifugals comes from associating the various parts of the formula to features of the sugar cake in a centrifugal. To make these associations, consider first the small gaps that exist between the individual crystals of sugar in the centrifugal basket. These gaps or interstices form a channel (or capillary) through which the mother liquor flows. It is reasonable to assume that the radius of the channel through the sugar is roughly proportional to the mean aperture of the crystals forming the channel. Second, the length of the capillary is proportional to the thickness of the layer of sugar cake in the centrifugal basket. Finally, the acceleration due to gravity, g, is replaced by the centripetal acceleration G generated in the centrifugal, t is the spin time, and μ is the viscosity of the mother liquor being purged from the sugar cake.

Using these associations it is now possible to write equation (14.1) as

initial purging rate depends on

$$\frac{(\text{mean aperture of sugar})^2 \times (\text{centrifuge } G) \times (\text{spin time})}{(\text{mother liquor viscosity}) \times (\text{sugar cake thickness})} \tag{14.2}$$

This simple expression has two uses. First, it indicates those properties of the massecuite and the centrifugal that affect the initial purging rate of the molasses. Second, it allows a simple comparison to be made between one centrifugal application and another. For example, one centrifugal may have a higher G value, a thicker sugar cake, and a shorter spin time than another. If the numerical value of equation (14.2) is calculated for both cases, a better initial purging rate would be expected from the centrifugal with the higher value. Alternatively, the effects of having a larger mean aperture (MA) for the sugar can be assessed in terms of a corresponding reduction in spin time and/or a decrease in centrifugal G.

Equation (14.2) is approximate and ignores factors such as the crystal size coefficient of variation (CV). It deals solely with the initial phase I rate of purging of mother liquor from the massecuite. Once purging nears completion, the gaps between the crystals are no longer full of mother liquor. All that remains is a small amount of mother liquor on the surface of the crystals and in the interstices between adjacent crystals. This liquor is held in place by surface tension forces. The amount of retained liquor defines the final dryness that can be achieved in a centrifugal. A full analysis of the purging process in phase III is extremely difficult; however, an analogy similar to that used to derive equation (14.1) can be adapted to give a physical understanding of the properties of the centrifugal and massecuite that govern the final dryness.

Consider an isolated sugar crystal as a simple sphere of radius r covered in a thin layer of syrup of thickness ∂r held in place against the centrifugal force by a surface tension force or the equivalent surface energy γ [4]. The limiting thickness of the liquid coating on the crystal (i.e., the thickness after a long spin time) can be estimated for a given set of centrifuge conditions by using the principle of virtual work. Under the influence of the centrifugal force, the liquid around the crystal elongates in the direction of the applied G force (see Fig. 14.1). If this distorted shape is assumed to be a spheroid (i.e., the shape of the three-dimensional ellipse), the work done by the centrifugal force can be calculated as the centrifugal force multiplied by the average distance moved by the liquid layer. This work done on the liquid must be counterbalanced by an increase in the surface energy of the liquid surface. If this is not the case, some liquid will escape and the liquid-layer thickness ∂r will be reduced. This simple energy balance approach allows the limiting liquor thickness ∂r to be estimated. Figure 14.1 shows this arrangement.

Assuming that the major and minor axes of the spheroid in Figure 14.1 are such that a is only slightly larger than r, the surface area of the lower half of the spheroid can be approximated as

$$\pi r^2 + \pi a r$$

and the increase in the surface area caused by displacing the liquid layer by a distance ∂a is

$$\pi r \, \partial a$$

The increase in energy is the product of the liquid surface tension γ and the increase in the surface area of the liquid:

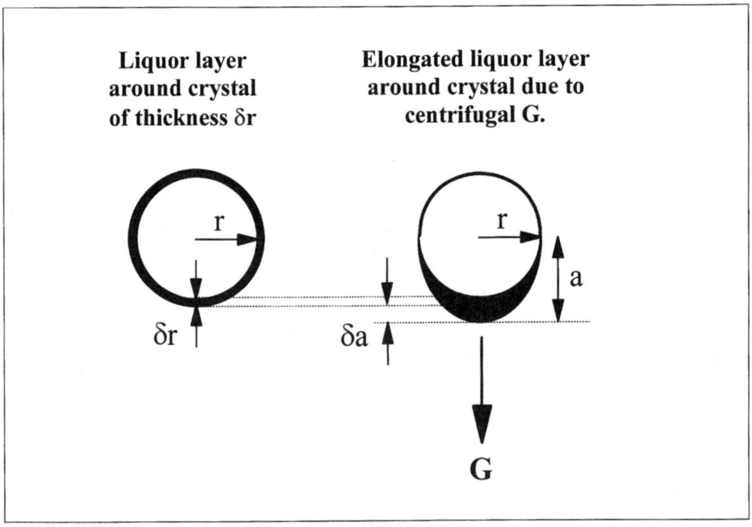

FIGURE 14.1 Liquor layer around a crystal.

$$\gamma \pi r \, \partial a$$

The work done by distorting the liquid layer from a sphere to a spheroid is given by the product of the mass of liquid and the average distance the liquid moves. For a liquid of density ρ and a centrifugal acceleration of G, the work done is

$$4k\pi r^2 \, \partial r \, \rho G \, \partial a$$

where k is a constant relating to the geometry. Equating the work done by displacement to the increase in surface energy gives

$$\partial r = \frac{\gamma}{4kr\rho G}$$

In this simple model the liquor layer of thickness ∂r is assumed to be present around each crystal. The total volume of liquid remaining in the sugar cake is a product of the number of crystals and the volume of liquid surrounding a single crystal. Expressing the liquor volume as a proportion of the total cake volume in the centrifugal with k' as the packing factor for crystals in the cake gives

$$\frac{3\gamma}{4\pi r^2 \rho G k k'} \tag{14.3}$$

Equation (14.3) incorporates the major physical factors affecting the final moisture. With typical values of sugar mean aperture $2r$ of 0.6 mm, liquor density ρ of 1350 kg/m^3, surface tension of saturated sugar solution γ 75 mN/m, a centrifugal G of 1000 times gravity, and taking the product kk' to be unity equation (14.3) gives a limiting liquid volume of about 1.5%. As this liquor is saturated syrup, the water content is around one-third of the total, giving a limiting moisture of approximately 0.5%. Although this is a plausible figure, the theory behind the derivation of equation (14.3) is approximate at best and equation (14.3) is presented to give the reader insight into the physical factors governing the operation of a centrifugal and not to derive accurate numeric answers. A more detailed analysis must consider the liquor trapped at the contact points between adjacent crystals (see Honig [1]).

Taking the liquor density as a constant and ignoring this and other constants, equation (14.3) can be rewritten as

final cake dryness depends on

$$\frac{\text{surface tension}}{(\text{mean aperture})^2 \times (\text{centrifuge } G)} \tag{14.4}$$

A larger crystal size, higher G, and reduced surface tension all improve the final dryness. A more complete and rigorous approach to the theoretical aspects of centrifugal filtration can be found in Refs. 2 and 3.

14.1.2 Practical Use of Theory

Equations (14.3) and (14.4) show the physical aspects of the centrifugal and massecuite, which affect the separation performance. The centrifugal force G and sugar MA appear in both equations (14.3) and (14.4). A larger G and larger MA mean a faster initial purging rate [equation (14.3)] and a lower final dryness [equation (14.4)]. Other aspects, such as cake thickness and spin time, appear only in equation (14.3) and consequently have less effect on overall separation performance. Although spin time and cake thickness undoubtedly affect separation performance, the stronger dependence on G, and particularly MA, is found in practice. Work published by Franzen [5] gives the expression

$$Z_E = \frac{Z^2 t_z}{h_c}$$

where Z_E is termed the spin intensity and is used to compare centrifugals. Z is the centrifugal force G, t_z the centrifuging time, and h_c the crystal layer thickness in the centrifuge basket. In this expression the spin intensity varies as the second power of the centrifugal G but only to the first power of spin time and cake thickness.

In earlier work Hugot [6] quotes the relationship

$$dn^2\theta = \text{constant}$$

as a measure to compare centrifugals where n is the basket rotational speed and θ is the drying time at spin speed and d is the basket diameter. Given that dn^2 is proportional to centrifugal G Hugot's expression reduces to the product of G and spin time, which is part of equation (14.2).

Finally, the importance of Equations (14.3 and 14.4) is their use in practical situations where the effects of changing the centrifugal settings or alterations to the feed massecuite are being considered. For example, if there is a need to increase the throughput and obtain maximum exhaustion from the crystallizers, changes might be considered to the process to produce a smaller MA crystal and reduce the mother liquor temperature to maximize exhaustion. These crystallizer changes would have a marked effect on centrifugal performance. The smaller MA and increased viscosity of the mother liquor resulting from the lower temperatures will affect the initial purging rates, and therefore feeding, of the centrifugal [Equation 14.2)]. Similarly the smaller MA will increase the amount of syrup remaining on the crystals and therefore impurity levels in the sugar [Equation (14.4)]. Overall, effectiveness of the centrifugal battery will decrease. Equations (14.2 and 14.4) can also be used to provide some guidance on the selection of different centrifugal types, such as the effects of large basket lip dimensions or low centrifugal G.

14.1.3 Centrifugal *G*

The acceleration generated by a centrifugal is expressed as a multiplication factor applied to the acceleration due to gravity. Gravitational acceleration is approximately 9.81 m/s^2, and therefore a centrifuge quoted as producing 1000G will produce an acceleration of 9810 m/s^2. G varies with the basket rotational speed, n (in rpm) and the basket diameter d according to the well-known expression

$$G = \begin{cases} \dfrac{dn^2}{1788} & \text{with } d \text{ in meters} \\[2ex] \dfrac{dn^2}{70{,}416} & \text{with } d \text{ in inches} \end{cases} \tag{14.5}$$

In practice, the sugar in a centrifugal experiences a range in G values even when operating at constant speed. For batch centrifugals it is the thickness of the sugar layer that leads to a range of G's. For continuous machines, where the sugar layer is normally very thin compared to the basket diameter, it is the conical basket shape that leads to a range of G's. For a batch basket with d_1 as the maximum diameter and d_2 as the minimum diameter of the sugar cake, the expressions above can be used if d is replaced by d_1, $(d_1 + d_2)/2$ and d_2 for maximum, mean, and minimum G, respectively. Table 14.1 shows the various G factors for typical centrifugal baskets. The difference between the maximum G at the basket wall and the mean G is quite marked, particularly for thick sugar cakes.

14.2 BATCH CENTRIFUGALS

In most refined sugar production a three-boiling process is adopted and this needs three stages of separation. The majority of affination magma and virtually all refined product sugar is processed with batch centrifugals at the final stage of separation. This section considers the mechanical, process, and electrical aspects of batch centrifugals. Recent development work on the continuous centrifugal is now allowing it to be applied to some affination and refined massecuites. Section 14.3 reviews these developments.

14.2.1 Mechanical Design

Figure 14.2 shows the sectional view of a batch centrifugal. Other designs of batch centrifugal are the same in principle but different in detail (see Section 14.2.3). The major component of the centrifugal is the basket. It is a drum typically 1.2 to 1.8 m (48 to 70 in.) in diameter, with the cylindrical portion being perforated.

TABLE 14.1 Maximum / Mean / Minimum Centrifugal Basket *G* Factors

Max Diameter (mm)[a] / Lip (mm)[b]	1200 / 178	1200 / 220	1400 / 220	1400 / 254	1600 / 220	1600 / 254	1800 / 254	1800 / 265
Spin 900 rpm								
Minimum *G*					525	495	585	575
Mean *G*	—		—		625	610	700	695
Maximum G					724	724	815	815
Spin 1000 rpm								
Minimum *G*	472	425	537	499	649	610	722	710
Mean *G*	571	548	660	641	772	752	864	858
Maximum *G*	671	671	783	783	894	894	1006	1006
Spin 1100 rpm								
Minimum *G*	571	514	649	603	785	739		
Mean *G*	691	663	798	775	934	911	—	
Maximum *G*	812	812	947	947	1082	1082		
Spin 1200 rpm								
Minimum *G*	679	612	773	718	934	879		
Mean *G*	882	789	950	922	1111	1084	—	
Maximum *G*	966	966	1127	1127	1288	1288		
Spin 1300 rpm								
Minimum *G*	797	718	907	843				
Mean *G*	966	926	1114	1083	—		—	
Maximum *G*	1134	1134	1323	1323				

[a]English equivalents: 1200 mm, 48 in.; 1400 mm, 55 in.; 1600 mm, 63 in.; 1800 mm, 72 in.
[b]English equivalents: 178 mm, 7 in.; 220 mm, 8.7 in.; 254 mm, 10 in.; 265 mm, 10.5.

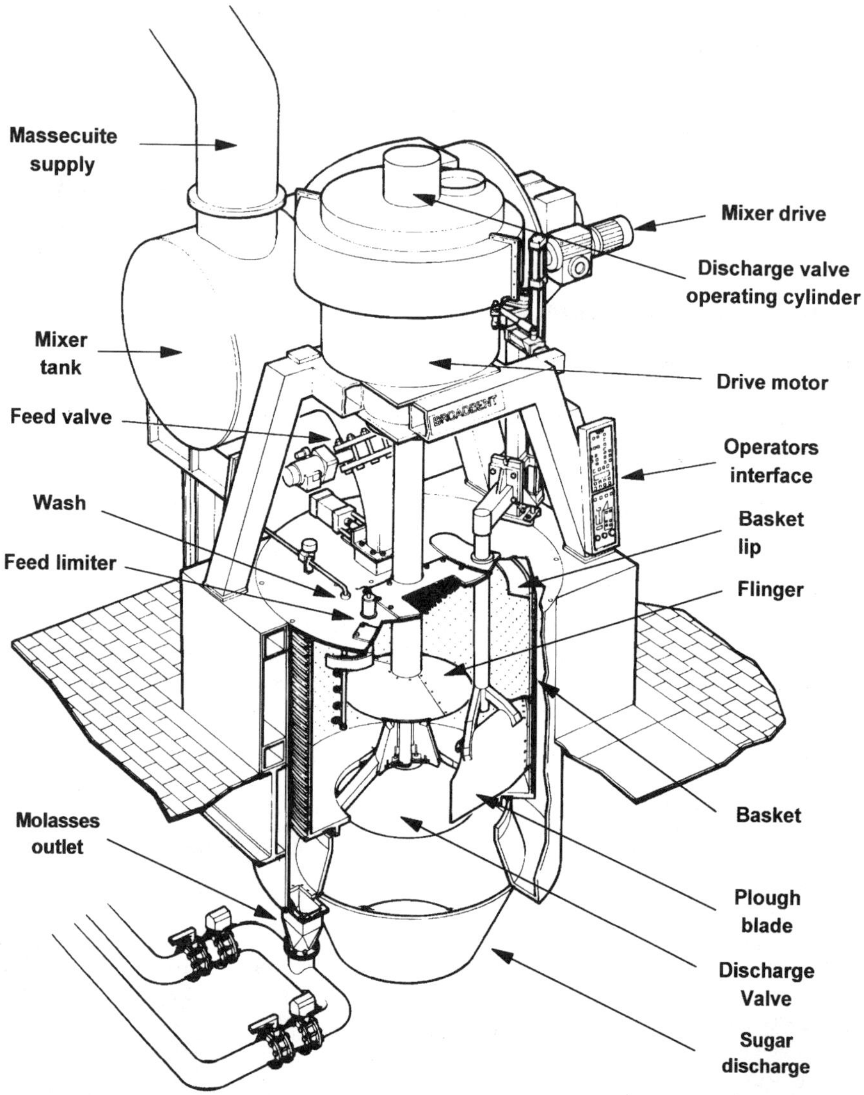

FIGURE 14.2 Batch centrifugal sectional view. (Courtesy of Broadbent Centrifuges Ltd. © 1997.)

Baskets are manufactured to widely used national and international standards (e.g., BS767, EN12547) from high-tensile steels, and designs are available using either a thick perforated shell or a thinner shell reinforced with high-tensile steel hoops. Baskets are subject to fatigue loading of typically millions of stress cycles during an average basket life, and the quality of manufacture and nondestructive testing (NDT) is of paramount importance. Repair or modification should only be undertaken by those aware of the loading regime and life requirements (see Section 14.4).

The cylindrical shell of the basket can contain several hundred drilled perforations 3 to 6 mm ($\frac{1}{8}$ to $\frac{1}{4}$ in.) in diameter. The top of the basket has a solid lip that fixes the maximum thickness of massecuite, typical lip dimensions are 170 to 250 mm (7 to 10 in.). The basket bottom contains large openings through which the washed and dried sugar can be discharged. The openings in the basket bottom incorporates a valve mechanism that is closed for feeding and spinning, and opened just before the sugar is discharged from the basket. A discharge mechanism (normally called the *scraper* or *plough*) is pneumatically operated and comprises a shaped blade that is used to plough or scrape the dried sugar off the basket wall and allow it to fall through the open valve in the bottom of the basket onto a conveying system below.

The basket is attached to a spindle that extends up from the basket bottom and connects to an electric drive motor mounted above on the top of the centrifuge support structure. Motor sizes are normally in the range 100 to 500 kW (135 to 650 hp). A mechanical brake is mounted below the motor, and in the design shown in Figure 14.2, the entire rotating assembly (comprising basket, spindle, brake, and motor) is supported on bearings just below the motor.

In an ideal world the feed to the centrifuge produces a perfectly balanced load in the basket. In a practical world a degree of nonuniformity in the basket load must be tolerated and the design of any batch centrifugal must accommodate the associated out-of-balance (OOB) forces and resulting vibration. This is accomplished by some form of flexible suspension, which supports the rotating assembly (comprising motor, spindle, and basket) in the centrifugal support frame. The effect of OOB is likely to be the only mechanical aspect of the batch centrifugal to intrude on normal automatic operation. For this reason these effects are considered in more depth below.

The ability of the centrifugal to handle OOB loads is dependent on the mechanical design and the integrity of the structural support on which the centrifugal is mounted. The flexible suspension and the rotating assembly comprise a pendulum, with a natural frequency of swinging (or resonant frequency) normally in the range 90 to 120 oscillations per minute. It is important to feed the centrifugal at a speed in excess of this frequency (i.e., above 90 to 120 rpm) to avoid the potential of large OOB loads. This is a result of the normal behavior of any resonant system to undergo a phase change at the resonant frequency. Below the resonant frequency any OOB load will cause the basket to swing in a way that accentuates the OOB. Above the resonant frequency the opposite occurs and the centrifugal tends to become self-balancing and is inherently more stable. Stability is enhanced further by the use of additional frictional damping to restrain the rotating assembly from swinging, although this increases the transmission of OOB forces to the surrounding structure of the building. The most adverse condition in a batch centrifugal is the presence of excessive amounts of free liquor on the surface of the cake at high speeds. Under certain conditions the liquor surface can break up into waves, and a very large out of balance can occur rapidly, which in extreme cases may lead to damage of the centrifugal. Such events are very rare and occur only with very poor massecuites (i.e., very small MA, false grain or excess liquor) or with blocked screens. Every effort should be made by centrifugal operators to avoid such conditions (see Section 14.2.4).

14.2.2 Screens

The perforated basket shell supports the filter screens. Typically, the screen comprises two or more meshes, with the inner filter or working screen being made of stainless steel or brass and containing slots or holes about 0.5 mm (0.02 in.) wide with a total open area of about 20%. One or two coarse stainless steel wire meshes (e.g., 4 and/or 7 mesh) are fitted between the working screen and the basket shell. Their function is to act as a spacer to allow the liquor to flow behind the working screen to the nearest perforation before escaping to the outer casing and to provide mechanical support for the thin working screen.

Figure 14.3 shows a small section of screen 10 mm wide with holes 0.55 mm in diameter. It might be thought necessary to have hole sizes much smaller than the sugar MA to avoid large losses. In practice, only the crystals immediately adjacent to a screen hole have a chance of being lost. The remaining crystals tend to form an arch that spans the hole, avoiding further losses. It is not uncommon for the average crystal size to be similar or below the screen hole diameter. As the sugar crystals do not move relative to the screens, there is virtually no wear of the screen by the sugar and no damage to the crystals. Some wear or damage can result from operation of the discharge mechanism, and it is not uncommon for this to be the reason for changing the screens in batch centrifugals.

14.2.3 Alternative Designs of Batch Centrifugals

The mechanical design of centrifugals differs among manufacturers. In this section we highlight some of the significant mechanical aspects in other batch centrifugals, which differ from those described above.

Figure 14.2 shows a solid spindle connecting the basket and motor. This is not the most common design and requires the use of a purpose-built electric drive motor to suit the electrical and mechanical duty. The majority of batch centrifugals have some form of flexible coupling just above the suspension. The use of a coupling allows a standard electric drive motor to be used which requires greater headroom and restricts the available methods of operating the discharge valve. The absence of flexible coupling in the spindle allows the valve mechanism to be incorporated into the top of the drive motor, as shown in Figure 14.2. The alternative design with a flexible coupling in the shaft has the discharge valve opening mechanism located on the top of the basket outer casing.

FIGURE 14.3 Batch centrifugal screen: 0.55-mm holes, sugar MA 0.6 mm.

Certain designs, particularly the smaller ones, feed directly onto the basket bottom. This approach has the advantage of allowing easier access to the screens, as no flinger is required (see Fig. 14.2). However, it also has the disadvantage that the massecuite must flow further to distribute itself evenly up and down the basket.

Various designs of plough mechanism have been developed. Some require the basket to be constrained from swinging on its suspension so that the plough mechanism can approach close to the basket wall and remove the majority of the sugar. Other designs use techniques to remove virtually all the sugar without the need to constrain the basket. The latter method is to be preferred, particularly from the point of view of cleanliness and yield of sugar from the centrifugal.

14.2.4 Process

A typical process cycle, starting with an empty basket at feed speed, would comprise the following steps:

1. *Feeding.* The feed valve opens and massecuite flows into the basket. The level of massecuite fed into the basket is controlled by a feed limiter (typically, a mechanical switch or some form of proximity switch utilizing ultrasonic or other noncontacting techniques) to sense the presence of the massecuite surface. Typical basket speeds during feeding are 150 to 250 rpm, with typical feed times of 12 to 25 s.
2. *Acceleration.* The centrifugal drive system accelerates the basket up to the set spin speed. This speed can be fixed by the electrical supply frequency or may be adjustable, depending on the sophistication of the drive used. Typical maximum speeds are 1200 to 1500 rpm for a 1.2-m (48-in.) basket, giving a centrifugal force 980 to 1530 times that of gravity. Average acceleration times are in the region of 40 s, depending on the basket size and maximum speed.
3. *Wash.* During acceleration a water or syrup wash is normally applied to the sugar. Washes are applied for 10 to 20 s (depending on the sugar and wash jets, etc.) and the wash usually ends at or near the point at which the basket reaches full spin speed.
4. *Spin.* Full spin speed is maintained for the chosen spin time. This can vary quite widely depending on the grade of sugar being produced. Typical figures for refined white sugars are 15 to 30 s.
5. *Deceleration.* After the spin time is complete, the basket is decelerated to plough speed (50 to 60 rpm). The method of deceleration varies according to the drive system employed. Deceleration times are broadly the same as acceleration times (e.g., 40 s). Substantial recovery of the energy supplied to accelerate the centrifugal is essential if an economical power demand is to be achieved.
6. *Plough.* Once the basket has reached ploughing speed, the basket discharge valve is opened and the mechanical plough/scraper blade is moved slowly into the sugar at the top of the basket. Once the layer of sugar has been fully penetrated, the plough blade is moved downward to cut the remainder of the sugar out of the basket. The sugar falls out of the discharge valve in the basket bottom onto a conveyor system below (see Fig. 14.2). The time to complete the ploughing operation is typically 20 to 30 s.
7. *Accelerate/basket wash.* Once ploughing is complete, the discharge valve closes and the basket is accelerated to feed speed in 5 to 10 s. It is normal to wash the basket wall during this acceleration period to maintain a clean working screen for the next batch. For plough systems that remove all the sugar from the screen reliably, it is not necessary to wash the screen every cycle.
8. The cycle repeats.

A typical batch centrifugal cycle as laid out above requires approximately 150 s and repeats at the rate of 24 charges per hour. Centrifugals with smaller drives more commonly cycle at 18 to 20 charges per hour.

In the remainder of this section we consider the use of batch centrifugals on high-grade massecuites or affination magma. Figure 14.4 shows a typical cycle for an inverter driven (i.e., variable-speed) centrifugal operating on refined white sugar. Both the basket speed and the maximum G force in the basket against time are shown. The individual elements described above are marked on the graph. Each of the individual elements of the time cycle is considered in more depth below. Of the six elements only the washing and spinning (drying) are fully productive in terms of making good sugar. The acceleration period is partly productive, as some filtration occurs during acceleration and the wash period normally begins during acceleration. The feeding, deceleration, and ploughing elements are necessary but do not contribute *directly* to separating the crystals from the massecuite. All could theoretically be reduced without compromising the quality of the sugar. It is therefore possible to consider all but washing and spinning as nonproductive time in the centrifuge cycle. Ideally, this nonproductive time should be as short as possible (i.e., fast deceleration, fast feeding, and ploughing). For the example shown in Figure 14.4, the productive and nonproductive elements are of about equal length at 86 s. The ratio of productive to non-productive cycle time is a useful measure of batch centrifugal processing efficiency.

a. Feeding. To maximize the potential performance of a batch centrifugal the basket must be fed with a uniform layer of massecuite. Failure to provide a uniform layer results in:

- *Reduced capacity.* The basket volume is not fully utilized.
- *Poor washing.* Thinner areas are washed more than thick areas.
- *Nonuniform drying.* The thinner areas dry more than the thick areas.

Even if the full capacity of the basket is not being employed (e.g., 75% fill only) it remains important to get as uniform a layer as possible. One cause of nonuniform feeding is feeding with a basket speed that is either too fast or too slow. If the speed is too fast, the massecuite purges too rapidly on hitting the screen and is unable to flow to the remainder

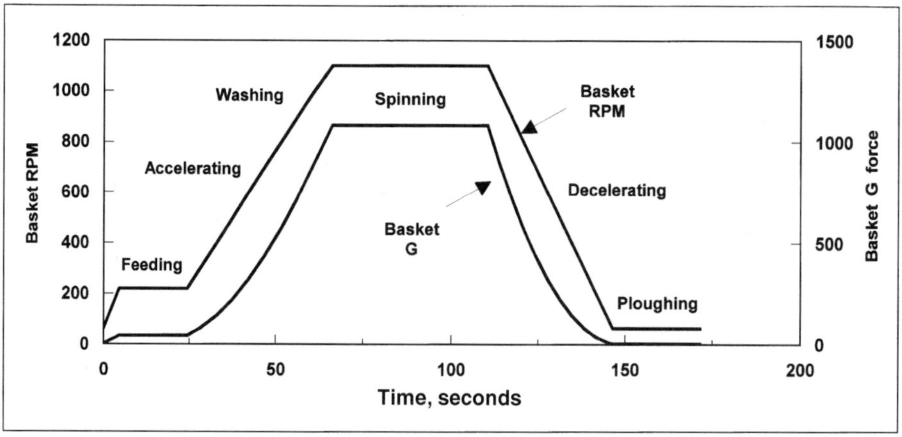

FIGURE 14.4 Typical batch centrifugal process cycle: basket rpm and g as a function of time, cycling rate 21 charges per hour.

of the basket. If the feed speed is too low, the massecuite tends to slump toward the bottom of the basket. Ideal feed speeds vary between massecuites and basket sizes. As a guide, good feeding usually results between 20 and 70*G*.

Feed rate also affects distribution of the massecuite in the basket. If the rate is too slow, the massecuite will not flow around the basket to give a uniform layer, causing reduced basket capacity and out-of-balance loads. If the feed rate is too high, the basket may, in extreme cases, become flooded and the massecuite may move in a wave and cause large out-of-balance loads to occur at feed speed. The ideal feed times are normally in the range 12 to 25 s, depending on the massecuite, basket size, and feed speed.

The volumetric capacity of a basket is

$$V = \pi h\ (ld - l^2) \tag{14.6}$$

where V is the basket volumetric capacity, h the basket internal height, l the depth of the basket lip, and d the basket diameter. In using this expression the basket internal diameter should take account of the thickness used by the backing and working screens. Typical screen thicknesses are 3 to 5 mm in total. During the 10 to 25 s when massecuite is being fed into the basket, a proportion of the mother liquor is spun off, allowing additional feed to be added to the basket. The magnitude of this effect depends on the massecuite and the feed time. It is not uncommon for these effects to allow 10% more feed to be added to the basket.

The uniformity of the feed massecuite is important. A local mixer tank containing a volume equivalent to two or three centrifugal charges for each centrifugal in the battery is common practice. The mixer acts to limit the rate of massecuite variation to a point where the controls fitted to most batch centrifugals can adjust the feed valve setting to maintain automatic cycling without operator intervention. If batch crystallizers are used, it is often necessary to oversee the centrifugal during the sudden change in viscosity that can occur when changing from one crystallizer to another. Advanced control schemes are now being developed to remove the requirement for operator intervention. Continuous crystallization tends to reduce these sudden changes in viscosity and makes the task of fully automatic operation simpler.

b. Accelerating and Washing. Washing is arguably the most important part of the centrifugal process cycle. It is the ability of the batch centrifugal to combine excellent washing with good drying and the absence of crystal damage that accounts for its widespread use in the sugar industry. Wash is normally applied partway through the acceleration after the majority of the mother liquor has been purged from the cake. In the past the general practice was to use two washes, the first during acceleration, with the second following after a delay of a few seconds. The single wash is now more common, and work reported by Stowe [7] indicates that there is no benefit from using two washes. The use of a single wash allows the time cycle to be reduced by a time equal to the delay between the two washes.

The application of wash water has two main beneficial effects, (1) to increase the purity of the sugar and (2) to reduce its final moisture. These effects are achieved by the wash water diluting the remaining mother liquor adhering to the sugar crystals. This dilution reduces the viscosity and allows the mother liquor to flow off the sugar crystals more easily [see equation (14.2)]. The result is an increase in sugar purity and a reduction in the sugar moisture. From this it will be clear that washing is an important part of centrifugal operation.

It is possible to separate the flow of initial mother liquor from the massecuite that is spun off during the early part of the acceleration from the high-purity liquor spun off during the washing of the sugar at the end of acceleration. This is achieved by using a classification valve on the liquor discharge line from the outer casing. This feature is sometimes used in the production of refined sugars where the higher purity wash syrup (and any screen wash liquor) is sent forward to the R1 pan and the lower purity mother liquor is returned to the R2 pan. Splitting the syrup in this way improves the energy efficiency and yield of the overall crystallisation process.

As the wash water flows through the cake it becomes progressively more saturated with sugar. After the first 50 to 75 mms (2 to 3 in.) the water is near saturation and this "clean" near-saturated solution displaces the massecuite mother liquor [8]. Washing therefore has the unwanted effect of dissolving sugar. The level of dissolution depends on the temperature and the quantity of wash water according to the expression (after Charles [7])

$$S = 64.397 + 0.07251t + 0.0020569t^2 - 9.035 \times 10^{-6}\, t^3 \tag{14.7}$$

where S is the % w/w sugar in a saturated solution and t is the temperature of the solution in °C. The weight of sugar that dissolves in a unit mass of water is then

$$M = \frac{S}{100 - S} \tag{14.8}$$

where M is the mass of sugar melting in unit mass of water and S is given by equation (14.7).

Normal wash quantities for affination are in the region of 1% w/w of the massecuite fed to the centrifugal and nearer 2% for refined product production. Taking as an example a 1300-kg load and 2.0% w/w wash, a total of 26 kg of water would be used. If the temperature is 50°C (122°F), equations (14.7) and (14.8) give the mass dissolved sugar as 67 kg. If the massecuite contained 55% sugar cryystals, this amounts to a loss of 9.4% of the sugar crystals in the charge. In practice, the amount of sugar dissolved is less than these figures suggest by perhaps 20 to 30%. This occurs for two reasons: because (1) the wash water does not become fully saturated as it passes through the cake, and (2) the wash mixes with the supersaturated syrup remaining on the crystal and no dissolution starts until the mixture falls below saturation. Nevertheless, the losses due to dissolving the sugar can be large. It is prudent to ensure that the centrifugal wash system provides a uniform wash, with the lowest wash quantities consistent with the required sugar quality and centrifugal cake thickness. Overwashing makes the centrifugal station easy to run but is wasteful in terms of evaporative energy consumption and recirculating sugar load in the refinery. Tests to confirm correct operation of the wash system are generally worth the effort. For a more detailed review of the performance and costs savings that can be achieved, see Ref. 9.

The uniformity of the wash application is most important. If one section of the sugar is washed less than the rest (e.g., badly adjusted/blocked wash jets or uneven distribution of massecuite in the basket), either a proportion of the discharged sugar will be badly washed and therefore substandard, or additional dissolution losses will occur in regions where the sugar layer is thin. The correct approach is to ensure even washing of a uniform cake with minimum wash water use to ensure the correct sugar quality.

It is informative to consider where the ash and color impurities are located. The impurities are found in two places: in the mother liquor adhering to the sugar crystals, and within the crystals themselves. The second type of impurity is termed the *occluded impurity* (or *occluded ash*) and cannot be removed by washing. Further information on color and its removal by washing is given in Ref. 10.

c. Spinning. Final drying of the sugar takes place during spinning. The approximate variation of moisture with spin time and crystal MA/CV for white sugar production is shown in Figure 14.5. As would be expected from the discussion in Section 14.1, the benefits of a longer spin diminish as the time increases. The final moisture is determined by crystal size (MA/CV), liquor surface tension, and basket *G* [see Equation 14.4)].

d. Decelerating. Deceleration is nonproductive time. No further drying takes place during deceleration (except in extreme cases of very low spin times). Deceleration times should be as short as possible within the limits set by the drive system components.

e. Ploughing. The plough mechanism should remove as much sugar from the basket as possible to maximize throughput and avoid the need for a large amount of screen washing prior to recharging the centrifugal. Some plough designs allow virtually all the sugar in the basket to be discharged. Others leave a thin residual bed of sugar of 1 to 2.5 mm (0.04 to 0.1 in.) on the screen after ploughing. If the screen is fully cleaned by the plough mechanism, the screen may require washing only periodically (e.g., every three to five batches) to maintain a clean filter screen. If a residual bed is left, it must be washed off the screen prior to beginning the next cycle. If it is assumed that a residual bed of 1.5 mm (0.06 in.) remains on the screen, then for a basket with a 180-mm (7-in.) lip, this corresponds to a recycle load on the crystallizer of 1% of the centrifugal station output. A more detailed assessment of the benefits of full removal of sugar during ploughing is given in Ref. 11.

f. Mass Balance: Affination. A useful way of looking at the process performance of any centrifugal is through a mass balance over the centrifuge. What flows in must flow out, and the mass balance is instructive in seeing where it goes. An example is shown in Figure 14.6. The flows marked are those for a battery of batch centrifugals on affination duty. To optimize the centrifugal for a particular refinery, it is useful, although time consuming, to produce the mass balance.

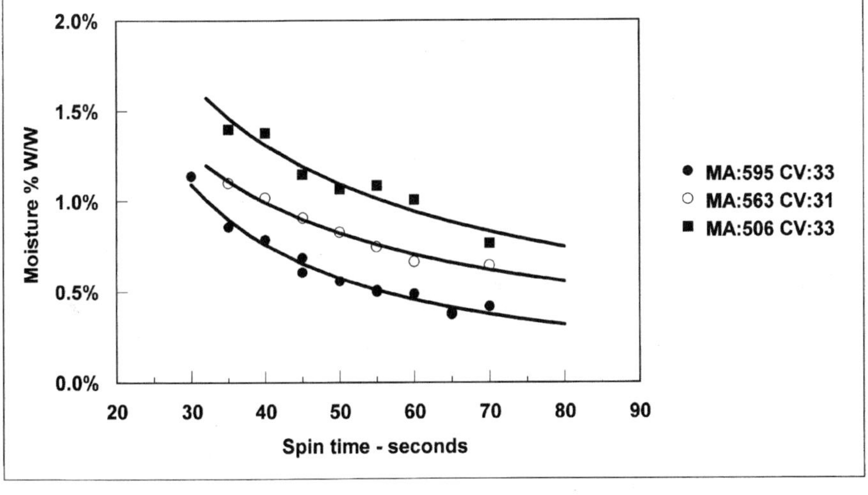

FIGURE 14.5 Variation of sugar moisture with spin time and crystal size. (From Ref. 9.)

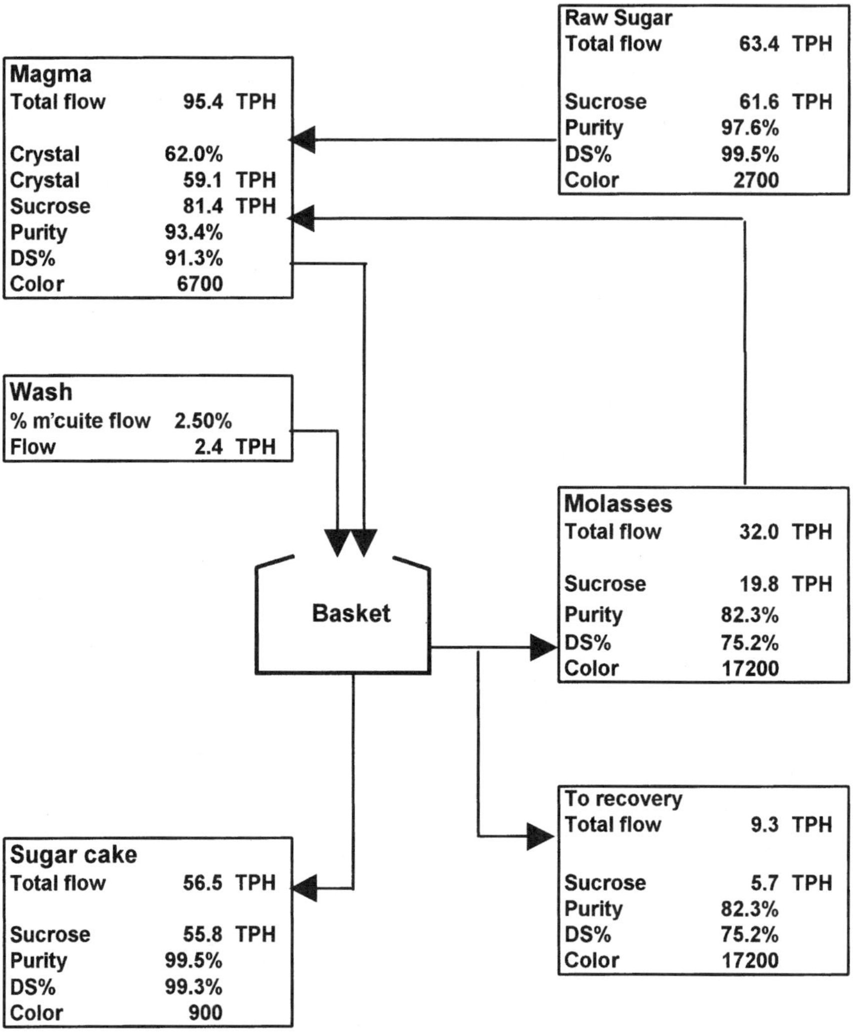

FIGURE 14.6 Typical mass balance: affination.

Prior to entering the centrifugal, the raw sugar is mingled with a proportion of the molasses to produce magma with a high crystal content. During the mingling process the sugar crystals come into contact with one another, and the high-color surface layers become softened prior to removal. In this example about 4% of the sugar is dissolved in the magmatizing process and 5.6% in the centrifugals. Overall 90.6% of the raw sugar crystals end up in the sugar cake and the color of the raw sugar is reduced from 2700 ICU to 900 ICU. The sugar discharged from the affination centrifugals is dissolved, and therefore a low moisture content is not required. Short spin times can be used (e.g., 10 s) and high cycling rates of up to 30 charges/h can be achieved. Continuous centrifugals are also used to process affination magma (see Section 14.3.5).

g. Mass Balance: Refined Product. Figure 14.7 shows a typical mass balance for the first refined product. In this example 8.4% of the sugar crystals in the massecuite (3.6 tonnes/hr) are lost into the molasses and the refined sugar is being discharged at 0.7% w/w moisture and at a color of 15 ICU.

14.2.5 Electrical Design

Batch centrifugals require complex controls to undertake the movements and actions required to process each batch. To minimize energy consumption, sophisticated electronic drive systems are now commonplace. In this section we provide an overview of these areas.

a. Controls. Modern batch centrifugal are controlled by programmable logic controllers (PLCs) which control the centrifugal in accordance with process parameters entered through an operator's interface. Modern interfaces usually provide additional information on throughput, centrifugal vibration levels, maintenance information, fault history, and so on,

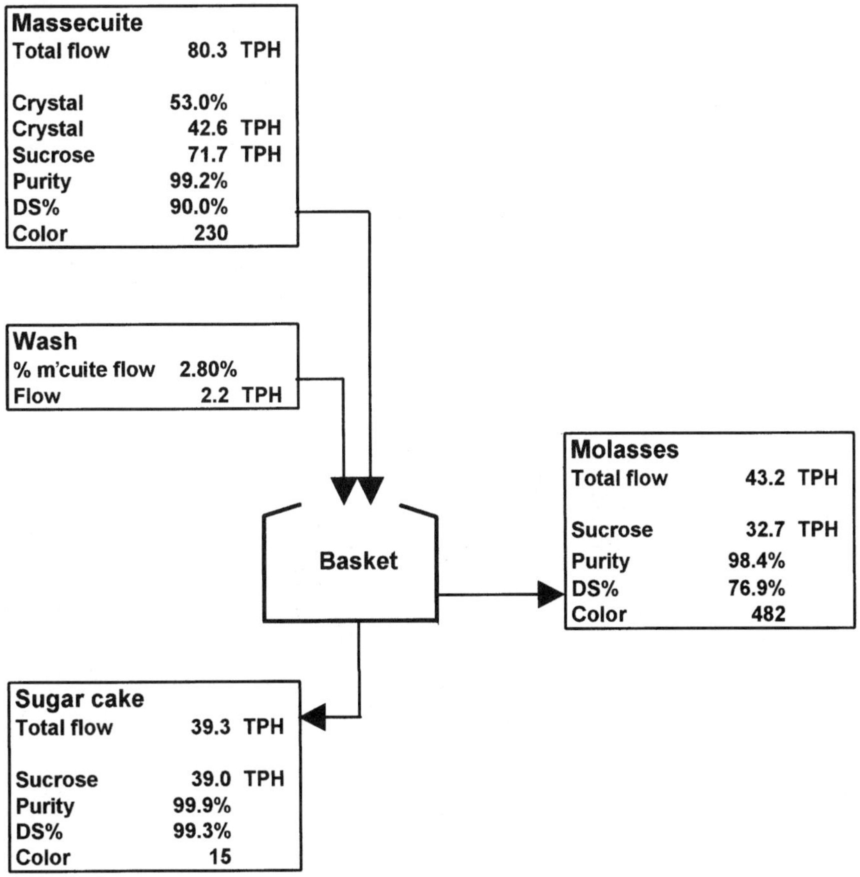

FIGURE 14.7 Typical mass balance: refined product.

and provide communication facilities to allow data exchange and a degree of remote operation from a plant control computer or distributed control system (DCS). As batch centrifugal developments continue, the possibility of true remote operation without operator interventions is now near at hand, with the necessary controls such as adaptive feed controllers and analog basket fill sensors already available. The control system plays a vital part in maintaining safe operation of the centrifugal, and any modifications to the control program therefore need to be undertaken with great care. Any urge to defeat or disable interlocks provided by the control system should be resisted.

b. Drives. Centrifugal drives come in a variety of guises. Pole changing and dc motors are still in common use, with electronic variable frequency supplies becoming more popular. Figure 14.8 shows the power requirement and basket speed as a function of time for a 1000-kg (2200-lb) centrifugal. The drive motor is a three-speed pole-changing induction motor, and Figure 14.8 shows the characteristic large power peaks that occur when switching from one winding to the next, the negative peaks during deceleration giving a degree of regeneration.

Figure 14.9 shows a cycle for a similar centrifugal fitted with an electronic variable-speed drive. In this case there are no sharp peaks, and the drive is considerably more energy-efficient, as it able to regenerate more of the energy back into the electrical supply during braking. The energy consumed by the three-speed design in Figure 14.8 is about 2.3 kWh/tonne of massecuite processed, and the equivalent number for the inverter-driven centrifugal in Figure 14.9 is 0.93 kWh/tonne.

Larger batch centrifugals are available only with electronic variable-speed drives. Small and intermediate-size centrifugals are available with either pole-changing or variable-speed drives. Table 14.2 summarises the current availability.

Three basic types of variable-speed drive are in use: pulse-width modulation inverters (PWMs), current source inverters (CSIs) and dc drives. All are complex and require careful

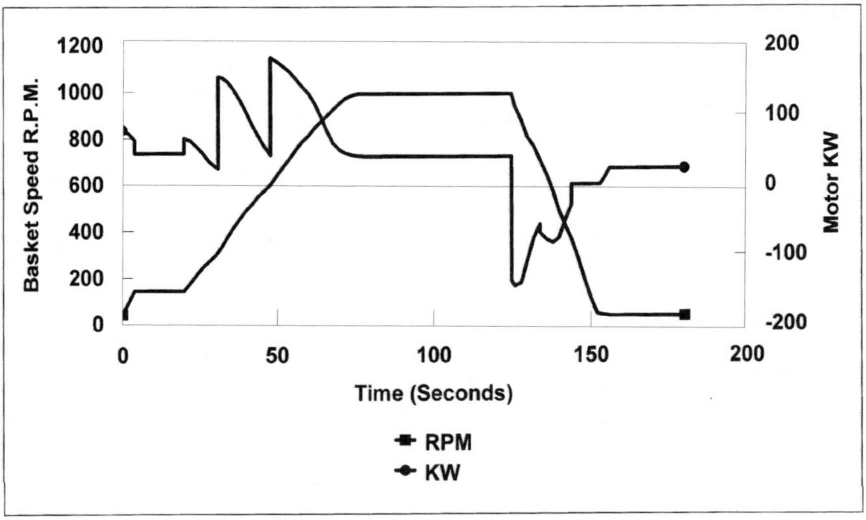

FIGURE 14.8 Batch centrifugal process cycle: 1000-kg pole-changing three-speed drive motor. Regenerative kilowatts shown as negative; cycling rate, 20 charges per hour.

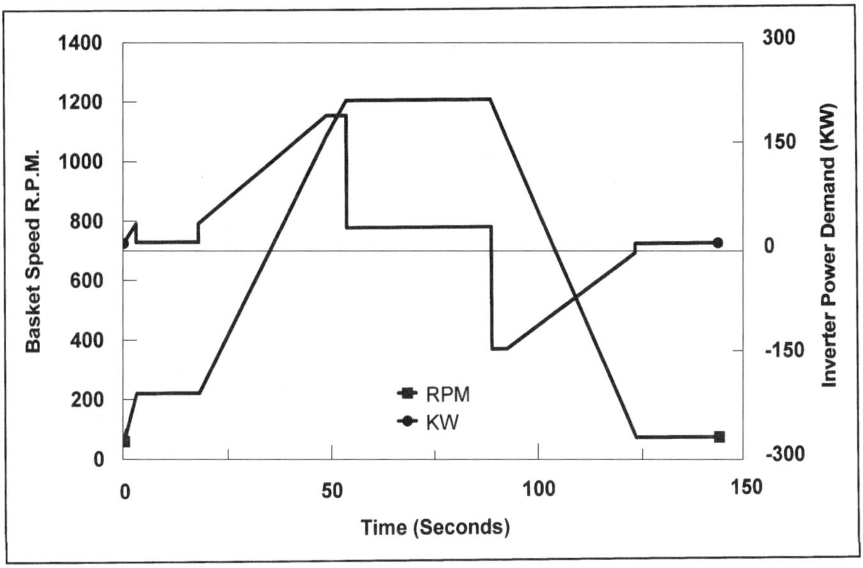

FIGURE 14.9 Batch centrifugal process cycle: variable-speed inverter-driven motor. Regenerative kVA shown as negative; cycling rate, 20 charges per hour.

installation and a reasonable factory power supply to operate reliably. Several factors that should be considered when selecting a drive include:

1. Impact of harmonics on the factory supply (see Section 14.2.5d)
2. Energy consumption (see Section 14.2.5c)
3. Tolerance to supply voltage and frequency variations
4. Installation requirements (e.g., air conditioning, filtered cooling air, space)
5. Reliability and service support
6. Capital cost
7. Process benefits of variable spin speed and so on.
8. Electric cable lengths, routing, and screening

Electronic variable-speed drives excel at factors 2 and 7, while pole-changing drives provide benefits in the remaining categories: 1, 3, 4, 5, 6, and 8. Often, the most important criterion is energy consumption (2), and as a result, electronic variable-speed drives, al-

TABLE 14.2 Availability of Drive Systems for Batch Centrifugals

Basket Capacity kg	Two-Speed Pole-Changing Drive	Three-Speed Pole-Changing Drive	Variable-Speed Electronic Drive
Up to 1000	Yes	Yes	Yes
1000–1300	—	Yes	Yes
1300–2500	—	—	Yes

though of higher capital cost, are common. Reference 11 provides further details on the relative merits of inverter drive systems.

c. Energy Consumption. In broad terms the specific energy consumption (i.e., kWh per tonne of massecuite processed) for a batch centrifugal depends primarily on the sophistication of the drive system. The simplest drive system, no longer used, is a single-speed motor. Additional sophistication comes, in order of merit, from two-speed, three-speed, and variable-speed drives (which is, in effect, a many-speed pole-changing drive). Table 14.3 shows a summary of the energy requirement of each general type of drive. The main reason for having additional sophistication (and therefore cost) is to recover the energy in the rotating mass (basket, sugar, motor, etc.) during deceleration from spin speed to plough speed. This may be offset against the higher capital cost of a variable speed drive when making a lifetime economic study of total costs.

Table 14.3 refers to a centrifugal with a basket of 1.22 × 0.91 × 0.18 m (48 × 36 × 7 in.) operating with a total cycle time of 164 s (22 charges/h) with a spin time of 25 seconds at a maximum speed of 1200 rpm (980*G*). The mean loss rate (kW) is the average power consumed by the centrifugal over a cycle. The relative peak power demand gives the size of the maximum peak kVA demand as a proportion of the peak demand of the variable-speed drive. The lower the figure, the better. The minimum power factor indicates the level of reactive kVA demanded by the drive type. A low figure means that cable sizes and switchgear must be more highly rated, and additional generator capacity will be required to supply the centrifugal. An ideal power factor would be 1.

It is interesting to consider the continuous centrifugal. The power consumption is strongly dependent on the basket size and speed. For a 1.2-m (48-in.) basket processing 8 tonnes/h of final recovery massecuite at 2000 rpm the power consumption is about 5 kWh/tonne [12]. For such a continuous centrifugal the starting power peaks are large (250 to 300 kVA) and are greater than a batch centrifugal processing the same quantity of massecuite, although the peaks occur only at startup (e.g., once per week). High-grade continuous centrifugals processing A-grade massecuites operate at lower speeds and consequently have a lower power demand, similar to that of a modern variable-speed batch centrifugal.

Finally, it is important to note that often the largest energy demand associated with a centrifugal operation is the evaporation of the wash water added in the centrifugal—not the drive motor power demand. For example, the saving obtained by using electronic drives for a battery of centrifugals processing 30 tonnes/h is around 1 kWh/tonne, or 30 kWh of electrical energy. With a wash rate of 2.5% w/w on massecuite the battery will use 750 kg of water per hour, and 30 kWh (100 MJ) of thermal energy is sufficient to evaporate 50 kg

TABLE 14.3 Summary of Drive Types

Drive Type	Mean Loss Rate (kW)	Energy (kWh / tonne massecuite)	Relative Peak (kVA) Demand	Minimum Power Factor
Single Speed	57	3	1.9	0.45
Two-speed pole-changing				
Reverse plough	47	2.5	1.3	0.55
Forward plough	45	2.4	1.3	0.55
Three-speed pole-changing	38	2	1.2	0.6
Electronic variable-speed				
CSI	15	0.9	1	0.25
PWM	15	0.9	1	0.9

of water, so a reduction in wash consumption from 750 kg to 700 kg provides the same energy saving as changing to electronic drives. The detailed economics will depend on the relative costs of electrical power and steam.

d. Sectional Drive Systems and Harmonics. Sectional drives are a special case of standard pulse-width-modulation (PWM) inverters. A sectional inverter drive uses a common input converter and common storage capacitor for all the centrifugals in the battery, with each centrifugal having its own PWM output inverter. This greatly simplifies the overall drive system. Figure 14.10 shows these components as part of a typical sectional PWM inverter arrangement for supplying five centrifugals.

It is sensible to consider sectional drives when the number of centrifuges is four or more. In addition to simplifying the overall drive system, sectional drives offer the significant benefits of lower capital cost and reduced harmonic impact on the main supply. Smaller benefits accrue from a slight increase in energy efficiency (1 to 2%) and simplified installation, with savings in supply switchgear and cables. The benefits of sectional drive systems generally increase as the number of centrifuges increase. It is essential to consider the possible sequence of operation of the centrifugals within the battery and possible failure modes, to ensure that a sectional drive is sized adequately and suitable for the duty expected.

All inverter drives generate harmonic currents which can create interference in nearby cables, telephone lines, and so on. In addition, the harmonic currents interact with the supply impedance to produce harmonic voltages that can disrupt other electrical equipment attached to the supply network and increase losses in the supply equipment, particularly supply transformers, where a down rating of 15% is not uncommon. For centrifugal drives on cyclic duty, the PWM inverter produces the lowest kVA demand and the lowest harmonics, but these may require further reduction depending on site conditions. The use of a PWM sectional drive rather than individual inverters reduces the flow of harmonics to

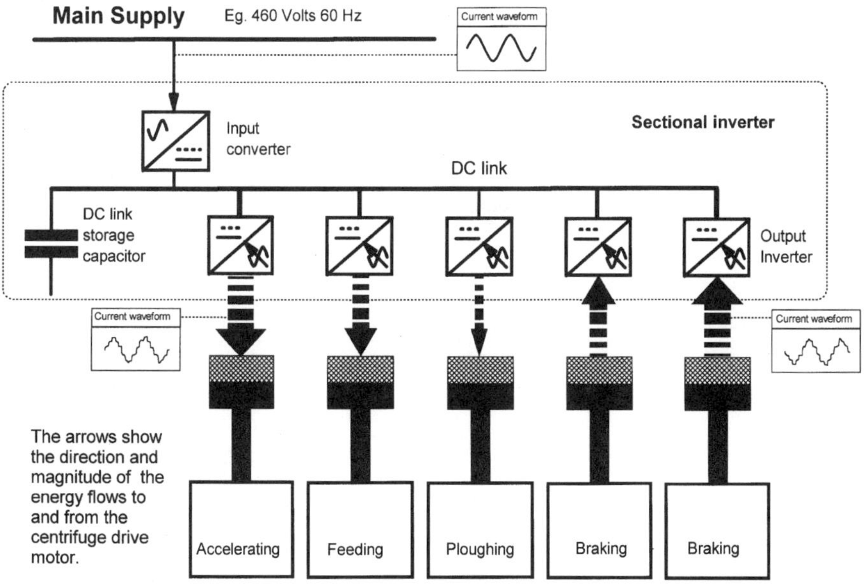

FIGURE 14.10 Sectional inverter drive system for batch centrifugals.

the supply mains by allowing the energy regenerated by one centrifugal drive motor during braking to be fed into another which is accelerating without passing through the factory electrical supply system, as shown in Figure 14.10. This internal transfer of energy via the dc link means that the associated harmonic currents do not appear on the factory electrical supply; they are contained within the drive itself. The reduction in harmonics achievable by this approach can be 40% or more. The reduction depends on site conditions, such as the supply transformer, the main generator, and the centrifugal process cycle. There are also several methods for reducing further the harmonics generated by means of inductors, phase-shifting, and electronic switching devices.

There are many ways to implement sectional drives for batch centrifugals, depending on site requirements and other factors; all give the benefits outlined above. The most advanced inverter drives in the late 1990s are fitted with active front ends. These use the latest semiconductor technology (IGBTs) together with advanced control electronics which nearly eliminate harmonic current on the main supply. This technology can be applied to both individual inverter drives and sectional drive systems. The use of such systems is likely to grow over the coming years as costs fall and the limits imposed on supply harmonics by regulatory bodies are reduced.

14.3 CONTINUOUS CENTRIFUGALS

Continuous centrifugals became widely used in the processing of low-grade recovery massecuites in the 1950s. Prior to this, batch centrifugals were used predominantly with spin times extending to several minutes. In Table 14.4 we compare some of the general aspects of batch and continuous centrifugals on low-grade massecuites. The importance of each of these categories depends on the particular installation and process requirement; nevertheless, it is clear that the continuous centrifugal offers significant advantages in certain areas.

There are several recent variants of the basic continuous centrifugal design, considered in more depth in Sections 14.3.2 to 14.3.4. However, many of the topics in Section 14.3.1 are generally applicable to all continuous centrifugals. The fundamental principles governing separation in a continuous centrifugal are exactly the same as those of a batch centrifugal; filtration under G forces. However, there are major mechanical differences between the two.

14.3.1 Low-Grade Continuous Centrifugals

Figure 14.11 shows a sectional view of a continuous centrifugal for processing low-grade sugars. The centrifugal runs at a constant speed with a constant flow of feed and discharge.

TABLE 14.4 Characteristics of Batch and Continuous Centrifugals

Parameter	Continuous	Batch
Capacity / floor space	High	Medium
Electrical load	Constant	Variable
Power consumption	Medium–high	Low–medium
Staffing	Low	Medium
Maintenance	Low	Medium
Washing	Poor	Good
Crystal breakage	High	None
Dryness of sugar	Medium	High
Massecuite variation	Less sensitive	More sensitive
Capital cost	Low	High

Source: Data from Ref. 13.

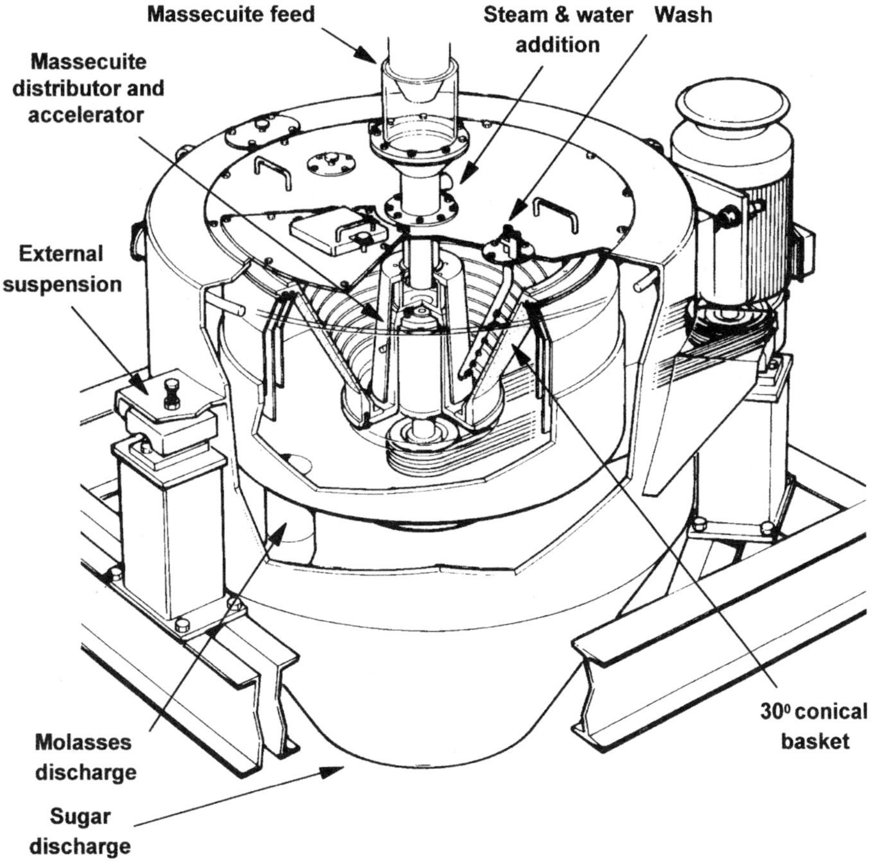

FIGURE 14.11 Continuous centrifugal sectional view. (Courtesy of Broadbent Centrifuges Ltd. © 1997.)

There is no requirement for complex controls, ploughing mechanisms, discharge valves, and so on, giving lower capital and maintenance costs. This simplicity is one of the principal advantages of the continuous centrifugal.

The basket is conical, with the sides at an angle of normally 28 to 34° from the vertical. The angle is chosen such that the sugar and massecuite will climb up the basket screen under the influence of the centrifugal force. The sugar is discharged over the top lip of the basket, whereas the molasses flows partway up the basket before passing through the filter screen and out of the inner casing via a molasses discharge pipe. There is a stationary baffle or seal arrangement close to the basket top lip to prevent remixing of the molasses and sugar within the case.

As the sugar slides out of the basket it is under the influence of a large centrifugal force, the residence time within the basket is short, and the sugar layer is correspondingly thin. Reference to equation (14.2) shows that the short residence time and thin sugar layer compensate for one another to a certain degree. Equation (14.2) also indicates that the smaller crystals (MA) and higher viscosities found in low-grade massecuites will have an adverse effect on the quality of the separation. Massecuite is fed into the continuous centrifugal through a static pipe and distributor/accelerator arrangement designed to provide a uniform layer of massecuite at or near the center of the narrow end of the conical basket (see Fig. 14.11).

The filter screen is supported on a backing mesh in a way similar to that of the batch centrifugal. The screens must be firmly fixed in the basket or they will be centrifuged out of the basket along with the sugar. As the sugar slides up the basket it wears the screen and damages the crystals. Further damage to the crystals occurs when they fly off the basket top lip and hit the outer casing. In many applications this is not a problem, as the resulting sugar will be remelted. Section 14.3.1b considers screens in more detail.

The drive mechanism is a simple belt drive. Basket speeds range between 1400 and 2200 rpm, with basket diameters in the range 1 to 1.5 m (39 to 59 in.) Maximum centrifugal G values are fixed by the basket material (almost universally stainless steels) at around $3000G$. Franzen [14] suggests typical G values of 2400 for low-grade applications, 1200 for affination, and 1500 for intermediate-grade applications.

A variety of washing techniques using water or steam are available in the continuous centrifugal. Figure 14.11 shows a wash pipe in the basket; in addition, water/steam injection points are provided in the centrifugal massecuite feed system. Continuous centrifugals are often used with water added via the wash jets, plus water added to the feed massecuite (sometimes known as *lubrication water*). It is also very common to add steam to heat the massecuite. The relative amounts of these additions depend largely on the massecuite being processed (see also Section 14.3.1c).

Continuous centrifugals are suspended on antivibration mounts. In the example shown in Figure 14.11, these are external to the case. The rate of feed to the centrifugal can be controlled either by a manual valve on the feed line or alternatively, by an automatic control valve linked to the main drive motor current. This setup can be used to maintain a reasonably constant throughput with varying massecuite conditions. This ability to handle variable massecuite with minimum operator intervention is one of the strengths of the continuous centrifugal. Grieg et al. [15] give useful details on the theory, performance, and characteristics of continuous centrifugals.

a. Alternative Designs of Continuous Centrifugals. Standard continuous machines vary only slightly in mechanical design among manufacturers. Some alternatives to the layout shown in Figure 14.11 are summarized below. Figure 14.11 shows the suspension external to the outer casing; some designs incorporate suspension mounts internal to the case. This leads to greater complexity but allows the outer casing to be solidly sealed to the building floor if required. For a given out-of-balance force, the internal suspension reduces the level of vibration transmitted to the outer casing but increases the amplitude of vibration at the basket.

Feed arrangements come in several variants. The center feed shown in Figure 14.11 can include a variety of features (some patented) to improve the heat transfer from the steam injected. Some designs use a feed arrangement where the feed is deposited directly into the bottom of the basket by an off-center static pipe. This arrangement has the advantage of mechanically simplicity but is less effective at providing a uniform massecuite layer on the basket.

b. Screens. A fundamental difference between the batch and continuous centrifugal is the way the sugar behaves on the screens. Once a batch machine has been fed, the sugar remains stationary relative to the screen until discharged at low speed, whereas in a continuous centrifugal there is always relative motion between the sugar and the screen. Continuous centrifugal screens are more important in defining process performance than are screens in a batch centrifugal. Only a tiny number of crystals are anywhere near the slots or holes in the screen of a batch centrifugal. However, in a continuous centrifugal a large proportion of the sugar will come into contact with the slots as the crystals slide up the screen. The minimum slot dimension in the continuous screen must be significantly less than the particle size (MA) of the sugar; otherwise, the losses through the screen become excessive. For final recovery duties, any loss of sugar through the screen is a loss to the refinery.

Figure 14.12 shows the calculated magnitude of this effect for screen slot sizes of 0.04 mm (0.0016 in.), 0.06 mm (0.0024 in.), and 0.09 mm (0.0035 in.) and sugar coefficient of variation (CV) values of 30% and 40% as a function of the sugar mean aperture (MA). The calculation assumes that the particle size distribution is Gaussian (characterized by MA/CV values) and that any crystal in the size distribution that is less than the screen slot size will be lost. For example, Figure 14.12 shows that a 2.3% loss is expected for a massecuite containing crystals with a MA of 0.3 mm (0.012 in.) and a CV of 40% on a screen with 0.06-mm (0.0024 in.) slots.

Any wash or steam addition further increases these losses by dissolution. Small openings and large crystals are necessary to minimize losses. If the screen slot size is reduced, the screen will have a smaller total open area, which restricts the flow of mother liquor through the screen [16]. This in turn reduces the throughput of the centrifugal. To get reasonable throughput plus low losses, it is necessary to have a large open-area screen with a small slot size, particularly for final recovery massecuites. Such screens have less metal in them and tend to wear faster. Typical screens contain slots of 0.04 to 0.13 mm (0.0016 to 0.0051 in.) and 6 to 22% open area, although many other combinations incorporating holes rather than slots are commercially available.

Screens are manufactured from nickel–chrome or stainless steel. Nickel–chrome screens are made by chemically etching the slots in a thin nickel plate, then electroplating a hard-chrome surface onto the screen. Stainless steel screens are manufactured by cutting each slot with a laser in a process pioneered in Australia. This is a costly but yields a screen made

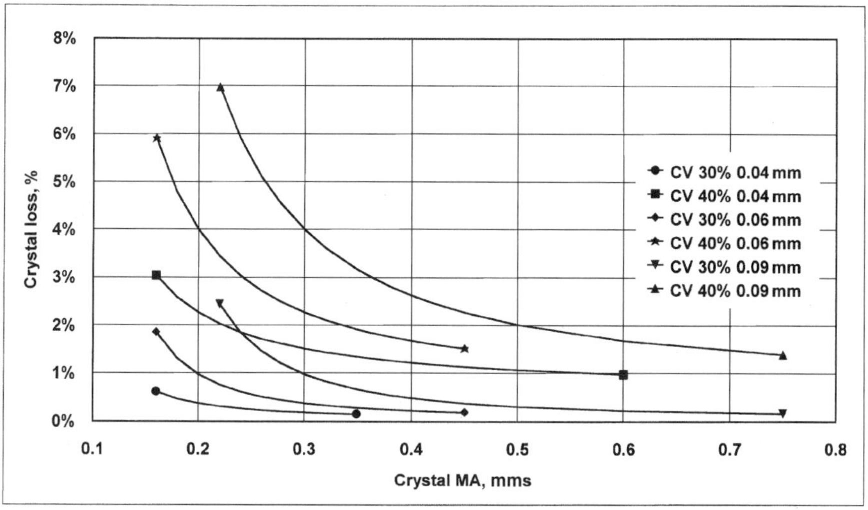

FIGURE 14.12 Calculated crystal loss for 0.04-, 0.06-, and 0.09-mm screens as a function of MA and CV. Graph shows maximum possible loss of crystal assuming no crystal breakage on the screen.

from hard-wearing solid stainless. The life of a stainless screen has been reported to be over three times that of a nickel–chrome equivalent [12]. In some installations the additional cost of these screens can be justified by the longer life; however, mechanical damage to the screen resulting from tramp material passing through the centrifugal can reduce this life significantly.

Sections of unused and worn screens are shown in Figures 14.13 to 14.15. Figure 14.13 shows a portion of unworn screen manufactured by Stork Veco with a slot size of 0.06 by 2.2 mm and a large open area of 22%. Figure 14.14 shows a similar type of screen having a slot size of 0.09 by 2.4 mm and an open area of 14%. Figure 14.15 is the same screen as shown in Figure 14.14 after use. The wear on the surface and enlargement of the slot size as a result of this wear are clear.

It is important to check regularly (e.g., weekly) for excessive wear or damage on the screen. Monitoring the average molasses purity rise over time from a continuous centrifugal is one method of monitoring screen condition and can be used to plan screen changes. A purity rise of 1 to 2% over a short time indicates a worn screen, as does the appearance of small crystals in the molasses. An alternative method is to stop the centrifugal and inspect the screen condition and slot width. Any screen with mechanical damage or slots more than 30% oversized should be replaced. Screen lives of 3 to 6 month can be expected. To achieve the longest life it is important that the backing screen should support the working screen uniformly at many points. Any high spots in the working screen are rapidly worn away.

c. Purity Rise and Purging Efficiency. The separation performance of most continuous centrifugals is measured in terms of the purity rise and the purging efficiency. The term *molasses purity rise* refers to the increase in purity caused by sugar being lost through the

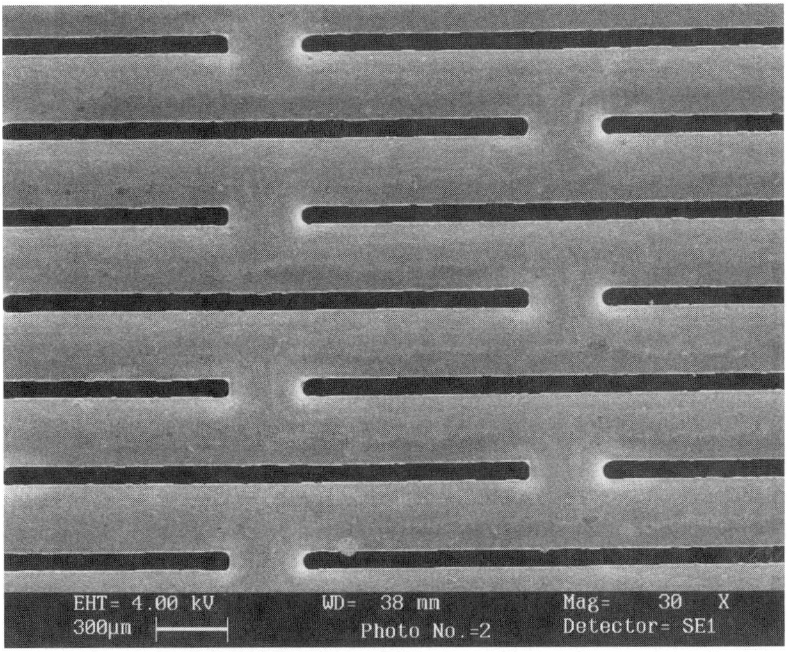

FIGURE 14.13 Unworn 0.06-mm slots.

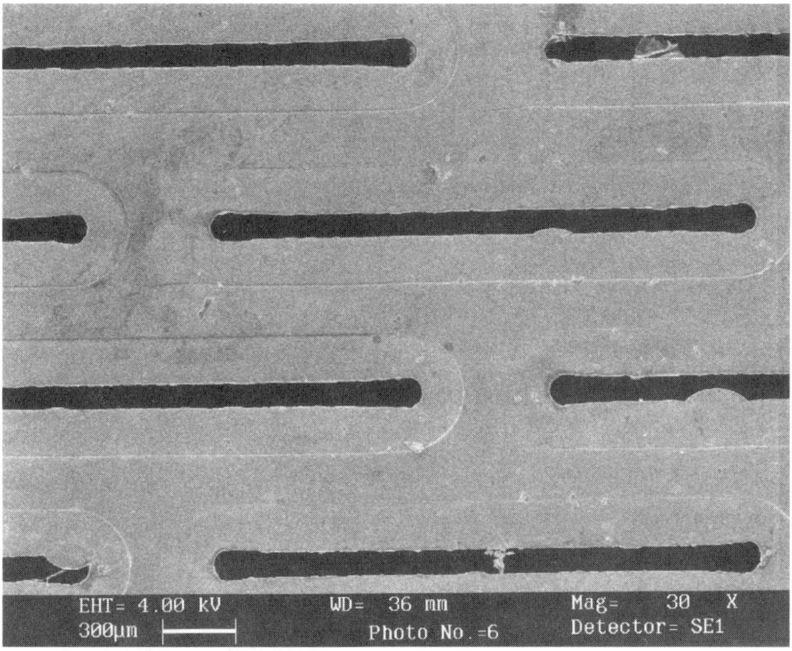

FIGURE 14.14 Unworn 0.09-mm slots.

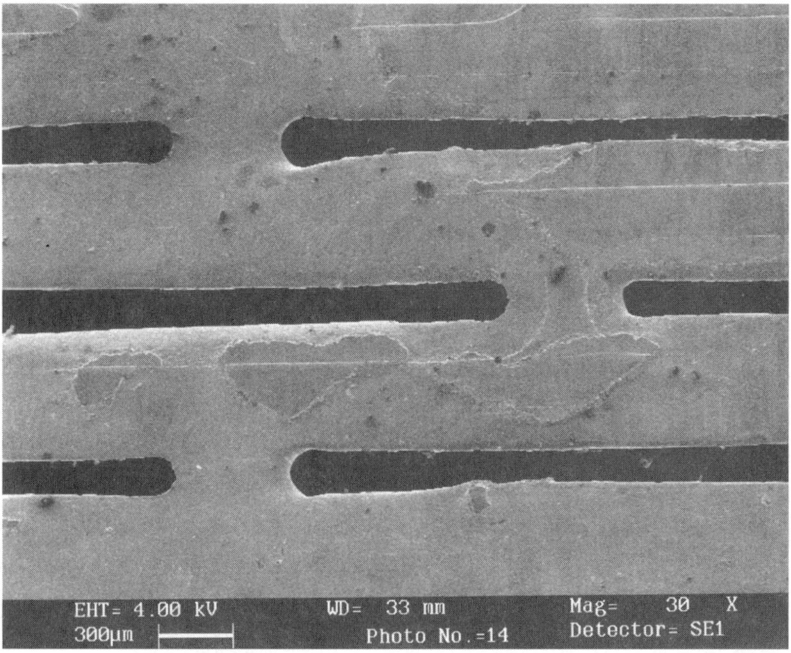

FIGURE 14.15 Worn 0.09-mm slots.

screen into the molasses. As a result, the molasses purity is higher than the massecuite mother liquor purity. Purity rises can be calculated from the sugar loss and the massecuite characteristics. For a typical massecuite with a crystal content of 40%, purity of 60%, Brix/DS of 96%, and mother liquor purity of 31.4, the purity rise and molasses purity are shown in Figure 14.16.

Figure 14.17 shows purity rise against throughput for a typical recovery massecuite [17]. The data show a slight reduction in purity rise as the throughput increases. This is to be expected as a thicker layer of sugar on the screen (higher throughput) means that a smaller proportion of the crystals come into direct contact with the screen slots, so the opportunity for losses is reduced. Although the level of purity rise varies with process conditions, values are generally in the range 0.5 to 3.

Purging efficiency PE is defined as [18]:

$$PE = \frac{100(P_s - P_m)(100 - P_{ml})}{(P_s - P_{ml})(100 - P_m)} \tag{14.9}$$

where P_s is the sugar purity and P_m and P_{ml} are the massecuite and molasses purity, respectively. Inspection of equation (14.9) shows that for a sugar purity of 100%, the purging efficiency is also 100%. PE is the percentage of nonsugars in the massecuite that end up in the molasses. The goal is to have all the nonsugars in the molasses (PE = 100%); however, if the crystals contain an element of nonsugars (color), a PE of 100% cannot be achieved irrespective of centrifugal performance. An alternative and simpler measure is the purity of the sugar P_s. For low-grade final recovery applications, the more important measure is the purity rise, as this is related directly to the factory sugar loss. For higher-grade applications, the emphasis is more on sugar purity. To get the best performance for both sugar purity and purity rise, it is necessary to perform a two-stage separation. This technique, known as *double curing* or *double purging*, is often applied to C sugar (see Ref. 19 and Section 14.3.4).

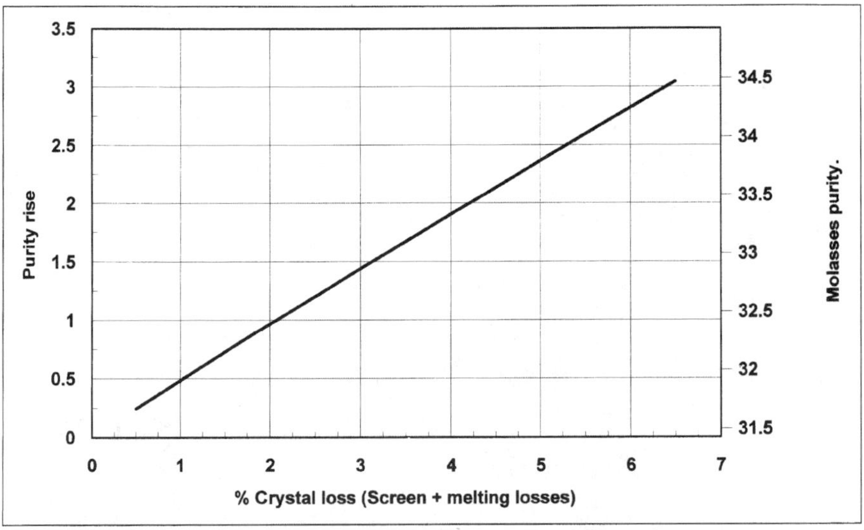

FIGURE 14.16 Purity rise and molasses purity as a function of crystal loss. Massecuite properties: crystal content, 40%; purity, 60%. Brix / DS, 96%; mother liquor purity, 31.4%.

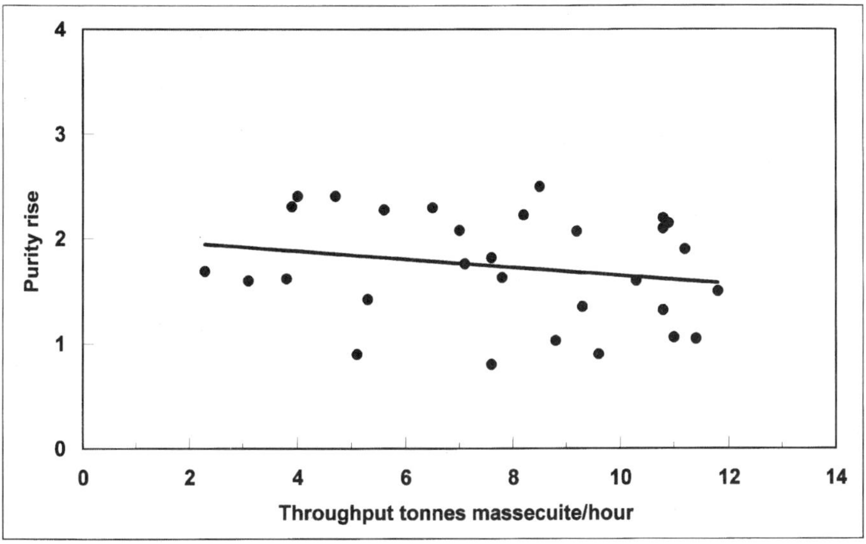

FIGURE 14.17 Purity rise versus throughput recovery massecuite. Basket diameter, 1.22 m; speed, 2000 rpm; wash/steam to massecuite ratio, 4%.

Purging efficiency falls as throughput increases, as expected from equation (14.2). Higher throughput means thicker cakes and therefore worse separation performance.

d. Color Line. As the massecuite moves along the basket wall, the separation of the mother liquor from the sugar produces a visible color change. Initially, the massecuite is brown, and as the mother liquor is purged, the color changes to a lighter brown. This color change, which can be seen on the basket wall, is called the *color line.* It acts as a useful guide to correct operation of the continuous centrifugal. It is not a sharp line but rather a change in shade from dark to light brown over 50 to 75 mm (2 to 3 in.) or so. It is normal practice to keep the color line between one-third and halfway up the basket wall. All other things being equal, Table 14.5 summarizes the effects of having a high or a low color line.

The high color line means that there is a lot of mother liquor (i.e., higher throughput) and it is therefore not being purged from the sugar until later in the basket (purging efficiency lower). With a higher throughput the crystal layer on the screen is thicker and less will be lost through the slots (i.e., purity rise lower). The opposite applies for a low color line. A stationary color line indicates a stable set of feed conditions and uniform massecuite properties. To see the line, it is useful to use a stroboscope or bright light. (N.B.: If it is necessary to remove any safety covers from the centrifugal to see the basket, this must be done only after all necessary safety precautions have been taken.)

TABLE 14.5 Effects of Low and High Color Lines

Parameter	Low Color Line	High Color Line
Purging efficiency	Higher	Lower
Molasses purity rise	Higher	Lower
Throughput	Lower	Higher

e. Steam and Water Addition. Washing as an aid to improved purging efficiency must take place below the color line, prior to purging of the majority of the mother liquor. Washing above the color line is usually wasteful, as the crystals are dispersed on the screen and the wash may not come into intimate contact with the sugar. Washing close to the basket lip can result in some of the wash liquor escaping with the sugar, giving increased moisture. Improved sugar quality can be achieved by applying wash directly to the purged crystals immediately above the color line via a spray bar. This will aid removal of the remaining layer of mother liquor adhering to the crystals at the cost of dissolving more of the exposed crystals.

Typical wash quantities are 3 to 7% of massecuite processed. Sugar crystal losses due to washing are less for a continuous than for a batch centrifugal, as the wash water does not come into contact with the crystals long enough to become saturated, so the losses are less, but equally, the effectiveness of the wash is also less. Nevertheless, it is important not to overwash the sugar, or increased sugar losses and evaporation energy demands will result.

It is clear from equation (14.2) that the viscosity is a significant factor in determining process performance. A lower viscosity allows better separation, and heating the massecuite by 10°C (20°F) reduces its viscosity by half [13]. As the massecuite fed to a centrifugal is supersaturated, there is a benefit in heating the feed to the centrifugal to the point where the supersaturation is reduced to zero. At this point the viscosity will be at its lowest value consistent with not dissolving any of the sugar, and the centrifugal will be working at its optimum operating point. The degree of heating required depends on the degree of supersaturation. Typically, a 5 to 10°C (10 to 20°F) is used.

Heating can take place external to the centrifugal using some form of in-line heater, such as steam or water pipes. Care must be exercised to avoid heating the massecuite locally adjacent to the heating elements to the point where the mother liquor is no longer saturated. An alternative and more common approach is to inject steam into the accelerator cone of the centrifugal (see Fig. 14.11). The steam heats the massecuite directly but also condenses and dilutes the massecuite, reducing the supersaturation. Only very limited dissolution of the crystals occurs, because the steam addition takes place very near the basket and there is insufficient time available for full saturation of the condensate to take place. Tests by Trojan [16] show that the typical massecuite–steam contact time is 1 to 2 s and a purity rise of no more than 0.1% occurs when heating the massecuite from 53°C up to a maximum of 70°C. The resulting viscosity reduction allowed the capacity of the test centrifugal to be increased by a factor of 3 for a temperature rise of 10°C. It is generally necessary to perform tests to find the optimum washing and heating conditions for a given application.

f. Mass Balance: Recovery Massecuite. Figure 14.18 shows the mass balance for a continuous centrifugal on final recovery duty. In this example 2.1 tonnes of sucrose per hour is lost with the molasses. This is a loss to the factory, as no further sucrose recovery takes place on the molasses.

g. Electrical Design. The simplicity of the basic continuous centrifugal extends to the electrical drive and controls. Other than basic startup and shutdown steps, the controls generally incorporate feedback from the motor current to control the feed valve to maintain a constant throughput. The high-speed continuous centrifugals require large amounts of energy to overcome the windage drag on the spinning basket and to supply the kinetic energy to the discharged sugar and molasses. None of this energy can be recovered, and consequently, the power consumed per tonne of massecuite processed is large, about 3 to 5 kWh/tonne. This energy consumption is significantly higher than that of any batch centrifugal. See Section 14.2.5c for additional information on energy efficiency.

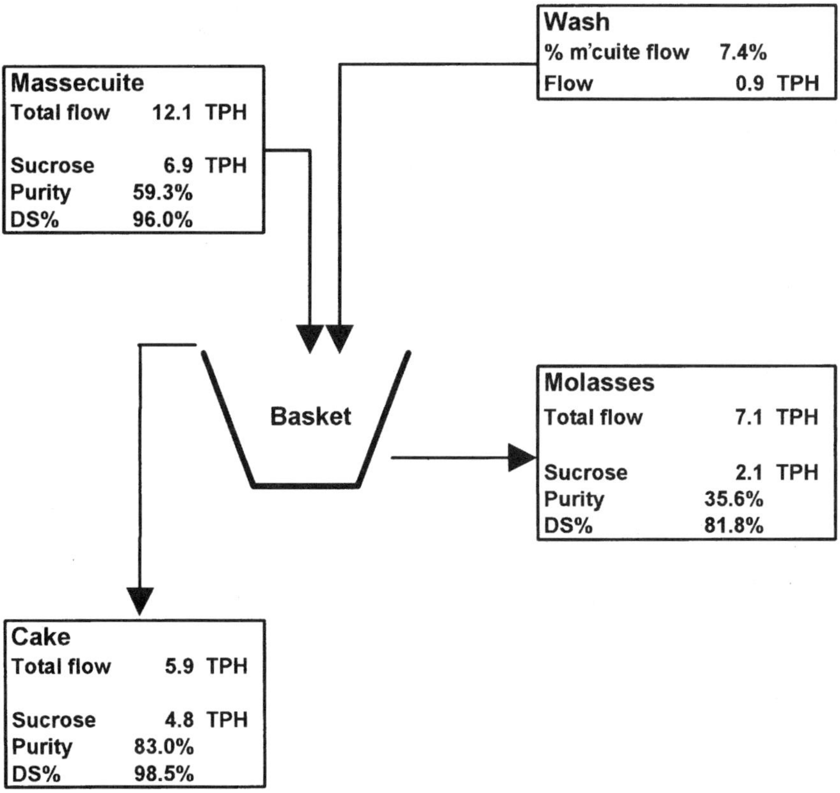

FIGURE 14.18 Typical mass balance: final recovery massecuite.

Figure 14.19 shows the test results for motor current against throughput for a 1220 mm diameter centrifugal running at 2000 rpm. The motor current can clearly be used as a measure of the throughput. For these tests the intercept of 45 A is the power consumed by the centrifugal with no massecuite feed. This is the power necessary to overcome the windage and bearing friction at 2000 rpm and to supply the drive motor magnetizing losses.

14.3.2 Magma Mixing Centrifugal

The sugar crystals from a continuous centrifugal are usually mixed with syrup to produce a magma for further processing. A magma mixer is used or, more simply, syrup is added directly to the sugar conveyor beneath the centrifugal. Another approach is to use a magma mixing centrifugal. These are based on a standard design such as that shown in Figure 14.11 and contain additional parts to allow mixing within the centrifugal casing. Syrup is pumped through jets (similar to wash jets) in the top of the centrifugal casing, where it mixes with the sugar being discharged. The mixing process is helped by the high-speed motion of the sugar as it leaves the lip of the basket. Unlike the standard crystal discharge continuous centrifugal, the sugar is discharged via a pipe as a magma.

The advantages are a saving in space and capital equipment; separate mixers, tanks, pumps, and so on, are not required. This leads to a clean, simple installation. In addition,

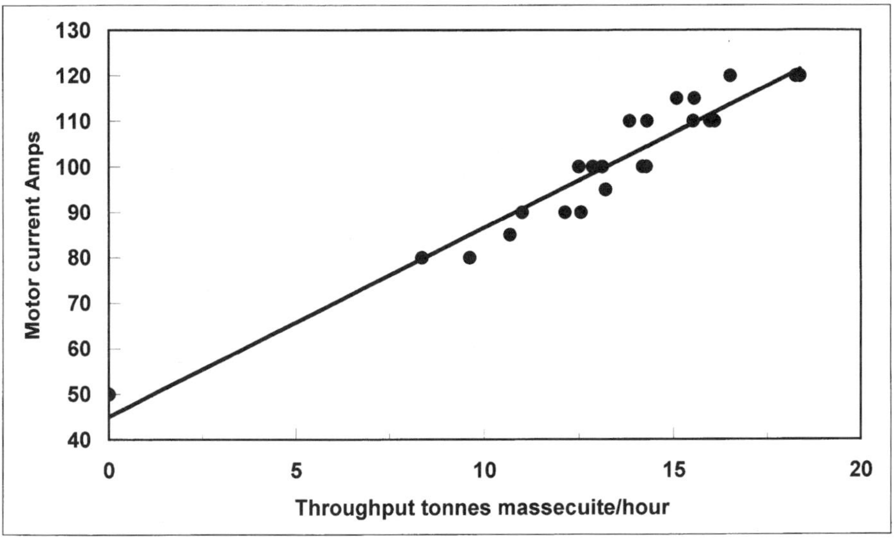

FIGURE 14.19 Motor current versus throughput: recovery massecuite. 1.22-m-diameer basket; 2000 rpm; massecuite temperature, 55 to 58°C; wash to massecuite ratio, 1.5 to 3%.

as no mixer drive is needed, there is a small energy saving [14], the mixing energy coming instead from the kinetic energy of the discharged sugar that would otherwise be lost in a nonmixing centrifugal. The disadvantages are twofold. First, a pump is often required to provide high-pressure syrup for the spray jets in the centrifugal. This pump is an additional item of equipment which consumes power and reduces the energy saving. Second, it is difficult to check the separation performance of the magma mixing centrifugal (depending on the design), as samples of centrifuged sugar must be obtained prior to the mixing operation. There can also be a tendency to set the centrifugal up to produce a lower-Brix magma. This is done to avoid potential blockage problems with high-Brix viscous magmas. A lower-Brix magma carries with it the penalty of sugar dissolution and a higher evaporative energy demand. Generally, the use of separate tanks and mixers is simpler to control but takes up more space and uses more equipment. The electrical controls on magma centrifugals are generally more complex and incorporate additional control loops to match the syrup addition to the throughput and to stop the machine in the event of loss of the syrup supply. See Ref. 20 for additional information on the design on mixing centrifugals. Most centrifugal manufacturers provide magma versions of their continuous centrifugals, and many provide kits to modify an existing standard low-grade continuous centrifugal to provide the magma mixing features.

14.3.3 Melting Centrifugal

The melting centrifugal can be considered as a further development of the magma mixing design, and broadly speaking it has the same advantages/disadvantages as those of the magma centrifugal. In melting versions the sugar is totally dissolved inside the outer case of the centrifugal by the addition of water or syrup. For the sugar to melt fully, it must remain in contact with the water/syrup for a suitable length of time (depending on the temperature, level of liquor saturation, and agitation). This is achieved by using an internal trough in which the magma collects. Agitation is provided by both the kinetic energy of the discharged sugar [14] and water/syrup jets, with the fully dissolved syrup overflowing to the centrifugal outlet. The syrup is normally discharged via a pipe and flange connection in the bottom of the centrifugal case. See Ref. 20 for additional information. As with magma centrifugals, the electrical controls on melting centrifugals incorporate additional control loops to match the syrup addition to the throughput, or alternatively, to the Brix of the discharged syrup. Additional interlocks are generally provided to stop the machine in the event of loss of the water/syrup supply.

14.3.4 Double Purging

Double purging is a technique used to reduce the recirculation of impurities back to a higher-grade station while minimizing the loss of sucrose to the molasses stream. As the name indicates, two stages of separation are used. Figure 14.20 shows a typical double-purge arrangement on recovery massecuite using two standard continuous centrifugals and ancillary equipment.

The first separation stage is set up to avoid sugar losses by using little or no wash, thereby avoiding losses through dissolution. High *G* is used to obtain reasonable liquor removal. The resulting low-purity molasses is discarded with minimal sugar losses. The sugar discharged from the first centrifugal is mingled with syrup to produce a high-crystal-content

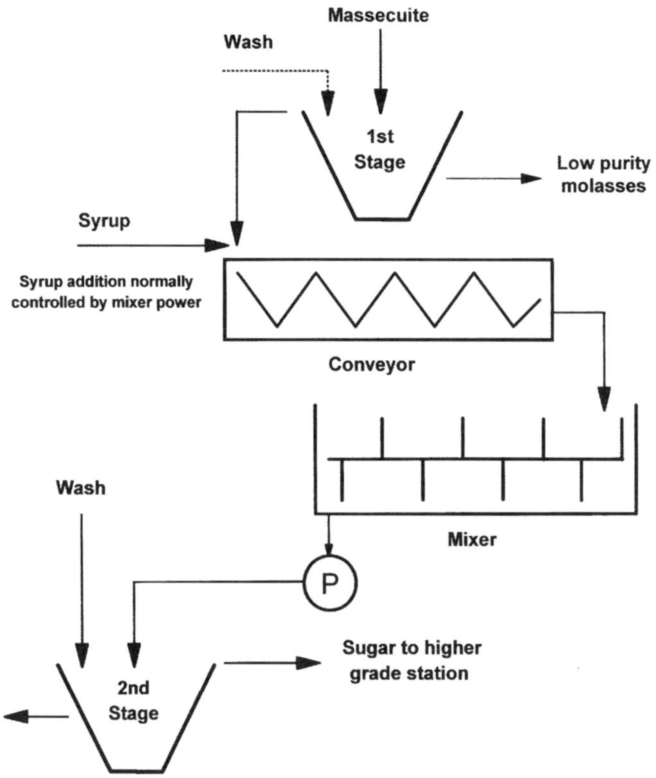

FIGURE 14.20 Typical double-purge arrangement.

magma. During the mingling process, the crystals come into contact with one another, which encourages removal of the low-purity liquor layer adhering to the crystal surface.

The second separation stage is set up to produce high-purity sugar by washing. By optimizing the first separation stage for a low-purity rise and the second stage for good sugar purity, a more efficient overall separation is achieved than is possible from a single separation stage. The benefits are a lower sugar content in the recovery molasses and a higher-purity sugar as feed for the higher-grade stations. See Ref. 19 for a comparison of single and double purging of C-massecuite in raw sugar production.

Since the early 1990s several manufacturers have developed continuous centrifugals which perform two separation stages within a single unit. Figure 14.21 shows a typical schematic arrangement of such a double-purge centrifugal. Most double-purge designs employ a separate basket for each stage. The first stage operates as a magma mixing type (see Section 14.3.2), with the discharged magma directly feeding the lower second-stage basket via pipework internal to the centrifugal case.

The benefits of double-purging centrifugals are a large saving of floor space, no requirement for mixers and conveyors, and a reduction in power demand, as the energy for mixing the magma comes in part from the kinetic energy imparted to the sugar from the first-stage separation rather than from a separate mixer drive motor. The penalties are some additional mechanical complexity within the centrifugal, the need for a pressurised syrup supply for the magma mixing stage (as opposed to a standard mixer), and some difficulty in assessing the quality of the sugar as it passes from the first to the second stage. Franzen [14] compares the electrical energy requirement of a double-purging arrangement similar

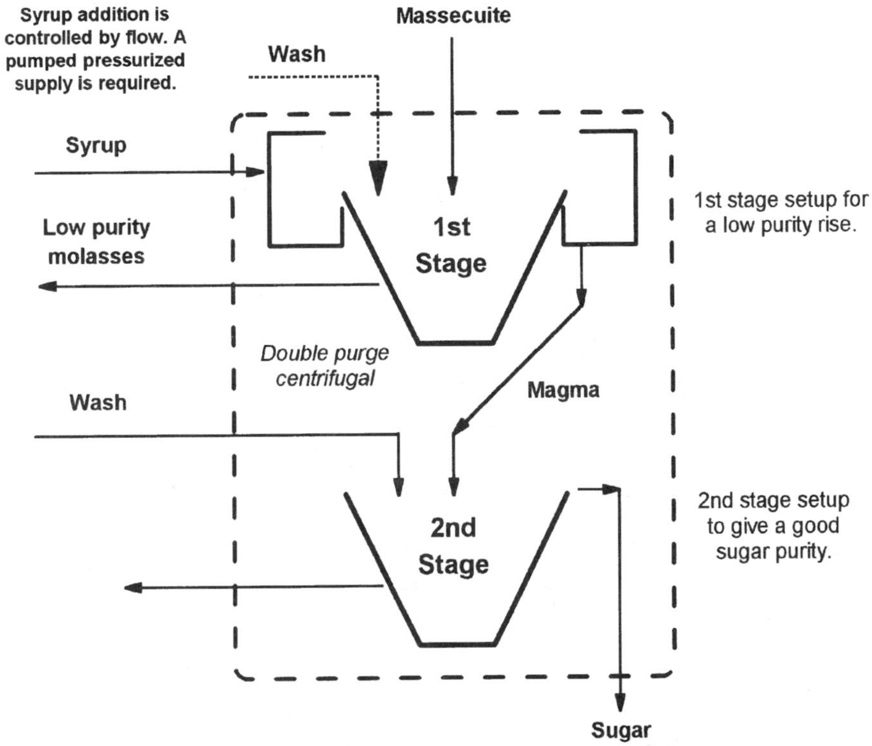

FIGURE 14.21 Schematic diagram of a double-purge centrifugal.

to that shown in Figure 14.20 to a double-purge design similar to Figure 14.21 and concludes that there is an energy saving of 20 to 30%, given equivalent throughputs and sugar qualities.

The double-purge centrifugal (Fig. 14.21) performs the same task as two standard centrifugals plus a mixer (Fig. 14.20), and both approaches are used to improve sugar purity in low-grade applications [21]. When considering the relative energy consumption between the two approaches, the largest element of power consumption is normally the centrifugal drive motors. For a system optimized with regard to centrifugal size, the power requirement of the centrifugal drives will be virtually identical whichever approach is adopted, and the differences will be governed primarily by differences in the ancillary equipment, pumps, mixers, conveyors, and so on, required by each approach.

a. Double-Purge Designs. There are three basic design of double-purging centrifugals. Two employ double baskets. One of these, manufactured by Dorr Oliver, suspends both baskets from a single overdriven spindle similar to that of a batch machine. This approach has the benefit of a simple single motor drive. The relative *G* between the two stages of separation is fixed by the mechanical geometry of the design, which may lead to a degree of inflexibility. An alternative design provided by BMA and others is based on mounting two continuous centrifugals one above the other, each with its own drive motor, which provides flexibility in basket speeds but requires two drives and associated switchgear and controls. The third design, by Hein Lehmann and shown in Figure 14.22, performs the two separation stages on a single basket. The first stage of separation takes place on the lower part of the basket and the second stage of the upper part. A mingling device separates the two parts, giving a simple overall design with fixed relative *G* between the two separation stages. As the first stage of separation takes place on the smaller-diameter part of the basket, the *G* is low and this design is unsuitable for double-purging viscous recovery or C massecuites. This design is, however, suitable for double-purging remelt or B massecuites. Figure 14.22 also shows a lump strainer on the massecuite feed line. The centrifugal shown is arranged to melt the sugar prior to discharge.

14.3.5 High-Grade Continuous Centrifugals

Since 1950 continuous centrifugals have been used to replace batch centrifugals for processing of some recovery massecuites. This is a result of the lower capital and maintenance costs of the continuous design combined with adequate process performance. If the continuous centrifugal is to be used successfully as an alternative to the established batch centrifugal on refined white sugar and affination, it must be able to provide (1) improved washing to produce high-purity low-color sugars, (2) good sugar dryness, and (3) no significant sugar crystal breakage (refined product application).

In the past, continuous centrifugals have generally been optimized for the production of low-grade sugars. Although various manufacturers have worked on continuous machines for high-grade applications since the 1960s they have not found widespread application. To succeed in the production of a refined product, the optimization process needed to be taken further. After revisiting the fundamental problems and requirements in 1988, a team in Australia (jointly by Bureau of Sugar Experiment Stations and NQEA Pty) made progress in the optimization of continuous machines for high-grade sugars. In this section we cover briefly some of the work done and the results obtained on affination and refined sugars. See Refs. 22 and 23 for further background information.

a. Mechanical Design. The basic design of the current high-grade continuous centrifugals is very similar in concept to the proven low-grade designs. Figure 14.23 shows a sectional view of a high-grade design. There are three main mechanical changes from the

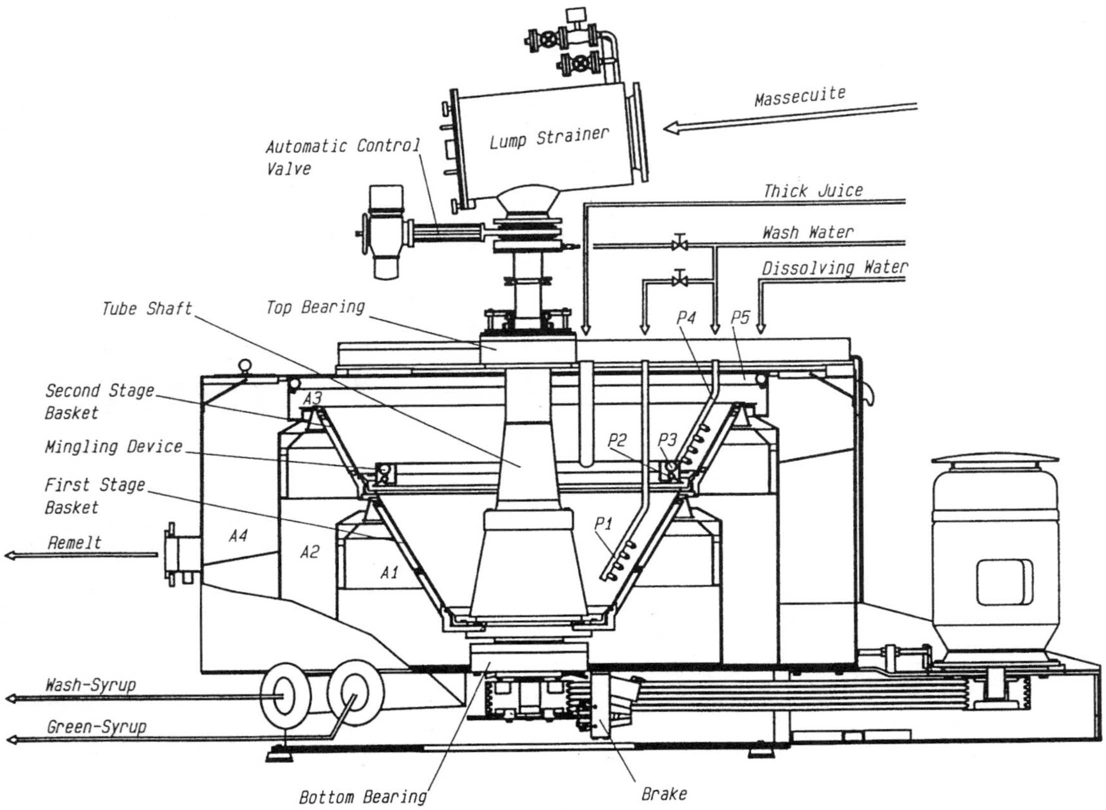

FIGURE 14.22 Example of a double-purge centrifugal. (Courtesy of Hein Lehmann GmbH.)

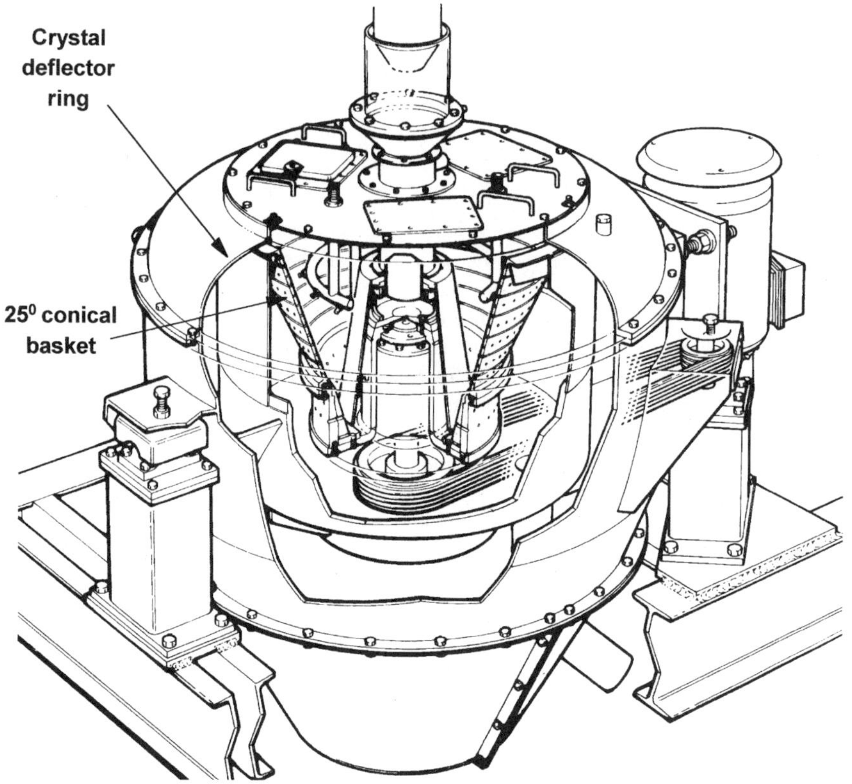

FIGURE 14.23 Sectional view of a high-grade continuous centrifugal. (Courtesy of Broadbent Centrifuges Ltd. © 1997.)

low-grade design. First, the basket angle is modified to suit the massecuite condition found with high-grade sugars. Typically, a basket angle of 30 to 40° is used for low grades, whereas a high-grade continuous basket has an angle in the range 24 to 26°. The reason for this change is to maximize the time the sugar crystals remain in the centrifuge basket; the longer the crystals stay in the basket, the better the separation [see equation (14.2)].

The second change is to reduce the basket rotational speed from the 1800 to 2200 rpm used in low-grade designs to 700 to 1000 rpm for the high-grade design. This speed reduction further increases the residence time in the basket, but it has only a limited direct effect on the separation performance, as the *G* in the basket has also been reduced, which counteracts the increased residence time. The benefit of the speed reduction comes from allowing the increased residence time to be used for washing. Washing is effective only if the contact between the sugar crystals and the wash liquor occurs for a reasonable length of time (in batch centrifugals it is about 15 to 25 s).

The third important design point is the incorporation of a crystal deflector ring within the centrifugal case (see Fig. 14.23). When the washed and dried crystals leave the basket, they slide round the crystal deflector ring, slowing up as they do so. This gentle reduction in speed greatly reduces the crystal damage compared to that in normal low-grade continuous centrifugal design, where the crystals fly out of the basket at high speed and strike the vertical outer casing wall violently. In addition, the reduced operating speed of the high-grade design also contributes significantly to the avoidance of crystal damage. The mechanical design of the crystal deflector ring is important to achieve minimum crystal breakage and to ensure that it is not worn away by the high volume of sugar crystals passing over it. See Ref. 22 for more information. Other aspects of the mechanical design of high-grade continuous centrifugals are very similar to those of the low-grade machine (see Section 14.3.1).

b. Process Performance. High-grade sugars are generally larger and more uniform in size than low-grade sugars and have to be separated from less viscous mother liquor. Equation (14.2) shows that both of these factors aid separation. These benefits are in addition to those resulting from increased residence time.

To maximize the performance of the high-grade continuous centrifugal it is important to ensure that the feed viscosity is minimal by having the highest possible massecuite temperature consistent with not redissolving the sugar crystals. If the massecuite is well heated and crystal size (MA) is good (e.g., >0.5 mm or 0.02 in.) and the crystal CV is low (ideally, 30 or below), the process performance is good, as shown in Table 14.6 and Figure 14.24. One shortcoming of the high-grade design is that the moisture of the sugar discharge is higher than that of a batch centrifugal on the same grade of sugar (see Table 14.6 and Figure 14.24). However, as the feed to a high-grade continuous centrifugal is often subject to additional heating prior to centrifuging, the discharge sugar is hotter, which aids subsequent drying. If the energy requirement to heat the sugar is added to the energy requirement of the centrifugal itself, the high-grade design is less energy efficient than a batch centrifugal. If the drying energy is excluded, the high-grade design is slightly more energy efficient than a modern inverter-driven batch centrifugal. The high moisture of the discharged sugar can lead to lump formation; however, correct operation with respect to washing and maintaining low moisture will largely overcome the formation of lumps in the discharged sugar. Lumps,

TABLE 14.6 Summary of Process Performance

Parameter	High Grade	Old Batch	New Batch	Comment
Crystal MA (mm)	0.62	0.66	0.66	Measured after centrifugation
Crystal (CV)	30	30	30	Measured after centrifugation
% Fines	1.5	1.22	1.22	Mass % less than 0.25 mm after centrifugation
% Massecuite crystal content	52.4	52.4	52.4	Same for both batch and high-grade continuous
Discharged sugar moisture (% w / w)	1–2	0.75	0.75	
Wash consumption (% massecuite)	3–6	2–4	2–3	
Power consumption (kWh / tonnes mass)	0.8	2.2	0.9	Figures ignores the energy required to dry the sugar
Throughput per centrifugal tonne mass	30–40	17	22	

Source: Data based on Ref. 23 and NQEA–Broadbent test work.

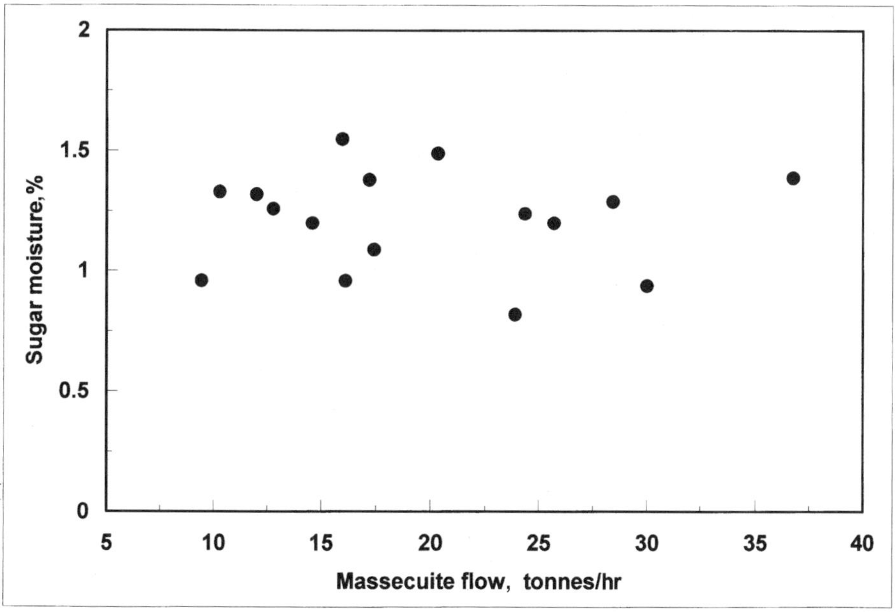

FIGURE 14.24 Sugar moisture versus massecuite flow rate for a high-grade continuous centrifugal.

if formed, comprise smaller sugar crystals with a higher proportion of syrup than that found in the bulk of the sugar.

Table 14.6 shows some general results [23]. Included in the table are comparative batch centrifugal results for an old two-speed 48 × 30 design and a larger modern inverter variable-speed design. Figure 14.24 shows the range of moistures obtained with a typical high-grade continuous centrifugal.

Results published by Greig et al. [24] give sugar colors from the high-grade continuous equivalent to those from batch machines for second and third refined sugars and approximately 25% higher than the batch machine for first refined sugar. Further development is considered necessary before the high-grade continuous centrifugal will gain wide acceptance on refined sugar.

The high-grade continuous centrifugal can perform a similar function to the batch centrifugal on affination. Raw sugar may contain tramp material which should be removed prior to entering a high-grade centrifugal to avoid screen damage or blockage problems. Figure 14.25 is an example of a mass balance for the affination duty, which shows the low wash water requirement of 1.3% of massecuite flow and the high sugar moisture content of 1.3% (DS% 98.7).

Typically, the amount of wash water used in a high-grade is more than that used in an equivalent batch centrifugal. This would imply higher losses through dissolution of the sugar. In practice, the wash liquor in the high-grade is in contact with the sugar for only a limited time and does not become fully saturated with sugar. In contrast, the wash liquor in a batch machine approaches saturation during its passage through the thick sugar layer in the basket. Broadly speaking, the sugar losses due to washing are similar, but evaporation energy demands are higher for the continuous centrifugal. The screens in a high-grade

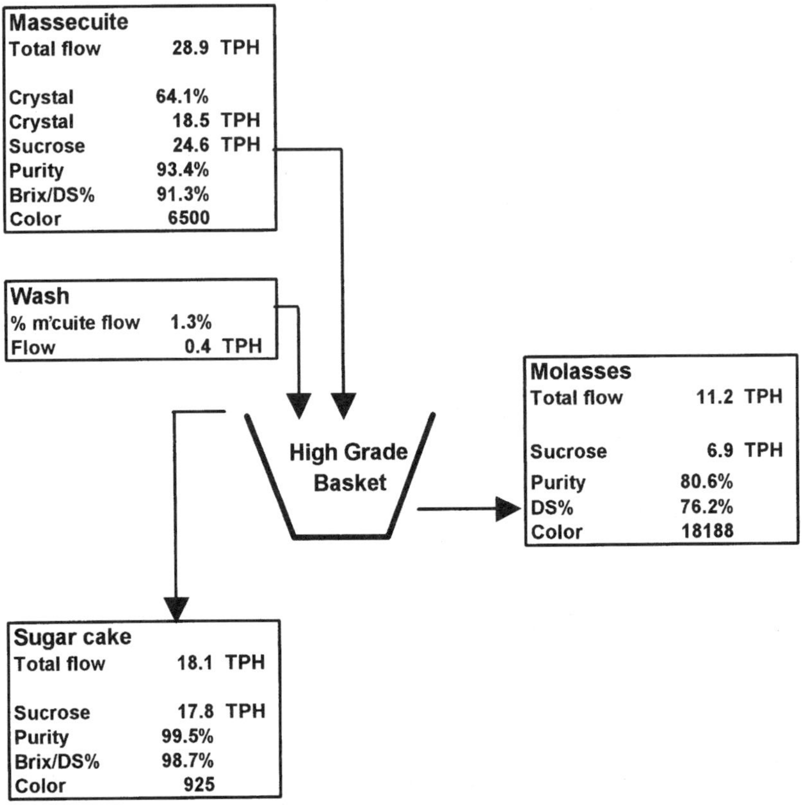

FIGURE 14.25 Affination mass balance for a high-grade continuous centrifugal.

centrifugal are similar to those in a low-grade design, with the exception of the slot size, which due to the larger crystal size being processed, can be increased to allow better purging (e.g., 0.15 mm or 0.006 in.). Other aspects of the use and maintenance of a high-grade centrifugal are similar to those of the low-grade continuous centrifugal described in Section 14.3.1.

c. Electrical Design. The control system on a high-grade continuous centrifugal is significantly more complex than that used on a normal low-grade continuous centrifugal. In addition to the automatic feed-rate control available for the low-grade continuous centrifugal (see Section 14.3.1g), the high-grade controls generally incorporate additional control loops for massecuite temperature and wash and steam addition. The controls for a high-grade continuous centrifugal are similar in complexity to those on a batch centrifugal. The centrifugal drive differs little from the standard low-grade design other than requiring a smaller motor as a result of the lower basket speed. If the additional energy requirement to dry the wetter discharged sugar is ignored, the energy efficiency of the high-grade centrifugal is slightly better than that of a modern inverter-driven batch centrifugal.

14.4 SAFETY AND MAINTENANCE

Failure to install, maintain, or operate any item of moving mechanical equipment correctly can constitute a severe hazard. Centrifugals are usually installed for a period of 20 to 30 years before replacement and may not undergo a total mechanical stripdown for 10 to 20 years. It is therefore most important to install the centrifugals on suitable foundations and undertake maintenance, and inspection in accordance with the manufacturer's instructions and good engineering practice.

The amount of stored energy in any centrifugal basket when spinning is large, with typical values being equivalent to a 1-tonne car traveling at 450 km/h (280 mph). The long-term effects of corrosion (which may not have been quantified at the time of initial installation) plus the cyclic loading makes it of paramount importance that the baskets be inspected regularly in accordance with the manufacturer's instructions, usually at least every 12 months.

When inspecting a basket, key areas to consider are the perforations and welds, together with the dimensions of the shell and hoops (if fitted). Crack detection methods can be employed, such as dye penetrant (DP), magnetic particle (MPI), or ultrasonic methods for basket hoops. Magnetic techniques are only applicable to ferromagnetic materials, which excludes many stainless steels. Dye penetrant techniques can only detect cracks that have not become filled with molasses, so careful and complete steaming and drying of the basket is necessary before using DP techniques.

Controls on centrifugals are designed to provide safe operation and the sensors, interlocks, and control program should not be bypassed or modified without a full and complete understanding of the safety implications and the effects on the complete centrifugal. Reference to the manufacturer is strongly recommended. Given good maintenance a centrifugal will operate safely for many years, as have thousands of centrifugals in many sugar mills and refineries throughout the world.

REFERENCES

1. P. Honig, ed., *Principles of Sugar Technology,* Vol. III, Elsevier, Amsterdam, 1963.
2. L. Svarovsky, ed., *Solid–Liquid Separation,* 3rd ed., Butterworth, Woburn, MA, 1990.
3. J. M. Coulson, J. F. Richardson, J. R. Backhurst, and J. H. Harker, *Chemical Engineering,* Vol. 2, 4th ed., Pergamon Press, Elmsford, NY, 1991.
4. D. Tabor, *Gases, Liquids, and Solids,* Penguin Library of Physical Sciences, New York, 1969.
5. P. Franzen, *Proc. 28th Conf. Austral. Soc. Sugar Beet Technol.,* New Orleans, LA, Mar. 1995.
6. E. Hugot, *Handbook of Cane Sugar Engineering,* 3rd ed., Elsevier, New York, 1986.
7. J. C. P. Chen and C. C. Chou, *Cane Sugar Handbook,* 12th ed., Wiley, 1993, Chap. 5.
8. H. Eichhorn, *Zuckerindustrie,* 3, 4, and 7, 1994.
9. P. D. Thompson and G. C. Grimwood, *Int. Sugar J.,* Dec. 1996.
10. P. W. van der Poel, J. L. M. Struijs, J. P. M. Vriends, and A. A. W. Marijnissen, *Int. Sugar J.,* 89(1060), 1987.
11. B. St. C. Moor and M. S. Greenfield, *Proc. S. Afr. Sugar Technol. Assoc.,* June 1998.
12. P. G. Atherton et al., *Proc. Austral. Soc. Cane Sugar Technol.,* 1992, pp. 205–212.
13. W. Hunter, *High Viscosity Liquid Separation in the Sugar Industry,* Tate & Lyle Refineries, Greenock, Scotland, 1978.
14. P. Franzen, *Int. Sugar J.,* 93(1116), 1991.

15. C. R. Greig, E. T. White, and C. R. Murry, *Proc. Austral. Soc. Sugar Cane Technol.*, 1992, pp. 289–295.
16. G. Trojan, *Int. Sugar J.*, 76:99–102, 1974.
17. P. Sahadeo, *Proc. S. Afr. Sugar Technol. Assoc.*, June 1992.
18. J. C. P. Chen and C. C. Chou, *Cane Sugar Handbook,* 12th ed., Wiley, New York, 1993, Chap. 8.
19. L. M. S. A. Jullienne, *Proc. S. Afr. Sugar Technol. Assoc.*, June 1989.
20. T. D. Milner, *Proc. Hawaii. Sugar Technol. Conf.*, 1991.
21. E. D. Bosse, *Proc. Am. Soc. Sugar Beet Technol.*, Anaheim, CA, Mar. 3–6, 1993.
22. P. D. Thompson and G. C. Grimwood, *Zuckerindustrie,* 122(10):777–780, 1997.
23. L. K. Kirby, C. R. Greig, P. G. Atherton, E. T. White, and C. R. Murry, *Int. Sugar J.*, 92(1104), 1990.
24. C. R. Grieg, A. L. Schinkel, R. L. Watts, T. H. Dalrymple, and P. D. Scroope, Paper 667, *Proc. Sugar Ind. Technol. Conf.*, 1994.

CHAPTER 15

Refined Sugar Drying, Conditioning and Storage*

INTRODUCTION

The final unit operation in the production of refined sugar is drying and conditioning. Thus as a last opportunity to enhance the quality of the product, one would expect this process to be well understood and optimized. Unfortunately, the opposite is more often the case—familiarity with the standard drying equipment and the apparent simplicity of the process have often conspired to exclude drying from the sugar technology spotlight. In 1952, Crawford [12] described sugar drying theory and practice as "a reasonably confused conglomeration of ideas and methods which suggests that the basic facts about drying and cooling sugar are not clearly understood." In 1957, Lyle [30] lamented that drying and cooling "are among the least satisfactory of our processes."

Much work has been done since then (as the list of references at the end of the chapter testifies), including a seminar on the subject of drying, conditioning, and storage at the 1960 meeting of the Sugar Industry Technologists (SIT) [51]. Recent years have seen drying receive some serious attention, most notably in Australia, while conditioning has undergone scrutiny in Australia and South Africa. However, that we yet have much to learn is perhaps demonstrated by two additional SIT seminars devoted to these topics in 1994 [52] and 1996 [53], and by Tait's [54] exasperated reference to sugar conditioning in 1996 as a "mystic art."

The field of refined sugar drying, conditioning, and storage is therefore one in which sound theoretical foundations prop up a range of postulations and speculations. It is one in which operational decisions are still often made without clear understanding, and new equipment is still frequently "designed" by rule of thumb. In this chapter we set out to negotiate this minefield by providing an overview of the theory of drying and conditioning, considering some of the practical aspects of plant design and operation, touching on some of the important issues in white sugar storage, and describing the most commonly used types of equipment.

* By Dave Meadows.

15.1 GENERAL CONCEPTS

When sugar is discharged from the centrifugals it contains between 0.6 and 1.5% moisture [30], which exists in two forms:

1. *Inherent moisture* [42]: water occluded within the crystal structure. It is uncertain whether this water is in the form of occluded liquor or chemically bound as sucrose hydrates or hydrated polysaccharides forming part of the sucrose complex. There is to date no evidence of the migration of this moisture to the crystal surface, so as far as drying processes are concerned, it may be considered to be inert.
2. *Surface moisture:* water on the crystal surface, which exists as a saturated (at about 75 Brix) or even initially undersaturated syrup film of a purity less than or equal to that of the crystal. It is this moisture that participates in drying processes.

The moisture content of the sugar must be reduced for two main reasons: (1) to improve the handling characteristics of the sugar (to ensure that it becomes and remains free-flowing) and in particular to reduce its propensity to cake (set or harden) in storage or transit (discussed in more detail later in the chapter) and (2) to reduce the likelihood of microbiological or chemical degradation (loss of sucrose or color formation), which also affect product quality. As a guideline in preventing such deterioration, sugar should be dried to below a threshold moisture as determined by a ratio known as the *safety factor* (f), where

$$f = \frac{\text{moisture (\%)}}{100 - \text{polarization}}$$

Hugot [27] asserts that if f is greater than 0.3, the sugar will deteriorate rapidly. However, there is no general agreement on the safe limiting value of f. In India, a value of 0.22 is used for white or refined sugars, with 0.20 recommended for raws. South African practice is to use 0.23 for all grades. Hugot suggests that sugar be kept below 0.25.

A modification of the safety factor known as the *dilution indicator* (DI) is in general use in Australia:

$$\text{DI} = \frac{100 \times \text{moisture (\%)}}{100 - [\text{polarization} + \text{moisture (\%)}]}$$

This is, in effect, an expression of the moisture percent nonpolization, and it is considered that a DI value greater than 50 (which corresponds to a safety factor of 0.33) indicates considerable risk of deterioration [27].

White or refined sugar is typically dried in conventional driers to a moisture content (as determined by oven drying) of between 0.02 and 0.03%. This is normally accompanied by some cooling of the sugar, and further cooling may be achieved by the use of dedicated coolers.

15.2 DRYING

15.2.1 Drying and Cooling Equipment

a. Rotary Cascade Driers and Coolers. The workhorse of refined sugar drying has certainly been the rotary cascade drier, historically (if misleadingly) referred to as the *granulator* (Fig. 15.1). This device comprises a nearly horizontal rotating cylindrical drum run-

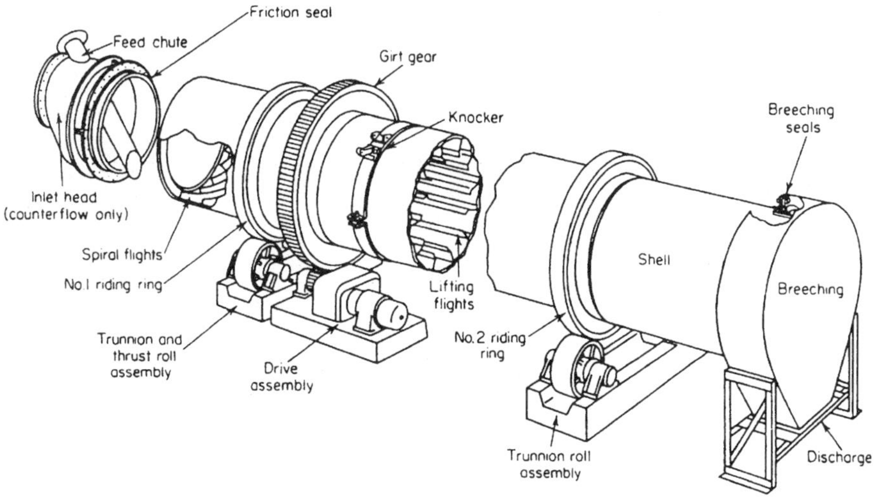

FIGURE 15.1 Typical rotary cascade drier. (From Ref. 38.)

ning on steel rollers or cradles. The drum is driven by a gear ring or chain sprocket on its outside. The rollers run on two path rings fastened to the outside of the drum. An alternative drive system drives the drum directly via rubber wheels on which the drum rests. In either system, thrust rollers are provided to prevent axial movement. The inside surface of the drum is fitted with flights or lifters that fill with sugar, lift it, and pour it out as the drum rotates (Fig. 15.2). As the drum is sloped toward the discharge end, the sugar moves in progressive lifts and falls toward the discharge. Air is blown or drawn axially through the drum, passing through the falling curtain of sugar to achieve the drying. Countercurrent movement of air and sugar is most common, but co-current operation is practiced in some

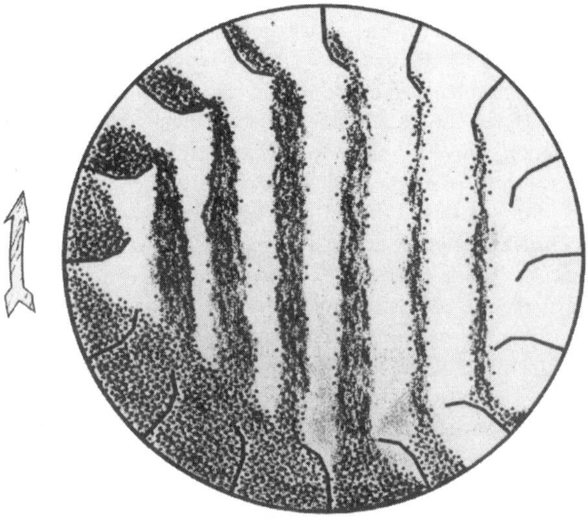

FIGURE 15.2 Cascade drier operation.

cases. The standard arrangement has the fan after the drier, which has the advantage of maintaining the drier slightly below atmospheric pressure, so that sugar dust is not expelled through any of the seals (an environmental nuisance and explosive hazard). Where the air is to be heated, a heater may be installed in the duct supplying air to the drier, or in the stationary hood at the end of the drum.

Early drier designs achieved their requirement for heating by means of heating jackets on the outside of the drums. These had to be supplied by steam and have condensate removed via glanded rotary valves using flexible piping. These were complex, maintenance intensive, leaked, and were generally unsatisfactory [30]. A later variation was the use of a large-diameter steam-heated pipe along the axis of the drum. Apart from the disadvantage of sugar buildup on this pipe, it created a gap in the sugar curtain below it, equal to its width, through which air could pass without contacting sugar. The use of finned-tube steam radiators at the drier entrance has proved far more successful and is almost universal. Very recently, some installations have experimented with electrical heating elements in the airstream, but usually only as a "top-up" to a radiator. Electrical elements do present the risk of very high-temperature surfaces providing a source of ignition to sugar dust.

Designs of lifters or flights vary widely, from individual scoops to axial plates several meters in length, usually with sawtooth edges. The flights should be designed to carry all the sugar held up in the drier, preventing any sugar progressing by rolling along the floor of the drum without being lifted, a motion sometimes known as *kilning*. In addition, the spacing and design of the flights should be such that the curtain of falling sugar across the drum cross section is as nearly continuous as possible.

In his 1936 volume, Tromp [56] described the use of rubber hammers mounted on the periphery of the rotating drum to dislodge moist sugar that had caked inside the drum. These rotated with the drum or were operated by cams. More recent practice has dispensed with these. In general, caked sugar in the drum must be removed manually. Some units in which the slope of the drum is adjustable do, however, allow a slight reversal of the slope (toward the sugar feed end) to allow any caked sugar to be washed out into a collection system provided. However, caked sugar rarely happened in a well-designed granulator.

Control of the residence time of sugar in a cascade drier is achieved by adjusting the rotational speed or (if possible) the slope of the drum. Some driers are designed with adjustable slopes, either by raising/lowering one end or by mounting the entire drier on a seesaw frame [56]. Many modern drier installations make use of variable-speed drives, allowing the drum speed to be adjusted at will; older units usually require changing of the drive pulleys to effect a step change in speed.

The main disadvantage of cascade driers is the damage that may be done to crystals by the tumbling motion. There is undoubtedly some attrition, or at least scratching, of the crystal surfaces in this type of drier, causing some loss of the "sparkle" or "luster" of the crystals. A further cause of this is that in countercurrent flow, any fine dust coming off the sugar as it dries will be carried by the airstream to the wetter sugar, where it may adhere to the wet surfaces of the crystals.

Cascade driers may be used as sugar coolers by eliminating the air heater or even employing an air chiller. A very common arrangement is to use two rotary cascade units installed one above the other, with the sugar passing in series through first the top and then the bottom unit, the former acting as a drier (with heated air) and the latter as a cooler. Tromp [56] erroneously suggests that in a "good installation" the two units should share the drying duty equally. In fact, drying will certainly take place in both the dryer and the cooler, but due both to the higher initial sugar moisture and the higher temperature operation, the lion's share of the moisture removal will take place in the dryer. Some careful optimization *is* necessary, however, in setting the air temperatures and flow rates to obtain maximum drying and cooling.

A modification of the standard rotary cascade unit is available to allow operation as a combined drier–cooler. This involves the use of a draft tube along the axis of the drum, extending from the discharge end to the interface between the drying and cooling sections. Cool air is then supplied to the discharge end of the drum in the normal way, and this air travels the full length of the drum before being removed at the sugar feed end. Hot air is supplied along the draft tube, and the drying section therefore operates at an air temperature resulting from the mixing of the hot and cool airstreams. This system is energy efficient in that it reuses the heat released in the cooling stage to enhance drying.

An interesting variation on the cascade drier is the *multitube drier–cooler* (supplied by FCB), which consists of 12 minidrums, each with lifters and employing the cascade principle, arranged in two concentric rings of six each, set into a tubeplate at each end and rotating together. The sugar passes along the inner drying drums and then returns along the outer cooling drums before exiting the unit. The suppliers justify the added complexity on the basis of:

1. Reduced crystal damage and dust formation due to very short falls of the crystals in the minidrums
2. Low space requirement to achieve both drying and cooling
3. The use of countercurrent cooling and co-current drying in a single unit
4. Low drive power consumption due to the balanced load of the product around the axis of rotation (a standard cascade unit requires power input to lift the sugar up one side of the drum)
5. Reduced mainenance cost because of less accessory equipment (e.g., motors, dust collectors) used

b. Rotary Louver Driers and Coolers. The main competitor to the cascade drier for the last 50 or so years has been the rotary louver drier (Fig. 15.3). This comprises a cylindrical drum, as for the cascade unit, but containing an inner shell made up of overlapping tangential louvers. Each louver is connected to the outer drum by a radial baffle, thus dividing the annulus between inner and outer shells into individual air passages. Air entering these passages must pass out via the gaps between the louvers into the inner shell. The diameter

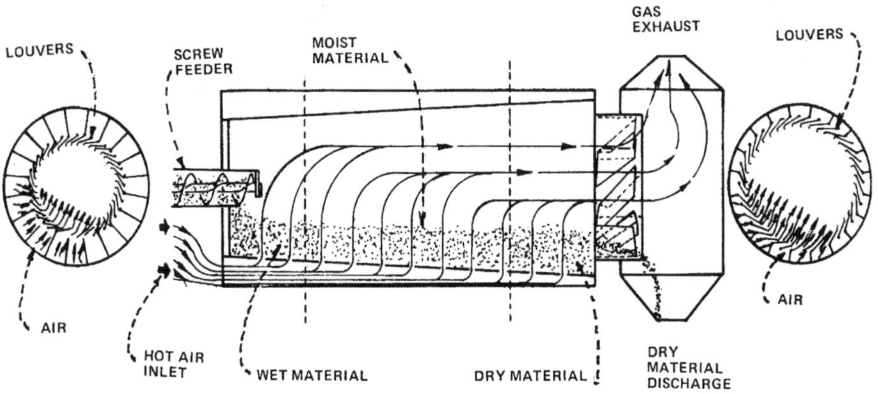

FIGURE 15.3 Rotary louver drier. (From Ref. 38.)

of the inner louver shell gradually increases toward the discharge end. This causes the bed of sugar that forms on the louvers to progress toward the discharge as it slides or rolls down the louvers with the drum rotation. There is therefore no need to slope the drum, and the outer drum is perfectly horizontal. This type of drier has most accurately been described as a "mechanically assisted semifluidized bed." Figure 15.4 is a cross section demonstrating the principle of operation.

Air is supplied by fans to the air channels at the end of the drum via a specially designed duct, which is sized to supply air only to those channels that lie under the bed of sugar. The diverging louver shell serves to maximize airflow through the feed end of the drier, where the sugar is wettest, and offers the greatest resistance to airflow, while permitting lower airflows to the discharge (dry sugar) end. The suppliers also claim that the air supply duct is designed so that "the greatest volume is presented to the centre of the bed where the bed is thickest and where the resistance is also greatest" [16].

Air removal from the drier is from a stationary hood at the discharge end via an exhaust fan, although some units have air removal via a pipe exiting the drier at the feed end, and in some cases, air is drawn from both ends. This type of drier therefore requires multiple fans, having one or more supplying the air (usually passing first through a radiator-type heater) in addition to the exhaust unit, and these must be balanced by setting dampers in the air ducting.

The main advantage of the louver drier is that it handles the product more gently than does the cascade drier, therefore reducing damage to the crystals. These units may therefore be expected to produce less dust and a more sparkling sugar. More intimate contact between air and sugar is also claimed, but when comparing with a well-designed and operated cascade unit, it is difficult to see why this should be so. Some classification of sugar crystals may occur in the bed of sugar, caused by the combined fluidization and rolling motion.

The retention time of sugar in the drier is adjusted by adjusting the bed depth. This is done by adjusting the weir arrangement at the discharge end of the drier. The weir

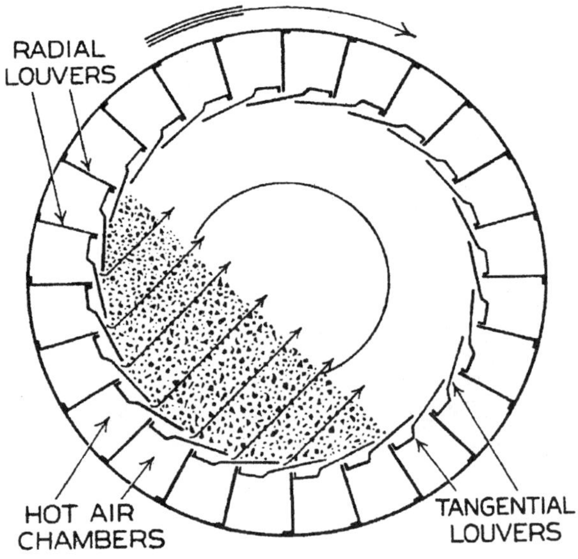

FIGURE 15.4 Louver drier operation. (From Ref. 16.)

comprises either a series of removable rings of varying inner diameter, or a series of vanes, the angle of which is adjustable. The rotational speed of the drum may also be varied (by variable-speed drive or by changing pulleys, as before). Care should be exercised in adjusting this parameter, as too high a speed will result in thinning of the bed at the feed end, causing insufficient airflow through the sugar at the discharge, while too slow a speed will cause the sugar to build up at the feed, preventing sufficient air from passing through the bed at this point.

The rotary louver may, of course, be used as a sugar cooler without modification, but its design also facilitates ease of use as a combined drier–cooler. This is because the drum may be divided into two sections along its length, with the air channels in the two sections separated. Heated air may then be supplied to the first section, to provide rapid drying, while ambient or chilled air is supplied to the latter section to cool the dried sugar.

c. Fluidized-Bed Driers and Coolers. A device that has found more recent acceptance for drying refined sugar is the fluidized-bed unit. Various designs are available, but all operate on the principle of a bed of sugar being fluidized by a stream of air. The simplest designs consist of a vessel containing a distributor plate above a plenum chamber into which the air is blown. The sugar is introduced onto the plate at one end and it forms a bed, which, being fluidized, flows across the plate to exit at the other end. Above the bed, the sides of the vessel may diverge to reduce the velocity and allow disengagement of fines. Whether or not this is the case, a dust collection system is necessary downstream of the unit. Figure 15.5 is a diagram of a simple unit installed at Gledhow factory, South Africa in 1977. This was operated successfully, but it proved necessary to install baffles in the unit to aid plug flow.

The key element in the design of these fluidized-bed units is the distributor plate, as it is this device which ensures even distribution of air across the bed and therefore even fluidization and proper flow patterns. Designs of distributor plates vary from fairly sophisticated bubble caps, which physically prevent weeping of sugar through the plate, to simple perforated plates, which rely on the velocity of the fluidizing air to prevent weeping. Some plates have apertures that converge in the direction of airflow, so that a crystal falling into an aperture will encounter a diverging opening, which will prevent blockage.

A problem encountered with earlier fluidized-bed units was that the minimum fluidization velocity for wet sugar is about an order of magnitude higher than that for dry

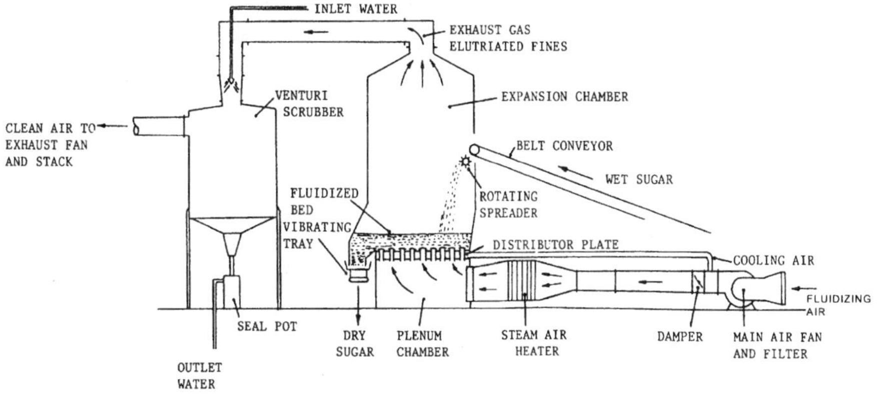

FIGURE 15.5 Simple fluidized-bed drier. (From Ref. 20.)

sugar. This meant that the bed would not fluidize evenly, with very poor fluidization at the feed end. It has therefore become standard for suppliers of these driers to compartmentalize the plenum chamber and sometimes the bed itself, using adjustable weirs that allow the sugar to overflow from one compartment to the next. This allows the air supply to each section of the bed to be tailored to the fluidization needs of the sugar in that section. Even then, the Stork fluidized-bed driers are provided with a rake system in the first two compartments to keep the sugar in motion "when fluidization of the sugar is not yet effective due to its high moisture content" [25]. Figure 15.6 is an example of a BMA unit with compartmentalized plenum chamber and bed.

The variation of minimum fluidization velocity with sugar moisture makes fluidized-bed drier operation sensitive to feed sugar moisture. This effect is reduced, but not eliminated, by the compartmentalization referred to above; clearly, a batch of excessively wet feed sugar would stop fluidization at the feed end unless accompanied by an increase in the

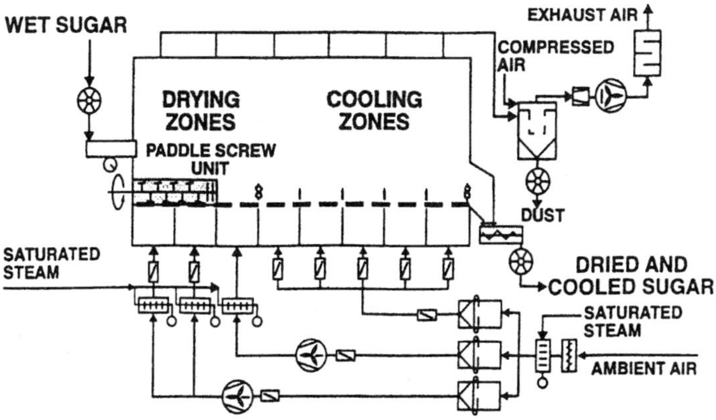

Schematic

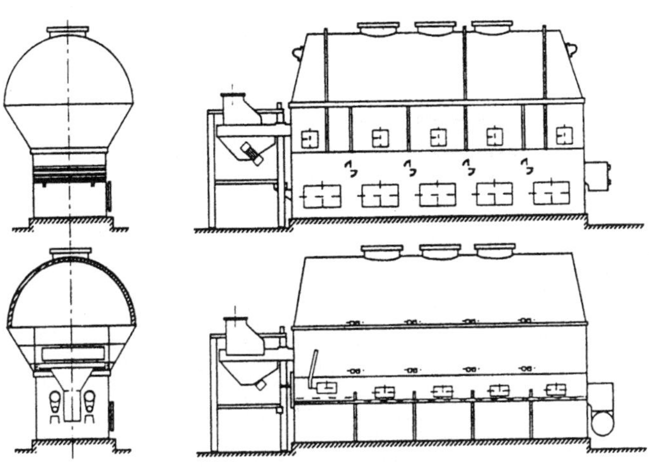

Front and side view

FIGURE 15.6 Modern compartmentalized fluidized-bed drier. (From Ref. 5.)

rate of fluidizing air. If operation is reasonably steady with respect to feed sugar moisture, this problem will not occur, but where large fluctuations in moisture are anticipated, it may be necessary to recirculate some of the dry sugar to the inlet to counter the effect of such fluctuations.

Various designs exist for breaking up lumps at the entrance to the drier, such as a grid of vibrating rods or a row of rotating paddles. Figure 15.7 shows such a paddle arrangement, as installed in BMA fluidized-bed driers. If the sugar is properly fluidized, crystal damage in the drier should be minimized, as the principle of fluidization is that each crystal is surrounded and supported by a layer of fluidizing air.

The compartmentalized fluidized-bed design lends itself very readily to use as a cooler or combination drier–cooler, as each of the plenum chamber compartments may be supplied with air at a different temperature using separate fans or dedicated heaters. BMA even employs a technique called *temperature swing,* in which the second compartment of a drier is supplied with hot and cold air alternately. This, it is claimed [4], provides rapid drying without heating the sugar excessively.

An alternative to the standard fluidized-bed drier is the *vibrating fluidized-bed* unit. This piece of equipment is similar to the ordinary fluidized bed in shape and appearance, but imparts additional motion to the sugar bed by vibrating either the entire unit or the plenum chamber and bed section. In some units, the distributor plate has a slight slope so that vibration-induced motion is toward the discharge. Although not the intended mode of operation, it is thus possible to operate the unit sub-fluidized, (i.e., with insufficient air to fluidize the sugar). Such operations are frequently encountered in India and Pakistan, where a thin layer of sugar moves across the distributor plate due to vibration only, and satisfactory results are obtained as long as the throughput is kept low (the layer is kept thin).

The advantage of the vibrating unit is that in imparting additional energy to the sugar crystals, fluidization is achieved more easily. By lowering the minimum fluidization velocity, the vibrating unit should cope more easily with wet feed sugar. Suppliers also claim that the vibration aids in turning over the bed and ensuring uniformity of bed temperature. Another small benefit is that any fine sugar that weeps through the distributor plate will be conveyed along the floor of the plenum chamber and may be removed from the end if the unit is sloped. The disadvantages are additional moving parts and increased power input.

Another possibility for cooling sugar in a fluidized bed is the use of cooling surfaces within the bed (e.g., hollow plates or tubes). These are usually supplied with cool or chilled

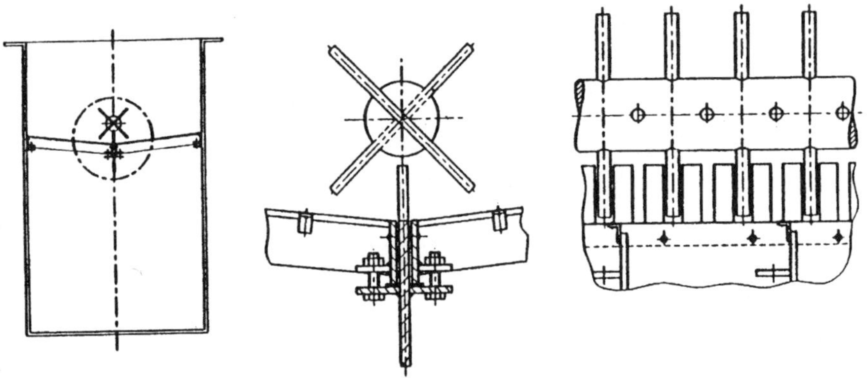

FIGURE 15.7 Paddle-type lump disintegrator. (From Ref. 4.)

water, thereby both increasing the temperature driving force and providing a heat sink. Care should be taken to ensure that the water used is not below the dew-point temperature of the fluidizing air, as moisture could condense on the cold surfaces and dampen the sugar.

d. Other Driers and Coolers. *Tray driers* are also in use on refined sugar, particularly in Europe. These have horizontal disk trays mounted on a vertical shaft in a cylindrical vessel supplied with heated air. Two major types exist. The first, and most common, has slots in the trays through which the sugar falls from tray to tray (a drop of about 150 mm). The rotating shaft (about 2 to 4 rpm) has scrapers that scrape the sugar toward the slots. In the second type, the shaft and disks rotate rapidly enough that the sugar is thrown outward and off the edge of the trays. The falling sugar is then caught in cones which deposit it near the axis of the tray beneath. Suppliers, particularly of the first type, claim reduced crystal damage due to the short falling distances between trays. In addition, retention times are high (up to 45 min) and drying is slow, which can be beneficial, as discussed later. These units may, of course, be used as coolers. *Plate hopper coolers* are mass-flow hoppers containing vertical water-cooled plates, past which the sugar flows by gravity. *Hollow flight screw coolers* (the most well-known being the Holoflite) are screw conveyors that cool the sugar by means of water passing through the hollow flights.

15.2.2 Drying Theory and Modelling

a. Drying Theory. Classical drying theory asserts that the drying of a solid takes place in two stages:

1. *Constant rate stage.* The surface of each particle is covered by a continuous film of moisture, and subject to environmental conditions remaining unchanged, evaporation proceeds at a constant rate.
2. *Falling rate stage.* The film of moisture becomes depleted and evaporation slows due to reduced surface area. Replenishment of the film occurs as moisture moves by vapor diffusion or capillary action from the interior of the particle to the surface, and this transport becomes rate limiting.

Figure 15.8 illustrates these two phases graphically.

Sugar, as a soluble material, is interesting in that it appears to follow the classical theory, but the mechanisms involved are quite different. The stages in the drying of sugar are:

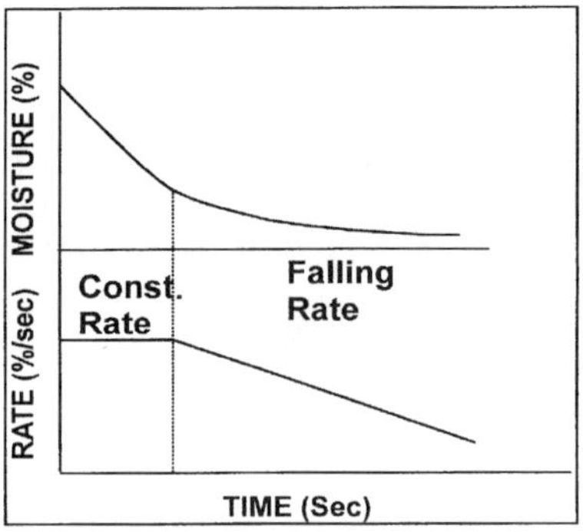

FIGURE 15.8 Classical drying phases.

1. *Pseudoconstant rate stage.* The film of syrup on the surface of the crystals is undersaturated and moisture evaporates freely at a nearly constant rate, governed by evaporative considerations.
2. *Falling rate stage.* The surface film has become sufficiently concentrated that evaporation is slowed significantly by the influence of the solute (Raoult's law). This effect is exacerbated by the fact that crystallization of sucrose then begins from the supersaturated film. Due to the speed with which moisture is removed from the film, a concentration gradient is established across the film and the crystallization process produces amorphous solid sugar in and on the surface of the film, which interferes with the evaporative process by trapping moisture. The higher the purity of the sugar, the faster the crystallization rate and therefore the greater the effect of this phenomenon.

Moisture available on the surface of the crystal (readily evaporated) is known as *free moisture,* while moisture that is associated with or trapped under amorphous sugar is known as *bound moisture* [42]. Conventional drying is concerned with removing free moisture plus a little of the bound moisture. From the preceding it is clear that three mechanisms are involved in the drying process:

1. *Evaporation* of moisture from the syrup film at a rate governed by the vapor pressure difference between the film and the bulk airstream.
2. *Diffusion* of water molecules through the surface film and across solid amorphous sugar barriers on the crystal surface, driven by concentration gradient.
3. *Crystallization* of sucrose molecules in the film onto the crystal surface and as amorphous sugar, effectively diluting the film and thereby making moisture available for diffusion and evaporation.

The conventional drying process is predominantly evaporation rate controlled, with the other two mechanisms coming into play toward the end of the drying operation, and then only in the case of higher-purity sugars. Control of drying is therefore the control of the syrup film vapor pressure (a function of sugar temperature and film concentration) relative to the partial pressure of water in the drying air (a function of air temperature and moisture content).

b. Modeling of Sugar Driers. Most of the fundamental modeling of the sugar drying process has been done in Australia. The foundation was laid by Crawford [12], who used psychrometric equations to model the simultaneous heat and mass transfer taking place in a sugar drier. He combined his results in a graphical relationship for raw sugar, relating outlet air temperature, inlet air humidity, air/sugar ratio, and overall moisture loss.

A simple model of the drying process was developed by Pope et al. [40], making some fairly broad assumptions and assuming an overall heat transfer coefficient of 30 $W/m^2 \cdot K$. However, Schinkel and Tait [44] developed the first complete model of the drying process based on the conditions in the film on the surface of each crystal. This model was further detailed by Tait et al. [55], who divided the drier up into incremental sections. Solution of the mass and energy balance equations was done iteratively for each section based on conditions in the adjacent sections, using the following relationships:

1. Sensible heat transfer between sugar and air:

$$Q_s = UA(T_s - T_a)$$

where Q_s is the sensible heat transferred (W), U the convective heat transfer coefficient

($W/m^2 \cdot K$), A the exposed surface area of the crystals (m^2), T_s the sugar temperature (°C), and T_a the air temperature (°C).

2. Latent heat absorbed by evaporation:

$$Q_e = EL$$

where Q_e is the latent heat absorbed (W), E the evaporation rate (kg/s), and L the latent heat of vaporization (J/kg).

3. Mass transfer rate by evaporation:

$$E = Mk_g A\ (p_s - p_a)$$

where E is the evaporation rate (kg/s), M the molar mass of water (18 kg/kmol), k_g the mass transfer coefficient ($kmol/s \cdot m^2 \cdot Pa$), p_s the vapor pressure of water in the surface film (Pa), and p_a the partial pressure of water in the air (Pa).

4. Crystallization rate from the surface film:

$$X = G \cdot SA \cdot \rho$$

where X is the crystallization rate (kg/s), G the crystal growth rate (m/s), SA the crystal surface area (m^2) (note that this is total rather than exposed surface area), and ρ the density of crystal sugar (1586 kg/m^3). The crystal growth rate (G) is a function of the sucrose/impurity/water ratios of the surface film and depends on empirically determined solubility and rate coefficients. The correlations used by Tait et al. [55] are given in the paper, but various alternatives are available in the vast array of material published on sugar crystallization.

It is important to note two assumptions made in this model:

1. There are no concentration or temperature gradients across the sugar crystals or the surface film. This means that no diffusion term is used.
2. The crystallization of sucrose is assumed to take place only on the crystal surface. The possibility of amorphous sugar formation in or on the film is not considered.

The authors fitted this model to plant data for raw and refined sugar drying in rotary cascade driers, using unheated countercurrent air, and deduced the following:

1. Close agreement with measured data can be obtained using a heat transfer coefficient of 300 $W/m^2 \cdot K$ and a mass transfer coefficient of 4×10^{-9} $kmol/s \cdot m^2 \cdot Pa$. This mass transfer coefficient seems reasonable, but the heat transfer coefficient appears orders of magnitude too high, based on work done elsewhere.
2. Only marginal improvement in drying may be obtained by increasing the exposed surface area of the crystals (improving the flight operation in the drier). This conclusion should be treated with some caution—it defies both operational and theoretical logic.
3. Overcooling of the sugar is deleterious to drying, as it removes the "energy necessary to evaporate" the moisture. In fact, what it does is reduce the vapor pressure of the water in the surface film. This is a valuable observation, as it means that when drying with cool air, increasing the air/sugar ratio may in fact retard drying. Also, in general, wetter sugar must be fed hotter to the drier for adequate drying to take place (and cooling of sugar during conveying before the drier is to be avoided).

The concept of operating plots is introduced for driers, showing an acceptable operating region on a plot of feed sugar moisture versus feed sugar temperature; an example is given in Figure 15.9. This model (which has become known as the *Schinkel model*) was improved by Shardlow et al. [46] by allowing a concentration gradient across the surface film and incorporating a diffusion term based on Fick's law of molecular diffusion. This is used to provide an adjusted water concentration in the film (at the film surface) using

$$\frac{W}{W_0} = \frac{p_a}{p_s} + \left(1 - \frac{p_a}{p_s}\right) \exp\,(-Dt)$$

where W is the time-dependent water concentration in the film (kg/m^3), W_0 the initial water concentration in the film (kg/m^3), D the molecular diffusion coefficient of water in the film (s^{-1}), and t the time (s). p_a and p_s are as before.

The Schinkel model was further altered by reducing the film crystallization rate to 40% of the calculated value in line with studies on stagnant molasses films carried out at the Australian Sugar Research Institute. In fitting the model to raw sugar drier data, a heat transfer coefficient (HTC) of 3.6 W/m^2 · K, a mass transfer coefficient (MTC) of 4.4×10^{-10} kmol/s · m^2 · Pa, and a diffusion coefficient of 0.001 s^{-1} were obtained. The HTC and MTC values agree very well with those obtained in test work on refined sugar in South Africa (not published).

The model above also incorporated a transport model for a countercurrent rotary cascade drier, drawing on earlier work done by Steindl et al. [50]. This was a detailed geometric analysis of crystal travel, incorporating factors such as air drag and particle bounce, and was validated on a pilot and factory scale. The model also allowed prediction of the point at which kilning (sugar rolling along the bottom of the drum without being lifted) would begin. The Shardlow model was proven in a most courageous way, by using it as the basis of a feedforward, PLC-based control system for a raw sugar drier at Proserpine factory in Australia.

Some published heat transfer coefficient data for dedicated sugar coolers are:

1. Petri [39] found, for a fluidized-bed cooler containing water-cooled plates, an average HTC of 142 W/m^2 · K (based on plate surface area), in an experimental range of 82 to 160 W/m^2 · K. These values have been confirmed in South Africa on a pilot unit (1 tonne/h) using water-cooled tubes.

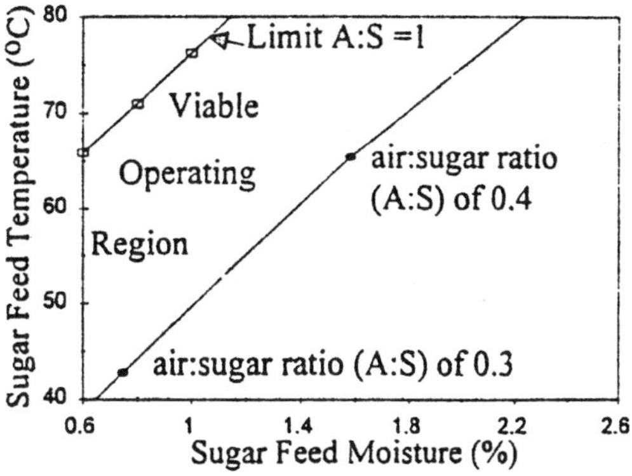

FIGURE 15.9 Example of a drier operating plot. (From Ref. 55.)

2. Manley [51] reports an HTC of 62 $W/m^2 \cdot K$ for a water-cooled hollow-flight screw. For a similar unit, McGimpsey [51] recorded a value of 75 $W/m^2 \cdot K$.

15.2.3 Drier Design

Without having to use a detailed model of the drying process, some generalized design equations for a rotary cascade drier are available. The holdup of a rotary drier varies with the feed rate, the number of flights, the shell diameter, and the air rate. Friedman and Marshall [21] give a correlation for holdup as a percentage of drier volume (for zero air-flow):

$$H_v = \frac{25.7F}{Sn^{0.9}D}$$

where H_v is the product volume % drier volume, F the specific feed rate ($m^3/s \cdot m^2$ drier cross section), S the slope (m/m), n the rotational speed (s^{-1}), and D the diameter (m). With airflow, the figure above will alter due to air drag (increase for countercurrent and decrease for co-current). A typical value of 3% is quoted for H_v when working with a drier slope of 0.1 m/m.

In calculating the heat transfer in a cascade drier, difficulty is often encountered in estimating the exposed particle surface area for heat transfer. The overall surface area of granulated sugar is in the range 7.7 to 9.4 m^2/kg [8]. A common approach, described by Miller et al. [33] and Sharples et al. [47], is to express the heat transfer as

$$Q = UaV(T_s - T_a)$$

where a is the exposed surface area per unit drier volume (m^2/m^3) and V the drier volume (m^3). Q, U, T_s and T_a are as before.

A correlation is then used for the term Ua:

$$Ua = \frac{20(N - 1)G^{0.67}}{D}$$

where N is the number of flights and G the specific airflow ($kg/s \cdot m^2$ drier cross section). Ua and D are as before. This is applicable for airflows in the range 0.37 to 1.86 $kg/s \cdot m^2$ and was derived using pilot driers (200 and 300 mm in diameter) with rotational speeds of 5 to 35 rpm and numbers of flights from 6 to 16. An alternative estimation of Ua is that derived by Saeman [43], who, on trials with sand in a pilot drier, found Ua to be a function of solids holdup. His graphical relationship is given in Figure 15.10.

The design of the flights for a rotary cascade drier is a key element of the design, as it is these that determine the quality of contact that is achieved between the air and the sugar. Hodgson and Keast [24] wrote the definitive paper on the subject, in which they used a geometric approach to estimate the capacity of flights. They found that two-sided flights were inherently superior to single-sided flights (see Fig. 15.11), with an optimum included angle of 105°, and that the use of serrations improved the continuity of the sugar curtain across the full width of the drier. Further norms for flight design [38] are that flight width should be about $\frac{1}{12}$ the drum diameter and the number of flights should be equal to six to nine times the drum diameter (in meters).

Table 15.1 provides rule-of-thumb ranges for rotary cascade drier design and operation. Little design information has been published on rotary louver driers. They would typically

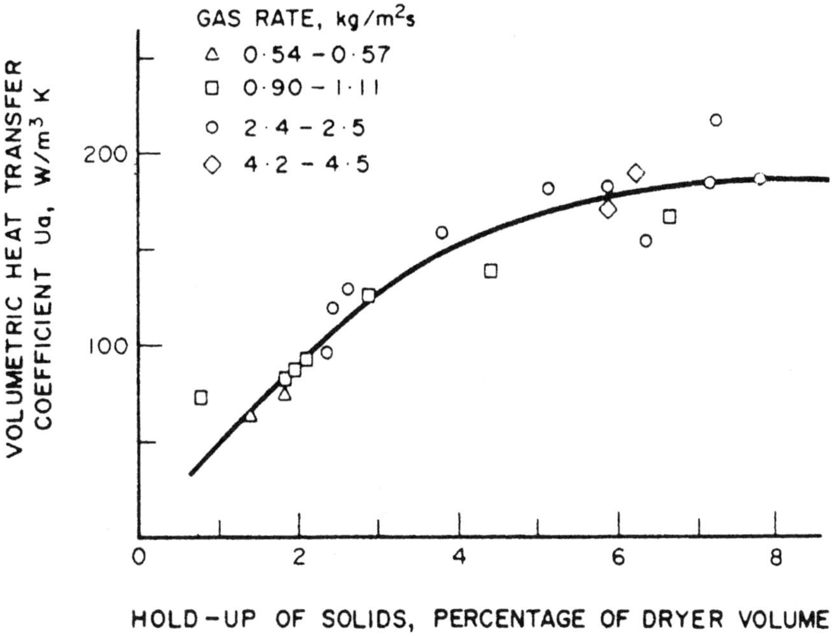

FIGURE 15.10 Correlation of volumetric HTC with holdup. (From Ref. 43.)

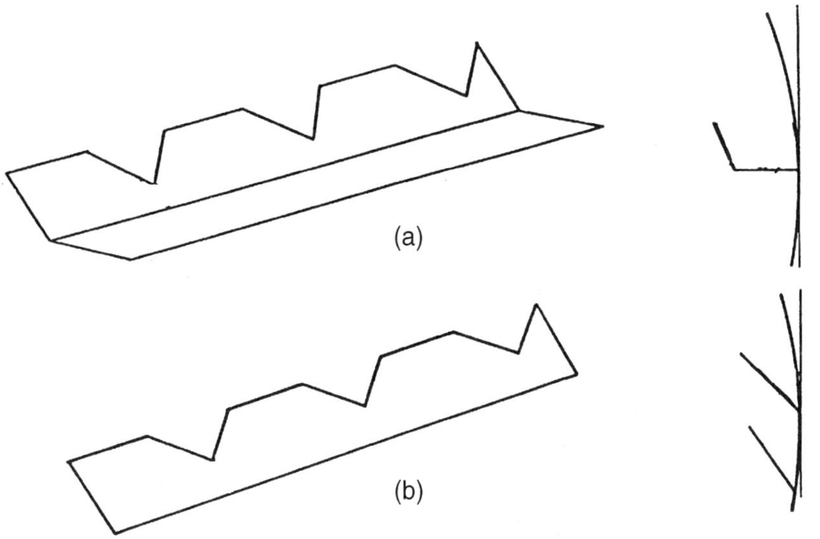

FIGURE 15.11 Typical cascade drier flights: (*a*) double-sided; (*b*) single-sided. (From Ref. 24.)

TABLE 15.1 Rotary Cascade Drier Design Parameters

Specific volume (m^3/tonne · h)	1.5
Holdup (m^3 sugar/m^3 drier volume)	<0.08
Slope (m/m)	0.02–0.08
Rotational speed (rpm)	3–6
Retention time (min)	5–10
Air velocity (m/s)	<1.25
Air/sugar mass ratio (kg/kg)	0.33–1.0
Specific drive power consumption (kWh/tonne · h)	0.8–1.0
Specific fan power consumption (kWh/tonne · h)	0.3–1.0

have a specific volume requirement in the range 1.0 to 1.25 m^3/(tonne · h), an air/sugar ratio toward the top end of the cascade range, and rotational speed and retention time in a range similar to the above. The Link-Belt Company (supplier of the original Roto-Louvre drier) recommended as operating norms a rotational speed between 1 and 4 rpm and a superficial velocity of air up through the sugar bed of between 0.6 and 1.5 m/s.

Table 15.2 is a set of design information for fluidized-bed driers, as published by suppliers. Twaite and Randall [58] report the use of fluidized-bed coolers at British Sugar and provide a performance comparison among five units. They report a minimum fluidization velocity of 0.5 to 0.6 m/s and recommend a freeboard velocity no higher than 0.4 m/s. They also achieved success with a vibrating fluidized-bed unit on a sticky speciality sugar.

15.2.4 Drier Operation

In all drier types, performance depends on:

- Airflow rate (or more accurately, air/sugar ratio), temperature, and relative humidity (and absolute moisture content, but dehumidification is seldom practiced in conventional drying)
- Sugar quality (including moisture content) and temperature
- Retention time

Retention time may be varied in different ways, depending on the drier type:

- In fluidized-bed units, by adjusting the bed height
- In rotary louver units by adjusting the speed or adjusting the bed depth (modifying the outlet ring weir)
- In cascade units by adjusting the speed or the slope, where provided for

TABLE 15.2 Fluidized Bed Drier Design Parameters

Required screen area (m^2/tonne · h)	0.5
Specific power consumption (kWh/tonne · h)	3–5
Retention time (min)	10
Airflow (N · m^3/sugar)	2500
Air velocity (m/s)	1.1–1.3
Bed height (mm)	300–400
Air pressure drop (kPa)	3.5

Since the slope dictates the number of times each sugar crystal will be lifted and dropped through the airstream before exiting the drum, this parameter determines the *effective* residence time (the period in contact with the air). Altering the speed, on the other hand, changes only the period spent by the sugar in the flights or lifters (not the period in contact with the air). This may be effective in increasing the throughput of the drier (reducing sugar holdup) but will have little effect on drying effectiveness. In general:

1. The speed of a cascade drier should be adjusted to alter the maximum throughput of the unit (quantity of sugar that can be transported from feed to discharge without kilning).
2. The slope of a cascade drier should be adjusted to alter the achievable moisture removal, or the final moisture of the product sugar.

Table 15.3 illustrates the sensitivity of the drying operation to various operating variables. The table was generated using a model that simulates the behavior of a cascade drier in countercurrent mode. The base case is of sugar at 50°C and 1% moisture, dried using air at 50°C and an air/sugar ratio of 0.6, to produce product sugar at 38.7°C and 0.12% moisture. Clearly, the absolute results are specific to this application, but the relative effects of the operating variables are informative:

- Increased residence time has the expected effect in reducing product sugar moisture, but if the drier is already operating at its design capacity, this adjustment is not available to an operator.
- Large changes in airflow rate (air/sugar ratio) are required to achieve relatively small improvements in product moisture.
- Increases in either sugar or air temperature have the greatest potential effect on product moisture. In the countercurrent arrangement, hotter sugar achieves this with a smaller undesirable increase in product sugar temperature than that of hotter air.

These results illustrate the fact that drying is energy intensive and that increasing the energy supplied to the process via the sugar or the air will have a positive effect on evaporation. Of these, supplying energy via the sugar is the more efficient, as it is not dependent on heat transfer efficiency. For this reason, feeding cool, wet sugar to the drier is a recipe for disaster, regardless of the other operating variables. Any significant increase in feed sugar moisture should, as far as possible, be balanced by an increase in its temperature.

Further evidence of the energy-dependent nature of drying is given by the following illustration. If the previous base case was changed so that unheated air at 30°C was used, the product moisture would rise to 0.25%. If, then, an attempt was made to counter this

TABLE 15.3 Sensitivity of Drying Operation to Operating Variables

		Change in Product	
	Percent Increase	Moisture (%)	Temperature (°C)
Residence time	10	−10	No change
Airflow	30	−11	+1.7
Air temperature	20	−27	+4.3
Sugar temperature	10	−25	+2.8

by increasing the airflow by 30%, the surprising result would be that the product moisture would *increase* to 0.26%! Despite the fact that the increase in airflow reduces the moisture content in the air along the drier, thereby enhancing evaporation, this is outweighed by the fact that it also cools the sugar, carrying away the energy vital to the drying process.

Schwer [45] recommends an air temperature around 65°C, while Farag [19] suggests as high as 100°C. Caution is advised, however, as excessively high drying temperatures will tend to dry the sugar too fast, causing the crystallization of amorphous sugar on the surface of the syrup film. This "case hardening" will retard further drying and conditioning and entrap moisture that may cause caking problems at a later stage. High-temperature drying also results in higher dust levels, from the overdried sugar as well as the amorphous surface sugar itself, as it is abraded from the crystals.

Optimum operation of driers is a complicated business—a far cry from the traditional "let the drier get on with it" approach. Figure 15.12 is a typical set of moisture and temperature profiles along the length of a countercurrent cascade drier. If we accept that our objective in drying is dry, cool sugar, it may be surprising to see that in the latter part of the drier, we actually heat the sugar up after it has cooled evaporatively. This is necessary to enhance evaporation sufficiently to obtain our target sugar moisture. However, the operating parameters for this example are far from optimized. This drier might benefit from hotter sugar and cooler air, or a different air/sugar ratio, or even possibly co-current operation. For a particular drier, therefore, careful optimization of the operating variables has the potential to yield a drier, cooler sugar without equipment change.

A study of drier operation should also consider the preceding unit operations in the process. Farag [19] notes the importance of good centrifugal operation and remarks that it is more energy efficient to remove moisture by physical means in the centrifugals than by phase change in the driers. In general, inadequate drying of the sugar may be remedied by more centrifugal washing, not less, as a water (or pure sucrose solution) film is far easier to evaporate than a molasses film. Neilson and Blankenbach [36] stress how drying can be aided by boiling a better (more regular and conglomerate-free) crystal and suggest that pan automation and the installation of pan stirrers may be justified on the basis of savings on drying and conditioning equipment.

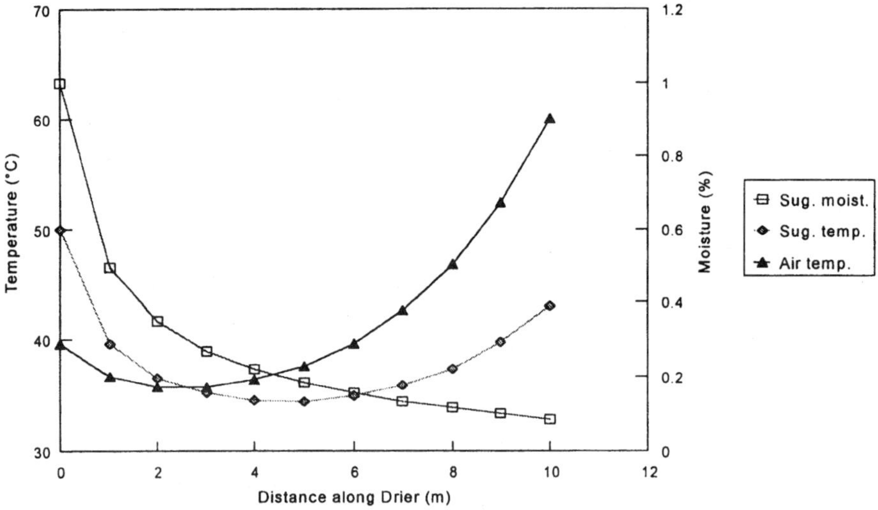

FIGURE 15.12 Typical countercurrent drier profiles.

a. Automation. Few sugar driers are highly automated. Adjustment of the airflow through a drier is typically done manually and infrequently. The standard level of automation is a control loop regulating the steam flow to the air heater radiator, the set point of which is set manually. Figure 15.13 shows a BMA fluidized-bed system at the next level of complexity, in which the air temperature is regulated on the basis of the sugar temperature.

The moisture of the outlet sugar is often measured on-line, by temperature (a change in moisture is often accompanied by a change in temperature), conductivity, or near-infrared (NIR) reflectance. Baird and Beatts [2] report success with the latter, typically obtaining correlation coefficients greater than 0.95 on dry sugar, while using conductivity on the wet feed sugar. Wright and Johns [59] used the traditional operator's method of watching for wet sugar (i.e., watching for a change in the dust cloud at the end of the drier) as a basis to develop a more sophisticated measuring system. They used a light attenuation meter to sense the level of the dust in the drier discharge hood as a measure of product sugar moisture. In all of these cases, however, control intervention was manual.

The Australian method of moisture and temperature control is to add water to the sugar before the drier, to make use of the evaporative cooling in the drier (Australian practice is not to heat the air). Shardlow et al. [46] have fully automated a drier on this basis, using measurements of sugar temperature and moisture (conductivity and NIR) into and out of the drier, as well as drier speed, sugar rate, and air temperature, flow, and humidity. These are fed into a drier model (discussed previously) running on a programmable logic controller (PLC) and used to control the water addition rate to the feed sugar. Figure 15.14 is a diagram of their system, and Figure 15.15 is a sample of the quality of control. This is a bold step in the right direction with regard to the application of science to drier operation.

15.3 CONDITIONING

15.3.1 Sugar and Its Environment

a. Equilibrium Relative Humidity. After white sugar has been dried in a conventional drier to a moisture content of around 0.03% (as determined by oven drying), it still contains typically 0.05 to 0.08% moisture in total (as determined by Karl Fischer titration; see later),

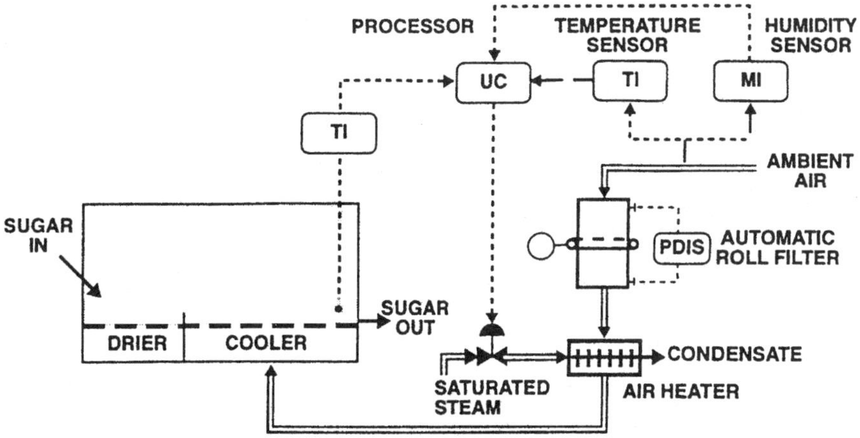

FIGURE 15.13 BMA fluidized-bed drier control system. (From Ref. 5.)

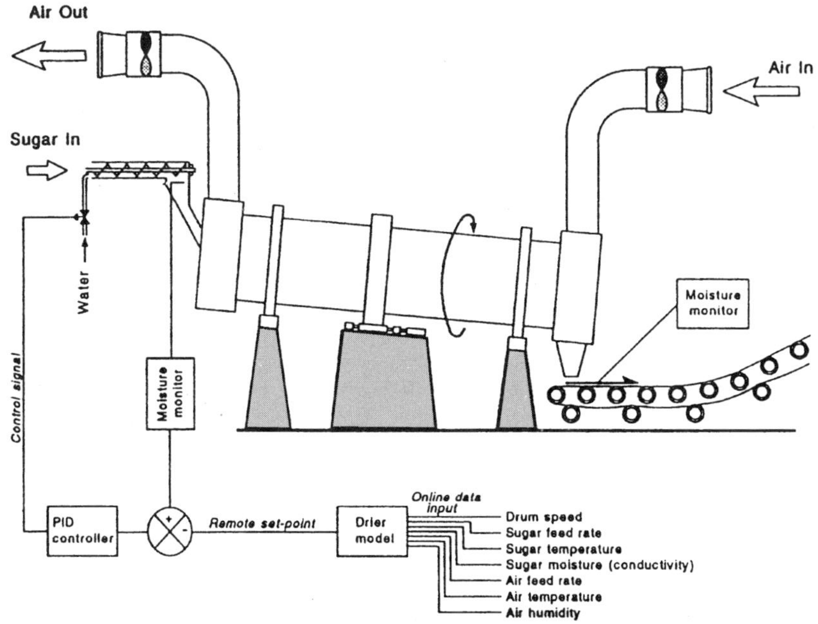

FIGURE 15.14 Complete feedforward drier control system. (From Ref. 46.)

most of which is present as bound moisture. It is the task of the conditioning process to remove as much as possible of this bound moisture, chiefly to prevent caking of the sugar in storage or transit. Note that moisture that is trapped between crystals during the formation of conglomerates is categorized as bound moisture. Despite the fact that it is no longer on the surface of the conglomerate crystal, it appears to play a part in caking and may therefore not be classed as inherent moisture.

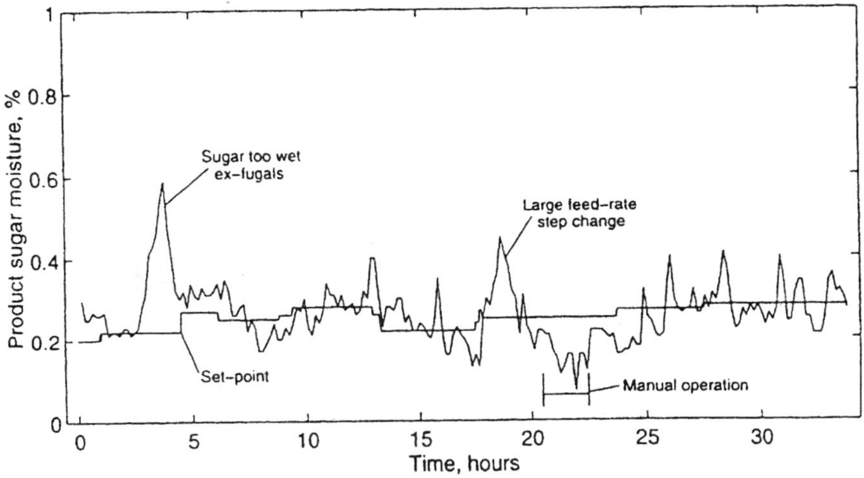

FIGURE 15.15 Sample controlled drier trend. (From Ref. 46.)

The question obviously arises: Why would "dry" sugar cake? Being an hygroscopic material, sugar will either absorb moisture from, or release it to the environment, depending on the composition of the crystal surface film (as glucose and fructose, for example, are more hygroscopic than sucrose) and the moisture content (relative humidity) of the surrounding air. Figure 15.16 shows the sorption isotherms for white sugar of different qualities; this demonstrates how increased surface film (smaller crystal size) increases the hygroscopicity of the sugar. Mikus and Budicek [32] recommend that to avoid hygroscopicity problems, sugar to be stored should have invert levels below 0.01%. The equilibrium relative humidity (ERH) of a particular sugar is the RH at which it will neither lose nor absorb moisture. The ERH of pure sucrose is 86%, which means that any sugar exposed to 86% RH will eventually dissolve. The ERH values of refined sugars usually lie between 65 and 75%, with 76% the upper limit (Lu et al. [29]).

Authors differ on what is a "safe" RH for sugar storage, but a common rule of thumb used by most refiners is that an RH below 60% is acceptable. The situation is complicated, however, by the fact that the ERH varies according to the moisture level of the sugar itself. Since impurities are concentrated in the surface film, drying of the crystal will further concentrate these impurities and therefore lower the ERH. What is disturbing is the rate at which refined sugar absorbs moisture in a high-humidity environment. Chapman [8] cites a moisture gain of 0.05% per hour, while South African tests have shown rates as high as 0.18% per hour. Some technologists believe that the ERH of refined sugar also depends on the crystal size and crystal distribution of the sugar.

b. Caking. Caking (or setting or hardening) is a phenomenon in which sugar crystals give up bound moisture, resulting in supersaturation at the crystal surface and consequent crystallization. At points of contact between crystals this surface crystallization causes intercrystalline bridging. The sugar then ceases to be free-flowing and is referred to as *caked.* Caking may be anything from soft, friable lumps, through surface crusting to rock-hard setting. A U.S. refinery at one stage purchased a coal crusher to allow reprocessing of badly caked sugar in the 5-lb bag, before finding solutions to their caking problems!

In practice, caking usually occurs due to a change in the relative humidity of the air in contact with the sugar, due commonly to temperature gradients. This may be illustrated by considering a warm mass of sugar cooling down at its boundaries. As the interstitial air

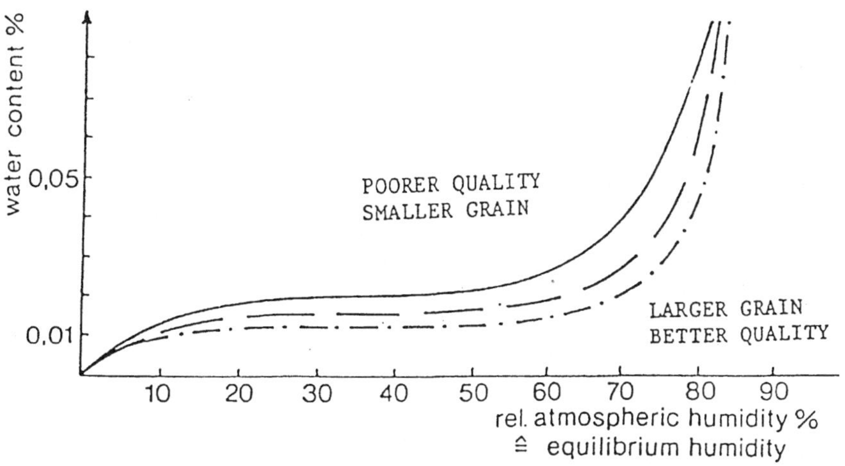

FIGURE 15.16 Sorption isotherms for white sugar. (From Ref. 53.)

cools down, its relative humidity rises beyond the equilibrium relative humidity of the sugar, and the sugar absorbs moisture. This absorption lowers the partial pressure of water at the boundaries, and moisture migrates to these regions from the warm center. Sugar at the boundaries thus undergoes surface dissolution as it continues to absorb moisture. It therefore either becomes progressively damper, or, more typically, warms up again as ambient conditions change, gives up surface moisture, recrystallizes, and consequently cakes. In the warm inner regions, the relative humidity has dropped, so moisture on the crystals evaporates, causing surface crystallization and, possibly, additional caking. This is why caking at the surface of a sugar mass is usually in the form of a thin, hard shell, while the center forms a large, relatively soft lump. The compression of sugar that takes place in bulk stores or stacks of bagged sugar will tend to increase the likelihood of caking by increasing the bulk density and therefore bringing more crystal surfaces into contact with one another, creating more contact points.

The moisture participating in caking has two possible sources: It comes from either the surrounding air (or the interstitial air entrapped as sugar is poured) or from the sugar crystals themselves. Caking due to the former, dubbed *deliquescent caking* by Chen and Chou [9], is avoidable only by preventing contact with moist air. Caking due to the latter, dubbed *efflorescent caking*, is preventable by sufficient drying of the sugar. Sugar conditioning should therefore aim to remove any remaining free moisture and reduce the bound moisture content of a refined sugar to the point where it will not cake. This is done, in practice, by exposing the sugar to low-humidity air in conditioning silos for extended periods. However, as Bagster [1] emphasizes, it is not possible to make a noncaking sugar; given sufficiently adverse conditions, any sugar will cake. Conditioning is at all times a practical compromise. A sensible conditioning objective is that offered by Bruijn et al. [6], that the moisture should be reduced "to the point where no serious caking will occur when the sugar is subjected to temperature (or humidity) gradients slightly in excess of those expected in practice."

The problem of moisture migration due to temperature gradients is most pronounced when sugar is in transit. Sugar transported into an area of low RH (or sold into homes that are centrally heated and therefore at low RH) are particularly susceptible to moisture loss and caking. Sugar dispatched warm, when coming into contact with a cold environment (through the steel sides of a truck or container, or directly in the case of a load of packed sugar on a flat-bed truck) will experience significant moisture migration, as described previously. Excell and Stone [18] described how bags of sugar on the floor and sides of shipping containers were found to be wet at their destination. Conditioning assists with this problem by minimizing the quantity of migratible moisture, but the best guarantee against moisture migration problems is to cool the sugar before packing or dispatch, to within 5 to 10°C of the average ambient temperature. In warm weather climates it is more practical to cool the sugar to a temperature of 5 to 10°C below the average ambient temperature.

c. Moisture and Caking Analysis. Residual moisture in sugar may be determined analytically in one of four ways:

1. Oven-drying for 3 h, with moisture loss assessed by mass loss
2. Oven drying for several days
3. Karl Fischer titration using methanol (as detailed by Bennett et al. [3])
4. Karl Fischer titration using formamide (method as above)

Method 1 measures free moisture and perhaps part of the bound moisture. Methods 2 and 3 measure free plus bound moisture, as methanol washes, but do not dissolve the crystals. Method 4 measures free, bound, and inherent moisture, as formamide dissolves the crystals. It could be argued that method 3 is the most appropriate, as inherent moisture does not cause caking, and variations in the ratios of inherent to bound moisture are not known. De Bruijn and Marijnissen [13] have described an analytical procedure for differentiating between the three types of moisture, using Karl Fischer (methanol) analysis, but grinding the crystals for total (including inherent) moisture determination. Various laboratory instruments are available for the rapid determination of residual moisture, most based on thermogravimetry (employing mechanisms such as microwave, infrared, or halogen/infrared). A different approach was employed in the Boonton meter, reported by Obst [37], which used radio-frequency absorption.

An accurate measure of the caking propensity of sugar is even more problematic. Bruijn et al. [6], Excell [17], and Ramphal [41] have reported the use of a number of standard tests to measure this property. Two examples are:

1. *Test-tube test.* Samples are taken in test tubes (150 mm long and 20 mm ID) which are then stoppered, and the lower 20 mm of the tube is immersed in a water bath, which is set at 10°C for 2 h and then at 40°C for 2 h. The test tube is then removed and the sugar poured carefully onto a flat surface. If the sugar flows freely and contains no lumps, the test is negative. The presence of any lumps or any signs of sugar adhering to the walls of the tube constitutes a positive result.
2. *Large-scale test.* Typical test units consist of a perspex cylinder, closed at the bottom by a hollow aluminum disc and at the top by a Perspex (polymethyl methacrylate) lid fitted with an O-ring. Each column (holding approximately 7 kg of sugar) is closed and then sheathed with a polystyrene insulating cylinder. Water at 5°C is then passed through the aluminum base for 2 h, followed by water at 40°C for 2 h. The sugar is then poured out in the same way as in the test-tube tests, and any lumping or adhesion observed. In this larger-scale test, it is possible to weigh any caked material in an attempt to quantify the seriousness of the problem.

While the effectiveness of conditioning should in fact be assessed according to a sugar's caking propensity, the foregoing tests are both subjective and qualitative. It is therefore usually necessary to rely on residual moisture (possibly in conjunction with caking test results) as an indication of the condition of the sugar.

15.3.2 Modeling the Conditioning Process

Conditioning is an "extension" of the drying process and has consequently received little attention as a separate process worthy of fundamental modeling. Rodgers and Lewis [42] complained in 1963 that storage (incorporating conditioning) "has scarcely been considered as one of the technical processes of sugar manufacture." However, regarding conditioning as an extrapolation of drying may lead to erroneous conclusions, as drying theory does not explain why the residual moisture removal frequently takes days to execute rather than minutes or hours. Figure 15.17 is a set of typical moisture curves for the conditioning process.

Various authors have recognized that the rate of conditioning has more to do with sucrose crystallization and water diffusion in the crystal surface film than with simple evaporative mechanisms. However, the first attempt at modeling these processes was done by Mikus and Budicek [32], who asserted the formation of a "crust" on the surface of the film. This crust is formed as the outer surface of the film exceeds the metastable limit during drying and nucleation begins to take place on the surface. This process continues until a continuous crust results. The authors calculated an average thickness of this crust (for a

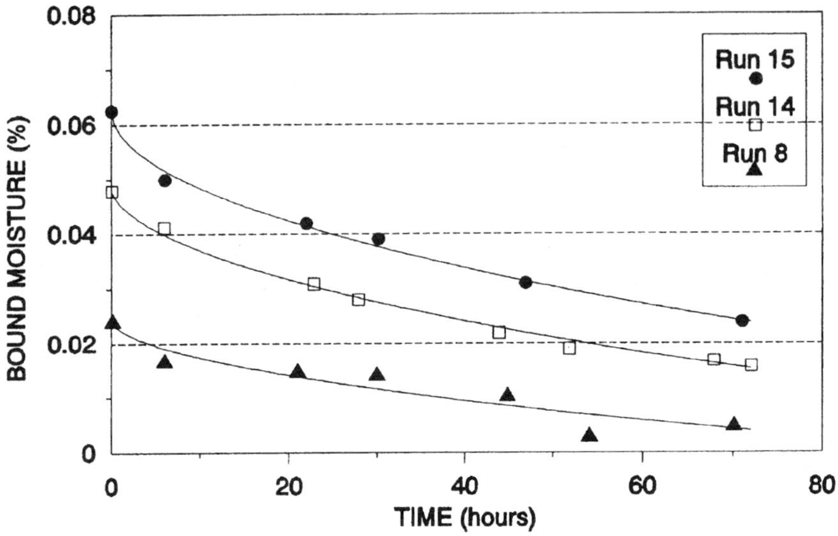

FIGURE 15.17 Typical moisture curves during conditioning. (From Ref. 31.)

crystal of assumed properties) and determined moisture loss rates based on rate equations for three processes: (1) crystallization of sucrose in the film, (2) diffusion of moisture across the solid crust, and (3) evaporation of moisture from the crust surface. The relationship between the rates was 4500:1:90000 for crystallization:diffusion:evaporation, clearly identifying diffusion as the rate-controlling step.

The concept of the solid surface layer was taken further by Meadows [31], who developed a complete model of the conditioning process. This too assumed the existence of a solid surface layer, which he termed an "amorphous" sugar layer. Figure 15.18 is his representation of a portion of crystal surface undergoing conditioning, from which four process mechanisms can be derived:

1. Crystallization of sucrose molecules out of the supersaturated surface film, effectively diluting the film
2. Evaporation of moisture at the interface between the film and the amorphous shell, governed by the equilibrium vapor pressure of the solution at its particular concentration, serving to concentrate the surface film
3. Vapor-phase diffusion of the moisture across the amorphous layer, which lowers the partial pressure of moisture at the film interface
4. Diffusion/convection of the moisture into the bulk interstitial air, which lowers the partial pressure of moisture at the outer surface of the amorphous layer

Calculation of typical rates for the processes yielded the relationship 6500:100,000:1 for crystallization:evaporation (incorporating bulk convection):diffusion, again identifying diffusion as the rate-controlling step (although the absolute rates were each an order of magnitude slower than those of Mikus and Budicek). These rate equations were then combined into an expression for the conditioning process:

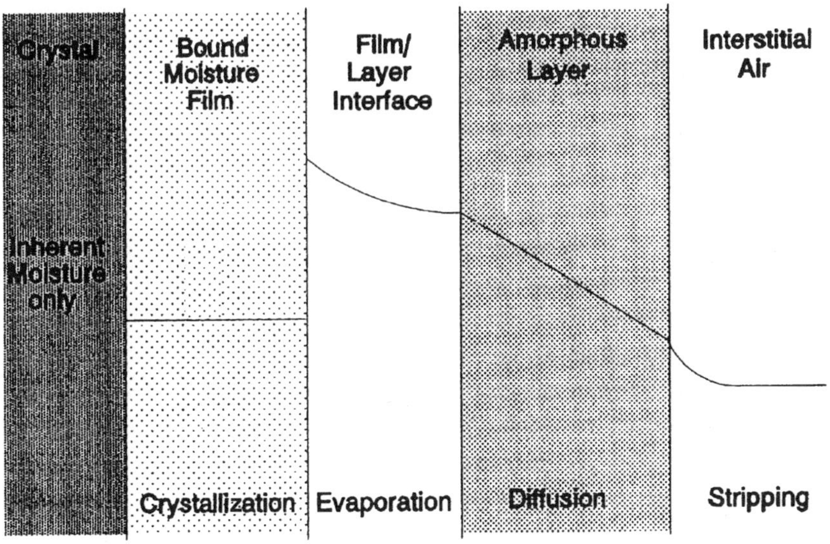

FIGURE 15.18 Portion of crystal surface undergoing conditioning. (From Ref. 31.)

$$M = K - (K^2 + M_i^2 - 2KM_i + 2kt)^{0.5}$$

where M is the moisture content of the film at time t (kg), M_i the initial moisture content of the film (kg), t the time of conditioning (s), and K and k are constants depending on the diffusivity and thickness of the amorphous layer, the initial mass of sucrose in the film, the saturation concentration of sucrose in the film, and the crystal surface area. This equation was successfully fitted to data from pilot-scale conditioning tests. The model rests on the assumptions of a continuous film of syrup on the surface of each crystal, a continuous amorphous layer on the surface of each crystal, and the crystallization of sucrose from the surface film taking place at the point of moisture removal (i.e., the amorphous layer) rather than the surface of the crystal itself. These assumptions limit the accuracy of the model, but it is at present the most complete published model available.

If the models above are accepted, all but the diffusion step in the conditioning process may be assumed to be instantaneous. The conditioning rate then depends on the saturated film vapor pressure (a function of temperature and purity) and the partial pressure of water in the bulk airstream (a function of temperature and moisture content). In this respect it is very similar to ordinary drying. However, it differs in that the rate also depends on the thickness and nature (diffusivity) of the amorphous sugar layer, which is a function of the sugar's drying history.

The most recent modeling of the conditioning process was done by Tait [54], who ignored the possibility of an amorphous layer or crust and modeled the conditioning process based on the effect on vapor pressure of a highly concentrated surface film. He constructed a steady state, as well as a "batch" model, although details of the models are not reported.

15.3.3 Conditioning Plant Operation

Much of the work on sugar conditioning in recent years has been done in South Africa, and a synopsis of their findings is useful at this point.

a. South African Experience. In 1966, Howes [26] presented observations on the conditioning process to the South African Sugar Technologists' Association (SASTA). He asserted that bound moisture removal took 7 to 10 days, and conventional wisdom at the time was that the moisture migrated from within the crystal itself. To aid the conditioning process, he recommended automation of pans, the installation of pan stirrers, and the boiling of a larger grain.

When bulk refined sugar transport was first considered in South Africa, climatic constraints made it clear that conditioning of the bulk sugar would be essential. Extensive pilot conditioning work was carried out, both prior to the commissioning of the first conditioning silo in 1982, to establish design and operating parameters, and in the years hence, to further optimize silo operation. Pilot work was begun by Bruijn et al. [6] in the early 1980s to obtain design data specific to South African conditions for the construction of the first silo at the Tongaat-Hulett sugar refinery. Both laboratory-scale and pilot-scale [40 kg of sugar per pilot silo) work was done, and the caking propensity of sugar was determined by means of test-tube and full-scale testers (full-scale being a device built to simulate a section through a railcar!). Conclusions from this work were:

1. Between 48 and 72 h was required for conditioning.
2. Larger crystals conditioned slower, usually due to high conglomerate counts.
3. Higher temperatures accelerated moisture loss, but only for the first 24 h.
4. Increased airflow rate increased initial drying rate but had no effect after 48 h.
5. "Conditioned" (noncaking) sugar had Karl Fischer (formamide) moistures between 0.04 and 0.05% (corresponding to oven moistures of 0.02 to 0.025%).

This work was repeated by Excell [17] prior to the construction of the second South African silo, using sugar from the proposed site (Noodsberg factory). The foregoing conclusions were corroborated, with the exception that airflow rate was found to have no effect on moisture removal rate. The phenomenon of moisture migration from warm to cool sugar was also demonstrated, and the equilibrium relative humidity of refined sugar was confirmed at 76%. Ramphal [41] carried out further pilot work in 1989 to examine the effect of temperature on conditioning rate. He found that reduction of conditioning temperature from 40°C to 32°C increased conditioning time by 25%. He also reiterated that 72 h was required for adequate conditioning.

b. Factors Affecting Conditioning. In a field where consensus is the exception, authors generally agree that conditioning is strongly dependent on temperature, as higher temperatures increase the driving force for conditioning. There appears to be a limiting moisture content of 0.007 to 0.008% (Karl Fischer methanol method), beyond which the sugar will not condition further. On this basis Stachenko et al. [49] recommend a temperature of 50 to 55°C, requiring a conditioning time of 40 to 48 h "for equilibration." Increasing the conditioning temperature has two constraints, however. Hotter product sugar is more prone to caking while cooling to ambient temperature, and excessive temperature gradients may cause caking in the conditioning silo itself. Of particular interest is the fact that after the first 24 h, the rates of moisture loss appear to equalize at all temperatures (subject to the lower moisture limit) (Rodgers and Lewis [42], Bagster [1]). This means that elevated temperatures are beneficial for the first 24 h only! No theoretical explanation for this phenomenon is offered, but it has been confirmed experimentally by many authors.

The temperature of the air is much less significant than the sugar temperature, as both the air/sugar ratios typical in conditioning and the low specific heat of air will result in the air equilibrating to the sugar temperature fairly rapidly as it passes through the silo. However, Ramphal [41] has found that large differentials between air and sugar temperature cause caking in the silo; South African practice is to keep any differential below 6°C.

As explained previously, conditioning is not limited by the rate of evaporation. For this reason, conditioning is independent of the airflow rate, which should be selected to achieve a uniform distribution of air throughout the sugar, without overloading the dust extraction system. Various authors (Rodgers and Lewis [42], Dowling [15], Schwer [45]) have reported insensitivity to the RH of the air as well, although the modeling of the process described previously suggests that lower RH should accelerate conditioning. Tests done by McGimpsey [51] corroborated this, finding that conditioning with drier air (20% RH instead of about 40%) halved the required conditioning time as well as producing a drier product.

Due to their larger specific surface area, smaller crystals condition faster than larger ones, making the larger crystals the determinants of the overall conditioning time (Chapman [8], Lu et al. [29]). However, for the same reason, smaller crystals cake more easily than large. Smaller crystals contain a greater percentage of water (available for migration and caking) at a given ERH. It is therefore necessary for conditioning to proceed to lower moistures for smaller crystals. Although neither large nor small crystals are universally advocated, authors agree that improved conditioning may be obtained by increasing the uniformity (lowering the CV value) of the grain. Sieving off of fines is also beneficial to conditioning and reduces caking propensity (Mikus and Budicek [32] mention removal of the fraction smaller than 0.3 mm). High levels of conglomerates have a definite deleterious effect on conditioning, due to the moisture that is trapped between the constituent crystals. Some authors maintain that it is only the large, star-shaped, "true" conglomerates that have any real effect.

In view of the diversity of influences on the conditioning process, there is no single optimum conditioning time for refined sugar. Estimates of the theoretical time requirement range from 36 h to over 100 h, while existing installations around the world operate at between 24 and 72 h. Kavan and Mikus [28] calculate a theoretical time requirement of 36 h, but Budicek and Mikus [7] recommend using twice this for conditioning plant. Most research, however, points to somewhere near 48 h as a reasonable practical compromise, given acceptable sugar quality and operating conditions. Bruijn et al. [6] point out, however, that "difficult" sugar (high bound moisture content, high conglomerate count, high CV) continues releasing appreciable amounts of moisture after 7 days of conditioning. There is, however, a growing conviction that "every little bit helps" in the conditioning of sugar—it is worth supplying conditioning air to bins and hoppers even if the sugar residence time is low. This reasoning is based on the fact that most of the moisture lost during conditioning is lost in the first 24 h.

The physical design of conditioning silos is dealt with in Section 15.4.2, but Table 15.4 contains a list of recommended operating norms for conditioning plant.

TABLE 15.4 Operating Norms for Conditioning Plant

Specific air rate ($m^3/h \cdot$ tonne sugar)	3[a]
Conditioning temperature (°C)	40–50
Minimum conditioning time (h)	24
Recommended conditioning time (h)	48
Air relative humidity (%)	10–20
Feed sugar moisture (% Karl Fischer)	<0.10
Product sugar moisture (% Karl Fischer)	<0.05
Sugar fines content (% through 300 μm)	<10
Postconditioning cooling target (°C)	<35

c. Influence of Other Processes. Other processes in the refining of sugar, in particular pan boiling and drying, have a significant effect on the efficiency of the conditioning process. The removal of free moisture from the sugar crystals in driers is believed to be the main determinant of the amount of bound moisture left trapped on the crystals. Sugar dried at high temperatures has been found to have a higher residual moisture content than sugar dried at lower temperatures. This may be attributed to two mechanisms. Drying that is too rapid has been found to lead to severe agglomeration of previously clean crystals. In other words, conglomerates may be formed in the drier rather than in the pan. The second mechanism is the formation of a "shell" of low-permeability, amorphous sucrose around the crystal when drying is too fast to allow the sucrose to crystallize on the crystal surface. The moisture trapped under this layer is then bound moisture and must first diffuse through the layer in order to be removed. Dowling [15] showed that sugar dried at higher temperatures (110°C) subsequently released greater quantities of bound moisture than that dried at lower temperatures (55°C), suggesting that increasing temperatures to accelerate drying should be done with great caution.

It should be noted that it is not possible to avoid the formation of bound moisture completely using conventional equipment (Stachenko et al. [49]), but slower drying is advocated. An interesting (but as yet unproven) approach to accelerating the rate of conditioning is the use of a preconditioning step, hinted at by several authors but discussed in detail by none. This involves holding the sugar in a bin for a period of several hours (e.g., 6 to 10 h) before conditioning and after drying. The bin is not aerated, and the theory is that the RH of the interstitial air will quickly equilibrate with the sugar. Thereafter, the amorphous sugar layer may experience partial redissolution in the saturated environment, allowing the sucrose to crystallize on the crystal surface rather than the film surface. Although this description is not particularly convincing, some refineries in North America hold their sugar in "preconditioning" bins for 6 to 10 h before conditioning. Also, Tait's [54] batch conditioning model describes a process whereby, by the use of a "wave" on high-humidity air, later conditioning can be accelerated so substantially that conditioning may be accomplished in 4 h! Tait refers to this as "relaxing" the sugar, and, while the claim of 4-h conditioning requires better substantiation, this certainly appears to present an avenue for further investigation.

Careful pan boiling addresses the cause rather than the symptoms of the conditioning problem. Formation of conglomerates in particular depends on the level of supersaturation as well as the degree of circulation. There is complete consensus that well-controlled pans with mechanical circulation (and well-prepared seed) make a vast contribution to proper conditioning by producing uniform crystals and reducing conglomerate counts. Rodgers and Lewis [42] report conglomerate counts as low as 15% (with accompanying photographs) and attribute this largely to mechanical circulation. Chapman [8] says that mechanical circulation reduces syrup inclusions (in conglomeration) by 33 to 50%. Stachenko et al. [49] found a drop in conditioning time from 4 or 5 days to 2 days after installing pan stirrers, and assert that increased capacity on the pan floor to allow slower boiling pays for itself by allowing a smaller conditioning plant. Dowling [15] even reported that the conditioning rate varied with the color of the liquor to the pans!

Good centrifugal washing can assist in drying and conditioning and reduce the likelihood of caking. Many refineries in North America have a lower limit on their centrifugal wash time to ensure that the syrup film is properly removed, even when a shorter wash would achieve the required sugar color. Their centrifugals are also set to trip if steam pressure drops below a lower limit, to eliminate any possibility of poor steamings producing unacceptable product. Chapman [8] notes that the conditioning process increases the bulk density of sugar (from about 770 kg/m^3 to about 870 kg/m^3). On this basis, he calculates a saving of 10% on packing materials as a result of conditioning.

d. Some Guidelines. Chen and Chou [9] list 11 tips for effective conditioning:

1. Store in "curing" bins or silos.
2. Boil good grain, free from conglomerates.
3. Boil and centrifuge sugar at relatively low temperatures to minimize residual dissolved sugar in the surface film.
4. Spin off as much water as is practically possible at the centrifuge.
5. Dry as slowly as practicable, with low air temperature.
6. Screen to remove conglomerates.
7. Place dry sugar in holding bins or silos for required time (16 to 24 h) with dry air percolation.
8. Recirculate sugar during storage in bins or silos, if practicable. If not, move sugar down about 2 m in height per day.
9. The sugar temperature should be less than 38°C before storage or packing.
10. Use dry air to ventilate sugar in conveyors after treatment.
11. Displace humid air from railcars before loading with bulk sugar.

15.4 STORAGE

Sugar storage takes place in one of four forms:

1. Bins and hoppers, for short-duration storage; capacity up to a few hundred tons
2. Short-term bulk storage silos, of tall cylindrical construction, frequently used for conditioning; typical capacity 1000 to 3000 tons
3. Long-term bulk storage silos, of low air/sugar ratio; typical capacity 20,000 to 30,000 tons
4. Bagged sugar warehouses

In this section we deal with design and operational considerations of the three types of bulk sugar storage and touches very briefly on bagged sugar storage issues.

15.4.1 Bagged Sugar Storage

Sugar packaging technology is an entire topic on its own and will not be dealt with in this chapter (a good summary is provided in Chapter 17 of the *Cane Sugar Handbook* [9]). However, some mention of the considerations surrounding the warehousing of bagged sugar is required. Chen and Chou [9] provide a list of guidelines for storage of sugar in packages:

1. Sugar temperature should not exceed 38°C.
2. Relative humidity should not exceed 60%.
3. Keep pile heights to a practical minimum.
4. Sugar should be stored separately from other products to avoid picking up moisture, odors, and so on.
5. Maintain limited inventories with rapid turnover.
6. Pack as much as possible for direct shipment.

7. In humid areas, use heat or air conditioning to control relative humidity and temperature changes.
8. Pack only conditioned sugar in moisturetight packaging.

Moisture containment and exclusion are achieved by the use of moisture-impermeable barriers in packing materials (bags, balers, or overwraps). This is an expensive option but is unavoidable in some cases (and climates), particularly when packing unconditioned white sugar. An additional point to remember is to ensure that pallets are dry. The Thames refinery in London dries any wet pallets (or new pallets—"green" wood contains high levels of moisture) before use. A moisture barrier between the bottom layer of bags and the pallet is sometimes used.

Bagged sugar warehouses should be built of masonry with sealed roofs and concrete or wood floors. For capacity calculations, Hugot [27] suggests a bulk density of 800 kg of sugar per cubic meter of stack (not allowing for corridors around the stacks). The main moisture exclusion practice is pressurizing the warehouse or store with conditioned air. Chen and Chou [9] report that a system installed to control RH to 65% in a warehouse 90 m long by 36 m wide employed four electric heaters and a single ventilating fan, and consumed 2000 kWh (period not given). The conditioning system may be run intermittently (only during periods of high ambient RH). Tight control of RH is not necessary; the minimum requirement is to keep the warehouse temperature a few degrees above ambient, with a minimum of 16°C, and to ensure good circulation or air around the stacks.

When stacking bagged refined sugar, it is advisable to give the sides of the stack a batter or taper of 20° for hessian bags and 7.5° for paper bags [27,30]. An example of a sophisticated automated bagged sugar storage system is that installed at the C & H refinery in Crockett, California. The automated storage and retrieval system (AS/RS) coordinates the palletizing, storage, and dispatch of about 700 different refined sugar products at a maximum receiving capacity of 76 pallets/h and a maximum dispatch rate of 126 pallets/h. The installation cost $24 million and was justified on the basis of reduced inventory requirements and reduced operating costs.

15.4.2 Bulk Sugar Storage

a. Design and Operation. Lyle [30] was of the opinion, in 1957, that "no refined sugar silo should be larger than one whose contents a shift manager can, on his own responsibility, decide to remelt"! Clearly, the sugar industry has transgressed his rule since then and the design and operation of bulk white sugar storage facilities has grown in importance.

Spencer and Meade [48] list eight desirable practices for preparing refined sugar for bulk handling, transport, and storage. The points were distilled from the discussion in the 1960 SIT Symposium on drying, cooling, and conditioning of sugar [51]:

1. Drying and primary cooling in conventional granulators
2. The addition of chemically dried air, using the Kathabar system, to the cooler–granulator
3. Additional cooling to give sugar of 35 to 45°C (the cooler discussed at the Symposium was the water-cooled Holo-flite cooler–screw conveyor)
4. Storage in bins that are properly insulated (Masonite, Fiberglass, air layer, or otherwise)
5. A slow movement of dry air through the stored sugar
6. "Maturing" or weathering the sugar by allowing it to stand in storage for several days
7. Moving the sugar from one storage bin to another to break up slight setting that may have occurred during the weathering period

8. Sweeping the contained air out of hopper cars (also out of bins and silos) with predried air before loading

The fundamental design consideration for bulk sugar storage is the angle of repose of sugar. For wet sugar, this may be as high as 55°, while estimates for dry white sugar are in the range 30 to 35° [9]. However, caution is advised in using angles of repose to design conical bottoms or offtakes, as experience with multiple-offtake silos has shown the flow angle of sugar to be much steeper, in some cases in excess of 70° (with the sugar at shallower angles remaining static). Lyle [30] relates an experience where the outlets from bulk storage bins, at 60° to the horizontal (51° in the valleys), were not steep enough, and the "sugar built up its own hopper in a hard crust." In large multiple-offtake silos, the bottom hoppers may be designed for the sugar flow angle, or, as has been done for a 25,000-tonne silo in Swaziland, a decision may be taken to allow the sugar to build up stationary cones between the hoppers, to be removed when the silo is emptied.

Chen and Chou [9] suggest a bulk density range of 785 to 830 kg/m^3 for conditioned white sugar. However, significant compaction of sugar occurs in a bulk store. Hugot [27] describes how sugar with a bulk density of 840 kg/m^3 compacts to 990 kg/m^3 at a depth of 7 m (by a combination of pressure and aligning of crystals as the sugar moves) and suggests using an average bulk density of 880 kg/m^3 for a deep silo.

Devices and arrangements for distributing and withdrawing sugar in bulk storage vary widely, from simple single chutes to screw conveyors that move within the sugar. An even distribution of the sugar entering a silo is important not only for maximum space utilization, but also because sugar pouring onto the top of a heap will classify as it rolls down. This may lead to sugar of differing crystal size distribution and bulk density at the center and periphery of the silo. This would have a deleterious effect on conditioning, as well as possibly causing problems with some of the sugar in meeting product specifications.

The stresses on the silo walls as the sugar moves downward are obviously a key element of the structural design of the storage vessel. Even withdrawal of sugar from the bottom of a silo is therefore not only a process consideration (to ensure good residence-time distributions), but must be ensured to prevent increased wall stresses due to differential movement of sugar at different points in the vessel. Figure 15.19 is a plot of the pressure on silo walls exerted by the sugar. The design should take account of the fact that should rod-type flow develop in the sugar, the dynamic stresses on the walls may be as high as 2.3 times the static stresses [8]. Epoxy-lined mild steel is fast becoming the material of choice for silo construction in Europe. Tate & Lyle's conditioning silos have epoxy-lined walls and stainless steel discharge cones. Gramercy refinery in the United States has had success with epoxy lining of a concrete silo. Lantic refinery lines the bottoms of the interior walls with stainless steel, leaving the rest troweled smooth. The benefit of self-passivating steels is reduced by the fact that the moving sugar abrades the passive layer. Mild steel can be used quite safely as a material of construction for conditioning silos, as the conditioned air protects it against corrosion.

There is some debate about the necessity of insulating bulk sugar storage equipment. This, of course, depends on capacity, sugar quality (degree of conditioning) and temperature, climatic conditions, and materials of construction. However, it must be said that it is highly undesirable to allow temperature gradients to exist in bulk sugar storage, as this causes moisture migration (discussed earlier), which may lead to widespread caking. Sugar of different temperatures or qualities should never be stored in contact with one another. One system for ensuring high-efficiency insulation is to blow temperature-controlled air through an annulus around the silo wall; this is quite commonly employed in colder cli-

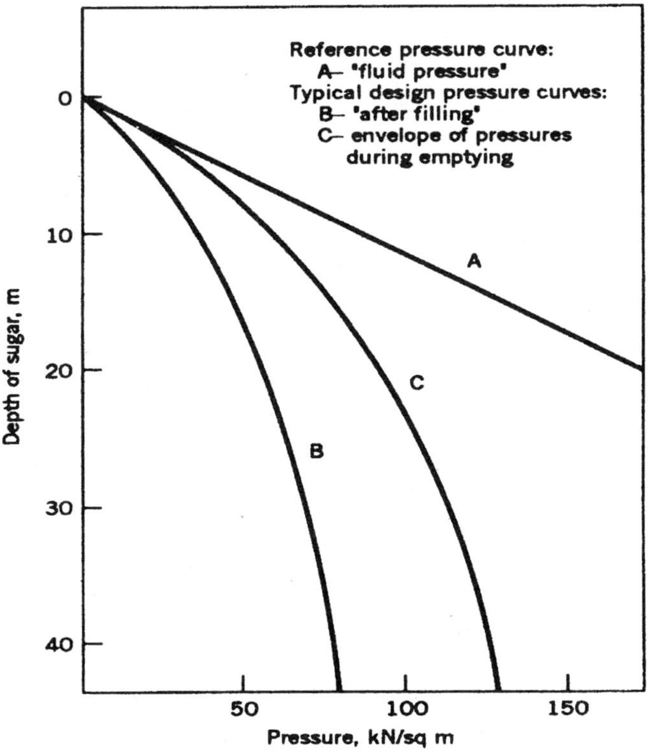

FIGURE 15.19 Sugar pressure on the walls of a silo. (From Ref. 9.)

mates. Sugar in bins or silos should be kept in motion as far as possible (certainly until well conditioned). When no outloading is taking place for extended periods, this involves recirculating the sugar, or, where capacity is available, transferring the load completely from one bin or silo to another (this ensures that no dead zones exist).

Mikus and Budicek [32] suggest four categories for sugar storage and comment on their operation as follows:

1. Nonventilated silos with unheated walls must be supplied with sugar that is conditioned, fines-free, and cooled.
2. Nonventilated silos with heated walls must also be supplied with conditioned sugar. The sugar must be of uniform quality and at the storage temperature and must not be allowed to dry out in the silo (e.g., at the surface).
3. Ventilated silos with unheated walls must have sufficient air that it can be used for both moisture removal and temperature gradient prevention. The air must be of "conditioning quality," and the quality of the sugar should not fluctuate.
4. Ventilated silos with heated walls require only that the air be sufficient to carry away excess moisture.

b. Typical Silos. An example of a simple bulk sugar store is the A-frame building, as shown in Figure 15.20. Economic designs of these stores do not have side walls much above 3 m (unless buttressed), due to the pressure exerted by the sugar on the walls. The sugar is

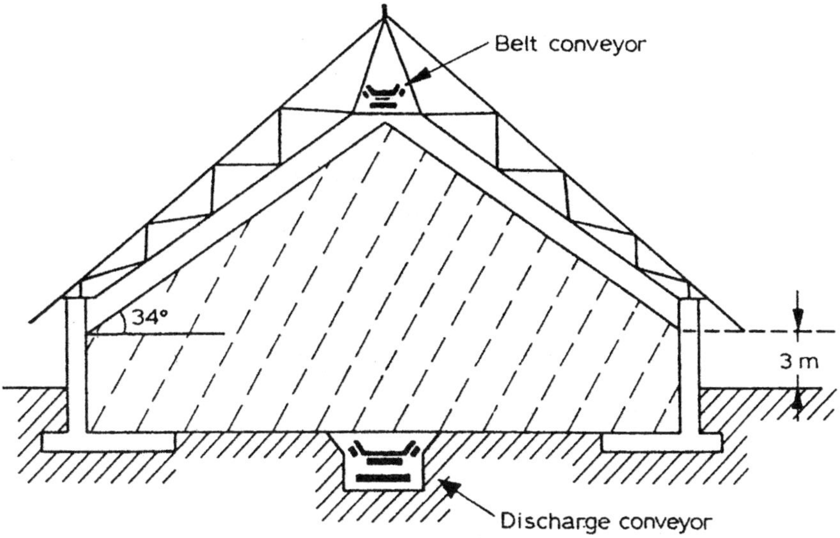

FIGURE 15.20 A-frame sugar silo. (From Ref. 27.)

supplied to the store by a conveyor running in the apex of the roof and is discharged by a throw-off carriage, a moving plough, or a moving conveyor. Recovery of sugar is usually by means of an underground belt, onto which the sugar discharges through slide gates in the floor. Sugar remaining along the sides of the store must be pushed to the centre manually. This is usually the cheapest method of bulk storage for large tonnages if space is not an issue. Its main disadvantages are that a large sugar surface is exposed to the air, the design makes controlled ventilation or air conditioning difficult and expensive, and true first in, first out operation is not achievable. The parabolic silo is a variation on the A-frame design, employing an arched shape, which, when full, has sugar pressing against the sides for about two-thirds of their height. Since the building is sealed, it may be air conditioned.

Vertical cylindrical silos are the cheapest option for smaller tonnages, occupy the least floor space, and may easily be air conditioned. The move in Europe is toward vertical steel silos, even for large units of capacity up to 30,000 tons. Figures 15.21 to 15.23 show three designs of cylindrical bulk silo, the first and second employing gravity feed and withdrawal (the second using multiple discharge hoppers). The third uses a moving-screw conveyor to remove the sugar. Most such mechanical methods of sugar recovery have the inherent disadvantage that they operate on a last in, first out system. Where mechanical withdrawal is used in a first in, first out silo (such as with the Starvrac extractor, a serrated screw rotating about the silo axis on the floor), the reclamation system is buried in the sugar, which makes mechanical failure disastrous.

15.4.3 Storage Issues

a. Air Conditioning. Air passage through the sugar is strongly recommended by most authors. In addition to minimizing the likelihood of caking by carrying away excess moisture, it eliminates problems that may be associated with localized moisture concentration, such as yeast activity. However, where the intention is not to condition the sugar, it is quite common practice to keep the RH of the air up rather than down. It is felt that by keeping the RH just below the ERH of the sugar, the sugar will not give up moisture in any signif-

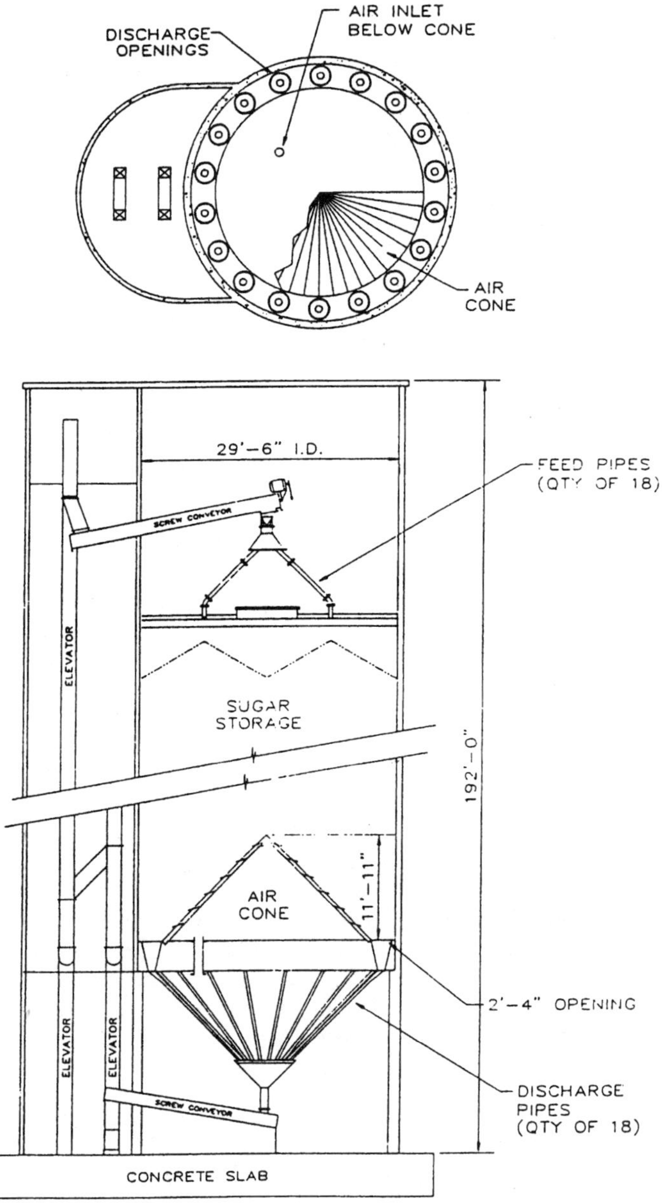

FIGURE 15.21 Vertical cylindrical silo. (From Ref. 51.)

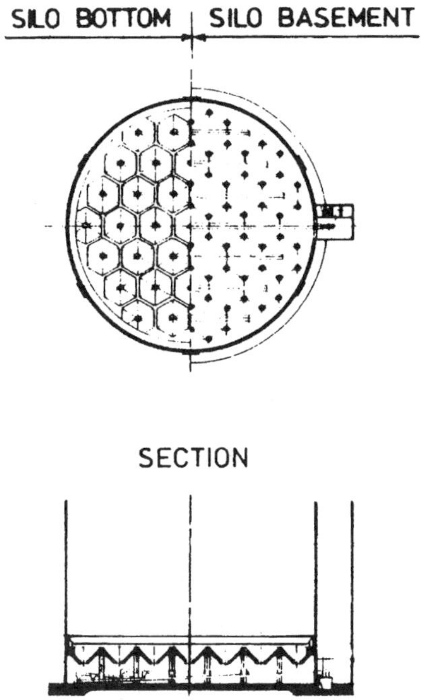

FIGURE 15.22 Cylindrical silo with multiple offtakes. (From Ref. 9.)

icant amounts. An RH of 60% is typical, but the Thames refinery had even better results by going as high as 70% (they had crusting of the sugar at 60%) [57]. The high RH has a secondary benefit of reducing dust problems and explosion risk.

The standard method of dehumified air production is to chill the air to a required dew point (required moisture content at saturation) and reheat it to the silo operating temperature, resulting in a specific relative humidity. It is often not necessary to reheat the air, as the rise in temperature across the blowers may be sufficient to reach the required temperature. Chemical dehumidification is growing in popularity. A neat system is the *desiccant wheel* concept (an example is the system supplied by Munters). In these units, the process air is dehumidified by being passed through a chemical desiccant, which is held in a slowly rotating frame. When the desiccant moves out of the process airstream, it moves into a regeneration stream, where the moisture is driven off by a stream of hot air. It is then returned by the rotation of the wheel into the process air.

b. Moisture Exclusion in Transit. Moisture exclusion from rail or road tankers is also important in preventing caking of sugar in transit. It is fairly common practice to blow conditioned air into trucks to dry them completely before filling. Some refiners use polyethylene "bonnets" under the lids of the trucks to prevent moisture ingress. A transportation method used in Europe is a container-sized, single-use plastic bag, which is inflated inside a container and then filled with bulk sugar. To empty, the bag is cut open, releasing uncontaminated sugar and leaving the container clean on removal of the bag.

c. Dust and Explosion. Sugar dust results from false grain generated in pans and/or crystal attrition or breakage during drying or handling. "Overdrying" of sugar will also exacerbate any dust problem. Apart from sugar loss, machine bearing surface wear, and

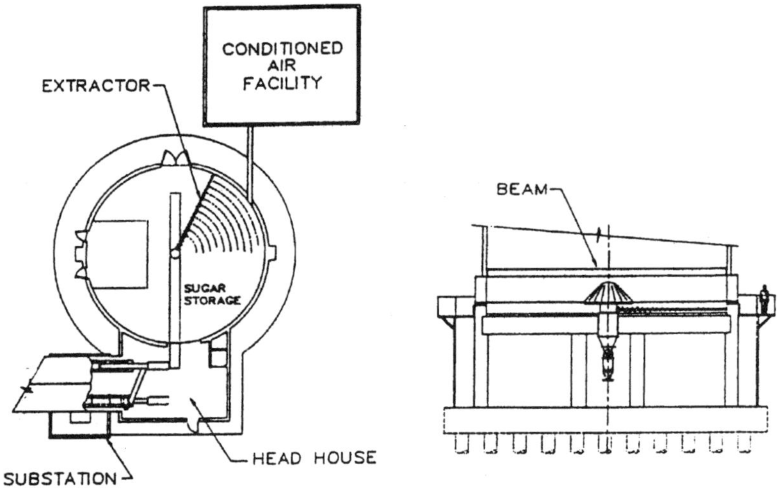

(a)

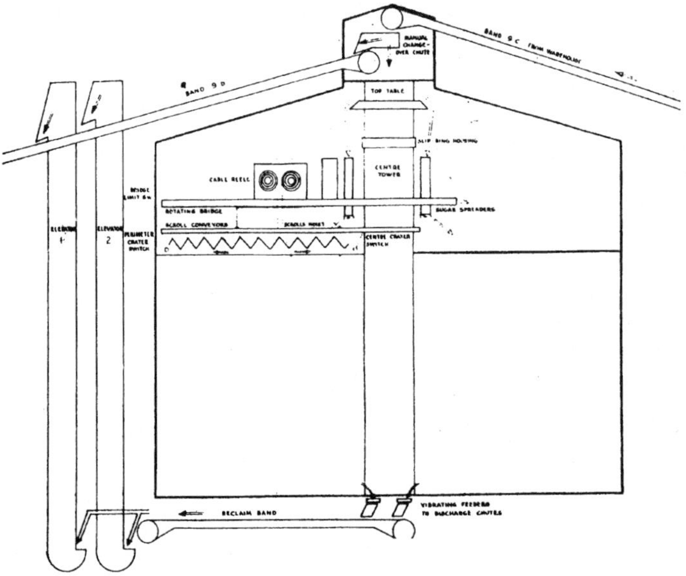

(b)

FIGURE 15.23 Mechanical offtake silos: (*a*) bottom offtake (from Ref. 51); (*b*) top offtake (from Ref. 57).

plant hygiene considerations, excessive dust also poses an explosion risk. Factors pertinent to the explosiveness of sugar dust are as follows:

1. Mikus and Budicek [32] quote a range of values for the minimum explosible concentration of sugar dust: from 6 to 19 g/m^3 (depending on particle size), with an average of 10 g/m^3. Other authors quote rather higher levels: Morden [34] refers to 40 g/m^3, while Hugot [27] quotes 60 g/m^3 for 0.1-mm particles, ranging down to 7 g/m^3 for the finest dust. Dust levels in sugar refineries are typically well below these levels, but the limit may be exceeded in the headspace of silos or bins. In addition, dust-laden surfaces contribute to any hazard, as an explosion will propagate by lifting dust from surfaces.
2. Estimates of minimum ignition temperature vary from 330 to 480°C [34], which means that ignition is unlikely to come from ordinary equipment surfaces in a sugar refinery. However, these temperatures are not especially high, which makes precautionary measures all the more important.
3. Sugar dust is categorized as explosion class St 1 (weak to moderate explosion), with a pressure rise rate of between 59 and 172 bar · m/s [34].
4. The risk of ignition through buildup of electrostatic charge depends on two variables. The powder resistivity of sugar dust is not well documented, but values have been quoted in the range 10^9 to 10^{12} $\Omega \cdot m$ [34], which puts it in the medium- to high-risk range (high-resistivity powders will tend to build up charge due to their electrical insulation properties). The minimum ignition energy of sugar dust was found by Morden [51] to be between 400 and 700 mJ [34], which indicates low sensitivity to ignition and low risk, but other sources cite values between 20 and 30 mJ [32,34], which represents moderate to high risk.
5. The maximum explosion pressure for sugar is approximately 9 bar [34]. However, in interconnected vessels, the propagation of an explosion from one to the other can cause some precompression of the dust cloud and therefore enhancement of the explosion pressure, to a value well above 9 bar.

The data above suggest that the refiner cannot afford to be complacent about explosion prevention and protection measures. Explosion prevention is largely the application of commonsense measures (though often at a price). The measures aim at avoiding the creation of a dust cloud or an ignition source (the "fuel" and "ignition" sides of the fire triangle), as it is normally accepted that the oxidant, oxygen, cannot be excluded. Common measures are:

- Providing dust extraction systems at all conveying transfer points
- Using the correct lighting installations for explosive environments in dusty areas
- Avoiding overloading of electrical reticulation
- Using suitably enclosed fuse boxes and switches (complying with the correct classification standard)
- Employing a practice of cleaning properly before welding or cutting
- Ensuring that a layer of dust is not allowed to build up in ducts

In addition, some factories (particularly in colder areas) apply a lower limit to the relative humidity of their conditioning air, to reduce the buildup of electrostatic charge that may occur under very dry conditions, with the associated ignition risk. The explosion protection of vessels may take one of three forms:

1. Providing suitable venting. Figure 15.24 is a nomograph for determining the required vent area for a vessel.
2. Providing a suppression system. This requires the early detection of an incipient explosion and the rapid injection of a suitable suppressant such as water, halon, or a range of powders.
3. Designing the vessel to contain the maximum pressure anticipated from an explosion.

d. Screening or Sieving. Particle size classification of some sort is employed routinely in almost all refineries. In its simplest form, screening is used to "scalp" off the coarsest fraction (lumps), which is often done using screens attached to the discharge end of rotary driers. At the other extreme, the refined sugar may be fully classified into a range of speciality products (often with esoteric names such as "brilliant," "fruit," or "Bakers's Special"). The classification is usually achieved using a wire mesh or perforated plate, through which particles smaller than the screen aperture may pass, while the larger fraction is carried over the surface. The screening is normally assisted by a gyratory or vibratory motion.

The most commonly used mechanical screening device is the vibrating or tapping type (of which the best known are supplied by Hummer, Rotex, or Sweco). These consist of flat screening frames, inclined about 35° to the horizontal, down which the sugar flows, while mechanical devices tap or vibrate the screen to ensure flow through the apertures. These units are usually provided with multiple parallel screens to allow the separation of several fractions at once. Older screening equipment was of the "bolter" type, which comprised a revolving casing slightly inclined to the horizontal, covered with a wire screen. A more modern alternative is the Rotex type, which has nearly horizontal screening surfaces rotated in an end-to-end and side-to-side motion and uses rubber balls held in compartments below the screening surfaces to bounce against the meshes and keep them clean. The most important operational parameters of a screening installation are:

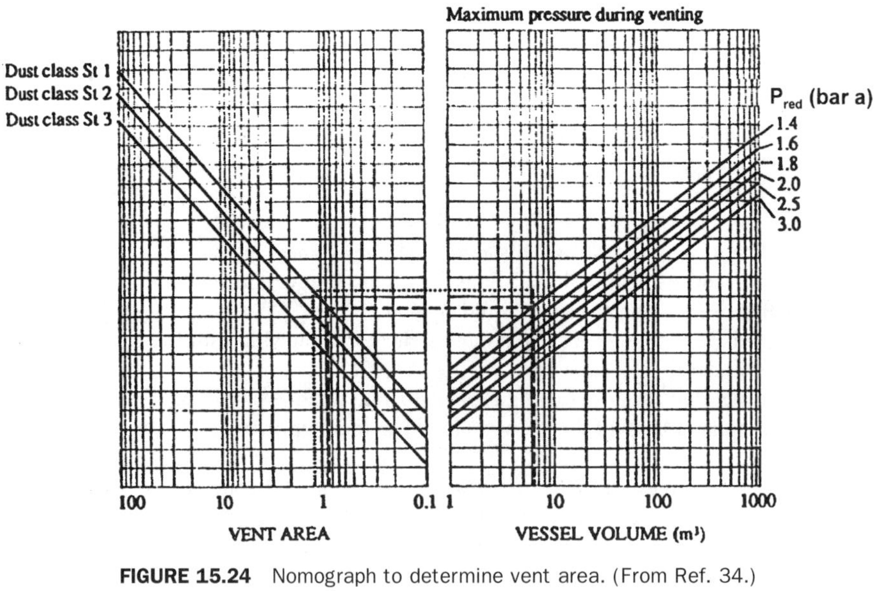

FIGURE 15.24 Nomograph to determine vent area. (From Ref. 34.)

1. Cleaning action, which should be effective without interfering with screening efficiency
2. Conveying action, which must ensure that the oversize fraction is conveyed off the screen at the desired rate
3. Feed-rate control, as too high a rate floods the screen, preventing some of the finer fraction from reaching the screen and therefore reducing efficiency

Screening practices vary widely from refinery to refinery, based on market requirements and local conditions. Many refiners believe that fines are a significant contributor to the caking propensity of sugar and screen all product sugar for this reason. The definition of "fines" also varies, being identified variously as that fraction passing through a 180-, a 250-, and a 300-μm screen, respectively, at three refineries in the United States. The part of the fines fraction that is superfluous to the requirements of the refinery for caster or other speciality sugar is remelted. One Canadian refinery boils all white sugar strikes to a minimum target of 85% on a 300-μm screen, to minimize caking propensity. Chapman [8] recommends that sugar to be stored in bulk in silos have fines below 300 μm removed.

Pilot work has been done in South Africa (unreported) and the United Kingdom to investigate the use of an airstream blowing through a falling curtain of sugar to classify the sugar. The South African work involved a series of louvers over which the sugar tumbled, while varying the air velocity resulted in different-sized fractions being entrained through the louvers. Lyle [30] reported the use of a wind-tunnel for sugar classification, but found that practical difficulties outweighed the high efficiencies obtained. A secondary benefit of screening is that it cools the sugar, often quite significantly. Yonkers refinery in New York reports a temperature drop across screening of 11°C.

e. Dust Removal and Collection. It is necessary to draw off sugar dust from conveyor transfer points and the filling points of bins and silos. This is normally done by means of hoods over the transfer points and ducting carrying the dust pneumatically to central collection equipment. In addition, the air passing through sugar driers carries away a dust fraction that must be collected (fluidized-bed driers are often used as a primary dust removal device). If an airborne classification system is to be used (as described above), this too requires a dust collection device.

Fabric filters consist of a battery of textile "socks" through which the air is blown. These are sometimes fitted with arrangements to dislodge the dust at regular intervals. The main disadvantage of these units is high-pressure drop due to progressive blinding, which causes a gradual reduction in airflow through the dust removal system and a loss in collection efficiency. It is also imperative that moisture (and vapor) be excluded from the system, as this would cause the cloth to blind. Electrostatic separators are highly efficient separators (above 99%) that use a very high dc voltage to carry the dust particles to the walls of narrow flow channels. These are, however, expensive and pose an explosive risk by providing a source of ignition in contact with a necessarily dusty environment. Other dry separators are available, such as standard cyclones (tangential entry and axial exit), axial-flow cyclones (axial entry and exit, with vanes in the flow path to impart spin), and baffle or louver-type devices. These offer trouble-free operation but do not provide the high efficiencies usually required.

Wet separators include irrigated cyclones (tangential or axial), irrigated fans, scrubbers, and baffle or louver systems. These systems usually achieve high efficiencies, above 95% for a combination of water sprays with cyclonic separation. A commonly encountered device is the Rotoclone irrigated fan, in which water is sprayed into the eye of the fan and is dispersed into the path of the dust-laden air entering the fan casing. Lyle [30] reports efficiencies of 96% for the Rotoclone; the device also has the advantage of combining the duties of a fan and a dust separation system in a single unit. Installations in South Africa have had success with banks of irrigated axial-flow cyclones in parallel (each unit no more than 100 mm in diameter). Baird and Beatts [2] experimented with a dust collection system integrated into the headbox of a rotary cascade drier and reported good results. The unit comprised a pack of chevron plates (more typically used for entrainment prevention in evaporators or pans), providing a tortuous path through which the air had to flow, combined with water sprays to keep the surfaces of the plates wet.

The major disadvantage of all wet dust collection systems is that they result in a well-aerated dilute sugar solution, which provides an ideal microbiological environment. Bacteria are washed out of the air, so it is impossible to keep the system sterile. In addition, efficient cooling of the water (or sweetwater) by the air in the separator eliminates temperature as a reliable means of microbiological control. The wet surfaces of a cyclonic separator, for example, are ideal sites for the proliferation of *Leuconostoc mesenteriodes.* There are several possible mechanisms of control:

1. Keeping the temperature as high as possible in any collection sumps or tanks (use steam sparging if necessary) to avoid degradation outside the separator itself.
2. In a recirculating system, keeping the pH up by adding lime (if fine spray nozzles are used, this may not be possible, due to the possibility of blockage).
3. Using a biocide either continuously or in shock doses.
4. Regularly washing out the separator, sump, or tank unit with biocide, lime, or steam.

f. Hoppers, Chutes, and Transfer Points. The primary design consideration in the design of hoppers is to ensure efficient and complete emptying without bridging or rat-holing. To guarantee zero holdup of sugar, the sides of an inverted-pyramid-shaped hopper should be between 75° (Lyle [30]) and 77° (Hugot [27]), which will result in 70 to 72° in the corners or valleys. An angle of 65° is acceptable for conical hoppers [27]. Practical constraints often make these ideals impossible, but designs using shallower angles should give consideration to systems for removing sugar that may hang up. Possible solutions include intermittent air sparging into the hopper to loosen the sugar and the use of membrane walls that can be flexed by a pulse of air behind the membrane to dislodge any buildup. Chutes feeding conveyors should be designed along the following guidelines:

1. The sugar must strike the conveyor vertically or in the direction of conveyor movement.
2. The chute must be as short as possible.
3. Valley angles in the chute must be kept above 60°.
4. A system for dust removal must be provided at the point of impact of the sugar.

If at all possible, steps should be taken to minimize the impact of the sugar on the conveyor (or bin or sugar pile). Various "shoe-type" devices exist for use at the bottom of the chute, to absorb or deflect the momentum of the falling sugar. An example is shown in Figure 15.25, where a plate with a parabolic cutout at the bottom of the chute feeds the sugar gently onto the conveyor, minimizing dust formation and belt wear. Another option is to choke-feed the sugar from the chute onto the conveyor, using plates with semicircular cutouts or rubber flaps to control the sugar flow. An ingenious chute design is the Cleveland Cascade chute, which has a series of minicones in the chute, arranged one under the other, with the axis of each inclined slightly to the vertical in the opposite direction to the cone above and below it. In this way, the sugar cascades through the chute in short, gentle

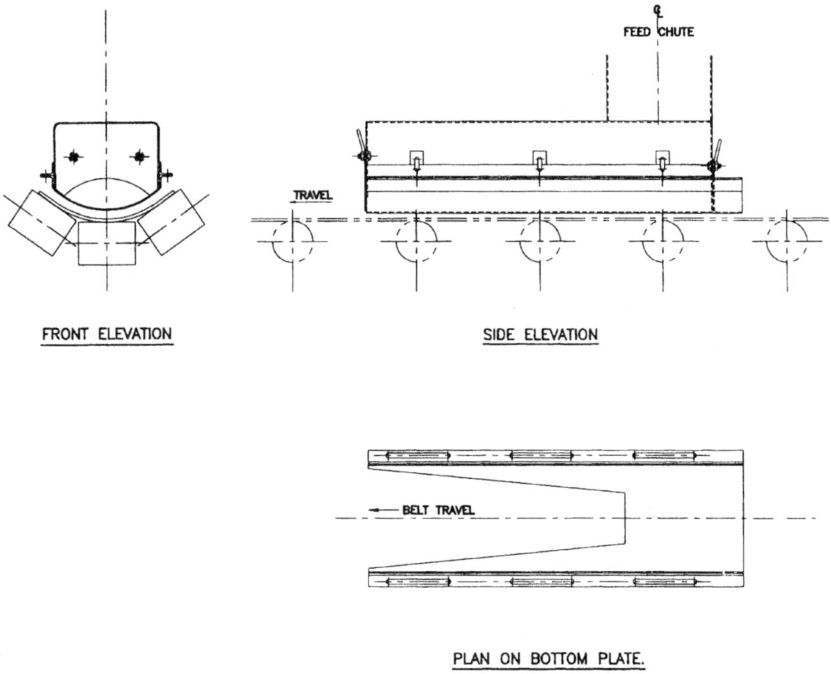

FIGURE 15.25 Chute "shoe" for impact minimization.

transfers from cone to cone. This unit is ideal for long chutes that would otherwise result in very high velocities.

g. Color Formation. When stored for long periods, refined sugar will experience a gradual increase in color. The rate at which this occurs is strongly dependent on storage temperature. Chapman [8] collected color increase data from four sites, and the results are summarized in Figure 15.26. Experience elsewhere indicates that Chapman's figures are on the high side (South African studies have shown color increases of anywhere between 10 and 100% over the period of a year), but the relationship with temperature is clearly demonstrated. The curve fitted to the data neglects the three Plaistow wharf (15 days) outliers. These points represent the color increase rate during the first 15 days of storage, but this rapid color increase during the initial storage period was not confirmed at the other three sites studied. However, South African work has found color increase to be more rapid during the first month of storage, after which the rate reduces. This may be a consequence of storing non- or improperly conditioned sugar, which experiences high rates of color increase while it conditions. Muro et al. [35] report a linear increase in color with time. De Mancilha [14] found that color formation rate increased logarithmically with temperature (the curve fitted to Chapman's data has both linear and logarithmic terms).

Chapman recommends storage below 20°C to minimize color formation, while Clarke et al. [11] suggest storage below 30°C and recommend further that sugar entering storage should be below 45°C. De Mancilha carried out tests in which it was found that "up to 35°C, browning was negligible." Most authors recognize that sugars differ in their suscep-

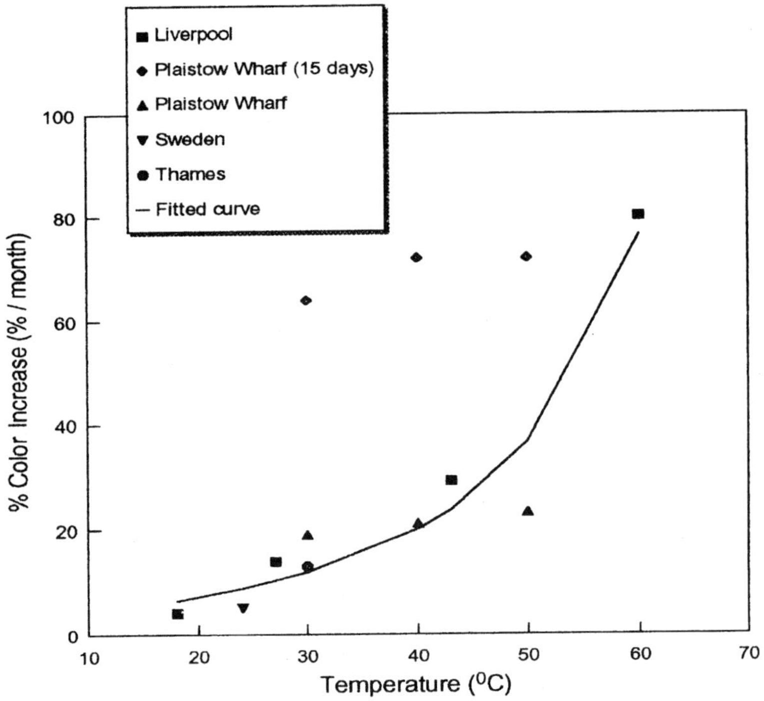

FIGURE 15.26 Rates of sugar color increases in storage. (From Ref. 8.)

tibility to color increase and their sensitivity to temperature. Clarke et al. and Chapman comment that beet and cane sugars may behave differently (Chapman mentions the contribution of amino acids). Cheng et al. [10] found that "colour development was more marked in sugars manufactured by a carbonatation process than by a sulphitation process."

Other factors that affect color development in storage are ash and moisture. Moisture and air are necessary for most color-forming reactions. A test carried out in South Africa measured color increase in identical sugar samples stored under identical conditions, except that half of the samples were packed in plastic (moisture-impermeable) bags and the other half in paper bags. The plastic-packed sugar increased in color from 32 ICUMSA units to 44 over a period of 11 months, while the sugar packed in paper increased from 32 units to 68 units in the same period.

Some ash components (e.g., iron) can catalyze color-forming reactions, but in general a higher ash content indicates a lower-quality sugar, which will have higher moisture and color levels as well. The distribution of color and ash is also important, as a sugar with these components concentrated in the surface film will develop color faster than one in which the distribution is throughout the crystal. This is a strong argument in favour of thorough washing of the sugar in the centrifugals.

h. Sugar Storage Overview. Mikus and Budicek [32] have compiled a superb overview of the entire subject of refined sugar storage, incorporating a most useful section on conditioning. They provide tables of physical and chemical properties of sugar, as well as summaries of published data pertaining to sugar storage. In addition, their list of references is the most exhaustive available on the topic. This work is highly recommended to those who wish to obtain a fuller understanding of this area of sugar technology.

15.4.4 Conveying

The topic of conveying must be introduced with Lyle's wry comment that "there is no satisfactory conveyor for refined sugar" [30]. However, sugar may be conveyed by a wide variety of methods, some of which are discussed below.

a. Screw Conveyors. Screw conveyors have a central shaft running along the axis of a U-shaped trough, with a screw or helical ribbon attached to the shaft. The sugar is pushed along the trough by the action of the screw and may be discharged at the end or through a hole in the bottom. The greatest disadvantage of this type of conveyor is that the sugar crystals may be ground between the screw and the shell, although Spencer and Meade [48] claim experimental evidence to show that little crystal breakage occurs if the trough is filled to less than one-third of its capacity. In any event, the sugar level in a screw conveyor should be below the level of the shaft, particularly if the sugar is wet. Another characteristic of the screw conveyor is mixing, which may be used to advantage in mingling or coating operations, for which the helical-ribbon type is particularly suited. Tromp [56] gives the capacity of a screw conveyor as:

$$Q = \frac{\pi d^2}{4} \mathrm{Cv} \rho p N$$

where Q is the conveying rate (mass per unit time), d the outside diameter of screw or ribbon, Cv the coefficient of fullness and slip, ρ the bulk density of the sugar, p the pitch of screw or ribbon, and N the rotational speed of the screw. An average Cv value for a screw filled to just below the shaft is given as 0.2.

b. Grasshopper Conveyors. Grasshopper conveyors are simple troughs supported from below on a series of links or flat springs. The trough moves in a reciprocating motion, forward and upward on the forward stroke and backward and downward on the return stroke. Since this action is rapid, the sugar is thrown forward with each stroke, to land at a position farther advanced along the trough. The supporting bands are inclined backward (about 30° to the vertical) and the stroke is short (typically, 20 to 30 mm). Grasshopper conveyors are simple, easy to operate, and clean, may be quite long, and can be loaded in heaps (as is the case below batch centrifugals) as the grasshopper action quickly levels out the sugar. Grasshoppers may be discharged at the end or out of a hole in the bottom of the trough. Disadvantages are high maintenance due to the considerable vibration that they generate and possible classification of the sugar in the conveyor. To determine the capacity of a grasshopper conveyor, it is necessary to estimate the average travel of the sugar crystals per stroke, after which it is a relatively simple matter to multiply by the frequency and the filled cross-sectional area of the trough. Tromp [56] suggests that the trough should not be filled to a depth greater than 150 mm and also proposes the use of a coefficient of slip (suggested value 0.5) in calculating conveyor capacity. The effect of this factor is to reduce the capacity by half to account for slippage. Variations on the grasshopper conveyor are:

- *Twin conveyors:* overcome the vibration problem of a grasshopper by oscillating two identical units 180° out of phase.

- *Vibrator conveyors:* operate on the same principle, but using high-frequency vibration. These can be enclosed (dust free) but consume a lot of power and certainly classify the sugar.
- *Slip-stick conveyors:* recent innovation, operating by means of a slow forward stroke followed by a rapid return stroke, both in the horizontal plane (along the conveyor axis). They rely on the sugar slipping on the return stroke to advance the sugar along the conveyor.

c. Belt Conveyors. Belt conveyors are continuous belts consisting of several plies of fabric impregnated, bonded, and coated with rubber. These run on rollers called *idlers,* spaced at intervals of about 1 to 1.5 m for the top, load-carrying strand and much larger spacings for the lower, empty strand (enough to keep the sag within acceptable limits). The top idlers are usually in sets of three or five, with the center ones horizontal and the outer ones inclined at about 20° to the horizontal, thus forming a trough. Troughing the belt more than doubles its capacity and minimizes the likelihood of spillage over the edges. Troughing idlers should be pitched closely or the load will be disturbed as it passes over a set of idlers, causing flattening out of the load and reduction in belt capacity. Return idlers may be single tubes or multiple narrow wheels if the material is sticky. The capacity of a belt conveyor varies with the square of the belt width.

Belt conveyors may be inclined, but Lyle [30] does not recommend more than 22.5° for dry sugar. Hugot [27] prefers a limit of 20° and recommends 16°. A problem with belt conveyors is ensuring the return strand is clear of adhering sugar, which is crushed by the return idlers and creates dust and mess. Many installations use brushes or scrapers after the head pulley, but these are not always 100% effective, present a risk of product contamination, and add to any dust problem.

Feed onto a belt conveyor should always be done on the center line of the belt, as asymmetric loading will cause the belt to track to one side, resulting in spillage and belt wear. Discharge from belts is usually over the head pulley, but where intermediate discharge is required, a throw-off carriage may be used (ploughs on belts are not usually successful). A throw-off carriage is a frame running on rails along the conveyor, so that it may be moved to the desired discharge point. It contains two pulleys around which the belt is looped, and the load is discharged down chutes on either side of the belt. These devices are commonly employed to distribute sugar in a large storage bin or silo. An alternative is to use a shuttle conveyor, which is half the length of the silo, mounted on wheels, and can be driven in either direction. The shuttle is positioned to carry the sugar from the main belt discharge at the silo midpoint to the desired discharge position.

A recent innovation in belt conveying is a system that uses specially positioned idlers to roll the edges of the belt over to form a closed sheath that is kept closed by idlers above as well as below the conveyor (an example is the U-CON conveyor [23]). These can transport sugar up steep slopes, apparently even vertically. The belt can load and discharge in a conventional open form and close only for the inclined portion of its travel.

Steel belts offer an alternative to rubber belts, as the rolled steel surface is easier to scrape or wash and material can be ploughed off cleanly. However, these belts are either flat or slightly troughed and are therefore of lower capacity than rubber belts of the same width. Steel belts may be inclined, unless the material is lumpy, as lumps will tend to roll backward down the conveyor and fall over the edge, as the troughing is insufficient to hold them.

An interesting variation on the belt conveyor is the *air-supported belt* (Aerobelt being the best known). This belt has no idlers and travels over a deck plate of perforated steel, through which air is blown, supporting the belt on an air cushion. The air-supported belt minimizes friction, thus minimizing power consumption, greatly reduces maintenance requirements, and is clean and neat. A *thrower* is a very short (about 1 m) belt conveyor comprising a head and a tail drum and two troughing wheels between them. The belt is driven very fast (over 10 m/s) and the sugar dropped onto it is thrown up to 20 m in a fairly neat stream, which can be controlled further by the use of a target plate. These devices are useful for distributing bulk sugar evenly into storage (such as in trimming the holds of ships), but will obviously cause abrasion of the sugar.

d. Bucket Elevators. Bucket elevators consist of chains or a belt to which buckets or scoops are attached. These run over sprockets or pulleys at each end and may be inclined, but are most often vertical. They are usually encased in sheeting to control the dust that they generate, and the material is fed into the boot, which is the bottom of the casing, to be scooped up by the buckets. The feed chute should be sized to prevent loading of the elevator at a greater rate than the buckets can remove, to avoid jamming.

Bucket elevators may be low or high speed in type. High-speed units have short-link chains or belts. Of these, belts are quiet and cheap but are not suitable for sticky materials that build up on the belt and the head and tail pulleys. High-speed elevators discharge by throwing the sugar out of the bucket tangentially as it goes over the head pulley, and so are only suitable for sugars that discharge fairly readily. Wet, sticky sugars will compact into the bucket as it picks up its load and will not discharge rapidly enough at the head. For these products, a slow-speed unit may be more suitable. These typically use longer-link chains and discharge by dropping the sugar into a chute. This may be achieved using a deflector on each bucket to deflect the sugar falling from the bucket above, by tucking the return strand of the chain under the head pulley, out of the way of the sugar chute, or by inclining the last portion of the elevator before discharging.

The major disadvantage of the bucket elevator is that some grinding of the sugar and consequent dust production are unavoidable. Buildup of sugar inside the conveyor casing may also present problems. A less frequently occurring but more serious problem is the potential for failure of one of the chain links, as proper lubrication of these links is not practical. Link or pin failure may result in the entire elevator dropping down and wedging in the casing, becoming what Lyle [30] refers to as a "major mess-up."

Gravity bucket conveyors have buckets that hang between pivots, so that the bucket always hangs the right way up, regardless of the orientation of the chain. The conveyor may therefore be horizontal or vertical, or change from one to the other. Buckets are emptied by engaging the buckets with a hook and tipping them over. Problems with these conveyors are in controlling swinging of the buckets, and that sugar tends to get into the pivot bearings and prevent buckets from hanging correctly, thus tipping their loads at the wrong time.

e. Other Conveyors

- *Drag chain or drag cable elevators or conveyors:* a chain or cable with plates or washers at intervals, which drag the material up a closely fitting casing, chute, or pipe. The Redler conveyor is probably best known, consisting of a single chain of rakelike links, and is most commonly used for coal. The Floveyor is a cable-and-disk version running in a pipe and is quite successful on dry sugar.
- *Skip hoists:* large, single-bucket (skip) elevators that carry the load to the top, tip, and then reverse to return the skip to the bottom, in a time sequence.
- *Plate, tray, or apron conveyors:* overlapping steel plates that form a continuous band or trough. These are, however, expensive and maintenance intensive and have largely been replaced.

A range of conveying mechanisms are available for conveying bagged sugar, including cranes, slings, grabs, swing tray elevators, slat conveyors, rubber apron conveyors, and gravity roller conveyors. The details of these will not be discussed here.

f. Run-Back. All elevators or inclined conveyors must be equipped with a mechanism to prevent their running backward in the event of motor failure or shaft breakage, and these mechanisms must be inspected regularly. At best, runback will result in mess and inconvenience; at worst, it will lead to serious and expensive damage (a large bucket elevator running back may reach such speed that it destroys itself).

g. Pneumatic Conveying. A completely different approach to sugar conveying is pneumatic transport. An effective option is dense-phase pneumatic conveying in which the sugar is conveyed in batches at much lower velocities than dilute phase, using much less air and consequently with little crystal breakage. The successful use of this system at the Hulett refinery in South Africa was reported by Gelling [22]. The advantages were that it was clean, sealed (contamination-free), space-saving, safe, quiet, and required little maintenance. Disadvantages are the need for clean, dry, filtered air, the need for a pressure vessel, the requirement for good dust collection, and the fact that chokes can be difficult to clear. The system is cost-effective for long or complex routes, but cannot compete over short, straight transfers.

h. Magnets. Magnets should be fitted in the path of the product sugar to remove any steel contaminants such as nuts, bolts, shavings, or metallic scale. The most common types of magnet and their uses are as follows:

1. Fixed magnets are positioned just above or just penetrating the sugar stream on the conveyor. They must be cleaned manually.
2. Target or impingement magnets are fitted so that the stream of sugar, usually coming over the headshaft of a conveyor, hits the surface of the magnet. This ensures better likelihood of pickup of contaminants, but presents the risk of material being swept off the magnet by the sugar. These must also be cleaned manually.
3. Magnetic conveyors have a conveyor with steel slats that passes between the magnet and the sugar stream, and the magnetic material is conveyed sideways and dumped.
4. Magnetic drums are positioned at the end of a conveyor so that the sugar passes over the drum. As the drum rotates, the portion of its surface in the non-sugar-carrying position is deenergized, dropping the material that it has picked up.

REFERENCES

1. D. F. Bagster, *Int. Sugar J.*, 71:263–267, 298–302, 1970.
2. J. C. Baird and R. M. Beatts *Proc. Austral. Soc. Sugar Cane Technol.*, 1989, pp. 237–245.
3. R. G. Bennett, R. E. Runeckles, and H. M. Thompson, *Int. Sugar J.*, 66:109–113, 1964.
4. E. D. Bosse, *Proc. 50th Congr. Sugar Ind. Technol.*, 1991, pp. 171–185.
5. E. D. Bosse, *Proc. 56th Congr. Sugar Ind. Technol.*, 1997.
6. J. Bruijn, B. S. Purchase, and A. B. Ravno, *Int. Sugar J.*, 84:361–365, 1982.
7. L. Budicek and O. Mikus, *Listy Cukrov.*, 96:202–207, 1980.
8. F. M. Chapman, *Proc. 20th Conf. Brit. Sugar Corp.*, 1970.
9. J. C. P. Chen and C. C. Chou, *Cane Sugar Handbook*, 12th ed., Wiley, New York, 1993, pp. 499–523.

10. H.-T. Cheng, W.-F. Lin, and C.-R. Wang, in *Maillard Reaction in Foods and Nutrition,* Am. Chem. Soc. Symp. Ser., 215:91–102, 1983.
11. M. A. Clarke, M. A. Godshall, R. S. Blanco, and X. M. Miranda, short report to CITS Scientific Committee presented by G. Mantovani, 1992.
12. W. R. Crawford, *Proc. 19th Conf. Queensl. Soc. Sugar Cane Technol.,* 1952, pp. 75–81.
13. J. M. De Bruijn and A. A. W. Marijnissen, *Proc. Conf. Sugar Process. Res.,* 1996.
14. F. A. S. De Mancilha, *Bol. Tec. Copersucar,* spec. ed., Oct. 1987.
15. J. F. Dowling, *Proc 25th Congr. Sugar Ind. Technol.,* 1966, pp. 34–59.
16. J. L. Erisman, *Ind. Eng, Chem.,* 30(9):996–997, 1938.
17. T. L. Excell, *Proc. 58th Congr. S. Afr. Sugar Technol. Assoc.,* 1984, pp. 56–60.
18. T. L. Excell and V. C. Stone, Proc. 63rd Congr. S. Afr. Sugar Technol. Assoc., 1989, pp. 68–72.
19. S. Farag, *J. Austral. Soc. Sugar Beet Technol.,* 20(3):207–216, 1979.
20. J. R. Fitzgerald, K. Taylor, and G. W. Bestwick, *Proc. S. Afr. Sugar Technol. Assoc.,* 1980, pp. 52–55.
21. S. J. Friedman and W. R. Marshall, *Chem. Eng. Prog.,* 45:482, 573, 1949.
22. B. Gelling, *Proc. 55th Congr. Sugar Ind. Technol.,* 1996, pp. 327–335.
23. S. F. Hansen, *Proc. 55th Congr. Sugar Ind. Technol.,* 1996, pp. 313–326.
24. M. C. J. Hodgson and W. J. Keast, *Proc. Austral. Soc. Sugar Cane Technol.,* 1984, pp. 211–218.
25. D. Hoks and E. Elfrink, *Zuckerindustrie,* 118(6):465–468, 1993.
26. A. M. Howes, *Proc 40th Congr. S. Afr. Sugar Technol. Assoc.,* 1966, pp. 214–219.
27. E. Hugot, *Handbook of Cane Sugar Engineering,* 2nd ed., Elsevier, Amsterdam, 1972.
28. V. Kavan and O. Mikus, *Listy Cukrov.,* 98:152–160, 1982.
29. C. J. Lu, E. H. Hsu, and H. C. Tseng, *Proc. 16th Congr. Int. Soc. Sugar Cane Technol.,* 1977, pp. 2599–2610.
30. O. Lyle, *Technology for Sugar Refinery Workers, 3rd ed., Chapman & Hall, London, 1970.*
31. D. M. Meadows, *Proc. 53rd Congr. Sugar Ind. Technol.,* 1994, pp. 388–404.
32. O. Mikus and L. Budicek, *Sugar Technol. Rev.,* 13:53–129, 1986.
33. C. O. Miller, B. A. Smith, and W. H. Schuette, *Trans. AIChE,* 38:841, 1942.
34. K. Morden, *Int. Sugar J.,* 96:(1142):48–55, 1994.
35. M. Muro, A. Kosiavkin, and J. A. Garcia, *Centro Azucar* (Univ. Las Villas, Cuba), 1(3, Sept.–Dec.):pp. 37–84, 1974.
36. A. P. Neilson and W. W. Blankenbach, *Proc. 23rd Congr. Sugar Ind. Technol.,* 1964, pp. 109–114.
37. E. F. Obst *Will the Boonton Moisture Analyser Predict the Tendency of a Sugar to Cake?,* American Sugar Company, Baltimore Refinery, 1966.
38. R. H. Perry and C. H. Chilton, *Chemical Engineers' Handbook,* 5th ed., McGraw-Hill, New York, 1973.
39. P. H. Petri, *Proc. 19th Congr. Sugar Ind. Technol.,* 1960, pp.119–136.
40. G. Pope, E. Taske, T. Wieden, and G. A. Brotherton, *Proc. Austral. Soc. Sugar Cane Technol.,* 1993, pp. 167–173.
41. R. R. Ramphal, *Proc. 63rd Congr. S. Afr. Sugar Technol. Assoc.,* 1989, pp. 64–67.
42. T. Rodgers and C. L. Lewis, *Int. Sugar J.,* 64:359–362, 1962; 65:12–16, 43–45, 80–83, 1963.
43. W. C. Saeman, *Chem. Eng. Prog.,* 58:49–56, 1962.
44. A. L. Schinkel, and P. J. Tait, *Proc. Austral. Soc. Sugar Cane Technol.,* 1994, pp. 229–306.
45. F. W. Schwer, *Proc. 23rd Congr. Sugar Ind. Technol.,* 1964, pp. 115–129.

46. P. J. Shardlow, P. G. Wright, and L. J. *Proc. Austral. Soc. Sugar Cane Technol.*, 1996, pp. 368–375.
47. K. Sharples, P. G. Gliken, and R. Warne, *Trans. IchE,* 42:T275, 1964.
48. E. F. Spencer G. P. Meade, *Cane Sugar Handbook,* 9th ed., Wiley, New York, 1963.
49. S. Stachenko, W. A. R. Allen, and J. A. Swan, *Proc. 25th Congr. Sugar Ind. Technol.*, 1966, pp. 75–122.
50. R. J. Steindl, M. Sheehan, I. T. Cameron, *Proc. Austral. Soc. Sugar Cane Technol.*, 1994, pp. 287–294.
51. Symposium: Drying, cooling and conditioning of granulated sugar, *Proc. 19th Congr. Sugar Ind. Technol.,* 1960, pp. 159–195.
52. Symposium: Refined sugar storage systems, *Proc. 53rd Congr. Sugar Ind. Technol.*, 1994, pp. 155–211.
53. Symposium: Drying and conditioning of white sugar, *Proc. 55th Congr. Sugar Ind. Technol.*, 1996, pp. 201–294.
54. P. J. Tait, *Proc. 55th Congr. Sugar Ind. Technol.*, 1996, pp. 212–220.
55. P. J. Tait, A. L. Schinkel, and C. R. Grieg, *Proc. 9th Int. Dry. Symp.,* 1994, pp. 203–211.
56. L. A. Tromp, *Machinery and Equipment of the Cane Sugar Factory,* Norman Rodger, 1936.
57. E. C. Tubb, *Proc. 27th Congr. Sugar Ind. Technol.*, 1968, pp. 28–43.
58. N. R. Twaite and A. J. Randall, *Int. Sugar J.*, 89:130–135, 1987.
59. P. G. Wright and L. J. Johns, *Proc. Austral. Soc. Sugar Cane Technol.*, 1993, pp. 174–179.

CHAPTER 16

Packaging, Warehousing, and Shipping of Refined Products*

INTRODUCTION

The last three operations in sugar refining—packaging, warehousing, and shipping—are very important ones in several aspects. First, they account for a significant proportion of the total refining cost. The packaging operation is very labor intensive since this activity may, in a typical refinery, cover a dozen or more packaging lines located in various distant areas of the refinery. In addition, one of the largest losses of sugar in a refinery is in product giveaway. Highly sophisticated weight measurement devices and skilled technicians are needed to minimize the size of this economic loss. Again, in the packaging operation, container material is invariably a very high cost item in sugar refinery operations. The warehousing department also tends to be labor intensive since it involves handling the packaged goods several times between the time the products are palletized and the time the same products are shipped, sometimes weeks later. Second, these three activities are the most critical in terms of customer acceptance of the product. An attractive package design free of defects often is a deciding factor for the selection by the consumer of one brand of sugar over another. Finally, it is of utmost importance to pay close attention to product safety. The use of scalping screens, magnets, and metal detectors is an essential part of any well-designed packaging system. The adherence to good manufacturing practices, including an excellent pest control program, is all part of the packaging and warehousing activities.

16.1 PACKAGING

16.1.1 Typical Packaging Line Arrangement

Packaging lines must not only package a product, but must also include provisions to ensure product safety and minimize product giveaway. Consumer packaging lines include a primary

* By Jean-Paul Merle.

container that contains the product and a secondary container that collects primary containers in various unit quantities: 8 units of 5 lb, 12 units of 4 lb, 16 units of 2 lb, and 24 units of 1 lb, being the most common sizes in the United States. Recent trends have been to limit the weight of secondary containers to less than 50 lbs because of lifting injuries. Industrial packaging lines include bags of various types in unit weights of 25, 50, and 100 lbs. The latter weight is becoming less and less common, again because of lifting injuries. In addition, the trend toward very large sizes is increasing rapidly; the 2000-lb supersack, or tote bag, is the most commonly used in such applications.

A typical packaging line setup is shown in Figure 16.1. The product to be packaged, if free flowing, passes through a vibrating stainless steel scalping screen with a typical opening of $\frac{1}{8}$ in., where sugar lumps and other materials are eliminated. The product passing through the scalping screen next passes over a rare-earth magnet prior to being packaged. The primary container or industrial bag passes over a check weigher that will reject over- and underweight products. Next, the package enters a metal detector that, depending on the product packaged and the size of the package, can detect metal contaminants down to 1 or 2 mm in size. Consumer lines usually have individual metal detectors, while industrial bags pass through a common metal detector on their way to the warehouse's palletizers. The individual packages are then coded using ink-jet coders. These coders are fully automated, provide a clearly legible code, and require a minimum of attention. Following the coder, the consumer packages are packed in a case or bundle that is also coded. The industrial coded bags are conveyed directly to the warehouse.

16.1.2 Products to Be Packaged

The various grades of sugar produced by a typical sugar refinery generally fall into the following categories: (1) white granulated sugar, (2) powdered sugar, (3) soft brown sugar, and (4) other special products. White granulated sugar is packaged in sizes varying from a few ounces to a 2000 pound supersack. Soft and powdered sugars are packed in containers from one pound or 500 grams to 100 pound or 50 kilograms. Depending on the type of

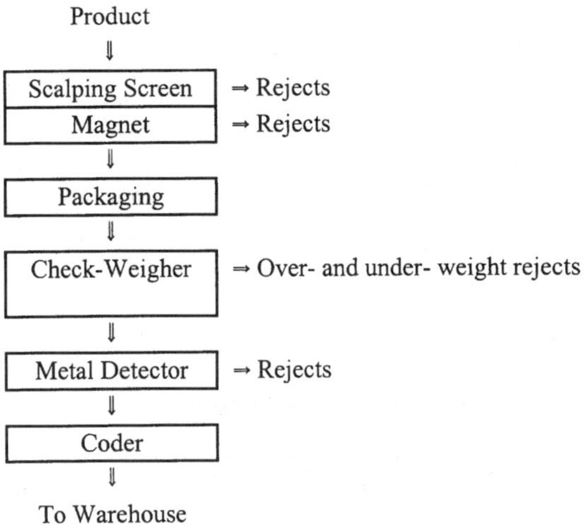

FIGURE 16.1 Typical packaging line arrangement.

sugar and its physical characteristics, the container material used varies from a carton with or without a liner to polyethylene bags to lined or multiwall paper bags. Furthermore, for the consumer products, the packages are either cased or bundled. As mentioned earlier, container cost is a significant item in total refining cost and care and judgement must be exercised in deciding on the mil thickness of the particular type of container material. While conservation of the product integrity is of primary importance, attention must also be paid to the packaging material basis weight.

16.1.3 Packaging Equipment

Over the past several decades, packaging equipment has evolved from a series of chain, gear, cam and other positive-acting mechanical and electric drives to the new solid-state electronic microprocessor-controlled units. The new state-of-the-art drives and controls have greatly improved the operating efficiency of the packaging machines but have also created a demand for highly skilled technicians to maintain and troubleshoot very sophisticated systems. The considerable progress in the electronic field has also led to more completely automated machines that require less operating labor. This progress has also contributed to better weight control systems with significant reduction in the amount of product giveaway. Sugar refiners package three basic products (granulated, soft brown, and powdered sugar) in various container types in unit weights varying from a few grams to 2000 pounds to satisfy a large number of customers (Table 16.1).

a. Types of Packaging Machines. In general, there are two basic types of machines that package goods for the industrial and consumer markets. Consumer-grade products, which vary in size from the packet to 10 lb or 5 kg, are usually packaged in continuous motion at high output rates, except for powdered and soft sugars, which run at slower intermittent motions. Packaging machines for industrial-grade products are almost invariably of intermittent motion and operate at much lower output rates. The Bosch model H-PLDBR of German manufacture and the Swiss SIG machine are examples of high-speed, continuous-motion packaging machines, whereas the Bemis 7115D, of U.S. manufacture, is a common unit used for industrial-size packages. Packaging machines are also differentiated by the type of containers used (i.e., whether preformed or of the form–fill–seal type), by the method of filling (volumetric or weight), and by various types of seals, such as glued, sewn, valved, or heat-sealed. Generally, consumer-grade packaging machines include a sys-

TABLE 16.1 Container Types and Sizes

Container Type	Product	Container Sizes
Packet	Granulated	$\frac{1}{7}$ oz, $\frac{1}{8}$ oz, $\frac{1}{9}$ oz, $\frac{1}{10}$ oz, 4 g, 6 g
Carton	Granulated	1 lb, 2 lb, 1 kg, 4 lb
	Soft brown	1 lb
	Powdered	1 lb
Polyethylene	Soft brown	2 lb, 4 lb, 2 kg
	Powdered	2 lb, 4 lb, 2 kg
Paper pocket	Granulated	4 lb, 5 lb, 10 lb, 2 kg, 5 kg
Paper bag	Granulated	25 lb, 50 lb, 100 lb, 20 kg, 50 kg
	Soft brown	25 lb, 50 lb, 100 lb, 50 kg
	Powdered	25 lb, 50 lb, 100 lb, 50 kg
Supersack	Granulated	2000 lb, 1000 kg

tem to form the container, fill using a combination volumetric-trim weight, and have a glue sealing system. Industrial-grade units generally use preformed bags, batch in the product through a scale, and use a heat-sealed system.

b. Container Feeding Systems. Packaging machines producing sewn bags usually do not have a bag feeding system. Bags for filling are manually placed on the filler by the operator. More modern units featuring heat-sealed closure for paper bags include a system where the bags are picked up automatically, usually by vacuum, and fed into the packaging machine for filling. For consumer-size containers, paper pockets or polyethylene pouches are formed at the packaging unit from a roll feeding the unit continuously. Preformed cartons are manually stacked by the operator and fed automatically into the packaging machine for filling.

c. Filling Systems. Depending on the type of sugar handled, the filling system can be by electrically driven augers controlled by a signal from load cells, by air to blow the product into the bag as in the case of powdered sugar, or by dumps from gross weighers. In the case of granulated and powdered sugars, some method is utilized to compact or densify the product to limit the size of the container and reduce material cost. The method used to achieve this can be a mechanically driven shaker or an electric vibrating probe that dips into the container.

d. Weighing Systems. There are three general methods by which the quantity of sugar delivered to a packaging container is controlled:

1. Volumetric, with provision for manual or automatic adjustment for variations in bulk density of the material. The Model 406 Velocitron uses this system.
2. A combination of volumetric and weight wherein an initial volume is dumped into the package, after which the package passes over a check weigher to verify the weight and send a signal to a secondary addition system that meters, usually by an electrically driven auger, the quantity of product necessary to make up the weight to the label weight. The Bosch Model H-PLDBR has such a system.
3. A scale with a load cell that batches the weight of product to be delivered to the container. This system usually also includes a trim final product addition. The Bemis Model 7115 D is an example of this type of weight control.

There is another weighing system that has gained popularity in recent years and is capable of delivering a very accurate weight control. It is a combination-type weigher, a version of which is manufactured by Ishida. In this weigher, the product falls onto a circular vibrating dispersion table that conveys the sugar to the radial troughs, which in turn vibrate to convey the product into the pool hoppers. The weigh/drive units discharge the product in the pool hoppers into the weight hoppers below if they are empty or require more product; otherwise, the pool hoppers hold the product until the next cycle. The load cells in the weight/drive units measure the weight of product in the weigh hoppers, and the results are sent to the calculation unit. The calculation unit selects the combination of weigh hoppers whose total weight is closest to the target weight with the minimum of product giveaway. The calculation unit signals the drive units to discharge the products in the weigh hoppers selected. The product dumped from the selected hopper falls through the discharge chute and out to the packaging machine. The cycle then repeats itself. The arrangement is shown in Figure 16.2. The weighing system is capable of coming as close as 0.2 g to the label weight of 907 g. Scale performance is continuously displayed in terms of X-bar charts, average weights, standard deviation, and so on.

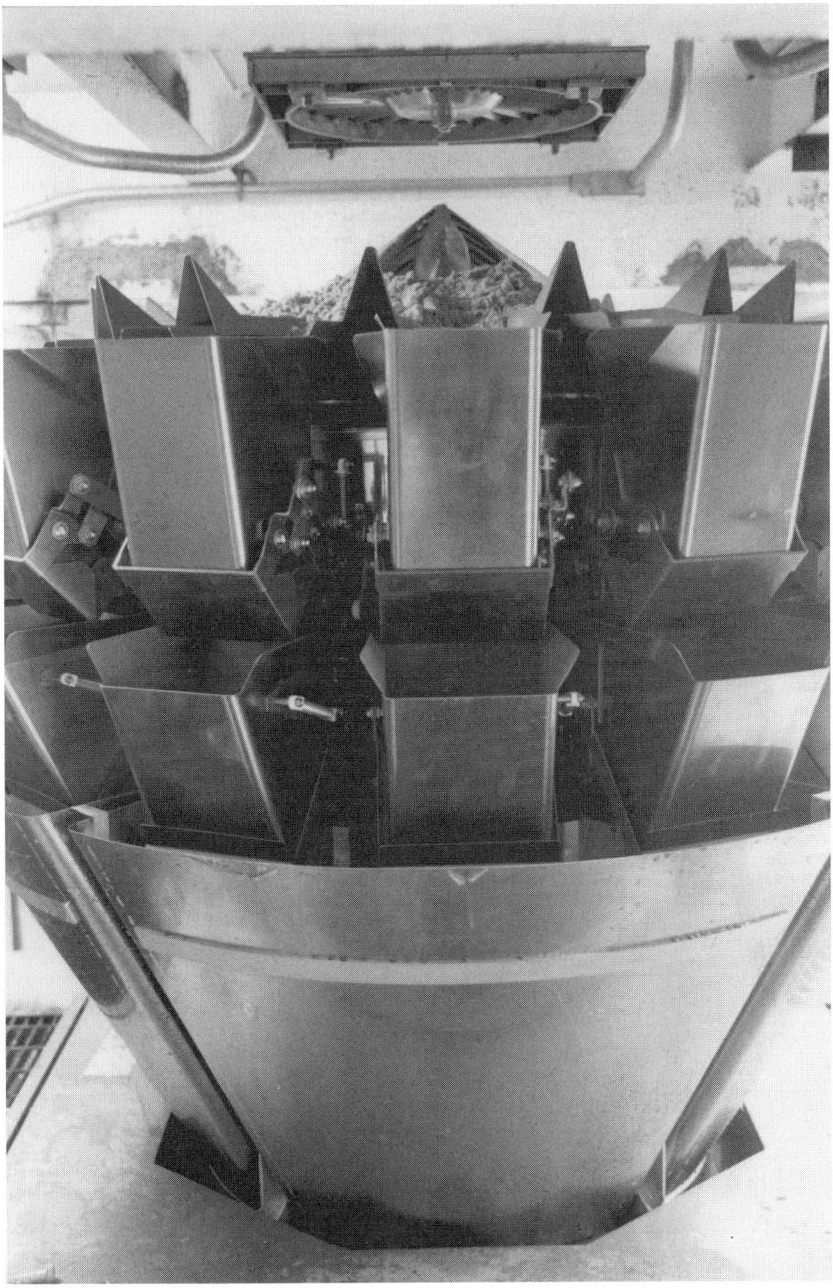

FIGURE 16.2 Combination weigher.

e. Sealing of Containers. There are four types of closing systems used: (1) self-closing valve, (2) sewn, (3) glued, and (4) heat-sealed, the last two systems being the most common. The heat-sealed closure type applies to polyethylene bags and to larger paper bags that have preapplied adhesive. In the latter application, it is important to establish a regular cleaning schedule for the heater bars; otherwise, wrinkles in the closure area may lead to product leakage.

f. Speed. In continuous-type packaging machines used for consumer-grade products, line speeds up to 220 packages per minute are quite achievable. For intermittent-type industrial packers, line speeds vary between 8 and 135 packages per minute for bag sizes between 4 and 100 lb. In between these two extreme values are several packaging units that deliver 30 to 40 containers a minute. Plastic pouch form–fill–seal units are good examples of the latter speed range.

16.1.4 Selection of Packaging Equipment

There are several important considerations in the selection of packaging equipment. The grade of sugar to be packaged is necessarily the primary consideration. Some sugars, including powdered and soft brown, are not free-flowing and special designs are required to overcome this lack of product mobility. Typically, various types of flakers for soft sugars are used wherein the product is forced mechanically through openings of various sizes. Products such as soft sugars that contain a high level of moisture and have the crystals coated with a thin film of syrup can present more problems downstream of the flakers, where the product tends to stick to surfaces and build up a crust that can eventually break off and be packaged with the product. In the selection of the packaging equipment, care must be taken to provide for adequate sloping surfaces and specially designed liner materials that will minimize the buildup. Furthermore, product buildup on surfaces necessitates frequent cleaning and results in lost time and poor equipment utilization. For powdered sugar, densifiers are commonly used to deaerate the product prior to packaging it. Inadequate deareation can result in high container costs to case-sealed packages. A second important consideration is the packaging speed required. From a labor utilization standpoint, there is generally a clear economic advantage in being able to store a particular product in a bin and package it out as fast as possible. Again, from a labor utilization standpoint, the equipment selected should be automated as much as possible without unduly affecting the operating efficiency of the packaging machine. Packaging systems should be chosen incorporating electronic features that display statistical weight control, line speed, and other pertinent useful operating data. Location of the packaging equipment is another important factor to consider. Floor space is often at a premium in a manufacturing facility, and a compact unit generally offers an advantage. Next, a decision has to be made on the location of the equipment. If the product to be packaged is easy to convey, the packaging equipment can often be located remotely from the source of the product. In other cases, such as soft sugar, the preference would be to locate the packaging machine close to the source and avoid complicated and expensive conveying systems. There are many other factors to be analyzed in the selection of packaging equipment.

- Availability for spare parts
- Access to out-of-plant service to support the efficient operation of the equipment when in-house expertise becomes inadequate
- Price
- Ease and speed of changeover from one container size to another

- Good documentation, especially regarding electrical components and control
- Easy access to machine parts for repair and cleaning
- Frequency of cleaning required, mostly in glue closure applications
- Dust containment and removal

Finally, the equipment purchaser would make a wise investment in training operators and maintenance personnel thoroughly at the equipment supplier's facility prior to installation of the packaging line.

16.1.5 Statistical Weight Control

Knowledge of the National Institute of Standards and Technology (NIST) Handbook 133, *Checking the Net Contents of Packaged Goods* [1], together with the basic principles of descriptive statistics, is critical for a successful in-house package weight control program. Handbook 133 cites the cooperation between NIST and the states in having uniformity in weights and measure laws and methods of inspection. Two essential requirements must be met by all producers of packaged goods:

1. The average quantity of the contents of packages in a lot, shipment, or delivery must equal or exceed the quantity printed on the label.
2. The variation of individual package contents from the labeled quantity must not be "unreasonably large."

a. Control Charts. The purpose of statistical weight control is to give the customer the quantity of product printed on the package and to remain legal with Handbook 133 as applied by the weights and measures official in each county of the United States. The way to control the net weights at the packaging machines is to take a sample of packages in each time period, determine each net weight, plot the average, and compare it to the target printed on the average control chart. Then the decision is to increase, leave alone, or decrease the amount of product packed.

The target is set at the labeled weight for a *zero giveaway program.* The upper and lower control limits are set at a standard deviation of ± 3 of the averages calculated by dividing the standard deviation of individual weights by the square root of the sample size. The sample size is determined by the number of filling heads on the packaging machine. If it has only one filling head, the sample size is five consecutive primary packages taken off the line after the top is sealed. If it has multiple filling heads (from two to eight), the sample size is equal to the filling heads. If it has nine or more heads, only five consecutive packages are collected.

The operator is to select 1 minute at random in each 30-min time period, collect the correct number of consecutive packages off the line, determine the net weight of each package, and enter them on the control chart. Each average and range is plotted against its sampling time. If the average falls between the upper and lower control limits, no adjustment in the package input is necessary. However, should the average fall outside the limits, a correction is needed to bring the average close to the target and within the control limits. The technique is to make a change in the fill quantity equal to half the difference between the average and the target values.

b. Operator Training. A training course taught in 20 h over 5 days is a necessary step for a successful statistical weight control program. Subjects needed to be covered are basic descriptive statistical concepts, frequency distributions, variable control charts, sampling, inferential thinking, flowcharting, packaging quality, and good manufacturing practices. After the classroom experience, talking to the operators on the station to correct any misapplications, explaining new techniques, or updating new experiences among operators is a never-ending task.

16.1.6 Packaging GMPs

Good manufacturing practice (GMP) is a necessity with regard to packaging lines. A GMP packaging line should have the following components:

- An enclosed hopper or bin to hold sugar
- A rare-earth magnet
- A scalping screen
- A clean, dry, well-maintained packing unit
- A lot coder that clearly marks the package
- A postpackaging metal detector set to the appropriate sensitivity
- Belt conveyor lines and palletizers that are maintained and will not damage the product after packaging
- Station documentation of maintenance and checking of magnet and metal detectors
- Regular station audits
- Good personnel hygiene

a. Sugar Bin or Hopper. The sugar bin or hopper should be inspected for integrity regularly. There should be no overhead leaks of water or other contaminants into the bin or hopper. The bin should be made of a food-acceptable material or lined with a food-grade paint or epoxy. Aeration and dust collection systems should be inspected and maintained to prevent potential contamination.

b. Rare-Earth Magnet. A rare-earth magnet should be installed just before the packing unit. Rare-earth magnets are recommended over ceramic or other magnet types because of their greater intensity and efficiency. The purpose of the magnet is to trap metal fragments that are inevitable when scrolls, elevators, or bins are used. To this end, either galvanized iron or magnetic stainless steel equipment is recommended.

c. Scalping Screen. Scalping screens are used to remove and monitor for extraneous contaminants while allowing sugar to pass through. Scalping screens should be cleaned by a vacuum unit to eliminate the possibility of brush bristle contamination. If more thorough cleaning is needed, the screens should be removed.

d. Packing Unit. Maintaining the packing unit is a GMP issue. Regularly scheduled and documented maintenance should be performed on all packaging equipment. Packing units should be checked regularly for abnormal wear and the potential for lubricant contamination of the sugar. Excessive glue should be wiped from the equipment to assure that the product will not be contaminated by an excess of glue. There should be no moisture, water, or other potential contamination problems (such as peeling paint) over the packing unit. Safety rules for the packing unit should be documented and adhered to by all personnel.

e. Package Coder. After a package has been filled and sealed, there should be a lot code unit that labels each package with the pack unit, date, and time the sugar was packed. Packing codes are essential to identifying the lot of sugar if needed for recall purposes or quality problems. For GMP, packing coders should be checked and documented as to proper functioning.

f. Metal Detector. After a package has been sealed and coded, it should pass through a metal detector sensitive enough to pick up any metal contamination. Recommended sensitivities range from a 1- or 2-mm metal bead for 5-lb bags to a 3-mm metal bead for 100-lb bags. It should be noted that metal detectors are more sensitive to ferrous metal than to nonferrous metal.

g. Belt Lines and Palletizers. Belt lines and palletizers can cause bag damage. Damage can be in the form of excessive abrasion or punctures from belt seams or palletizing equipment. To prevent this type of damage, belt lines and palletizers should be checked regularly.

h. Documentation. Enough cannot be said about documentation. The GMP paper trail is a prerequisite of a good GMP program. If records are not kept, there is no proof that GMP exists. This could lead to a loss of business if a customer audits the facility. Documentation should take place at the packing unit. Preventive maintenance and repairs should be documented and kept readily available for the station operator and quality assurance group. Documentation should include magnet samples, scalping samples, metal detectors, and lot code function. Documentation of the packing unit can reduce downtime, protect product integrity, and increase productivity, and it provides additional evidence in case of litigation.

i. Packing Unit Audits. Packing units should be inspected regularly for potential critical or serious contamination points, such as, but not limited to, maintenance issues, metal detectors, magnets, scalping screens, and documentation in place. Audits are an additional quality check of the station and a necessary part of GMP.

j. Personnel Hygiene. Personnel who work at the packing unit should wear a hairnet and beard restraint if needed. Jewelry, such as watches, rings with stone insets, bracelets, and necklaces, should not be worn—both for equipment safety and to prevent contamination. Personnel should wash and dry hands regularly since they are working with a food product.

16.1.7 Container Material

Typically, crystalline sugar is no longer sold unpackaged at retail. Although package specifics vary from country to country to meet societal, sales, and consumer requirements, cost-effective product protection is the fundamental concern. High volumes and low profits make it difficult to cost-justify expensive packaging. As a result, sugar packaging has been, and continues to be simple, lightweight, and low in cost.

Since the cost of sugar lost to handling and in-transit damage is far less than the cost of packaging materials needed to reduce to near zero, in-transit and handling damage have traditionally been tolerated. More recently, customers working to improve quality within their operations have insisted on better package performance. Improved rail and truck loading procedures, including wall lining, dunnage, and the addition of stretch-wrap film for load protection, having been implemented as the lower-cost alternative to upgrading primary and secondary packaging. Plastic stretch film is now commonly used to stabilize loads and protect the product from dust and moisture. Where more protection is required by the customer, corrugated caps and trays are added to the pallet loads.

Sugar is most often packaged in one-way, nonreusable containers that frequently rely on product support for stacking strength. Sugar package sizes vary from individual paper packets containing a few grams, to woven-plastic bulk bags holding more than 1 ton. Package sizes for particular grades are determined by marketplace customer requirements. Primary containers for sugar include:

- Individual paper packets
- Folding cartons
- Gable-top "milk" cartons
- Plastic pouches
- Plastic-lined jute or woven polypropylene bags
- Multiwall paper bags
- Glass jars
- Large cans
- Metal drums
- Bulk bags

Secondary containers include:

- Corrugated cases
- Plastic stretch or shrink films
- Heavyweight papers

16.1.8 Selection of Packaging Material

Obviously, a package must contain and carry the product through storage, distribution, sale, and end use with minimal damage or leakage. Although sugar does not require grease or oxygen barriers, it can flow freely through small holes, take on odors, develop compaction set, and harden as a result of changes in relative humidity. With seasonal inventory buildups and subsequent storage at distribution centers, it is not uncommon for packaged sugar to be stored six months to one year before it is purchased by an end user. During that time the sugar is kept usable by protecting it from moisture, strong odors, and infestation (predominantly, insects).

a. Moisture Barrier. Crystalline sugar is somewhat hygroscopic. Under prolonged conditions of high relative humidity (in excess of 70%), the surface moisture of the sugar can exceed 1%. The surface of neighboring crystals is dissolved, and as the moisture evaporates, a solid mass forms. Fine crystal, powdered, and brown sugars require packaging with greater moisture protection. Plastic films or plastic-coated papers with lower water vapor transmission rates (WVTRs) are commonly used as barrier plies. The barrier can be placed next to product for special application where fiber contamination is an issue. However, where conditioned sugar cannot be used, the barrier ply is more often nested between paper plies, leaving paper next to the sugar for wicking off moisture in hot sugar as it cools. For special moisture protection, metallized films or foil laminations can be used. However, metal in the packaging renders metal detectors ineffective as a means of determining the presence of metal in the product. Low- and high-density polyethylene films are typically used where a moisture vapor barrier is required. Table 16.2 shows the approximate WVTRs required of a barrier film for sugar in normal commerce.

TABLE 16.2 WVTR Required of a Barrier Film for Sugar

Sugar	WVTR[a]
White	
Coarse	None
Granulated	None
Fine	0.4–1.3
Soft brown	0.45–0.75
Powdered	0.45–1.3
Agglomerated products	0.6–0.75

[a]g/100 in.^{2}/24 h at 100°F and 90% RH.

b. Odors. Crystalline sugar will absorb strong odors from the immediate environment. Since this characteristic is related to the surface area of the sugar, the finer sugars, with more surface area, are more susceptible to odor pickup. Although heavier papers and plastic films would reduce odor transference, most are not impermeable to odors and therefore do not guarantee that the sugar would be safe from long exposure to odors. Higher-priced packaging materials suitable as odor barriers are normally not cost-effective. Therefore, the more common approach is to instruct that sugar not be stored near odorous materials.

c. Migration and Abrasion. Governments often regulate and define packaging materials suitable for contact with food products and restrict substances that migrate or transfer from the packaging to the product. With sugar, migration of chemicals from packaging to product is a concern, but less so than with multicontent foods, particularly those with liquids, fats, and acids. Neither dry granular sugars nor brown sugar, with its slightly higher moisture and nonsugar content, are likely to encourage this kind of migration. Hence, glass, metal, or plastic containers with corrosive-resistant properties are not necessary. However, the sharp edges of sugar crystals can physically abrade packaging materials and there is some potential for the presence in the sugar of minute particles of packaging material (wax, paper fibers, adhesives, etc.) that have been abraded from the container. This is somewhat less so with brown sugar because moisture in the syrup dissolves sharp edges. Nevertheless, where sugar and packaging touch, it is important that the packaging materials be suitable for food contact.

16.1.9 Convenience Features

Cost considerations have limited the number and variety of convenience features used in sugar packaging. Nevertheless, where customer preferences or competitive advantages afford opportunities, convenience features have been added. Rigid containers of free-flowing sugars often include a pour spout. Some spouts are simple diecut holes that are pressed opened. Others include more elaborate, integral carton board designs (where the spout is formed by the consumer) and resealable spouts made of plastic or metal that are added to the carton during packaging. Plastic clips or plastic/paper-coated metal ties are sometimes included with smaller, flexible plastic pouches. After opening the package, the opened end can be gathered, twisted closed, and held in place with the clip or tie. A more sophisticated feature consists of an integral plastic "zipper" that allows the package to be resealed with sliding finger pressure. Multiwall bags frequently incorporate a string or paper band that when pulled opens the bag by unraveling stitching or tearing the paper. Consumer bags that are less than 12 kg in weight sometimes include a paper or plastic handle intended to make the bag easier to pick up.

16.1.10 Sizing of Bags

Customer requirements relating to rack storage of pallet loads and in-transit damage have led to standard-size pallets and loads that minimize package overhang or underhang. Packages that are placed in secondary shipping containers or wrapped into bundles for shipment are typically rigid, allowing the final load sizes to be accurately predicted. Large 10- to 50-kg) multiwall or plastic bags, being flexible, do not easily lend themselves to theoretical load sizing. For example, if the filled bag width needs to be reduced slightly to prevent pallet overhang, simply decreasing the empty bag's face width and increasing the gusset width (to make up for the lost volume) would leave the girth of the bag and the subsequent filled footprint substantially unchanged. Reducing the total girth would produce the desired narrower width but might also result in insufficient product headspace to fill and seal the bag properly. Increasing the length of the bag would afford the needed volume but produce a different pallet overhang. Final bag sizing frequently requires test filling of enough sample bags of various sizes to physically form and evaluate the theoretical load patterns.

16.1.11 Warehouse Retail Stores

To reduce in-store handling costs, some businesses have adopted strategies that present pallet load increments of products in rack storage at retail. Although presentation of the package is not expected to be as attractive as on a store shelf display, there is little opportunity or incentive for the retailer to culling or improve the package appearance at the point of sale. Therefore, special unit loads are employed designed to protect the product in rack storage, identify the package to the consumer, and allow convenient selection of individual packages directly from the pallet load. Typically, the packages are placed in corrugated trays. Load protection is afforded by tie sheets, corrugated pallet trays, and caps. The loads are stabilized with stretch wraps.

16.1.12 Graphics

Folding cartons are normally printed using offset lithography on clay-coated news or clay-coated solid unbleached sulfate boards. The process is suitable for four-color process printing as well as line work. Pockets or pouches made from bleached kraft paper or plastic rollstock are printed using the flexographic process, which is less sophisticated than offset lithography and better suited for line work. Therefore, these packages employ simplified artwork and special colors in lieu of screened colors. Multiwall bags, intended for industrial customers, are printed in one or two colors with greatly simplified designs on less sophisticated flexographic presses. Loose sugar from packing line or sifting from the packages can result in severe ink rub and package abrasion during conveying and shipment. Low-rub inks and overprinted lacquer or varnish can help to minimize ink rub and light scuffing. However, external package abrasions is best addressed by minimizing the amount of loose screen in contact with the outside of the package.

The graphics designs include the typical brand, product, weight, nutritional data, ingredients, and bar code information. Ingredient labeling for sugar is relatively uncomplicated. A list of ingredients such as cornstarch or maltodextrin (added to various powdered sugars), molasses and caramel color (added to some brown sugars), dyes and waxes (added to colored sugars), and invert sugar may be required in a declaration of ingredients, depending on prevailing laws. Primary package identification using bar codes, such as the UPC (Universal Product Code) and EAN (European Article Numbering), as well as shipping container symbols are in widespread use. Bar codes are used by manufacturers, distributors, and retailers to identify packages and collect data automatically.

16.1.13 Cost Considerations

Even though newer, more sophisticated packaging materials are available, sugar is still packaged in traditional, low-cost paper and flexible plastic. While the sugar packaging has remained substantially unchanged, the packaging equipment has not. The same cost-saving forces that work to retain traditional packaging materials have resulted in the installation of state-of-the-art sugar packaging equipment with accompanying promise of faster speeds coupled with reduced labor, giveaway, and packaging material waste. On predominantly manual packing lines, package nonconformities may be tolerated in that they can be detected and culled by the operators. However, automated packing lines, particularly high-speed lines, are more sensitive to the natural variations in container material. To prevent losses resulting from misformed packages or machine jams, it is prudent to develop and maintain an informed approach to container specifications.

16.1.14 Specifications

Certainly, the high annual cost of packaging materials (rivaling labor and energy) is in itself a justification for detailed specifications. But where understanding and control of container variability are critical to the particular packaging operation, it is even more important to identify and establish current container requirements that can be documented and communicated as formal, written container material specifications. The specifications must include targets and, more important, acceptable ranges for attributes that significantly affect the performance of the package during packing, shipping, handling, storage, and consumer use. Involving the packaging vendors in the specification-creating process will help to identify areas where the package requirements and the vendor's capabilities do not coincide. By resolving these differences in advance of container production, many emergency situations and losses resulting from nonperforming containers can be eliminated. Specifications should communicate all details currently understood to be important to the performance of the material and should be subject to revision as a result of evolving circumstances or processes.

16.1.15 Container Testing

Responsibility for conformance of container material rests with the packaging vendor and should be integral to vendor's process and procedures. Test results, process documentation, certification, and samples can be provided with each packaging material order. Still, even the most modest capability for on-site testing at the sugar packaging plant offers a number of advantages. The ability to accurately quantify tear, tensile, porosity, compression strength, moisture content, ink rub, slide angle, caliper, melting points, viscosities, and similar attributes permits quality audits of the materials received. When packaging-related problems arise, the ability to test attributes quickly in-house provides essential knowledge needed to address on-line problems. Long-term advantages of in-house testing include the ability to benchmark and track evolving changes in container materials and the evaluation of new materials and packaging concepts. Confirming that critical characteristics are within specification is particularly valuable when experimenting with packaging alternatives that offer potential container cost savings. As performance minimums are approached, the ability to tolerate nonconforming attributes diminishes. Quantification of the point at which an attribute contributes to package failure helps to identify the low-range minimum and high-range maximum that suggest a target specification.

16.1.16 Standardized Testing

The characteristics of the container materials used for packaging sugar can vary with ambient conditions. For example, wood pulp–based packaging, such as corrugated, carton board and multiwall papers, are hygroscopic. Since many of their properties vary with relative humidity and temperature, it is important that testing be conducted under standardized conditions. Use of the standards and testing procedures of established organizations and associations for packaging materials offers accepted criteria for reasoned decisions on package quality. Packaging requirements are likely to include bar codes. As with other package attributes, it is important to have the in-house capability to audit code quality. If bar codes are used internally in the packing, conveying, or warehousing operations, the ability to analyze bar code quality is indispensable. Laboratory instrument designed to verify and analyze the attributes of bag codes offer valuable insight into suspected code quality and scannability issues.

16.2 WAREHOUSING

16.2.1 Warehouse Conditions

Basic conditions are listed below with a brief explanation.

a. Humidity. White sugars packaged in paper containers should be stored in an atmosphere of less than 60% relative humidity. Between 60 and 75% RH, moisture pickup is sufficient to cause caking following exposure to drier conditions and loss of moisture. The degree of caking will depend on how much moisture the sugar absorbed during the exposure time. Pulverized and fine grain sugars are much more sensitive to high humidities. Above roughly 80% RH, sugar will absorb enough moisture to liquefy. Soft sugars, by contrast, need to retain their higher moisture level, so should be stored at 60 to 70% RH.

b. Temperature. Both white and soft sugars should be kept cool, 10 to 40°C for white and 5 to 25°C for soft sugars. At high temperatures, color will develop in both white and soft sugars. Moisture loss is more rapid from soft sugars at high temperatures if the RH is less than the recommended 60 to 70% range. Variations in temperature should be avoided as much as possible, particularly those caused by drafts of alternately cold and warm air through the day and night hours. Such changes accelerate the caking of sugars.

c. Odor. White sugars, particularly pulverized sugars and fine grains sugars, can pick up odors from the storage environment. For this reason, sugars should not be stored in an atmosphere where strong odors can be detected. Some sources of strong odors are garlic and spices, perfumed detergents, tires, odorous chemicals, and pesticides.

d. Stock Rotation. A system of first-in/first-out stock rotation must be used. Soft sugars and pulverized sugars have a finite shelf life and must be distributed to customers prior to their expiration dates.

e. Stacking Height. Granulated sugars can generally be stacked very high without caking consequences. Pulverized sugars, on the other hand, are sensitive to compression set and must not be stacked higher than the recommended height, preferably single pallets and no more than two pallets high. If it is necessary to stack pallets, corrugated board sheets must be placed between the loads to protect against pallet wood fragments and splinters.

f. Warehouse Inspections. Periodic inspections of outside warehouses by either the consignor or a contract inspection service are strongly recommended. Most well-managed warehouses will contract for this service on at least an annual basis. The results of these inspections should be used as a tool for improvement and should also be available for review by warehouse customers.

16.2.2 Automated Warehouse

A unique, modern, fully automated warehouse is in operation at one of the cane sugar refineries in the United States. The facility is known as the *automated storage and retrieval system* (AS/RS). As its name implies, products are automatically palletized, stored, and reclaimed for shipment. A flow diagram of the complete system is shown in Figure 16.3. Products that are cased, bundled, or bagged are read by bar code scanners before entering any of several palletizers. The scanner identifies the grade and packaged product and sends a signal to the palletizer to form the palletizing pattern for this particular product. For example, the palletizer may form a load of 50-lb soft sugar bags and immediately afterward reset the load pattern for 50-lb granulated sugar. All the products are palletized on special hardwood boards that bear a bar code on two of the four sides of the pallet. The pallet is a 55 by 46 by 1.5 in. hardwood slab. The bar code is later read by a scanner, which creates the history of this particular load in the computer. All the palletized products leaving the palletizers merge into a common conveyor that directs the loads to one of several stretch-

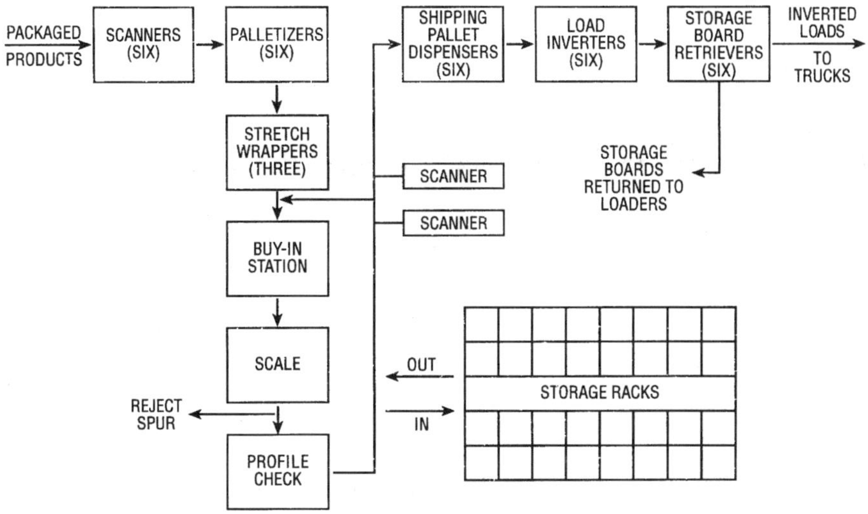

FIGURE 16.3 Automatic storage and retrieval system product flow diagram.

wrap machines. Following this operation, the load passes through a *buy-in station*, where the history of this particular load is created. Two separate scanners identify the bar code number of the board and the products to be stored. At this point the computer has already allocated a slot for the load in any of the slots in the storage racks. The load then passes over a platform scale that verifies whether the product weight is within the range that is expected for the particular load. Depending on the product, the weight varies between 1700 and 3200 lb. Loads that are outside the expected range are rejected. The last verification before storing the pallet is a profile check. The height and all four sides of the pallet are read by photoelectric lights to ascertain that the load is within specified dimensions in order to be stored in one of the racks. Failure of this profile check results in the load being rejected. The pallet then enters the storage racks and its location is stored in the computer. There are 15,600 locations available in the racks, which are comprised of six symmetrical aisles, 50 positions in length, 13 positions high, and two positions on each side of the crane. A section of the storage racks and input and output conveyors is shown in Figure 16.4. The double-deep storage scheme used in all racks means that a unit load will never be buried by more than one other load. The unit load is picked up by a shuttle telescoped out from the crane, the shuttle containing this unit is moved to the center of the crane aisle, and the crane proceeds to its selected destination. While the crane is moving horizontally at a speed of 350 ft/min, its carriage containing the load is concurrently being elevated at 95 ft/min. Hence, any location in the total rack is accessible in approximately 1 minute. At the same time that loads are being stored in the racks, the cranes also pick up loads for orders that are fed into the computer and bring them out of the storage racks. As a load is conveyed toward the shipping spurs, it is identified by a scanner to determine whether it holds products to be shipped or fresh products that have failed the weight or profile check. In the latter case, the load is diverted to the loop conveyor and later rejected. Otherwise, the load proceeds to the second scanner, which directs it to one of the six shipping spurs. At the shipping spur, a customer pallet is automatically placed on top of the load. The load then proceeds to an inverter that clamps and inverts it 180°, placing the customer's pallet beneath the unit load as shown in Figure 16.5. The load finally proceeds to the final stage, where the pallet is retrieved by a pallet collector and is routed back to a robotic stacker, where it is stacked with similar pallets and prepared for insertion in the palletizers for the next round trip. The load advances a few feet and is picked up by a fork truck operator and loaded into an adjacent truck. This is the first and only manual handling of a pallet.

16.3 PEST CONTROL IN THE WAREHOUSING OF SUGAR

Warehousing requirements for sugar are, for the most part, the same as for any food product. Desired conditions are listed below with a brief explanation. Because sugar is not a complete food and will not support insect or vermin growth, direct infestations are rarely a problem.

a. Lighting. Good lighting is important in areas where sugar is received, stored, or assembled for shipping. Adequate light allows for thorough inspections of products prior to shipment.

b. Floor Condition. The floor should be clean, smooth, and free of spilled materials or accumulations of dust and debris. There should be no holes or unsealed cracks, particularly at the floor wall junction. Any drains should be protected against the intrusion of both rodents and insects.

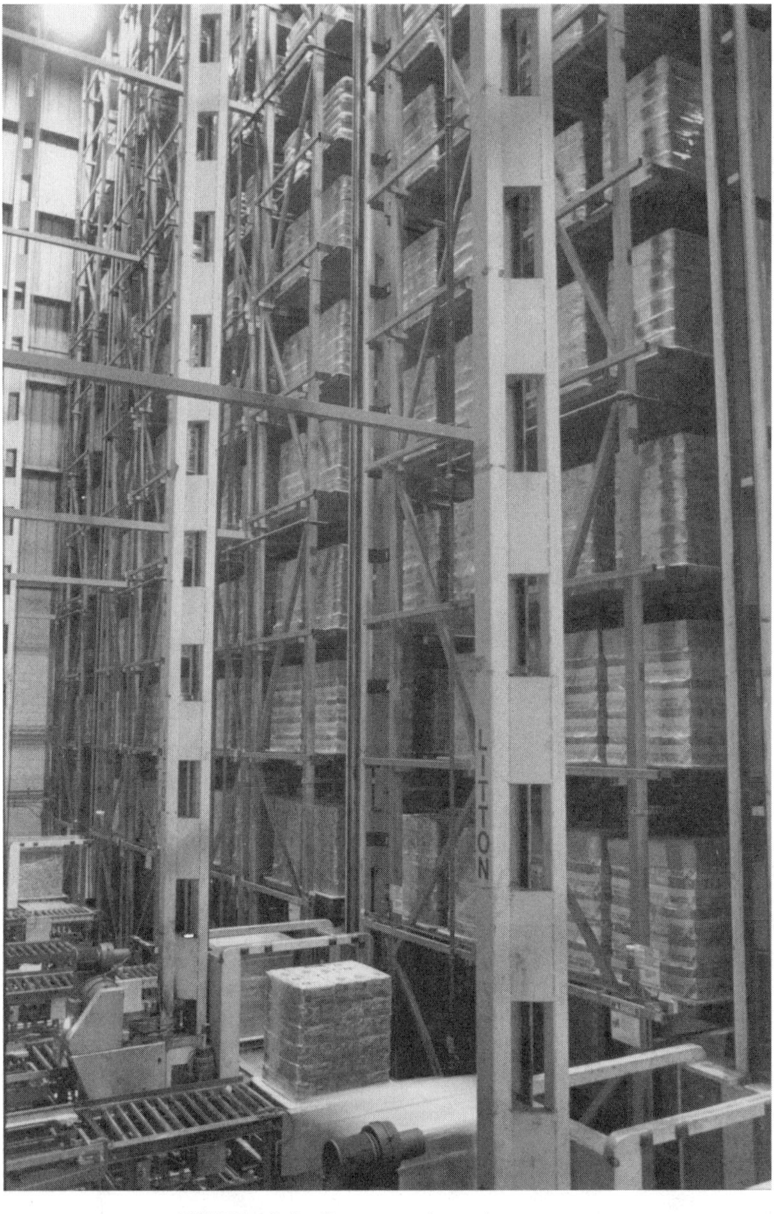

FIGURE 16.4 Storage racks and conveyors.

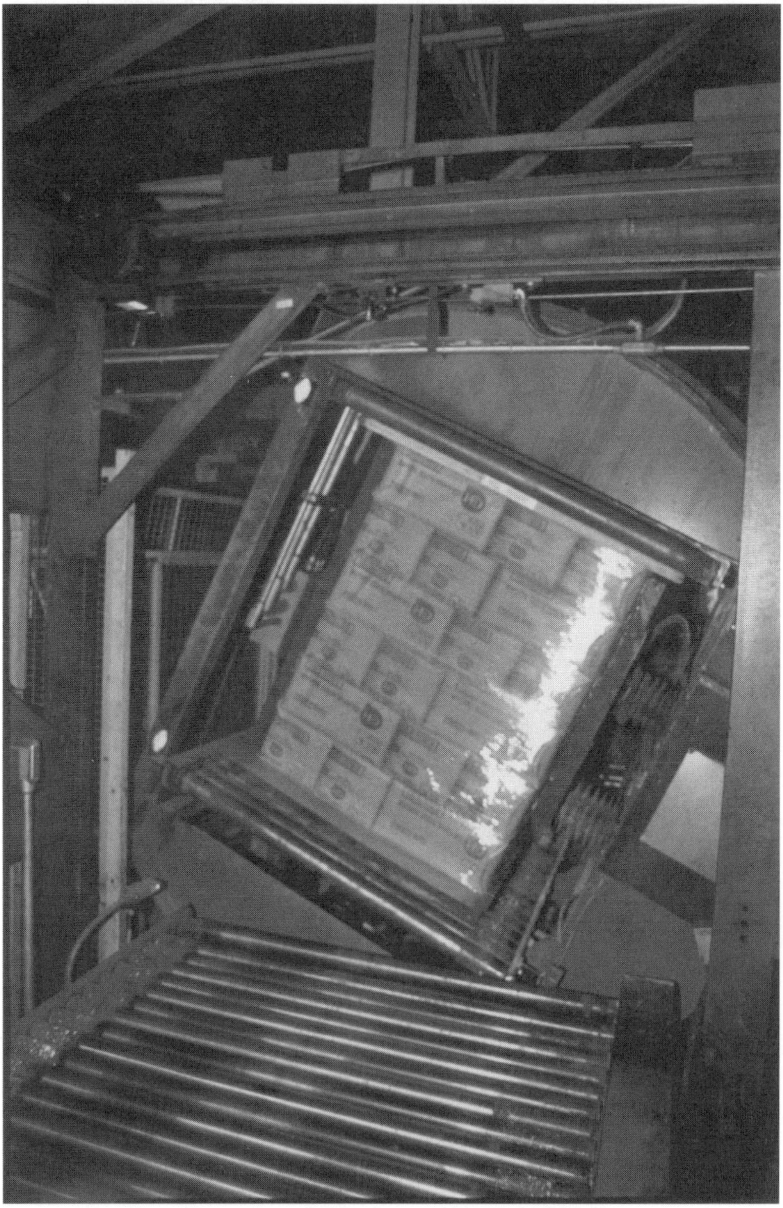

FIGURE 16.5 Unit load on customer's pallet.

c. Other Stored Products. Other dry food products, such as cereals, dried legumes, and dry pet foods, are frequently found in public warehouses. These foods will provide the necessary nutrients for insect and rodent vermin to reproduce. Although good warehouse practices will minimize the chances of infestations in these stored products, it is prudent to keep sugar separated from these foods. A well-run warehouse with these other food products present will use pheromone traps or other devices to monitor for the presence of various beetles and moths.

d. Package Damage and Spillage. Recovery of damaged packages is sometimes done in a separate area. Exposed product in damaged packages is much more vulnerable to insect infestation. Great care must be taken to ensure that these areas are kept clean and free from insect infestation. For the same reason, spilled product in storage areas must be cleaned up immediately.

e. Rodent Protection. Rodent protection has three important segments, exterior control, exclusion, and interior control and monitoring. The exterior of a warehouse should be free of grass or shrubbery that may harbor rodent burrows. Refuse containers must be kept closed when not in use. This will eliminate any edible refuse as a food source or rodent attractant. Since the exterior will be the primary source of any rodents, it is important to set up a perimeter line of bait stations to eliminate them before they can enter the building. Bait stations must be tamper resistant and must be secured to the ground. The bait used should be a single-feeding anticoagulant type. Bait stations must be properly identified and should have a label on the interior of the cover for recording the service history by the pest control technician. Each bait station should be numbered and shown on a map of the building and grounds.

Rodents and other vermin can be excluded by tightly closing doors and windows. Doors should be kept closed at all times when not in use. Any needed ventilation should be through screened openings. Self-leveling dock boards should be close fitting to prevent the entrance of rodents.

The interior must be protected and monitored by traps and/or glue boards. Traps can be either the snap or multiple-catch type. Snap traps should be of the expanded trigger type and can be baited with peanut butter if desired. Snap traps are cheap, effective, and easily checked during a warehouse perimeter inspection. Multiple-catch traps or covered glue boards are needed when perimeter inspections are less frequent. The inspection frequency should be such that any dead rodents do not become a source of flies. As with bait stations, the location of traps or glue boards must be recorded on a map of the building. Multiple-catch traps and glue boards must also be provided with labels to record the frequency of attention. Inspection records must show the location of rodent catches.

f. Floor–Wall Junction. Most warehouses with a poured concrete floor have a gap between the floor edge and the wall. Depending on the construction technique, this gap may be up to an inch wide. It is important that this gap be sealed and maintained. The soil under concrete floors can settle and provide a harborage space for insects and/or mice.

g. Perimeter Line. An 18- to 24-in-wide access path should be maintained around the entire perimeter. This path should be marked by a white or light-colored line, and preferably, the entire path should be painted a light color. This provides a clean surface where any insect activity or rodent excreta will be readily apparent. The path also provides access for inspecting stored material and regularly spaced perimeter traps or glue boards. No materials should be stored against the wall. Additionally, periodic aisles should be provided in large blocks of stored materials. This allows more complete inspection for possible infestations.

h. Moisture or Water Sources. There should be no source of moisture available to vermin within a warehouse. Examples of moisture sources are water-pipe, drainpipe, or roof-drainpipe leaks. Additionally, access to lavatories must be restricted by tight-fitting and self-closing doors.

i. Pest Control Contract Services. Unless the warehouse has an in-house pest control person, a licensed pest control service must be employed. Each visit by a pest control technician must be recorded on a report, and the findings and recommendations discussed with the warehouse manager. Copies of these reports should be forwarded to all warehouse customers.

j. Incoming Product Inspections. Warehouse practices should include the routine inspection of all incoming materials. This is to prevent the introduction of damaged or vermin-contaminated product into the warehouse. A portable black-light or ultraviolet-light source should be available to identify rodent urine spots or tracks on suspect materials. Some training and experience is needed with these lamps to be aware of their limitations and false diagnoses. Fluorescence by itself does not indicate rodent urine with 100% certainty.

16.4 PACKAGED SUGAR SHIPPING

16.4.1 Modes of Transport

Three types of transportation are commonly used in shipping packaged sugar: truck, intermodal, and rail boxcar. Truck transportation pertains to the use of trailers and tractors to transport product and is the most flexible and widely used means of delivering product to the customer. Truck tractors are used to transport "over the road" either trailers or containers mounted on a trailer chassis. Intermodal transportation consists of transportation by means other than over-the-road and can be either rail or water transport. It is used primarily when transferring products over longer distances. Intermodal is further broken down into TOFC (trailer on flat car, or piggyback) and COFC (container on flat car). TOFC uses flatbed railcars to transport either truck trailers or containers on truck trailer chassis. This method allows cheaper rail rates while retaining the flexibility of truck transport to and from the railyards with minimal handling. Articulated railcars carry either containers loaded one or two high or truck trailers on integrated five-car train units. The five-car units have tightly coupled cars and an advanced suspension system. The premise behind this system is that most rail transit damage is the result of two actions: (1) the slack action of conventional car couplings allow significant shock energy to be transmitted to the cargo, and (2) the sway action of conventional cars results in cargo motion and package abrasion. Both of these actions are minimized in articulated units. Articulated units are available in either conventional cars or well cars. Well cars allow the placement of containers within a well or depression and lower the center of gravity by approximately 3 ft. The effect of this is to further diminish the amount of sway during transit. Articulated cars are the mode of choice over long distances, due to a lower incidence of transit damage.

Rail boxcars are most often used in shipping long distances to customers or warehouses with rail service. The advantages are direct service and larger loads per shipping unit. As with any shipping unit, loading practices have a large effect on product condition at destination.

16.4.2 Cleaning and Inspection of Vehicles

Railcars or trucks are an extension of the warehouse and have similar requirements for cleanliness and physical condition. They should be mechanically sound, clean, and odor-free so that they will not cause damage to or compromise product quality. This is not always the case, and in these instances, a judgment must be made as to whether to clean and accept a vehicle for loading, or reject it. The following guidelines can help in making this judgment.

a. Mechanical Inspection of Vehicles. Any defects found with the running gear that could jeopardize a vehicle's safety should be reported to the carrier and the vehicle rejected. Vehicles with leaks in the roof or walls or damaged doors that cannot be closed tightly should be rejected. Holes in the inside wall liner should be inspected thoroughly. Food materials can enter wall cavities and serve as an infestation source. Weak spots in a vehicle's floor are a safety hazard to forktrucks and if detected should result in the vehicle's rejection. Nails or wood splinters on sidewalls and floors must be removed prior to loading. Since the interior of transportation vehicles is often dark, a good flashlight or other light source is a necessary inspection tool.

b. Cleanliness. Shipping vehicles are to be clean and free of dirt, dry residue, liquids, and insects. Vehicles not received in this condition must be rejected or cleaned thoroughly prior to loading. Railcars and trucks containing live insects should not be loaded without first eliminating them. Their presence indicates the possibility of an insect infestation in a wall or floor void or, at best, inadequate cleaning following the previous load. Some refrigerated trucks have been found harboring ant infestations in the insulation and fruit fly infestations in the refrigeration unit. Floor depressions should be clean and dry. Liquids are sometimes found in the movable bulkhead floor tracks or strap tie-down depressions of railcars. This liquid is invariably dirty, might be a toxic residue, and at the very least, can promote a breeding place for insects. It does not take much liquid in these depressions to contaminate sugar through a slip sheet placed directly on the floor.

c. Odor. There should be no significant odor in a transportation vehicle. When making a judgment as to the suitability of loading a vehicle with a slight odor, consider the following: any odor present may be stronger or weaker to someone else, depending on the nature of the odor and the person's nasal acuity. A vehicle that is being checked has possibly been open for awhile and the odor may not be as strong as it will be when the customer opens the door or when the temperature rises. Pulverized sugars are particularly susceptible to contamination by adsorbed odors.

16.4.3 Pallets and Slip Sheets

Due to the weight of a pallet of sugar, used No. 1 or new hardwood pallets are recommended. These can be made of oak and other hardwoods, including Douglas fir. They should have no unsound knots, splits, wane or white speck. Unsound knots are knots that have dropped out, leaving a hole and reducing pallet strength. Splits are cracks through the full thickness of a board. Wane is excessive bark on a board, which can reduce its strength. The extent of wane, as a defect, depends on its width, length, and location. White speck is defined as small inclusions of dried pitch. The sawing operation producing the boards results in their white appearance. White speck should be limited to the point where it does not affect the strength of the pallet board. In addition, pallets should be free of mold, decay, and noxious odors. It is best to use mildly used pallets because they do not have as much of a fresh wood odor as new pallets. Some other defects that should be observed when inspecting pallets are pinworm and grubworm holes. These holes may be permissible as long as they do not affect the structural strength of the pallet. Infestation of *Lyctus* (powder-post) beetles, termites and other wood-destroying insects in pallets should not be acceptable for storage or shipping sugar. Finally, it is prudent to check pallets for protruding nails, staples, or anything that might puncture a package.

To prevent packaged sugar from becoming contaminated with wood splinters, corrugated board is used between the pallet and the sugar. The board should be sized to cover the pallet completely with minimal overhang. Slip sheets are used for loads shipped with no pallet. For rail shipping on slip sheets only, 48 by 46 in. 60-point select fiberboard is recommended. For truck shipments, 42 by 54-in. 60-point select fiberboard is recommended.

16.4.4 Balancing Loads

When a container is accepted for loading, the load weight must be evenly distributed throughout the container. Moreover, the maximum weight on any axle must not exceed the state highway weight limits.

16.4.5 Bracing Loads

All shipments traveling by rail, either intermodal or boxcars, need to be braced to prevent loads from shifting during transport. There are three methods typically used to brace loads: dunnage and airbags, strapping, and skid plates. Dunnage typically consists of honeycombed corrugated fiberboard and is available in various thicknesses. It is used when there are 2 to 18 in. of space to fill. Filling this space secures the load and prevents load shifting. Although dunnage can be used alone, airbags should only be used with dunnage. The airbags used should be 4-ply 48 by 96 in. bladders. They should only be used if the space to be filled is no more than 18-in. Airbags should never be used alone because they have a tendency to pop out when the air expands, with an increase in elevation during transit.

Strapping is also used to secure loads in place. The straps are attached to the walls of the container by an adhesive or by built-up grip points. Generally, securing only the last four loads against tightly placed loads in the front is adequate to prevent load shifting.

Another way to prevent loads from shifting is by the use of a skid plate, thin 4 by 5 in. metal plate with punched spurs protruding from both sides. This plate is then set between the bottom of the pallet board and the container floor. Usually, only the last few loads have the skid plate placed under them. The skid plate spurs grab the pallet and floor surface during transport with the help of the weight of the pallet load. To protect the side of the pallet load tangent to the wall, corrugated fiberboard can be used to line the wall. This added protection, along with the load being stretch-wrapped, will prevent the packaging from being scuffed and soiled.

16.5 BULK DRY SUGAR SHIPPING

16.5.1 Modes of Transport

Bulk dry sugar is shipped in trucks, railcars, or semibulk bags of 1 tonne or more. There are two predominant types of bulk truck delivery systems: bottom dump and pneumatic.

Trailers for each of these types are virtually always constructed of aluminum, due to its light weight and freedom from rust or corrosion. A third type using collapsible tanks is used less frequently. Railcar shipments of bulk sugar are made in either Centerflow or Airslide railcars.

a. Pneumatic Trucks. Pneumatic trucks are equipped with a high-volume air blower and return-air filter system. In operation, air is blown continuously from the truck to a silo or tank and back to the truck filter system, which removes small quantities of carry-through sugar dust. The truck tank is pressurized. This allows the sugar to drop into the high-velocity airstream within a tube passing under the tank outlets. This tube is connected by hose to the receiving system, which carries it into the silo. The high velocity experienced with air conveyance results in some sugar crystal degradation. Depending on the end use of the sugar, this may or may not be important. Pneumatic trucks can also be used as bottom dump units by opening cover plates at the bottom of the pneumatic tube just below the outlets of the truck tank. The advantage of pneumatic trucks is their versatility and the ability to serve as a complete unloading system. Their tanks are heavier since they must withstand an unloading pressure of up to 15 $lb/in.^2$ and they must also carry a blower and filter system. Because of this extra weight, their cargoes are slightly smaller.

b. Bottom Dump Trucks. Bottom dump trucks consists of tanks with a conical bottom and outlet gate. They are emptied by gravity, with the sugar dropping into a screw conveyer or other takeaway device. Their simplicity results in lower maintenance costs and increased payload. Additionally, sugar crystal damage during the unloading is minimal.

c. Combination Trucks. A third type of bulk sugar truck transport, used less frequently, consists of regular truck trailers to which have been added a set of collapsible plastic-coated fabric tanks. The tanks are collapsed and raised to the ceiling by cables when not in use. This allows the trailer to be used for the transport of regular cargo. Prior to loading bulk sugar, outlet port covers in the floor are removed and the tank bottoms are lowered to the floor. The tanks are filled through ports in the top of the trailer and emptied by gravity through the outlet ports in the floor of the trailer. A circular pneumatic bladder in the bottom and lower walls of the tank is expanded to empty the sugar completely. This type of transportation system allows sugar transport one way and packaged cargo on the return trip. The disadvantage is the difficulty in inspecting the tanks for cleanliness and maintaining a watertight seal on the loading port hatches, located on the outside top of the trailer.

d. Supersacks. Supersacks are a recent development in the transportation of sugar. They consist of a strong woven plastic outer bag with a suspension loop at each corner, a plastic inner liner, a filling port at the top, and an outlet port at the bottom. They can be moved with a forktruck using either their suspension loops or a standard pallet. They are shipped in trucks, containers, or railcars. For intermediate volumes of sugar, they are more convenient than bagged sugar.

e. Bulk Sugar Railcars. Bulk sugar railcars are available in either Airslide or Centerflow types. Each of these two types must have a food-grade paint lining on the interior surface. There should be no loose or peeling liner. Airslide cars have a canvas belt on the bottom of each side to allow fluidizing the sugar with pressurized air during discharge. It is difficult to remove lumps of sugar that accumulate on the belt without sending a person inside to move the sugar out manually. When these cars are washed, particular care must be taken to ensure that the canvas belt is dried or mildew will develop on it. Centerflow cars have the advantage of simple construction and easy cleaning.

16.5.2 Inspection of Bulk Railcars Prior to Loading

The top hatches must be tight sealing to prevent the entrance of rain, dust, and insects. The gaskets must be intact, in correct position, glued in place, and show evidence of pressure contact on the entire circumference. Proper contact can be verified by placing a small piece of paper or other suitable material between the gasket and hatch coaming, latching the lid shut, and pulling on the paper. The degree of gasket seal can be judged by the force needed to remove the paper. Some Centerflow cars are equipped with trough hatches, which run most of the length of the car. These hatches are designed for loading cereal products. They do not always seal tightly and should be avoided for sugar loadings. No odor should be detectable in a bulk railcar. The most common source of odor is a fresh interior liner application that has not been cured properly or completely. Any car with a detectable odor should be rejected until the odor has dissipated either through additional cleaning or the complete curing of a new liner.

a. Cracks at Welds or in the Wall. Cracks are more evident on freshly cleaned cars. Significant cracks are often outlined by a dark rust stain. They are usually present at welds and represent the potential to admit rainwater.

b. Clean Interior. The interior of railcars received for the first time must be clean and free of all prior commodity residues. During use, a buildup of sugar on the wall will develop. If this buildup has been subjected to high humidity and has begun to get sticky, the car will need to be washed. Under normal circumstances, a slight buildup of dry sugar on the interior walls is acceptable if, upon return, any dislodged clingage is removed prior to reloading. Any car that has been returned with an open or unsealed hatch will without question need to be washed prior to loading. Any car that has had a break in service must be washed prior to resuming sugar service. The outlet slide gates of Centerflow cars have a spring-loaded wiper that can accumulate materials behind it. Care should be taken to ensure that this portion of the car has been cleaned prior to receiving it. Centerflow cars have a screened vent in one end. This vent should be inspected to ensure that it is intact. Otherwise, insects or airborne contaminants may enter.

16.5.3 Inspection of Bulk Sugar Trucks Prior to Loading

Bulk sugar trucks should be inspected for a clean exterior, tight-fitting dome covers, and a clean interior. A flashlight or other suitable light source will aid interior inspections. Slight clingage residues of clean sugar on the inside walls are acceptable as long as the sugar does not appear dirty or show evidence of moisture. Any vents should be equipped with a fine synthetic fiber air filter. There should be no holes or leaks in the tank. All openings to the sugar should be capable of accepting a tamper-evident seal. Any sugar clingage that has broken loose from the side during the return trip should be emptied prior to loading. As in railcars, no odor should be detectable in a bulk truck. No trucks should be loaded that do not have a guarantee of dedicated sugar service or that are not freshly washed.

16.6 BULK LIQUID SUGAR SHIPPING

16.6.1 Modes of Transport

Bulk liquid sugar is shipped in either tank trucks or railcar tankers. The majority of shipments are made to local destinations in trucks due to their versatility, quick delivery times, and the lack of rail service to numerous customers. Railcar shipments of liquid sugars are more economical over long distances. However, due to its water content and resultant extra freight, liquid sugars are seldom shipped over long distances.

a. Rail Tank Cars. Rail tank cars are designed for transporting many different liquid products, each with its own hazards and requirements. Tank cars are available in many different configurations of size, outlets, interior lining, dome style, and insulation. Tank cars are constructed of mild steel and coated on the interior with an epoxy liner. Railcars used for sugar must have a liner approved by the U.S. Food and Drug Administration for food contact use. Some railcar tanks are cylindrical with a horizontal bottom. Others are of double conical construction. The double conical tanks have better drainage since the bottom is sloped to the center. Some of the more modern cars have a stainless steel dome and hatch cover. The advantage of this lies in the elimination of rust and paint chipping, which frequently occurs in this area due to impact dings during loading and washing. Since the dome headspace in a railcar provides an unsubmerged surface for microbiological growth, it is important to keep it as small as possible. The gasket used on the hatch cover must be intact and made of food-grade rubber. Tank cars should be sized to carry a full load, consistent with the maximum rail gross weight requirement. There should be no more excess space than is necessary to protect against expansion. Due to car instability from sloshing during transit, there are rules governing the maximum headspace allowed in railcars.

Tanks cars have two types of outlet valve: top and bottom operated valves. Top-operated valves have the disadvantage of needing a packing gland at the valve stem top. These glands can be sources of microbiological contamination. Bottom-operated valves also have a packing gland but are usually submerged and under pressure, which minimizes the potential for product contamination. Stainless steel ball bottom-operated valves with stainless steel fittings are generally used for food products. These can have a built-in pocket around the ball, which is impossible to clean when the valve is wide open. When the valve is fully closed, any contaminated material in this pocket can enter the hole in the ball. For this reason it is important to clean the car with the valve partially closed to allow circulation through both the ball and the surrounding cavity.

b. Pressure Relief. Excessive pressure in a rail tank car can result from overfilling, with subsequent expansion due to a temperature rise, overpressuring while using air pressure to discharge the car, or the unwelcome possibility of gas generation from fermentation. For safety, rail tank cars are equipped with pressure relief vents. These can be either spring-locked poppet valves or frangible disks. Disks are preferred since they are easily cleaned and present less opportunity for microbiological growth.

c. Railcar Cleaning. Railcars used for liquid sugar are usually dedicated to either sugar or corn syrup use. This simplifies cleaning the tank since these products are water soluble and allow washing with hot sanitizing water only. The main concern is to use filtered hot water in a rotary spray head with enough pressure and volume to reach the ends. The washing time should be such that the drainage water is maintained at a temperature of at least 80°C for a minimum of 15 min. Usually, the first wash water is discarded to waste and then the water is recirculated for the remainder of the wash. On railcars with an internal ball valve outlet, the valve should be left partially closed to allow water circulation through the cavity surrounding the valve. The hatch cover inner surface must be cleaned manually. It is advisable to use 200-ppm chlorine bleach or a compatible sanitizing solution. The pressure relief system should also be cleaned using this sanitizing solution. After the hatch cover and pressure relief vent have been cleaned, they must be rinsed and protected against overhead contaminants while left open. Since it is difficult to dry a rail tank car thoroughly, it should be loaded within a short time of washing. This will minimize the development of microorganisms in any wash water residues.

d. Transit Time. Rail transport is much slower than truck transport. This results in potential microbiological problems because conditions are no longer available to protect the headspace from microbiological growth. The headspace of permanent liquid sugar storage tanks is protected by either internal ultraviolet lamps or filtered and sanitized forced-air ventilation. Either of these methods will prevent microbiological growth in the headspace. Their absence in railcars frequently results in low levels of microbiological contamination due to moisture condensation on unsubmerged surfaces. The low osmotic pressure in this condensate allows microbiological growth much more rapidly than in the product itself. This growth can be minimized by choosing a tank car, which by size and design allows the maximum weight consistent with the least tank headspace and the smallest access dome volume.

e. Railcar Loading. Railcars are typically loaded from the top using an articulated loading spout. They can also be loaded from the bottom using a hose. When loading from the top, care must be taken to protect the hatch opening from overhead contaminants. Railcars are usually filled to a specified volume using an outage gauge and weighed later during transit. As with all loadings, all car openings must have a tamper-evident seal applied with the seal numbers recorded.

f. Tank Trucks. Tank trucks are constructed of either aluminum or stainless steel. Stainless steel is preferred since aluminum has a porous surface and can pick up odors from a prior load. If not removed during the washing, these odors can be transferred to the sugar. Fruit concentrates are a particularly significant risk of odor contamination. With thorough washing or dedicated sugar service, aluminum tanks are entirely acceptable for sugar transport. Tank trucks are often equipped with a discharge pump or discharge manifold with pipes running to the front for use with a tractor-mounted pump. When washing the trailer, care must be taken to ensure that all legs of these pipes, the pump, and the discharge hoses are thoroughly washed. Truck tanks are prone to develop stress cracks after years of service. Covering cracks with a patch should be discouraged unless the patch underside is vented to the outside by a hole drilled into the tank body. If a crack is covered with a patch, the potential exists for a pocket of microbiological contaminants to develop if the patch should crack from the same stresses that caused the original crack.

g. Cleaning. Its prior load determines the type of washing a truck gets. Sugar and syrup residues are easily washed out and the truck sanitized using only hot water. Insoluble residues such as oils and fats will require the use of detergents followed by a hot water rinse. In all cases, the truck hoses, discharge manifold legs, and pump must be included in the wash-water stream. The wash water must be hot enough to provide a final exit temperature of not less than 80°C. This temperature should be maintained for a minimum of 10 min or as specified by the customer. Tank trailers are equipped with pressure/vacuum relief vents. These can be a source of microbiological contamination if not kept clean and must be included with each truck washing. In some cases they will need to be disassembled and soaked in a 200-ppm bleach solution followed by a thorough rinse to effect proper cleaning.

h. Loading and Unloading. Good sanitary practices require the loading of liquid tanks through a closed loading system. The most common is the direct connection of a loading hose using Cam-Lock fittings. Although a truck tank will be equipped with a breathing filter to allow the exit of displaced air, it is a good idea to loosen the tank hatch fastening devices while leaving the hatch cover closed. This will allow the escape of loading material if the tank should accidentally be filled beyond its capacity. The loading system should have a fine mesh screen system in place to remove any particulates that may be present. Regular cleaning of this screen with the findings documented will provide available information in a HACCP (Hazard Analysis and Critical Control Points) plan. During unloading, customer procedures must be followed regarding sampling and ensuring there is adequate space available in the receiving tank. The receiving fixture should be sanitized prior to connecting the hose, to avoid pumping contaminants into the tank with the load. A coarse mesh screen should be installed in the discharge line to safeguard against introducing particulate contaminants. The customer should also monitor this screen to document any particulate contaminants received with the load.

Acknowledgment

The author wishes to acknowledge the contribution to this chapter of the following people from the C & H Sugar Company: April Blackmore, Edward Hansen, Ralph Leporiere, and Jim Peterson.

REFERENCE

1. National Institute of Standards and Technology, *Checking the Net Contents of Packaged Goods,* Handbook 133, 3rd ed., NIST, Gaithersburg, MD, 1994.

CHAPTER 17

Remelt and Recovery House Operations

INTRODUCTION

Crystallization takes place in single-effect vacuum pans, where the syrup is evaporated until saturated with sugar. At this point seed grain is added to serve as nuclei for the sugar crystals, and more syrup is added as the water evaporates. The growth of the crystals continues until the pan is full. Given a skilled sugar boiler (or adequate instrumentation) the original crystals can be grown without the formation of additional crystals, so that when the pan is just full, the crystals are all of the desired size, and the crystals and syrup form a dense mass known as *massecuite*. The strike or contents of the pan is then discharged through a foot valve into a mixer or crystallizer.

The massecuite from the mixer or crystallizer is drawn into revolving machines called *centrifugals*. The cylindrical basket suspended on a spindle has perforated sides lined with wire cloth, inside which are metal sheets containing 400 to 600 perforations per square inch. The basket revolves at speeds from 1000 to 1800 rpm. The perforated lining retains the sugar crystals, which may be washed, if desired. The mother liquor, molasses, passes through the lining because of the centrifugal force exerted, and after the sugar is purged, it is cut down, leaving the centrifugal ready for another charge of massecuite. Modern installations are exclusively of the high-speed type, with automatic control of the entire cycle. Low grades may be purged by continuous centrifuges.

17.1 PROCESS ANALYSIS

17.1.1 Sugar Boiling Process

In the cane sugar industry the sucrose available in the syrup is crystallized in several stages, conducted at descending purities. The procedures followed in carrying out the pan operations constitute the pan boiling system, which may consist of two, three, or four stages. The

[*] By Chung Chi Chou.

last boiling, where the massecuite is cooled slowly before being purged in the centrifugals, requires the most time.

The first step is to make grain. It is preferable to make grain at a vacuum not exceeding 25 in. Hg at sea level, which means operating temperatures of 150 to 160°F, depending on purity. At those temperatures, the viscosity will be lower and the rate of crystallization growth faster. The pan is closed, vacuum is raised, and the full amount of injection water is turned on the condenser to give a fast rate of evaporation with the liquor, syrup, or molasses before graining. Steam is turned on after the heating surface is covered with liquid. The pan must not be loaded too high, since this will slow operations because of the effect of hydrostatic head. After boiling starts, feeding should be continuous to hold the level slightly above calendria closing off the feed before nucleation starts.

17.1.2 Full Pan Seeding

The best method of obtaining good seed grain is full pan seeding by adding at the proper moment the full amount of grain of predetermined size to equal the total number of grains in the finished strike. No grain is formed in the pan at any time, and the concentration must be held in the crystal growing or metastable phase. Seed is introduced as soon as the saturation point is reached, as indicated by instruments (Fig. 17.1).

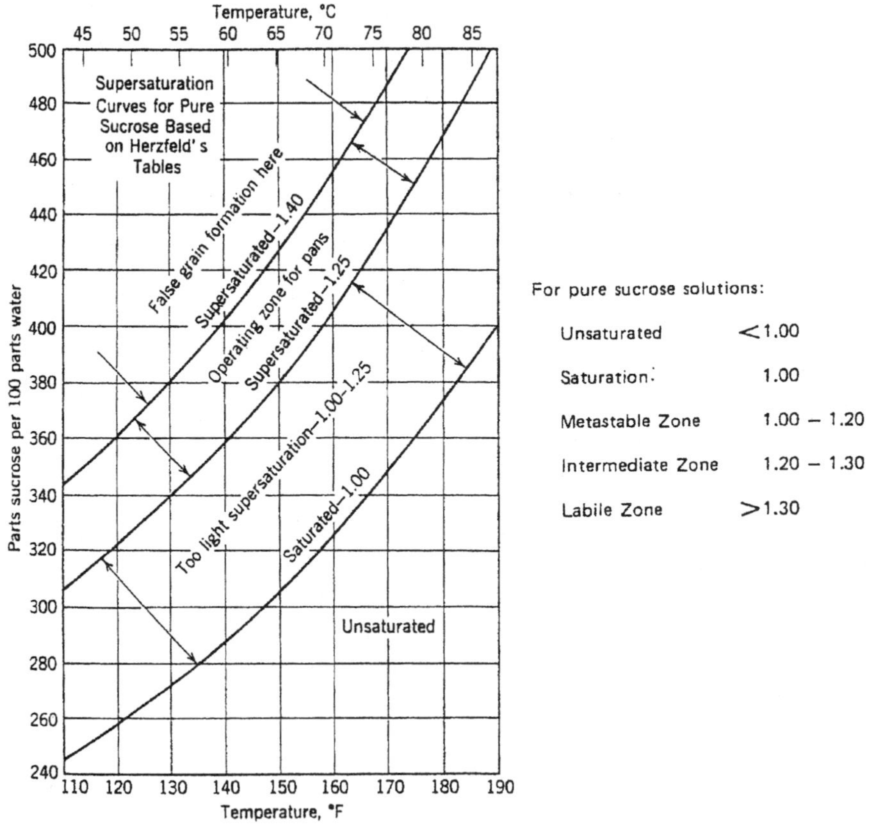

FIGURE 17.1 Supersaturation curves for pure sucrose.

17.1.3 False Grain and Conglomerates

When "shock" seeding without addition of seed is used, the formation of grain must be arrested, by increasing the pan temperature, by dilution, or both. Using this technique it is easy to overshoot the mark, dropping the mass below saturation and dissolving all the grain, which necessitates a fresh start. Even if the grain is not all destroyed, part of it may be, leaving an insufficient quantity and requiring additional nuclei to make up the loss. If the concentration is carried too high, false grain or smear will form and must be dissolved by dilution, preferably with water. Even before false grain appears, conglomeration may take place. *Conglomerates* are defined as groupings of a number of crystals that then grow together as one. Once formed, conglomerates will so remain to the end of the strike. These group crystals are objectionable because impurities and dirt lodge in the crevices, preventing proper washing and yielding a poor product of high color and low filterability.

17.2 LOW-GRADE CRYSTALLIZATION

17.2.1 Improvements in Boiling Low-Grade Strikes

Gillett in Hawaii developed some revolutionary new data that have had a great impact on the technology of low-grade boiling. It has been found necessary to raise the low-grade massecuite purity in order that enough sucrose be made available in the mother liquor to allow the crystal crop to grow. Lower massecuite purities did not give lower molasses purities but did result in poor-purging massecuite, lower-purity low-grade sugar, and extended curing time. The optimum low-grade massecuite purity has been determined between 65 and 67 apparent purity (AP). Massecuite purities exceeding 65 to 67 resulted in higher molasses purities. This is explained by the fact that more sucrose was available in the higher-purity massecuites than physical limitations would permit to be crystallized out [1]. In the measurement of AP, refractometer Brix is used for calculation of purity.

17.2.2 Crystal Size

The size of the crystal grown in the pan is important to the recovery process. Small grains will be dissolved in centrifugal washings and increases the purity of the syrup. This will decrease crystal yields and will also cause inboiling of the syrup, lowering operation efficiency and increasing energy cost. Referring to Gillett's article on low-grade sugar crystallization [1], Crockett's crystal size for final low-grade massecuite averages about 0.35 mm, for which the surface area is estimated to be 11.4 mm^2/mg of crystals. This relatively large average crystal gives higher-purity low-grade sugar (hence less recirculation of nonsugars), not only because of good drainage but also because of the smaller aggregate surface area, which carries less molasses film. It has been further determined that a low-grade sugar crystal size of approximately 0.35 mm (average) with a massecuite of this type gives optimum results with respect to low-grade centrifugal work.

17.2.3 Crystal Content

This deficiency in crystal concentration meant that the grain crop was "sparse," that considerable distance normally existed between suspended crystals, and that an unnecessary limitation was imposed with respect to the amount of crystal surface area available for adsorption of sucrose from the mother liquor. This contributes toward inferior results in a low-grade system. Other factors—large variations in crystal sizes of low-grade sugar, slow massecuite curing rates, high molasses purities, and poor purging massecuites—seemed to be manifestations of these grain-deficient massecuites. In the case of low-grade work, the viscosity of the mother liquor is high because of the lower solubility of sucrose in low-purity materials and the relatively high saturation that must be maintained to secure a reasonable rate of crystal growth that even when the volume of crystals suspended in the massecuite is relatively small, the massecuite still is comparatively viscous and stiff. As the crystal volume concentration increases, the massecuite tends to become less fluid [1].

17.2.4 Crystallization and Viscosity

Supersaturation brought about by decreased solubility is the driving force of crystallization in both the pan and in the crystallizers. In the pan boiling operation, solubility is reduced by evaporating water from solution, whereas in crystallizer conditioning, solubility is reduced by cooling the massecuite. The conditions of operation in the crystallizer are in direct contrast to the isothermal conditions that exist in the vacuum pan, although the mechanism of sucrose deposition is the same. The rate of crystallization is determined by the degree of supersaturation, temperature, crystal surface area, and nature and concentration of impurities. The viscosity is also influenced by the same factors, and the limit to which the massecuite can be cooled depends on the ability of the crystallizer to handle the material physically at high viscosities. Viscosity is therefore a dominating factor in the technology of the process.

17.2.5 Cooling of Crystallizers

Low-grade massecuites, after having been boiled to maximum workable consistency in the vacuum pan, are discharged to the crystallizers at a temperature of 65 to 70°C (149 to 158°F) and a supersaturation of approximately 1.20. The mother liquor of the massecuite cannot be exhausted of crystallizable sucrose at this temperature. To continue crystallization to the ultimate limit of exhaustion, therefore, it is necessary to lower the temperature of the massecuite progressively to the minimum at which it can be mechanically stirred, the limiting factor being viscosity. The progressive decrease in sucrose solubility resulting from diminishing temperature maintains the supersaturation necessary for crystal growth. Crystallization should be allowed to continue at the minimum temperature until a saturation temperature suitable for centrifugal processing is attained, which is approximately 131°F. The minimum temperature to which the massecuite may be lowered varies with the characteristics of the massecuites and the crystallizer capabilities. Usually it falls between 113 and 122°F with high-density massecuites, and above 95°F with lower-density massecuites.

17.2.6 Reheating in Crystallizers

Reheating in the crystallizer is accomplished by circulation of warm water through the cooling elements. The water temperature should not be more than 6°F above the saturation temperature of the mother liquor. Rapid reheating and immediate purging in the centrifugals at the desired temperature minimize increased molasses purity. On the other hand, if reheating is slow, losses will be much more serious. On the average, about a 13°F temperature increase was required to halve the viscosity of saturated molasses. If a massecuite containing a mother liquor viscosity of 1000 P is reheated 20°F, it would be quite easy to purge and would not suffer serious loss from re-solution.

17.3 BOILING SCHEME

The ultimate goals of recovery house remelt operations are (1) low blackstrap molasses polarization/dry solid (PD) purity, (2) minimum blackstrap volume produced, (3) low remelt color, (4) reduced chemical and microbiological sucrose loss, (5) minimum energy use, (6) maximum throughput with existing equipment, and (7) reduced staffing requirements, for both operation and maintenance. To achieve these goals, each company would devise the "best" boiling scheme based on the best available process technology in that company. The best technology includes chemistry of remelt, such as viscosity and rheology, solubility of constituents, and theory of molasses formation and exhaustion; and such chemical engineering aspects as material balance, energy balance, crystal population balance, and process control. Examples of the boiling scheme, as described in the remelt symposium at the 1987 Sugar Industry Technology technical meeting follow.

17.3.1 C & H Refinery, United States

At the C & H refinery, we boil four grades of remelt massecuite, as shown in Figure 17.2. A special remelt strike is boiled out of the syrup from the fourth granulated boiling, with the addition of some No. 4 liquor from the char house. (No. 4 liquor is an intermediate cut in the breaking of char filters from high to low purity.) The massecuite purity is about 90°. The sugar is remelted and sent back to the raw melter. The syrup is used as one of the basic ingredients in producing soft sugar.

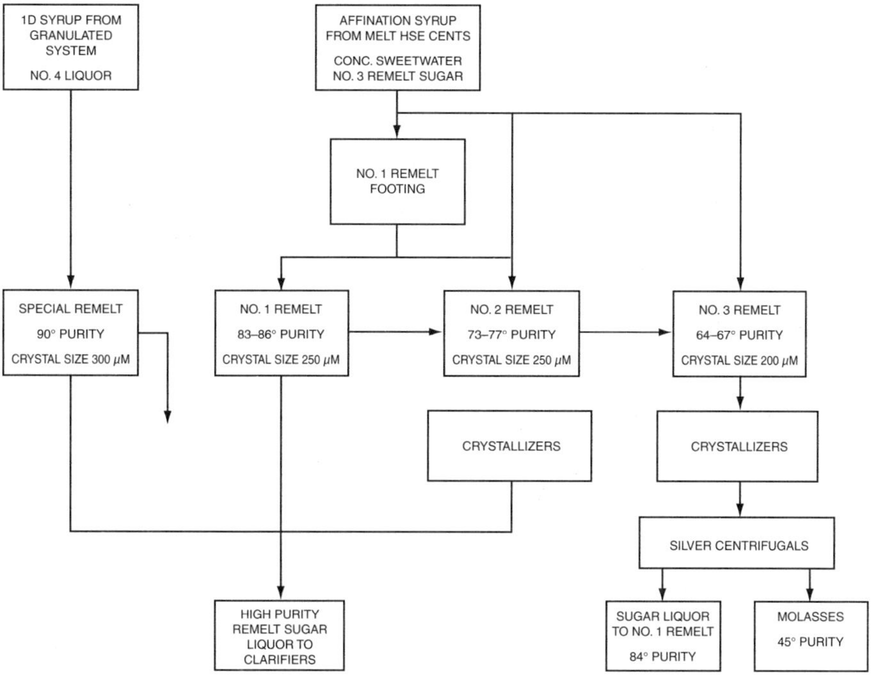

FIGURE 17.2 C & H Sugar Company remelt system flow diagram.

The other remelt strikes boiled are referred to as No. 1, No. 2, and No. 3 remelts. We recover sucrose and eliminate nonsucrose in a three-skip boiling scheme. The normal practice is to prepare a footing strike with sufficient seed for two strikes. The charge for this strike consists of 3000 gallons of affination which is concentrated at 4 in. Hg absolute. Before seeding, the absolute pressure is adjusted to 7 in. Hg. In practice, the seeding point is largely determined by the pan temperature at a controlled absolute pressure of 7 in. Hg. A slurry of 4 lb of fondant sugar in cool, saturated liquor is admitted to the pan. After seeding, the steam flow is cut back and the absolute pressure programmed to change to 9 in. Hg in 16 min. If the quality of the grain appears good, the absolute pressure is slowly reduced to 4 in. Hg and the affination feed started. The steam flow is raised as required and the strike boiled to the desired volume and consistency. A portion of the strike is then cut into another pan. The cut is boiled up with affination syrup or concentrated low-purity sweetwater and completed as a No. 1 remelt strike with a messecuite purity of about 84°. The seed pan is boiled up with No. 1 remelt syrup and completed as a No. 2 remelt strike with a massecuite purity of about 75°. No. 2 remelts are usually cured in a crystallizer for 8 to 12 h before purging. No. 1 remelts go directly to the centrifugals. The dissolved sugar from No. 1 and No. 2 remelts is returned to the raw melter.

The No. 3 remelt strike is charged with affination syrup and seeded with a slurry containing 11 lb of fondant sugar in cool saturated liquor. The strike is then finished with No. 2 remelt syrup to a calculated purity of 65 or 66°. No. 3 remelts are cured in crystallizers from 12 to 32 h, depending on equipment availability. We cool the messecuite to 50°C, which is the purging temperature. In practice, there is never sufficient time to cool the massecuite to the optimum temperature and then reheat it. The new continuous crystallizer, under installation, will remedy this situation. Massecuite will be cooled and held at 40°C for about 18 h, at average flow rates, and then reheated for purging. Dissolved No. 3 remelt sugar is returned to the remelt pans to be boiled into a No. 1 remelt strike, or may be used in the soft sugar system.

17.3.2 Refined Sugars Inc.

a. A/B Operation (No. 1 and No. 2 Remelt). After four years of operating the two boiling system with marginal success, we opted to try a modified three-boiling system. Our remelt facility is divided into two systems. The A/B system uses the same equipment to boil both A and B sugars. Scheduling is a function of the availability and purity of the various feed materials. The C system is dedicated to C boilings. Grain for the A/B system is first established on affination. Our aim is a final mass purity of 80. Since affination purities are generally higher, we normally back-boil A molasses to control the final purity. Each grain strike is cut once, thus producing a footing for either an A or a B boiling. The A boilings have a target for final mass purity of 80, and the B's, 70.

b. C Operation. The C system establishes its own grain on affination. The grain strike is normally boiled to a final mass purity of 60 to 65. The C grain strike will serve as footing for three C mass boilings, which have a final mass purity in the range 55 to 60. The C boilings are discharged to a receiver that feeds a continuous crystallizer. Under normal operating conditions, the retention time will be in excess of 20 h. The C mass is purged by continuous centrifugals with the C sugar, of approximately 88 purity, being melted and sent to affination.

17.4 EQUIPMENT AND MATERIAL

17.4.1 Vacuum Pans

The function of the vacuum pan is to produce satisfactory sugar crystals from syrup or molasses. The concentration of the feeds used in the pans is usually 60 to 65 Brix, and may reach 74 Brix in the refinery. High densities reduce the steam consumption and cut down the duration of the cycle, but too high a density may involve the danger of producing conglomerates and false grains.

In general, there are coil pans, which use live steam, and calandria pans, which use low-pressure exhaust steam. The coil pan has three main disadvantages: (1) it restricts steam economy by having to use live steam, (2) maintenance costs are high, and (3) mechanical circulators cannot be installed—only perforated steam coils provide limited added circulation. A calandria pan is a specially designed single-effect evaporator with large-diameter short tubes and a large downtake for the circulation of heavy viscous massecuite elaborated in batches (Fig. 17.3). Syrup and molasses in certain proportions develop sugar crystals, beginning with the heating surface being covered just sufficiently to obtain circulation and finishing with a full load called a *strike.*

17.4.2 Mechanical Circulation

The advantages of mechanical circulation are (1) thorough and uniform distribution of feed; (2) control of densities by indications of an ammeter on the driving motor; (3) adequate movement of the mass, depending on controlled mechanical equipment; (4) minimum variations of temperature; (5) improved crystal formation (Figs. 17.4 and 17.5); (6) better performances with lower steam pressure and larger capacities; (7) relatively high circulation velocities maintained from the beginning to the end of the strike; (8) better color of products because of reduced heat injury; and (9) increase in massecuite output rate.

17.4.3 Crystallizers

The function of the crystallizer process is to reduce sucrose losses to molasses. The related processes preceding and following can have a major influence on the results obtained. In practice, the exhaustion of molasses involves four successive operational steps: (1) boiling a massecuite to maximum workable consistency in a vacuum pan, (2) cooling the massecuite in the crystallizer to crystallize the recoverable sucrose remaining in solution, (3) reheating the cooled massecuite to its saturation temperature to reduce viscosity, and (4) separating the crystals from the exhausted molasses in the centrifugal machines.

The mother liquor in final, low-purity massecuites cannot be exhausted of crystallizable sucrose adequately in the vacuum pan, primarily because of the sharply decreasing crystallization rate and the high viscosities encountered as the mother liquor approaches the state of exhaustion. Consequently, after the massecuite has been boiled to maximum workable consistency in the pan, it is discharged to the crystallizer, where crystallization in motion takes until the mother liquor becomes a substantially exhausted molasses. Crystallization is a subsidiary operation in which the massecuite is slowly stirred while cooling. Progressive lowering of temperature reduces solubility and forces crystallization to continue. Continuous stirring minimizes internal differences in temperature and supersaturation and thus reduces the danger of fresh nuclei forming.

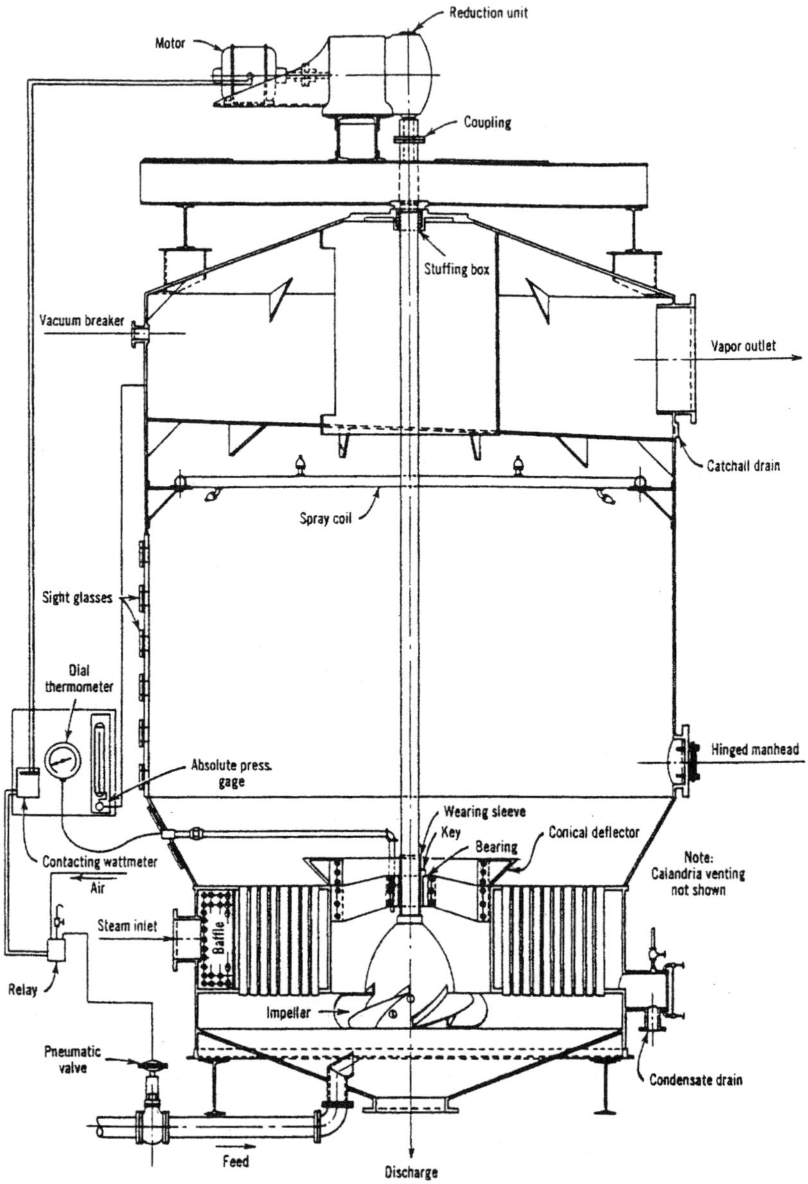

FIGURE 17.3 Calandria pan.

17.4.4 Centrifugal Machines

The massecuites from the vacuum pans or from the crystallizer go first to a mingler, generally a troughlike container with revolving arms to prevent crystals from settling. The crystals in the masscuite are separated from the surrounding molasses or syrup by centrifugal force in a centrifugal. Essentially, a centrifugal consists of a perforated drum or basket

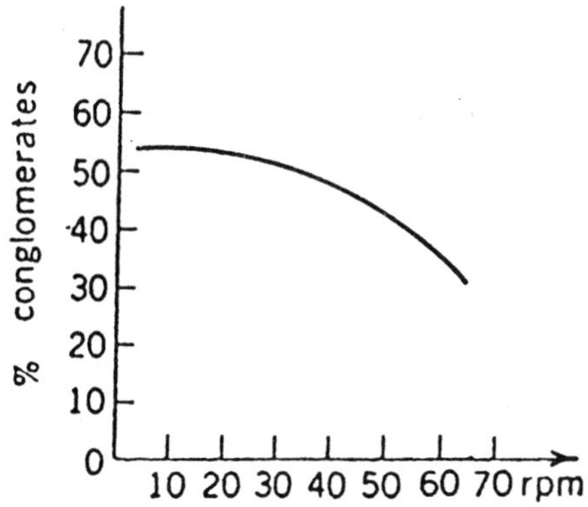

FIGURE 17.4 Stirrer influence on percentage of double crystal and conglomerate formation after graining time.

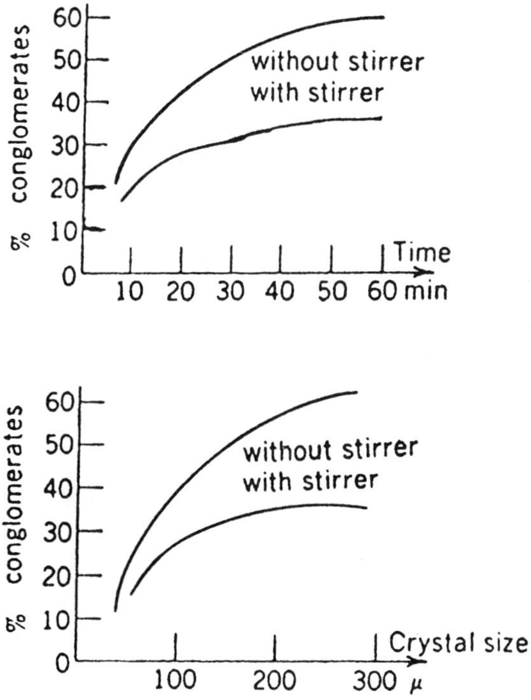

FIGURE 17.5 Percentage of double crystals and conglomerates in relation to time and crystal size.

revolving on a vertical shaft or axis called the *spindle*. The basket revolves within a metal casing or curb, which catches the molasses spun off by the centrifugal force. The basket has perforated vertical sides lined toward the center, first with brass-wire backing screen, and then with a perforated sheet known as a *lining*. The backing screen permits more rapid drainage of the molasses. Each battery of machines has its own mingler, and the charge of massecuite is fed to each machine through its own special spout or gate at the bottom of the mingler (Fig. 17.6).

a. Operation of Centrifugals. The machine is set in motion, the charge is immediately added by opening the mixer gate, and the massecuite rises in the basket because of the centrifugal force generated by the revolving basket. The massecuite distributes itself over the perforated lining, the molasses is expelled into the curb, and the crystals are retained. The spinning or purging continues until the sugar crystals are almost free of molasses, after which the crystals may be further purged of molasses by spraying the wall of sugar with a measured quantity of water. Included in the cycle of filling, purging, and washing is the

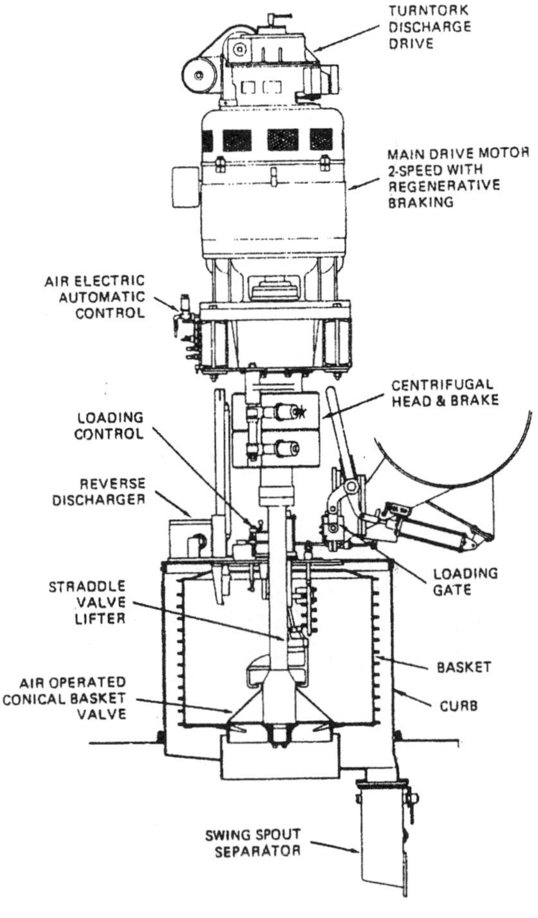

FIGURE 17.6 Centrifugal machine.

discharge or cutting down of the sugar, effected by opening the discharge valve, sometimes called the *cone* or *bell*, in the bottom of the machine. Then the plow is lowered into the machine, by hand or by automatic control, while the basket is revolving slowly, and the sugar discharges into a scroll conveyor underneath the row of machines.

b. Size and Capacity. Centrifugal baskets of modern machines are 48 in. with a depth of 30 or 36 in. or 54 in. in diameter with a depth of 40 in. Baskets of the high-speed machines of present-day installations revolve 1200, 1400, and even 2000 times a minutes. The introduction of mechanical dischargers or plows greatly increased the output, and self-discharging machines gave the capacity of about 130 tonnes per machine per 24 h.

c. Development. The advantages of continuous purging are evident (Fig. 17.7). It offers mechanical and electrical simplicity, low maintenance costs, low and constant power consumption, and low filtration resistance associated with liquid flow through a moving and uniform thin layer of sugar crystals. Application acceptance involves a wide range of low-grade massecuites in cane refineries and in cane and beet factories whose recovered sugar crystals are to be melted or mingled with syrup as magma for vacuum pan footings.

d. Centrifugal Design and Capacity. The massecuite is fed into the spinning conical basket, either by the side or at the center or at the center. The centrifugal force moves the massecuite up the wall of the cone over a perforated stainless steel screen, having the perforations approximately 0.005 in. As the massecuite is moving up the cone, the sugar crystals and the molasses are separated, and the sugar is washed with steam, water, or both. The molasses passes through the screen into a receiving trough while sugar is discharged over the rim into the annular space surrounding the molasses receiving compartment.

The continuous centrifugal of proper design, and associated with a proper feed system and massecuite fluidity, can offer dramatic comparative capacity advantage when processing the lower-grade massecuites generally, because of thin layer filtration and because lower

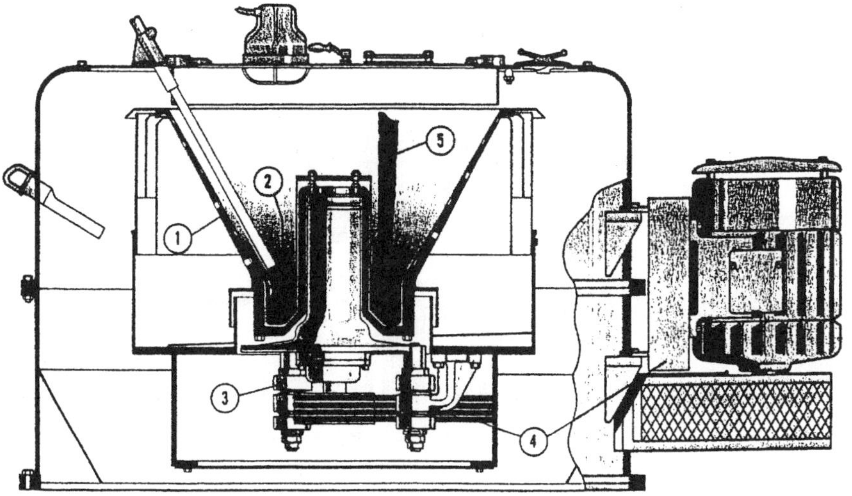

FIGURE 17.7 Continuous centrifugal. 1, Stainless steel basket; 2, loading bowl; 3, support; 4, drive; 5, massecuite side feed. (Courtesy of Western States Machine Co.)

sugar quality is tolerated. There are high-capacity machines for higher-viscosity cane sugar–producing areas. The higher-capacity operation has reduced the incidence of undesirable higher molasses purity, which is more often related to lower-capacity operations. Results provide smaller centrifugal plant requirements and higher sucrose recovery.

High-capacity results are related to (1) modification by enlargement or increased head of older restrictive feed systems; (2) pretreatment of massecuite by water (and possibly steam) in the feed system of the individual machine, to provide continuous high rate and constant flow of properly directed massecuite at reduced mother liquor viscosity; and (3) possible massecuite pretreatment by molasses in pan or crystallizer.

17.4.5 Double Purge Centrifuge

The Dorr–Oliver double-purge centrifuge is a continuous centrifuge that contains two baskets within one casing, giving the ability to subject the fillmass to a two-stage washing process within a single machine (Fig. 17.8). Differences in the top and bottom baskets of the machine are shown in Table 17.1.

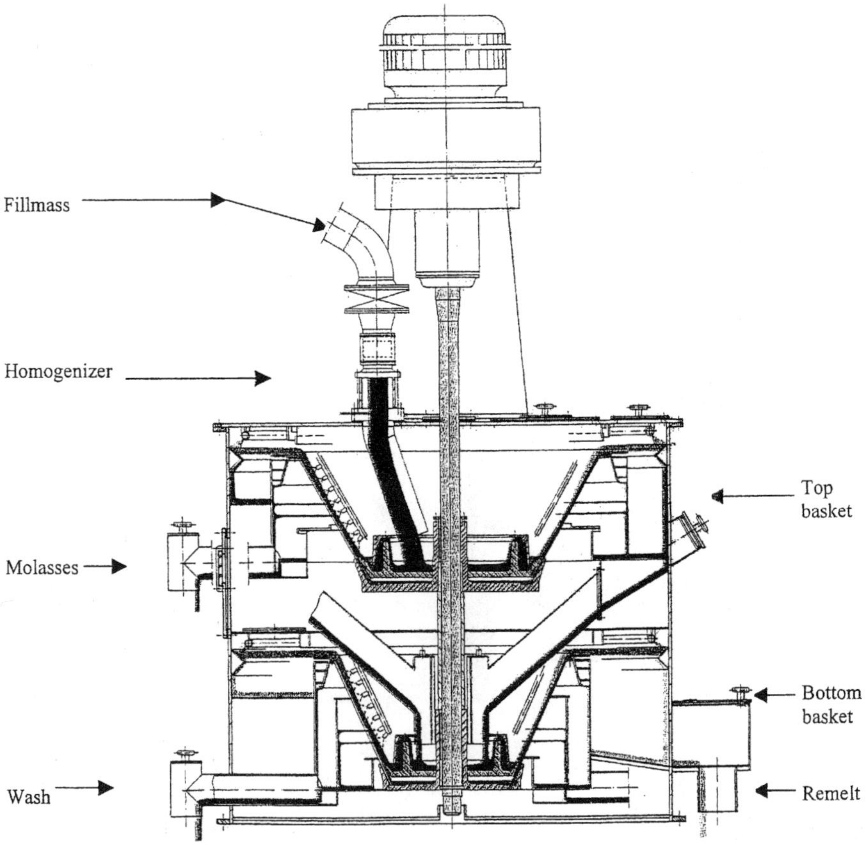

FIGURE 17.8 Double-purge machine.

TABLE 17.1 Double Purge Centrifuge Parameters

Parameter	Top Parameters Basket		Bottom Basket
Diameter (ft)			
Maximum	4.43		3.45
Minimum	2.33		1.78
Surface area (ft^2)	20.99		16.15
Angle of slope (deg)	30		25
Operating speed (rpm)		1750	
G-force		1797	
Sieve size (μm)		60	

a. Description of the Process. The low-grade fillmass is first fed into a homogenizer where the feed is mixed thoroughly with water and steam. This is in preparation for the separation of the syrup and crystals in the first basket. The fillmass is discharged from the homogenizer onto an accelerator pan where the fillmass is redirected evenly onto the basket surface. The separation that takes place in the first basket is without any steam or water addition. The syrup is discharged through the sieve of the first basket and sent to molasses tanks. The sugar crystals are discharged over the rim of the first basket and are immediately wetted with affination syrup by a spray ring. The sugar crystals mixed with the affination syrup is called *affination magma*. This magma flows into the second basket through two feed chutes. The magma flows into an accelerator pan, where it is distributed evenly onto the screen of the second basket. Wash water is used to further separate the syrup from the sugar crystals and also to adjust the color of the sugar. The syrup is discharged as wash and green syrup. These two streams combine into one common stream and is reused in the recovery process. The sugar is discharged from the second basket and is diluted with condensate by a spray ring. This is called *remelt liquor*.

REFERENCE

1. E. C. Gillett, *Low Grade Sugar Crystallization*, California and Hawaiian Sugar Refining Corp. Ltd., Crockett, CA 1948.

CHAPTER 18

Application of Membrane Technology in Sugar Manufacturing*

INTRODUCTION

The interest of sugar producers in using membranes either in lieu of or in addition to the existing, more traditional unit operations dates back to the 1970s. The developments until 1982 were well summarized in a detailed review by the DDS group [1]. Despite the number of tested and proposed applications, no commercial units were installed by that time. In the mid-1980s in a reaction to the expected escalation of energy prices, the interest turned to reverse osmosis for concentration of dilute sugar solutions, primarily the beet raw juice [2], and later, refinery sweetwaters [3] and products of molasses desugarization [4,5]. Increasingly higher requirements on sugar quality in the developing countries, most notably perhaps those of soft-drink firms and confectionery manufacturers, combined with the prohibitive cost of construction of new conventional refineries has led since the early 1990s to testing of membranes as a part of novel simplified and, potentially less costly ways of refining raw sugar or for production of refined quality sugar directly in the mills [6–11]. The interest of soft-drink producers in membrane applications is understandable in view of the high correlation [12] between the negative sensory characteristics (unrefined, burned flavor) of the sugar against its turbidity, a characteristic most directly amenable to drastic improvement with ultrafiltration or microfiltration membranes. In the sugar beet industry the interest turned to the possibility of using membranes to replace the traditional, very effective, yet costly juice purification process [13].

The prospects in the sugar industry are undoubtedly being bolstered by the rapidly multiplying commercial applications in related industries, notably the corn sweeteners, fruit juices, and pulp and paper industries. Dextrose clarification, starch concentration, fructose syrup sterilization, caustic soda recovery, and others [14–16] are now used by the corn processors to reduce the use of diatomaceous earths, enhance the performance of ion-exchange decolorization systems, and reduce or eliminate carbon posttreatment. Ultrafiltra-

* By Michael Saska.

tion is reportedly used for decolorization and biological oxygen demand reduction of, and recovery of, lignosulfonates from wood pulping liquors [17–20].

18.1 FUNDAMENTALS AND NOMENCLATURE OF MEMBRANE SEPARATIONS

Membranes are semipermeable barriers, permeable to most but not all components of the feed, and in this respect, not different from conventional filter media. As indicated in Figure 18.1, they do, however, target for retention particles 10 to 10,000 times smaller than conventional filtration and that imparts to membranes and membrane filtration a number of characteristics as to their construction and operation that make them justifiably a distinct field of fluid separations.

As with conventional filters, the pore size of the membrane and the porosity of the solid deposit (filter cake) are of the same order of magnitude as the target particles. In a simplified way, the effect of the membrane pore size and the size of the target particles on the flow ($L/m^2 \cdot h$) across a membrane can be adequately described by the Hagen–Poiseulle equation for flow through cylindrical pores of diameter d and surface density n (1/unit area),

$$F = \frac{nd^4P}{128\eta l}$$

where P is the pressure difference across the membrane (transmembrane pressure), η the fluid viscosity, and l the sum of the thickness of the membrane and that of any solid deposits on the membrane surface. It is readily evident that all other parameters being the same, the

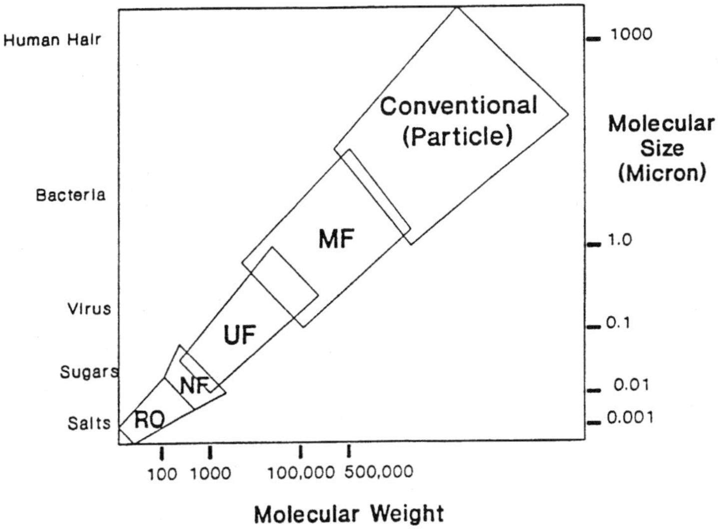

FIGURE 18.1 Classification of membranes based on their molecular weight cutoff (MWCO). RO, reverse osmosis; NF, nanofiltration; UF, ultrafiltration; MF, microfiltration. (From Ref. 25.)

filtration rate through a membrane would be 10^4 to 10^8 lower than for a filter with a conventional mesh size and therefore totally impractical for any industrial application. Practical filtration rates can only be achieved by minimizing the combined thickness of the filtration medium, with the membrane configured as a 1 to 10-μm thin layer on top of a more porous support and typically, cross-flow (or tangential) flow of the liquid with respect to the solid surface (Figure 18.2). Systems do exist where the high shear at the membrane surface is generated by vibrating or rotating the membranes or elements in their proximity [21–23], but their applications so far appear limited primarily to smaller, higher-value applications. An exception may be the rotational CR system supplied by ABB of Sweden to a number of Scandinavian pulp and paper mills [22].

According to the size of the target particles, membrane applications are conveniently divided (Fig. 18.1) into microfiltration (MF), ultrafiltration (UF), nanofiltration (NF), and reverse osmosis (RO) and characterized either by their molecular weight cutoff (MWCO) rating (molecular weight of particles rejected by the membrane with a 90% or higher efficiency) or for microfiltration membranes by the absolute size in micrometers of the particles rejected. Clarification or turbidity removal is the main effect of MF and UF of sugar liquors, while the lower range of UF and the upper range of NF membranes (MWCO 1000 to 5000) produces in addition to complete turbidity removal a significant decolorization effect and a measurable purity increase. Nanofiltration membranes can separate salts from sugars and can, at least in principle, be used for de-ashing of sugar liquors. Reverse osmosis membranes retain all molecules, with the exception of the solvent, water, and present an option to evaporation for dilute solutions.

The large number of membrane manufacturers and suppliers precludes a complete review, but a partial list of those known to have been involved in testing of sugar-related applications is given in Table 18.1. A more complete list can be found in the literature [24,25]. Both polymeric and inorganic membranes are currently marketed, each having specific advantages and drawbacks.

A number of materials are used in manufacturing polymeric membranes [26], those suitable for sugar industry applications are limited to the more-temperature-tolerant ones [e.g., polysulfone (PS), polyethersulfone (PES), polyvinylidene fluoride (PVDF), polytetrafluoroethylene (PTFE), and others]. Frequently, the porous support and the thin skin layer are made from different materials, and for higher flux and reduced fouling characteristics the surface may be chemically modified. The process temperature of a polymeric membrane element is limited rather than by the material of the membrane itself by the thermal stability of the adhesives used to attach the membrane sheets, hollow fibers, or tubes together or to

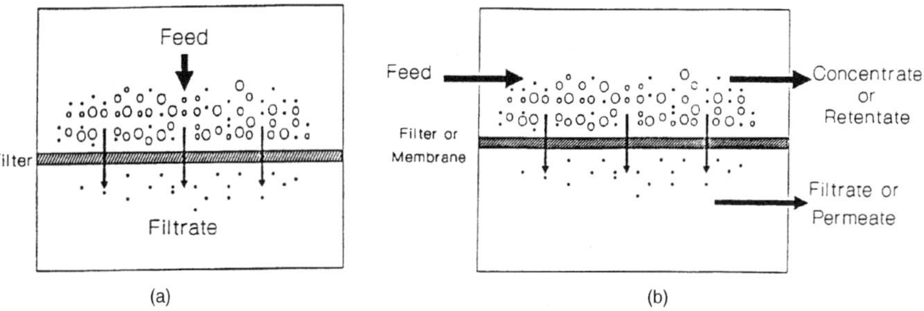

FIGURE 18.2 Principles of dead-end filtration (*a*) and tangentional (cross-flow) filtration (*b*). In the cross-flow design, concentrate (retentate) is fully or partially recirculated at a high flow rate along the membrane surface. (From Ref. 25.)

TABLE 18.1 Membrane Manufacturers and Suppliers with Known Programs Related to Sugar Industry Applications

Company	Manufacturer or Supplier	Membrane Type[a]
Applexion, Des Plaines, IL, and Epone, France	S	Ceramic (UF, MF), carbon (UF, MF), poly / SW (NF, UF)
Ceramem, Waltham, MA	M, S	Ceramic (UF, MF)
Dow / Film Tec, Minneapolis, MN	M, S	Poly / HF, poly / SW, (RO, NF, UF, MF)
Graver Separations, Glasgow, DE	M. S	Stainless steel (UF, MF)
Koch, Wilmington, MA	M, S	Poly / HF, poly / SW (UF, MF), carbon (UF / MF)
Membrane Products, Charlotte, NC	M, S	Poly / SW (RO, NF)
Niro Hudson, Hudson, WI	S	Poly / SW (RO, NF, UF, MF), ceramic (UF, MF)
Osmonics / Desalination, Minneapolis, MN, and Escondido, CA	M, S	Poly / SW (RO, NF, UF, MF)
Pall Filtron, Northborough, MA	M, S	Poly(UF, MF)
PCI / Membrane System, Milford, OH	M, S	Poly / HF
U.S. Filter / Membralox, Warrandale, PA	M, S	Ceramic (UF, MF)

[a]Poly, polymeric, organic; SW, spiral wound; HF, hollow fibers, tubular; RO, reverse osmosis; NF, nanofiltration; UF, ultrafiltration; MF, microfiltration.

the central permeate tube (spiral wound membranes) or to the end plugs (hollow fibers and tubes). Figure 18.3 illustrates the typical arrangements. As all polymers yield under stress (pressure and elevated temperature), certain degree of irreversible compaction, leading to a flux decline, is to be expected for new elements. Operating temperatures of up to 95°C [27] with Koch spiral polymeric and 99°C with Dow tubular polymeric membranes [13] were reported.

In production of inorganic membranes, the thin membrane, typically a 1 to 5-μm layer or ZrO_2, TiO_2, or α-alumina is fused to a porous rigid support, usually in the form of multichannel monoliths of extruded ceramics, α-alumina, carbon, or stainless steel. The monoliths are held in the housing with polymeric gaskets, which may in some applications limit the operating temperature, although gasket materials that can withstand temperatures of 130°C or more are available. Ceramic membranes are very resistant to chemical attack and mechanical wear, but are brittle, and mechanical shocks must be avoided. Porous stainless steel–based membranes are all-welded, with no seals or gaskets, and fully resistant to thermal and mechanical shocks.

Unless it is compensated for by increasing the effective transmembrane pressure, the filtration rate declines during the operation because of buildup of the filter cake layer on the surface of, or penetration of particles within the pores of, the membrane, or plugging of the channels with fibers and course particles. The potential for plugging is largest for spiral-wound membranes, where the feed flow meanders through the mesh of the feed spacer, but even with the straight-channel geometry (tubes and hollow fibers) the potential for plugging exists at the inlet to the elements. In most applications pretreatment is needed and bag filters or self-cleaning filters may be appropriate. Microfiltration can be a cost-effective pretreatment for nanofiltration and reverse osmosis applications. Cleaning of the membranes is usually best accomplished with dilute sodium hydroxide or sodium hypochlorite, but occasional cleaning with an acid (e.g., hydrochloric, nitric, citric, oxalic, etc.)

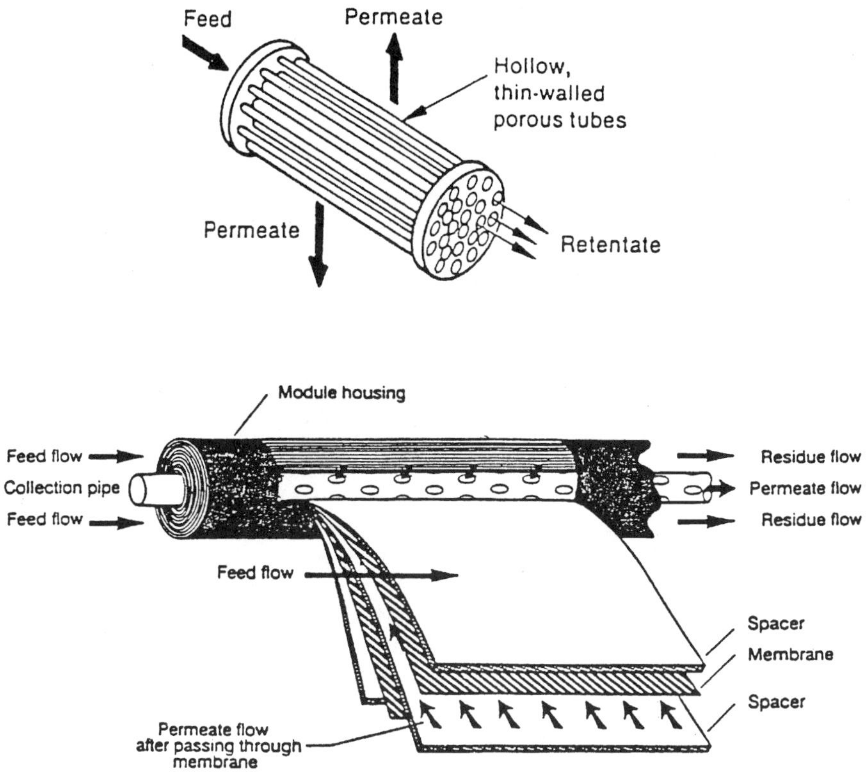

FIGURE 18.3 Membrane configurations.

may be required to remove inorganic foulants. To prevent damage to the membrane, the pH limit and chlorine tolerance provided by the manufacturer must not be exceeded.

Membranes make up some 30 to 70% of the total capital cost of the system (Table 18.2). Obviously, when considering the choice of the membrane, the filtration rates and expected lifetime of the membranes must be factored in. For example, for ultrafiltration of clarified juice (Table 18.3) the tests show that filtration rates for membranes of comparable

TABLE 18.2 Estimated Capital Cost for a 500-gal / min Two-Stage Dextrose Clarification System with Ceramic Membranes

Item	Capital Cost	
	Amount	% of Total
Membranes	$1,413,000	68
Housings	353,000	17
Pumps	161,000	8
Controls, valves, pipes, and fittings	100,000	5
Clean-in-place (CIP) system	50,000	2

Source: Adapted from Ref. 28.

TABLE 18.3 Membrane Cost, Estimated Production Rate for Ultrafiltration of Clarified Juice, Estimated Lifetime, and Calculated Membrane Cost per Lifetime of Resin and 1 m^3 of Clear Juice

Membrane Type	Cost ($/m²)	Filtration Rate (L/m²·h)	Estimated Lifetime (y)[a]	Cost ($/m³)
Ceramic	1200–1600	250–300	5–8	0.17–0.44
Polymeric spiral wound	70–120	70–100	1–2	0.12–0.59

[a]Five-month processing season per year.

pore size, at comparable processing conditions, are some three to four times lower for spiral polymeric membranes than the multichannel ceramics, probably because of the higher hydrophobicity of the surface and thicker skin layer of the polymeric membranes. With all the factors included, the estimated capital cost of the membranes will contribute to the overall cost by \$0.12 to \$0.59 per 1 m^3 of the juice, or about \$1 to \$5 per ton of sugar.

For other applications the advantages of one type over another may be more pronounced. Tubular polymeric or ceramic membranes with large-diameter channels may be required for handling feeds with a high concentration of suspended solids, such as the filtrate from conventional rotary vacuum mud filters. Hybrid polymeric–ceramic systems may be advantageous, with polymeric membranes installed in the initial stages of the system (low concentration factor) and a wide-channel-diameter ceramic membrane for the final concentration in the latter stage of a multistage system.

A typical industrial ultrafiltration or microfiltration system consists of two to five feed-and-bleed stages (Fig. 18.4), where the retentate from one stage becomes the feed for the downstream stage and the permeates from each stage are combined. Within each stage the total membrane surface is determined by the number of elements placed in series, and the pressure drop across an element limits the number that can be operated in series.

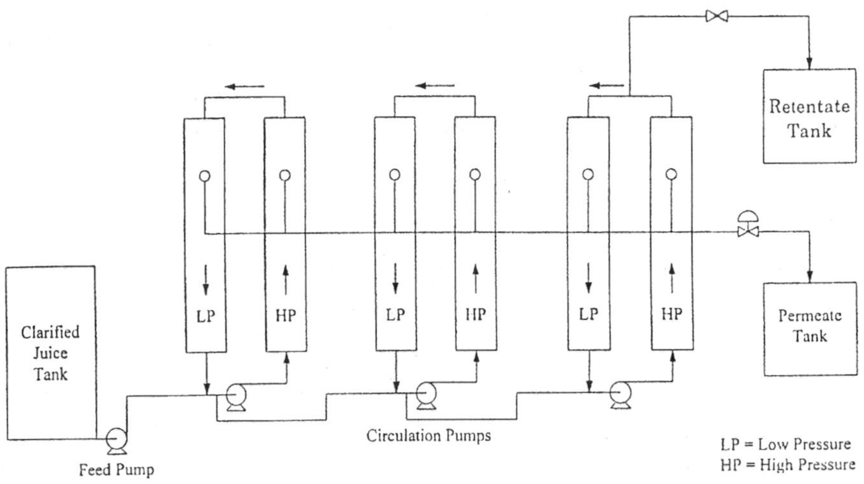

FIGURE 18.4 Three-stage feed and bleed continuous system. (From Ref. 8.)

Two modules in series, each about 1 m long, are typically used for the 2- to 4-mm-channel-diameter ceramic membranes (Fig. 18.4).

18.2 CLARIFICATION WITH MEMBRANES

The startup in September 1994 of the 940-m^2 UF installation in the Puunene, Hawaii mill [8] marked the first large-scale commercial membrane project in the sugar industry. The system, a part of the Applexion's NAP or New Applexion Process [6,7], fed with the conventional clarified juice (Fig. 18.5), produces a 1 to 3-NTU (nephelometric turbidity units) juice (100 to 300 NTU is typical for conventional clarified juice), low enough for softening with ion-exchange resins and production of a super very low color (SVLC) sugar. The following results were obtained in the first year of operation [8]: sugar pol 99.45, sugar color 600 ICU, clarified juice purity increase 0.65%, increase in boiling house recovery 0.8%, 50 to 90% less chemical use in evaporator cleaning, and shortening of boiling time by 15 to 30 min for A and B strikes, respectively. No data concerning the removal of polysaccharides were given, but in tests with a Louisiana and Florida clarified juice, both starch and dextran concentrations were reduced substantially (Table 18.4) with the 300,000 MWCO ceramic membranes [29] and eliminated completely with a 50,000 MWCO spiral polymeric membrane [27].

In the design of a UF system for clarified juice, some flexibility exists as to the concentration of the juice to be filtered, and it would be desirable, if warranted by the filtration rates, to filter the juice at higher concentrations as the volume to be handled (curve V in Fig. 18.6) is only about 60% at 25 and 40% at 35 Brix of the clarified juice volume. The effect on the filtration rate of the juice concentration is significant, and it may be estimated, at least for pure sucrose solutions for which accurate physical data are available, from the Hagen–Poiseulle equation realizing that the dry solids–based flux G is

$$\frac{G = F \times \text{density} \times \text{Brix}}{100}$$

Obviously, the volumetric flux F (Fig. 18.6) declines with increasing concentration, but the dry solids flux G is expected, at least based on this simplistic approach, to pass through a broad maximum at around 30 Brix. This prompted suggestions that UF should be done on a intermediate syrup at 20 to 40 Brix after the first- or second-effect evaporator [30–32]. The experimental results with cane juice do not always support this expectation, as precipitation of sparingly soluble calcium salts at higher sucrose concentrations may reduce the actual filtration rates.

While in the NAP process, a very high quality SVLC sugar with 300 to 500 ICU color is first to be produced from ultrafiltered and preferably, softened clarified juice and then remelted and decolorized with a standard ion exchange process, a direct white sugar process (without remelting) was proposed [9,33] with a combination of microfiltration of clarified juice followed by decolorization of the syrup with a polymeric adsorbent. The 0.2-μm MWCO polyethersulfone-based membranes were tubular type, with each capillary a hollow tube of 1.5 or 3 mm internal diameter, and a supplier's recommended operating temperature of 80°C [34]. Operating temperatures of 90 to 99°C were reported with the identical membranes for raw beet juice microfiltration [13,34,35]. The sugar produced in this process was reported [33] to be less than 35 ICU color and produced at a cost, including labor, chemicals for cleaning the membranes and regeneration of the adsorbent, energy, maintenance, and

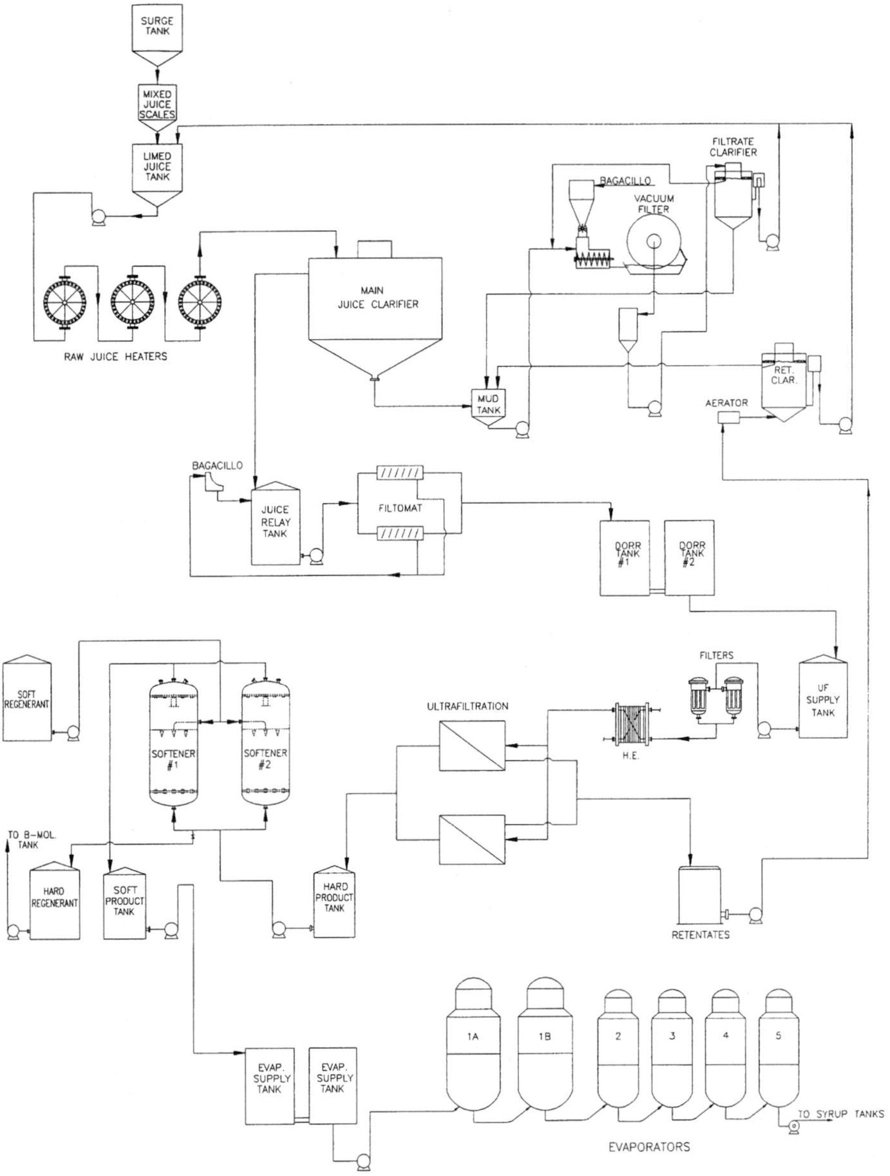

FIGURE 18.5 SVLC process with ultrafiltration of clarified juice and ion-exchange softening of the ultrafiltered juice installed at the Puunene, Hawaii mill. The retentate from the UF membranes is clarified in a flotation clarifier and returned to the mixed juice tank.

TABLE 18.4 Ultrafiltration of Clarified Juice, Average Performance of Five 20-h Tests[a]

Parameter	Feed	Permeate	% Change
Brix	14.2	14.2	0
Turbidity (NTU)	240	1	−99.6
Color (ICU)	11,600	11,400	−1.7
Starch (ppm / Brix)	560	300	−46
Dextran (ppm / Brix)	620	140	−77

Source: Data from Ref. 29.

[a]Membranes: Kerasep, Rhône-Poulenc (19-channel, 300,000 MWCO), batch tests. TMP 3.2–4 bars; 90°C velocity, 6.5 m / s.

replacement of membranes and adsorbent, of less than $40 per ton of sugar. A prototype system with a nominal permeate capacity equivalent to 1000 to 1500 tonnes of cane a day was installed in 1997 at the Florida Crystals' Okeelanta factory [36]. The original hollow fiber membranes have since then been reportedly replaced with stainless steel membranes [37].

Filtrate from rotary vacuum mud filters used in the raw sugar mills is usually too turbid to pass on directly to the evaporators and is usually recycled back to the clarifiers, increasing the load on the clarifiers and sucrose loss. An option to clarify the turbid filtrate with membranes and avoid the recirculation is noted in Fig. 18.7. Large-channel ceramic

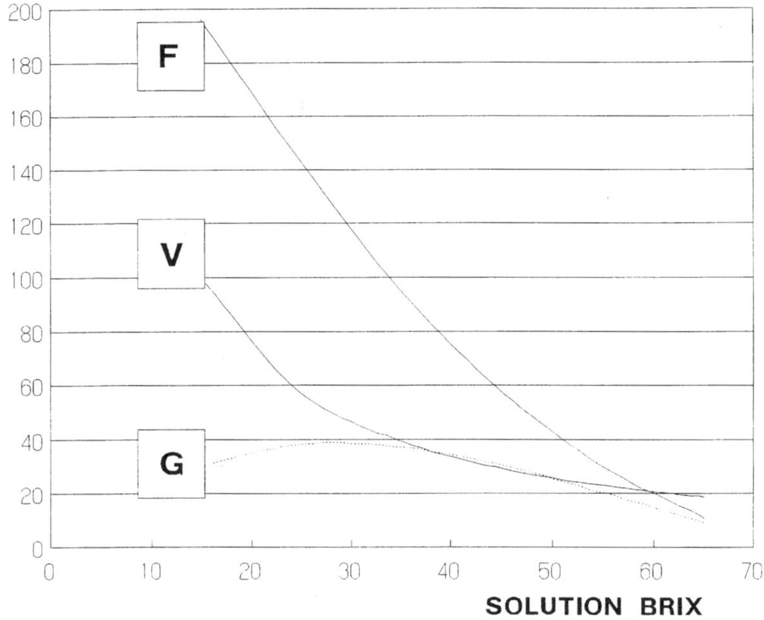

FIGURE 18.6 Theoretical effect of solution concentration on the (relative) volumetric flux *F* (permeate volume / unit area / time), dry solids flux *G (mass dry solids in permeate / unit area / time), and volume of the feed V.* Calculated from the Hagen–Poiseulle equation using the tabulated values of sucrose solutions viscosities at 80°C. The solids flux *G* goes through a broad maximum at around 30 Brix concentration.

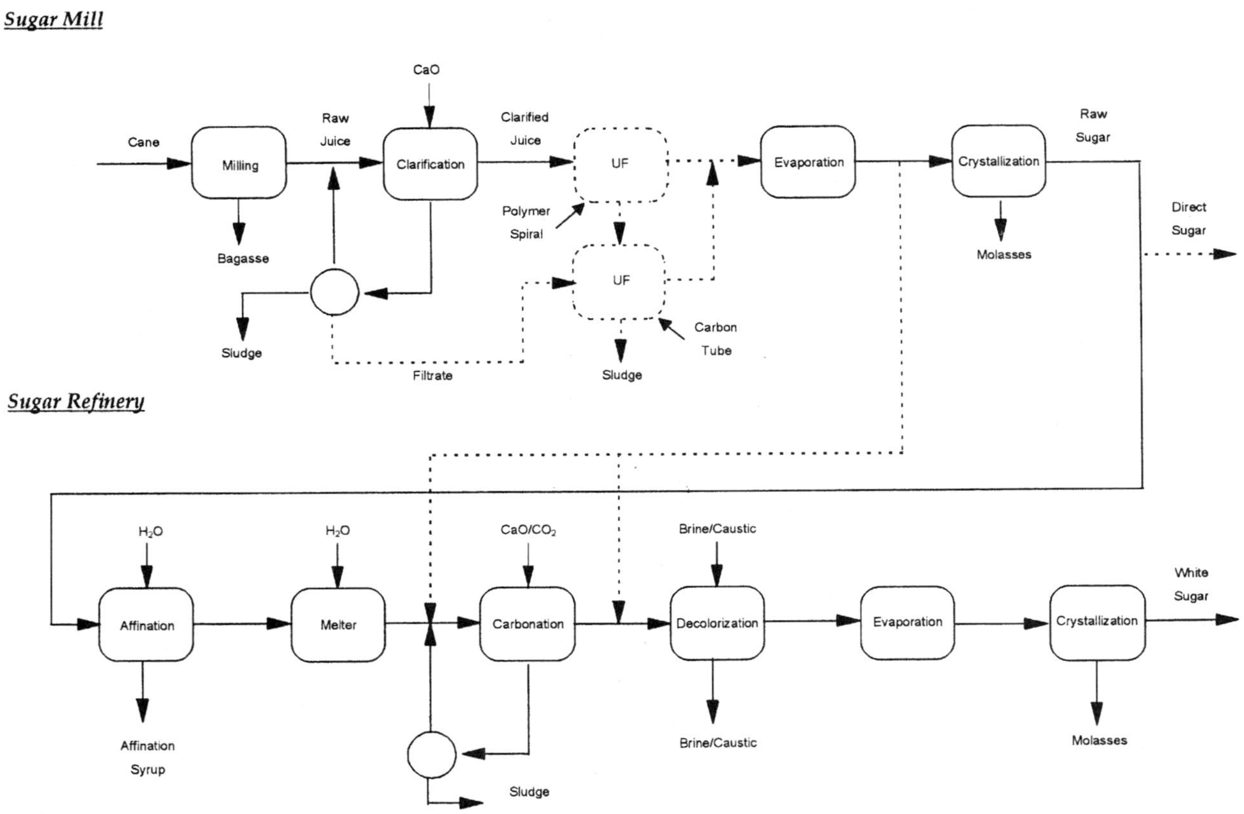

FIGURE 18.7 Conceptual process flow in a sugar mill and a refinery with membrane filtration. Conventional process is outlined with solid lines; dashed lines indicate the alternate processing of clarified juice and rotary vacuum filtrate with ultrafiltration membranes and the simplified refining process of the turbidity-free raw sugar. (From Ref. 27.)

membranes were tested [38] for this application and were found to produce turbidity-free juice at a rate of 100 to 300 L/m^2 · h. Unless removed in a pretreatment filter, fine bagacillo particles may cause plugging of the channels. With bigger diameter channels, the potential for plugging is expected to be less and $\frac{1}{2}$ inch diameter polymeric tubular membranes were reported [39,40] to produce fluxes in the 110 to 160 L/m^2 · h range on a clarifier underflow without any prefiltration. Although perhaps less so with polymeric than inorganic surfaces, wear of the membrane layer with abrasive suspended solids is a concern, and careful SEM examination of the membrane surface after extended testing needs to be performed to detect possible signs of abrasion.

Although the performance of ceramic membranes for ultrafiltration of clarified juice is well proven, their capital cost may in some circumstances prove prohibitive. Mill tests with spiral polymeric membranes by the Koch Membrane group [27] indicated that despite the lower fluxes and expected shorter lifetime (Table 18.3), the spiral membranes may be cost competitive. An economic analysis (Fig. 18.8) showed a favorable ROI for even a modest 1% sugar yield improvement. The analysis further assumed a 15% throughput increase, from the increase in capacity of the clarifiers because of the membrane treatment of the filtrate (no filtrate recirculation as indicated in Fig. 18.7) and faster sugar boiling. The 15 to 20% shortening of the boiling times for ultrafiltered liquors was documented in a number of pilot tests [27,41] and reported as well from the full-scale juice UF system [8]. No credit was taken in the analysis for improvements in the sugar quality and reduced refining costs.

A process for producing liquid sucrose or invert syrup for industrial uses with spiral polymeric ultrafiltration membranes is marketed by AmCane International [42]. The combination of the cane separation technology and absence of the traditional liming clarification produces a high-purity low-color juice that is clarified by ultrafiltration and further decolorized and demineralized with ion-exchange resins.

Membrane filtration of B or C molasses or equivalent refinery streams is difficult, although the effects of the filtration (turbidity removal, viscosity reduction) are relatively higher than with higher-purity streams. In a refinery stream of 60 purity, the viscosity was measured before and after ultrafiltration [29] at a range of concentrations. The reduction

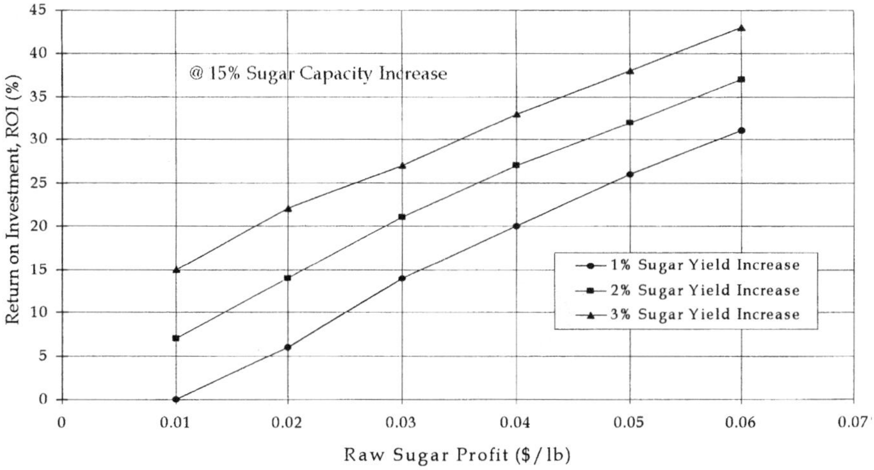

FIGURE 18.8 Estimated return on investment (ROI) from ultrafiltration of clarified juice with spiral polymeric membranes and ultrafiltration of mud filtrate.

in the Brix range pertinent to sugar boiling was found to average around 1000 cP or 40%, corresponding approximately to a 2 Brix difference in concentration. Presumably, ultrafiltration of this product would permit boiling the massecuite to a 2 Brix higher concentration, without increasing its viscosity in comparison with unfiltered liquor, and it was estimated that a 5 to 7% gain in crystal yield would result. The filtration rates for the low-grade products diluted to 30 to 50 Brix are relatively low, typically in the range 20 to 80 $L/m^2 \cdot h$ with the ceramic 300,000 MWCO or 0.1-μm membranes (Figs. 18.9 and 18.10) and should be, in any design considerations, anticipated to fluctuate considerably in day-to-day operations as their composition changes [43]. On the dry solids basis, the filtration rates are 15 to 20 kg dry solids $(DS)/m^2 \cdot h$, some two to three times lower than those for ultrafiltration of clarified juice (ca. 45 kg $DS/m^2 \cdot h$) with the same membranes. This corresponds to the higher non sugar content (15% non sugars on DS in clarified juice vs. 40 to 60% in B and C molasses) but is also likely effected because of the degradation in the prolonged low-grade boiling and reduced solubility of calcium salts at the 30 to 50 Brix concentrations employed in membrane testing of low-grade streams. A striking example of what was interpreted as irreversible effects of crystallization or agglomeration was noted in tests with ultrafiltration of a Louisiana B molasses [29], where the flux was found drastically different for a freshly diluted molasses and a diluted molasses that was kept cold after dilution and prior to the UF testing (Fig. 18.9).

Ultrafiltration and microfiltration of affination syrup was studied [44] as a potential way to reduce the cost associated with sugar recovery from the high-viscosity stream that may in the conventional refining add up to 40% to the overall refining cost. The decolorization rates (Table 18.5) compare quite well with those found for clarified juice (Fig. 18.11) with identical ceramic membranes. The color of the sugar boiled from the respective permeates was reduced by 34 to 69% in comparison with the control (i.e., sugar boiled from untreated affination syrup).

18.3 DECOLORIZATION WITH MEMBRANES

The decolorization effect of a membrane depends on its pore size (MWCO), origin of the feed, and operating conditions, primarily the temperature and concentration factor (recov-

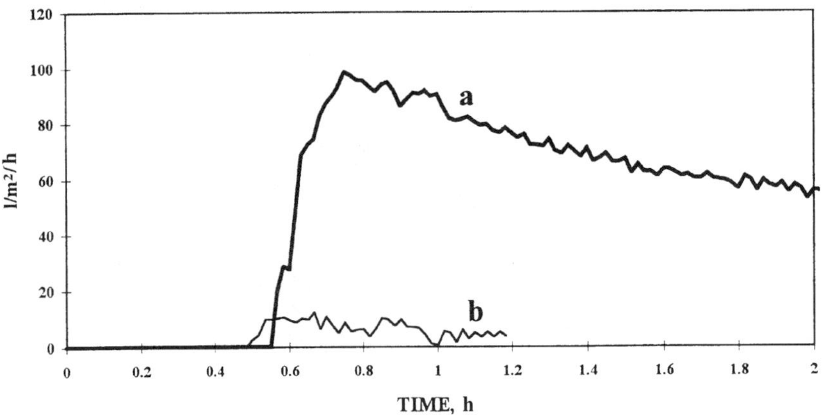

FIGURE 18.9 Ultrafiltration of a Louisiana B molasses diluted to 40 Brix. Membranes; Kerasep 300,000 MWCO (Rhône-Poulenc); transmembrane pressure, 3 bars; temperature, 90°C, recirculation velocity, 5 m / s. Curve a, freshly diluted molasses, curve b, diluted molasses kept at 10°C for 24 h prior to ultrafiltration.

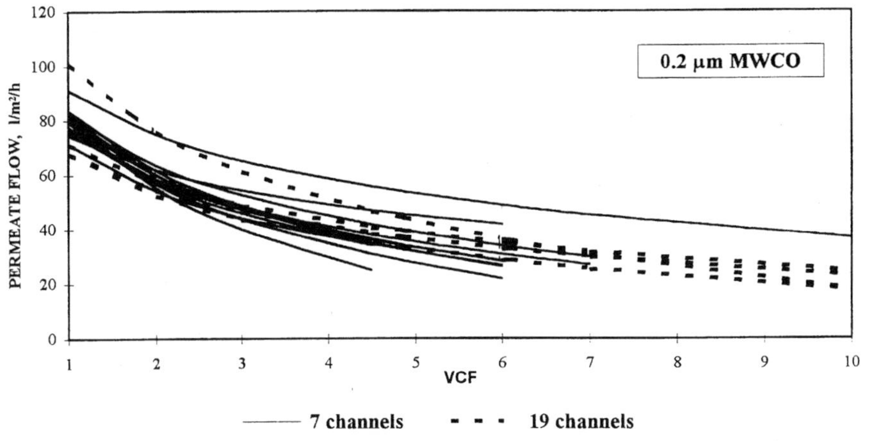

FIGURE 18.10 Microfiltration of a C molasses diluted to 40 Brix. Membrane, Kerasep 0.2 μm; transmembrane pressure, 1 bar; temperature, 85°C, recirculation velocity, 5 m/s.

ery). For the cane juice with 12,000 to 20,000 ICU color (the common range for Louisiana clarified juice) and a series of polymeric and Al_2O_3/ZrO_2 ceramic membranes, the average decolorization effect (1—color of mixed product/juice color) varied [38] from about 15% or so for ceramic 300,000 MWCO membranes to over 70% for 2500 MWCO thin-film polymeric membranes (Fig. 18.11). This is somewhat lower than found earlier [31] in brief tests with higher-color South African clarified juices. The tests concerning the crystallization properties of the permeates from a series of membranes indicated that the membrane effect on the sugar quality for membranes with MWCO > 5000 comes primarily from the viscosity reduction of the crystallizing liquors, and its effect on the efficiency of the centrifugation with respect to the removal of the liquor from the surface of the crystal. The heat and mass transfer coefficients appear enhanced, resulting, characteristically for A strikes, in a 20 to 30% boiling time reduction with 300,000 MWCO membranes. A 20% increase in rate of crystallization of cane syrup was reported from laboratory study of a ultrafiltration membrane [45]. For membranes with MWCO > 10,000, 40 to 60% whole-color reduction with respect to (membrane) untreated sugar is typical, while the laboratory double-affined color of 250 to 350 ICU was found nearly identical for untreated and membrane-treated sugars at MWCO > 5000. Only for membranes with MWCO < 5000 was the affined-color reduced, typically to 50 to 70 ICU.

TABLE 18.5 Ultrafiltration and Microfiltration of Affination Syrup and Crystallization of Filtration Products[a]

Membrane Type (MWCO)	Decolorization (%)	Turbidity Removal (%)	Sugar Color (ICU)
Control	—	—	1090
15,000	40	96	335
300,000	21	95	400
0.1 μm	7	96	720

Source: Data from Ref. 44.
[a]Feed color 24,000 ICU, 45 Brix. Membrane: Kerasep, Rhône-Poulenc.

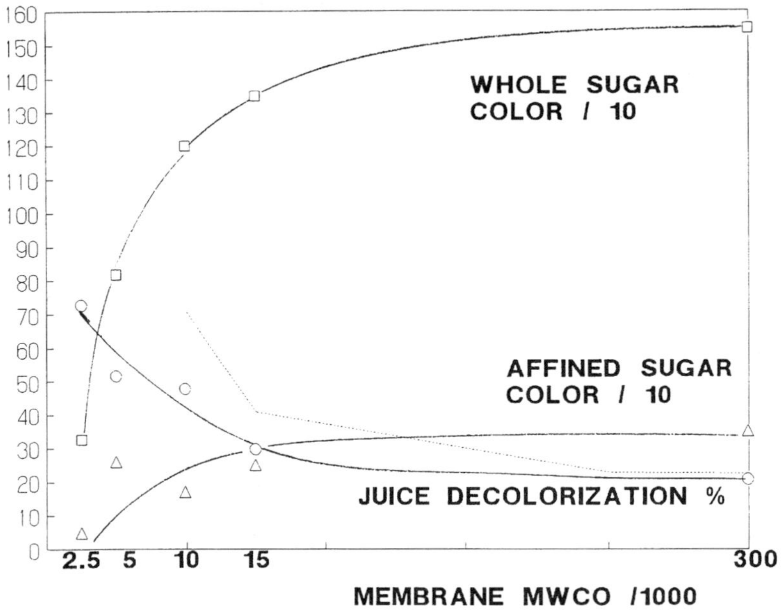

FIGURE 18.11 Decolorization of clarified juice with ultrafiltration membranes and color of the whole and affined sugar boiled from the ultrafiltered juice. 15,000 and 300,000 MWCO membranes Kerasep (Rhône-Poulenc); 2500, 4000, and 10,000 Osmonics/Desalination G-10, G-20, and PW spiral polymeric membranes. (From Ref. 38.) Dashed line, juice decolorization effect with Carbosep (Rhône-Poulenc) inorganic membranes (31).

A direct white sugar process [11] was described that would use a combination of, first, ceramic UF membranes (300,000 MWCO or 0.1 μm) for clarification, followed by 2000 to 3500 MWCO spiral polymeric membranes for decolorization of the clarified juice. The product quality (color, ash, and invert content) was reconfirmed in more recent larger scale-tests [41]. Additional testing indicated that direct filtration, with only a 100-μm coarse prefilter, without the use of ceramic membranes, with polymeric 2000 to 3500 MWCO membranes is feasible, giving a permeate of the same quality albeit at a reduced filtration rate. The total additional cost from operating the two in-series membrane systems was estimated at only $13 to $24 per tonne of sugar, depending on the length of the processing season. Indications are that even better results (affined color $<$ 50 ICU) can be obtained if special precautions were taken to eliminate any postmembrane (evaporation and sugar boiling) generation of color. Those colorants are likely to be of high molecular weight and particularly detrimental to the color of the final product.

Simultaneous clarification and partial decolorization of either whole or affined cane raw sugar as an alternative to chemical clarification (phosphatation or carbonatation) and decolorization with bone char has been studied as well [46–48], although it appears at a disadvantage compared with membrane treatment of clarified juice, because of the additional dilution (to 30 to 50 Brix) required to secure reasonably high filtration rates. Although it is well established that the size (molecular weight) of the colorants inside the sugar crystal tends to be higher than in the corresponding mother liquor [49,50], which would favor higher color rejection (decolorization effect) for raw sugar liquors than with the cane juice, the amount of the "internal" colorants is only some 20% of the total (Fig. 18.11) and the overall decolorization effect may not be expected to be effected substantially. A 70 to 85% decolorization was obtained [46] in tests with 2500 MWCO membranes (Fig. 18.12).

In principle, the effective size of the colorant molecule can be increased by a chemical reaction with, or at least an electrostatic coupling to, a color-scavenging agent, a cationic flocculant with or without the presence of a nascent calcium phosphate surface [51,52]. Typically, 25 to 35% color removal rates are reported from phosphatation refineries, and as high as 70% when enhanced with the color precipitant [51], typically prior to liming and phosphoric acid addition. Pretreatment prior to filtration with ultrafiltration and microfiltration membranes of a 50 Brix, 3000 ICU color raw sugar liquor with a cationic quaternary ammonium color precipitant, liming, and phosphatation gave total decolorization rates of 20 to 60%, depending on the membrane pore size [53,54] and filtration rates in the range 20 to 60 $L/m^2 \cdot h$. The overall decolorization effect is equal to, but not greater than, the sum effect of the two operations performed separately, (i.e., flocculation–liming–phosphatation) followed with conventional filtration and membrane treatment of the untreated raw sugar liquor. Similarily, accounting for some scatter in the data, the filtration flux did not appear enhanced by the pretreatment. It is doubtful that the colorant–flocculant–calcium phosphate aggregate can survive the high shear conditions in the recirculation pumps of the membrane system.

Decolorization of spent brine from ion-exchange decolorization resins was tested [55] with tubular 400 MWCO SelRO (Membrane Products Kiryat Wizmann) nanofiltration membranes with a total carbon elimination slightly in excess of 80%. More recently, color rejections of 98 to 99% were reported [56] with spiral polymeric membranes of around

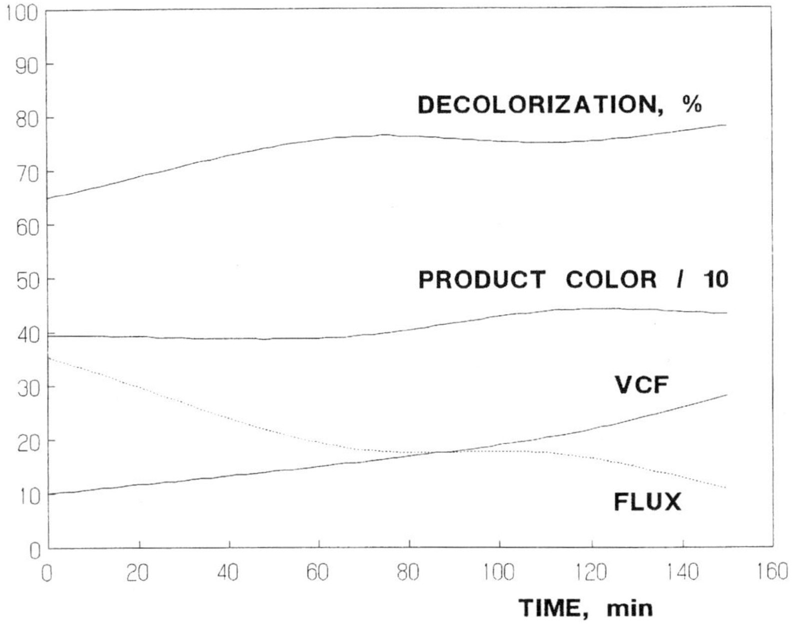

FIGURE 18.12 Decolorization of melted raw sugar liquor. Membrane, Osmonics/Desalination GH2540C1390, 2500 MWCO, 83-mil feed spacer; feed concentration, 31 Brix; transmembrane pressure, 300 psi; recirculation rate, 6 gal/min; final concentration factor, 2.8. (From Ref. 43.) Flux in $L/m^2 \cdot h$.

1000 MWCO and an industrial system was installed at the Marseille refinery in 1997 [57]. The purification and recycling of the spent brine are expected to save annually 1100 tonnes of NaCl and 140 tonnes of NaOH, and its estimated payback time is 3 years. Treatment of the nanofiltration retentate with fungi is being tested to further reduce the waste disposal load. Spiral polymeric membranes with a 2500 MWCO [46] as well as ceramic nanofiltration membranes [58] were also tested for brine recovery.

18.4 REVERSE OSMOSIS CONCENTRATION OF SUGAR SOLUTIONS

Concentration of dilute solutions by reverse osmosis (RO) is an established option to vacuum evaporation in many industries and has been tested on sugar beet juices [2,59], refinery char waste waters [3,60], and dilute products from molasses desugarization process [4,5]. Because of the osmotic pressure limitations, RO of sugar solutions is limited to about 30 Brix final concentration, and less for high-salinity streams, such as the waste product from molasses desugarization. The latter appears the only large-scale commercial application to date [5].

REFERENCES

1. W. K. Nielsen, S. Kristensen, and R. F. Madsen, *Sugar Technol. Rev.*, 9:59–117, 1982.
2. S. E. Bichsel and A. M. Sandre, *Int. Sugar J.*, 84(1005):266, 1982.
3. K. W. Lee, G. Keblish, and C. C. Chou, Paper 591, *Proc. Sugar Ind. Technol. Conf.*, 1989.
4. M. Saska. and J. deLataillade, Paper 653, *Proc. Sugar. Ind. Technol. Conf.*, 1994.
5. *Concentration of the Raffinate Fraction from Beet Molasses Desugarization Chromatographic Separators,* Bulletin, NIRO Filtration, Inc., Hudson, WI, 1997.
6. D. Hervé, X. Lancrenon and F. Rousset, *Sugar Azucar,* May 1995
7. R. J. Kwok, X. Lancrenon, and M. A. Theoleyre, U.S. patent 5,554,227, Sept. 10, 1996.
8. R. J. Kwok, *Int. Sugar J.*, 98(1173):490, 1996.
9. J. P. Monclin, U. S. patent 5,468,300, Nov. 21, 1995.
10. J. P. Monclin, U. S. patent 5,468,301, Nov. 21, 1995.
11. M. Saska, A. S. Deckherr, and C. E. LeRenard, *Sugar J.*, Nov.–Dec. 1995, pp. 19–21, 29–31.
12. D. W. Bena, G. Radko, and J. Kuntz, *Int. Sugar J.*, 95(1139):459–472, 1993.
13. S., Galt, K. B. McReynolds, and J. P. Monclin, *Proc. Workshop Products Sugarbeet Sugarcane,* Helsinky, Finland, Aug. 10–11, 1994.
14. R. L. Simms, *Corn Refin. Assoc. Sci. Conf.*, 1990.
15. Koch Membrane Systems, Inc., Wilmington, MA, private communication.
16. X. Lancrenon, M. A. Theoleyre, and G. Kientz, *Int. Sugar J.*, 96(1149):365–367, 1994.
17. J. Wagner and X. S. Xing, *TAPPI Proc. Environ. Conf.*, 1989, pp. 89–93.
18. J. L. Gaddis, D. S. Fong, C. H. Tay, R. G. Urbantas, and D. Lingren, *TAPPI J.*, Sept. 1991, pp 121–124.
19. H. Barnier, A. Maurel, and M. Pichon, *Pap. Puu,* July 1987, pp. 581–583.
20. Georgia-Pacific Corp., Bellingham, WA, private communication.
21. *Pallsep TM VMF Vibrating Membrane Filter System,* Pall Filtron Corp., Northborough, MA, 1996.
22. E, Dahlquist, *Fluid/Part. J.*, 5(4):189–192, Dec. 1992.
23. *Introducing Pacesetter,* Publ. P490, Rev., Membrex, Inc., Fairfield, NJ, Apr. 1990.
24. W. S. W. Ho and K. Sirkar, eds., *Membrane Handbook,* Van Nostrand Reinhold, New York, 1992.

25. C. A. Merlo, W. W. Rose, and N. L. Ewing, *Membrane Filtration Handbook/Selection Guide,* National Food Processors Association, Dublin, CA, 1993.
26. T. Matsuura, *Synthetic Membranes and Membrane Separation Processes,* CRC Press, Boca Raton, FL, 1994.
27. M. Saska, J. McArdle, and A. Eringis, *Proc. 28th Annu. Meet. American Soc. Sugar Cane Technol., St. Petersburg Beach, FL,* June 17–19, 1998.
28. N. Singh, and M. Cheryan, *J. Food Eng.,* 37:57–67, 1998.
29. M. Saska, *Proc. 23rd Congr. Int. Soc Sugar Cane Technol.,* 1996, pp. 147–161.
30. R. J. Kwok *Proc. Workshop Sep. Process. Sugar Ind.,* New Orleans, LA, April 18–19, 1996.
31. M. N. Patel, Tech. Rep. 1623, Sugar Milling Research Institute, Durban, South Africa, Mar. 13, 1992.
32. S. Kishihara, H. Tamaki, S. Fuji, and M. Komoto, *J. Membr. Sci.,* 41:103–114, 1989.
33. J. P. Monclin and S. C. Willett, *Proc. SPRI Workshop Sep. Process. Sugar Ind.,* M. Clarke, ed., 1996, pp. 16–28.
34. *FilmTec Membranes,* Food and Dairy Product Bulletin, Dow Chemical Company, Midland, MI 1993.
35. V. Kochergin, *Proc. Conf. Sugar Process. Res.,* Savannah, GA, Mar. 1998.
36. S. J. Clarke, private communication.
37. S. Wittwer, in Proc. Symposium of Advanced Technology for Raw Sugar and Cane and Beet Refined Sugar Production, New Orleans, LA, Sept. 9–10, 1999.
38. M. Saska, *Proc. American Soc. Sugar Cane Technol. Meet.,* Baton Rouge, LA, Jan. 30–31, 1997.
39. T. J. Tyndall, in Proc. Symposium of Advanced Technology for Raw Sugar and Cane and Beet Refined Sugar Production, New Orleans, LA, Sept. 9–10, 1999
40. J. F. Alvarez and C. Baez-Smith, in Proc. Symposium of Advanced Technology for Raw Sugar and Cane and Beet Refined Sugar Production, New Orleans, LA, Sept. 9–10, 1999.
41. M. Saska and Y. Bathany, unpublished.
42. *AmCane Jucana Juice: Production Process,* AmCane International, Inc., St. Paul, MN, 1996.
43. V. Kochergin, *Membrane and Chromatographic Processes in the Sugar Industry,* Practical Short Course on New Developments in Membrane Technology, Food Protein R&D Center, Texas A&M University, Houston, TX, Mar. 1998.
44. S. Cartier, M. A. Theoleyre, and M. Decloux, Paper 711, *Proc. Sugar Ind. Technol. Conf.,* Montreal, 1996.
45. S. Kishihara H. Tamaki, N. Wakiuchi, and S. Fuji, *Int. Sugar J.,* 95(1135):273–277, 1993.
46. M. Saska, and H. S. Chang, Paper 728, *Proc. Sugar Ind. Technol. Conf.,* 1998.
47. P. Punidadas, M. Decloux, and G. Trystram, *Ind. Aliment. Agric.,* July/August 1990, pp. 615–623.
48. M. Decloux, E. B. Messaoud, and M. L. Lameloise, *Ind. Aliment. Agric.,* July–August 1992, pp. 495–502.
49. M. Saska and Y. Oubrahim, *Sugar J.,* June 1987, pp, 22–25.
50. M. A. Godshall, M. A. Clarke, X. M. Miranda, and R. S. Blanco, *Conf. Proc. Sugar Process. Res.,* New Orleans, LA, Sept. 27–29, 1992.
51. M. Moodley, *Proc. S. Afr. Sugar Technol. Assoc.,* June 1993, 155–159.
52. J. R. Elvin, Decolorization and Clarification of Sugar Solutions using Colour Precipitants and Flocculants. Tate and Lyle Process Technology, 1996.
53. M. A. Theoleyre and S. Boudoin, European patent 9308826, July 19, 1993; Australian patent 9467551, January 27, 1995.
54. S. Cartier M. A. Theoleyre, and M. Decloux, *Proc. Conf. Sugar Process. Res.,* New Orleans, LA, 1996.

55. D. M. Meadows and S. Wadley, *Proc. S. Afr. Sugar Technol. Assoc.*, June 1992, pp. 159–165.
56. M. A. Theoleyre, S. Cartier, and M. Decloux, *Proc. Sugar Process. Res. Inst. Conf.*, New Orleans, LA, 1998.
57. F. Verhaeghe, R. Malgoyre, M. A. Theoleyre, and S. Cartier, Paper 737, *Proc. 57th Meet. Sugar Ind. Technol.*, 1998.
58. S. McGrath, Paper 744, *Proc. 57th Meet. Sugar Ind. Technol.*, 1998.
59. W. Capelin, *Int. Sugar J.*, 84(1007):323, 1982.
60. C. C. Chou and H. C. Weber, Paper 566, *Proc. Sugar Ind. Technol.*, 1988.

PART III

Refinery Design and Process Control

CHAPTER 19

Refinery Design Criteria*

INTRODUCTION

To meet the challenge of the twenty-first century, a new refinery should be designed with the following goals:

1. Staffing of no more than four operators per shift from raw sugar receipt to sugar conditioning, screening, including boiler operation, electricity generation, and wastewater treatment. However, operation for packaging of sugar is not included. This is because the number of operators depend greatly on the company's product mix. Simplification of refinery processes is also needed in order to achieve the goal.

2. A sucrose loss of no more than 0.55% on raw sugar receipt. *Sucrose loss* is defined as sucrose in raw sugar less sucrose in product shipped and sucrose carried to blackstrap molasses. Packaging over-fill is also included in the sucrose loss. A 0.2% of excess sucrose loss would cost a company as much as $250,000, with a refining capacity of 1000 tonnes/day operating 300 days/yr, $440/tonne of raw sugar.

3. A total energy use of 70 L of No. 6 oil per tonne of raw sugar processed operating 250 days/yr with weekly shutdown. This is equivalent to 3136 MJ/tonne, or 134.85 MBtu/hundred weight or 2973 MBtu/tonne of sugar. For a refinery running continuously for 340 days/yr, the total energy consumption should be significantly less than 60 L/tonne. The target should be 0.5 tonne of steam per tonne of raw sugar.

4. For optimum process efficiency, the raw sugar polarization should average 98.3. Attempts should be made to avoid purchasing raw sugar with polarization less than 97.5.

5. Sucrose carried to blackstrap molasses should not exceed 1% of sucrose in raw sugar processed. To achieve this goal, the blackstrap molasses purity should be less than 45%; nonsucrose content eliminated by the refining process should be at least 5%, and some low-purity products should be incorporated in the overall product mix.

6. The maintainance cost should not exceed 20% of total production cost.

* By Chung Chi Chou.

In the following sections we enumerate plant designing requirements to achieve the goals mentioned above.

19.1 pH CONTROL

Without proper pH control and technical set points established at various stages of the process, it would not be unusual to experience a sucrose loss of over 1.0% of raw sugar melted. The magnitude is substantial; an increase of 0.5% sucrose loss (i.e., from 0.5% to 1.0%) would cost a refinery with 300,000 tonnes/yr capacity approximately $750,000/yr at a raw sugar price of $485/tonne in the U.S. market.

19.2 ON-LINE CONTINUOUS TESTING INSTRUMENTS

In addition to temperature and density controls at various points in the refining operation, the following are essential to maintaining process and operating efficiency. All instruments listed below are commercially available:

1. Colorimeters for remelt liquor, melt liquor, press-filtered carbonated liquor, and fine liquor to the pans
2. Turbidity meters for press-filtered liquor
3. Purity analyzers for remelt liquor and affination syrup
4. Sugar detectors with minimal use of hazardous chemicals and waste generation should be acquired for both condensing water and sewer discharge
5. Continuous on-line moisture analyzer for granulated sugar products
6. Continuous on-line color measurement of granulated products

19.3 PROCESS CONTROL AND SPECIFICATIONS

There are over 25 points in the refining process for control of the temperature, pH, density, color, turbidity, conductivity, purity, and vacuum pressure. The control points and critical limit for each unit operation/process must be specified. This would ensure that the refinery will perform at top effectiveness at optimum conditions and produce consistent high-quality product, minimizing the need for laboratory testing.

A dynamic matrix controller (DMC) employed in a popular distributed control system (DCS) should be sufficient to limit staffing to five operators or fewer per shift between the raw sugar warehouse and sugar drying and storage. The essential point to remember is that the key to the success of any control system is based on sensing/measuring devices. Therefore, it is of utmost importance to evaluate and select carefully the sensing and/or measuring devices.

19.4 ENERGY USE AND WATER CONSUMPTION

Presently, the total energy use of a "best"-run refinery with carbonation facility is about 3200 MJ per tonne of raw sugar processed. The level of 75 L of No. 6 oil is equivalent to 3500 MJ per tonne of raw sugar processed. The average total energy use in the sugar refining industry is about 4200 MJ/tonne of raw sugar with weekly shutdown.

The total refinery water use, excluding water for the condensers, of a "best"-run refinery is about 200% of the weight of raw sugar melted; an average refinery probably runs at

285% of raw sugar melted. The more water is used, the higher the energy required for evaporation and the higher the sucrose loss, which would result in a poor sugar yield. The keys to minimizing energy use and water consumption are prudent process selections and a good utility conservation program. A heat train to recycle energy resources should be an integral part of the turnkey design system. All the vapor should be reused whenever possible to the limit of the second law of thermodynamics. For reasonable process efficiency, the polarization for raw sugar process should be less than 98.2.

19.5 PROCESS SELECTION AND DESIGN

19.5.1 Raw Sugar Warehouse

For versatility in handling different raw sugars, a tunnel conveyor spanning the full length of the warehouse with at least 30 floor openings should be built. Raw sugars of various origins should be strategically placed to allow blending of raw sugars with automatically controlled floor openings. Any low polar (less than 97.5) raw sugar with a high moisture content should be blended and processed immediately to avoid inversion of sucrose even before the sugar leaves the warehouse.

19.5.2 Affination Station

Mingling syrup Brix should be at about 72, with a maximum magma temperature of 42°C. The ratio of mingling syrup to raw sugar should be controlled by the wattage of the motor on the mingler. Belt weigher control is maintenance intensive and should be avoided. Affination control can be based on color, ash, and/or purity.

19.5.3 Melting Operation

It is recommended that the premelter be set for 72 Brix and a temperature of 70°C. The main melter should discharge the liquor at a temperature of 75°C and a minimum of 68 Brix for maximum energy conservation. Normally, melting is accomplished by recycling vapors from the evaporators.

19.5.4 Decolorization Process

For the most effective operation, the decolorization process chosen should be designed to give a fine liquor color of 150 maximum and an average of 120 color (ICUMSA method at 420 nm). The best choice for a decolorization scheme is carbonatation followed by ion exchange or granular carbon treatment. However, the ion-exchange method is not environmental friendly with respect to waste disposal. Granular carbon is known to have high capital and operating costs, intensive maintenance, high sugar loss, and low sulfate removal. However, the process is environmentally sound. The choice of either process should be based on the economical and environmental considerations, for the present and future.

a. Flue Gas Treatment. The stack flue gas should be cooled to a maximum of 45°C. This is followed by a separate scrubber with a soda ash solution to remove most SO_2 contaminants. The quantity of water needed for the cooling water is usually seven times that of the water (with soda ash) used for the scrubber. The gas compressor selected should be such that the temperature increase across the compressor be limited to 25°C. At any rate, the gas entering the A saturators should not exceed 80°C, to minimize both chemical and physical sucrose losses and to avoid environmental problems.

b. Carbonation. The retention time at the A saturators should be 40 min and at the B saturator 20 min; therefore, two A saturators in parallel are needed. Some refineries are designed for four saturators, one as a spare for either A or B should the process be run for more than a month. If the refinery is to operate more than 300 days/yr, the preferred mode of operation is to run 18 days consecutively and shut down for 3 days for maximum plant efficiency.

To avoid a sudden change in quality, remelt liquor is ratioed/metered into the washed sugar liquor. pH values should be around 9.6 and 8.3 for A and B saturators, respectively, when measured at 20°C. Some refineries minimize sucrose loss by operating A saturators at 75°C and reheating the liquor to 85°C prior to entering the B saturators.

Membrane press filtration should be used for desweetening the carbonate cake to reduce both the polar and percent moisture content in the cake. Low moisture levels (less than 30%) make it feasible for the cake to be used as a raw material for other industries. It goes without saying that a low-percent-moisture cake would save on freight costs. It should be reiterated that carbonation is considered the best decolorizing process by many sugar technologists, assuming that the disposal of the carbonate cake is not a major problem.

c. Phosphatation. If the phosphatation process is selected, Tate & Lyle process technology's service is recommended. This is because T&LPT is considered to be the best in the sugar industry in the design of a phosphatation system.

d. Granular Carbon Versus Ion Exchange. For a refinery with a 1000-tonne/day capacity, in a granular carbon configuration, eight columns 10 ft deep by 42 ft high, each with a column carrying 2400 ft^3 of granular carbon and a carbon regeneration kiln with much peripheral equipment are required. In an ion-exchange decolorization scheme, only three columns (each 8 ft deep by 24 ft high) are required. The more equipment necessary to run the operation, the more staffing and maintenance are expected, resulting in higher refining operating costs. However, a pulse bed granulator carbon system is preferred. Only two columns would be needed.

The major disadvantage of the ion-exchange resin process is the disposal of the dark-color sodium chloride produced in the regeneration of the resin and possibly waste resin itself in the future. One advantage of the ion-exchange process is that it can remove up to 80% of the sulfate in the liquor. Sulfate normally causes pan scaling, thus resulting in potential product contamination, loss of refining capacity due to time spent to remove the scale, high energy use, and high sucrose loss.

The operating and capital costs of the ion-exchange process are about one-half and one-third of that of the granular carbon system, respectively. One disadvantage of the granular carbon system is the high sucrose loss associated with the process; the average sucrose loss is probably 0.04% on an operating day and as much as 0.3% during weekend shutdown periods. Regardless of which technology is chosen, polishing filters are required before and after the decolorization process.

19.6 EVAPORATION

At minimal, a two-effect evaporator should be installed to produce a fine liquor (going to the pans) of at least 75 Brix for energy conservation. The evaporating system should be designed to have excess vapors for reuse in vapor melting, heating of processing streams and air to the granulators, heating for the pan boiling, and so on. Most refineries use vapors from the evaporator for sweetwater evaporation. A single-effect thin-film evaporator may cause product quality problems due to potential overburning of the sugar liquor. When overburning takes place, the resulting carbonaceous matter needs to be removed by polishing the filter again to avoid contamination of the sugar products.

19.7 SWEETWATER EVAPORATOR

Many continuous refining processes, such as decolorization, press filtration, and dust collection from sugar drying, conveying, and packaging generate low-density sweetwater. Water balance and the ability to manage other continuous refinery processes in an orderly manner are put at risk when upsets and downtimes occur in the raw sugar melting process which would stop the orderly consumption of this low-density sweetwater. It is for this reason that most well-run refineries have a sweetwater evaporator to concentrate low-density high-colored material and manage the imbalance that occurs.

Without a proper evaporating system to manage water imbalance resulting from equipment failure or operator error, the following are likely to occur:

1. Sweetwater and syrup containing a significant quantity of sucrose are allowed to accumulate on the lowest elevation in the refinery. Floors exposed to flooding will have to be replaced in time. Flooded floors are hazardous to employees.
2. Continuous operations such as adsorbent/decolorization column desweetening and revivification will have to be interrupted. Without orderly regeneration of the adsorbent, the quality of the final product and adsorbent itself will suffer.
3. Large quantities of sucrose from sweetwater and syrups were dumped to the sewer. Not only will charges for sucrose losses be severe, but the charges from the municipality to process the excess biological oxygen demand (BOD) material will be expensive.
4. The refining output (capacity) and product quality would suffer, resulting in loss of sale and customer dissatisfaction.
5. The general sanitation of the facility deteriorated over time since the spills could not be cleaned immediately. This is critical for maintenance of good manufacturing practices.
6. Ultimately, light-density materials, such as sweetwater, have to be concentrated in the remelt boiling equipment. Orderly remelt boiling is time consuming and can best be accomplished with a one-strike boiling time of about 4 to 6 h. When it is necessary to handle sweetwater, the performance of the recovery house (e.g., yield) will be affected severely.

19.8 WHITE SUGAR BOILING

A heat train through the vacuum pans, heat exchangers, evaporators, and other refinery heating systems should be designed to recover a minimum of 80% of condensate with a good-enough quality to be reused at the boiler house. Over 60% of the energy use in a refinery is consumed in the crystallization of sugar. A twenty-first century refinery should employ continuous sugar boiling coupled with a vacuum cooling crystallizer to bring the total energy use below 70 L of No. 6 oil per 100 lb of raw sugar melted. The design of the entire system should be such that both conglomeration and coefficient of variation of sugar crystal is minimized.

With fine liquor at a color of 120, under normal conditions, No. 4 sugar should be sufficiently low in color to go forward to the final product without reprocessing. No. 4 syrup purity is normally too high to be sent to the remelt house; therefore, it would be more advantageous to boil one additional strike and remelt No. 5 sugar so that it can be combined with press-filtered carbonated liquor before the polishing presses. Vacuum pans should be of the short-tube design with agitators to circulate the massecuite effectively. Additionally, an automatic seeding system should be installed at the pan floor to achieve a minimal staffing requirement.

19.9 RECOVERY SUGAR BOILING

Affination syrup, No. 4 or No. 5 syrup, and the concentrated excess sweetwater should be processed in a three-stage boiling scheme. Sugar from the first strike should be sent forward to washed sugar liquor. Sugar from the second strike can also be sent forward—however, only after it has been double-purged. Final sugar can either be sent forward or used as a footing for the second strike after it has been double-purged. Double purging can be achieved efficiently using the double-purging centrifugal machines commonly used in beet factories in Europe. This scheme is particularly suitable for a fully automated refinery. The conventional double Einwurf system requires frequent operator attention and is maintenance intensive. The system works quite well for a conventional refinery where labor is readily available, but it would not be conducive for a modern fully automated refinery to meet the challenges of the twenty-first century.

19.10 SUGAR CONDITIONING

Because of the weather conditions where most beet processors are located, and the purity and nature of the impurities present in the beet sugar, sugar conditioning techniques for beet sugar manufacturing may not be applicable for cane sugar conditioning. Hot air entering the granulators should not exceed 90°C. The temperature of the sugar discharged from the granulators should not be greater than 45°C. It is recommended that the sugar from granulators be stored in holding bins for at least 5 h but no more than 7 h before a screening operation to remove fine and coarse sugars. To prevent caking in bulk storage, the screened sugar should be placed in conditioning bins for at least 12 h with conditioned air blowing through it. If a silo is used, sugar should be moved out from the bottom and recycled to the top of the silo during the shutdown day at 6-h intervals.

19.11 WASTEWATER TREATMENT

For a modern refinery, treatment of streams having high BOD, high temperature, and/or high solids content will be required on site. For oxidation-ditch-type BOD treatment to be successful, cooling of hot streams such as contaminated condensates for optimum microbiological activity will be necessary. Prior separation of any large quantity of suspended solids entering the BOD treatment area will probably be cost-effective. The key is to minimize the number of streams connected to the wastewater treatment system.

19.12 QUALIFICATION OF PROCESS OPERATOR

To operate a fully automated refinery effectively, process operators could have a combination chemical, electrical, and mechanical engineering bachelor's degree with a B average. A corporation can mitigate risk through organization design. Safety, environmental assurance, product quality, profitability, and customer service level are fixed by the type of organization management created. This level of discipline training may be perceived as an usually high skill level; however, it is not. As corporations meet their obligation to coexit on a friendly basis with the community and to make a profit, the need to minimize and eliminate operator error becomes obvious. Operator error is responsible for as much as one-half of serious incidents affecting operations, safety, or environmental pollution. Given complex processes and systems, operators must be competent to detect small aberrations and act before serious incidents or personal injuries occur. Selection and training of process operators should begin as soon as possible. Only competent and responsible operators can ensure highest process efficiency and enhance the profitability of corporations.

19.13 THE BASIC PRINCIPLES OF CANE SUGAR REFINING*

1. Sugar refining is a series of separations.
2. Once a separation has been made, good must not be added to bad, nor bad to good.
3. The best possible degree of separation must be secured from each separation.
4. The quality standards laid down must be achieved, but not appreciably exceeded.
5. Syrups should be stored cool and neutral, or slightly alkaline.
6. Heating of syrups and masses should be done as late and as fast as possible.
7. Stock in process should be as small as possible.
8. All processes should be carried out as fast as possible.
9. There should be as few processes as possible.
10. Each process should be as simple as possible.
11. Process units should be as large as is economical.
12. The addition of water to any sugar product should be held to a minimum.
13. It is better to thicken up lights be adding sugar than by evaporating water.
14. Sugar that has been crystallized out must not be redissolved.
15. The addition of ash to any part of the process makes molasses and loses money.
16. Reprocessing causes sugar destruction and incurs costs twice over.
17. Dry sugar should be subjected to the least number of water drops.
18. It is no use being technically correct if the result is losing money.
19. The cleaner the plant, the better the result.
20. A bad plant run by good operators is preferable to a good plant run by bad operators.

*Courtesy of C&H Sugar Co., Crockett, CA.

CHAPTER 20

Process Selection*

INTRODUCTION

As the twentieth century draws to a close, we are seeing an unmistakable trend toward allowing sugar prices to rise or fall with unfettered market forces. Trade liberalization is likely to make the market even more competitive and responsive to change over the next decade. This reshaping of the dynamics of international commerce has resulted in major dislocations in the sugar industry that are being felt at the present time. In addition, the costs of fuel and labor have changed dramatically over recent decades, not only in absolute terms but also relative to the price of sugar. Resulting pressures—to do better with fewer resources—have provided the impetus for a new conceptual framework of sugar technology and bold proposals for change that not so long ago would have seemed unthinkable.

Some of these innovations incorporate new and sophisticated techniques that can properly be described as revolutionary. In several cases, what is new is not a concept but rather, the availability of advanced materials and control systems, the latter in large measure a consequence of the computer revolution that is transforming the world. Someone designing a new factory or refinery today is faced with unprecedented choices, but at the same time cannot afford to choose wrongly. The configurations and very appearances of the installations of the future may be unrecognizable to many of us today.

20.1 SEPARATION STRATEGIES

Separation processes are at the very core of the sugar industry. We start with a complex mixture and want to end up with very pure material, in essence 100% sugar. Few, if any, other foodstuffs can even approach this goal, because crystallization is such a highly effective purification operation. As we know from thermodynamics, to go from an impure condition of high disorder to a pure one of high order requires an input of energy. However, energy consumption is only one of the criteria that determine the outcome when we make a selection among separation process. Other issues to be addressed are capital investment

* By Richard Riffer.

outlays, operating costs, labor requirements, and environmental issues. In addition, we want to be able to monitor our operations with rapid, automated, accurate, and nonpolluting procedures, using robust, reliable instrumentation.

Every operation we examine will show a balance sheet of advantages and disadvantages. It should not be surprising that the relative merits of any given unit process from the perspective of industrialized countries such as Japan and Germany are likely to differ from those seen from a developing economy. Some examples of refining schemes collected from published literature in representative cane refineries are shown in Figure 20.1. In several cases changes have been introduced since the data for this 1990 survey were compiled.

The choices to be reviewed here are: (1) clarification vs. filtration, (2) phosphatation vs. carbonation (alternatively called carbonatation); (3) granular carbon vs. bone char, (4) ion-exchange resin vs. carbonaceous adsorbents, (5) powdered vs. granular activated carbon, (6) the use of blends of granular carbon and bone char, (7) the use of color precipitants

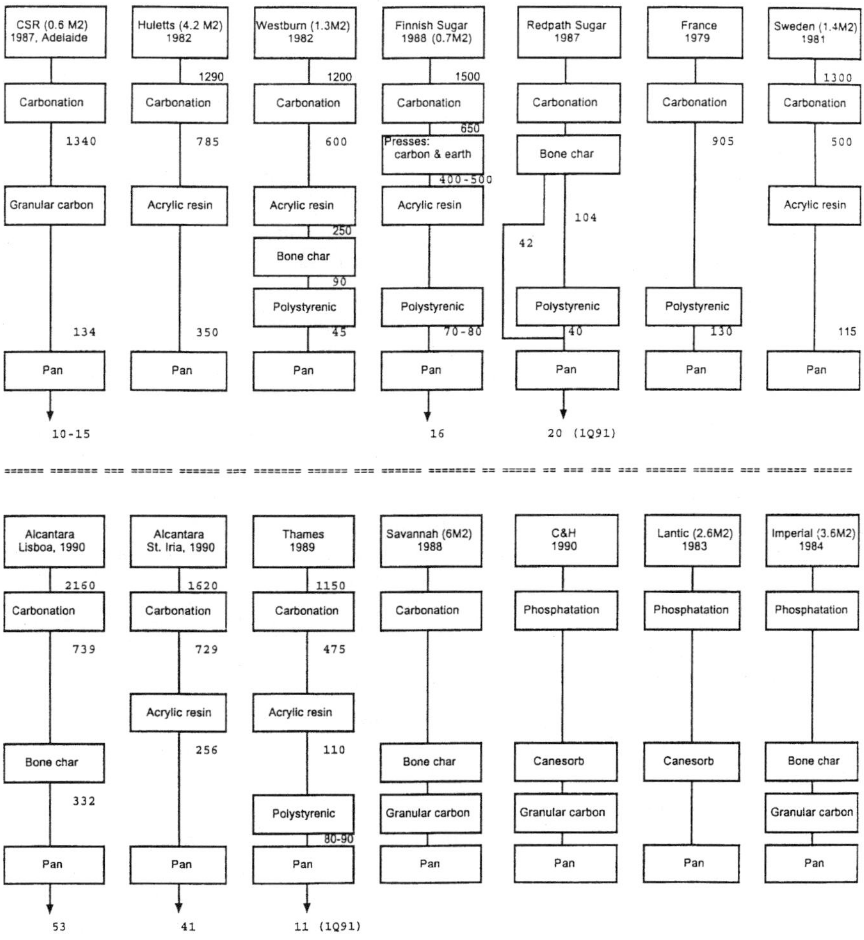

FIGURE 20.1 Decolorization schemes (colors in ICUMSA units).

and chemical decolorizers, (8) membrane processes vs. traditional separation techniques, and (9) chromatographic separations vs. traditional procedures. Items 1, 2, and 7 are generally considered as primary decolorization/turbidity removal processes. Items 3 to 6 are classified as secondary decolorization processes. Items 8 and 9 represent future trends in refining techniques.

20.2 CLARIFICATION VERSUS FILTRATION

Pressure filtration using diatomaceous earth is, of course, a basic operation for removing particulates in many applications. The particles that are to be removed must obviously be larger than the screen openings, where they are to be mechanically caught. In the case of sugar, however, we immediately run into trouble. The smallest openings in the filtration medium—typically about 0.5 μm—are larger than roughly one-third of the suspended solids in raw sugar that contribute to turbidity.

The size distribution of particulates in sugar samples is a continuum beginning well below 0.1 μm, but a demarcation line is sometimes drawn at 0.45 μm to distinguish colloidal matter from settleable solids. This concept is pure contrivance, an arbitrary cutoff point with no basis in nature. On the other hand, because of the relationship between particle size and sedimentation rate, a distinction can be drawn between species that do or do not settle spontaneously within a time frame that is useful for human endeavors—that is, minutes rather than days or weeks. Since such a point lies somewhere between particles of radius 1.0 μm and those of 0.1 μm, the 0.45 figure appears to be a workable compromise.

Note that this analysis deals with gravitational forces opposed by frictional forces dependent on the viscosity of the medium and has nothing to do with zeta potentials. However, to the degree that precipitation contributes to the separation across the filter, the particles in question would have to display potentials that are conducive to aggregation, or at the very least not unfavorable.

In contrast, clarification under optimal performance is equivalent to a 0.1-μm filtration, removing material that might require tens of hours to settle without intervention. For this reason, the brilliance achieved in clarified liquors can never be duplicated by ordinary mechanical discharge Sweetland or similar leaf-type systems. Suspensoids much smaller than 0.1 μm that survive clarification are not likely to settle within a finite time—or more correctly stated, within a time frame useful for human enterprise.

Although filtration has an important role to play in the refinery, this should preferably be limited to a polish treatment, such as for pan feed liquors, rather than a principal upstream treatment that precedes decolorization. Filtration that follows a clarification operation to protect the decolorization adsorbent from carryover is of course very different from filtration as an alternative to clarification.

Although filtration as a concept is easily visualized, theoretical analyses can be quite complex, not only because one is dealing with continuous dynamic change, but also because the fluid is typically multicomponent in nature. Such a process lends itself to mathematical treatment, which, however, is beyond the scope of this chapter. In the case of sugar liquors, temperature fluctuations significantly affect fluid viscosity and hence performance. Several grades of diatomaceous earth are available with different flow characteristics.

Apart from clarity considerations described above, pressure filtration stations are labor intensive and difficult to keep clean. In addition, diatomaceous earth is costly and its dust poses health concerns. If employees are required to wear dust masks for extended periods, this does not make for a comfortable work environment.

It is possible to burn off the organics in a process filter cake and reuse the material, but whether this is cost-effective will depend on such factors as fuel and capital costs for a regeneration station, as well as the costs for disposing of once-used filter aid.

Before a decision is made to regenerate kieselguhr, the economics should be looked at very carefully. Reused material is likely to perform less well, and inorganics such as calcium salts in the recycled fraction cannot be incinerated off and would require acid treatment, adding another layer of complexity to recovery schemes. If the recycle portion is substantial, this reduced quality material will inevitably adversely affect the performance of a blend with virgin material.

In a sense, the notion of a "choice" between clarification and filtration is not appropriate because the two processes do not accomplish the same end, despite very significant overlap. However, before clarification became established as a general unit process in sugar refining, its superiority to pressure filtration had to be demonstrated to the satisfaction of numerous plant managers responsible for technical decisions. Within that context, the idea of selecting one operation in preference to another was very real and legitimate.

20.3 PHOSPHATATION VERSUS CARBONATION

Although these two clarification methods are roughly equivalent in accomplishing the removal of particulates down to a diameter of about 0.1 μm, their modes of action are somewhat different. In phosphatation the principal mechanism is flocculation, whereas carbonation acts primarily by the inclusion of impurities within the calcium carbonate crystal system. In effect, a filter aid is made in situ. A "choice" exists for cane refineries only, because beet factories use carbonation exclusively.

A definitive comparison between the two processes has been made by Bennett [1]. He compared phosphatation at 0.03% P_2O_5 (without polyacrylamide) with carbonation at 0.6% CaO. Bennett's findings can be summarized as follows, with this author's added remarks shown in brackets:

Color removal. Removal is similar. However, phosphatation refineries typically report 25–40% decolorization, whereas the figure for carbonatation refineries is 30–50%. [The fraction removed is believed to comprise colored species associated or bonded to high-molecular-weight material, especially polysaccharides. Also, it should be remembered that the purpose of clarification is not only to remove suspended particles, but also to remove color. Any such decolorization reduces the burden on adsorbents later in the process, but such colorants *can* be removed subsequently, whereas microscopic suspended solids cannot.]

Starch. Phosphatation is far more effective because there is no mechanism to link starch to the calcium carbonate crystal system whereas it *can* be incorporated into the flocculated phosphate mass. [This is an important consideration, because starch is often one of the most troublesome macromolecular components in raw sugar.]

Calcium and magnesium. Carbonation is much more effective. At low levels of P_2O_5, phosphatated liquors may show a *gain* in calcium, magnesium, and anions as a result of leakage of finely divided calcium phosphate into defecated liquor.

Sulfate and total anions, including organics. Carbonation is much more effective. It should not be surprising that the process effectively removes any anion forming a sparingly soluble calcium salt. [It is for this reason that vacuum pan scale is not a problem in the refinery with carbonation.]

Phosphate. Levels in treated liquor are similar [but bear in mind that one of the two processes involves addition of phosphate].

TAP (total alcohol precipitate). This is an indication of colloidal material together with its associated inorganic components, precipitated in 75% ethanol in pH 3.5 acetate buffer. In general, carbonation is considerably more effective. [However, this fraction is not well defined, and variability among raw liquors of different origins is difficult to assess.]

Turbidity, measured as an attenuation index at 900 nm. Phosphatation performed slightly better in the Bennett study [but it should be pointed out that the method used to monitor turbidity is insensitive and the result is sensitive to instrument geometry. Also note the inconsistency here between removal of "colloids" as TAP and "turbidity." Using the definitions applied here, these two terms are clearly not referring to the same constituents]. Since both processes achieve almost complete removal of turbidity, it is not particularly useful to focus on slight differences in performance.

Protein. The two methods appear to be equivalent [but protein levels are generally very low in sugar liquors, and this fraction is not well characterized].

These technical considerations alone do not suffice to permit a choice between the two defecation processes. Carbonation requires a higher capital investment but has a lower operating cost. It also generates far more waste for disposal than the phosphatation option and requires higher maintenance costs. The use of flotation polymers and color precipitants markedly enhances the phosphatation process but at added operating cost. Also despite the fact that carbonation uses more extreme pHs and temperatures, it is phosphatation that results in a higher sugar loss. Of course, phosphatation is a lower-pH process, and despite the fact that phosphoric acid and lime may be added in either order, the common practice of adding acid first certainly creates a transient low-pH environment.

Interestingly, Bennett tried a mixed process of phosphatation followed by carbonation. One would expect to get the best of both worlds, but what he in fact obtained was a costly system in which the impurities could be neither filtered off nor floated away.

At the time of this writing, the ratio of cane refineries in the world using phosphatation to those using carbonation is roughly 2:1, but this statistic may show some sensitivity to geographical area and to the age of the installation. Waste disposal requirements are likely to carry increasing weight for new installations.

20.4 GRANULAR CARBON VERSUS BONE CHAR

One difference here that appears to be of growing importance to consumers has nothing to do with performance. Bone char is, of course, of animal origin, a fact of concern to a segment of the population, although no animal material is present in the sugar itself. If the customer's concern has a religious component, it may be helpful to know that cattle bones only are used in this application, never the bones of other animals. Customer inquiries on this topic have become increasingly frequent in recent years. Granular carbon used in the sugar industry is of course made from coal, although softer vegetable carbons made from agricultural waste are used in other applications.

Both bone char and granular carbon are highly porous materials in which most of the surface area is within the honeycombed interior of the particles. The pore size in such systems can be described in terms of fractals. The surface area within a single bone char filter (1200 ft^3) is about 1900 square miles, roughly the size of the state of Delaware. Granular carbon has a considerably higher surface area per unit weight than bone char. Therefore, the decolorization capacity is much higher. Large surfaces are required for decolorization because colorants are adsorbed as a monomolecular layer.

The adsorptive forces are relatively nonspecific. Although the adsorptive forces on carbonaceous materials are fairly weak, typically less than 5 kcal/mol, a large colorant molecules might be bound at a multiplicity of sites. Consequently, most color cannot be washed off at the end of a cycle, and regeneration requires thermal degradation of the adsorbed material. Since most colorants are weakly acidic, a portion can be desorbed with strong alkali, the basis of the "lye test" to assess adsorbent quality after regeneration.

Not everything picked up is colored, of course, starting with sugar itself, which is present in sugar liquor at a level of more than 600,000 ppm, compared to a much weaker concentration of each of the various components of the highly heterogeneous nonsugar fraction. We notice the decolorization primarily because color measurement is how we assess the adsorption and indeed how we monitor the entire refining process. If we could measure, for example, the weight of the adsorbent before and after the liquor cycle and obtain a complete analysis of the material adsorbed, we might come to a very different conclusion about what exactly it is that the adsorbent is picking up.

Colorants do, however, have structural features that provide an edge in getting adsorbed. As already noted, most are weakly acidic, and the acid function provides a means of electrostatic linkage to the adsorbent, a mechanism lacking in neutral sugar. A portion of the colorant fraction is high-molecular-weight material, formed either via polymerization or by complexation with large molecules, especially polysaccharides. Large species can bond to an adsorbent at multiple sites, improving the enthalpy gain. Large complex molecules may also contain nonpolar regions that are available for hydrophobic bonding. In addition, such substances may be only marginally soluble and hence easily removed from solution. Polymeric dark thermal degradation products have graphitelike structural features that would be expected to have an enhanced affinity toward carbon, which is itself graphitelike.

Micropores within the range of about 100 to 400 Å are useful for sugar liquor decolorization. This represents a small fraction of the overall porosity. Most colorant molecules are considerably smaller than 100 Å, but the relatively large pore-size requirement is a consequence of their diffusion characteristics: The hydrated molecules are bulky and in their viscous sugar liquor environment move sluggishly toward the sites where they are ultimately adsorbed. Since such diffusion is the slow or rate-determining step in decolorization, performance can usually be improved by reducing the flow rate, which in effect increases contact time. With a constant adsorbent quality and uniform feed liquor, the kinetics but not the thermodynamics can be manipulated to control the outcome—assuming that there are no other constraints, such as limited decolorization capacity. The porosity of the adsorbent declines with use, and the fraction of pores in the useful range declines.

A low-purity stream, such as soft liquor, very rapidly exhausts the de-ashing capacity of bone char, whereas the decolorization capacity is depleted much more slowly. The implication is that adsorption sites in carbon regions remain active after the calcium phosphate surface is saturated. Furthermore, the capacity of bone char to remove polymeric color decreases faster during a cycle than its capacity for monomeric color, a phenomenon that may be rooted to both kinetics and statistical mechanics.

Granular carbon has a much larger decolorization capacity than does bone char and requires a shorter contact time, so a much smaller plant is possible. Service cycles are much longer—weeks rather than days. The burn rate refers to usage levels, which are about 0.5 to 0.8% on melt for carbon, 5 to 15% for bone char. On the other hand, bone char but not carbon removes a portion of the inorganic impurities: namely, a significant fraction of the divalent and multivalent ions. In the refinery, the most important species in this category are calcium, magnesium, and sulfate. Removal lessens the severity of scale buildup on evaporative surfaces during subsequent processing. Some potentially troublesome cations, such as ferric iron, may be present in the form of stable complexes that are not adsorbed on either carbonaceous substrates or even ion-exchange resins. Silica may be present as uncharged SiO_2 or as colloidal material—also not easily removable and hence a candidate for subsequent scale formation.

The ash removal function in a granular carbon refinery may be allocated to ion-exchange resin, but if the refining scheme does not include a designated operation specific for de-ashing, the granulated sugar produced will have a slightly higher ash content. Ash levels in sugar products are most commonly determined by conductivity measurements, which are, however, sensitive to the makeup of the inorganic fraction. Correlation to the more reliable but slower and more labor-intensive gravimetric measurement is generally good to only one significant figure.

Unlike the case for the adsorbed organic material, the adsorbed inorganics on bone char can be removed by a washing cycle. This means that the sweetwater will be low in purity and will have to be diverted to waste at a higher Brix than washings from granular carbon. However, granular carbon needs a larger amount of water for desweetening per unit volume of carbon. Both carbon and bone char systems are difficult to automate, unlike the case for ion-exchange resin.

Animal charcoal is of course mostly bone (i.e., calcium phosphate), supporting the carbon component (about 10%). Like teeth, bone can be demineralized at low pH. The carbon, which is mostly amorphous but contains graphitelike zones, is distributed throughout the char structure. As indicated above, most colorants are anionic at refining pHs and are adsorbed at cationic calcium sites. Adding calcium chloride to a liquor tends to saturate phosphate sites to produce more cationic surface and typically improves decolorization performance. However, use of calcium chloride may cause problems in other parts of the refining process. On the other hand, a high sulfate level is detrimental because these small ions compete with colorant for available cationic sites and "win" because of their much higher mobility.

Granular carbon is kilned at 880 to 1000°C compared to 500 to 600°C for bone char. Burning in the absence of oxygen, sometimes called *retorting,* refers to the pyrolysis of adsorbed organics: that is, thermally degrading them to small volatile molecules. *Roasting,* by contrast, involves heating in the presence of a small amount of air to oxidize such substances. Retention time in the furnace is an important control parameter. Char losses due to discarded dust and heavies are about 0.3 to 0.5% per cycle, which means that the material will last for about 200 cycles, or roughly four years. Losses per cycle are much higher for carbon, about 4%, so that the adsorbent lasts far fewer cycles, about 25. However, since each cycle is much longer, 3 to 6 weeks as opposed to 4 to 5 days for char, the life is about 2 years.

The pressure drop through a bed of carbon or bone char is described by the Poiseuille equation,

$$P = \frac{k\mu LQ}{g\beta^2 A}$$

where k is a proportionality constant dependent on particle shape and packing density, μ the viscosity, L the bed depth, Q the flow rate, g the conversion factor between pounds mass and pounds force, β the average particle diameter, and A the cross-sectional area of the bed. The viscosity is, of course, sensitive to percent solids and temperature. Since the particle diameter appears in the equation *squared,* pressure drop is highly susceptible to fines. If particle-size uniformity is poor, *channeling* takes place; that is, the fluid takes the path of least resistance through the bed, bypassing a portion of the adsorbent. This is apt to be accompanied by a decline in performance and an increase in the amount of sweetwater generated. Note also that the *shape* of the column holding a given volume of adsorbent—its height and cross section—affect the pressure drop.

Some, but not all, users of bone char kiln it before first use. This practice depends largely on the route used to inject new char into the service stock and does not serve to somehow "activate" the char.

Good-quality service stock bone char will maintain the pH of the feed liquor, but the use of unbuffered carbon results in a pH drop, with accompanying sugar loss through inversion. Hence in cane sugar refining, magnesite—an alkaline magnesium carbonate—is used as a buffering agent for carbon. It is best to use a buffering agent that is integrated into the carbon itself, to avoid stratification. In the beet industry, no such agent is needed because the process is weakly alkaline and off-liquors are maintained at pHs greater than 7. The amount of sucrose permanently adsorbed on carbon—2 to 6% of the carbon weight—is much higher than in the case of bone char—about 0.4%—but, of course, the cycles are much longer. Clearly, the proper way to evaluate the sugar loss to the adsorbent or indeed that accompanying any operation is as a percentage of the sugar processed.

The quality of bone char or granular carbon may be monitored by specific gravity measurement. Heavy material is indicative of incomplete regeneration, with pores remaining partially occluded with organic material. On the other hand, overburning will soften the carbon, and bone char can sinter, resulting in irreversible damage to its crystal structure. In the case of bone char, the increase in specific gravity can also stem from permanent blockage of micropores with inorganics, such as calcium sulfate. It is common practice to remove such material from the active service stock by using a specific gravity separator.

20.5 ION EXCHANGE RESIN VERSUS CARBONACEOUS ADSORBENTS

A new sugar refinery built anywhere in the world today—whether in an industrialized country or a developing economy—would very likely not be designed with a char house. Present-day bone char manufacture is almost entirely in support of the established facilities that contain older bone char installations. Such systems typically contain elaborate conveying, washing, and kilning components that represent large capital investments made in the distant past. As long as these assemblages can be made to continue functioning, there may be little incentive to invest in the very significant capital outlay of an alternative system.

This situation could change if the market for bone char were to become sufficiently small that traditional sources of supply became uneconomical or unreliable. The difficulty in automating carbonaceous systems is also a cost factor, particularly where labor is costly and in short supply. But by and large investment in upgrading is inextricably linked to the general health of the sugar industry.

The case of granular carbon is somewhat different. There are relatively few bone char suppliers in the world and char is used for little else than the sugar industry. In the case of carbon, by contrast, less than half the production is devoted to sugar refining. Thus availability of carbon over the long term is more securely assured. Furthermore, because carbon has about 10 times the decolorization capacity of bone char, a much smaller plant is possible. On the other hand, because unlike bone char, carbon does not remove ash, it should be used in conjunction with a complementary operation that does perform this function. Carbon also enjoys some niche applications, such as odor control and polish decolorization, although these functions are likely to be accomplished by powdered rather than granular material.

The application of ion-exchange resins to supplement or replace carbon-type adsorbents is by no means a recent innovation. Industrial uses of synthetic ion-exchange systems became increasingly common in the post–World War II period, but the development of macroporous—*macroreticular,* to use the term introduced by the developer—resins by Rohm & Haas Co. in 1960 had a profound effect on the progress of ion exchange in the sugar industry. These substances contain pores of diameter larger than 400 Å, near the ideal range for decolorizing sugar liquors and at the same time for resisting irreversible fouling. The smaller pored traditional resins are often called *gel resins* to distinguish them from the large-pore type.

Decolorizing resins are usually, but not always, strongly basic resins in their chloride forms. This means that their functionality is analogous to a strong quaternary base, in contrast to the weakly basic types, which are functionally like immobilized amines. In addition to pore-size considerations already mentioned, they can be categorized by the physical structure of the resin "backbone" that supports the ion-exchange sites. *Acrylic resins* have a high affinity for high-molecular-weight aliphatic color, aren't particularly good for phenolic color, and are more resistant to irreversible fouling than their *polystyrene resin* counterparts. The latter type behaves similarly to activated carbon and is good for removal of phenolics, since it itself has an aromatic structure. (To readers unfamiliar with organic chemistry terminology, the use of *aromatic* in this context has nothing to do with fragrance but refers to a benzenoid molecular structure. *Aliphatic,* used earlier, means paraffinlike, or nonaromatic.)

Resin decolorization systems are sometimes characterized in terms of ion-exchange capacity, which although valid for such applications as softening and demineralization, is inappropriate for color removal for polystyrenic resin since much if not most decolorization occurs on the matrix or resin backbone, not necessarily at the ion-exchange sites. Thus it is not surprising that styrene and acrylic resins behave very differently in sugar liquor decolorization despite their both having strong base functionality. Nevertheless, some users of resin decolorization systems use fallacious ion-exchange capacity measurements to monitor the resin's decolorizing potential.

Matrix adsorption depends on backbone surface area, the polarity of the matrix, and of course the hydrophobic character and molecular size of the colorant. Adsorption is of the van der Waals type rather than stoichiometric. Adsorption at ion-exchange sites, in contrast, depends on the number, nature, and distribution of anionic functions on the colorant molecules, the nature of the ion-exchange sites, and the pH and ionic strength of the liquor. The overall controlling factor is accessibility of colorant to both types of retention sites, determined by the size and shape of the colorant in relation to the pore-size distribution in the hydrated resin.

Some advantages of resin systems are:

- They are regenerated chemically rather than thermally; hence costly fossil fuels are not required.
- They require a much shorter contact time than do carbonaceous systems; hence a much smaller volume of adsorbent is needed.
- Unlike unbuffered carbon, use does not result in a pH drop.
- Water requirements are much lower.
- They can be regenerated in place rather than having to be transported to a remote regeneration facility. At the same time, resins have good hydraulic characteristics.
- The sweetwater volume is relatively low and of high purity.
- Unlike bone char, they have a high color capacity.
- Equipment cost is relatively low and automation is easy.
- Unlike granular carbon, makeup requirements are low.

- Sugar losses are lower.
- Resin can be economically exhausted to a much higher percent of its total capacity than can carbon.
- Resins display much more favorable adsorption isotherms than do carbons in the lighter color ranges. Therefore, they can be used in cases where the amounts of carbonaceous adsorbents required would be economically prohibitive.

On the other hand, resin as an adsorbent is more costly than carbon or bone char per unit volume: There are high initial and makeup costs. Potential environmental risk is to water rather than to air. Resin also has the potential to impart fishy off-odors to product, particularly if not used properly. (The odor is removable—ironically, perhaps—with activated carbon.) In addition, it may present a problem in the future to dispose of highly dark color regeneration effluent discharge. A typical resin cycle length is 24 h, but this can be adjusted to any reasonable period desired by system design parameters.

The de-ashing capability of strong base resin is very different from that of bone char. Both can remove multivalent anions, such as sulfate and phosphate, but resin cannot remove cations such as calcium and magnesium. On the other hand, a separate softening or cation-removing function can be accomplished using cationic resins, either of the strong acid or weak acid type. Both gel and macroporous resins are used; the gel type has a greater capacity but is less resistant to attrition. Of course, this is offset in part because it requires less frequent regeneration.

For decolorization applications, resin is generally used in the *chloride* or *salt form*. The *hydroxide* or *free-base form* is typically used for polish decolorization in conjunction with de-ashing, usually in a mixed bed with an acid resin for softening or removal of cations. Exposure to cation resin in the hydrogen form in the absence of anion resin in the hydroxide form is a potential source of inversion losses. However, because the hydroxide form is relatively unstable at typical operating temperatures, the liquor is generally cooled to about 45°C for such processing. Thermal degradation follows the mechanism of decomposition of quaternary ammonium hydroxides, producing volatile amines with objectionable odors and—depending on the resin structure—alcohols and alkenes.

Other sources of reduced resin life are oxidation, by dissolved air or the presence of trace metal catalysts (which produce peroxides and free radicals), irreversible fouling referred to above, mechanical attrition, and osmotic shock. This last-mentioned category refers to stresses on the physical structure of the resin beads that result from relatively rapid transitions from an environment of high osmotic pressure, such as high-Brix process liquors or high-salt regeneration media, to one of low osmotic pressure, such as wash water.

There are a number of specialized resin applications in use in the sugar industry. However, our focus here has been on those that correspond to the principal functions of bone char and granular carbon—decolorization and de-ashing.

20.6 POWDERED VERSUS GRANULAR ACTIVATED CARBON

The principal advantage in using powdered carbon rather than the granular material is that no significant capital investment is required. The disadvantage is, of course, the operating cost associated with using a substance on a throwaway basis. The carbon—smaller than 300 μm or about a U.S. No. 50 standard sieve—is used as a precoat and/or ad-mix on a filter press, with a contact time of less than several seconds. Usage levels are commonly expressed as a fraction of kieselguhr consumption. At high usage levels, flow rates can be severely impeded.

Powdered materials have very favorable kinetics characteristics—that is, rates of adsorption—compared to granular substances. A short contact time is sufficient because of the high surface area per unit mass that is characteristic of very fine particles. It is the same phenomenon that makes powdered sugars susceptible to moisture and odor pickup.

Powders tend to generate nuisance dusts and are difficult to handle. In addition, it is often difficult to predict refinery performance of such substances on the basis of batch trials in the laboratory. Typically, powdered carbon is used to make a small improvement in color, not as a principal decolorizer.

Ecosorb, a product of Graver Chemical Co., is an anion resin-activated carbon powder composite used similarly to remove low levels of color and odor. The resin component is available in the bicarbonate form to resist pH drop. Ecosorb is more costly than powdered carbon but has superior hydraulic properties—that is, higher flow rates with lower accompanying pressure drops. Some of these composites have a very high moisture content, about 60%, and may present microbiological problems.

20.7 USE OF BLENDS OF GRANULAR CARBON AND BONE CHAR

After the introduction of granular carbon as a regenerable adsorbent in the 1950s, efforts were made to use it in conjunction with bone char, to combine its superior decolorization properties with the de-ashing capability of bone char. Because the kilning temperature requirements were very different (carbon could not be regenerated below 900°C and bone char could not above 600°C) early attempts consisted of separating the two components and regenerating them separately, then remixing them for each service cycle. However, this approach was never commercially successful.

In 1982, Calgon Corp. came up with a compromise solution: the granular carbon could be regenerated successfully at the lower bone char kiln temperature if held in the furnace for a much longer time, about twice as long as had been customary. Although it was recognized at an early stage that both components would suffer initial pore blockage and surface loss, this was expected to occur in the smallest pores that were not effective in sugar decolorization.

Calgon's patented *Canesorb process* in a typical application consisted of adding 20% granular carbon to an existing bone char system. This approximately doubled the capacity of the system; that is, liquor cycles were about twice as long as formerly. Naturally, this was a very attractive prospect to the cane sugar industry, and the process was widely adopted in North America in the mid-1980s. Note that the patented innovation was a process rather than a new kind of carbon.

By the early 1990s, however, most users had discontinued the process. The principal problem was that color loading of the blend had to be carefully controlled, because both components were being regenerated at their limits. Overloading was resulting in irreversible activity loss as a consequence of incomplete regeneration. At the same time, there was no mechanism to eliminate the inactive carbon from the blend, and this fraction continued to be recycled without benefit. An additional problem was that bone char is a harder substance that tends to degrade the carbon physically by grinding as the dry blend is conveyed. Furthermore, differences in specific gravity between the components would tend to stratify the mixture, particularly when fluidized. The partial replacement of bone char with granular carbon may result in loss of some de-ashing capacity and some buffer capacity, resulting in increased sucrose inversion.

The decline in service stock activity as a result of incomplete regeneration is gradual and masked in part by the infusion of high-quality makeup. The problems associated with irreversible fouling of carbon may not be noticed until they have reached serious proportions.

20.8 USE OF COLOR PRECIPITANTS AND CHEMICAL DECOLORIZERS

The topic of color precipitants is described more fully in Chapter 6. These substances require little capital investment but result in relatively high operating costs. Use is typically as an adjunct to phosphatation clarifier systems, although they can also be used with a carbonation process.

Bleaching-type decolorization, using peroxide or hypochlorite, is little used. Such treatments are costly and may destroy sugar. *Sulfite* is often described as an inhibitor of color formation rather than a bleach for existing color, although the bleaching action of sulfur dioxide has been known at least since Roman times. Color discharged by added sulfite can be regained by hydrolysis, unlike the case for peroxide and hypochlorite, which destroy color by cleaving the conjugated bond system.

The use of added sulfite in the United States is more common in the beet industry. There is considerable toxicological concern about sulfite residuals, however. About 1 million Americans, primarily severely ill asthmatics dependent on corticosteroid medication, are very sensitive to even low levels of sulfite. As a result, processed foods in the United States are required to display sulfite contents on their labels if appreciable levels (over 10 ppm) are present.

20.9 MEMBRANE PROCESSES VERSUS TRADITIONAL SEPARATION TECHNIQUES

Membrane processes offer the prospect of various innovations, including those focused on decolorization. Advances in materials science have overcome some of the limitations encountered in the past, although a primary obstacle in the case of reverse osmosis has not been merely lack of technological progress but rather, an immutable law of physics. Recent novel applications treating molasses and cane juice offer considerable promise for economic membrane utilization, which heretofore has rarely progressed beyond the pilot plant.

One might expect sugar decolorization using membranes to be easily achievable. The considerable difference in size between a sugar molecule, about 0.001 μm, and for example most melter liquor color bodies, about 0.004 to 0.2 μm, would suggest a natural application for this technology, which is after all a partition based primarily on molecular size and only secondarily on other attributes, such as polarity or functionality.

The reality, however, is very different. Reverse osmosis (RO) has never become a common unit operation in the sugar factory or refinery, not only because of the high capital costs of installation but also because membrane materials developed for primary use in other fields, such as water treatment, either could not withstand the required higher temperatures for sugar applications or else rapidly became fouled with nonsugar debris, especially polysaccharides. Such substances in sugar streams can also increase viscosity, adding to handling problems unrelated to osmotic pressure.

The economic impetus that has driven RO research in the sugar industry has not been decolorization but rather, the advantage of achieving a concentration without a change of state and the fuel expenditure associated with the high enthalpy of water vaporization—or, more simply, savings of steam. Even if the stability and fouling difficulties were resolved, the role of RO in concentrating dilute sugar streams is subject to a serious theoretical limitation—the requirement to mechanically oppose enormous osmotic pressures and at the same time maintain a reasonable throughput flux with increasing Brix.

The ultrafiltration (UF) alternative reduces the fouling problem substantially simply by using a more porous medium. However, with pore diameters larger than a hydrated sugar molecule, UF cannot be used as an alternative to conventional evaporation. Applications have thus focused on cleanup of juice, sugar liquor, or molasses for additional processing.

Over the past 25 years, cellulosic and polyamide UF membranes have gradually given way to new materials that could tolerate temperatures up to 90°C in continuous service, such as polyethersulfones, polyvinylpyrolidines, and similar substances. More recently, organic membrane materials have to some degree been supplanted by ceramics, which are far more stable and better suited to elevated-temperature processing. Oxides used for ceramic materials are highly resistant to degradation over a wide range of temperatures and environments.

The pore size of ceramic UF membranes that have been tested in various sugar applications is typically within the range of 0.02 to 0.2 μm, which might be expected to remove at least a portion of the higher-molecular-weight color fraction, either polymerized material or smaller colorant species complexed or covalently bonded to polysaccharide or protein.

Recent tests run on clarified juice and molasses at Hawaiian Commercial & Sugar Co. in Puunene using a Rhône Poulenc pilot plant did not monitor color but instead, focused on turbidity, reporting NTU (i.e., 90° light scatter) normalized on Brix. This was not an oversight but rather an implicit recognition that decolorization was not a primary goal of the tests. Removal of turbidity was essentially total. Such processing thus might be viewed as a pretreatment for subsequent decolorization, such as by chromatographic separation or by a secondary nanofiltration (NF) membrane treatment.

The prospects for economic membrane utilization in the sugar industry no longer appear to be constrained by the state of materials science, which was the case 25 years ago. Recent progress suggests that membranes are most likely to be used successfully in the future in conjunction with other unit operations, including a second membrane process, to accomplish a specific task. Decolorization is only one of several objectives of the novel membrane applications that have been reported in the recent past. Some of these innovations offer the prospect of dramatic change in our industry, in ways that could not have been imagined when we took our first tentative steps with reverse osmosis just a generation ago.

Figures 20.2 and 20.3 illustrate differences in filtration among conventional filtration, ultrafiltration, nanofiltration, and reverse osmosis filtration.

20.10 CHROMATOGRAPHIC SEPARATIONS VERSUS TRADITIONAL PROCEDURES

Chromatographic procedures use adsorbents such as ion-exchange resins but differ from traditional adsorbent uses in an important respect: Rather than separate unwanted minor impurities from the principal material of interest, in our case sugar, chromatography partitions the feed mixture into its components. We can see that such an application would be most useful for separating blends containing significant amounts of several constituents. This is clearly *not* the cause for raw sugar liquor, which is typically about 99% pure sucrose and in which the remaining 1% may contain dozens, if not hundreds, of components.

Sugar industry products that might be candidates for such treatment are low-purity syrups and of course molasses. The technique can be used to recover sugar from such media using some type of recycle loop. The maximum sucrose recovery from such a system can be computed using an infinite series model [2]. In the broader world of sweeteners, chro-

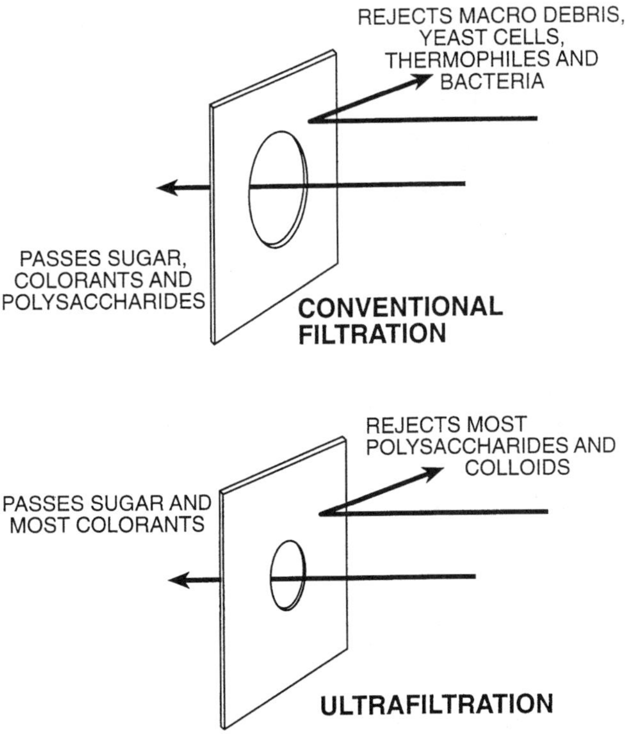

FIGURE 20.2 Conventional filtration (kieselguhr) versus ultrafiltration.

matographic separation is a well-established operation for production of high-fructose corn syrups.

The term *ion exclusion* is sometimes used interchangeably with chromatographic separation, but there is an important distinction. Ion exclusion belongs to a subset in which an inorganic component, such as the ash fraction of molasses, is rejected from porous resin beads that preferentially retard the passage of sugars from the bed. What all chromatographic separations have in common is a column, usually vertical, of some sort of adsorbent of small granules that interact differentially with the components of a solution as the fluid moves downward in intimate contact. For industrial applications, continuous operation typically means a cyclic procedure in which the several components are collected in turn at the bottom of the column. The products can be directed to the proper streams either by a timer or by direct detection of the particular component. Batch operations, in contrast, may be difficult to integrate into a process that is otherwise continuous.

One of the characteristics of chromatography is that if high degrees of separation are desired, the solutes are likely to be collected at high dilution. Thus there exists a trade-off between cost and degree of separation which will depend on the value of the recovered material. The economics of recovery of, for example, streptomycin, might be different from that for sugar.

Recent research and development work in this area for the sugar industry has been most active in the beet sector, where there have been dramatic successes in molasses desug-

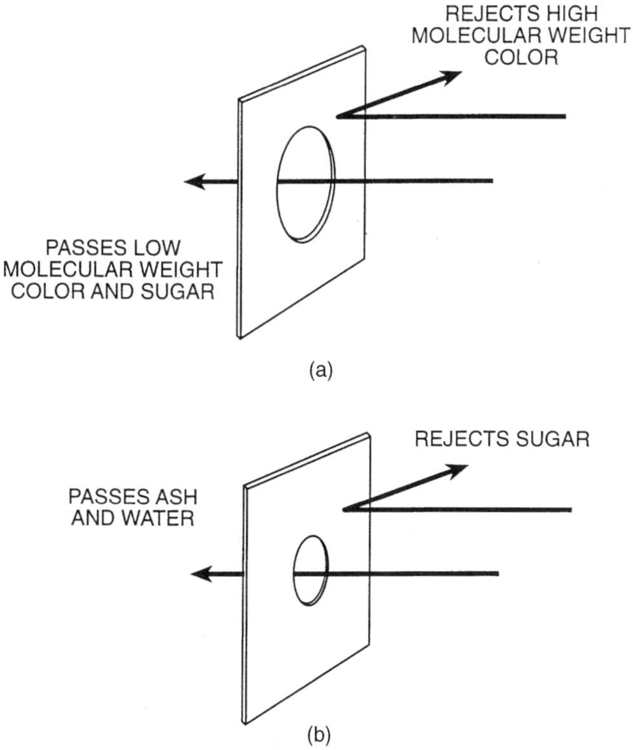

FIGURE 20.3 Nanofiltration (*a*) versus reverse osmosis (*b*).

arization and purification of diffusion juice in the recent past. Cane molasses is much more difficult to handle. Unlike its beet counterpart, it has a very significant invert content, a much higher concentration of suspended solids, and much higher levels of color and hardness. The composition of the ash fraction is less favorable. Because blackstrap is a less than perfect feedstock, researchers began looking to earlier and earlier processing stages, even to the very cane juice itself. A commercial plant at Hawaiian Commercial & Sugar Co. on Maui treats clarified juice by ultrafiltration, then softens it on resin. The proposed second phase will be the chromatographic recovery of sucrose from the B molasses.

The use of membrane methods to clean up chromatography feedstocks or partially concentrate product streams seems a natural application for this technology. Unlike the case for membrane technology, where improvements in materials science have been paramount in development, recent innovation in chromatographic separations stems largely from the availability of advanced control systems.

Some separation processes will continue to be difficult, however dramatic our technological progress. We may have learned in recent decades how to concentrate dilute solutions without the energy expenditure associated with a change of phase, and we may have learned how to make better use of waste heat. But no matter how clever we are, or how hard we work, our accomplishments will always be limited by the laws of physics. Our yields and efficiencies can never be 100%, however hard we try. Evaporator surfaces will always differ locally from the bulk of the solution being evaporated, with all the implications that difference entails, and high-Brix sugar solutions will always display high viscosities and high osmotic pressure.

REFERENCES

1. M. C. Bennet, *Proc. Tech. Session Cane Sugar Refin. Res.*, 1972.
2. M. Kearney, Paper 603, *Proc. Sugar Ind. Technol.*, Vancouver, 1990.

CHAPTER 21

Instrumentation For Process Control*

INTRODUCTION

The instrumentation used in the sugar processing industry is varied, usually specific for the task under consideration, has been used for a long period of time, and is generally tested under actual operating conditions before gaining acceptance. The reasons for using instrumentation in operation of a sugar processing facility are numerous and the same as any processing plant using modern methods of operation to produce a salable product. In the sugar refinery the reasons are highlighted by the important fact that the material produced is a direct consumption food. An out-of-spec chemical in another industry is no great loss, whereas a problem with refined sugar production can be detrimental to human life since most refined sugar products are manufactured for consumption as is and will not be processed further.

Automation is achieved by using instrumentation in an effective manner to achieve control and manage a processing plant. Instrumentation has its benefits in tight process control, achieving energy savings, and reducing the need for expensive labor. Usually, instrumentation will give a rise in production capabilities by better use of existing equipment and processes. In many cases, automation through instrumentation will produce a consistent product, which is better than material produced by manual means. In addition, instrumentation will give the user the ability to track and record the processing with a documented trail for problem analysis as well as complaint investigation. Today's world has found a real use for the computer and programmable controller in managing processes to produce refined sugar.

21.1 TYPES OF INSTRUMENTS

For purposes of discussion, process instrumentation will be divided into two general types:

1. *Pneumatic.* These are devices that use a compressed gas (normally, air) as the source of power and the means for achieving control of a process. Pneumatic controllers and

---

* By Walter Simoneaux, Sr.

control valves have been used in industry, and in particular the sugar refining industry, for more than 40 years. Many factories still used the pneumatic controller and control valve as the basis for individual control loops throughout the plant.

2. *Electronic.* These are devices that use electrical current of some type to sense and effect control of a process. The electricity used can be either ac or dc or a combination. In many process control applications, the process sensor is electronic in operation and the final control element is a pneumatic control valve.

Each type of instrumentation possesses properties that make it a good or poor choice for use in measuring or controlling a sugar refining process. The two basic types of instrumentation are characterized in Table 21.1.

21.1.1 Pneumatic Instruments

The price of pneumatic instruments is generally less than that of an electronic instrument doing the same job. Pneumatic controllers and sensing devices are easy to install from a practical standpoint. Since most pneumatics use compressed air as the source of power, piping and regulating air is the main concern when installing this equipment. Pneumatic controllers use very little electrical power. A chart-drive motor and/or case light are the power users in a pneumatic control loop. The plant air compressor uses electrical power, but the amount of air consumed by a pneumatic controller is very small in the overall air usage picture. Pneumatic instruments can tolerate harsh conditions such as heat, humidity, and line voltage fluctuations. Signal conversion is necessary for using the output of a pneumatic controller to communicate with a computer or programmable controller. Generally speaking, pneumatic instrumentation problems are diagnosed in a straightforward manner using a pressure gauge to watch the reaction of the process controller or control valve as the air is increased or bled down. Usually, cleaning a small orifice or changing a diaphragm are the problems encountered. Pneumatic systems demand a clean and dry air supply, and

TABLE 21.1 Pneumatic Versus Electronic Instrumentation

Pneumatic	Electronic
Initial and maintenance costs are reasonable	High initial cost
Ease of installation	Requires conduit and shielded wire
Low electrical power consumption	No compressed air required
Tolerates harsh plant conditions	Temperature humidity limitations
Signal conversion for programmable controller interface	Able to communicate directly with programmable controller
Immune to electromagnetic interface	Shielding and separate conduit runs
Speed of response is mechanical	Senses with the speed of light
Lubrication may be required	No lubrication necessary
Fail-safe condition easily arranged	Battery backup or spring return
Requires a compressed gas for power	Stable power is necessary
Transmission distance is limited	Can span long distances easily
Excellent in spark-free atmospheres	Dangerous in flammable conditions
Troubleshooting via pressure gauge	Expensive and specific test equipment
Efficiency limited to mechanical means	Efficiency is excellent
Firesafe	Can operate in cold environments
Submersible in pure form	Exposure to water is to be avoided
Has specific task limitations	Is the best choice in certain applications

many control problems can be avoided by making clean air a major priority. Loss of air supply shuts down pneumatic control systems, and critical first-line instruments need backup or emergency air supplies. A wise choice of fail-safe conditions for a control valve can keep problems to a minimum. Signal lines for air to control valves do not suffer from electrical interference, but this air transmission cannot be done effectively over long distances. Long lengths of air line affect time lag in control and lack of response. Most pneumatic instruments and control elements are mechanical devices and suffer some time lag regarding indication and actual control movement. Environmentally, the pneumatic controller and valve are spark-free in dangerous atmospheres.

21.1.2 Electronic Instruments

Electric instrumentation is generally high in initial cost and has a high maintenance load over the useful life span. In the pure application, electronic instruments use no compressed air or gas, which lends itself to areas where air is not available. Electronics do not perform well in damp areas and are affected by voltage fluctuations but normally are very accurate and perform their duties with the speed of light. Electronic instruments can usually communicate with a computer or programmable controller directly and eliminate the need for signal conversion. Test equipment can be costly and technicians usually require extensive training for diagnosis and repair. Signal transmission is possible over long distances, but installation of this capability can be expensive. Fail-safe conditions may warrant the need for battery backup or spring return on critical control valves.

The pneumatic controller in today's world has an industry-wide acceptance. This controller has a standard output and this output is normally in the ranged 3 to 15 psi. This standard output is recognized as a benchmark in instrumentation, and most suppliers have accepted this standard. Some specific applications use 10 to 50 psi as the output, but 3 to 15 psi remains the dominant output provided by equipment vendors.

The electronic controller has a similar role in today's automation. The electronic controller has a standard output rated at 4 to 20 mA and is recognized by most producers as the normal output for control. There are more outputs offered in the electronic controllers input or outputs, such as 0 to 1 V, 0 to 10 V, and 0 to 100 mV. There also exists more equipment for signal conditioning, that is, converting from almost any input to any desired output by using an appropriate signal altering device.

Almost every area in a modern sugar refinery uses some type of instrumentation as the method of control. These areas may be called different names in different plants but instrumentation is vital to every facet of plant operations. A list of plant areas and control functions is given in Table 21.2 and represents a basic sugar refinery operation.

The information in Table 21.2 gives us many ideas for using automation through instrumentation. These areas use many types of ideas to accomplish a task. Listing the parameters on which most of the control in a sugar plant is effected can encompass the majority of modern refinery instrumentation. These parameters are (1) level control, (2) density control, (3) pH control, (4) flow control, and (5) temperature control. A discussion that details these parameters should cover almost all major refinery areas and the associated instrumentation used to accomplish process control. Sugar process is done primarily in the liquid state. The starting product is usually a solid, as is the final product. In most cases, however, the actual processing is in a liquid state, which requires much more control.

21.2 LEVEL CONTROL

Level monitoring and control is a control function of great importance since the bulk of refined sugar processing is done in the liquid state. From the mingler to the vacuum pan

TABLE 21.2 Refinery Areas Using Automation

A. Raw sugar to melt house
   1. Melt scale control and totalization
   2. Mingling syrup dosing control

B. Raw sugar mingler
   1. Mingler density control
   2. Mingling syrup temperature control
   3. Mingling syrup storage tank level control

C. Affined sugar to melt house
   1. Melter density control
   2. Melter temperature control
   3. pH control (if required)
   4. Melter level control

D. Clarification–defecation
   1. Temperature control of feed to clarifier(s)
   2. Flow rate control
   3. Various tank level control
   4. pH control
   5. Chemical dosing proportional to flow

E. Filtration
   1. Temperature control of press feed liquors
   2. Level control in various tanks serving press department
   3. Inventory and dosing of filter aids
   4. Flow rate monitoring and control
   5. Pressure control of feed to presses

F. Decolorization, ion exchange, and bone char
   1. Temperature control of feed materials to columns
   2. Level monitoring and control
   3. Pressure regulation
   4. Flow monitoring and control
   5. pH control
   6. Flame safeguard of regeneration equipment

G. Pans and evaporators
   1. Level monitoring and control of pans, storage tank, mixers, and evaporators
   2. Steam pressure monitoring and control (makeup)
   3. Computer control of pan and evaporator operations

H. Power house
   1. Steam–water level monitoring and control
   2. Fuel monitoring, inventory, control of combustion
   3. Flow control of liquid and gases
   4. Compressed air generation and regulation
   5. Generate steam and water balances, trending

I. Plant services
   1. Inventory of process supplies: filter aid, chemicals, etc.
   2. Inventory of liquid products: sugar and molasses
   3. Reconciliation of input versus output regarding raw and refined sugars
   4. Stock-in-process inventory
   5. Refinery cost data

liquid is the state of material being processed. After the pans the material state changes from liquid to solid and back again as the extraction proceeds to a final molasses point. Ways to measure level and devices to control level vary in simplicity and manner of operation.

21.2.1 Bubbler Tube (Air Purge)

The bubbler tube method of level measurement is used largely in nonpressurized vessels and open tanks. The source of power is a 20-psi regulated air supply. The equipment used is a differential pressure (DP) cell whose low side is connected to the atmosphere and whose high side is connected to the liquid head of the system. Figure 21.1 is a view of the components of a DP cell. The cell measures the differential or difference between the pressure applied to both sides of a diaphragm chamber. This difference is converted to some value (usually by electronic means) and can be used to effect control. The *liquid head* of a fluid in a tank is the product of the height of the fluid (h) multiplied by the density (d) of the

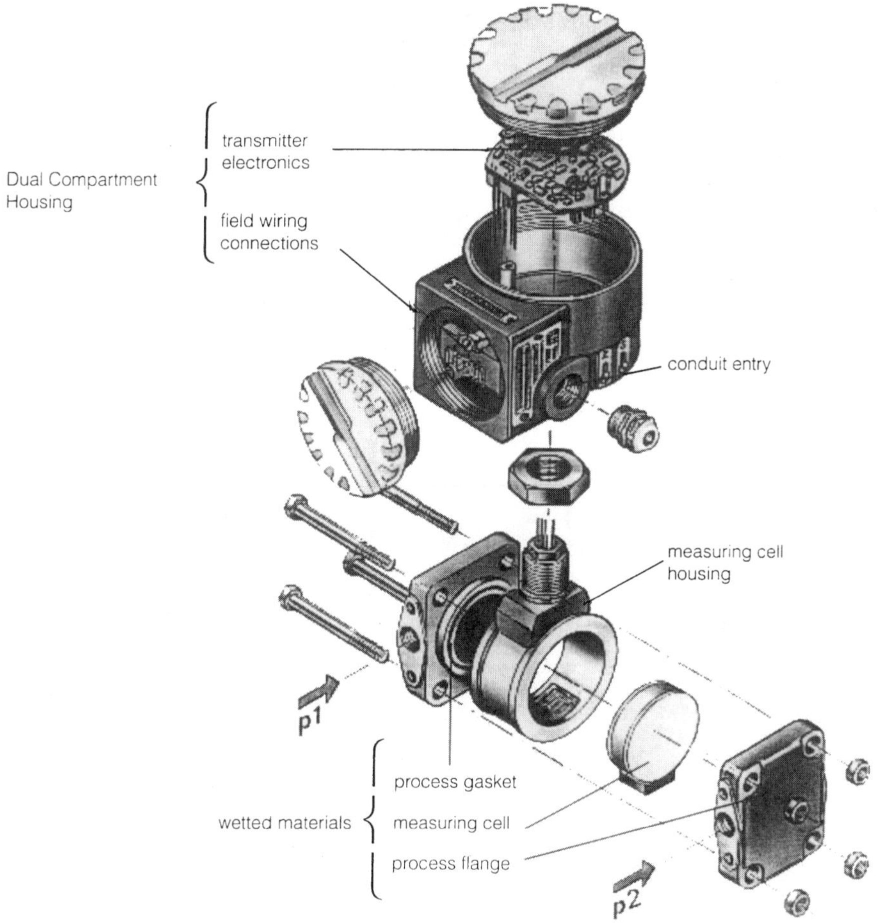

FIGURE 21.1 DP cell.

fluid. These head numbers are usually expressed in small quantities such as "inches of water." Figure 21.2 is a simple bubbler tube used to measure level. Air is fed to the system by a rotometer and allowed to bubble out of the end of the standpipe into the liquid. Since the formula is: head = density × height, keeping one of the parameters constant will define the second parameter automatically. When used to measure liquids of constant density, the head is the level of the fluid.

When the level of the system is held constant, the head is the density of the fluid. This device can be used to measure density or level by choosing which unit value to become constant. When the air is bubbled into the tube, the liquid over the end of the pipe creates a backpressure, and this pressure is measured by the DP cell. The cell has been referenced to some standard (e.g., 0 to 50 in. water) and can generate an output signal (3 to 15 psi or 4 to 20 mA) and be used for control purposes. For level applications, the density of the fluid must be constant (within reasonable amounts). For density applications, the level of fluid in the chamber must be held constant for the system to be accurate. In density applications, a water reference can be used for constant calibration of the apparatus. Figure 21.2 is a simple level control application. The liquid is constant in density, so any head changes are a result of liquid-level changes. This system has advantages when measuring levels of corrosive liquids, oils, and so on, by a prudent choice of materials for the impulse line in the liquid. The system is simple to install, and altering the depth of the impulse line

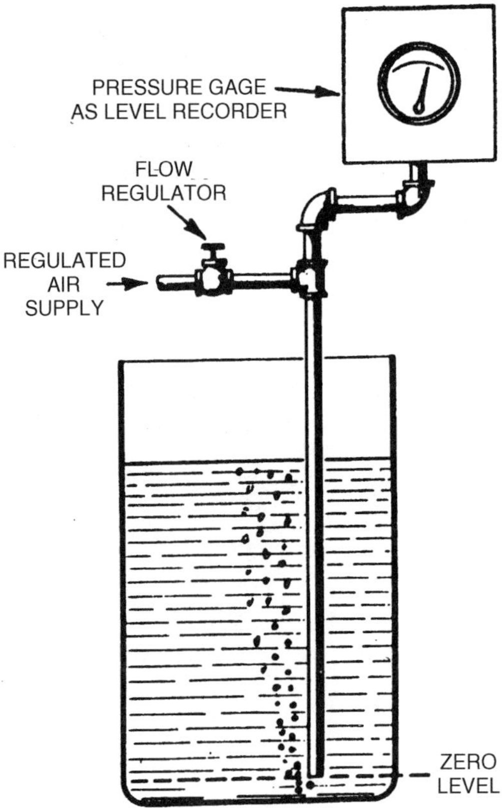

FIGURE 21.2 Level by bubbler tube.

or controller set point easily changes the set point. The most expensive part of the system is the DP cell, which can be used as the controller by choosing the input range of the cell very carefully. *Important:* This system cannot tolerate leaks in the air piping after the rotometer! Any air leaks are recognized as a change in head and generate errors. Suspended solids in the liquid have little effect since the air purges out and has a self-cleaning effect.

21.2.2 Electrical Capacitance Probe

The electrical capacitance probe uses the property that water or aqueous sugar solutions will conduct electricity. Today's *radio-frequency impedance* capacitance probes use a small-diameter stainless rod ($\frac{1}{2}$ to $\frac{5}{8}$ in., depending on the application) encased in a protective plastic (Teflon). This coated rod is immersed to the full depth of the liquid. The tank is usually made from steel, but any material can be used for the vessel. If the material used does not provide a ground connection, it is necessary to insert a ground rod in the tank. The principle involved is measuring the capacitance between two plates of a system immersed in a liquid with a constant dielectric constant. The formula for RF impedance is

$$C = \frac{kA}{d}$$

where C is the capacitance, k a dielectric constant, A the capacitor plate area, and d the distance between capacitor plates.

In a typical capacitance level probe installation, the probe is connected to the tank in a manner to demonstrate a grounding contact between the probe and the tank's metal walls. This will make the distance between the probe and the wall a constant. It is assumed that the dielectric constant of the liquid is a constant. Therefore, the capacitance becomes equal to the level of liquid covering the plates of the capacitor (probe and wall). This relationship is linear, so in a round or square tank the capacitance becomes the level using electronics and the output is 4 to 20 mA after the probe is zeroed and spanned in the actual process liquid being measured. Probes should be calibrated in the actual process conditions so that the small effect of temperature can be controlled. Probes can be used as continuous or point sensing, and lengths from 1 to 25 ft are common. Applications include open or closed tanks, pressure or conditions, semiliquids such as massecuites, corrosive liquids, or liquids with suspended solids such as mud waters. Figure 21.3 illustrates the principle of measuring level using capacitance. The capacitance probe works well in vacuum applications.

21.2.3 Differential Pressure Cell

DP cell makes use of the fact that a standing column of liquid will produce a pressure equal to the height of the liquid multiplied by the density of the liquid. If the liquid has a constant density, this pressure can be directly related to level. The two chambers of the cell are shown in Figure 21.1. When the highest column of liquid is connected to one chamber, this side is called the high side, representing a higher-pressure area. The opposite chamber is then defined as the low side. A DP cell measures the difference in pressures in the chambers, and the output of the cell represents this differential. The output of a DP cell can be in the range 3 to 15 psi, 4 to 20 mA, or some common output signal. Figure 21.4 represents a typical DP cell installation with the cell output used to control the level in the tank. In open or unpressurized vessels the low side is left open and referenced to air. In pressure or vacuum applications, the both chambers are connected to the vessel. The DP cell is the choice for

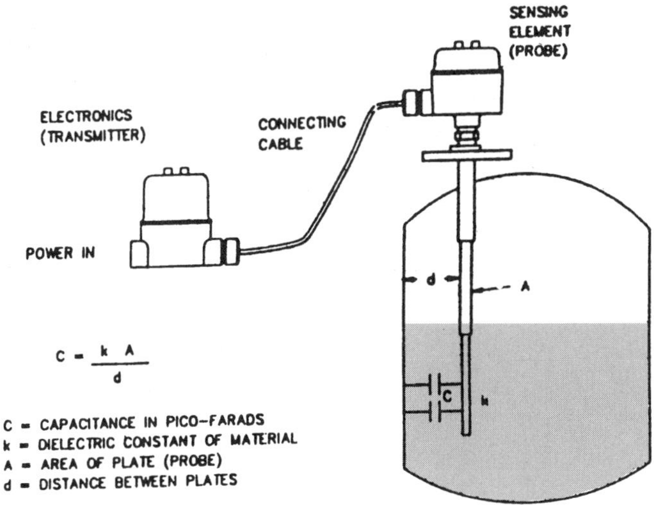

FIGURE 21.3 Level by capacitance.

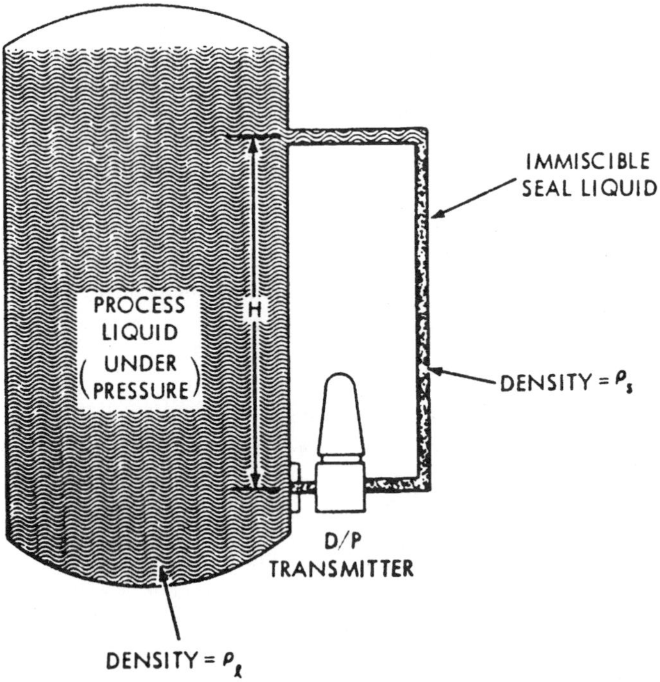

FIGURE 21.4 Level measurement in a pressurized vessel.

systems producing small changes in pressure, small changes in multiple pressure inputs, and applications where the differential pressure must be measured and some math function performed on the differential, such as orifice plates and venturi flow systems.

21.2.4 Ultrasonic Level Devices

High-frequency sound waves can be used to measure level in liquids and granular materials. This system uses high-frequency sound waves, which are beamed at a target (liquid level) by a transponder, and the echoes from the liquid are reflected back so that the transponder can absorb the wave. The electronics calculated the time of flight against the calibration standards and can accurately determine level. Usually, the system is a noncontact type of unit mounted above the vessel and aimed at the liquid or solids within the vessel.

Figure 21.5 is a typical installation for measuring level by ultrasonic means. In this application the reflector has a calibration device so that the instrument constantly has a reference. Ultrasonic level detection should be evaluated carefully with the following precautions:

1. Variable-density liquids may produce false echoes.
2. Liquids with moderate to dense foam will give variable results, which are usually in error.
3. Agitation by natural or manual means will cause unpredictable and inconsistent results.
4. Vapors from the liquid service tend to confuse the echo and produce errors.

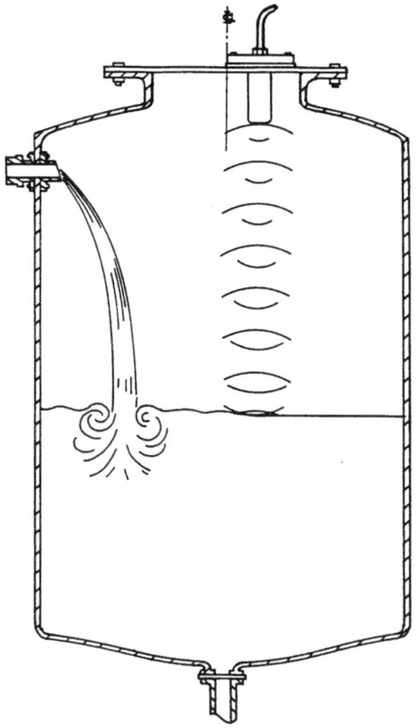

FIGURE 21.5 Level by ultrasonic sensor.

5. Temperature changes in the liquid will give sound reflections that are constantly changing and give rise to inconsistent results.
6. The transponder unit mounted above the tank has definite temperature limits.
7. Ultrasonic devices have defined cone angles, usually 7 to 8°, and restrict its use in tall, narrow tanks with internal bracing or long mixer shafts.
8. Cooling/heating of the vapor space tends to confuse the echo and generates unpredictable results.

Figure 21.6 demonstrates the principle of bouncing a sound wave off a liquid surface with the transponder mounted above the open tank. This unit has a reference reflector for constant calibration on line.

21.2.5 Conductivity Probe

The principle of a liquid being able to conduct electricity can be used to measure level. Insulated rod(s) are installed in a vessel and the point liquid touching/not touching the rod can be used to indicate a level in the vessel. Usually, this type of level device is point specific and is generally not a continuous application. The liquid in the vessel must conduct electricity. In most applications, one electrode is used and the touch/no touch action of the liquid on electrode is used to control a relay, which can sound an alarm, start or stop a pump, or open or close a valve(s). This type of point control makes for fast on–off operation

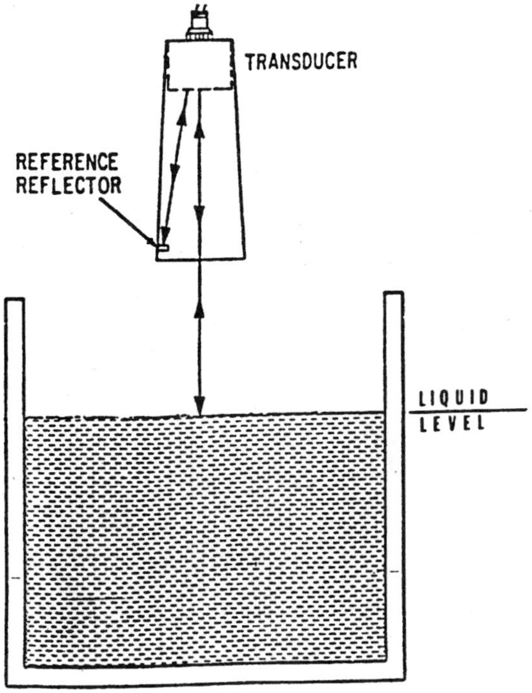

FIGURE 21.6 Ultrasonic level device.

and is detrimental to relays, pumps, and control mechanisms. Two electrodes whose tips are spaced at different levels give the system a *zone of control.* This will eliminate the jerky on–off control in favor of timely, spaced starts and stops.

Figure 21.7 illustrates a conductivity level control loop in a closed vessel, which may be under pressure or vacuum. Figure 21.8 demonstrates electrode placement for maintaining the level within a span for control purposes. In either case the advantages of liquid conductivity are:

- It can be used for interface measurement between conductive and nonconductive liquids.
- The apparatus has a low cost and a relative ease of installation.
- The design is simple and straightforward.
- There are no moving parts.

Limitations are:

- Liquids must be conductive.
- Problems are associated with liquids, which tend to coat the electrodes.
- Electrolytic corrosion is a harmful side effect.
- Ac voltage can be used to reduce corrosion but sparking may occur.

21.2.6 Vibration or Tuning Fork Devices

A vibration device uses an exposed tip, which is made to vibrate at a certain frequency, and uses the liquid being measured to dampen this controlled vibration. These devices are almost entirely point-level devices and do not have the capability of measuring level in a continuous range. The newest tuning fork instruments vibrate at 400-Hz frequency, and liquid covering the forks activates the electrical switches, which indicate point level. Thick, viscous materials such as centrifugal molasses, pan footings, or magma are handled by this type of device. Temperature and density have very little effect on the probe. The main disadvantages are point-specific applications and lack of a range of level, but if the purpose is to keep the tank full or prevent it from going empty, the unit has a useful role. Figure 21.9 illustrates a tuning fork in use as a level sensor.

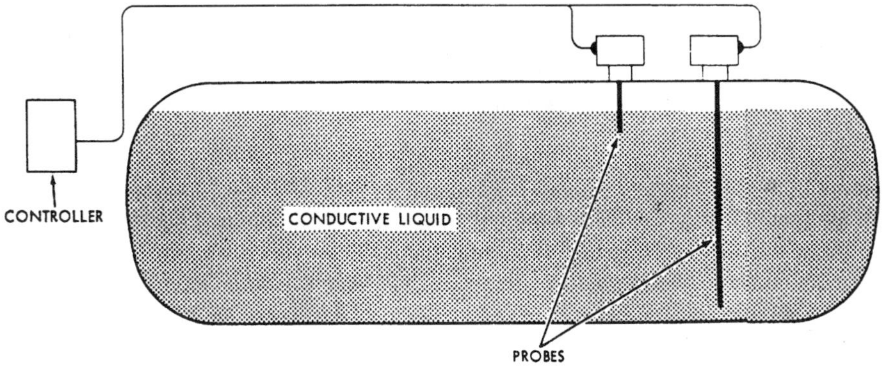

FIGURE 21.7 Conductivity level in a closed vessel.

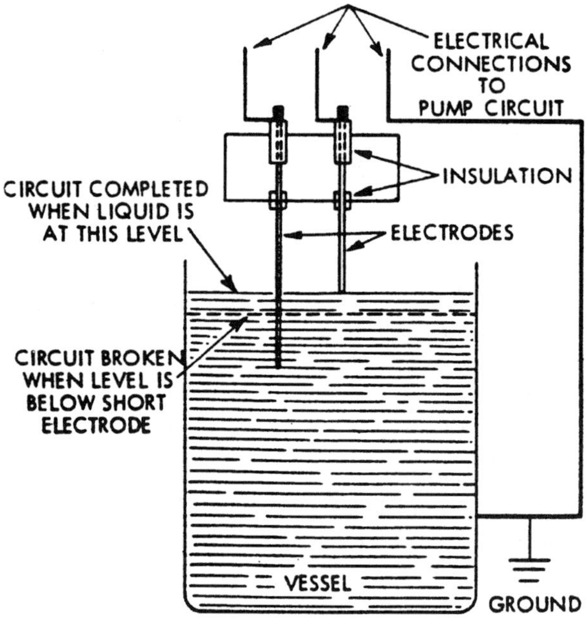

FIGURE 21.8 Conductivity level application.

21.2.7 Resistance Tape

Another means of measuring continuous level is the resistance tape method. Figure 21.10 illustrates the basic mechanism of level measurement by this method. A resistance wire is wrapped in a uniform spiral around the full length of a nonconducting strip. The assembly is encased in a plastic tube that is sealed at the bottom. Inside the tube is an electrical conductor. A slight pressure keeps the tube away from the resistance wire for the length of tube in the vapor space. When liquid rises and covers a portion of the tube, the hydrostatic head pressure squeezes the tube and the electrical conductor inside the tube shorts out the wire up to the liquid level, resulting in a lower resistance in the circuit. The higher the level, the lower the resistance. The outer envelope must be flexible and therefore is subject to breakage by abrasion, but using a stilling well the full depth of the tank can minimize this problem.

Temperature will affect the outer envelope, and this type of instrumentation has a defined temperature limit for its application. The interior of the envelope is a separate chamber, and condensation in the chamber must be avoided since any collection of moisture will short out the strip at unwanted spots. The inner chamber usually has a controlled atmosphere with moisture-absorbing desiccant purging this airspace. Foam does not have a hydrostatic head and the tape usually does not read foam. Any undesirable shorting of windings will usually make the entire strip unusable. This type of level device has applications in the refinery, where conditions are constant and temperatures are ambient and stable. Water-level control is one specific application.

21.2.8 Nuclear Level Devices

Modern science has given us new and exciting applications for nuclear energy. In level applications, a radioactive source is attached to a vessel and the receiving device is usually

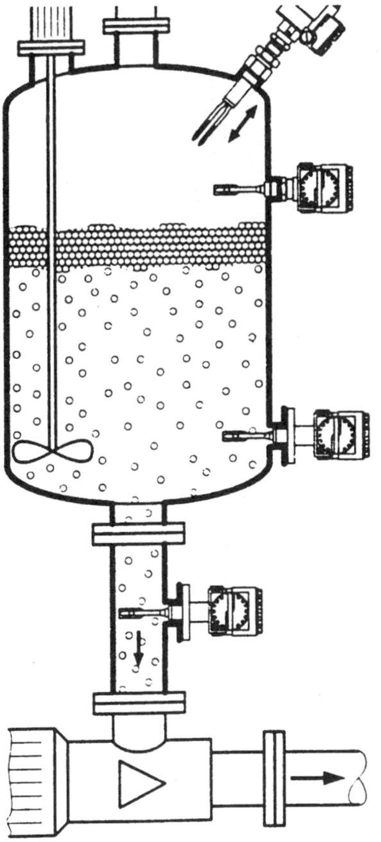

FIGURE 21.9 Tuning fork level sensor.

mounted on the opposite side of the tank. The detector can measure level by the absorption change influenced when the level changes in the tank. The instrument is nonintrusive and is mounted on the sides or bottom of the vessel. These systems involve the use of gamma radiation. The system usually consists of three parts: (1) a source (^{137}Cs), (2) a detector, and (3) a transmitter. The source and detector are mounted on the vessel across from one another, while the transmitter can be mounted in a convenient location. As the level in the vessel rises, the amount of radiation reaching the detector grows smaller and smaller as the material absorbs the gamma rays being emitted. These systems work well in harsh conditions and are unaffected by temperature, pressure, abrasion, vapors, dust, shock, or vibration. This device has a negative perception in the sugar refining industry because of the word *nuclear*. There is a licensing requirement and leak checks by the Nuclear Regulatory Commission are mandatory. Purchase cost is generally higher than for contact-level devices. Figure 21.11 illustrates an application of a continuous-level device using nuclear power. Figure 21.12 shows that use of this type of device can be point or continuous in its operation by choosing the detector carefully.

21.3 DENSITY CONTROL

Sugar refining makes use of the principle of impurity separation. This separation is usually done in a liquid state. Liquids in a sugar refinery are varied but there must be some method

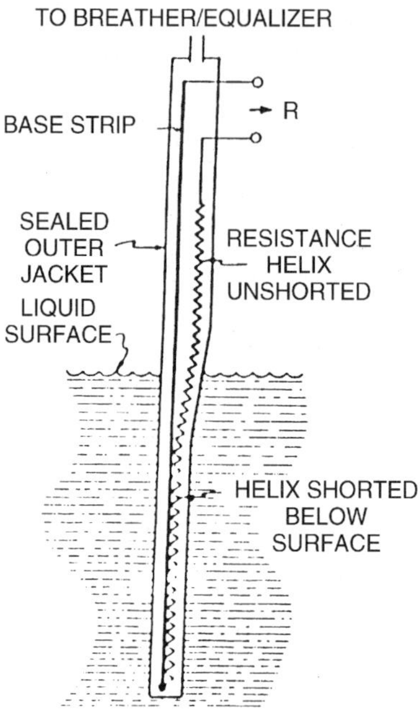

FIGURE 21.10 Level measurement by resistance tape.

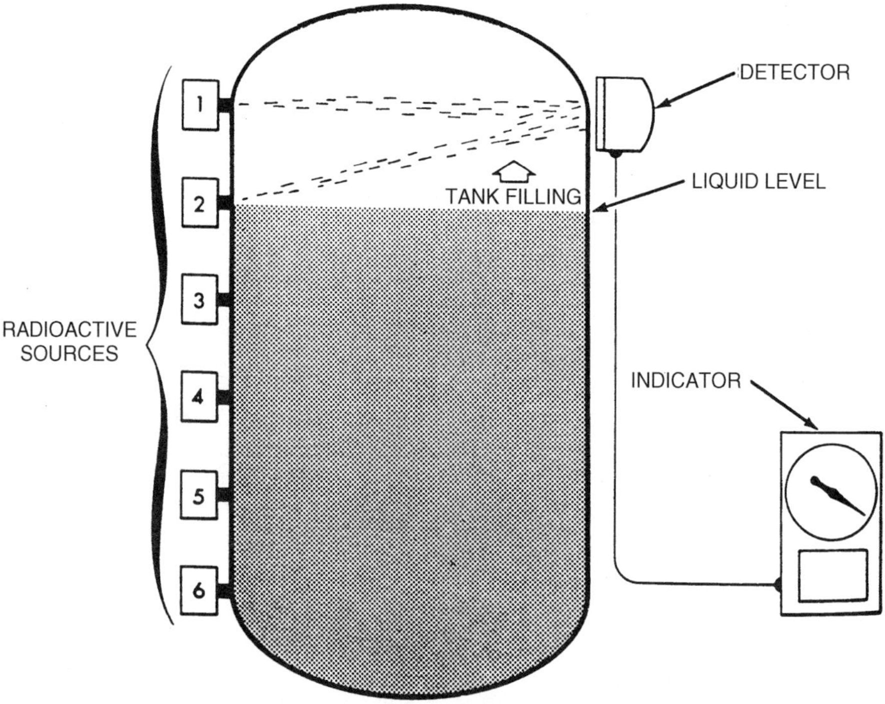

FIGURE 21.11 Continuous level by nuclear device.

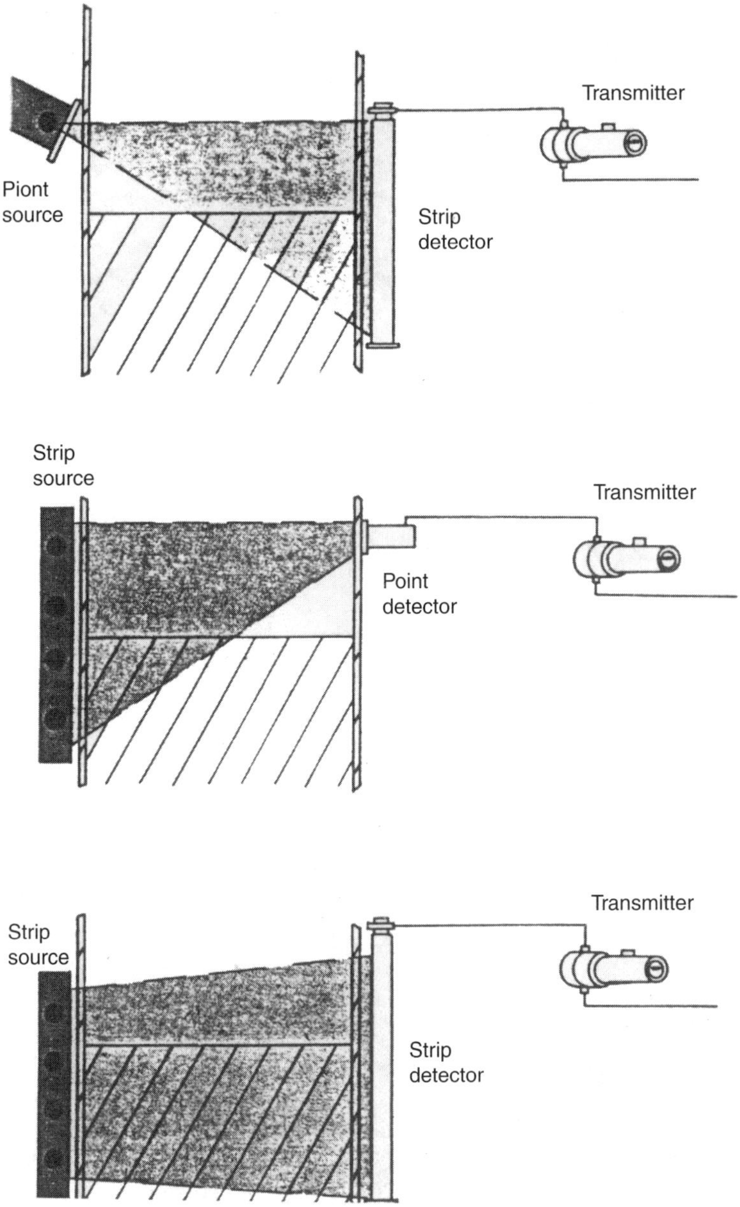

FIGURE 21.12 Nuclear level device.

of measuring and controlling density. Measuring density is important since many of the processes used have defined density parameters. After measuring the density, the means and methods of control become important for process control. Liquids whose density is left to chance will cause big problems in processing. Too light a density will require more energy and time to process, usually requires more tankage for a given refining rate, and is subject to microbial growth. Too heavy a density causes chokes, and in addition to damaging equipment, eventually will slow or stop production. Density is usually measured in two ways with good accuracy.

21.3.1 Nuclear Density Gauges

The equipment used to measure density by a nuclear device is very similar to the equipment used to measure level. The sensor is usually mounted to a pipe filled with the liquid in question. The amount flowing past the sensor is constant and the gamma radiation absorbed by the liquid is attributed to density changes. The same precautions are necessary as mentioned in the level applications. Figure 21.13 illustrates a nuclear device mounted on a pipe, used to measure density. Figure 21.14 is a nuclear density device mounted on a vacuum pan.

21.3.2 Vibrating U-Tube Devices

The vibrating U-tube device has been used for many years to measure density in various liquids. Sugar solutions work well in a U-tube device. The tube is constructed using small-bore tubing ($\frac{1}{2}$ in.) and is made to vibrate at a set frequency by an oscillator. The vibrations of the tube filled with water are considered as the reference point. When a sugar solution is passed through the tube, the amplitude of the vibration is changed, and this offset from reference is sensed and converted electrically to units representing density. Figure 21.15 illustrates the device, which uses controlled vibration to measure density. Precautions to be taken with this equipment are:

1. Solutions should be free of entrained air since the device will measure the air bubbles as a change in density.
2. Solutions should be free of suspended particles since these particles will show up as density changes.
3. External vibrations are to be avoided for obvious reasons.
4. Sugar solutions should be free of undissolved crystals.

21.4 pH CONTROL

pH measurement and control is important to the orderly operation of a sugar refinery. Many of the processes have defined pH ranges for optimum performance. In many cases, the actual reaction (e.g., carbonation) is dependent on a given pH limit. Controlling sucrose inversion in light-density sugar solutions is supplemented by pH adjustments. Liquid sugar production and storage requirements are benefited by optimum pH control. Two electrodes are used to measure pH. The *glass electrode* (sometimes called the *measuring electrode*) has a sensitive glass tip which measures the hydrogen-ion concentration in the solution under test. A small voltage (millivolts) is produced between the electrode tip and the hydrogen ions in solutions. This potential rises or falls with the rise and fall of hydrogen-ion concentration in the solution. The *reference electrode* provides a constant potential against which

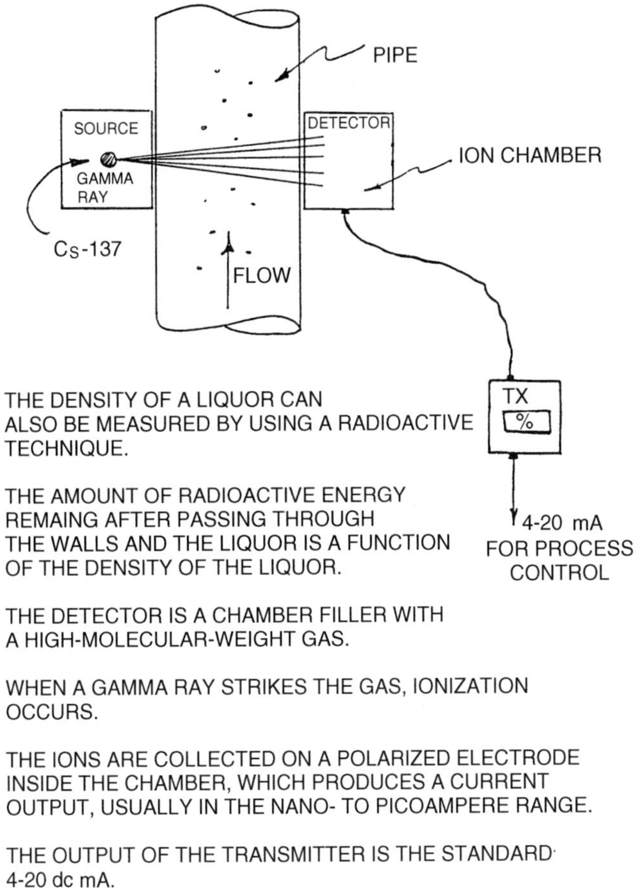

FIGURE 21.13 Flow measurement by nuclear device.

the varying potential of the glass electrode is compared. The reference electrode is filled with 0.1 *N* KCl solution, and this electrolyte is allowed to "leak" into the solution. The system is completed with a thermocompensator used to correct the potential measurement for temperature fluctuations. Figure 21.16 illustrates a combination pH electrode, with the measuring and reference electrodes packaged as one unit. At the bottom is the liquid junction where the electrolyte will escape into the solution being tested. Figure 21.17 illustrates a pH measuring and control system with output contacts, remote electrode wash, and the standard 4 to 20-mA control signal to drive a recorder or controller.

21.5 FLOW CONTROL

Flow is the parameter used to rate speed when processing in the liquid state. Flow must be measured and controlled carefully for a sugar refinery to operate safely. The major flows in a sugar refinery are of liquid and gases. Instrumentation used to measure flow is determined by process conditions such as liquid type, temperature of liquid or gas, whether the liquid

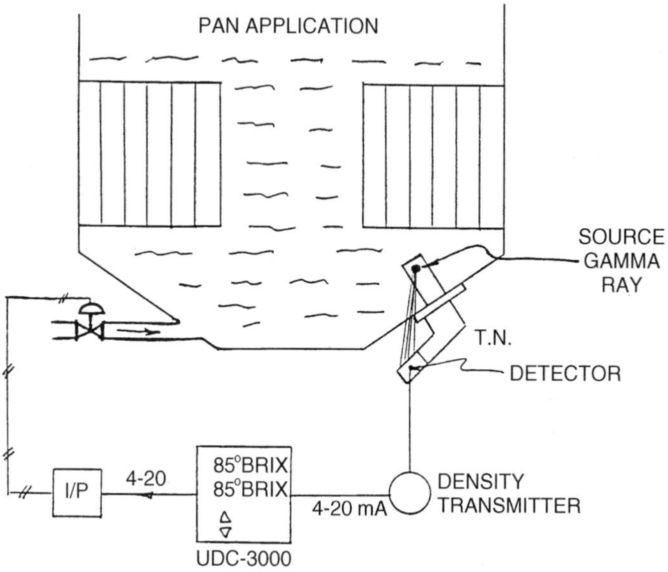

FIGURE 21.14 Nuclear density device.

is clean or dirty, accuracy desired, and costs associated with purchasing, installing, and maintaining flow measurement equipment. The devices used to measure, calculate, estimate, or compute flow are as varied as the types of liquids or gases under study. Some of the devices used to measure other parameters can also be used to measure flow. For example, the differential pressure cell used to measure level can easily be adapted to measure flow. Means or methods used to determine flow are discussed in the following sections.

21.5.1 Orifice Plate and Differential Pressure Cell

The orifice plate is a thin, circular metal plate with a sharp-edged hole in it. The plate is held in place by pipe flanges with taps to measure the pressure on either side of the plate. The flow path has a constriction placed so as to reduce the flow cross section. The pressure drop between the full pipe section and the constricted area is mathematically transformed to yield the flow rate. Figure 21.18 is an illustration of the orifice plate installed in a pipeline and depicting the pressure profile across the plate as flow is occurring. When flow passes through the hole in the orifice plate, its speed is increased and the increase is seen as a drop in pressure across the plate. This pressure drop across an orifice plate is the basis for using this device to measure flow. There exists a relationship such that the flow across the plate is proportional to the square root of the differential pressure across the plate. These differential pressures are small units such as 0 to 200 or 0 to 100 in. water. The square root is extracted electrically, and modern installations use smart transmitters to measure the pressures, calculate the square root, and display the flow in linear units. Figure 21.19 is a flow-measuring setup using the orifice plate and differential pressure cell. The orifice plate as used for flow measurement of liquid or gases has limitations:

1. The bore of the orifice plate must be chosen carefully so as to range the instrument properly.

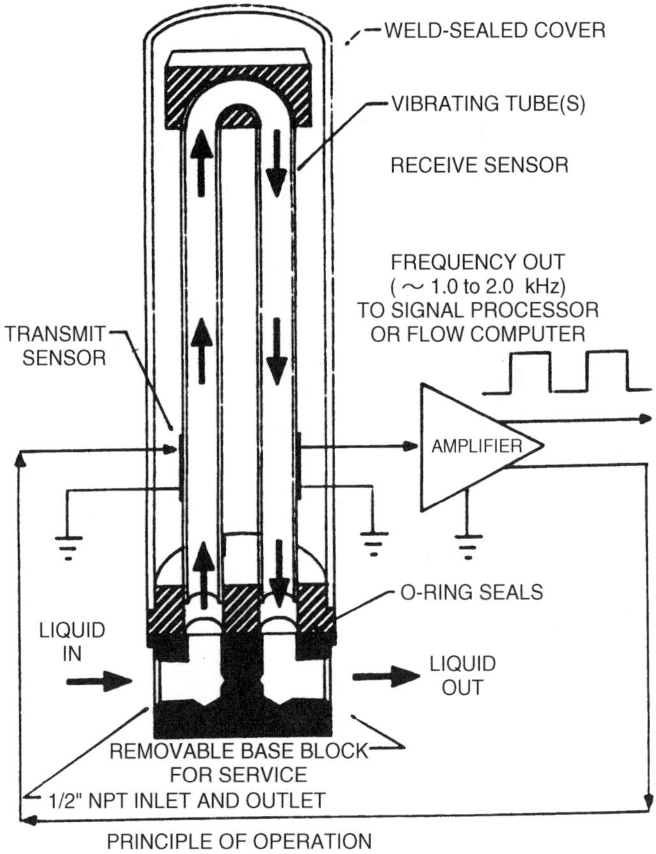

FIGURE 21.15 Vibrating U-tube for density measurement.

2. The liquids or gases must be relatively clean since abrasive flow will destroy the bore of the plate.
3. The plate must be inspected periodically to ensure proper configuration of the bore.
4. Impulse lines from the flange taps to the DP cell must not be allowed to clog, leak, or freeze.
5. Flow systems using this device usually require more pumping power since there is a permanent pressure drop in the system.

21.5.2 Magnetic Flowmeter

The magnetic flowmeter has been around for many years. It is a very reliable method of measuring liquid flows. This device offers obstructionless flow measurement and does not create a pressure loss across the flowmeter. The instrument's body contains a pair of externally energized magnetic coils and exploits the conductivity of the flowing liquid as it cuts across the magnetic field created by these magnetic coils. Two electrodes placed in a plane perpendicular to the coils pick up the tiny voltage produced by electromagnetic induction (e.g., 0.0015 mV/gal). This flow device is the most widely used flow technology today. The

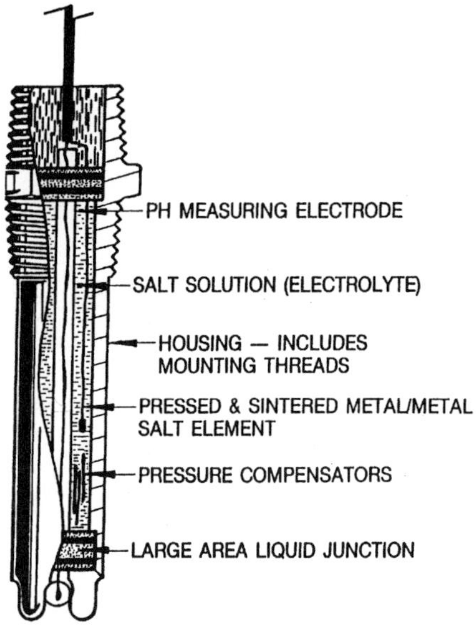

FIGURE 21.16 Combination pH electrode.

measured fluid must be electrically conductive and nonmagnetic. Magmeters operate totally outside the pipeline and are useful for corrosive chemicals, sewage, or liquids, which contain suspended solids. Carefully choosing the tube liner and maintaining flow velocity will reduce coating of the electrodes with conductive residues which can cause erroneous voltages. Figure 21.20 shows the electromagnetic field produced by the coils, with the flow plane cutting

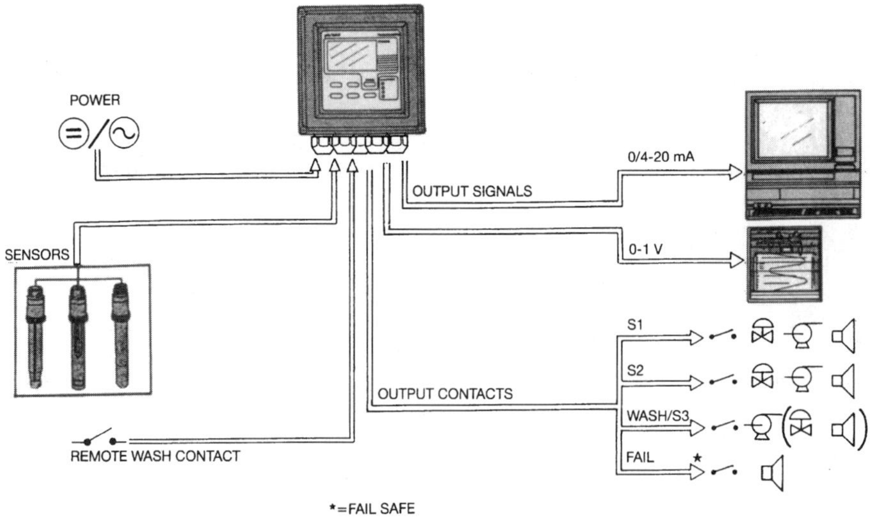

FIGURE 21.17 pH control loop.

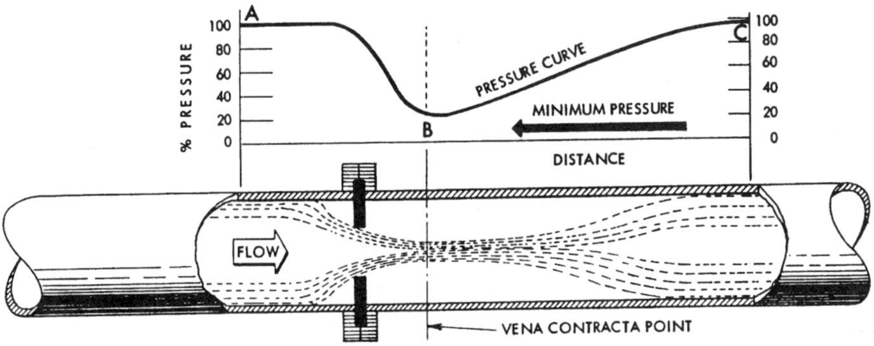

FIGURE 21.18 Flow profile across orifice plate.

across the electrodes. There is no pressure loss with a magnetic flowmeter since the flow tube is the same size as the pipe. This flow-measuring device produces a linear signal that does not require any type of calculation.

21.5.3 Ultrasonic Flowmeters

The two types of ultrasonic flowmeters are Doppler and time-of-flight. *Doppler flowmeters* use a beam of ultrasonic energy and receive reflections of this beam from bubbles or entrained particles. Since the particles are moving, the frequencies of the reflections differ from the original signal, and the difference is proportional to the flow velocity. Particles (bubbles

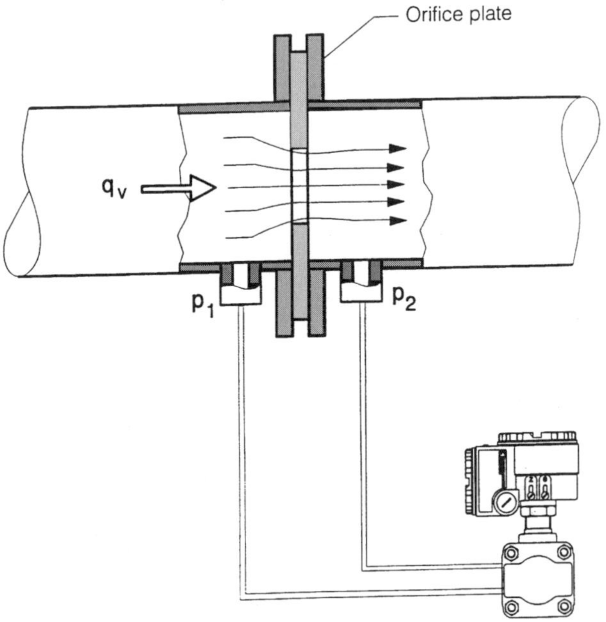

FIGURE 21.19 Flow measurement using orifice plate and DP cell.

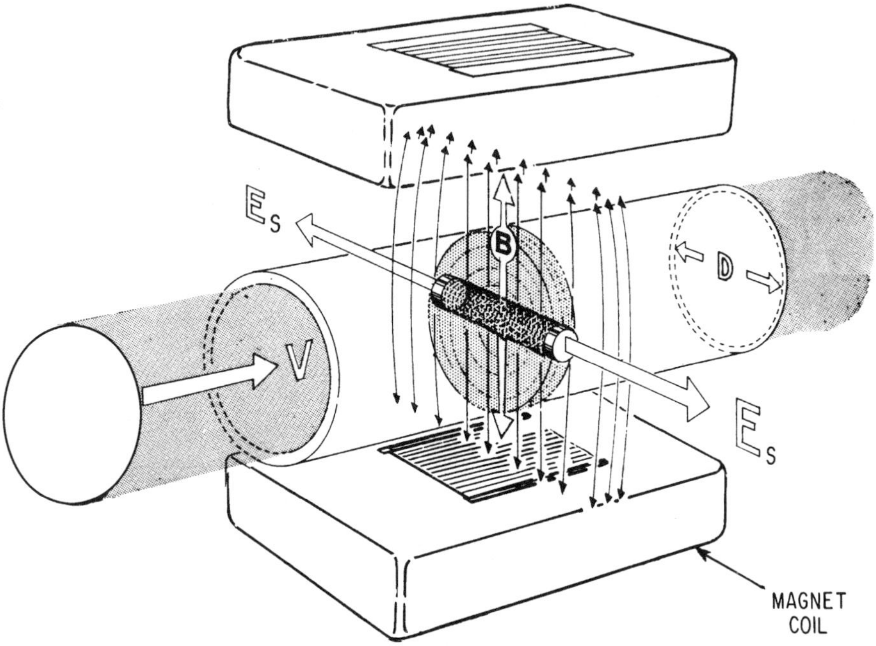

FIGURE 21.20 Magnetic flowmeter.

or suspended matter) must be present in sufficient amounts to reflect the energy to the transceiver. *Time-of-flight flowmeters* beam ultrasonic energy across the pipe and measure the time required for the beam to traverse the pipe. Pulses of ultrasound are sent upstream and downstream, and the time difference between the two is calculated as the flow velocity. These types of meters are strapped to the outside of the pipe. Usually, these devices can measure flow in either direction. These flowmeters are subject to error and their accuracy limits their application. Figure 21.21 illustrates a Doppler application and Figure 21.22 illustrates a time-of-flight installation.

21.5.4 Venturi Flowmeter

The venturi flowmeter is an obstruction type of flow-measuring device. A device that is gentle to a flow because its sides constrict and widen over the length of the body is inserted into a pipe to measure the flowing stream. Flow is derived from measuring pressures at the front and after maximum constriction, then extracting the square root of the differential pressure. Figure 21.23 is a pictorial view of a venturi flowmeter. Limitations are that liquids should be clean and there is a pressure loss associated with any obstruction in the flow.

21.5.5 Turbine Meter

The turbine meter is used to measure liquids and consists of a rotor mounted on a shaft. The flow passing the rotor causes it to rotate and drive a counting device. Particles tend to damage and limit the life of the rotating parts. As this is a mechanical device, there is a

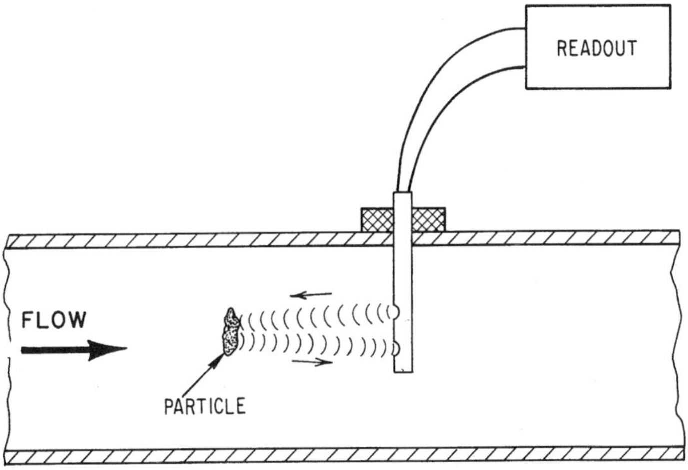

FIGURE 21.21 Doppler flowmeter.

defined life span due to wear, but the low cost of this device lends itself to many clean water applications. Figure 21.24 illustrates a propeller flowmeter.

21.5.6 Coriolis Mass Flowmeter

Mass flow rate metering is the preferred fluid flow for critical control applications. In a sugar refinery this device will express flow and control in mass units which will give a close look at the process operation to gauge bottlenecks and rate the processing. Most refineries have an expressed processing rate, which is usually different from the day-to-day operation rate. Mass flow can determine the actual rate in real-time values. A combination of flow tube vibration and deflection due to mass flow results in a twisting of the flow tubes within the meter, as illustrated in Figure 21.25. The degree of twist is directly proportional to the mass flow rate. Sensors mounted on the tubes measure the amount of twist or deflection and the sensor signal is scaled, integrated, and filtered to produce the standard control output relative to mass flow (Fig. 21.26). Coriolis mass flow is unaffected by the fluid's temperature, density, pressure, and viscosity and is insensitive to material buildup or coatings that change the cross-sectional area of the meter's tubing. These immunities lend this device to unlimited applications in all processing industries. Pressure drop is the limiting factor in applications, as is the size of tubes available, which restrict applications to smaller ranges. Because these flowmeters are very accurate, their use has been applied to tight-control uses such as chemical processes based on mass flow balances, managing expensive fluids, and custody transfers. Table 21.3 compares flowmeters and the costs associated with purchase, operation, and repairs over the life of the meter. The Coriolis mass flowmeter is the premier flow measurement technology today.

21.6 TEMPERATURE CONTROL

Temperature is probably the most frequent process measurement made in a modern sugar refinery. Many of the procedures, processes, and methods used in making refined sugar and

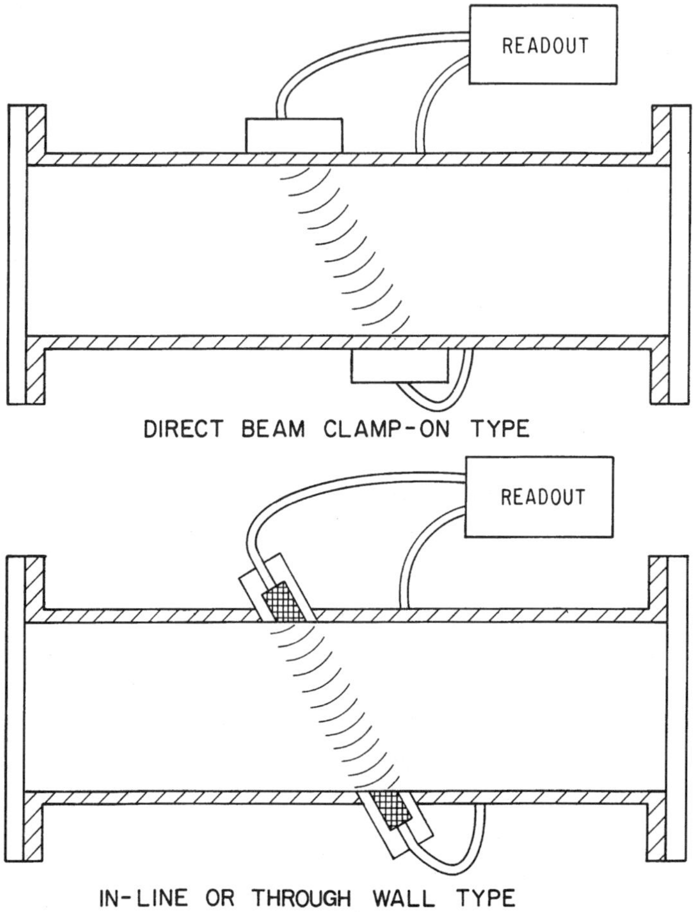

FIGURE 21.22 Time-of-flight flowmeter.

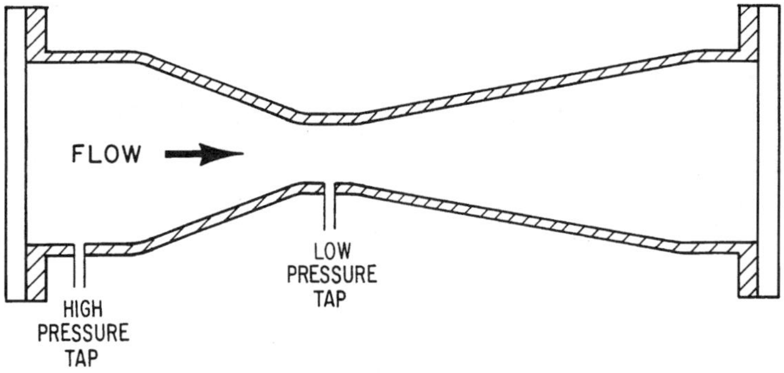

FIGURE 21.23 Venturi flowmeter.

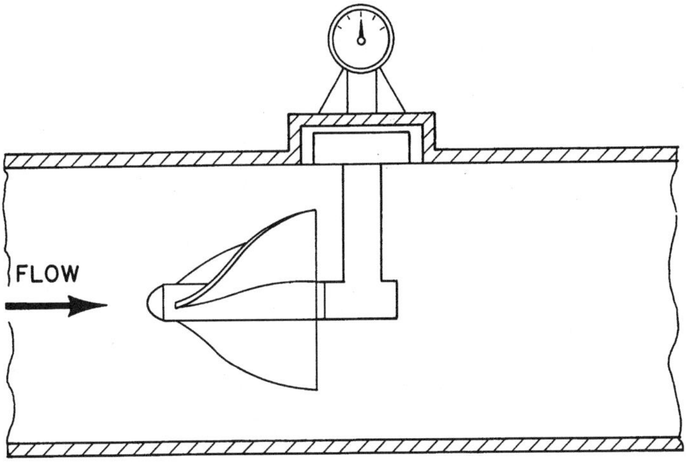

FIGURE 21.24 Propeller flowmeter.

liquid sugars require measuring and controlling temperature at many points in a plant. Auxiliary support functions such as steam generation require temperature measurement and regulation. In many areas of the country, temperatures at various times of the year have a definite impact on sugar processing.

21.6.1 Thermocouple

The thermocouple is a device used to measure a change in temperature. The thermocouple works on the principle that when two dissimilar metals are joined and heat is applied, a small dc voltage is generated at the free ends, proportional to the temperature difference between the free ends and the ends where the metals are joined. The end where the dissimilar metals are joined, sometimes called the *hot junction,* is exposed to the process. Usually, this joint is made by fusion welding to make a permanent connection. The other end of the joined metals, called the *cold junction,* is connected to the instrument, which will read the voltage and convert it to a scale or temperature display.

The cold junction temperature, which will be different from that at the hot ends, must be zeroed out for the instrument to read the true hot junction temperature. This zeroing, called *cold junction compensation,* exists on all thermocouple types. The thermocouple is usually encased in a thermowell to protect it from the harsh conditions of the process. The thermocouple is connected to the instrument with extension wire, which is made of the same combination of metals as those of the thermocouple itself. There are many types of thermocouples, based on the metals involved. Listed in Table 21.4 are the insulation colors for the three most common types of thermocouples. This demonstrates the ease with which thermocouples and thermocouple extension wire can be recognized. The outer insulation color of extension wire indicates the type, as does the insulation color on the positive wire. In all cases the negative wire insulation color is red. Constantan is an alloy consisting of 45% nickel and 55% copper; Chromel, an alloy of 90% nickel and 10% chromium; Alumel, an alloy consisting of 94% nickel, 2.5% manganese, 2% aluminum, 1% silica, and 0.5% iron.

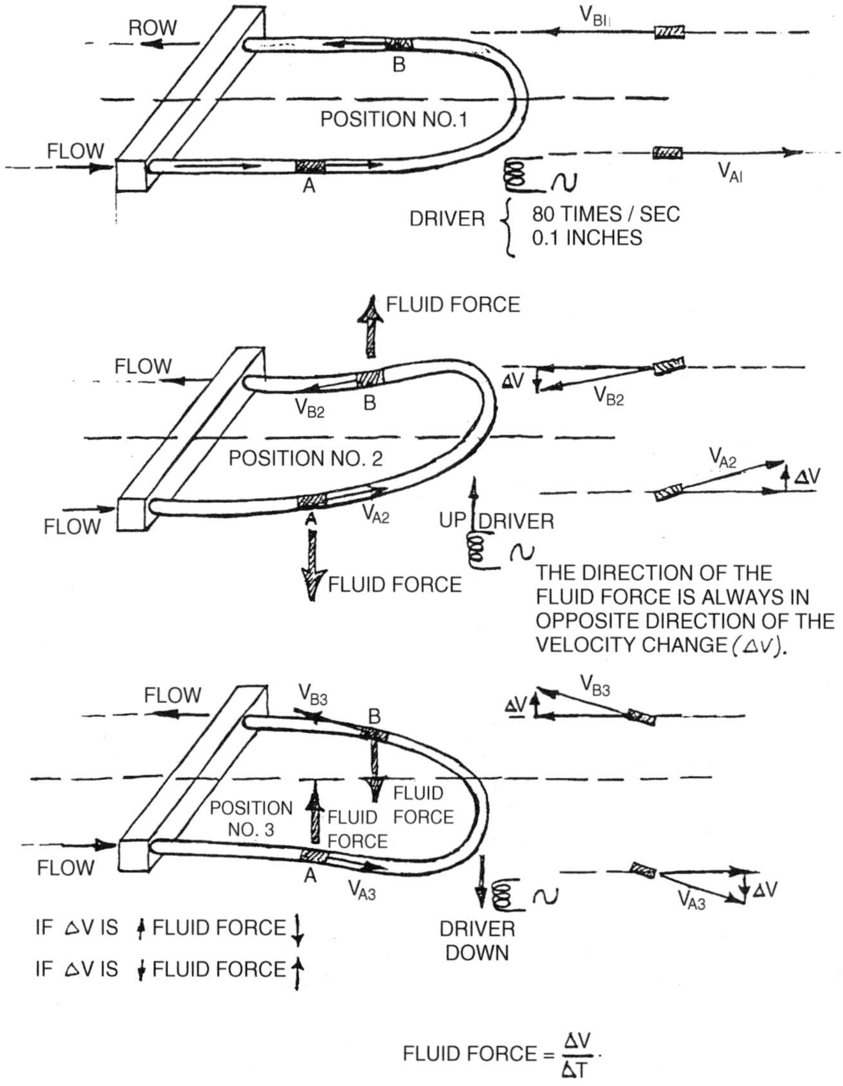

FIGURE 21.25 Coriolis mass flowmeter.

21.6.2 Reference Temperature Detector

The reference temperature detector (RTD) is another device used to measure a change in temperature. The RTD is composed of a resistance bulb and a receiving instrument, which converts a change in electrical resistance to a temperature reading. The temperature-sensitive portion consists of platinum, nickel, or copper wire, wound on a silver or copper core, and encased to prevent damage. The resistance of the wire used in the winding is directly related to temperature. As with the thermocouple, the RTD is usually set in a thermowell for protection from the harsh conditions of the process. The RTD is very accurate in the moderate temperature range, but the thermocouple has the accuracy advantage in the higher-temperature ranges. Cost as a requirement gives the nod to the thermocouple as a first

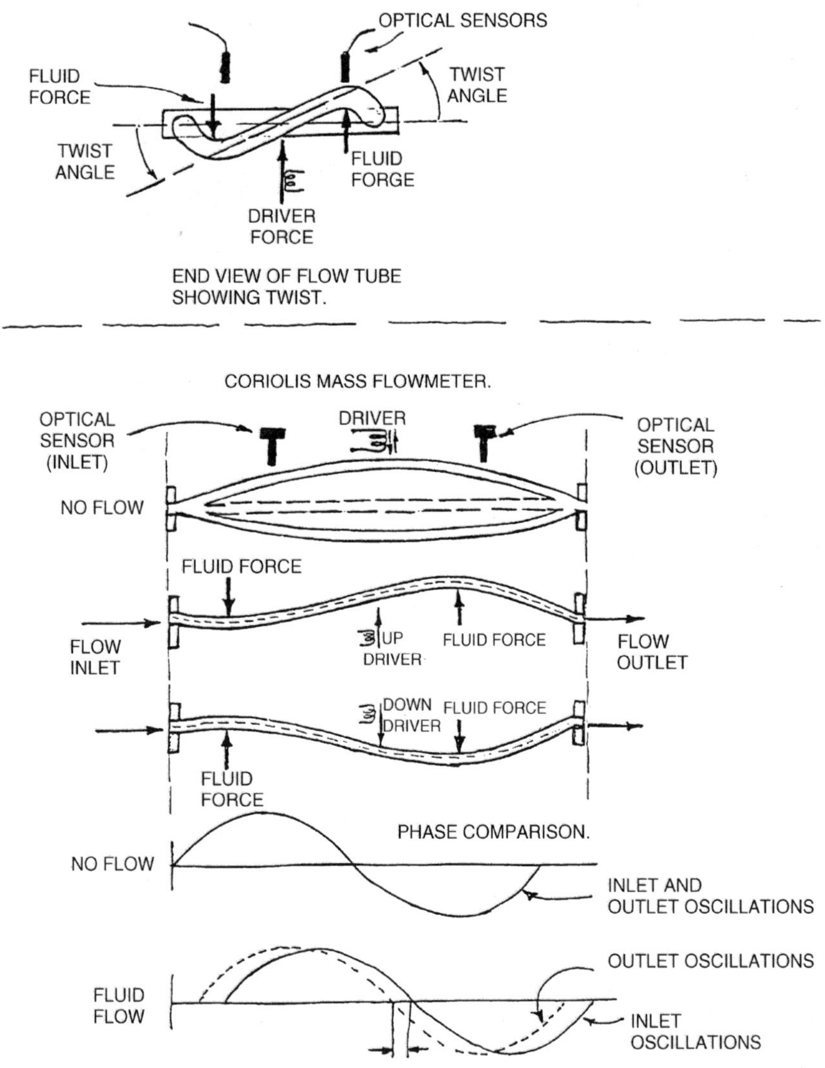

FIGURE 21.26 Coriolis mass flow sensors.

choice, but the purchase of thermowell, extension wire, and reference junctions makes the two types of temperature-measuring devices about equal for many applications. The expenditures for measuring various parameters in industry today are distributed as follows: pressure, 13%; level, 15%; temperature, 29%; flow, 43%. Density is not mentioned, but the ideas are valid and show approximate numbers for how money is spent to measure a process.

21.7 PROCESS CONTROLLER

The process controller is the device, which receives the measurement of process variable and used this information to generate a control signal, which is usually fed to some type of control valve. The controller can be either electronic or pneumatic or as is usually the case,

TABLE 21.3 Comparison of Flowmeter

	Flowmeter Type								
Parameters[a]	Coriolis	Magnetic	Orifice	Positive Displacement	Turbine	Ultrasonic-Doppler	Ultrasonic-TOF	Variable Area	Vortex
Meter cost (5)	6,500	3,000	1,400	3,500	2,100	2,500	4,000	1,500	2,000
Pumping cost ($)									
1 year	1,744	—	1,182	710	710	—	—	850	686
10 years	17,740	—	11,820	7,100	7,100	—	—	8,500	6,860
Mean time between failures (yr)	10	10	5	2	2	5	5	2	5
Number of repairs in 10 years	1	1	2	5	5	2	2	5	2
Cost to repair (% of meter cost)	100	20	200	200	200	200	50	100	200
10-year cost to repair ($)	6,500	600	2,800	7,000	4,200	5,000	2,000	1,500	4,000
Meter % accuracy (0–100 gal / min)	0.15R	1.0R	1.0FS	0.2R	0.25R	5.0FS	1.0R	2.0R	1.0R
Error (gal / min at 50% scale)	0.75	1	10	1	1.25	10	1	10	5
Yearly loss at $0.01 / gal ($)	3,942	5,256	52,560	5,256	6,570	52,560	5,256	52,560	26,280
Loss in 10 years at $0.01 / gal ($)	39,420	52,560	525,600	52,560	65,700	525,600	52,560	525,600	262,800
Total costs, 10 years ($)	64,240	56,160	547,540	70,160	79,100	533,100	58,560	537,100	275,660

[a] Meter cost is based on a 2-in. meter size, quantity 1. Pumping cost is based on Krohne published data for unrecovered pressure loss, pump efficiency of 80% and electricity cost of $0.05 / kWh. Cost to repair is based on published data or relative complexity, obstructions, moving parts, and experience. Yearly loss and loss in 10 years are estimated amounts of money overspent or undercollected due to meter inaccuracy. Calculations based on product cost of $0.01 / gal. In "Meter % accuracy" row, R stands for rated, FS for full scale. Pumping costs are based on 1990 estimates; all other numbers based on March 1997 calculations.

TABLE 21.4 Thermocouple Specifications

Thermocouple Type (Overall Insulation Color)	Wire Positive Metal (Insulation Color)	Wire Negative Metal (Insulation Color)
Type T	Copper	Constantan
(blue)	(blue)	(red)
Type J	Iron	Constantan
(black)	(white)	(red)
Type K Chromel–Alumel		
(yellow)	(yellow)	(red)

a combination of the two types. The controller is the heart of process control since this device does the actual monitoring and control of a process variable. The *control loop* consists of the sensor (level device, thermocouple, flowmeter, and density cell); the controller, which will produce the control signal; and the control valve, which is the final element and actually does the control in the loop. A typical level controller would use a capacitance probe to sense the level and send this signal to the controller. The controller will compare the actual level to some preset level chosen for reference and generate a signal to the control valve to take some action to maintain the preset level. The actual level is called the *process variable* (PV). The PV is the actual condition as it exists while the sensor is scanning the process. The *set point* (SP) is the value one wants to control the process parameter. It is the purpose of the controller to try to make the set point and process variable identical or as close as possible. In some cases the sensor and controller are in the same instrument. A DP cell can use air to measure levels (bubbler tube) and generates the 3 to 15 psi control signal for a pneumatic valve. In this case the set point has been defined very closely for sensor application. Usually, the controller is an analog device panel mounted in the control room performing a single function. A digital controller can be programmed to make calculations and generate the control signal for the valve. The digital controller can handle scaling, complex equations, floating-point arithmetic, and many changing variables.

21.8 ACTUATORS AND CONTROL VALVES

The final element in the control loop is the control valve. The type of valve one uses to effect control is based on past experience, cost, close control needed, distance from the controller, availability of air or electricity, physical location, temperature, fail-safe position, code requirements, and so on. The majority of actuators in the sugar refining industry are pneumatic in operation. Many plants use computer control, which dictates electronic measurement, electronic control, and an electronic output signal. Usually, this electronic signal is converted to pneumatic at the control valve. When an electronic signal is converted to a pneumatic signal, an I/P converter is used (current to pneumatic). When a pneumatic signal is converted to an electronic signal, a P/I is used (pneumatic to current). Usually, the signals involved are 4 to 20 mA and 3 to 15 psi gauge pressure.

An actuator is used to provide the power to stroke the control valve. The pneumatic actuator must provide enough torque to drive the control valve under the most severe conditions. This power is provided by air pressure. The pneumatic actuator consists of a diaphragm clamped between two metal plates. On one side of the diaphragm is a chamber into which air pressure (3 to 15 psi) is admitted. The plates are connected to a shaft, which is physically joined to the stem of the control valve. The diaphragm is 16 in. in diameter in a large valve, and when 15 psi is applied over the surface area of the diaphragm (area =

201 in.^{2}) there is a force of approximately 3000 lb acting on the shaft and transmitted to the valve stem. When the air pressure is released from the diaphragm chamber, the valve is returned to its original position by a large spring, which has been placed against the plate and on the side opposite the air chamber. Valves, which are moved when air is applied, are defined as *direct acting.* Valves, which are moved when air is released, are defined as *reverse acting.* The choice of actuator action is usually dictated by process conditions; fail-safe needs insurance mandates and past operational experiences. Double-acting valve actuators are springless and use air pressure for movement in either direction. Connected to the actuator is the control valve, the workhorse of process control. The valve is the final element and is directly exposed to the conditions of the process. The choice of control valve type is mandated by cost, power source, torque at the valve stem, failure mode, valve size, frequency of operation, speed of operation, and control desired. The conditions of the material being controlled have a heavy bearing on the choice of control valve. Valve noise is another factor in making a choice of control valves. Figure 21.27 is a view of a pneumatic actuator and associated control valve. In this case the actuator is stroked by admitting air pressure to the port marked with an arrow. The air pressure over the area of the diaphragm will force shaft movement downward and will close this type of valve (weir type). The return spring is mounted under the plate, and when air is removed from the chamber, the valve will stroke

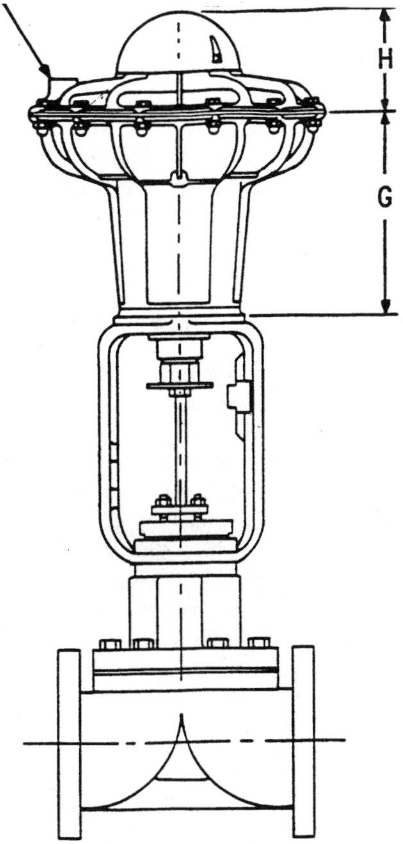

FIGURE 21.27 Actuator plus control valve.

upward and the valve will open. The H section is the air chamber and the G section is the spring chamber. Figure 21.28 depicts three types of actuators controlling a weir valve. In some cases it is desirable to match the control signal with the valve opening and therefore match the valve to the signal. Figure 21.29 displays the action of control signal and valve position. In this application the valve is an equal-percentage single-ported valve. Usually, this type of valve is sized in terms of flow capacity. It is interesting to note that when the control signal is 50% (9 psi) the valve will be half-open and carry half its designed capacity. This is useful when the control valve's capacity is chosen and the material being controlled is metered for some purpose. An example of this would be metering steam used for making up the low-pressure main in the refinery. The process control valve is usually equipped with a positioner. This device is mounted on the valve and has it movement mechanism attached to the valve stem. The positioner will track valve stem movement and provide feedback to the controller to ensure valve position as dictated by the process controller. Figure 21.30 is a view of a process control valve with a stem-mounted positioner. Figure 21.31 illustrates a double-ported control valve stem connected to a pneumatic actuator with the two chambers for the air and spring.

21.9 EVAPORATOR AND PAN CONTROL

Most refineries, if not all, use a vacuum pan for the boiling of the crystal sugar termed as production. Since almost all of the actual processing in a refinery is in the liquid state, there

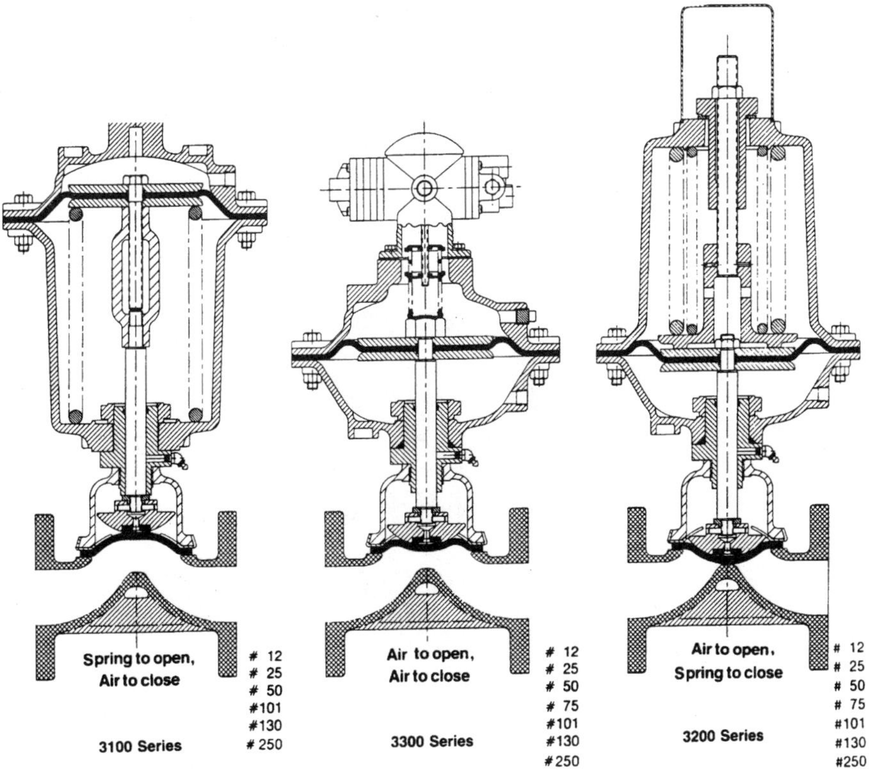

FIGURE 21.28 Control valves.

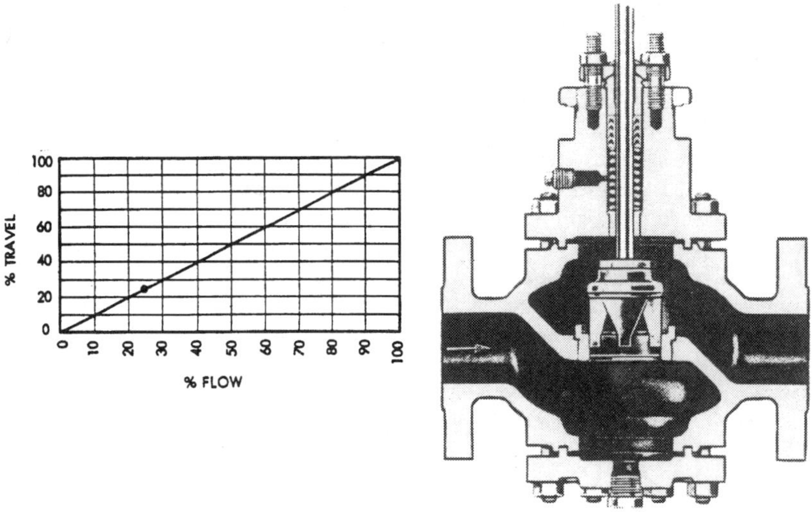

FIGURE 21.29 Equal percentage port control value.

FIGURE 21.30 Pneumatic control valve with positioner.

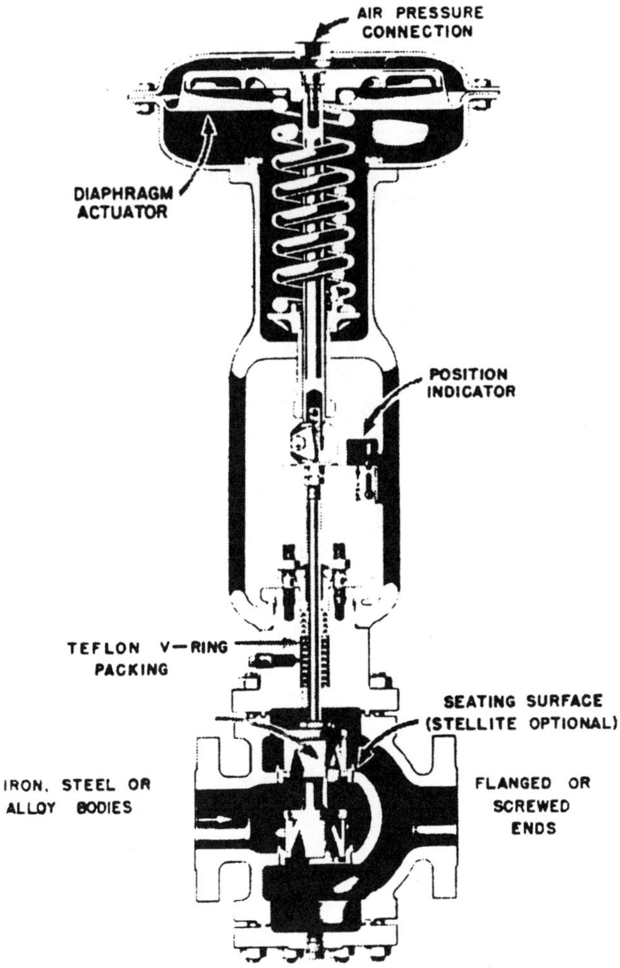

FIGURE 21.31 Double-ported control valve.

is usually a drop in density from melting raw to pan feed liquors because of dilution. The multiple-effect evaporator is used to increase the density of pan feed liquor, and the vacuum pan crystallizes this liquor under controlled conditions. These two operations require instrumentation to sense and control these processes. Many of the process parameters we have discussed are controlled in a single application. Figure 21.32 details the control scheme in an evaporator using the older technology of bubbler for level in feed and density and the DP cell for level under operating conditions. Many evaporators operate with this type of instrument today. Parameters under control are level, pressure (vacuum), density, flow, and temperature by virtue of vacuum control. Figure 21.33 illustrates evaporator control by digital means with a computer as the device to direct control functions. Most sensors are high-tech and electronic, which lends itself to direct digital control. The computer is constantly scanning the evaporator set for deviation and correcting these errors in milliseconds. In many actual cases, there is no human attendant for this evaporator and the computer

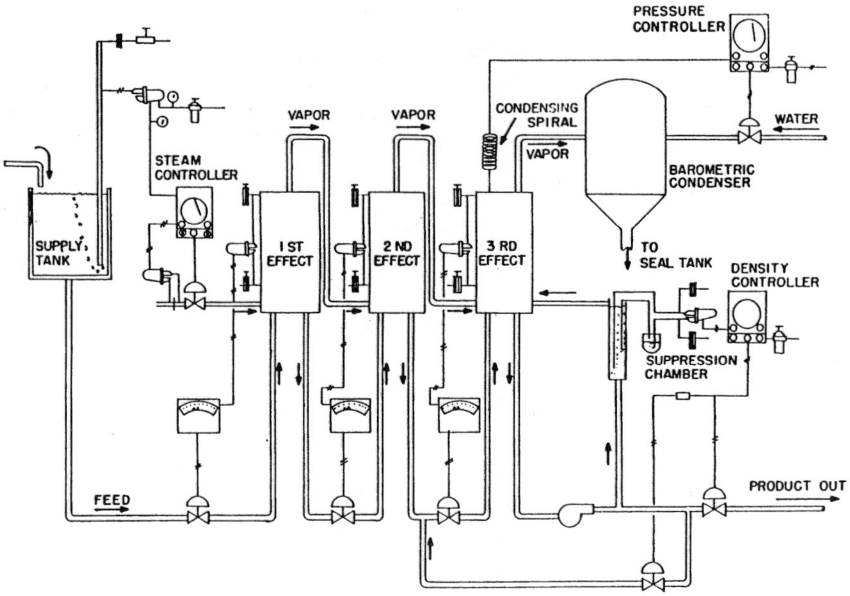

FIGURE 21.32 Multiple-effect evaporator with pneumatic instrumentation.

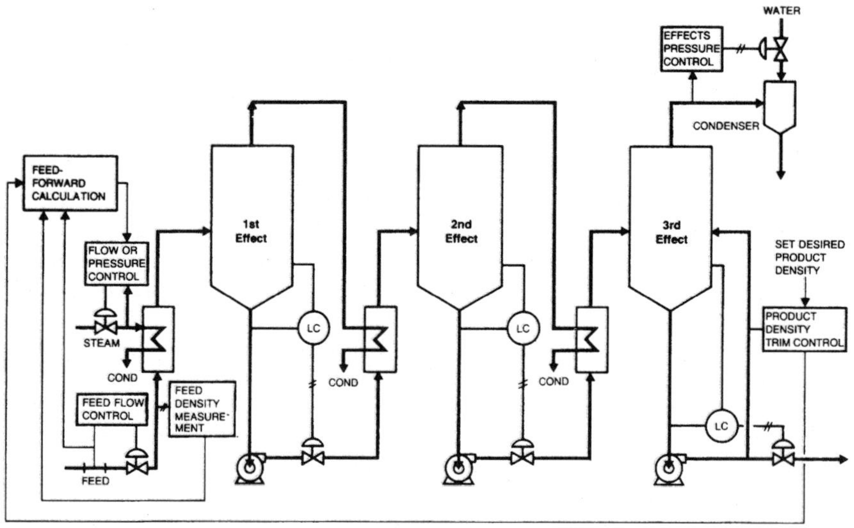

FIGURE 21.33 Multiple-effect computer control.

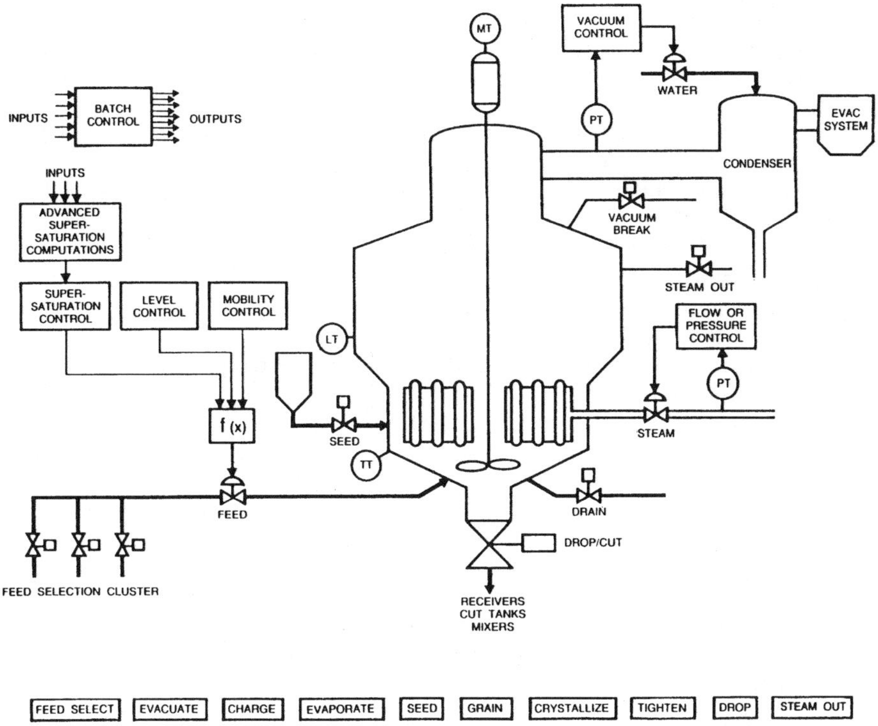

FIGURE 21.34 Vacuum pan control by instrumentation.

does all the supervision and control. The vacuum pan is an ideal device to automate with instrumentation. The process is straightforward and is repeated many times per day. Figure 21.34 depicts a vacuum pan with modern instrumentation. Once again the computer or programmable controller is operating the pan with little or no human interface. All the parameters are being controlled: density, pressure, level, flow, and temperature. Vacuum pan automation has become standard in the sugar refining industry.

The sugar processing industry has undergone significant changes in a very short period of time. Every aspect of the industry is using instrumentation in new and exciting ways to produce a better product with consistent quality for a cheaper price. Instrumentation will continue to play an ever-increasing role in sugar processing. No matter what the task to be handled, using basic instrumentation as a method of controlling a process will be around for many years to come.

REFERENCES

This chapter is based on information from the following sources.

1. Technical information bulletin, TI 11P/24/ae/08.94, Endress & Hauser, p. 6.
2. *Fundamentals of Industrial Instrumentation,* Honeywell, Inc., May 1979, p. 45.
3. E. L. Schuler, *Sensors,* Aug. 1989, pp. 45–51.
4. S. Parker, *Chem. Process.,* 1996 Fluid Flow Annual, pp. 67–71.

5. S. Omidbakhsh, *Eng. Dig.*, Feb. 1988, pp. 16–18.
6. W. H. Nagel, *Ind. Proc. Control,* Feb. 1986, pp. 57–60.
7. *Instrumentation Training Course,* Vol. 1, *Pneumatic Instruments,* Howard W. Sams, Indianapolis, IN, 1978.
8. G. G. Moyer, *Honeywell Food Processing Instruments,* AB-8, Industrial Division, Honeywell, Minneapolis, Minn, 1967.
9. *Instrumentation Training Course,* Vol. 2, *Electronic Instruments,* Howard W. Sams, Indianapolis, IN, 1978.
10. Technical information bulletin, TI 014F/24/ae, Endress & Hauser, p. 2.
11. A. D. Ehrenfried, *Sensors,* Aug. 1989, pp. 13–21.
12. C. H. Hoeppner, *Chem. Eng.*, Oct. 1984, pp. 71–78.
13. R. A. Pineiro, Honeywell, Inc. Sugar seminar, pp. 2–20.
14. J. Yokogawa, *General Specifications, EXA PH400-Series GS 12B6B2-U-H,* 5th ed., Newman, GA, Apr. 1995, p. 3.
15. Texas Nuclear Corp., *Control Eng.*, Feb. 1988, p. 101.
16. F. W. Kirk and N. R. Rinboi, *Instrumentation,* 3rd ed., American Technical Publishers, Homewood, IL, 1975, pp. 48–296.
17. *Dynatrol,* Bulletin J-67DA, CL-10TY Series, Automation Products, Inc., Houston, TX.
18. D. Creter, J. Samagond, and J. Beahm, *Flow Control,* Apr. 1997, pp. 12–49.
19. J. Pomroy, *Instrumention and Control Systems,* Mar. 1994, pp. 61–68.
20. R. Toft, *Sensors,* Aug. 1987, pp. 14–20.
21. J. Pumroy, *Chem. Eng.*, May 1996, pp. 94–102.
22. G. P. Trearchis, *Sugar J.*, Oct. 1968, pp. 28–32.
23. D. Reif, *Flow Control,* May 1997, pp. 31–35.
24. W. Ulanski, *Flow Control,* Mar. 1997, pp. 12–20.

CHAPTER 22

Operational Computers*

INTRODUCTION

The history of operational computers used in industrial control applications began with two discrete origins. On the one side were applications to automate electrical control functions such as starting and stopping motors. These electrical control applications influenced both the design of the equipment and the programming methods. This led to the development of the programmable logic controller (PLC). On the other side were developments to improve process automation through analog instruments such as level and flow controllers. These new devices were developed with a strong resemblance to the equipment they replaced and the configuration practices used by instrument technicians. This path yielded the distributed control system (DCS).

In the early years, the differences were quite distinctive; PLCs were the choice for electrical control applications, and DCSs were for process instrumentation. Today, however, those distinctions no longer exist or at most, are difficult to discern. The functions once limited to PLCs are now incorporated in DCSs, and vice versa. This is natural considering the nature of control theory and the development of computer technology. Throughout this chapter, they are referred to interchangeably as operational computers. Since both are derived from computers, a look at their core is in order.

22.1 COMPUTER ELEMENTS

Computers have four basic sections as described below.

22.1.1 Input

Data are received in several forms by the computer and converted into a machine-readable language in order to perform the program instructions. Inputs may be received through a keyboard, disk drive, modem, touch screen, and so on. Two fundamental methods are used

*By Ray Burke.

for inputting (and outputting) the actual data. First is the serial method, where data flow one bit at a time through a pair of wires and parallel where data are passed one word (16 bits) at a time by separating each bit out on a bundle of conductors.

22.1.2 Processing

The work center of the computer is the central processing unit (CPU). It is here that all operations within the computer are performed according to the programs. The flow of all information to or from the computer is directed within the CPU. The arithmetic–logic unit (ALU), control unit, and primary storage are located within the CPU. All inputs received will be processed by the CPU to determine what action should be taken, such as sending an output to a printer, video display, disk drive, and so on.

The rate at which instructions are processed is referred to the *clock speed.* A precision quartz clock or oscillator generates pulses that step the processor through its tasks. The design of the processor is the limiting factor as to how fast instructions are processed. Generally, improvements in computer technology lead to faster processor speeds.

22.1.3 Storage

Instructions and information to be processed by the computer are stored in some form of memory. Registers are the memory in the CPU holding information being processed. Dynamic random access memory (DRAM) is an array of memory chips that hold the operating system and program. Read-only memory (ROM) contains permanent information that can only be read. Memory is measured in bytes, usually increments of 1 million bytes.

Magnetic media have been the predominate method of storing programs and data replacing the use of punch tape and punch cards. Reel-to-reel, hard disks, and floppy disks are the most common types used. Optical disks (CD-ROM's) are increasing in popularity due to the large data capacity and integrity of data. Most are read only, although systems are available that are capable of writing output to an optical disk. Data on tape systems are read sequentially; that is, all data from the beginning of the tape must be read to locate the data desired. Disk systems are random access. The location of a file or data is found in the file allocation table. The desired file is accessed by the read head moving directly to the sector containing the target file.

22.1.4 Output

The product of the computing operations is the output. Its form is dictated by the program. Outputs may be seen on video displays, network interface devices, printers, or BCD displays. The need for peripheral equipment to facilitate communications and operations is common to most computers. Peripherals may be defined as specialized computer components that support the activities of the CPU. They may be either input or output devices (Table 22.1).

22.2 OPERATING SYSTEM

The operating system is a group of supervisory programs that maximize effective use of the hardware. These programs establish the type of video display, access to printers, disks, and input devices, and the keyboard language. Operating systems are chosen based on some desired characteristic (i.e., multitasking, multiuser) or because of the vendor's requirement for specific hardware.

TABLE 22.1 Common Computer Peripherals

Input	Output
Auxiliary computer	Auxiliary computer
Badge reader (magnetic)	Binary-coded decimal display
Bar code reader	Bar code printer
Card reader (punch or magnetic)	Card punch
Drawing board	Monitor
Disk (input)	Disk (output)
Keyboard	Output simulator
Magnetic tape	Magnetic tape
Personal computer	Personal computer
Programmable logic controller	Programmable logic controller
Digital control system	Digital control system
Mouse	Video projector

22.3 LANGUAGE

For any computer to perform work, a link must be established between the computer and the task. The language is a software program that provides instructions for the computer to accomplish useful work. There are many languages available today, but some are more popular than others with particular computer manufacturers. Most of the languages were designed for a specific application but may be used to perform other tasks. Following are some of the more common software languages:

- *ALGOL (Algorithmic Language):* a noninteractive, high-level programming language designed to facilitate programming of algorithms
- *APL (A Programming Language):* an interactive programming language based on the use of array and matrix elements
- *BASIC (Beginner's All-purpose Symbolic Instruction Code):* an interactive, high-level programming language that is easy to learn
- *C:* a general purpose, low-level programming language used to write major numerical text-processing and database programs
- *COBOL (Common Business Oriented Language):* a high-level noninteractive programming language designed for business applications
- *FORTRAN (Formula Translation):* a high-level noninteractive programming language particularly suited for scientific and mathematical applications
- *Pascal:* a high-level interactive programming language designed for reliability and efficiency in programs
- *PL/1 (Programming Language 1):* a high-level noninteractive programming language designed to incorporate the best features of a variety of languages

22.4 PROGRAMMING

Despite their immense complexity, computers can be reduced to their most basic components, which are simply electronic switches that are either in the on or off position. The most complex instructions are all accomplished by turning those switches either on or off with a numeric instructions consisting of only 0's and 1's, known as the *binary system.* Binary is a base 2 system that uses only the symbols 0 and 1. The total number of symbols is equal to the base (2), and the largest-valued symbol is one less than the base; $1 < 2$.

With only two symbols, the binary system is well adapted to use in computers, where the switchlike operations have two states: on or off. These are indicated by two voltage levels: usually, +3 V (on or logical 1) and 0 V (off or logical 0), which can be distinguished by a digital circuit. Logical arguments can be either true (1) or false (0). A number greater than the largest-valued symbol will be given a weighted value for each position. The weighted values for the binary system from right to left are the powers of 2 (l, 2, 4, 8, 16, 32, etc.).

22.5 OPERATIONAL COMPUTERS

The speed of operation in an operational computer refers to the amount of time it requires to perform a certain function. The execution of the program logic in the operational computer may be very fast; however, if the output controls a pneumatic valve that must be pressurized to respond, the actual speed of operation will be slower. The reliability of operational computers has become an almost forgotten asset. With improved manufacturing of components, thorough checkout procedures, and on-line backups, the operational computers available today are highly reliable when properly installed in the right application. Unlike most mechanical equipment, the longer a computer operates, the more likely it is to continue operating. Electronic components typically fail early in their life cycle.

Accessibility to the information contained within the operational computer's memory is a major advantage. The information is available for a multitude of noncontrol uses such as preventive maintenance. Actual run time can be calculated for a piece of equipment to schedule maintenance or replacement, thus avoiding a loss of production. Performance calculations can be made in real time and present more accurate accounting of production costs. Troubleshooting with the operational computer reduces the time required to locate faulty devices through the use of process graphic displays. Modifications to the process for changes in the sugar or changes to tune a piece of equipment to improve its performance can be accomplished quickly, accurately, and unilaterally through the use of operational computers.

22.5.1 Programmable Logic Controllers

A programmable logic controller is a general-purpose, computer-based electronic device containing a microprocessor that may be programmed to control a variety of machines or processes. The programs may be altered or reconfigured easily to accommodate new jobs or changes in the existing process.

22.5.2 Distributed Control System

A distributed control system is also a general-purpose, computer-based electronic device that contains a microprocessor which may be programmed to control a variety of processes. The programs may be altered or reconfigured to accommodate a new process or a change in the existing process. The concept of distributed control is to parcel out control functions to individual controllers rather than perform them in a single computer (centralized control).

22.5.3 Common Sections of PLCs and DCSs

Most PLCs and DCSs are common in many respects as to their functions and performance. They may be either a rack-mounted system or a single stand-alone unit. Each has four major components: (1) a central processing unit, (2) an input section, (3) an output section, and (4) a power supply.

a. Central Processing Unit (CPU). The central processing unit is the brains of the operational computer and directs all input/output (I/O) functions to the proper address while scanning the program logic continuously. All math functions and logic decisions are processed in the CPU. The memory used for both the program and data storage is also located in the CPU.

b. Input and Output (I/O) Sections. The input and output sections of an operational computer may handle either discrete or analog signals, although most units have the capability to handle both. These sections usually consist of a mounting base with modules that plug into it. Modules are the link between the processor and the outside world. All signals received or sent will pass through a module or data link connections to the data highway or programming unit. Since the I/O is the link to the outside world, it is also an entry point for unwanted noise and harmful power surges. Common practice is to place an optoisolator in each I/O circuit to protect the electronic components. The optoisolator contains a light-emitting diode (LED) on the signal end and a phototransistor on the receiving end. The signal is transmitted as a pulse of light from the LED through an air gap to the phototransistor, where the light pulse is converted back to an electrical signal. The air gap provides a high level of electrical isolation which protects the electronic circuits.

Two basic categories of inputs and outputs are (1) discrete and (2) analog. The simpler and less expensive PLCs are designed to handle discrete 1/0 only, while DCSs and higher-level PLCs have the ability to handle both. The latter generally are capable of exchanging information with other operational computers, host computers, or other intelligent devices.

Discrete I/O includes simple on–off functions and refers to the logical state of a device, which is either on (logical 1) or off (logical 0). Common discrete input devices are:

- Pushbuttons
- Selector switches
- Level switches
- Proximity switches
- Relay contacts

Common discrete output devices are:

- Lights
- Relays (coils)
- Solenoid valves
- Horns
- Motor starters

c. Analog I/O. Analog I/O refers to a changing, variable signal. The signal may be any numeric value ranging between the zero and maximum signal. Analog input signals typically are derived from process variables. Analog outputs are generally signals sent to a control element or indicating device. The distinction between an analog and a discrete signal is sometimes fuzzy, such as an analog signal that drives an electric actuator.

Analog Inputs. An analog input signals will usually be received in a continuous form. A transducer or other analog input field device converts the process variable to a proportional electrical signal. The transmitter is calibrated to generate an arbitrary zero value at some low process value and has a span value to define the maximum process value it will transmit. The signal is sent to the input module of the operational computer either as a current or voltage loop. The advantage of a current loop is that the signal is not attenuated (diminished) by the resistance of the conductor cables. On the subject of cables, common practice is to use twisted, shielded pairs of copper wire for voltage and current loops. For thermocouple devices, the cost of running thermocouple wire (to avoid cold junction potential) must be compared to using a thermocouple transmitter and running standard copper signal cable. Where a pneumatic transmitters is in use, a pneumatic to current transducer can be used to convert the pressure signal to a current signal which is digitally converted by the operational computer. Other types of transducer accept millivolt signals from pH meters or thermocouples. A recent development in transmitters is the "smart" type, which transmits digitized signals directly to the operational computer using frequency pulses instead of a current loop.

Common analog input signal forms are:

- 4 to 20 mA
- 0 to +1 V dc
- 0 to +5 V dc
- 0 to +10 V dc
- 1 to +5 V dc
- 5 V dc + 10 V dc

The analog signal is converted or digitized into an equivalent binary number by an analog-to-digital converter (ADC) device. The divisions of the parts of the input signal are referred to as the *resolution*. The resolution of the module will determine how many bits of data are used by the ADC.

Common analog input devices are:

- Potentiometers
- Pressure transducers
- Flow transducers
- Temperature transducers

Analog Outputs. Analog outputs are used to control field devices that respond to continuous voltage or current signals. The analog output module converts the binary output value to an analog signal understood by the field devices. This is accomplished by a digital-to-analog converter (DAC), which transforms the digital value to an analog signal. Generally, output signals are expressed as percent to correspond to the operation of final control devices (valves, dampers), and the magnitude is proportional to the maximum voltage or current value. A transducer or converter receives the current or voltage signal from the output module to adjust the final control device. Since most final control devices (valves and dampers) are pneumatically operated, the most common transducer is the current to pneumatic.

Common Analog Output Signal Forms are:

- 4 to 20 mA
- 0 to +1 V dc
- 0 to +5 V dc
- 0 to +10 V dc
- 1 to +5 V dc
- 5 V dc + 10 V dc

Common analog output devices are:

- Meters
- Valves
- Chart recorders
- Motor drives

d. Power Supply. The power supply is an essential part of the operational computer, and without its proper functioning, the electronic equipment will not operate. Microprocessor components require low dc voltage to operate and are sensitive to both voltage fluctuations and power surges. The power supply provides a stable source of electricity at the proper voltages. The power supply usually provides power for both the operational computer components (at 1 and 5 V dc) and for the I/O devices (at 24 or 125 V dc or ac).

22.5.4 Programming

Programming is providing coded instructions for the operational computer to perform a task. The program defines where the inputs come from, what is done to them, and where the outputs go. There are a multitude of programming methods usually determined by the manufacturer and the particular piece of hardware. Two of the common methods are described below.

a. Ladder Logic. Ladder is a programming method that represents a system of relays, switches, lamps, and so on. It also shows the sequence of events as the program is executed. Each element is represented symbolically on a diagram that resembles a ladder. Each step of the program is known as a rung. Ladder drawings are similar to electrical wiring drawings. The functions shown on a ladder logic diagram are initiated by inputs (on the left side) and generate specific control output responses (on the right side).

Examples of ladder logic functions are:

- Timers
- Integer move
- Jump
- Compare
- Add
- Counters
- Master control relay
- Integer
- Drums
- Subtract

b. Function Code Programming. Function code programming is a method that logically represents the operation of a control system. Most control systems are continuous (i.e., there is no set starting or ending point, as compared to ladder programming). It is easier to think of these continuous systems as loops. Function code programming shows the relationship between the components of the control loop: the inputs, control actions, and the outputs. Each function code is represented symbolically on a diagram that resembles a instrument control drawing.

Examples of function codes are:

- PID (proportional integral derivative)
- Multiply
- Square root
- Hand/automatic station
- High/low select
- Sum
- Divide
- Function generator
- And (Boolean logic)

22.5.5 Safety

Safety should be a major consideration in the control of any piece of equipment. One must anticipate all the possible control failures and guard against them. This also includes the fail positions of valves and dampers, interlocks, and permissive steps. The I/O simulator is a very useful tool to prevent accidents involved with new installations, both for checking out the system and for training personnel. Ensure safety during initial startup by closing block valves if possible, wiring centrifugals with empty mixers or using water instead of liquor.

There may also be a requirement to adhere to certain codes, such as National Fire Protection Association (NFPA) and National Electric Code (NEC), imposed by a governmental agency or insurance underwriter. These requirements often address functional standards for interlocks and start sequences as well as listing acceptable hardware. If colored graphics are used to display critical information, be aware that some operators may be color blind. Also, audible alarms may not be discernible in a high-noise area, so it may be necessary to supplement them with flashing lights.

22.5.6 Training

Training is essential to maximize the benefits of an operational computer system. This includes the design engineers, maintenance technicians, and the operators. Training may be conducted either off-site at the vendor's facility or on-site, preferably with the newly installed equipment. For generic training or very popular systems, instruction may be offered at technical institutions and third parties. The simulator used for system checkout makes an excellent training tool.

22.5.7 Security

While operational computers and their associated networks enhance the dissemination of information and allow for timely control adjustments to respond to changing operating conditions, consideration must be given to who has access to the information and who has rights to make changes. These options are typically available with the control software, but the user must consider how they are to be used. Backups and documentation are also important factors of system security. A major advantage of operational computers is that the programming or configuration can be changed. Therefore, it is equally important to maintain a system to document changes. Although systems are very reliable, consideration must be given to the possibility of a crash or failure. Two important factors are where the backup software and documentation will be stored and how the person using it will know that it contains the latest changes.

22.5.8 Benefits

There are considerable advantages to using operational computers. A short list of some of the most obvious advantages would read:

- Comparatively lower installation cost
- Unlimited applications throughout the process
- Versatility for adapting to change in control schemes
- Interface capability between operational computers and other computers
- Future expandability of individual operational computers and the entire system

22.5.9 Planning

If you don't need a plan, any path will do! Every successful project involving operational computers must begin with a thorough plan. Planning is simply starting where you are, deciding where you want to go, and laying out the steps to get there. Planing allows you to separate the overall project into small, systematic steps. It involves buying equipment in the short term that will be useful in the long term. Sometimes the process must be reengineered to support changes in equipment; planning allows those changes to be made with the future goals considered.

a. Planning Guidance. Separate the short- and long-term goals for the entire control project. Short-term goals may be operational computer installations at individual sites. They may be isolated as stand-alone applications for immediate installation and fast justification to management of their usefulness. This also allows operators to develop experience with the new equipment. Short-term projects are useful to develop experience in the engineering/design/maintenance team necessary for larger projects. Long-term goals may be an entire network, complete with supervisory control and data acquisition (SCADA) capabilities and generating management reports. Most systems are gradually built upon as the necessity for changes and availability of funds dictate. The familiarity that you develop with operational computers and the peripheral equipment will determine how smoothly and timely your installation progresses.

Once the need for a system has been justified, consider which manufacturer is best suited for your process. The manufacturer's sales engineers are helpful in determining the best application of their equipment to your process. Magazines, books, technical articles, and vendor mail-outs are good sources of information. Trade shows are an excellent way to see most of the current equipment as well as to discuss your needs with knowledgeable manufacturers' representatives. You may also want to consult in-house personnel about vendors used in related equipment or in other departments.

b. Justification. The remote interface capability of most operational computers increases the efficiency of supervisor's control of the process. With reliable controls and current information, the ability to make correct decisions regarding process changes from remote locations enhances the supervisor's management of the system. Another objective may be job displacement where cost reductions justify the project. The increased productivity realized by using operational computers are attributed to many things, including (1) maximum equipment efficiency, (2) reduced down time, and (3) continuous process control adjustments.

The application of an operational computer system to any process is a major undertaking and should be regarded as such. Decisions will have a direct impact on the future capabilities of the complete control system. Currently, many sugar companies are involved in some type of control system upgrading or are studying their options prior to undertaking such a project. Some considerations that help in choosing a system are as follows:

1. Define the individual areas of control and their position in a network.
2. Consider operations that you may eventually consider for possible data-gathering interfaces or remote controls in the future.
3. Develop a list of all specifications, with as much detail as possible, concerning the control required for separate applications and the information that will be extracted from the operational computer system to be exported to other computers for display or data storage.
4. Estimate the amount of I/O necessary for each application.

Separate processes into those suitable for the use of PLCs, DCSs, or computers, and those that are not. If it does not apply, it will be hard to justify. (Don't dock a rowboat with a radar!). We have a responsibility to provide ample justification to install any new control system. This justification may come in the form of increased production, reduction in the labor force, enhanced control, or replacement of obsolete and hard-to-maintain equipment. Lay out a diagram of the complete control scheme, placing equipment in its physical relationship with the process. Form a rough list of all hardware that may be necessary for your computer control scheme.

22.5.10 Initial Cost

The initial cost of an operational computer installation is usually high due to the many factors involved with any new installation. To program an operational computer, you must have a programming unit, preferably a video unit, and the appropriate software. A compatible printer or plotter is necessary to produce a hard copy of the program to facilitate debugging, troubleshooting, and ultimately, the system documentation. Simulators are necessary to check out the program safely, ensuring correct I/O sequencing of the program logic prior to actual equipment application. Since most people are not practiced programmers by trade, training for the particular manufacture's equipment will be necessary. Additional initial costs incurred may be the use of contractors to program, install, debug, and document the new installation. The task of cost justification may not be easy to measure in the case of an operational computer installation. Sometimes the indirect costs of an operational computer installation add up to more than the direct costs. When considering the cost justification for the project, always keep in mind the overall effect that the conversion to operational computers may have on the entire process.

22.6 CHOOSING AN OPERATIONAL COMPUTER SYSTEM OR CONTROL NETWORK

When dealing with vendors, "Let the buyer beware." This statement relates to the fact that you are buying into something relatively new for you. The vendors know that once committed to a system, you will have to stick with it or lose a portion of your initial investment. It is wise not to make any commitments for equipment purchases without first considering the long-term relationship that will be established with the vendor. Some key points to consider in selecting a vendor are:

- What stage is the equipment in? (Beta testing or near the end of its cycle)
- What is the commitment to support the purchased equipment?
- What is the commitment to upgrades? (Forward compatibility)
- What is the compatibility with other equipment?

Other general considerations are the existing operational computers or computer equipment within your process and whether this equipment may be expanded. Determine the interface capabilities the new equipment may have with other manufacturers' equipment. Consider "open architecture" when selecting equipment and software; this will help to avoid being locked into one vendor that cannot meet all your requirements.

Be familiar with the terminology associated with operational computers and computers. If a term comes up during a discussion that you are not familiar with, ask for an explanation. Do not let the vendor use their experience with certain terms to intimidate you.

Give each vendor all the information required for the overall control scheme that may eventually be implemented. You might consider asking for bids to prepare the specifications for the overall system before soliciting bids on the project. Let the vendor describe how their equipment could satisfy your control scheme and exactly what equipment would be necessary.

Ask about systems already in place and if possible, visit them to view firsthand what is being controlled. Talk to the people using the equipment.

Vendors will often speak of the rapid advance of technology in this equipment's field. There may be promises made involving these advances that should be made on paper as part of the purchasing agreement. Reduce the number of vendors to two or three and discuss their individual pro's and con's with upper management. After reviewing the literature available on each vendor's equipment, try forming a plan for their equipment's application to your process.

22.6.1 System Considerations

After reducing the field of vendors considered to handle your application, here are a few more considerations, other than operating hardware, before the purchase of a system. Backup hardware is critical, especially in an interfaced control scheme where there is a dependence between operational computers. If the vendor is not stocking the parts to support your hardware, there is a need for in-house stocking at an additional expense to the user.

Interfaces between operational computers and computer equipment will require cabling. The cost for this installation will usually be a one-time expense. It is very critical that this installation be done properly. The reliability of the entire communication network depends on cable integrity. Consider nonconductive (i.e., fiber optic) cables for the system backbone, especially the portion running through the plant where inductive fields are encountered. For an equipment conversion project, the existing wiring to solenoids, starters, and so on, may be reused if it is in good condition. New applications require all new wiring, and the integrity of these wires will directly affect both the amounts of startup time required and the overall reliability of the system.

Documentation for the operating software must be provided with the hardware system purchased. Any additional software required for systems control will require additional documentation supplied by the programmer. The engineering work required for each application will determine how quickly an installation goes on line. Knowing the details of an operation will diminish the amount of debugging necessary when starting up a system.

When a complete system is purchased, the supplier may furnish personnel to assist in the startup. This may not be the case if small pieces of hardware are purchased individually unless the customer requests assistance, usually for a fee. Most companies will provide access to a support group of people who are familiar with your equipment. This group will serve as your contact for technical assistance for most operational problems encountered. Some companies also have a 24-h hot line or modem hookup to lend assistance at any time. Training courses on almost all equipment sold are usually offered at either the manufacturer's facility or on site. Some of the courses are included in the price of the initial hardware purchase, and some are offered for a fee.

22.6.2 Startup

After the hardware has been purchased and installed properly in its prescribed environment, it is time to startup the system. Upon arrival, all equipment should be checked visually for shipping damage and completeness of the order. The cable integrity should be checked thoroughly to ensure that any startup problems encountered will not be compounded by faulty cabling. With the manufacturer's representative and the system manual, the system is powered up. Usually, it is best to correct each problem as it is encountered. Ensure that adequate plant personnel are available to make corrections outside the manufacturer's responsibility.

CHAPTER 23

Automation of a Sugar Refinery*

INTRODUCTION

Automatic control was introduced in cane sugar refining in the early 1950s. At that time, individual equipment was operated through closed control loops from a cubic control panel installed near the machinery using large-scale pneumatic instruments.

With the development and introduction of electric instruments in the 1960s, measuring accuracy, response, and controllability were greatly improved. Moreover, the miniaturization of instruments made it possible to quadruple the number of controllers and instruments accommodated in a single panel.

The introduction of electronic instruments in the late 1960s further improved response time and control accuracy. At the same time, the faceplate of instruments became smaller, so that a panel could accommodate roughly eight times the conventional number of instruments in the early 1960s, while the panel depth increased fivefold.

The input and output signal of electronic instruments was standardized worldwide at 4 to 20 mA or 1 to 5 V, leading to rapid adoption of automatic control systems in many fields. From that time on, efforts were directed at concentrating all control functions in the control room and it eventually became possible to operate all processes from panels concentrated in one or two central control rooms (CCRs). In our refinery, the number of operators could be greatly reduced, to 20 per shift.

In the late 1990s, these electronic instruments continue to play an important role in all fields of industry and have proven highly valuable. The first recommendation to plants planning to automate, therefore, is to start by adopting electronic instruments.

Furthermore, the progress of electronics enabled implementation of an interface with high-level computers and changing the set point of the analog controller by command from the process computer, thereby making it possible to control the flow rate, level, pH, temperature, and so on, according to a target value through control of manipulators using the output signals of the controller. As a result, factory automation (FA) was also adopted in the sugar refining process.

*By Naotsugu Mera.

With the development of small computers and rapid advances in both hardware and software in the mid-1980s, the distributed control system (DCS) based on a digital rather than an analog control system became the mainstream.

In the 10 years since 1985, most sugar refiners in Japan adopted DCS, implemented all control systems in a single CCR, and greatly reduced personnel expenses by running all processes with four to six operators per shift operating five to seven cathode ray tubes (CRTs).

An older analog control system with 800 to 1200 input signal points and 400 to 600 control loops can be replaced by five to nine central processing units (CPUs) and five to seven CRT operator desks in a DCS, although this depends on the size of the plant. One drawback with DCS that needs improvement is that as long as conventional detectors and manipulators are used, the accuracy of these instruments determines the overall quality of control. Also, all process data are printed out directly on logging typewriters linked to computers.

Many sugar refineries all over the world are still using analog control systems, and some countries are even now in the process of introducing analog systems. The best way of automating refineries will be discussed later. At this point, the shift from an analog system to DCS will be taken up. The life of analog instruments generally is about 15 years. It is desirable to make the change to DCS when the time for replacing these instruments has come. By that time, a factory should have established an optimum production technology that assures stable operation of the entire process and designed an automation concept based on this technology.

DCS is not cheap by any means but is believed to offer considerable advantages in that it greatly reduces labor costs. Personnel made superfluous by the introduction of DCS can be shifted to R&D or other departments, and all in all, investments in DCS can be recovered in a short period of time, also considering mounting repair costs arising from continuous use of superannuated analog instruments.

A typical system configuration is shown in Figure 23.1, and a process flow sheet is shown in Figure 23.2.

23.1 INTRODUCING A NEW CONTROL SYSTEM

The introduction of a control system has the following two objectives: (1) to ensure production quantity and quality by stable operation of the entire process, while fully utilizing the production capacity of a factory, and (2) to reduce production costs by cutting back labor expenses through introduction of automatic equipment. In either case, the conditions of the formula given below must be satisfied if a cost advantage is to be derived from introducing an automatic control system.

$$\frac{\log[R/(R - i \times I]}{\log(1 + i)} \leq \text{payback period}$$

where R is the overall cost reduction, i the interest rate, and I the total investment.

The depreciation period for sugar manufacturing equipment in Japan is 13 years, but for an investment to be profitable, the number of years obtained with the formula above should generally be less than 4 years. Note that R should not only include the reduction in labor costs, but also the amounts representing improvement in quality and factory recovery yield, energy savings, and so on.

SYSTEM CONFIGURATION

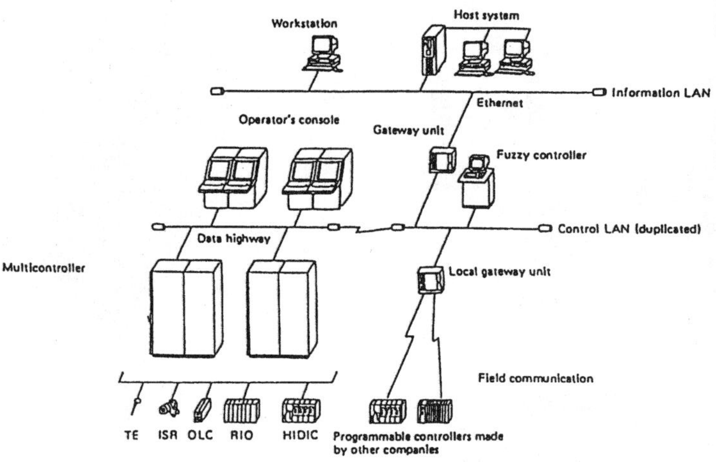

RIO	Remote I/O
HIDIC	Hitachi programmable controller HIDIC
ISR	Intelligent transmitter
OLC	One-loop controller
TE	Temperature detection end

more open CIM system.

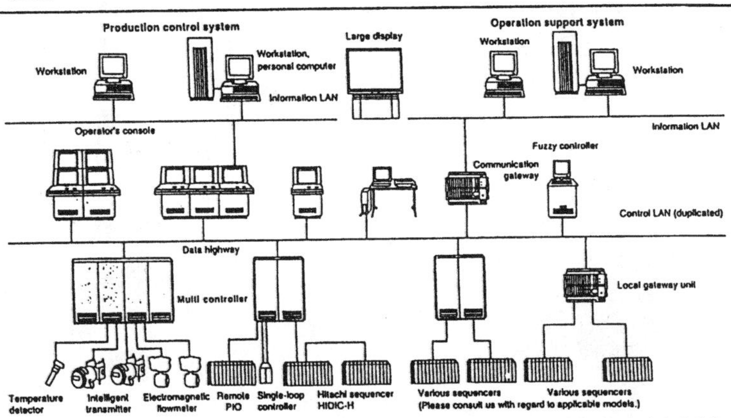

Operator's console (POC): Monitors plant operations, and configures/manages operation system.
Multi controller (MLC): Presides over plant operations. Various intelligent sensors, sequencers or like instruments can be connected by the intelligent PI/O method. Plant information is collectively manageable.
Fuzzy controller: Conducts fuzzy inference in real time.
Local gateway unit: Serves as an interface with the sequencers of different brands.

Gateway unit: Provides interfacing for linking with the information LAN. Besides Ethernet* (IEEE 802.3), computers made by different companies can be interfaced.
Workstation: Diverse workstations including HITAC 2050/3050 series are connectable.
Host computer: Process computer HIDIC, general-purpose computer HITAC and other computers are applicable.
Large display: Enlarges screen display of operator's console, workstation, industrial TV set, etc.

* Ethernet is a registered trademark of Xerox Corp.

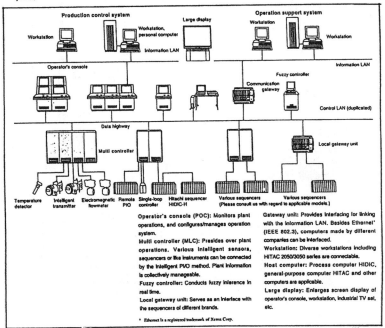

FIGURE 23.1 System configuration.

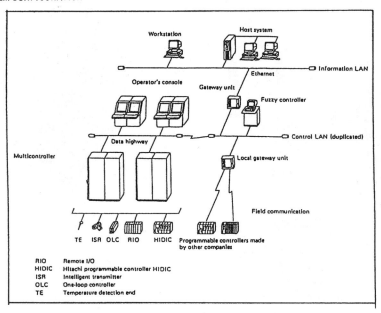

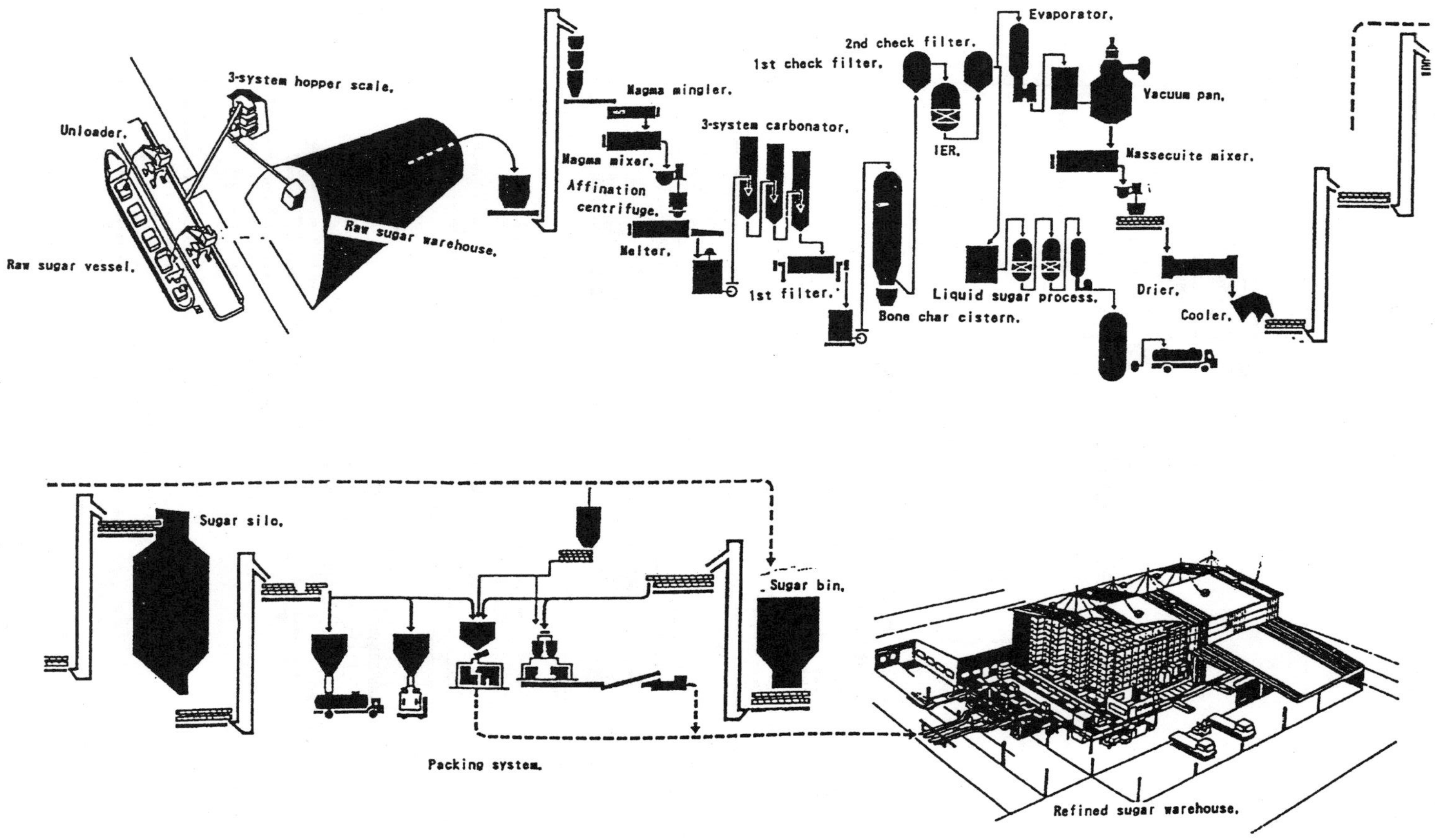

FIGURE 23.2 Process flow sheet.

Once a company has decided to introduce an automation system based on the considerations above, it will be necessary to define a control concept based on corporate management policies as well as the core points of the control system.

In the absence of data for determining the control parameters, it will be necessary to acquire data on process characteristics, time constants, and so on. through actual manual operation, express the characteristics of each factory in numerical terms, and reflect these in the software.

23.2 CONTROL CONCEPTS

The control concept and its aims are at least as important as the control system (hardware and software) selected. In theory, the entire process of a sugar refinery can be run by feedforward control by converting the daily melting weight according to the production plan into weight per hour and forcing it into the process every hour. However, this is not economical because additional intermediate tanks, and other equipment must be installed and additional facility capacity provided for postprocessing steps. On the other hand, if the feedback control technique is employed for all processes, there is a danger of not reaching the production target.

For that reason, combined control taking advantage of both methods is selected as the basic control concept. Closed control loops are made feedback loops in principle, but for the processes, it is desirable to use feedback control from affination to the evaporator and feedforward control from the vacuum pan onward. Accordingly, the boiling schedule is precisely planned for the massecuite weight to attain the planned production value, and to ensure that the necessary quantity of evaporated or fine liquor is obtained, preprocessing to the raw sugar stage is run by feedback control. In addition, using feedforward control for the processes from the massecuite mixer and centrifuge to the sugar silos will ensure that the production plan is met.

In this way, it is possible to design machinery and equipment with capacities to match the daily melting weight. From the operational aspect, the tank capacity can be reduced because the flow rate of each process can be roughly fixed. At the same time, load variations affecting utilities such as steam, electric power, and water can be reduced, resulting in more efficient use of energy. However, it is necessary to use somewhat larger evaporated or fine liquor tanks which act as an interface between the feedforward and feedback control in order to stabilize process variations.

With the signal level of instruments having been standardaized internationally, there are no constraints with regard to the choice of instruments, but to keep the installation uniform, maintain interchangeability of parts and make maintenance easier, it is desirable to use products of the same manufacturer if at all possible. This point should be kept in mind, especially if a future computer linkup is planned.

23.3 PROCESS CONTROL SYSTEMS

The control concept of each process is described below based on the process flow sheet introduced in Figure 23.1. The descriptions are designed to be applicable both to analog systems and computer systems.

23.3.1 Raw Sugar Handling

a. Raw Sugar Unloading. Imported bulk raw sugar is unloaded directly from a cargo vessel at a private wharf by gantry crane. Raw sugar is transported on an inclined belt conveyor to the three-stage hopper scale, where it is weighed automatically and the weight recorded. The weighed value determined here represents the volume of the business transaction and the reference weight for import customs tariff. The belt conveyors are matched to the unloading capacity and are interlock controlled. They are started sequentially last to first. When stopping, the first one is stopped and the remainder after the necessary conveying time. The three-stage hopper scale is also included in the interlock system.

b. Three-Stage Hopper Scale. The weighing hopper converts the weight into electrical signals by load cell, prints the gross weight and tare weight on the register, and calculates and prints out the net weight automatically. The scale must be legally certified because it determines the volume of the transaction. A factory that wants to unload 4000 tonnes of raw sugar daily at 7.5 tonnes per batch will need two hopper scale systems. The two systems are operated alternately under fully automatic control by a sequencer. Accuracy is 1/1000.

23.3.2 Raw Sugar Warehousing

a. Raw Sugar Receiving. The raw sugar warehouse is a semicylindrical concrete building suited for the repose angle of raw sugar. Raw sugar is transported to the top on an inclined conveyor and dropped from a traveling reversible conveyor (shuttle conveyor) moving back and forth on the ceiling to form a mountain in a desired place on the floor.

Filterability and decolorizability of raw sugar vary by country of origin. For this reason, mountains are created by country to make controlled mixture possible. In this way, a uniform raw sugar quality is obtained before the refining process. Raw sugar from large producing countries such as Australia is classified by production region. If raw sugar piles cannot be made separately due to a lack of space, sugars of different characteristics are placed next to each other to equalize these characteristics by overlapping the mountains at the base.

The position of these piles can be determined by remote controlling the shuttle conveyor. The shuttle conveyor is the first to be started for the unloading processes such as transport and weighing, and the last to be stopped. This sequence can be controlled sufficiently through independent local control and does not require linkage to the process computer. An importing country should have a warehousing capacity equivalent to a 30- to 50-day supply.

b. Raw Sugar Dewarehousing. While the raw sugar quality in refineries using domestic raw sugar is nearly uniform, refineries using imports have to deal with many different types of raw sugar and therefore must adapt their production accordingly. That is, it is necessary to mix two or three types of raw sugar before sending the mixture to the refinery process in order to reduce the need for chemicals used to obtain uniform filterability and decolorizability, which drive up production costs.

In the center of the warehouse floor are 10 to 20 outlet gate lines. The gates under the sugar mountains separated by country of origin open for the desired mixing ratio of raw sugar. The mixing ratio is monitored by an Industrial Camera and Television Set (ITV) installed in the CCR, and opening and closing of the gates can be freely adjusted by remote control.

The mixing ratio should be determined according to ash content and reducing sugar content rather than color value. For this, it is important to analyze samples from each cargo vessel and to confirm filterability and decolorizability experimentally. Mixing ratio control is very important for process control since it has a major effect on purification cost and recovery yield, and therefore on production cost.

c. Weighing Raw Sugar at the Process Inlet. Raw sugar is weighed at the process inlet on a three-stage hopper scale, which is fully automated, like the hopper scale used for import unloading. Since it is known that it takes about 1 minute to weigh one batch (hopper loading, clump, weighing, printout, unloading), equipment capacity can be determined by reverse calculation. The weighed value is converted to electric signals, fed directly to the host computer for addition, and the gross, tare, and net weights are stored in the register one batch at a time.

The added-up melted weight can be called up at a fixed time every day or at random for output on the CRT or printer to check the production progress. The three-stage hopper scales are checked monthly for accuracy and undergo official inspection and certification twice a year.

23.3.3 Affination

a. Magma Mingling and Heating. The raw sugar weighed on the three-stage hopper scale is sent through the sugar bin to the belt weigher for continuous weighing. The amount of green syrup added to the mingler is determined based on this weight. This amount greatly influences the affination recovery or factory yield.

The percentage of green syrup added is controlled automatically according to the magma mobility specified. It is determined considering raw sugar quality and green syrup purity.

The set value is determined by the weight. To make 1 m^3 of magma with a Brix of 93, Higashi-Nihon Sugar add 0.52 tonne of green syrup to 1 tonne of raw sugar, based on a green syrup with a Brix of 80 to 82 and a purity of 70 to 72. The magma is heated indirectly to 55°C in the downstream magma mixer to increase affination efficiency. To agitate and heat it, its temperature is maintained at a fixed level by controlling the amount of hot water flowing through the spiral screw tube at uniform temperature (90°C). The affination centrifuge capacity and affination recovery are at their optimum at 55°C.

The magma mixer is also level controlled. It controls the amount of raw sugar fed to the mingler so that magma is always above the heating coil and coloration due to scorching is prevented. Similarly, the distributor for the affination centrifuge is also temperature and level controlled. The level is controlled by adjusting the degree of opening of the magma mixer outlet gate.

b. Affination Centrifuge. One affination centrifuge is designed to hold 1 ton of magma and is driven by a 1500-rpm dc motor via a thyristor that converts alternating to direct current. All centrifuges, including the affination centrifuge, are run in groups by pitch time control to maintain the current near a fixed level (Fig. 23.3). The groups are operated by DCS, and fully automatic and unmanned operation of the centrifuges is made possible by using sequence control. Controlled variables are rpm, charge volume, spray water, and unloading. The amount of spray water is controlled by a digital timer based on the correlation between the color value of affinated sugar, syrup purity, and affination recovery.

23.3.4 Melting and Raw Liquor Brix Control

The melting system consists of a screw conveyor located below the affination centrifuge, a premelter, a cush-cush strainer, and a main melter. The washed sugar weight of one affi-

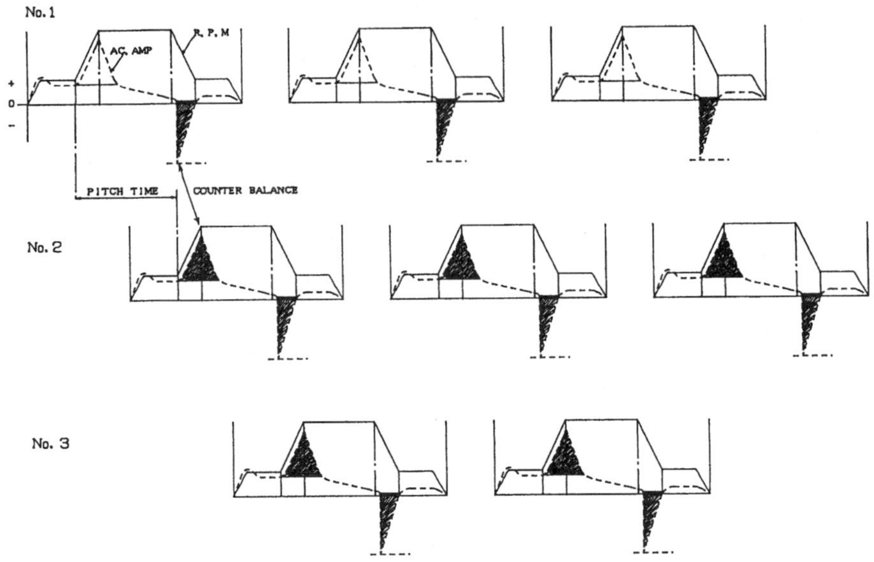

FIGURE 23.3 Pitch time control.

nation centrifuge basket is nearly uniform, so a fixed amount of sweetwater is added to make the Brix of the solution 67° on receipt of the unloading end signal from the sequencer. Dissolution is continued by agitation in the premelter. In the main melter, the Brix of the raw liquor is controlled at precisely 65 by adding 60°C hot water. Since remelt liquor is returned to the main melter from the low-grade recovery line, Brix and level control are required.

The Brix number can be measured to an accuracy of 0.5 with a differential pressure (DP) cell. The main melter level is controlled by feeding the Brix number back to the pitch time controller and restricting the raw sugar feed.

23.3.5 Raw Liquor Color Control

Refined granulated sugar in Japan has a color value (CV) equivalent to ICUMSA CV 4 to 7. To meet this value, the raw liquor CV is maintained at 1500 and activated granular carbon is added as an auxiliary decolorizer and to prevent flocking. This process takes place automatically in a sequence determined jointly by the interrupt timer and charge timer.

23.3.6 Carbonation

a. Lime Milk Treatment. $Ca(OH)_2$ consumption generally is 8 to 10 kg/tonne of raw sugar. Higashi-Nihon Sugar's processing capacity is 1050 tonnes of raw sugar per day. Accordingly, 10 tonnes of $Ca(OH)_2$ are sent to the 55-m^3 lime milk adjusting tank every day. There the lime milk is dissolved under flow control in 40 m^3 of hot water and stored in the lime milk storage tank (85 m^3), thereby always ensuring the necessary supply of $Ca(OH)_2$ at a uniform concentration (Baumé, Brix).

Both the adjusting and storage tanks are equipped with agitators to prevent the lime from settling. The carbonator is always supplied at a constant rate with more than the necessary amount of lime milk, and the excess is constantly recycled to the storage tank to prevent clogging of pipes.

b. Lime Milk Mixing. Because raw liquor and lime milk are kept at a known concentration, the amount of lime milk to be added to raw liquor can be determined by taking the raw liquor flow rate and multiplying it with a coefficient. The lime milk mixing ratio is calculated by computer based on filterability and decolorization in the first filter to control the amount of lime milk into the in-line mixer.

c. Carbonator pH Control. The same control principle can be applied to a two- and three-tower system. Boiler flue gas scrubbed in a gas scrubber is used as CO_2 gas. Its CO_2 content is only 10%, because the boiler is set to low-oxygen combustion to save energy. The CO_2 gas volume to the carbonation process is controlled to a specified value based on the pH value measured in each tower. The pH of each tower is determined by the filterability of the first filter, that is, by the $CaCO_3$ formation potential. The pH value in the first and second towers is fixed, but the third tower requires a slight adjustment. The excess CO_2 gas is recycled to the suction side of the CO_2 gas pump to reduce the power consumption of the pump.

23.3.7 First Filter

a. Distribution of Carbonated Liquor to First Filter. The first filter for removal of $CaCO_3$ generated in the carbonator normally consists of several units. It is therefore necessary to distribute the carbonated liquor in equal amounts to the filter units to prevent them from entering regeneration at the same time. To this end, the total carbonated liquor volume is measured, divided by the number of filter units active in the first filter by computer, and fed automatically at a uniform rate to each filter unit. The internal pressure of the filter increases over time, so the filter inlet valve opening is adjusted by computer at all times to obtain a constant flow rate. The point when the filter enters regeneration is determined by time under good filtering conditions and by internal pressure difference under poor conditions.

b. First-Filter Control. With constant flow control being performed, regeneration of filters takes place at nearly fixed intervals. One cycle is sequentially controlled on a time basis. That is, precoating, filtering, and regeneration (sluicing → lime cake removal by backpressure → cake ejection → backwashing → heating) can easily be sequentially controlled by DCS. Higashi-Nihon Sugar aims at a decolorization rate of 60 to 65% in the carbonator. Decolorization can be improved by increasing $Ca(OH)_2$ consumption, but there is no point in raising it above 10 kg/tonne of raw sugar.

23.4 PURIFICATION

Generally speaking, the following three systems are available for adsorption purification: (1) a pulse-bed adsorber using activated granular carbon, (2) a bone char cistern, or (3) a blow-up system using activated powder carbon. All three systems can be operated automatically by time-based sequential control, although methods differ among equipment manufacturers. Described here is the bone char cistern, which requires large-scale equipment and is rather complex in operation.

23.4.1 Bone Char Cistern

a. Purification. The bone char system normally involves 15 to 30 towers because it requires a holdup volume of bone char that is nearly equivalent to one day's melted weight. Higashi-Nihon Sugar uses a 16-tower cistern, of which 10 are normally used in the purification process under sequential control, and the others are for regeneration or at rest.

The flow rate is controlled by the DCS, as in the case of the first filter, to distribute the liquor equally to the towers in use. The end of the purification cycle is determined by a timer. The time set on the timer is that at which the brown liquor value reaches 0.5 (in the Stammer unit), which is usually 90 to 120 h.

b. Regeneration. All processes, from sweetening-off 1 and 2, washing 1 and 2, air forced out, char discharge, to settling, are operated sequentially by DCS.

The liquor at the end of sweetening-off 1 is recovered at 35° Brix as purified liquor, and the rest, down to 0° Brix, as sweetwater at the end of the sweetening-off 2 process.

For the purpose above, Higashi-Nihon Sugar has developed a high-precision fine-volume sucrose detector (patented) called the zero Brix detector. Char is settled with water to prevent bridging inside the tower. Dry char is weighed on a high-temperature belt weigher and supplied uniformly to each tower. Char weight must be kept constant to ensure a uniform flow to each tower.

High-specific-gravity bone char with decreased color matter and ash removability is discharged automatically and continuously to maintain char quality. The regenerated bone char is measured continuously for pH, mean aperture, sand content, and specific gravity to determine the makeup ratio of new char.

c. Kiln and Furnace. The maximum temperature inside the kiln is controlled at 550°C for bone char regeneration. The highest temperature the char reaches (hearth) is used to control the fuel supply to the furnace. The amount of air required for combustion is determined by proportional control based on the fuel–air ratio. To prevent overburning of char and to save energy, the temperature is controlled at 550°C and the percentage of O_2 kept as low as possible, normally at about 1.0 to 1.5% by feedback of O_2 from the flue gas.

d. First-Check Filter. As the bone char cistern liquor begins to flow, some powder char particles are mixed in with the brown liquor. To remove these particles and check char leakage, ceramic filters are used. The filter life cycle is 24 to 48 h, so it is usually not necessary to equalize the inlet flow. One filter cycle consists of preheating, precoating (kieselguhr), filtration, remaining liquor drain, sweetening-off, cake removal, and backwashing. The endpoint of filtration is determined by pressure, but for the other steps, sequential time-base control is sufficient.

23.5 ION-EXCHANGE RESIN FOR DECOLORIZATION

23.5.1 Ion-Exchange Resin

Recent ion-exchange resin (IER) systems consist of three towers, two of which are normally in the liquor decolorization cycle for 16 h and one in regeneration for 8 h based on cyclic sequential operation by DCS. One operation cycle, which includes preheating, sweetening-on 1 and 2, decolorization, sweetening-off 1 and 2, washing 1, regeneration 1 (spent NaCl), 2 (new NaCl), and washing 2, is controlled sequentially and cyclically by DCS.

The brown liquor treatment volume in this process is fixed at a value (= resin volume × 40 − 50) at which the average decolorized liquor color value (CV) in one cycle is 0.1

St, CV. The color value is measured at all times by an in-line spectrophotometer. Treatment is continued for as long as the color value is maintained at CV 0.1, to reduce regeneration cost.

The resin life is about 600 cycles, but it is made to undergo heavy regeneration every 50 cycles to maintain its decolorizing ability. IER decolorized liquor is called *fine liquor*. The color value is very important because it greatly influences factory recovery. This is especially true for refineries producing liquid sugar from fine liquor.

23.5.2 Second-Check Filter

The second-check filter is used to catch the crushed resin in the fine liquor. Like the first-check filter, it is sequence-controlled by DCS, but its cycle time is longer, normally 72 to 96 h.

23.6 EVAPORATOR

The liquor is concentrated in the evaporator to Brix 72° to save energy and to maintain the product quality by stabilizing the Brix of the evaporated liquor. A thin-film falling-type multieffect evaporator is most widely used, but the calandria type can be used as well, since the control method is basically the same. Whichever is used, the purpose of the evaporator is to keep the outlet Brix constant. For this, a coefficient representing the evaporator inlet fine liquor (Brix $\times$ flow $= K$) is calculated and supplied as input signal to the steam flow controller for control of the steam flow so as to obtain an evaporated liquor with a Brix of $72 \pm 1°$. This control method can also be applied to equipment provided with MVR.

The evaporator is self-balancing. No control of temperature, pressure, or other parameter is required, except for final-stage vacuum control which is done by controlling the injection water volume. All these variables are controlled by DCS.

23.7 VACUUM PAN BOILING

23.7.1 High-Grade Massecuite Boiling

It is recommended to use a program-controlled intermittent boiling process. In addition, several programs should be available for each type of sugar to be able to cope with changes in liquor quality.

A typical program used at Higashi-Nihon Sugar is shown in Figure 23.4. The control steps, including air exhaust (to create vacuum), liquor induction, evaporation, supersaturation control, slurry seeding, crystal growing 1 and 2, mobility control, massecuite tighten, boiling end, massecuite discharge, and washing can be performed by DCS or a combination of analog controller and programmer.

The supersaturation ratio (SSR) is calculated from the vapor and liquor temperatures. SSR is used only for determining the seeding point, which is at SSR = 1.1 to 1.2. After seeding, boiling proceeds controlled by the level and mobility program. The slurry seeding system was developed by Higashi-Nihon Sugar (patented) and is currently being used with great success in combination with DCS. Mobility is detected by a pilot mixer and controlled intermittently (liquor feed on–off) for crystal growing.

In low-mobility massecuite boiling, continuous control is possible, but intermittent control is recommended to obtain a uniform grain size and to minimize the coefficient of variation in crystal growing. With an analog programmer, the number of intermittent control steps is limited, but with DCS, any number of intermittent steps can be accommodated.

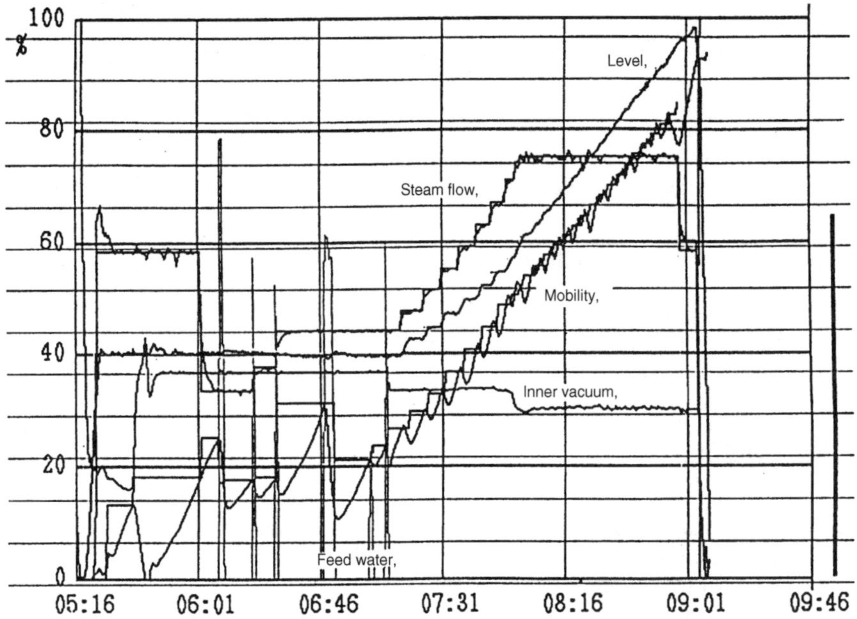

FIGURE 23.4 Typical vacuum pan boiling program.

Vacuum is regulated by controlling water injection to the barometric condenser and air induction into the save-all. To ensure high-quality crystals, it is desirable to install independent vacuum pump and condenser units. The desired massecuite purity in mixed boiling can be obtained by presetting the induction sequence and volume of feed liquor.

23.7.2 Low-Grade Massecuite Boiling

Because the magma seeding method is used for low-grade and final massecuite boiling, all process steps can be controlled by a mobility and level program. Mobility can be measured by changes in the stirrer motor current. Other processes are the same as for high-grade massecuite boiling, but better results can be obtained if the number of intermittent control steps is increased and the mobility range expanded. If the massecuite is too sticky/viscous in final boiling, surfactants for food use can be added.

23.7.3 Massecuite Mixer

The massecuite mixer volume is made slightly larger than the vacuum pan volume. Generally, the centrifugal capacity is made larger than the boiling capacity, so that when one centrifuge breaks down, separation can be continued with the remaining operative machines. Consequently, the massecuite mixer does not require level control but is equipped with a level indicator and high–low level annunciators. Discharging of the vacuum pan is triggered by a low-level signal. The massecuite mixer is provided with a hot-water pipe for dilution to prevent tightening (hardening) of massecuite.

23.7.4 Distributing Mixer (Submixer)

The submixer is used to raise the massecuite temperature to the optimum level immediately before centrifuging and to charge massecuite to the centrifuge at a constant rate. The heated massecuite is temperature controlled by 60°C hot water to prevent crystal dissolution. The massecuite temperature and volume charged into the centrifuge are kept at a stable level by level controlling the charging volume from the massecuite mixer to the distributing mixer.

23.7.5 Final Massecuite Crystallizer

Besides the boiling vacuum pan, the final massecuite process is equipped with a crystallizer to promote cohesion of sucrose to the sugar crystals and thereby increase the crystal volume. One crystallizer cycle is 72 h. The cycle comprises the sequence steps of massecuite (1) cooling (8 h), (2) constant-temperature maintenance = crystal growing (48 h), and (3) heating up (16 h), all of which can be controlled by a programmed sequence.

Only this final massecuite usually is double purged. The final sugar separated in the continuous centrifuge in the first stage is turned into magma by addition of green syrup, and again separated in a batch centrifuge in the second stage to obtain final sugar.

23.8 CENTRIFUGE

Most batch centrifuges nowadays are sequentially controlled according to the manufacturers' own standards and therefore are not dealt with here. If pitch-time-controlled dc motors are used to save energy and equalize electric load variations, power consumption can be reduced by making acceleration of centrifuges coincide timewise with braking of other machines. That is, the power consumed in acceleration can be supplemented by the power generated in braking. This overlapping of acceleration and braking cycles is shown in Figure 23.5.

23.9 SUGAR DRIER

The sugar product transportation system, including the drier and cooler, is interlocked for continuous operation with the centrifuges. It is therefore sufficient to control only the hot-air temperature at the outlet of the aerofine heater, which supplies hot air to the drier. Part of the exhaust air (not saturated with moisture) from the sugar cooler is induced as air for the drier via the dehumidification system. The air induction ratio is controlled based on the sugar moisture content at the drier outlet.

The drier may stop when centrifuge operation is interrupted. In such instances, the air intake and exhaust fans operate without interruption to prevent entry of foreign matter at startup and shutdown.

23.10 SUGAR COOLER

The sugar cooler, both the rotary type and the fluidized-bed type, uses dehumidified, sterilized, and refrigerated air. The relative humidity (RH) of the cooling air needs to be adjusted to keep it constant throughout the year. During the rainy season when the specified relative humidity cannot be maintained, RH can be reduced by heating the air. The drier–cooler system is shown in Figure 23.6.

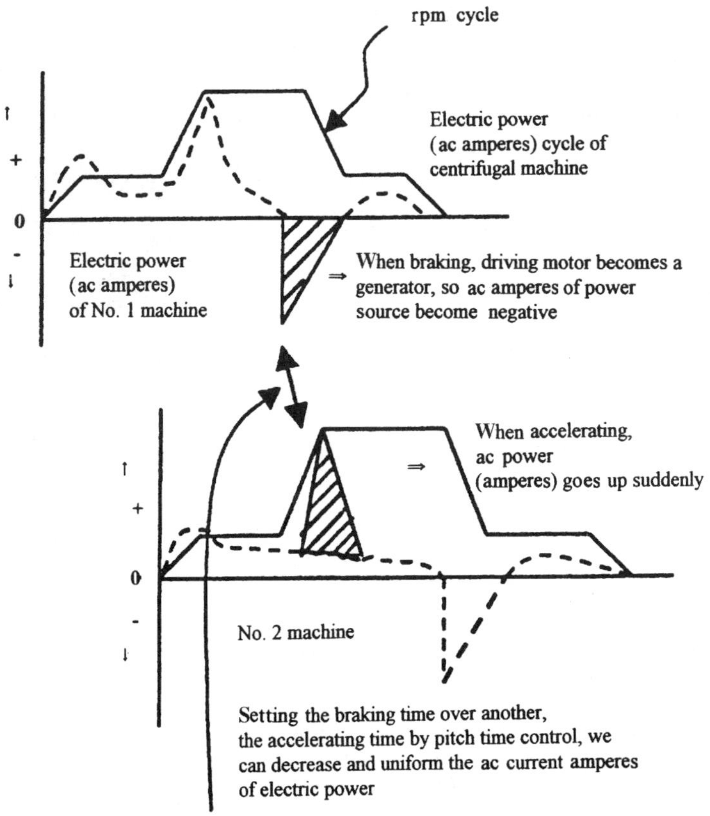

FIGURE 23.5 Pitch time control and electric power saving.

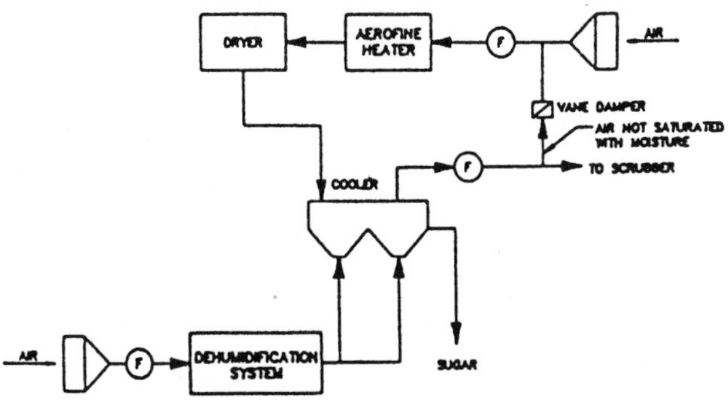

FIGURE 23.6 Drier–cooler system.

23.11 PRODUCT WEIGHING AND TRANSPORT LINE

Japanese sugar refiners produce more than 10 types of crystal sugar. The various sugar products are each weighed on a belt scale and the measured weights are added up by computer. These values are used for rough checking against the production plan. The transportation line from the production process to the sugar bin or sugar silo is sequentially operated (interlock start) to prevent spilling of sugar along the line. That is, the vibrating conveyers are started last to first, and stopped first to last.

12.12 SUGAR STORAGE

Soft sugar does not require aging, but hard crystal sugar is aged in silos for 48 h. For this, dehumidified, sterilized, and cooled clean air is injected from the silo bottom by a special air humidity control system that maintains air temperature and humidity at constant levels throughout the year to prevent caking (lumping).

To prevent condensation in the silo due to the difference in temperatures inside and outside the silo, silos are installed in enclosed buildings, where the temperature is controlled by steam-heated air conditioning. Temperature and humidity are controlled by computer.

23.13 PACKING MACHINE

Bulk lorries (10 to 20 tonnes) are weighed on the truck scale. Aluminum containers (1 and 2 tonnes), flexible containers (500 to 1000 kg), paper bags (20 to 30 kg), polyethylene bags (1 kg), and so on, are weighed and packed automatically by packing machines. The packing process is controlled by a sequencer linked with the weighing machine. One-kilogram polyethylene bags are packed 20 to a paper bag by the auto-sacker.

23.14 PALLETIZING MACHINE

Bags and gallon cans are stacked on pallets in units of 1.2 tonnes by a palletizer. The palletizer, equipped with a bag counter and sequencer, is fully automated for unmanned operation.

Sixty 20-kg bags or forty 30-kg bags can be stacked on one pallet, each bag in an alternate direction to the next. The palletizer is programmed to handle both 20-kg and 30-kg bags.

23.15 PRODUCT WAREHOUSE

An automatic warehouse control system enables fully automatic and unmanned warehousing and dewarehousing of product pallets. All processes, including stock control, are computerized. The warehouse is always stocked with a five-day supply, with a delivery allowance taken into account. Higashi-Nihon Sugar uses approximately 40 different kinds of packaging bags, warehousing and dewarehousing of which is controlled by computer on a first-in, first-out basis.

23.16 LIQUID SUGAR PROCESS

Sugar refineries usually produce two kinds of liquid sugar, liquid invert (LI) and liquid sucrose (LS), for which the production methods differ. Therefore, the IER system consists of a cation and a monobed (anion–cation mixture) line. Both of these lines are used for production of LI, but only the monobed tower is used for LS production. In the following stage, LI is concentrated to 76 Brix and LS to 67 Brix in the evaporator. The raw liquor used here is decolorized IER (anion) fine liquor from the refinery process.

All of these processes are controlled by DCS. Variables such as the inversion ratio, Brix, pH, ash content, and so on, are checked by detectors and controlled by computer to the required product specifications.

The quality of liquid sugar deteriorates within a short time. For this reason, flexible production control is required to cope with variations in demand.

23.17 UTILITY SYSTEM

Utility systems, including boilers, turbines, electric generators, process water treatment, and wastewater treatment, operate fully automatically under interactive control by exclusive high-speed computers. The control systems developed by the manufacturers of the utility equipment can be used as they are simply by adapting their control concept to the software in use.

23.18 SYSTEM MAINTENANCE

Controlling all processes by computer requires high-level technology on the part of computer hardware and software, machinery and equipment, detectors, and manipulators, and high-level technical skills on the part of operators. With the exception of the computer operating system, it is recommended that the software be created in-house to protect company know-how. In this way, software maintenance and improvements can also be done

FIGURE 23.7 Central control room.

in-house, thereby reducing costs. Hardware maintenance is performed by the computer manufacturer under a maintenance contract.

Mechanical and electrical engineers handle maintenance of detectors and manipulators. Computer engineers handle software maintenance. All work day shifts but are on call to deal with operational trouble at night. All control systems are monitored in the Central Control room shown in Figure 23.7.

23.19 MANPOWER

Higashi-Nihon Sugar has nearly 80 employees in the sugar refinery. The refinery operates in four shifts manned by seven workers each, a total of 28. Of the seven workers per shift, five are assigned to the refining process and two to utilities. In addition, there are six computer engineers and seven mechanical and electrical engineers.

CHAPTER 24

Integration of Raw and Refined Sugar Operations

INTRODUCTION

For the purpose of this chapter a *refinery* is defined as a site for production of refined sugar with all the facilities for drying, conditioning, packaging, and other processes. *Raw sugar* would then be defined as the raw material for this process. Cane to refined sugar is a continuum of processes, traditionally broken into two parts, with raw sugar as the transition material. The older terms for a refinery attached to a raw mill are *back-end* or *white-end*, but the term *annexed refinery*, introduced by Todd [1], is now preferred and is used throughout this chapter. Production of white sugars by single crystallization after enhanced juice purification by sulfitation and/or carbonatation and/or syrup clarification is not considered here. Two distinct crystallization operations are implicit in this discussion.

This chapter is an expended version of a paper presented at the 1997 Sugar Industry Technologists meeting in Montreal. The assumption is made that the reader is familiar with standard sugar industry terminology and unit operations. These are described in detail in other chapters of this handbook and in the *Cane Sugar Handbook* [2]. This chapter is intended to give an overview of the issues involved in this type of refinery, not as a prescription for operating a specific refinery.

The financial arrangements between the parties are not considered and the emphasis is on the relationships between raw and refining technologies. Several stand-alone refineries have recently been built, and some new annexed refineries are soon to come on line. Annexed refineries are usually smaller than stand-alone refineries and produce a smaller range of products, in part due to the lack of necessity of handling the lower-purity streams which are sent back to the raw operation.

The climatic conditions for optimum cane growth are probably not the optimum for drying, conditioning, and storage of refined sugar. The latter are the refinery operations

* By Stephen J. Clarke.

most adversely affected by high temperatures and humidity, and special care must be taken in setting up refining operations under these conditions.

Emphasis in the refinery operation on product quality, together with the costs of finishing equipment, dictate that annexed refineries cannot be at all mills. An optimally located mill for the refinery should be found. However, if the refining capacity of this operation becomes greater than the raw production capacity, there is the danger of imbalance in the raw boiling house due to recovery demands. The technology most simply or logically set up may not match the commercial demands of the system in terms of the product quality, which cannot be compromised.

Raw sugar quality plays an important role in determining the ease and efficiency of operation of this type of refinery. The range of quality of the raw sugar can be substantial and is often a measure of the relationship between the raw producer and the refiner. There is a wide range of such relationships between raw sugar producers and refiners, and these may be grouped as follows:

1. *No contact.* The raw sugar is produced at a location far from the refinery and there is no technical contact between the raw sugar producers and the refiners. Often, the raw sugar is marketed through a central organization and the identity of the raw sugar from individual mills is lost. There may be generalized feedback on raw sugar quality, but this would not be of much use to individual mills. The raw sugar may be many months old before being processed by the refinery.
2. *Some contact.* The raw sugar mill knows the identity of the refinery and there would be useful feedback on raw sugar quality. However, the production staff of the refinery would often have little contact with the mill processing staff, and decision making on mill operations would not involve the refinery. The time between production of the raw and its being refined may be days or months.
3. *Close contact.* This may be defined as common ownership but different locations for the raw and refined operations. The time between production of the raw and its being refined may again be days or months. The common ownership would normally lead to decisions being made on process technology involving both mill and refinery personnel.
4. *Intimate contact.* This is the case of an annexed refinery where both operations share, to varying degrees, common facilities. Changes in the raw sugar operation can have an immediate impact on the refinery. This is especially important if the refinery shares common water and steam systems with the mill. Some newer annexed operations are less tied as far as utilities are concerned, but changes in raw sugar quality would still have an immediate impact. This group of refineries is the subject dealt with in this chapter.

24.1 MILL–REFINERY RELATIONSHIPS

There is no standard relationship between the raw mill and its annexed refinery. Of the many local factors that have to be taken into account, some but not all are listed below and the more important are described in some detail in later sections.

1. *Refined sugar the only commercial product, or both refined and raw sugar commercial products.* Some mill / refinery combinations have been designed with refined sugar as the

sole product. In some of these, part of the raw sugar production is stored for later processing when the mill is not operating. No raw sugar is sold to an external refiner. When a refinery is attached to an existing mill, it is often, at least initially, to refine only a fraction of the raw sugar produced. In this case the raw sugar produced by the mill can have three separate destinations: the attached refinery in crop, the attached refinery out of crop, and an external refinery.

2. *Operation of the refinery on stored raw sugar in the out-of-crop period for the raw sugar operation.* The distinction between crop-only operation and this case is important since, although it may be the same refinery, in the out-of-crop period it has to cope with sugar from the warehouse rather than straight from the mill. The well-washed raw sugar sent directly, without drying, to the refinery melter will be quite different from the raw sugar suitable for storage over a period of up to 10 or 11 months. This requires that the raw sugar boiling house be set up to produce more than one quality of raw sugar—one suitable for immediate refining and the other suitable for storage. This is also the case if raw sugar is sold to a third party, but is less critical in this case since the third part is most likely to be a large stand-alone refinery with the capability to handle a wide range of raw sugars rather than the smaller annexed refinery with limited facilities. Further, the recovery operation has to be on its own rather than part of the much larger mill boiling house in the crop.
3. *Refinery to process sugar from other raw sugar mills.* If there is more than one raw sugar mill in the area, it is not cost-effective to operate an annexed refinery at each mill. One refinery would be designed to cope with processing as much of the raw sugar as required. This could result in the daily refinery output being larger than the raw sugar output for the mill involved. Alternatively, the annexed refinery would process all its mill raws on the run and the other raws out of crop. The issues in item 2 then become very significant since the annexed refinery has to become a stand-alone refinery out of crop. Variations in raw sugar quality between the raw producers could become significant.
4. *Raw sugar quality.* Refinery operations are usually most efficient at a fairly narrow range of raw sugar quality. In theory, the annexed refinery should have an advantage, but this is not necessarily the case. The refinery has to handle whatever the raw factory is producing at the time, with good weather conditions and fresh cane or with deteriorating cane under poor weather conditions. If the cane has been damaged by a freeze or a hurricane, the pressure is on the raw operation to get the cane through the system as quickly as possible. This can lead to reduced sugar quality due to inferior juice quality and rushed boiling house operations. The mill would prefer to operate at maximum throughput under conditions when the refinery throughput would be slowed due to inferior sugar quality.

   Upsets and irregular operations are regarded as part of life in a traditional raw sugar factory but not in a refinery. Steady and consistent operation is crucial in a refinery operation. Perhaps this trait will pass over to the raw factory operation.

   Since the raw sugar is to be melted immediately, it might be assumed that the factors necessary for good raw sugar storage do not apply. Other than the moisture content, this is not the case. Other raw sugar quality factors, such as ash and crystal size distribution, are measures of the operational performance of the raw process, especially in sugar boiling. Raw sugar crystal quality must be considered as the primary factor in being able to operate the refinery efficiently.

   If there is common ownership, the penalties normally incurred by failure to meet quality parameters may not be applicable. Even if the financial impact is not computed, the general guidelines for high-quality raw sugar should be met to avoid process problems in the refinery and to ensure that refined sugar quality specifications are met.

5. *Degree of shared facilities between the mill and refinery, including energy supply.* The annexed refinery will depend, to varying degrees, on the raw mill for electric power, steam, and process water (condensate). The wastewater treatment system may also be shared and emphasis should be placed on optimization of both water and energy use. Pinch technology can be applied to both with significant benefits [3]. The degree of connectedness between the raw mill and refinery can be at varying levels, often depending on the history of the system rather than on a logically developed plan. There are several levels of intimacy between the raw mill and the refinery, for example:
   a. Shared buildings with shared distribution systems for electric power, steam, and water, shared injection pumps, automation systems, and waste treatment facilities
   b. Separate buildings with shared electrical and steam and water sources but separate substations and distribution systems, common waste treatment, and so on
   c. Common laboratory and material accounting systems (dealt with in more detail in Section 24.3)
   d. Common operating and support personnel

   Each situation is different and the following is an example of one case where the connection is very close. The refinery shares a building with the raw mill and is dependent on the mill for steam and electric power and sends the final run-off, clarifier scums, and any excess sweetwater to the mill during the grinding season. In the out-of-crop operation, the mill boiling house is used for recovery operations. In the past this dependency situation has meant that the refinery could be shut down if a bagasse carrier failed and the boilers went down. With the new cogeneration facility, this is much less of a problem. This facility also supplies steam at a steady rate out of crop, much preferable to the oil-fired package boiler used previously. This more reliable and less expensive energy source still requires mill equipment for operation in the out-of-crop period. Since it is essential to return the clean first condensate to the cogeneration boilers, low-pressure steam from the cogeneration system is used only in the raw factory preevaporators. All refinery operations (and all other mill operations) are run on vapor from these preevaporators.

   If the annexed refinery is energy independent of the mill, especially if at some distance, operations are not so dependent, and it may be necessary for this refinery to have a separate recovery house. Some of the synergies of operation could be lost if close contact is not maintained.

6. *Refinery added to mill or refinery and mill designed as combined plant.* Many currently operating annexed refineries started life as raw sugar only operations and have since been converted. This type of retrofit is often awkward, with both plant layout unsuitable and equipment capacity mismatched. For example, there are installations with the air intakes for the refined sugar dryers downwind of the bagasse boilers. The location of the melter may be close to the raw sugar centrifugals at some distance from the refinery. It is, of course, simpler and cheaper to pump melt than to convey raw sugar but operational control becomes more difficult. Boiling houses may be separated, again leading to more complicated process control than could be achieved.

   Newer complete installations have the systems laid out at the beginning even if the construction of the refinery takes place after the raw operation. The areas of greatest interaction and synergy between the raw and refined operations are in the boiling house, and these should be as integrated as possible, including the automation system.

24.2 CAPACITIES, SEASONAL OPERATION, AND MATERIAL BALANCE

Annexed refineries are usually of lower capacity than stand-alone refineries. Annual production capacities, as true annexed refineries running in parallel with the milling operation, are determined by the length of the harvesting season. Longer seasons, for example in Colombia, are very desirable.

A large raw mill grinding 25,000 tonnes of cane for 150 days could produce approximately 400,000 tonnes of refined sugar in this period. Very large and problematic storage silos for the refined sugar would be required to provide the market with sugar year round. The alternative is to operate a smaller refinery almost year round but in two different modes, as described previously. This does not give the process and energy efficiencies inherent in true annexed refineries. The same production of refined sugar could be achieved in 330 days of mill or refinery operation with a cane throughput of 11,500 tonnes/day. The equipment required is much less and there is no requirement for long-term storage of refined sugar.

Generalized material balances for a raw sugar operation alone and with an annexed refinery processing all the raw sugar is shown in Table 24.1. These balances ignore the operations of the extraction system and the clarifier mud filtration system since they are assumed to be the same for the two cases. It is also assumed that the refinery phosphatation scums are returned to the raw process and that the final white runoff is also returned to the raw process. Reasonable values for undetermined losses are assumed in the raw and refined sugar operations. The conclusions are that the raw factory boiling house operation

TABLE 24.1 Production of Raw and Refined Sugar[a]

Raw Sugar

10,000 tonnes cane
↓
1206 tonnes sucrose into boiling house
→ 360 tonnes molasses at 35.0 purity and 80.0 Brix (101 tonnes sucrose)
→ 18 tonnes sucrose in undertermined loss (1.5%)
→ 1100 tonnes raw sugar at 98.8 polarization (1087 tonnes sucrose)

Refined Sugar

10,000 tonnes cane
↓
1206 tonnes sucrose into boiling house from cane
↓ + 140 tonnes sucrose into raw boiling house from refinery return
1346 tonnes total sucrose in raw boiling house
→ 404 tonnes molasses at 35.0 purity and 80.0 Brix (113 tonnes sucrose)
→ 20 tonnes sucrose in undetermined loss (1.5%)
→ 1228 tonnes raw sugar at 98.8 polarization to refinery melter (1213 tonnes sucrose)
  → 140 tonnes sucrose to return to raw house
  → 12 tonnes sucrose in undetermined loss (1.0%)
  → 1060 tonnes refined sugar

[a]Sucrose in bagasse and filter cake ignored.

become more complex with return of high-purity material from the refinery and increased molasses production. Optimization of the raw boiling house is discussed later.

24.3 MILL–REFINERY COORDINATION

For good performance by the joint facility it is important that the management be properly coordinated, especially where technical decisions are involved. Since there is so much overlap between the raw sugar boiling house and the refinery boiling house, the question arises as to whether there should be a single person with responsibility for both. Similar considerations apply to the management of the services required: steam, water, electricity, compressed air, instrumentation, and so on.

An integrated waste treatment system is desirable but can be difficult to design and operate when a refinery is added to the mill. Many wastewater handling systems are quite inefficient and can be operated to handle a greater volume of material than is normal practice. To achieve this, the losses from the operations have to be reduced.

Perhaps the most sensitive issue is the accounting for the sugar produced by the mill and sent to the refinery and the returns from the refinery to the mill. Failure in this area can lead to uncertainty, at best, in the efficiencies of the two operations and, at worst, can cause problems between operating personnel. Of particular concern are the returns to the raw factory; these quantities are small compared with the material flow in the raw operation but have a large impact on refinery recovery. Accurate weight measurements for all streams are essential if the efficiencies of both operations are to be determined to the necessary precision.

All scales must be calibrated frequently and the composition of all streams determined, preferably by analysis of samples collected by automatic sampling systems. The location of the raw sugar scales must be such that vibration and condensation are minimized. Since the raw sugar to be refined does not leave the factory, there is a tendency to locate the scales in a convenient place between the centrifugals and the melter, without questioning whether this is a good location for efficient scale operation. Scales used to monitor the transfer of material to and from the mill and refinery must be reliable and should be under the control of the single chief chemist and maintained by the common instrument group. Without good data it is quite possible to transfer inefficiencies from the mill to the refinery, and vice versa.

Physical losses in the refinery tend to be more significant due to the higher Brix and purity and the costs involved in reaching this stage in process. Consideration should be given to sump pumps to return refinery spillages to the raw operation. The ability to return refinery material easily to the raw house can lead to complacency with regard to losses in the refinery. Sudden returns to the raw operation of large volumes of refinery material can seriously upset the raw boiling house, especially if it is operating at capacity.

The overall losses in the total process are distributed between the cane extraction plant, the raw process, and the refinery. Targets should be set for all process flows and monitored continuously if steady and efficient operation is to be achieved.

A single chief chemist (a misnomer and better described as the "sugar accountant") with responsibility for both raw and refinery production data is probably the best solution. Such a person may be hard to find since detailed knowledge of both operations is required. The confidence of the production staffs of both operations is necessary for this chemist to operate efficiently. Refined sugar quality control and customer relations are a separate area and should not be the direct responsibility of this chief chemist.

A single laboratory organization is important for the chief chemist to operate efficiently. Separate areas within the laboratory must, of course, be designated for such measurements as incoming cane quality and refined sugar quality. Many analyses are common and can be run by the same personnel (e.g., massecuite and molasses analyses). Standardization of analytical methods is important to avoid such problems as the raw laboratory and refinery laboratory using different methods for raw sugar ash determination. If the mill/refinery is large, it is appropriate to have satellite laboratories at various sites (e.g., refinery filter operations where Brix measurement during desweetening is important, or any site where pH control is important).

As the operation becomes larger, it is important that as much data as possible be collected from on-line sensors. There are many areas in the processes where the impact of one process parameter on a larger operation is suspected but not documented. Modern data mining techniques should be used for statistical evaluation and control of process.

24.4 INTEGRATION OF TECHNOLOGY

The challenge in coordinating raw and refinery operations is to choose the appropriate technology for both. Standard and well-understood technologies exist for all raw and refining unit operations, and the tendency is to choose the safe option. However, applying the best currently available technology to a completely new combined facility would result in a factory quite unlike anything now operating.

Adding a refinery to an existing mill imposes many constraints on the refinery design. Changes must be made in both equipment and operations in the mill to achieve the optimum operation. Simply to require the new refinery to accept the raw sugar as traditionally produced is not acceptable. The choices are more complex if the refinery is also to operate on stored raw sugar after the cane processing is completed.

It is important to resolve the different constraints of cane processing and raw sugar refining. In the first case the demand is for throughput and recovery. If the raw sugar polarization is 98.2 rather than 99.0, this is not a problem from the perspective of the conventional raw sugar producer. The constraint on the refiner is product quality, with recovery being an important second factor. The refinery's efficiency is maximized at a consistent (and high) sugar quality.

In a raw sugar without a refinery or refining only a fraction of its raw sugar, the bottlenecks to cane throughput are in the raw factory. If there are problems in the refinery, the extra sugar can go to the warehouse. If all the raw sugar must be refined and there is little or no intermediate storage, a problem in the refinery can slow or stop the mill.

It is not possible to eliminate any of the unit operations in the raw factory. Traditional cane processing has developed a rather minimalist approach to achieve a (barely?) acceptable raw sugar quality. There is therefore more flexibility and opportunity for improvement in the raw factory than in the refinery. Traditionally, raw sugar polarization has been in the range 96 to 98; it is now possible, with the appropriate equipment but without major changes in process, to produce raw sugar with 99 polarization (i.e., a reduction of 50 to 75% in the non-sugar loading to the refinery). The raw operation must produce a consistent high-quality raw to reduce the investment necessary in the refinery to handle the periods when raw sugar quality is below optimum.

A review of the unit operations in the raw factory suggests that the areas for improvement in raw sugar quality are in clarification and crystallization. The unit operations in both raw and refining processes that affect each other are italicized in Table 24.2.

TABLE 24.2 Synergy between Raw and Refined Operations

Extraction
Clarification and filtration
Evaporation
Crystallization
Centrifugation
Raw sugar drying and storage
Affination
Clarification, purification, and filtration
Decolorization
Crystallization and centrifugation
Drying, conditioning, etc.

Standard clarification techniques work well if the cane is fresh and good operating practices are maintained. Enzymatic processing aids, such as amylase and dextranase, can be of great value if used appropriately and not indiscriminately. Syrup and filtrate clarification have both been found to be of value in improving raw sugar quality.

Membrane filtration of clarified juice has the greatest potential for improving raw sugar quality [4]. It is well established that membrane filtration of clarified juice results in higher-quality raw sugar [5]. The long-term reliability and operating costs of these systems remains uncertain. Also, the emphasis has been on the high quality of permeate, neglecting the very inferior retentate from which sugar must be recovered if yields are to be maintained. All purification systems generate low-purity streams, and the handling of these often presents more challenges than the purification system itself. Proper application of this technology requires a level of sophistication and automation not generally found in raw sugar operations.

Membrane technology may be especially useful under conditions of poor cane quality and consequent inferior clarification. However, when the membrane filtration system is needed the most is the time when it is most difficult to operate. There are problems of maintenance of acceptable throughput at reasonable cost under poor cane quality conditions.

The high capital and operating costs (especially the electric power requirements) of membrane filtration systems and the only modest premiums available from selling this superior raw sugar to outside refiners may result in this technology only being applied where the major financial benefits are derived from lower-cost operation of annexed refineries. One of the advantages of annexed refineries is the lower cost of electric power based on bagasse as fuel. The high power demand of membrane systems could stretch the overall requirements for power beyond that available from conventional bagasse-fired systems.

Modification of the raw factory crystallization scheme is standard practice for improving raw sugar quality. If all the raw sugar produced is not refined immediately, there is a case for modification of the boiling scheme to produce raw sugars of different quality. For example, the raw sugar produced for sale would be obtained by a different route than the raw sugar to be refined. This approach can cause problems if some of the sugar in the warehouse is to be returned to the refinery for out-of-crop refining.

The boiling scheme used in a mill/refinery which refines only one-third of production is shown in Figure 24.1. The most significant effect (in terms of ash and color in raw sugar) is not returning the excess C magma to the syrup and thereby to the A massecuite. The recycling in this system requires additional pan capacity since the A molasses produced with the washed A sugar has a significantly higher purity. When producing washed A sugar, the A molasses purity is 73 to 75; when sending this sugar to the warehouse and reducing washing, the A molasses purity decreases to 68 to 70. In the latter case (refinery not operating), the reported undetermined loss decreases by 10 to 20%. Whether this is an accounting problem remains unknown. If performed properly, these enhancements to raw house operation do not lead to reduced recovery and should lead to increased recovery.

In annexed refineries it is common practice to return the lower-purity refinery run-offs to the raw boiling house. The raw boiling house capacity is much greater than that required only for recovery operations. In crop this is not a problem but can become a serious problem in out-of-crop operations. In the example for the boiling scheme, the solids return from the refinery is only 2 to 4% of the solids flow in the raw factory. However, out of crop, the boiling house vacuum pans are of such a size that it is only possible to boil a low-grade strike every 2 to 3 days. Keeping the intermediate material at less than optimum Brix for extended periods leads to significant losses.

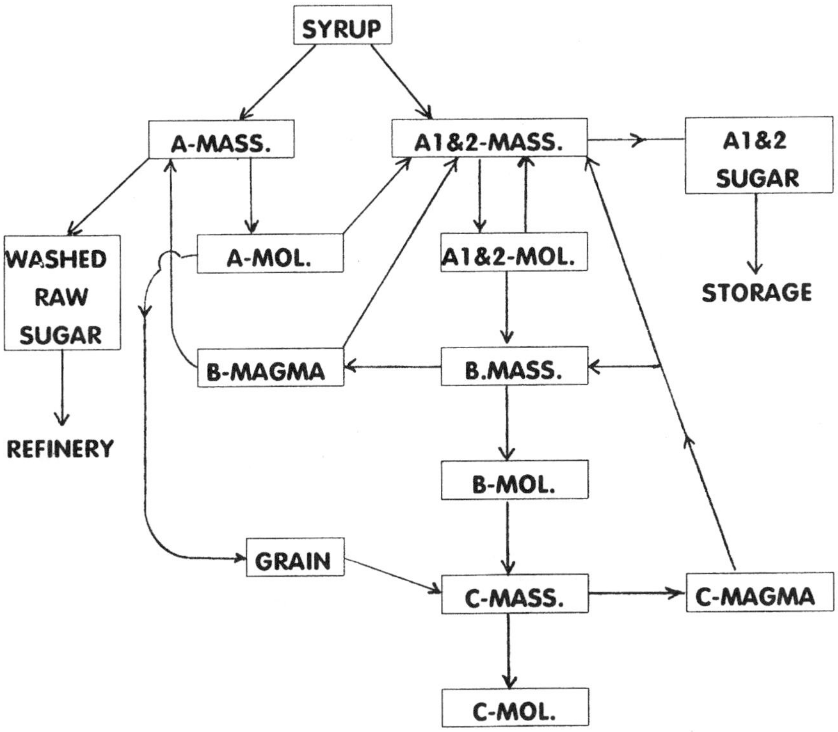

FIGURE 24.1 Boiling scheme for refining raw sugar.

The challenge is to optimize the performance of raw sugar production and to minimize refinery requirements. In a combined operation it may be inappropriate to draw a demarcation line between raw sugar production and refining. Certain operations cannot be removed: crystallization, centrifugation, drying, and conditioning. For the other operations, the question becomes the quality of "raw" sugar that must be produced for these stages not to be necessary.

With standard raw sugar, affination is essential to remove color and ash. Additional washing of raw sugar during crop operations is equivalent to affination. To avoid affination in out-of-crop operations, a stable raw sugar of low color and ash is required. The extra washing of the sugar to achieve this requires that the sugar be dried. (Is this a transfer of capital charges from the refinery to the mill?) A lack of affination capability restricts the flexibility of being able to refine lower-quality raw sugars from outside sources.

When the two operations are physically linked, standard practice is to use the same centrifugals, conveyors, and so on, for the affined sugar as used for the washed A sugar in crop. The centrifugals are therefore not idle out of crop. This advantage would not be available to a refinery at a distance from the raw house.

Many of the same constraints apply to the elimination of clarification in the refinery, with the additional requirement of low turbidity in the raw sugar. The microbiological requirements for refined sugar require filtration of the liquor before crystallization. Membrane filtration for raw sugar production is probably the only reliable technique available to produce raw sugar suitable for refining without clarification. However, it should be kept in mind that although turbidity may be completely removed from the clarified juice by membrane filtration, precipitation can occur during evaporation and crystallization, and these materials (typically, calcium sulfate and aconitate in high ash juices) are carried through into the raw sugar. Softening of the membrane filtered juice would eliminate most of the problem but is another example of the transfer of cost from the refinery to the raw house. The decision to eliminate clarification by membrane filtration must be based on a detailed analysis of each case, taking into account the probability of having to refine some below specification raw sugar.

Elimination of decolorization is even more problematic. This would be dependent on producing a "raw" sugar of less than 150 ICU. A technique for achieving this has been described [6]. This approach involves recrystallization of A sugar, after clarification of the liquor by phosphatation, to give the intermediate sugar. This sugar is dried and shipped to the refinery, where the only operations would be filtration and crystallization. This is basic sugar technology with no chemical treatment and minimal waste generation in the refinery. Concerns are for the high energy demands and potential losses during crystallization and centrifugation.

This approach is another example of transferring the cost of refining to the raw operation. With a common ownership this may be quite justifiable. Providing that the vacuum pan and centrifugal capacity is available and that the cost of energy is low, this would appear to be an interesting approach, especially in environmentally sensitive areas.

Each situation must be evaluated on its merits, and there can be no single best approach to the operation of linked raw sugar mills and refineries. We will certainly be seeing more interesting developments in the years to come.

REFERENCES

1. M. Todd, *Int. Sugar J.*, 99:379–384, 1997.
2. J. C. P. Chen and C. C. Chou, *Cane Sugar Handbook*, 12th ed., Wiley, New York, 1993.

3. Saving energy in refineries, *Sugar Ind. Technol. Symp.*, 1997.
4. Separation processes in the sugar industry, *Proc. Sugar Process. Res. Inst. Workshop,* M. A. Clarke, ed., 1996.
5. M. Saska et al., *Sugar J.,* Nov.–Dec. 1995, pp. 29–31.
6. M. Player and P. Fields, International patent WO 96 25,2552; 125:224986p, 1996.

CHAPTER 25

Off-Crop Sugar Refining for a Back-End Refinery*

INTRODUCTION

This chapter covers utilization of a *back-end refinery,* a refinery attached to a raw cane sugar factory, during the traditional off-crop period. There is a wide range of relationships between cane raw sugar producers and refiners, and this has been well described by Clarke [1]. In the South African context, a back-end refinery is one that is physically integrated with a cane raw sugar factory. The two plants are under the same roof and share all services, including labor and management. This would fall under the *intimate contact* category described by Clarke. The contribution of back-end refineries can be quite significant; in South Africa about 45% of the refined sugar originates from one large, central refinery, while the other 55% comes from back-end refineries.

The feasibility of traditional back-end refining is dependent on the length of the season, since this affects plant and capital utilization. The season lasts between 30 and 40 weeks in southern Africa, and this has had a positive impact on local back-end refining. Despite this, the possibility of extending the use of back-end refineries to the off-crop has been considered as far back as 1983. It was, however, only in 1993 that full-scale off-crop refining was done in a South African factory for the first time. Since then three other back-end refineries have operated during the off-crop for periods ranging from 3 to 10 weeks, and involving sugar tonnages of 13,000 to 65,000 tonnes. The process is thus only about five years old and many aspects need further optimization. Furthermore, because the author's involvement is limited to the South African region, the information given here is based primarily on conditions as found in that region.

Back-end refining has a number of advantages over a stand-alone refinery, many of which arise from facilities already available from the raw sugar factory, such as energy from bagasse; shared vapor and condensate systems; good control of raw sugar quality, which eliminates affination; and the use of the raw house equipment to handle the last refinery jet, thus removing the need for a recovery house. Many of these advantages are lost, or at

*By G. R. E. (Raoul) Lionnet

least reduced, with off-crop refining. In addition, other considerations, such as the isolation of steam lines, safety procedures, importing and/or storage of raw sugar, and energy balances, for example, now come into play. These new costs and the advantages that are now lost must be compensated by increased refined sugar productions and by better plant utilizations. Careful studies are therefore necessary. Some of the areas that have been found to be important include plant modifications and integration, energy considerations, material handling, processing, and performance.

25.1 PLANT MODIFICATIONS

Generally, the raw house and back-end refinery are well integrated, particularly with respect to steam and condensate piping. Obviously, those lines that are not utilized during off-crop refining need to be isolated, on a nonpermanent basis, for maintenance, safety, and energy loss considerations. Depending on the conditions, other modifications (e.g., to boilers, storage/conveying systems, etc.) may be needed.

Plant modifications fall under two main categories. The first involves the temporary isolation of piping and equipment not utilized during off-crop refining; the second concerns modifications needed to facilitate operations during off-crop refining. A description of the isolation of pipes and equipment for off-crop refining is given by Schorn [2]. The modifications required can be quite complex since a number of objectives must be achieved. They must not be permanent—it is necessary for personnel to be able to establish quickly whether or not the line or piece of equipment is isolated; changing from one mode to the other should not be too costly in terms of time, labor, or money; and at all times, safety measures must be satisfied. These comments apply particularly to steam, vapor, and condensate lines and equipment, where energy losses must also be prevented. It can take up to 3 days to isolate the plant.

Plant modifications to facilitate operation during off-crop refining depend very much on local conditions. They can be major if modifications to a boiler, for example, are needed or fairly minor if sweetwater has to be rerouted. Some of the areas where plant modifications have been done for off-crop refining are:

- Piping for jet 4 as new exhaustion schemes were investigated (described by Moodley et al. [3])
- Massecuite gutters to accommodate recovery massecuites in the raw house
- Piping and plant modifications for the storage of jets or refinery final molasses
- Piping and tanks for handling sweetwater, which would normally be utilized in the raw house
- Modifications to exhaust steam lines
- Minor modifications to condensate systems to use condensate as boiler feedwater
- Minor modifications to vacuum lines

Major plant modification (e.g., to boilers or concerning the storage of raw sugar) may be needed. These aspects are covered in specific sections.

25.2 ENERGY CONSIDERATIONS

Energy is one of the most important considerations for off-crop refining. Bagasse is not normally available during the off-crop. If it is possible to produce excess bagasse in sufficient quantities during the crop, the possibility of storing it can be assessed. Storage and conveying costs need to be considered against the cost of burning a fossil fuel. Another possibility is to store part of the bagasse during the crop, replacing it by burning coal. A problem particular to some local cane sugar factories is that often, only one of the boilers has been designed to burn fossil fuel, coal in this case—usually it is also a small boiler. This has been a bottleneck during off-crop refining at some back-end refineries; bagasse storage then becomes attractive.

Steam and vapor productions also need careful considerations. The approach adopted has been to feed exhaust steam into one or two of the raw house evaporators, to evaporate water, thus producing vapor 1 (V1) and condensate. Vapor 1 is used in the pans and for heating duties, while the condensate is used as boiler feedwater. The following points have been noted.

- Even when potable water has been used in the evaporators, the concentration effect has resulted in the need for *blowdown* at least once a day. This causes heat losses and increased requirements for clean water.
- Efforts must be made to recover as much condensate as possible to improve thermal efficiency. If makeup water has to be used, being at ambient temperature, it will tend to reduce the efficiency.
- Condensate from pans has been used in the melter. In general, all condensates should be tested for sugar and utilized optimally.
- Experience has shown that thermal efficiency is very dependent on steady operation during off-crop refining. A major problem, particularly during the first period of off-crop refining, has been the supply of raw sugar. Running out of raw sugar or even unsteady input tonnages and varying quality have been identified as major causes for poor thermal efficiencies. Generally, steady and consistent conditions are crucial in refinery operation [1].
- Heat losses during off-crop refining tend to be underestimated by factory personnel.
- Normal back-end refining allows the factory to control the quality of the raw sugar feed to the refinery. This control may be lost with off-crop refining, particularly if raw sugar is imported. Raw sugar quality will affect energy by influencing the number of refinery boilings that can be bagged and may even affect the need for an affination station.

The simplified flow diagram in Figure 25.1 shows the tonnages of water and vapor for a 20-tonne/h back-end refinery, bagging four boilings, with affination and a recovery house. It is possible to estimate the steam required for various options if the following assumptions are made.

- Exhaust steam (119°C, 190 kPa absolute) is used to boil water in one evaporator vessel to produce V1 (112°C, 150 kPa absolute).
- All evaporation (evaporators and all pans) is single-effect.
- An additional amount of V1 is needed for heating, losses, and so on, and this is equivalent to 15% of the quantity of V1 calculated from the theoretical balance.

Three cases are now compared. The first is for 20 tonnes of refined sugar per hour, using carbonatation and ion exchange, without an affination station, without a recovery house, and bagging four boilings. Then, just under 23 tonnes of steam are required per hour. Addition of recovery house operations increases the steam requirement to 24 tonnes /h and, with the further addition of an affination station, the steam requirement is now 25.4 tonnes/h. These results are summarized in Table 25.1.

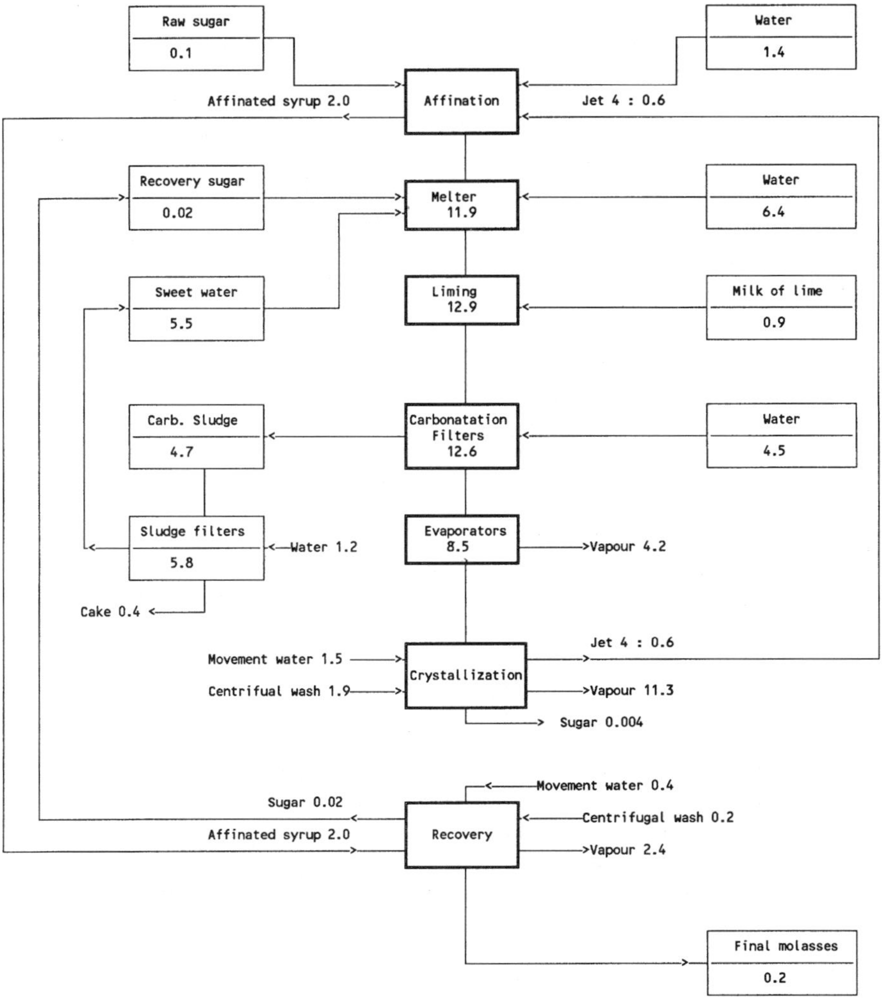

FIGURE 25.1 Simplified flow diagram showing hourly tonnages of water and vapor. The numbers are tons of water or vapor in the material.

TABLE 25.1 Steam Requirements in Back-End Refining for 20 Tonnes of Refined Sugar per Hour

Conditions	Tonnes Steam per Hour	Percent
No recovery house and no affination	22.8	100 (base figure)
Recovery house and no affination	24.0	105
Recovery house and affination	25.4	111

25.3 MATERIAL SUPPLY AND STORAGE

Materials such as bagasse, which affect the energy requirements, have been covered in Section 25.2. Off-crop refining also creates specific requirements with respect to other streams, particularly raw sugar. The supply of raw sugar can cause serious problems if it is not steady and at the required tonnage. Back-end refineries that do not have raw sugar storage facilities and thus rely on transport can have supply problems which will then affect thermal efficiency, sucrose losses, and production targets. If off-crop refining is to be adopted, raw sugar storage and conveying facilities need to be installed. The type of raw sugar stored in South Africa is of the *very high polarization* (VHP) quality, with a polarization of 99.3%, a moisture content of 0.11%, and an ICUMSA color of about 1500 units. With this type of sugar, there has been no evidence of deterioration in quality with storage, and handling has been without problems. To avoid affination, a stable supply of raw sugar of low color and ash is required.

Water supplies for off-crop refining need no special attention, and pan condensers can be fed from the usual system. Condensates should be used optimally with respect to energy. Makeup water for generating vapor 1, in the raw house evaporators, is a special off-crop requirement, and water quality becomes relevant. If the silica content is high, for example, blow-down will be frequent. The production of large volumes of sweetwater should be avoided as far as possible since this will result in both sucrose and thermal losses. Locally, effluent was not a problem during off-crop refining, the usual systems coping well.

The possibility of storing refinery or recovery house materials can be a practical option. Under South African conditions only lower-purity recovery house materials have been stored successfully. Generally, the stream should be of about 70 Brix and cooled to 50°C before being stored in conventional molasses tanks or in any suitable vessel. This material can then be returned to the raw house, usually to A molasses, during the next crop. No adverse effects have been noted but it must be stressed that the quality of the material in terms of the type and concentrations of the impurities is critical here. This is discussed further in Section 25.4.

Experience has shown that it is important to end the normal crushing season with as little material as possible left over as stock in the entire factory. Material in stock can complicate plant utilization during off-crop refining or can result in thermal efficiency problems if it has to be treated.

25.4 PROCESSING

The absence of recovery house where many refinery streams can be readily absorbed and processed is the main problem as far as processing is concerned. Raw house equipment, which is now idle, has to be used to process the various refinery products. The capacity of most of the raw house vessels is normally too large for the refinery streams [1], and this can cause processing difficulties. Performance can be reduced because of losses caused by long retention times but can be enhanced in the case of molasses exhaustion.

Off-crop refining has a different impact on each of the unit operations used in refining. During normal back-end refining, raw sugar quality can usually be controlled, and in South Africa, affination stations are not required. If the import raw sugar quality is low, affination may be necessary during off-crop refining, which would then have an effect on energy, equipment, ease of processing, and thus costs. A process flow sheet for affination is shown in Figure 25.2. All the jet 4 is used, with water, to produce a magma which is then centrifuged and the affination syrup fed to the recovery house. The process is adjusted to let the affination syrup reach a preset quality, in terms of purity and/or color, before being sent

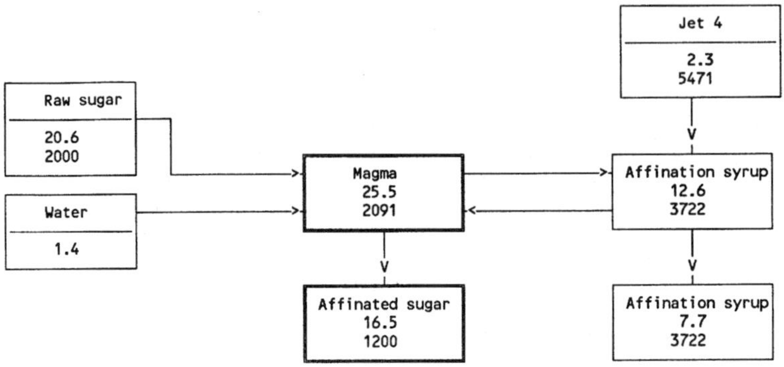

FIGURE 25.2 Affination flow sheet. The numbers are tonnes of product and color.

to the recovery house. This preset quality depends mostly on the inlet raw-sugar purity. For example, the affination syrup is removed and sent to the recovery house when raw sugar purity and color are about 96 and 4300, respectively. If affination can be avoided, jet 4 is sent directly to the recovery house.

Decolorization, filtration, and evaporation are not much affected by off-crop refining. Particular attention needs to be paid to energy-saving procedures, and all processes should be controlled to produce the minimum amount of sidestreams since the raw house is not available to handle these. The same comments apply to the refinery pan house, where movement of water, particularly, needs to be controlled. A typical South African situation is shown in Figure 25.3, where mass and color (ICUMSA units) balances are shown for a back-end refinery producing 20 tonnes of refined sugar per hour, using carbonatation and ion exchange, and bagging four boilings. A simplified flow sheet for a recovery house is shown in Figure 25.4.

Recovery house operation is one area that needs careful consideration during off-crop refining, particularly with respect to the following points.

- The refinery stream to be processed in the recovery house is usually of a small hourly tonnage, and the raw house equipment that is available for recovery purposes will normally be too large for that tonnage. This results in long retention times and can cause losses and operational problems.
- Liquor quality, in terms of the type of the impurities present, becomes important; oligosaccharides, for example 1-kestose, 6-kestose, and neo-kestose, cause severe crystal deformation and slow down the crystallization rate. This has been well investigated by Morel du Boil [4–6]. During normal back-end refining, these effects are reduced since the refinery returns are added to a much larger volume of raw house syrup. This is now not possible and the presence or absence of such impurities may well have an overriding effect on the operations in the recovery house.
- If the off-crop refining period is short, the possibility of storing the refinery returns should be investigated. Work in South Africa has shown that refinery molasses may be stored as long as the Brix is above about 70 and the temperature below 50°C. Apart for some crystallization, no problems were experienced for a storage period of up to 6 weeks [3].
- Operational flexibility in the raw house is an advantage. This depends on plant layout and on the cost of modifications to piping, tanks, vessels, and so on. One back-end refinery in South Africa has used this flexibility to achieve remarkably good recoveries [3].

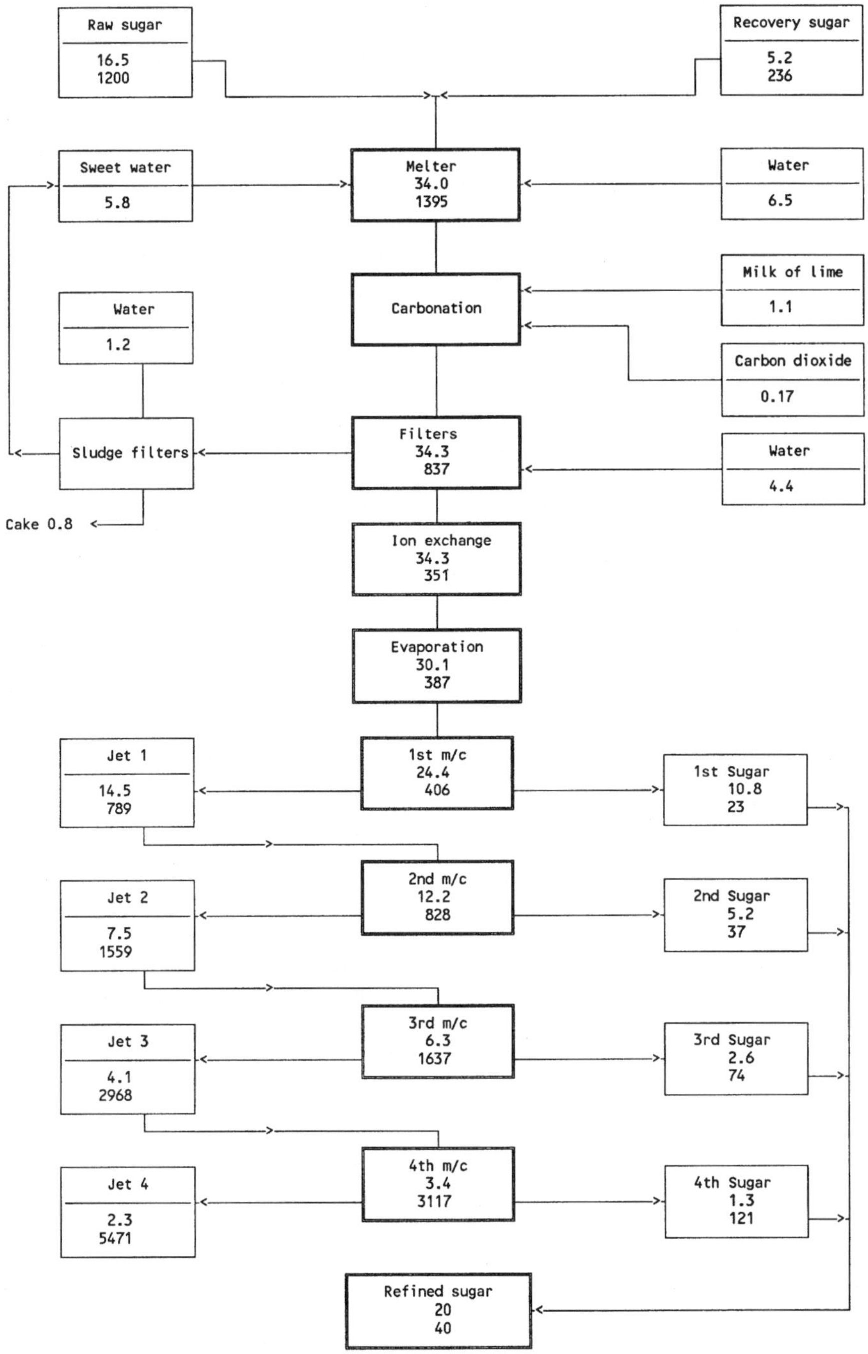

FIGURE 25.3 Simplified back-end refining flow sheet. Numbers are tonnes of product and color.

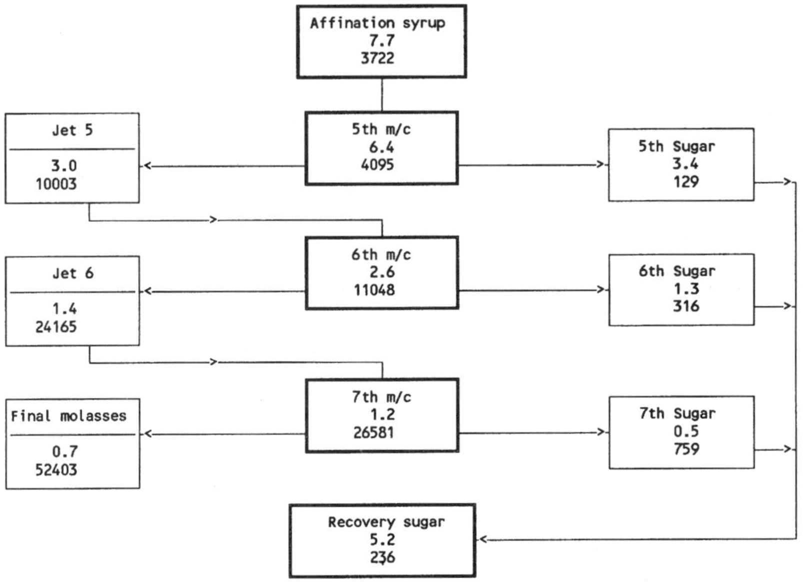

FIGURE 25.4 Simplified recovery house flow sheet. Numbers are tonnes of product and color.

The actual operation of an off-crop refining recovery house is more complicated than indicated in Figure 25.4, since this is where the capacities, liquor quality, and operational flexibilities mentioned previously come into play. A good description of various alternatives has been given by Moodley et al. [3]. One of the options described by these authors has been the use of a C-massecuite continuous pan to boil jets 6 and 7, which achieved good exhaustions. It should be noted here that the levels of oligosaccharides were low at that refinery.

Generally, the raw house batch pans can be used in the recovery house. Various boiling schemes have been tried, depending on local conditions. Fifth and sixth recovery massecuites have generally been cured as quickly as possible after striking, using A-batch machines. Seventh massecuites have been through four continuous C crystallizers and then cured in continuous C machines, with minimum amounts of steam or wash water. If the massecuites cool too quickly, pipe blockages can occur and steam injection points are essential. Experience has shown that the properties of the recovery streams are very dependent on the concentrations of oligosaccharides present in them and on viscosity. Most of the abovementioned operations were carried out with relative ease with low concentrations of oligosaccharides and low viscosities.

As far as performance is concerned, there is no basic reason for poorer results during off-crop refining, particularly with innovative use of raw house equipment. The main factors that have affected performance have been found to be the following.

- *Steady operation.* As was the case with energy, stop/start operations affect the process in terms of temperature, retention time, and general efficiency. A steady raw sugar supply is essential.

- *Raw sugar quality.* This affects decolorization directly and, as discussed previously, affects the operations in the recovery house. Erratic raw sugar quality is also undesirable.
- *Limitations from raw house equipment.* These might result in less-than-optimum operations.

REFERENCES

1. S. J. Clarke, *Proc. Sugar Ind. Technol. Meet.*, 1997.
2. P. M. Schorn, *S. Afr. Sugar Technol. Assoc.*, 69:177–180, 1995.
3. M. Moodley, P. M. Schorn, and I. Singh, *S. Afr. Sugar Technol. Assoc.*, 70:189–194, 1996.
4. P. G. Morel du Boil, *S. Afr. Sugar Technol. Assoc.*, 59:33–38, 1985.
5. P. G. Morel du Boil, *S. Afr. Sugar Technol. Assoc.*, 65:171–178, 1991.
6. P. G. Morel du Boil, *Int. Sugar J.*, 99(1179B, March):102–106, 1991.

CHAPTER 26

Energy Conservation for Sugar Refining*

INTRODUCTION

The management of energy resources did not begin with the oil embargo of the 1970s. Energy-intensive companies such as the sugar industry and the chemical industry were concerned with costs and availability of fuels for energy long before the embargo. It simply brought the need to manage energy into sharp focus. The energy manager of today has to understand the generation and use of steam and electricity, heat recovery, computer control, economics, and environmental concerns. The energy manager must also have the management skills to direct plants to more efficient use of energy resources. Oliver Lyles [1, p. 441] gave us three rules for energy conservation in 1957: (1) prevent the physical escape or loss of heat from the plant; (2) reduce the work required; and (3) reuse the heat in the process.

26.1 ENERGY MANAGEMENT

The best way to conserve energy in any organization is through a good energy management program. For energy management to be effective, the simplistic housekeeping approach must be replaced by a program that can focus on the total energy picture. An effective program is one that can adapt to changing energy prices, government regulations and deregulations, and changing environmental regulations.

The multidivisional corporation should organize energy activities on a corporate and plant basis. Corporations should take advantage of the volume of energy purchased at their various sites around the country or world. With the upcoming deregulation of electrical utilities and natural gas transportation, increased volumes in an increasingly buyer-friendly market should net significant savings. A plant planning to enter into an energy management program should have three objectives: (1) to establish clear energy goals, (2) to structure the program to achieve the goals, and (3) to give the manager authority to implement the program.

* By Joseph C. Tillman, Jr.

The selection of an energy manager is very important. In most cases, the person can be selected from existing technical personnel. Seldom will energy management be the only position this technical person holds. However, it is very important to pick a person who fits the framework of the three objectives mentioned previously. The structure of the program should not be modified to fit a particular person. The energy manager's position is best filled by a dynamic person who is self-motivated and able to make good decisions without relying on day-to-day guidance from management.

For this program to be ongoing, the plant energy manager needs to establish an energy audit program for the plant. An energy audit should identify where a plant uses energy and identify energy conservation opportunities. (In this book, *plant* refers to the refinery, packaging house, warehouse, and all storage and support facilities.) There are three basic types of energy audits:

1. *Walk-through audit.* This consists of a visual inspection of the plant. It should be done by an outside firm or someone not associated with the plant. In the case of multifacility corporations, energy managers from the other plants would be a good choice, and the least costly. This type of audit should be done monthly. The purpose of this audit is to look at operations and maintenance activities that could save energy. It should also serve as an opportunity to collect information about idle equipment, instruments, meters, and recorders that are not working. This additional information can be used in future audits. The participants in this audit should be looking for specific things, such as: leaks (steam, water, air, syrup, etc.), improper tuned boilers (O_2, CO_2), loose or missing insulation, and equipment running when not needed.

2. *Miniaudit.* This audit will verify that the Btu/hundredweight that is set as a goal is achieved. All measuring and recording devices needed to determine the actual Btu/hundredweight will have to be calibrated and working properly before this audit can take place. An audit of this nature should be done by an engineering or energy consultant firm. This audit should be done quarterly. In the miniaudit, usually no effort is made to separate the process from packaging. The end result of this audit will be to determine if it is economically feasible to make changes to achieve the desired or a better Btu/hundredweight goal.

3. *Maxiaudit (energy balance).* This audit will look at the energy used in each process of the plant. It will look at raw sugar through drying and cooling as well as energy consumption in the packaging house and warehouse. The audit will also look at the lighting, heating, and air-conditioning loads. There may be some type of computer modeling associated with this type of audit. These audits are generally very expensive and usually need the expertise of an outside firm. The need for such an audit should be reviewed thoroughly before entering into an agreement with an engineering firm or energy consultant. The maxiaudit should be done once per year. The energy balance audit should be analyzed thoroughly for the following:

- Can waste heat be recovered?
- Can new, more efficient equipment be justified?
- Can the process be modified or changed?
- Can throughput be increased with the same equipment?
- Can controls be upgraded to improve efficiency?

Any improvements that are made in the plant may require retraining of the operations and maintenance personnel. Training should take place while the improvements are being made so that the operators and mechanics will be comfortable in their new roles. The plant energy manager should conduct unannounced surveys at night and on weekends.

The plant manager, working with the energy manager, should develop a program that will ensure that energy conservation is a continuing effort. The success of such a program will depend on many things, such as:

- Having clear achievable energy goals
- Having the support of upper management
- Ensuring that recording equipment is calibrated and working properly every day
- Changing charts on a regular basis, including weekends
- Promoting employee involvement and commitment
- Keeping the entire organization focused on the energy goal
- Evaluating the program:
  - Review progress in energy savings.
  - Evaluate original goals.
  - Change program if necessary.
  - Revise goals as necessary.

It is the responsibility of the energy manager to see that the recorded data are collected, evaluated, and put into clear concise reports. The energy goals should be a part of these reports. The reports must be received on a regular basis by a committee comprised of the plant manager, energy manager, and department heads. Arrangements should be made so that hourly employees can be part of this process. Their insight is often very different from management's. The task of this committee is to measure where the company is, based on the goals and determine where the company is going. The committee should be able to account for the total energy consumption, energy costs, and the Btu/hundredweight. The reports mentioned previously should contain all the charts, graphs, and data to account for the energy used. The following items should be included:

- Million Btu (MMBtu) each day (startup, operating, shutdown, and down)
- Melt each day
- MMBtu/hundredweight
- MMBtu content of each fuel burned
- MMBtu of electricity generated and purchased
- Percent of each fuel used (including electricity)
- Total MMBtu for the period (weekly is adequate)
- Total melt for the period

From the melt for the period and the MMBtu for the period, calculate the MMBtu/hundredweight. Use the MMBtu (*of the steam only*) and the enthalpy of your steam to calculate the pounds of steam per pounds of melt:

$$\frac{\text{total MMBtu}}{\text{enthalpy (Btu/lb)}} = \text{total lb of steam}$$

$$\frac{\text{total lb of steam}}{\text{total lb of melt}} = \frac{\text{lb of steam}}{\text{lb of melt}}$$

Enthalpy is defined here as the total energy of a substance. For water, the total heat is from 32°F to its boiling point (based on water at sea level).

26.2 ENERGY

To understand energy, specifically the conservation of energy, we must first know a few facts about energy.

- Energy is defined as the ability or capacity to do work.
- The law of conservation of energy states that energy cannot be created or destroyed; it may be transformed from one form to another, but the total amount of energy never changes.
- As work is done, the original form of energy is diminished by the amount of work done and the energy is transformed to another form of energy.

In this section we look at two basic forms of energy: steam and electricity. In a refinery, steam is the major form of energy that is used, accounting for approximately 90% of the energy. Electricity accounts for approximately 10%. Energy sources are listed in Table 26.1.

Oliver Lyle handed down three basic principles for the conservation of energy:

1. Prevent the physical escape or loss of heat from the plant.
2. Reduce the amount of work required.
3. Reuse the heat in the process.

The first time that I was exposed to these concepts was at the Cane Sugar Short Course in 1980. They appeared in the notes of W. N. G. Buckner, who credited Lyles with these principles. These three basic principles should be required knowledge of all employees. In the balance of this chapter we look at these three principles and how they relate to a sugar refinery.

1. The physical escape or loss of heat from the plant may occur in the following ways:
   - Radiation
   - Steam traps
   - Steam leaks
   - Blowoff to atmosphere
   - Air leak
   - Water leaks
   - Condensate loss
2. The amount of work required may be reduced in the following ways:
   - Lower the process temperature.

TABLE 26.1 Sources of Energy Used in a Sugar Refinery and Their Btu Values[a]

Natural gas	1000 Btu / ft^3
No. 2 oil	140,000 Btu / gal
No. 6 oil	150,000 Btu / gal
Coal (bituminous)	12,000 to 13,500 Btu / lb
Electricity	3413 Btu / kWh[b]

[a]These are average values and may vary slightly from region to region.

[b]1 kWh = 3413 Btu.

- Minimize recirculation of char liquors.
- Keep Brix as high as possible.
- Lower the electrical loads.
- Run the plant at the highest possible rate.

3. The heat in the process may be reused by:
   - Process heaters
   - Multiple-effect evaporators
   - Flash blowdown steam

26.2.1 Steam Generation and Boiler Efficiency

Steam for a sugar refinery is generated in boilers that range in pressure from 100 psig saturated to 1200 psig superheated. The size and pressure of the boiler used are determined by the steam loads in the refinery, and whether or not electricity is being generated at the refinery. For distribution throughout the refinery, the steam will be reduced in pressure by turbines or pressure reducing valves (PRVs).

26.2.2 Feedwater (Makeup)

It takes 1 lb of feedwater to make 1 lb of steam. Feedwater is supplied at the same rate that steam is used. Dissolved solids are removed from the makeup water by softeners or by demineralization and evaporation, depending on the pressure of the boiler in which the water is used. Dissolved solids in water are measured in parts per million (ppm). The higher the boiler pressure is, the purer the water must be. High-pressure boilers require a lower (ppm) than do lower-pressure boilers. The purified makeup water passes through a deaerator to remove any oxygen. It is then heated and mixed with returning condensate from the process. Forty to 60% of the steam sent to the process is returned as condensate. The amount of condensate that is returned will depend on the condition of the evaporator and pan steam chests. Under no circumstances should condensate that contains any trace of sugar be returned to the boiler.

26.2.3 Typical Industrial Boiler Efficiency

In a sugar refinery, because of the nature of the pan operations, efficiency is lower than in most other industries. Sugar refinery boilers will be in the range 80 to 88%. Losses are summarized in Table 26.2.

TABLE 26.2 Boiler Losses (%)

Dry gas loss	5.16
Loss due to moisture and or hydrogen in the fuel	4.36
Loss due to unburned fuel	0.50
Loss due to radiation	0.30
Loss due to moisture in air	0.13
Loss due to tube corrosion	1.50
Total losses	11.95
Efficiency	88.05

26.2.4 Determining Boiler Efficiency

a. Direct Method.

$$\frac{\text{output}}{\text{input}} = \frac{\text{heat added to the incoming feedwater}}{\text{heat input of the fuel}}$$

The following data are necessary to calculate efficiency using the direct method:

A. Steam flow (lb)
B. Steam pressure (psig)
C. Steam temperature (°F)
D. Steam enthalpy (Btu/lb)
E. Feedwater flow (lb)
F. Feedwater temperature (°F)
G. Feedwater enthalpy (Btu/lb)
H. Blowdown flow (lb)
I. Blowdown temperature (°F)
J. Blowdown enthalpy (Btu/lb)
K. Fuel flow (lb, ft^3, gal)
L. Heat content of fuel [Btu/(lb, ft^3, gal)]

$$\text{Efficiency} = \frac{(A) \times (D) + (H) \times (J) - (E) \times (G)}{K \times L}$$

The limitations of this method are:

- Accuracy of the flowmeters
- Accuracy of temperature and pressure meters
- Accuracy of fuel weight (solid fuels)
- Heat content of the fuel (measurement accuracy)

b. Heat Loss Method. In this method we calculate the individual losses and subtract them from 100%. This method is very laborious, and not often used in the sugar industry. The use of this method can be obtained from most good textbooks that deal with steam boilers. Although the results from this method are generally a little more accurate than those of the direct method, it is far more time consuming.

Two items within the operator's control that effect boiler efficiency are dry gas loss and unburned fuel loss. Dry gas loss, usually the largest single item affecting boiler efficiency, increases with higher exit gas temperature or excess air valves. Every 35 to 40° of increase in exit gas temperature will lower the boiler efficiency 1%. A 1% increase in excess air by itself only decreases efficiency by only 0.05%. However, increased excess air leads to higher exit gas temperature, because less heat is transferred to the boiler tubes. Thus excess air can have a twofold effect on boiler efficiency. Unburned fuel loses will vary with the ability of the sugar boilers to keep the pan cycles separated. Pan cycles are not always controlled by the sugar boiler, but sugar boilers and driers that work well together can often keep the pan cycles apart, which will result in smoother, more efficient boiler operations.

26.2.5 Excess Air Required for Combustion

The percent excess air and the oxygen requirements for various fuels are listed in Table 26.3. The most effective way to achieve maximum boiler efficiency in a day-to-day operation is to embark on an educational campaign for plant management, supervisory staff, operators, and maintenance personnel. If everyone knows and understands the economic impact of the operation variables on fuel cost, this can lead to significant fuel savings.

TABLE 26.3 Percent Excess Air and O_2 Required for Various Fuels[a]

Fuel	Lb Air / lb Fuel	% Excess Air	% Oxygen
Gas	18.5	12–15	2.5–3
Oil	15.6	10–15	2–3
Coal	13.2	15–20	2.5–3

[a]Our atmosphere is composed of 23% O_2 by weight (0.075 lb / ft^3).

26.2.6 Boiler Tune-up

Proper maintenance and adjustment of boilers can make a substantial difference in fuel consumption. Combustion efficiency and fouling deserve special attention if a boiler is to deliver peak performance.

26.2.7 Combustion Efficiency

Combustion efficiency is a measure of how effectively the heat content of a fuel is being transferred in the burning process. Factors such as fuel atomization, airflow, and combustion temperature affect combustion efficiency. The combustion efficiency of a boiler can be determined easily by using an inexpensive CO_2 absorption kit to measure carbon dioxide in the flue gases. Table 26.4 indicates the type of readings that can be expected.

26.2.8 Fouling

Fouling of the boiler's tubes by soot or scale acts as insulation and prevents necessary heat transfer. Proper water treatment to prevent scale formation is essential for efficient operation of the boiler. Similarly, the fire side of boiler tubes must be cleaned of any soot deposits when oil or coal is burned. A good indication of a fouling problem occurs when boiler stack temperatures exceed the saturated steam temperature at the boiler operating pressure by 150°F or more.

26.2.9 Keeping Boiler Tubes Clean (Water Side)

The prevention of scale formation, even on small boilers, can produce substantial energy savings. For an individual case, the potential saving depends on the scale thickness and its chemical composition. Check boiler tubes visually for scale while the boiler is shut down for maintenance. Operating symptoms that may be due to scale include reduced steam output, excessive fuel use, and increased stack temperature. If scale is present, consider modifying the feedwater treatment and/or the schedule of chemical additives. The cost of

TABLE 26.4 Combustion Efficiency Using % CO_2

Rating[a]	Gas	No. 2 Oil	No. 6 Oil and Coal
A	10% CO_2	12.8% CO_2	13.8% CO_2
B	9.0% CO_2	11.5% CO_2	13.0% CO_2
C	8.5% CO_2	10% CO_2	12.5% CO_2
D	8.0% or less	9.0% or less	12.0% or less

[a]A, excellent; B, good; C, fair; D, poor.

modification can vary widely, depending on such factors as the type of treatment facilities already available and the chemical problems present, if any. The advice of a consultant or of a vendor of water treatment chemicals can be helpful.

Figure 26.1 shows the energy that is lost for various scales and scale thickness. Figure 26.2 shows deposits that can form in tubes when improperly treated feedwater is supplied to a boiler. Note how the tube is starting to "egg shape" where the scale deposit is the heaviest. If this scale is not removed, a tube rupture will occur as shown in Figure 26.3.

26.2.10 Heat Recovery from Boiler Flue Gases

Boiler flue gases are emitted to the boiler stack at temperatures at least 100 to 150°F higher than the temperature of the generated steam. Obviously, recovering a portion of this heat will result in higher boiler efficiencies and reduced fuel consumption. Heat recovery can be accomplished by using either an economizer to heat the feedwater stream or an air preheater for the combustion air. Normally, adding an economizer is preferable to installing an air preheater on an existing boiler, although air preheaters should be given careful consideration in new installations.

Before going through the calculations to determine how much heat can be recovered from flue gases, the boiler should be operating close to optimum excess air levels. It does not make sense to use an expensive heat recovery system to correct for inefficiencies caused by improper boiler tune-up. The lowest temperature to which the flue gases can be cooled depends on the type of fuel used: for example, 250°F for natural gas, 300° for coal and low-sulfur-content fuel oils, and 350°F for high-sulfur-content fuel oils. These limits are set by the flue gas dew point, cold-end corrosion, and heat driving force considerations.

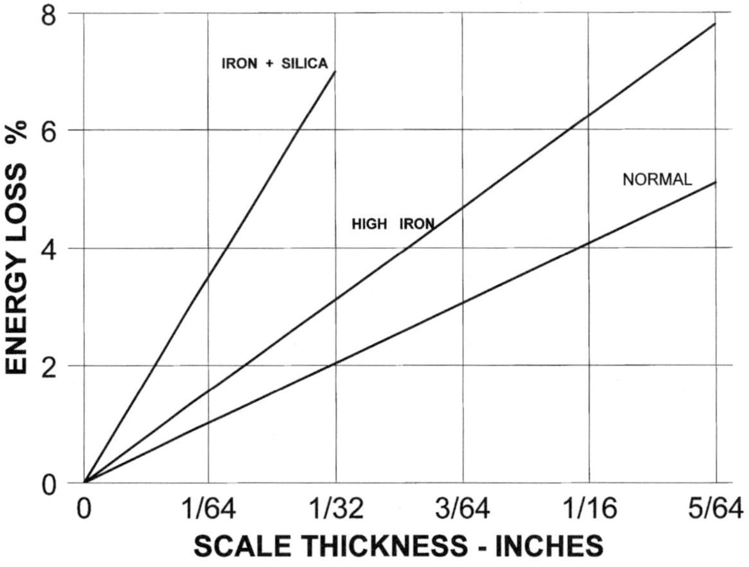

FIGURE 26.1 Energy loss from scale deposits in boiler tubes. (From Georgia Tech Energy Extension Service, *NBS Handbook 115,* Energy Tip 8.)

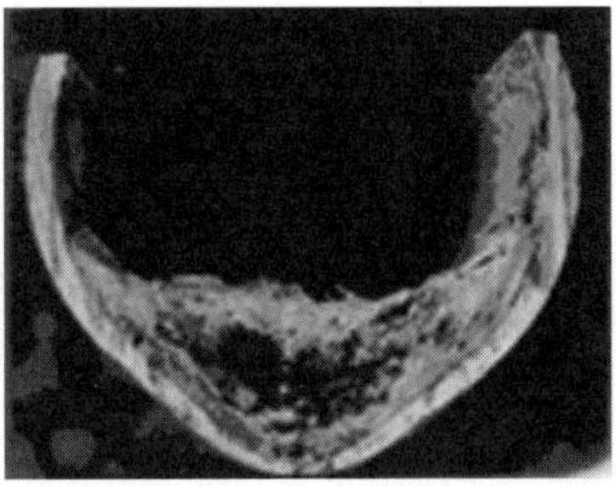

FIGURE 26.2 Internal tube deposits from normal scale. (Courtesy of Tom Gerrald.)

26.2.11 Flashing High-Pressure Condensate to Regenerate Low-Pressure Steam

High-pressure condensate can be flashed to lower-pressure steam. This is an alternative to throttling high-pressure steam to meet low-pressure steam needs. Flashing is particularly attractive when the return of condensate to the boiler is not economically feasible.

26.2.12 Minimizing Boiler Blowdown

Boiler blowdown is required to maintain the concentration of dissolved solids in the boiler water at acceptable levels. This operation is wasteful of energy since the purged stream is at the same temperature as the steam generated in the boiler. Minimizing blowdown rates reduces energy losses and improves boiler efficiency. Several measures are available to minimize blowdown, such as:

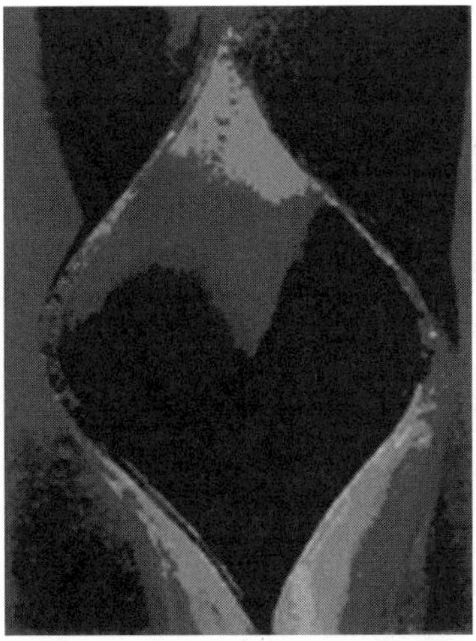

FIGURE 26.3 Wide-mouth rupture from a plugged tube. (Courtesy of Tom Gerrald.)

- For boilers operating between 0 and 300 psig, the American Boiler Manufacturers Association recommends a maximum total dissolved solids concentration of 3500 ppm. If the actual concentration in the blowdown stream is below this figure, it may be possible to decrease blowdown rates without causing fouling of the heat transfer surfaces.
- Chemical treatment of the makeup water lowers the feed total dissolved solids concentration and makes it possible to reduce blowdown rates without affecting the level of solids concentration in the purge stream.
- Use of antifoulants increases the maximum concentration of solids that can be tolerated in the boiler, thus allowing a reduction in blowdown rates. Care should be exercised because the maximum permissible level of solids, as well as the water treatment type and the antifoulant type, all depend on specific makeup water composition. Water treatment experts should be consulted to ensure that specified conditions will not cause excessive fouling.

26.2.13 Heat Recovery from Boiler Blowdown

Heat can be recovered from boiler blowdown by preheating the boiler makeup water with the use of a heat exchanger. This is done most conveniently in continuous blowdown systems. Figure 26.4 shows the heat that can be recovered from boiler blowdown.

1. Determine the method and quantity of boiler blowdown.
2. If continuous blowdown is not used, consider installation of a continuous blowdown system coupled with heat recovery.
3. Determine the savings due to heat recovery.
4. Determine the cost of installing a heat exchanger and continuous blowdown system if necessary.

26.3 PREVENTING THE LOSS OF HEAT

26.3.1 Insulating Bare Steam Lines

Bare steam lines are a constant source of wasted energy. Proper insulation can typically reduce energy loss by 95% and help ensure proper steam pressure to plant equipment. Figure 26.5 indicates approximate heat loss from bare pipe at the indicated pressures. Thermal conductivity or k factor is the amount of heat in Btu/h that will flow through a slab of material 1 ft by 1 ft by 1 in. thick when one surface is 1°F hotter than the other. The heat flow through insulation is measured in Btu/(ft^2-h) and is equal to k (temperature difference)/insulation thickness. The k is available for pipe insulation. The difference in the temperature of the pipe and the temperature of the surrounding air can vary depending on the area of the world you are in.

Example: With 4 in. of insulation, a temperature difference of 200°F, and a k factor of 0.05, the heat flow is 2.5 Btu/h.

26.3.2 Eliminate Steam Leaks

An obvious waste of energy occurs when steam leaks at pipe joints, valves, unions, and so on. Until the cost of energy skyrocketed, it was generally thought that small leaks should

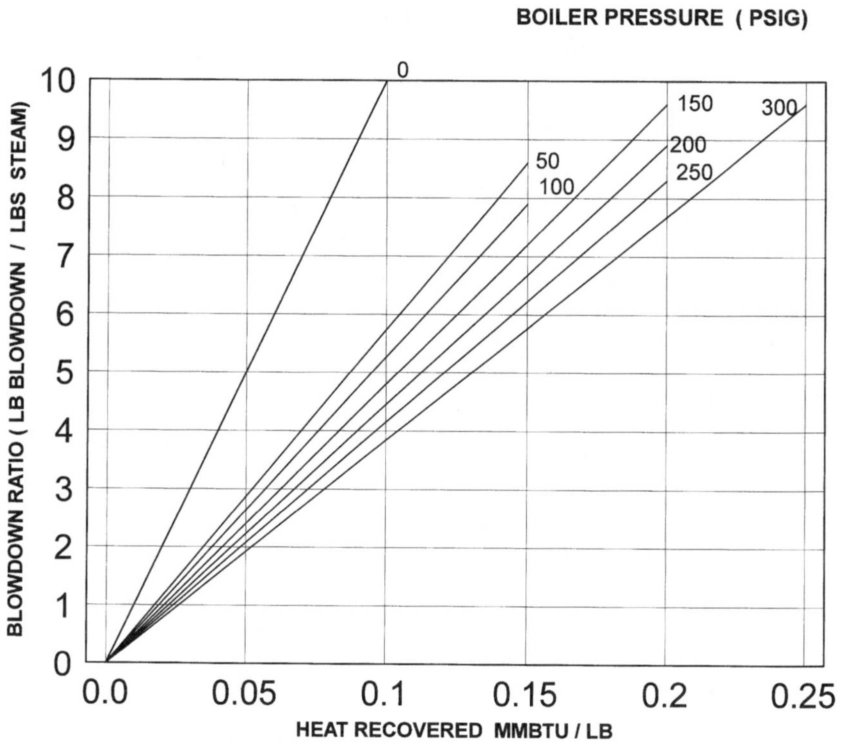

FIGURE 26.4 Heat recovered from blowdown. (From Georgia Tech Energy Extension Service, *NBS Handbook 115,* Energy Tip 23.)

be tolerated and that fixing them was not worth the time or cost. Table 26.5 shows the steam loss for holes of various sizes.

26.3.3 Inspecting and Repairing Steam Traps

Efficient operation of any steam system requires well-designed trapping that is inspected periodically and properly maintained. It is only in this way that condensate and air will be removed automatically as fast as they accumulate without wasting steam. Table 26.6 shows the energy loss per year through leaking traps. Initial inspections commonly reveal that as high as 7% of the traps in a system are leaking. It has been demonstrated that by careful maintenance and frequent inspection this can be reduced to 1% by following these guidelines:

1. Establish a program for the regular systematic inspection, testing, repairing, or replacing of steam traps. Include a reporting system to ensure thoroughness and to provide a means of establishing the continuing value of the program.
2. Inspection and testing on a regular basis should provide answers to the following questions:
   a. Is the trap removing all of the condensate and noncondensables?

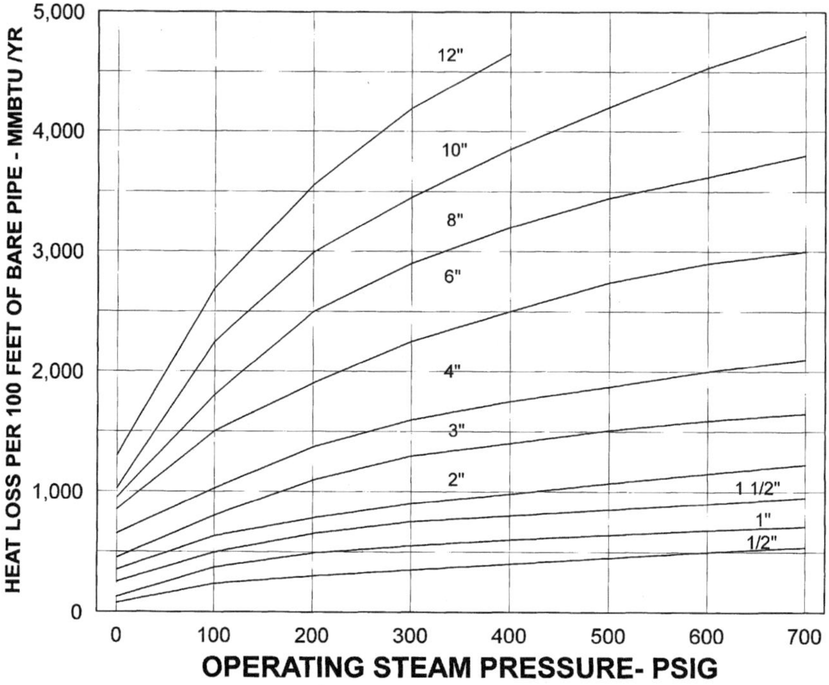

FIGURE 26.5 Heat loss per 100 ft of bare steam pipe. (From Georgia Tech Energy Extension Service, *NBS Handbook 115,* Energy Tip 4.)

b. Does it shut off tight after operation?
c. Is bypass, or separate discharge, closed and free of leaks?
d. Is frequency of discharge in an acceptable range? Too frequent discharge indicates possible undercapacity; too infrequent discharge indicates possible overcapacity inefficiency.

26.3.4 Eliminating Leaks in Compressed Air Lines

The cost of leaking compressed air is often considered insignificant. The following examples illustrate that appreciable energy savings can be realized by repairing leaking air lines.

TABLE 26.5 Loss of Steam (lb / h) Through Various Size Holes

Pressure	Hole Size (in.)				
(psig)	$\frac{1}{16}$	$\frac{1}{8}$	$\frac{1}{4}$	$\frac{1}{2}$	1
100	18.2	72.6	280.3	1161.3	4645.1
200	33.9	135.7	542.8	2171.1	
400	65.5	261.8	1847.7		
850	138.5	545.8			

Source: Power magazine, January 1979.

TABLE 26.6 Leaking Steam Trap Heat Loss[a]

Trap Orifice Diameter (in.)	Steam Pressure (psig) 15	30	50	100	150	300
$\frac{1}{8}$	160	235	355	610	910	1,180
$\frac{3}{16}$	370	525	805	1,400	2,050	2,625
$\frac{1}{4}$	655	875	1,425	2,410	3,640	4,335
$\frac{3}{8}$	1,470	1,970	3,220	5,430	8,200	10,500
$\frac{1}{2}$	2,625	3,765	5,720	10,600	14,565	18,300

Source: Data from *NBS Handbook 115,* Energy Tip 4.

[a]Based on 8,760 operating hours per year, steam discharging to atmospheric pressure.

Example: A complete inspection of a plant's compressed air system was conducted at the start of a regular monthly leak detection program. The air compressor discharge pressure was 100 psig. At a power cost of \$0.03 per kilowatthour, the cost of the leaks found in the compressed air system as shown in Table 26.7. The energy and cost savings possible by fixing compressed air system leaks can be estimated from Tables 26.7 and 26.8. For electrical costs other than those listed, use multiples of table values.

26.3.5 Installing Traffic Doors on Frequently Used Openings

Frequently used doors in heated warehouses and plants are often left open for traffic purposes during operating hours, regardless of weather conditions. This practice can lead to significant heat losses from a building during the heating season. These losses can be minimized by the installation of traffic doors on such openings, which confine heat and open only enough for passage.

Suggested Action: Estimate the annual heat loss for a specific case from Figure 26.6 and then determine the annual fuel cost savings that can be realized. Make an economic evaluation to determine if installation of a traffic door can be justified. Suppliers of transparent, overlapping plastic strip doors claim that these doors can eliminate 90 to 95% of the heat loss through the open area. The cost of such doors for openings of 60 to 100 ft^2 is in the range of \$300 to \$500, and a typical installation cost is \$75. In the southeastern United States, with a door open 5 to 10 h/day, payout times (before taxes) of 1 year or less are not uncommon.

TABLE 26.7 Cost of Air Leaks at 100 psig

Number of Leaks	Estimated Diameter (in.)	Air Lost per Year (ft^3)	Fuel Wasted (MMBtu/yr)	Cost at \$0.03/kWh
3	$\frac{1}{4}$	107,000,000	2,920	\$8,740.00
7	$\frac{1}{8}$	62,200,000	1,700	5,100.00
12	$\frac{1}{16}$	26,600,000	727	218.00
15	$\frac{1}{32}$	8,300,000	227	680.00
37		204,000,000	5,574	\$14,738.00

Source: Data from *NBS Handbook 115,* Energy Tip 13.

TABLE 26.8 Air Leaks at 100 psig

Hole Diameter (in.)	Air Lost per Year (ft^3)	Fuel Wasted (MMBtu / yr)	Cost at: $0.02 / kWh	$0.03 / kWh	$0.04 / kWh
$\frac{3}{8}$	79,900,000	21,900.0	$4370.00	$6560.00	$8740.00
$\frac{1}{4}$	35,500,000	972.0	1940.00	2920.00	3880.00
$\frac{1}{8}$	8,800,000	243.0	486.00	728.00	972.00
$\frac{1}{16}$	2,220,000	60.6	121.00	182.00	242.00
$\frac{1}{32}$	553,000	15.1	30.30	45.40	60.60

Source: Data from *NBS Handbook 115,* Energy Tip 13.

26.4 REDUCING THE WORK REQUIRED OF HEAT

26.4.1 Affination Station

A properly mingled magma will purge in the centrifugals with a minimum amount of wash water.* This can reduce the amount of affination syrup. Excess affination syrup must be boiled on the low-grade station. Centrifugals should be sequenced to prevent high electrical demand loads. The use of digital timers allows the setting of the exact amount of wash water required in the centrifugals. Wash water valves should be checked regularly for leaks.

Note: In a single-effect evaporator, it takes 1 lb of steam to evaporate 0.9 lb of water.

26.4.2 Brix Control

Example: Each of the 12 centrifugals uses 10 lb of water on the preflush and 15 lb of water for washing.

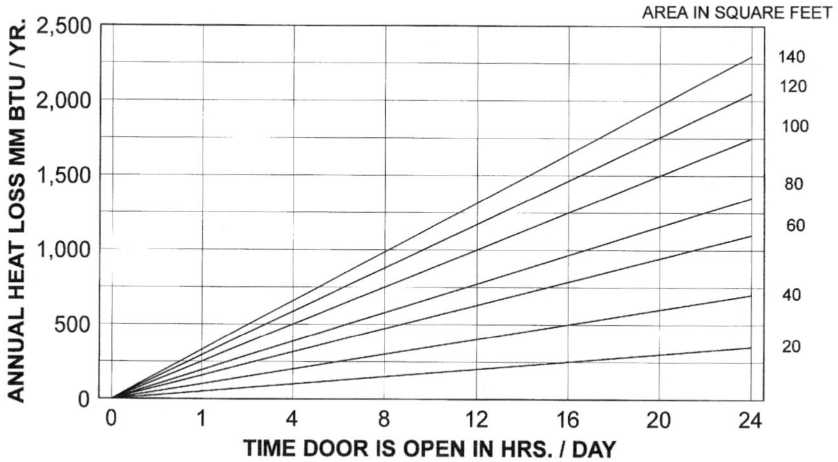

FIGURE 26.6 Heat loss through open doors. (From Georgia Tech Energy Extension Service, *NBS Handbook 115,* Energy Tip 5.)

*In Section 15 of Ref. 4, C. F. Stowe gives an example of too much wash water increasing the affination syrup solids by 12%.

$$12 \text{ centrifugals at 3-min cycles} = 5760 \text{ cycles/day}$$

If the wash water to each machine is increased by 1 lb on each machine:

$$(5760 \times 25) - (5760 \times 26) = 5760 \text{ lb/day} \times 230 \text{ days} = 1{,}324{,}800 \text{ lb/yr}$$

$$1{,}324{,}800 \text{ lb/yr } (1395 \text{ Btu/lb} - 198 \text{ Btu/lb})/0.86 \text{ (boiler efficiency)} = 1844 \text{ MMBtu/yr}$$

$$1844 \text{ MMBtu/yr} \times \$2.00/\text{MMBtu (coal)} = \$3688/\text{yr}$$

$$1844 \text{ MMBtu/yr} \times \$3.36/\text{MMBtu (oil)} = \$6194/\text{yr}$$

Example: Cost to lower Brix from 65 to 63. In a refinery where the melt is 700,000 tonnes/yr:

$$(700{,}000 \text{ tonnes}/0.63 \text{ Brix}) - (700{,}000 \text{ tonnes}/0.65 \text{ Brix}) = 34{,}188 \text{ tonnes water/yr}$$

It takes 1 lb of steam to evaporate 0.9 lb of water:

$$(34{,}188/0.9) = 37{,}987$$

For steam at $8.00/tonne:

$$(37{,}987 \text{ tonnes steam}) \times (\$8.00/\text{tonne}) = \$303{,}893/\text{yr}$$

26.4.3 Electrical Power

Electrical power is purchased, generated, or some combination of the two. Most refineries generate at least a portion of their power requirements. The turbogenerators used are usually backpressure or extraction units. A backpressure turbine receives superheated steam through the inlet nozzles, and exhausts at a lower pressure that is used in the process.

An extraction machine discharges steam at two pressures. The exhaust pressure, out the back of the turbine, will be the same as in the backpressure unit. There is also steam extracted at a point nearer the inlet valves that is higher than the exhaust pressure. This steam is also used in the process (Fig. 26.7). Steam that leaves a turbine will still have some superheat in it and must be cooled to saturation temperature for steam at the same pressure. The cost of electricity generated by refinery turbines is generally $0.02/kWH or less.* However, if the extraction or exhaust steam is blown to atmosphere, the cost of the electricity generated can be higher than that of purchased electricity.

26.4.4 Demand Charge

A portion of the plant's monthly electric bill from a utility company is a demand charge. A demand meter looks at the plant load every 15 min and registers the highest demand for the month, which is what the demand charge is based on. Some plants have gone to load shedding to keep demand cost down. One way to help keep the demand down is to make sure that the centrifugals are sequenced.

---

* This number is available from you company's powerhouse cost sheet, or may be calculated as follows:

$$[\text{fuel cost (\$)/MMBtu} \times \text{heat in the turbine (MMBtu/h)}] \times \text{(kilowatts generated by the turbine)}$$

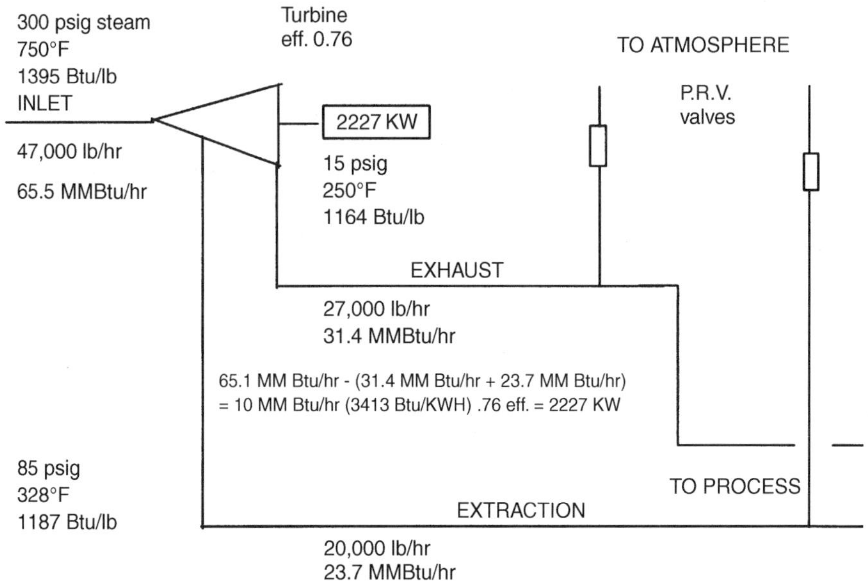

FIGURE 26.7 Turbine with extraction and exhaust.

26.4.5 Power Factor

The rate structure of many utility companies includes a power factor clause, which results in increased power cost when the power factor is below a specific level. Ninety-five percent of the plant electrical load is induction motors. A typical 10-hp 240-V ac motor will have a power factor of 86% (1 hp = 0.746 kW). The power factor can be improved to 97% with the use of capacitors.

$$\text{Power factor} = \frac{\text{kilowatts (kW)}}{\text{kilovolt amperes (kVA)}}$$

As an example of the use of the kilowatt multiplier table, assume that the original power factor is 70%, the original load is 100 kVA, and the improved factor is to be 95%. The load kW is

$$\text{kW} = 0.7 \times 100 = 70$$

From Table 26.9, the kilowatt multiplier is 0.691 for improving the power factor from 70% to 95%. The capacitor kilovars (kVAR) required is then found to be

$$\text{kVAR} = 70 \text{ kW} \times 0.691 = 48.4$$

For large motor control centers or substations, increasing the power factor with capacitors can result in using smaller cable and conduit, which will cost less to purchase and install. Power factor is defined as the cosine of the angle between active power, and apparent power (Fig. 26.8)

TABLE 26.9 Kilowatt Multipliers for Determining Capacitor Kilovars

Original Power Factor (cos)	Desired Improved Power Factor (cos)			
	95	90	85	80
60	1.004	0.849	0.713	0.583
62	0.937	0.732	0.646	0.516
64	0.872	0.717	0.581	0.451
66	0.809	0.634	0.518	0.388
68	0.749	0.594	0.453	0.328
70	0.691	0.536	0.400	0.270
72	0.635	0.480	0.344	0.214
74	0.580	0.425	0.289	0.159
76	0.526	0.371	0.235	0.105
78	0.473	0.318	0.182	0.052
80	0.421	0.266	0.130	0.000
82	0.369	0.214	0.078	
84	0.317	0.162	0.026	
86	0.264	0.109		
88	0.211	0.056		
90	0.155	0.000		
92	0.097			
94	0.034			

Source: Adapted from Editors of *Power* magazine (reviewed by J. D. Carpenter), *Power Handbook,* 2nd ed., McGraw-Hill, New York, 1983.

Example: A 100-kW load is operated at 0.86 power factor (100 kW/0.86 power factor = 116 kVA). If the power factor is improved to 0.96 (100 kW/0.96 power factor = 103 kVA). This will free up 13 kVA that can be used for other loads.

26.4.6 Motor Efficiency

$$\text{Efficiency} = \frac{[(746\text{ W/hp})\ (\text{hp})] \times 100}{\text{watts input}} \qquad (\text{watt} = \sqrt{3}\, ei \cos\theta)$$

High-efficiency motor savings:

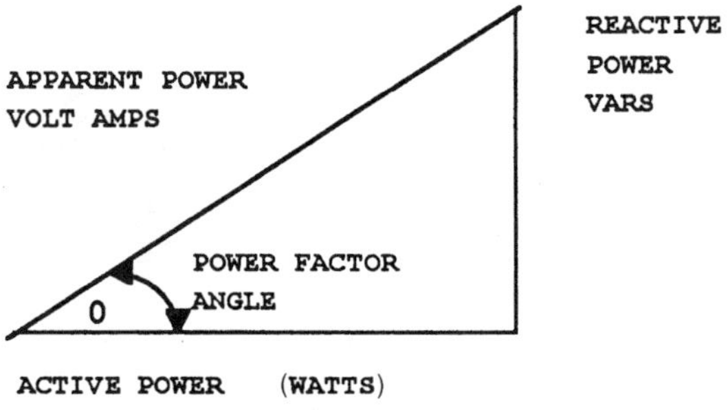

FIGURE 26.8 Power factor angle.

$$[(0.746 \text{ kW/hp})(\text{hp})(\$/\text{kWh})(\text{h/yr})][(100/\text{std }\%) - (100/\text{high }\%)]$$

$$[(0.746 \text{ kW/hp})(50 \text{ hp})(\$0.08/\text{kWh})(7200 \text{ h/yr})][(100/86) - (100/95)] = \$2366/\text{yr}$$

The use of energy-efficient motors to save energy and money can be broken down into three general categories: (1) in place of rewinding failed motors, (2) to retrofit an operating but inefficient motor for energy conservation, and (3) when a new motor is purchased. When making this decision, remember that the purchase cost of a new motor is a fraction of the operating cost over the life of the motor. To illustrate this, a 100-hp motor driving a forced-draft fan on a boiler operates 8000 h/yr, at full load, in a location where the cost of electricity is $0.07/kWh. For a standard efficiency motor (86.5% efficient) the cost to operate the motor is

$$(100 \text{ hp})\ (0.746 \text{ kW/hp})(\$0.07/\text{kWh})(8000 \text{ h/yr}) \times (100 \div 86.5) = \$48{,}296/\text{yr}$$

The annual cost of electricity being converted to heat is $6520 and $41,776 is the cost of electricity converted to mechanical energy.

In most cases, the cost of annual wasted energy exceeds the cost of the motor purchased. A high-efficiency motor 92% would cost $45,409/yr to operate, with $3633/yr losses. A premium efficiency motor 94.5% would cost $44,207/yr to operate, with $2431/yr losses. When projecting these numbers over the life of the motor, it is obvious that the purchase price of a motor is insignificant. With this in mind, the highest-efficiency motor should always be purchased. Table 26.10 indicates the importance of rewinding a failed motor to its original efficiency. When a decision is made to rewind a failed motor, use a qualified rewind shop. A good rewind shop can maintain the original efficiency of a motor; However, if the motor core has been damaged in the failure, it may not be possible to restore the motor to its original efficiency.

The payback for using a premium efficient motor in place of a failed standard efficiency motor is

$$\begin{aligned} \text{savings} &= (100 \text{ hp})(0.746 \text{ kW/hp})(\$0.07/\text{kWh})(8000 \text{ h/yr}) \\ &\quad \times (100 \div 86.5 - 100 \div 94.5) = \$4088/\text{yr} \end{aligned}$$

The payout is:

$$(\$3493 - \$2806) \div \$4088/\text{yr} = 0.168 \text{ yr}$$

This assumes that the labor to install either motor would be the same; However, if the motor base needs to be modified, this cost should be added to the difference in the motor cost.

TABLE 26.10 Comparison of Standard, High, and Premium Efficiency Motors

Motor Efficiency	Cost of Motor	Operating Cost per Year	% of Purchase	Cost of Wasted Energy	Energy Wasted Motor Cost (%)
Standard	$2,806	$48,296	1,721	$6,520	232
High	3,175	45,409	1,430	3,633	114
Premium	3,493	44,207	1,266	2,431	70

26.4.7 Converting to Energy-Efficient Fluorescent Lamps

One attractive alternative for energy conservation is replacing existing fluorescent lamps with the lower-wattage energy-efficient lamps. Generally, the new lamps are of lower light output and the light level will be reduced by about 5 to 11%. However, the wattage reduction will range from about 14 to 20% (see Table 26.11).

Example: An industrial area of 20,000 ft^2 is presently lighted to 85 footcandles (fc; the amount of direct light thrown by one candle on 1 ft^2 1 ft away) with 75-W standard fluorescent lamps (F96T12/CW). The number of two-lamp fixtures is 300. Mounting height is 20 ft. Replacing these lamps with 60-W energy-efficient lamps would yield the following savings:

- *Annual energy savings:* 9 kW × 10 h/day × 250 days/yr = 22,500 kWh
- *Annual cost savings:* 22.500 kWh at $0.04 = $900
- *Total initial cost (lamps and labor)*: $1938 + $1500 = $3438
- *Simple payback period:* $3438/900 = 3.8 yr

Maintained light level with new lamps is 75 fc. Establishing a policy of spot replacing lamps with energy-efficient ones yields a very early payback as follows:

- *Cost of lamps:* 600 lamps × ($3.23 − $3.16) = $42
- *Simple payback period:* $42/900 = 0.05 yr

Suggested Action: The following steps, in order, are suggested:

1. Determine if the slightly lower light level is acceptable. In most instances this is acceptable. If not, an improved maintenance program (i.e., washing fixtures more frequently) will boost the level with reduced wattage.
2. Immediately replace lamps operating beyond their efficient life or burned-out lamps with more energy-efficient lamps.
3. Survey existing lamps and fixtures and justify the replacement of as many of the standard lamps as possible with energy-efficient lamps.

26.4.8 Converting to More Efficient Light Sources

There is excellent savings potential in most plants by converting present lighting systems to more efficient light sources. A condensed comparison of light sources is shown in Table 26.12.

TABLE 26.11 Comparison of Lamp Types

Lamp[a]	Watts	Lumens	Length (in.)	Life (h)
Standard F40CW	40	3150	48	20,000+
Energy-efficient lamp	34–35	2800–3050	48	20,000+
Standard F96T12	75	6300	96	12,000+
Energy-efficient lamp	60	5400–6000	96	12,000+
Standard F96T12/CW/HO	110	9200	96	12,000+
Energy-efficient lamp	95–98	8500	96	12,000+

Source: Data from *Spectrum 9200 Lamp Catalog,* 22nd ed., General Electric.

[a]CW, cool white.

TABLE 26.12 Lumens per Watt

Light Source	Light (lumens) per Watt
Incandescent	17–22
Mercury	56–63
Fluorescent	67–83
Metal halide	80–115
High-pressure sodium	80–140

Source: Data from *Spectrum 9200 Lamp Catalog,* 22nd ed., General Electric.

26.4.9 Illumination from Using Skylights

In many plants the power required for lighting production areas, warehouses, or offices can be reduced significantly by installing skylights. Plants that are presently using skylights for building illumination have reported that electrical lighting is required only 20% of the time during daylight hours.

26.5 USING HEAT EMITTED IN THE PROCESS

26.5.1 Additional Effects to Reduce Evaporator Steam Cost

During the design of existing evaporation systems a great deal of attention was probably paid to selection of the optimum number of effects. However, as fuel prices continue to rise, and steam generation becomes more expensive, it may be economically attractive to add one or more effects to an existing evaporator. Each additional effect causes less steam to be consumed for a given amount of evaporation. Figure 26.9 can be used to estimate the annual steam costs for given evaporation rates, steam costs, and number of effects. The curves are based on 350 days at 24 h per operating days per year, and the assumption that $0.9N$ lb of H_2O is evaporated per pound of steam consumed (N = number of effects).

Suggested Action: Use Figure 26.9 to estimate the energy savings possible from increasing the number of effects in your evaporation system. The result may warrant a serious look at an evaporator upgrade.

26.5.2 Evaporation

Multiple effect is the reusing of the latent heat in vapors for successive evaporation.

Number of Effects	Water Removal/lb Steam
Single	0.9
Double	1.8
Triple	2.8

The original heat in the steam is split into two parts:

1. *Latent heat of vaporization:* the amount of heat added to a substance, which is at its boiling point, that will cause 1 lb of steam to be liberated from the substance (the latent

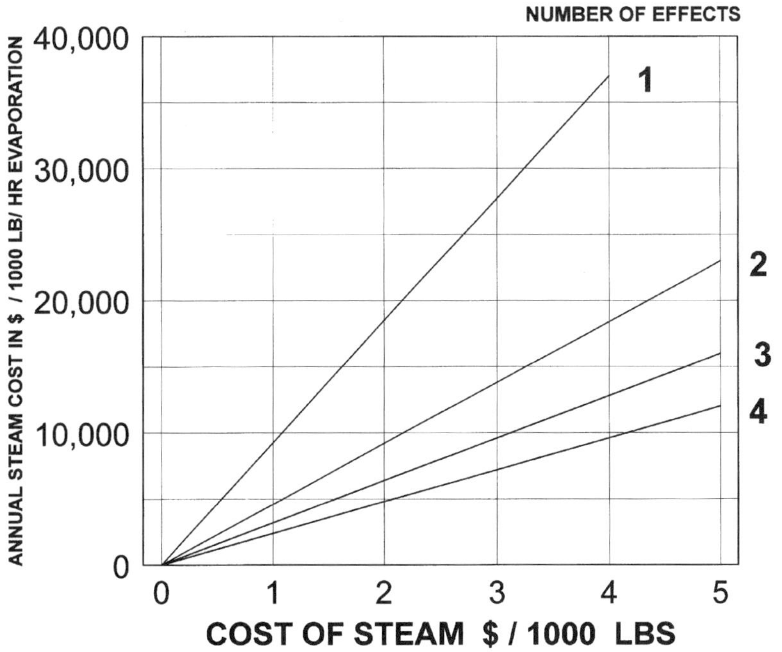

FIGURE 26.9 Multiple-effect evaporation. (From Georgia Tech Energy Extension Service, *NBS Handbook 115,* Energy Tip 16.)

heat of water is 970 Btu/lb). Latent heat is transferred to the product being treated and causes the formation of further steam by evaporating water from the product.

2. *Sensible heat:* the amount of heat required to raise a substance from its present temperature to its boiling point in Btu/lb (water has a sensible heat of 180 Btu/lb at 212°F). Sensible heat is retained in the condensate that forms when the original steam condenses.

In evaporation by steam, only the latent heat is transferred to the product being treated. This transferred heat is absorbed (1) as sensible heat to raise the product to its boiling point or (2) as latent heat for vaporization of water from the product.

The water vapor from the product is at a lower pressure than that of the original steam but holds all the heat of that steam, less the sensible heat left in the condensate, which is the sensible heat used to raise the product to its boiling point. Figure 26.10 shows an arrangement of boiling 20 Brix sweetwater in a vacuum pan to 65 Brix. The calculation associated with Figure 26.10 confirms that it takes 1 lb of steam to evaporate 0.9 lb of water in a single-effect evaporation system. Using the data in Figure 26.10, we can calculate the following:

1. Add sensible heat to raise temperature to (160 + 1.00) − 100 = 61.00°F.

$$[5400 \text{ lb/h})/0.20 \text{ Brix}] \times (61°\text{F} \times 0.91 \text{ Btu/°F-lb}) = 1.5 \text{ MMBtu/h}$$

2. Add latent heat to vaporize the water.

$$(1/0.20 \text{ Brix} - 1/0.65 \text{ Brix})(5400 \text{ lb/h})\ (946 \text{ Btu/lb}) = 17.70 \text{ MMBtu/h}$$

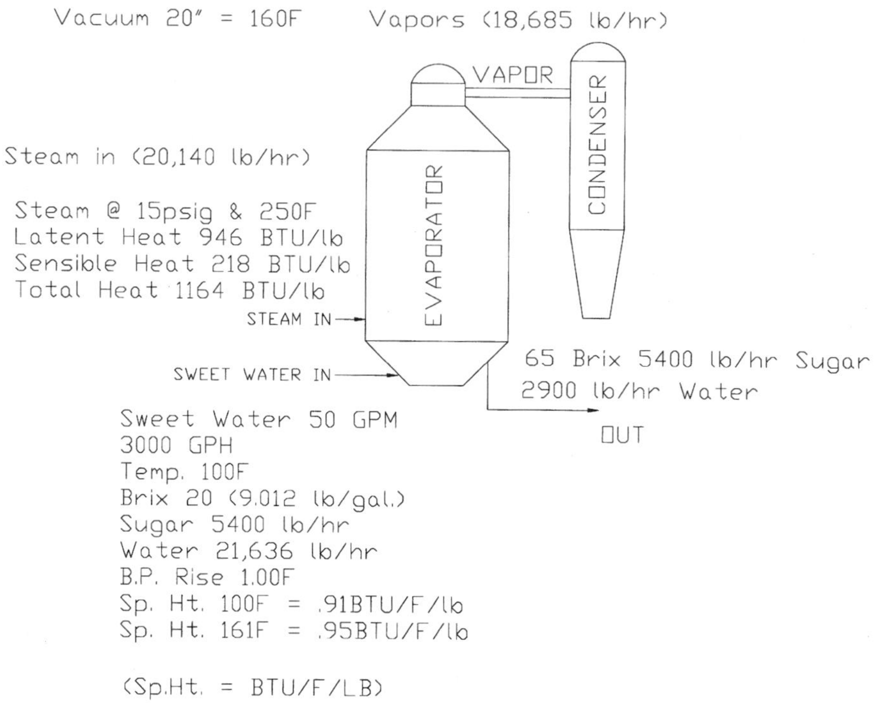

FIGURE 26.10 Evaporation using a vacuum pan.

3. Calculate the vapors evaporated.

$$(1/1.8 \text{ lb/gal} - 1/7.135 \text{ lb/gal})(5400 \text{ lb/h})(8.33 \text{ lb/gal}) = 18{,}685 \text{ lb/h}$$

4. Calculate the steam in.

$$(19.20 \text{ MMBtu/lb})/946 \text{ Btu/h}) = 20{,}295 \text{ lb/h}$$

5. Calculate the ratio of vapor to steam.

$$(18{,}685 \text{ lb/h})/(20{,}295 \text{ lb/h} = 0.92$$

26.5.3 Reusing Vapors for Heating

Improved energy economy can be achieved by using vapors from evaporators and pans for heating other process steams in place of primary steam. Thermal economy can be improved using this technique when a favorable steam to power ratio prevails and when a large surplus of vapor is unused and condensed.

ACKNOWLEDGMENT

This chapter is dedicated to my mentor in the sugar industry: Walter F. Oetgen, Jr., Chief Engineer, Savannah Sugar Refinery, 1954–1973.

REFERENCE

1. O. Lyle, *Technology of Sugar Refinery Workers*, 3rd ed., Chapman and Hall, Ltd., London, The Caxton and Holmesdale Press, Sevenoaks Kent, 1957.

CHAPTER 27

Technical Control and Sucrose Loss*

INTRODUCTION

Because of its continuous-flow process, large refineries, and 24-h operations, the sugar industry has always required a process control system to maintain efficient operation. In years past the chief chemist was the person responsible for overall technical operation. This person would oversee the laboratory operations, process control, quality control, raw materials, yield, customer service, and any technical problems or questions that might arise. Today in most sugar refineries, it is the technical manager or director who contributes to or oversees all technological functions.

We have all types of new quality programs, such as ISO 9000, quality cycles, statistical process control, and so on, and some of these programs can be reviewed in the *Proceedings of the Workshop on White Sugar Quality* [1]. None of these programs are addressed in this chapter, but each in its own way serves to improve quality control. What we discuss in this chapter are some of the basic controls required to refine sugar efficiently.

The quality level of sugar has been maintained at a high standard throughout the years and recognized as such by the food industry. Some of the major reasons for maintaining good quality control are:

- To assure the food industry of a safe product, FDA approved
- To ensure consumer satisfaction with food that tastes and looks good
- To maintain uniformity, thus allowing the end user to control his product
- To maintain production efficiency: less reprocess, loss, energy, etc.
- To provide a tiebreaker in sales with a product that is at least equal to competitors
- To encourage pride and motivation for the company

The technical department, laboratory, or whatever it may be called must generate meaningful data and reports pertaining to the present-day operation. Today, many of these reports are on-line via computer programs and data available to the operators in their control rooms. This allows for more rapid communications and hopefully, response to the

*By Joseph F. Dowling.

problems. However, with the greater amount of data available today, it is important that individual technical reports be presented in a manner so that management can see the most critical control data. Too much data can sometimes mask the small variances that become bigger problems later. One should maintain historical data to verify the feasibility of good performance and other factors. Technical targets should be established throughout the process to ensure arriving at a satisfactory performance. These targets should be related to the refineries' limitations, such as age of equipment, decolorizing capacity, and energy. In evaluating controls and targets, it is important to relate variances to the actual cost to the refinery. Understanding what poor performance is costing will lead to faster operating changes, equipment replacement, and so on. Technical control is a basic cost control.

This chapter is divided into three main sections:

1. *Process control:* basic analysis required to control raw sugar to salable product
2. *Quality control:* basic analysis required to meet standards and customer requirements
3. *Yield and loss control:* controls required to operate efficiently

27.1 PROCESS CONTROL

Control of the refinery process starts with knowledge of the raw sugar entering the refinery. Not all raw sugars will refine in the same manner. In the United States, as in most countries, the raw sugar is valued based on the polarization value. The raw sugar polarization tells the refiner the approximate sucrose content of the raw sugar. It should be remembered that the polarization is decreased by percent invert and increased by percent dextran. The percent moisture content of the raw sugar is important in that excessive moisture can lead to microbiological and chemical degradation. This moisture content is used in calculating a safety factor. Microorganisms that normally cause deterioration do not develop in high-density solutions. Thus, the ratio of nonsucrose (100-polarization) to the water is critical. It is generally felt that the safety factor should be at least below 0.3 to 0.25. The safety factor is calculated as follows:

$$\text{safety factor} = \frac{100 \text{ moisture}}{100 \text{ polarization}}$$

Generally in refining, the chemical analysis of the washed raw sugar is more important than analysis of the raw sugar because it is normally washed raw crystals that go directly to the raw melter. The raw sugar can be washed in the lab using saturated sucrose solution and a laboratory basket centrifuge or by the Domino raw sugar contract method [2].

The laboratory-washed raw sugar should then be analyzed for color, invert, and ash. The color will give you an idea of the color load to be put on the refinery and can also serve as a target color for the plant wash station. The lab-washed raw color can be called a theoretical wash, and one would expect the plant to perform at about a 95% efficiency, calculated by the following formula:

$$\% \text{ efficiently} = \frac{\text{raw color} - \text{plant-washed raw color}}{\text{raw color} - \text{lab-washed raw color}}$$

If the percent washed raw efficiency number becomes too high, the sugar would normally be overwashed and more sucrose would be sent to the recovery house. In reviewing the washed raw sugar color, it may be useful to evaluate how it reacts to color removal plus clarification by performing lab tests. Not all raw sugars react in the same manner, so performing a lab clarification test with varying amounts of phosphoric acid, decolorant, and so on, will make it possible to operate in the optimum range. The percent invert in the washed raw is important since this is the starting concentration that the refining process will probably add to. The carbonation process is an exception wherein the invert is destroyed. Any invert created in the refinery stream can add to color buildup, slowing of crystallization, and so on, and therefore invert should be maintained at a low concentration. Knowing the percent invert in the washed raw sugar can also help in calculating the invert gain across the refinery. It should always be remembered that any invert gain is sucrose loss.

The percent ash in the raw sugar and washed raw sugar is important in maintaining white sugar yields, finished product percent ash content, scaling in evaporators, heaters, and pans, remelt recovery, ion-exchange operations, and so on. Raw sugars are also analyzed as to grain size distribution, which will affect raw sugar washing characteristics in the centrifugals. Too much small grain could increase the number of crystals going through the centrifuge screens and the sucrose content of affination syrups or prevent syrups from washing the other crystals properly by forming tight layers. The dextran content of the raw sugar is determined to measure its effect not only on raw polarization but also on elongated crystals in the recovery house, filtration, and so on. Various filtration tests have been tried, but to date it is hard to find one that is suitable to predict how refinery filtration systems will react to various raw sugars.

Many technical papers on raw sugar quality characteristics have been presented through the years at Sugar Industry Technologists and Sugar Processing Research Institute workshops. A basic review should focus on the two papers relating to the Domino raw sugar quality standards [2] and the Sugar Processing Research Institute workshop [3].

Once the sugar is washed, it is normally dissolved in a melter which contains remelt sweetwater. All refineries generate various quantities and types of sweetwater, such as washing, of presses, decolorization column, and sweetening-off. Most refineries operate a recovery house where the affination and white sugar runoff syrups are crystallized to produce remelt sugars, which dissolve in the sweetwater along with the washed raw sugar. Some refineries clarify or try to purify the remelt sugars slightly before adding them back to the main stream. What is important here is that one know the chemical analysis of the sweetwater stream in order to know the load being put on the refining operation. The color and percent invert are important, since they will not only reflect the efficiency of the remelt recovery system but determine the color removal capacity required throughout the refining process. The pH on all sweetwaters should be maintained above 7.0 to minimize inversion of sucrose, and all sweetwaters should be used as quickly as possible to prevent microbiol destruction of sucrose. The raw sugar melter liquor should then be analyzed for Brix (too much water results in high energy), color, invert, and pH. In sugar refining pH is one of the most critical analysis performed in that too low pH will cause inversion and too high can cause color buildup and destruction.

If the next step in the refining operation is a clarification process, the Brix, color, invert, and pH should continue to be measured. A clarity test (turbidity) can be used at this point or after polishing. The percent color removal for this operation should be measured and an invert gain of less than 0.02% maintained.

Decolorization performed using bone char, granular carbon, or ion-exchange resins requires similar process control analysis, such as Brix, color, invert, and pH. Percent color removal targets should be established for whatever material is used, to be able to measure both that material's efficiency and the refinery efficiency. It is important to establish if granular carbon is being underburned, if there is resin fouling or resin bead breaking, added magnesite, and so on. The percent color removal and pH history plotted on a weekly or monthly basis will illustrate trends.

After decolorization the liquor is generally polished and evaporated just prior to feeding the white sugar pans. The pan feed liquor analysis should include Brix since one wishes to obtain a predetermined solids level by the more energy efficient previous evaporation stages. The color is important since it will determine the color of the granulated sugar. The pH should be above 7.0 to minimize invert gain during the boiling process, and a low percent invert content is required for higher granulated sugar yield.

Samples of the granulated sugar produced should be monitored for color, percent moisture, and grain size. A sieve analysis is performed to ensure the size distribution required by the sales demand and to monitor excessive dust or large grain that may have to be reprocessed. The percent moisture is generally monitored before conditioning of the sugar to evaluate drying and later performed on finished products when screened to various grain sizes. If one is producing liquid sugars directly from decolorized liquors via ion-exchange resins or other material, color, invert, and percent ash must be monitored for this stream.

A basic general shift process control report might look like the one shown in Table 27.1. These are the minimum data generally required to show flow through the main process stream. The remelt recovery house analysis for purity and other parameters would be handled separately. Most refineries now have on line Brix and pH measurements and limited color and ash recordings. Thus lab analysis for the most part is to monitor plant instrumentation and to act as auditors for the process. In some areas these analyses may not be required only once a shift (8 h), depending on instrumentation. However, lab data should be collected in a computer program and weekly summaries made available to management. Key trends should be followed and targets compared to performance. Being a continuous operation with large volumes, the main refining process can see longer-term damage result from short-time loss of control. The loss of a pH unit can result in a rapid increase in invert content over a $\frac{1}{2}$-h period, which will then take hours to dilute out of the refinery stream and return to a normal level.

Listed in Table 27.2 are the control guideline for refining processes for maximum efficiency.

27.2 QUALITY CONTROL

The first step in determining what quality control test to perform is to establish product quality standards that will meet customer needs. Through the years the sugar industry has

TABLE 27.1 Typical Control Analyses

Parameter	Brix	Color	% Invert	pH	Polarity
Raw sugar	—	4,000	—	—	98.2
Washed raw sugar	—	1,000	0.11	—	—
Lab washed raw sugar	—	850	—	—	—
% Efficiency wash	95.2	—	—	—	—
Sweetwater	20.0	3,000	0.40	7.4	—
Washed raw melter	67.0	1,800	0.19	7.2	—
Clarifier	66.5	950	0.22	7.4	—
Granular carbon	66.0	100	0.27	7.6	—
Pan liquor evaporator effluent	73.5	105	0.28	7.5	—

TABLE 27.2 Refining Process Control Guidelines

Parameter	Brix	Temperature (°F)	pH
Washed sugar liquor	67 ± 1	175 ± 5	7.9 ± 0.2
Affination syrup		125 ± 5	7.2 ± 0.1
Affination magma	90 ± 2	110 ± 5	
Water temperature for heating of affination syrup		160 ± 5	
Sweetwater for melting washed raw sugar		165 ± 5	7.5 min.
Centrifugal wash water		180 ± 5	7.0 min.
Low-purity sweetwater			7.5 min.
High-purity sweetwater			7.5 min.
A liquor to pan floor	74 ± 1	165 ± 5	7.8 min.
Nephelometer, 15 max.			
Color, 200 ICU max.			
3 syrup (jet)			7.5 min.
4 syrup (jet)			7.0 min.
White sugar fillmass	91 ± 1.5	155 max.	
Remelt fillmass		150 max.	
Final remelt fillmass		125 ± 5	
Purity, 55 ± 5			
Sucrose in sewer line			
30 ppm max.			
Sucrose in pans and evaporator's condensate			
7 ppm max.			
Sucrose in vapors from pans			
75 ppm			

had guidelines established by various industries, such as National Soft Drink Association, National Formulary, and National Canners Association. However, the U.S. sugar industry still varies in some of the terminology used for its products, methods of analysis, and specifications. Refineries normally have a product line of four to five categories: granulated, liquid, brown, molasses, and specialty.

27.2.1 Granulated Sugar

Granulated sugars are normally defined by the percent of crystals retained on a particular standard U.S. mesh screen (crystal size). Table 27.3 is an illustration of the most common grades of granulated sugar sold. To manufacture products within their grain size ranges, a refiner must screen crystallized sugar through various openings and separate to individual bins. The two most common grades are extra fine and fruit. *Extra fine* (sometimes called *fine*) has a grain size that generally falls between 20 and 100 U.S. mesh and is the largest-volume industrial granulated sugar sold today. Its size makes it ideal for bulk handling and less susceptible to caking. It is used primarily in the candy and baking industries. *Fruit* granulated sugar has a grain size distribution between 40 and 100 U.S. mesh. The smaller grain size and larger surface area makes this product popular in dry mixes for gelatin desserts, puddings, and drinks.

The courser granulated sugar is normally crystallized from the highest-purity sugar liquor. Therefore, sanding and course grade are the highest-purity crystalline product with the lowest color and best color stability. The first quality control test performed on finished granulated sugar should be a grain size analysis to be sure it complies with your internal

TABLE 27.3 Typical Analysis for Granulated Sugars

U.S. Screen Mesh	Sugar Grade: Coarse	Sanding	Extra Fine	Fruit	Baker's Special	Powdered 6×	Powdered 10×
% on U.S.:							
12	3	—	—	—	—	—	—
16	45–65	—	—	—	—	—	—
20	20–35	2–10	0–5	—	—	—	—
30	—	40–70	2–20	—	—	—	—
40	—	30–40	10–45	0–7	—	—	—
50	—	1–8	5–35	20–50	0–5	—	—
70	—	—	—	30–70	10–30	—	—
100	—	—	10–40	10–30	30–60	0–2	0–1
% through:							
100	—	—	0–8	0–10	10–30	—	—
140	—	—	—	—	5–20	—	—
200	—	—	—	—	—	88–100	94–100
Color (ICU)	15–30	20–35	25–50	25–50	25–50	25–50	25–50
% Ash, max.	0.015	0.015	0.025	0.030	0.035	0.030	0.030
% Moisture	0.04	0.04	0.05	0.05	0.05	0.50	0.50
% Starch	—	—	—	—	—	2.5–3.5	2.5–3.5

specifications and that of the customer. The percent moisture should be monitored to insure no future caking will take place if the product is handled properly. The color and percent ash should be randomly monitored.

An important test not normally publicized or standardized is the sediment test. All finished products should be evaluated for foreign contamination. This can be performed by visual examination of a spread out sample or by dissolving the sample and filtering through a sediment pad. Granulated sugar is one of the safest food ingredients available but outside contamination can cause major problems. A good review of the granulated sugar quality can be found in the *Proceedings of the Workshop on White Sugar Quality* [1].

27.2.2 Liquid Sugar

If liquid sugar is produced then the finished quality should be monitored for pH and ash content, color, solids, and invert. Typical analyses for liquid sugars are shown in Table 27.4. Microbiological testing becomes important with liquid sugars to control yeast, bacteria, and mold. One should test production samples after cooling, storage, and shipment. Since it is possible under the right conditions for a microbial population to grow in liquid sugar, proper sanitary procedures must be followed to ensure the quality of the product shipped. Standards have been developed by the American Bottlers of Carbonated Beverages and the National Canners Association and can be found in the *Cane Sugar Handbook* [4].

27.2.3 Brown Sugar

In the production of brown sugars one must monitor the color of the feed liquors to ensure proper color of the finished product. In both a boiled and a mingled brown, there is slightly more than 10% syrup on the outside of the crystals, which controls the color and flavor of the product. Both moisture and invert content are important in controlling the microbio-

TABLE 27.4 Typical Sugar Analyses

a. Liquid sugars

Parameter	Liquid Sucrose	Amber Sucrose	Liquid Invert	Total Invert
% Solids	67.0–67.9	67.0–67.7	76–77	71.5–73.5
% Invert	0.35 max.	0.45 max.	45–55	93 min.
% Sucrose	99.5 max.	99.4	55–45	7 max.
Color (ICU), max.	35	250	35	40
% Ash	0.04	0.15	0.05	0.05
pH	6.7–8.5	6.5–8.5	4.5–5.5	3.5–4.5

b. Brown sugars

	Boiled		Coated		Free-Flowing	
Parameter	Light	Dark	Light	Dark	Granulated	Powdered
% Sucrose	85–93	85–93	90–96	90—96	91–96	91–96
% Invert	1.5–4.5	1.5–5.0	2–5	2–5	2–6	2–6
% Ash	1–2	1–2.5	.3–1	.3–1	1–2	1–2
% Moisture	2–3.5	2–3.5	1–2.5	1–2.5	0.4 0.9	0.4–0.9
Color (ICU)	3000–6000	7000–11,000	3000–6000	7000–11,000	6000–8000	6000–8000

logical stability of the product and its drying tendency. Yeast will not readily grow in a low-moisture product, but if moisture and percent invert are too low, the brown sugar will dry out and go hard. Again in this area as in raw sugars, people related to some type of safety factor whereby percent moisture is related to nonsugars to ensure a stable product. Typical brown sugar analyses are shown in Table 27.4. General uses for sugar products and descriptions of how various chemical properties affect the products are given in *Sugar: A User's Guide to Sucrose* [5].

27.2.4 Quality Assurance

In today's laboratories, hazard analysis and critical control point system (HACCP) [6] programs in which the critical control points associated with a particular characteristic are identified have become very popular. For example, a program to ensure no metal contamination of the sugar can be drawn up by identifying the critical control points on a flowchart. For each control point (e.g., polishing filters, magnets, metal detectors), limits must be established that would initiate investigation and corrective action. All actions, including cleanings, standardizing, and so on, are recorded in logs along with all corrective actions. The program is audited by quality control personnel. The importance of a program of this nature is that everything in documented along with correction action by maintenance departments and others. Quality assurance is a never-ending job, and procedures installed to ensure quality must remain in place, be functioning, and be audited to obtain a top quality record.

27.3 YIELD AND LOSS

To reach the goal of a top-quality product at the lowest possible cost, it is necessary to obtain the maximum quantity of salable product from the raw sugar. This is normally referred to as *yield,* defined as

$$\text{yield} = \text{raw sucrose} - \text{sucrose loss} - \text{sucrose in molasses}$$

There are various ways to express and measure yield. It could be stated as a percentage of prime product related to raw sugar, such as 96% yield for a 98 polarization raw sugar. That is, 100 lb of that raw sugar would yield 96 lb of salable product (granulated sugar, liquid sugar solids, etc.). The reciprocal of this would be 1.0/0.96 = 104.0 and would mean that 104 lb of raw sugar was required to produce 100 lb of finished product.

One must establish a means of measuring yield in order to evaluate refinery performance. In doing this, we must decide which chemical properties will be balanced. The best yield measurement would be to convert everything to a sucrose base. That is, the sucrose content would have to be determined or calculated for the raw sugar. All finished products would have to be converted to a sucrose content or white sugar equivalent (WSE) basis. For brown sugars and low-purity syrups, the amount of granulated sugar (WSE) that could have been produced from these products must be calculated. The number of pounds of all salable products (no final molasses) on a WSE basis is divided by the pounds of raw sugar to give a WSE yield. This yield calculation is less affected by sales product mix, and the like, and is thus a truer measure of refinery performance. Every refinery is different in product mix, age, remelt recovery, and so on, and therefore should establish its own yield target. Another popular yield is the polarization yield, whereby everything is balanced based on polarization analysis.

To improve yield, the sucrose loss and sucrose in final molasses must be controlled. A typical sucrose and nonsucrose balance chart for yield calculation can be found in the 12th edition of the *Cane Sugar Handbook* [4].

27.3.1 Sugar Loss

Sugar losses may be measured by various methods such as polarization, total solids loss, sucrose loss, and so on. With today's improved methods of analysis, sucrose measurement is preferred. Sugar loss will normally have the greatest effect on yield. These losses can be divided into three main types: (1) chemical, (2) physical, and (3) microbiological.

a. Chemical Losses. Sucrose can become nonsucrose by chemical changes such as inversion or carmelization. During inversion, sucrose is converted to dextrose and levulose. Inversion will take place under acidic conditions (below 7.0 pH), and the amount of inversion depends on pH, temperature, and time. The lower the pH, the greater the inversion, and the higher the temperature, the faster the inversion. To minimize inversion, all liquids must be kept over 7.0 pH, process steps maintained at as low a temperature as desired and process time to a minimum. Prolonged storage of process liquors will add to inversion. The percent sucrose inverted for various pH, temperatures, and time can be found in the *Cane Sugar Handbook* [4]. Carmelization can take place at high temperatures and/or high pH and results in sucrose being changed into color compounds. One not only pays the price of removing the color but also that of sugar loss.

b. Physical Losses. These can be anything related to the actual physical loss of sucrose. They could be losses by spills or leaks going to the sewer, those from evaporator or vacuum pan entrainments or, most common, losses such as those caused by packaging overfill, scale error (receiving and shipping), analytical errors, accounting errors, and security.

c. Microbiological Losses. Bacteria and yeast in the refinery can invert or convert sucrose to decomposition products. They can also produce acids, which lower the pH and create more invert. Thus, monitoring pH and invert through the process will be a means of measuring and controlling microbiological losses. Because of their low density, sweetwaters are an ideal growth medium. Maintaining temperature above 165°F and Brix above 50 will inhibit microbiological activity.

27.3.2 Loss Control

We must be able to measure the losses in the refinery as best we can to be able to control them. Generally, a technical staff member should be assigned to loss control. In most refineries the quantity of sucrose discharged to the sewer is a major part of the loss. Thus, all sewer lines should be identified and tied into collection points where possible. The sewer effluent should be monitored on a continuous basis. The sugar concentration can be measured with such instruments as the Technicon AutoAnalyzer or YSI Analyzer at very low concentrations on a continuous basis. There should be an alarm system so that if there is a major spill, it will be detected as soon as possible. The amount of sugar leaving in press cakes, bagasse, and any other solid waste not exiting through a monitored sewer system should be measured.

Sugar can also be lost to absorbent systems. If char or granular carbon filters are not sweetened off properly, sugar will be burned in the furnace. Granular carbon, char, and powdered carbon will absorb sucrose that can not be washed out. The Bone Char Research

Group estimated the absorption to be 0.5% of the weight for bone char, 3.0% for granular carbon, and 5% for powdered carbon.

All packaging lines should be monitored for overfill. A simple loss report would be as follows:

Sucrose in raw sugar	98.10
Sucrose in finished products	96.12
Sucrose in final molasses	1.02
% Total loss	0.96
Sewer loss	0.41
Adsorbents	0.05
Packaging	0.04
Bagasse/press cakes	0.02
Unknown	0.43

All refineries will have some unknown losses due to analytical errors, shipping weights, raw sugar scale, sucrose destruction, and so on. However, a good refinery should be able to keep losses below 1% and identify 50% of the sucrose loss. The refineries loss data should be made available so that all personnel are aware of the performance. Loss control depends mainly on station operators' awareness.

Chen and Chou [4] describe the essential steps needed at each refining unit operation for sucrose loss control as follows:

- Incoming raw sugar
  1. Check conveyors for spillage.
  2. Check scales for cleanliness and accuracy.
  3. In sample taking and sample preparation, take all precautions to avoid evaporation of sample.
  4. Avoid excessive water spraying on conveyors.
  5. Melt high-moisture and/or dirty raw sugar first.
- Affination
  1. Optimize the operation of centrifugals.
  2. Control all materials within specified ranges.
  3. Be sure that no excessive water is used.
  4. Liquidate all low-Brix materials (affination syrup/sweetwater) for weekend shutdown.
  5. Apply biocide (bacteriocide) to sump.
- Clarification/filtration/evaporation
  1. Assure that all materials are within their respective ranges in temperature, pH, and Brix.
  2. Check leakage of filters.
  3. Check clarity of filtrate and filtrate receivers.
  4. Avoid excessive cake washing.
  5. Check polarization in mud against target.
  6. Check evaporators and pan entrainments.
  7. Check vacuum lines to avoid changes in vacuum.
  8. Check condensates (for calandria tube leakage).
  9. Apply biocide to low-Brix sumps.
  10. Avoid excessive use of water.

11. Minimize remelting/recycling materials.

- Decolorization
  1. Control optimum cycle of decolorization through char/carbon column.
  2. Observe optimum regeneration of char/carbon.
  3. Avoid channeling in the column for better sweetening-off and to use less water.
  4. For CAL carbon, add sufficient magnesite for proper alkalinity.
  5. Although sweetening-off for a CAL carbon column needs more time than char, keep it as short as possible.
- Pan boiling and crystallization
  1. Keep all materials within specified ranges in Brix, pH, and temperature.
  2. Control boiling time and temperature. These are important factors in sucrose destruction.
  3. Keep the grain from boilings as uniform as possible, with minimum conglomerates, to reduce remelting and provide better centrifugation.
  4. Check regularly for tube leakage in crystallizers.
- Purging station
  1. Do not overcharge.
  2. Do not overwash.
  3. Correct dripping of syrup to purged sugar.
  4. Check screen damage and fine grains in blackstrap.
- Drying Station
  1. Check rotoclones regularly to avoid flying sugar dust and/or the growth of microbes.
  2. Use bacteriocide wisely.
  3. Avoid spillage of sugar to floor.
  4. Keep sugar dryer/cooler in optimum operation for maximum output but less remelt.
- Packaging
  1. Check packages for overweights.
  2. Keep packing machines clean, with no spillage.
  3. Check packaging materials and package quality.
  4. Check all weighing scales.
  5. Avoid spillage from conveyors to floor.
  6. Keep all floors clean and dry.
  7. Handle properly all returns from bulk loads.
  8. Check sugar to silos from contamination and wet sugar.
  9. Cover all bins to avoid contamination or sugar dust loss.
  10. Check sugar dust collecting system to minimize physical loss.
  11. Control all remelts within specified ranges in pH, temperature, and Brix, and return to process without delay.
- Liquid sugar making and shipping
  1. Control all materials used in making liquid products within specified ranges in pH, temperature, and Brix.
  2. Check filters for leakage, dripping, and clarity of filtrate.
  3. Check sugar in water out from condensers.
  4. Check sugar in condensate from heater and evaporator.
  5. Follow strictly the procedure for inversion operations.
  6. Check weighing scales.
  7. Check pumps for leaks.
  8. Keep sump and all storage tanks free from microbe growth.

9. Assure the proper temperature for sufficient sterilization of trucks, cars, and storage tanks.

- Warehouse
  1. Keep the floor and ambient air clean.
  2. Check any damaged packages.
  3. Prevent package damage by nails or defects of pallets, conveyer-systems, forklift trucks, etc.
  4. Check quantity of loads in and loads out.
  5. Check weights of package trucks.
  6. Guard against possible pilferage.
  7. Maintain first-in, first-out principle.
- General precautions
  1. Maintain minimum stock-in-process.
  2. Exercise practical accuracy when taking stock-in-process and stock-in-warehouse.
  3. Clean up spillage and leakage promptly. When water is applied, use as little as possible.
  4. Avoid microbiological activity by regular cleaning and biocide application to holding tanks and sump.
  5. Report and correct any leakage of refinery equipment as soon as possible.

27.3.3 Recovery Operation: Molasses

Another way to lose sucrose is in final molasses. This is a by-product that has a much lower dollar value than the sucrose in granulated sugar, and other products. One must control both the volume of molasses produced and the total sugar (sucrose + invert). The amount of affination syrup produced while washing the raw sugar must be controlled so that the recovery house doesn't get overloaded. Affination syrup purity can be maintained in the low 80s while monitoring washed raw sugar color, chloride, and so on, to ensure proper washing but not overwashing.

Various boiling schemes for the remelt house (three boilings, two boilings, etc.) can be used, depending on the capacity of the station. Whatever the boiling scheme, targets should be established for strike purities and yields. The Sugar Juice Molasses (SJM) formula can be used to balance remelt boiling so that the lowest-purity final molasses is produced by back-boiling A or B molasses while still using all the feed liquor to the recovery station. In managing the remelt house, the amount of sucrose inverted while boiling and the amount of invert sent to the station should be minimized. Generally, 1 lb of invert will take 1 to 2 lb of sucrose to molasses. It is important to measure the total sugars in molasses since the sucrose could be inverted, giving a false low sucrose value and a false sense of recovery. In controlling the station the final molasses polarity or sucrose should be measured daily. The individual A strikes, B strikes, and so on, should have purities determined so that yield efficiency, and other parameters, can be monitored. The remelt sugar quality should also be measured daily for percent invert, color, and percent ash. Since this sugar normally feeds back into the process, we do not want to reintroduce too high a nonsugar content.

The recovery house can produce high-quality remelt sugar and low-sucrose molasses. This is not an easy task since the recovery house is affected by many variables. All raw sugars do not have the same nonsugar content or type. Sweetwaters fed to the station can vary, along with brown and white sugar runoff syrups. Rates of crystallization and extractability of various syrups will change. The recovery house could be the most technically demanding station to control.

27.3.4 Calculation of Yield

a. Raw Value. The U.S. Treasury Department requires that all sugar be described in terms of raw value for the calculation of tariffs and quotas. Raw value is defined as the equivalent of such articles in terms of raw sugar testing 96° by the polariscope. The Treasury Department's proclamation instructs that crystalline sugar be converted to raw value by multiplying pounds of sugar by a factor of 1.07, and this factor is reduced by 0.0175 for each degree of polarization under 100 with fractions of degrees prorated. The formula commonly used for calculation of raw value is raw value = (polarization −92)(1.75) + 93. Thus 100 lb of 100 polarization pure sugar = (100 − 92)(1.75) + 93 = 107 lb of raw value. Similarly, 100 lb of 96 polarization sugar = (96 − 92)(1.75) + 93 = 100 lb of raw value. The calculation above is based on the assumption that 107 lb of raw sugar of 96 polarization will produce 100 lb of refined sugar. In reality, the yield varies with process efficiency.

b. Sugar Output (% of Raw Sugar Melt)

	Example
P = polarization of raw sugar melted	98.1
S = sucrose loss (% of melt)	0.59
F = factor of safety of raw sugar	0.21
E = nonsucrose solids eliminated (% of nonsugar solids in melt)	10
B = P.D. purity of blackstrap	42
W = % water in blackstrap	28

$$P - S - \left[(100 - P) \times (1 - F) \times \left(\frac{100 - E}{100}\right) \times \left(\frac{B}{(100 - B)}\right)\right] = \text{sugar yield}$$

$$\% \text{ of melt} = 96.53\%$$

where

$$\begin{aligned} 100 - \text{polarization} &= \text{nonsucrose solid} + \text{moisture} \\ F &= \text{factor of safety of raw sugar} \\ &= \text{moisture}/(100 - \text{polarization}) \\ &= \text{moisture}/(\text{nonsucrose solid} + \text{moisture}) \\ 1 - F &= (\text{nonsucrose solid})/(\text{nonsucrose solid} + \text{moisture}) \\ E &= \text{nonsucrose solid eliminated, based on \% of nonsucrose solid in raw sugar} \\ \frac{100 - E}{100} &= \text{\% of nonsucrose solid left} \end{aligned}$$

c. Raw Sugar Melt Required for Estimated Net Refined Sugar

$$\frac{\text{Sugar production needed}}{\text{Sugar yield (\% of melt)}} = \text{raw sugar required}$$

For example, if 1,000,000 lb of white sugar production is needed, then

$$\text{raw sugar required} = \frac{1{,}000{,}000}{0.9653} = 1{,}035{,}947 \text{ lb}$$

d. Blackstrap Yield (% of Raw Sugar Melt (based on standard yield conversion formula)

$$(100 - P) \times (1 - F)\left(\frac{100 - E}{100}\right) \times \left(\frac{100}{100 - B}\right)\left(\frac{100}{100 - W}\right) = \text{syrup yield}$$

The factors are the same as in the sugar yield conversion formula.

$$\text{Molasses yield (\% of raw sugar melt)} = 3.23\%$$

e. Sucrose Carried by Nonsucrose Solids to Blackstrap Molasses

$$[(100 - P) \times (1 - F)] \times \left(\frac{100 - E}{100}\right) \times \left(\frac{B}{100 - B}\right) = \text{sucrose carried}$$

REFERENCES

1. *Proc. SPRI workshop White Sugar Quality,* New Orleans, LA, 1988.
2. E. Culp and A. Hageney, *Raw Sugar Quality Standards, Sugar Industry Technologists,* 1967, pp. 164–173.
3. *Proc. SPRI Workshop Raw Sugar Quality,* Savannah, GA, 1986.
4. J. C. P. Chen and C. C. Chou, eds., *Cane Sugar Handbook,* 12th ed., Wiley, New York, 1993.
5. Pennington and Baker, *Sugar: A Users Guide to Sucrose,* AVI Book, Van Nostrand Reinhold, New York, 1990.
6. *Hazard Analysis and Critical Control Point System,* Food Safety and Inspection Service Guidelines, U.S. Department of Agriculture, Washington, DC, Nov. 1989.

CHAPTER 28

Microbiological Control in Sugar Manufacturing and Refining*

28.1 CLASSIFICATION OF MICROORGANISMS

Microorganisms of importance to the sugar manufacturing process are classified as bacteria, yeast, or fungi. Viruses affect the sugarcane plant and sugar yields, but their presence does not impinge directly on the sugar manufacturing process. Molds and yeast commonally contaminate sugar and sugar containing solutions after manufacture. They are usually present as airborne contaminants. The majority of the bacteria found in the final sugar and sugar syrups are those that can survive production or refining processes. Initially, these organisms come into the factory from the field soil or as infectious agents in the sugarcane plant and are usually common soil contaminants of sugarcane fields. To facilitate an understanding of microbiology, please refer to glossary in chapter 1.

Bacteria are microscopic, single-celled organisms about 1 μm (10^{-6} M) in size. They are omnipresent, filling all available ecological niches. They are classified according to their growth temperature preferences and their ability to use gaseous oxygen in their metabolic processes. Many bacteria that survive the sugar production process are spore-formers. Spore-forming bacteria are extremely temperature resistant and can survive temperatures of 100°C (212°F) for more than 20 h. Aerobic thermophilic bacteria, classified in the sugar industry as the *flat sour group,* grow well between 45 and 60°C (113 and 140°F). Mesophilic bacteria have optimum growth temperatures of 20 to 45°C (68 to 113°F). Anaerobic thermophilic bacteria, referred to in the sugar industry as the *sulfur bacteria,* grow in the absence of oxygen and produce hydrogen- and sulfur-containing gases. Anaerobic mesophilic bacteria are commonly present in sugar. The most prevalent species are those in the lactic acid bacteria group.

Yeast are spherical, oval, or rod-shaped fungi with a cell size of about 10 μm which normally grow in a nonfilamentous form. Many yeasts are capable of growth utilizing oxygen, and some (the fermentative yeasts) use an alternative metabolism in the absence of oxygen, usually producing ethanol as a by-product. Some yeast can survive in high sugar

* By Donal F. Day.

concentrations (osmophilic yeast). Scarr [1] defined osmophilic yeasts as those that can grow at concentrations above 65° Brix. Most yeast species produce significant levels of the enzyme invertase. Yeasts, although temperature resistant (they survive temperatures of 60°C), usually do not survive the high temperatures of the sugar production process. Because of this, most yeasts found in sugar products are probably due to postproduction contamination. The optimum temperature for growth of yeast normally is between 20 and 35°C (68 and 95°F).

Filamentous fungi (molds) are minute, multicellular, generally saprophytic organisms, ranging in size from 10 μm to 1 mm. They generally require oxygen for growth and temperatures between 25 and 70°C (77 and 158°F). They are particularly prevalent in rotting piles of organic material such as bagasse. They can be serious plant pathogens. Sugarcane rusts are fungi. They are generally not a direct problem in the sugar factory but many cause human health hazards such as bagacilliosis, due to spore and toxin production.

Factors affecting microbial activity are as follows:

1. *Oxygen.* Microorganisms are classified on the basis of their oxygen requirements. Aerobic organisms require oxygen for growth; anaerobic organisms do not utilize oxygen, and in many instances oxygen is toxic to them. Some species of microorganisms are facultative, may utilize oxygen, or may utilize alternative methods of metabolism in the absence of oxygen. The fermentative yeasts are an example of this class of microorganism. Aerobic, anaerobic, oxygen-tolerant anaerobic bacteria, as well as oxidative and fermentative yeasts, can be found in both the sugar mill and sugar refinery environment. Fungi (molds) of importance to the sugar producer are generally aerobic.

2. *Moisture.* All microorganisms require some water for survival and growth. The requirement for moisture (free water) for microbial growth is normally expressed in terms of *water activity,* defined as the ratio of water vapor pressure of a solution to that of the vapor pressure of pure water at the same temperature and pressure. A dilute solution has a ratio close to 1. The lower the ratio, the higher the solids content of the solution. In general, as the Brix of a sugar solution increases, its water activity decreases. Most microorganisms require a water activity (a_w) of 0.9 to 1.0 for growth. Osmophilic and osmotolerant organisms can survive at water activity values as low as 0.6 to 0.7. It is for this reason that syrups must be maintained above 67° Brix to avoid microbial degradation. Table 28.1 shows the approximate minimum a_w values for the growth of microorganisms.

3. *Temperature.* Each microorganism has a specific temperature range for maximum growth. Psychrophiles grow generally between 4 and 20°C (39.2 and 68°F); mesophiles, 20 to 45°C (68 to 113°F), and thermophiles, 45 to 100°C (113 to 212°F). Extreme thermophiles are able to grow in boiling water. Bacteria of concern in sugar production are generally killed by temperatures above 60°C (140°F) for 15 min.

TABLE 28.1 Approximate Minimum a_w Values for the Growth of Microorganisms

Organisms	Minimum a_w
Most spoilage bacteria	0.91
Most spoilage yeasts	0.88
Most spoilage molds	0.80
Halophilic bacteria	0.75
Xerophilic molds	0.65
Osmophilic yeasts	0.60

4. *pH.* The pH also selects for the types of organisms that will grow in a sugar solution. Generally, fungi grow at pH values ranging from 4.5 to 7.0; yeast, 4.5 to 8.0; and bacteria 6.5 to 9.0. A major grouping of bacteria that grow in the acid range are the lactic acid bacteria, which grow across the range 4.0 to 8.0. This group includes the Leuconostoc, Lactobacilli, and Streptococci. The minimum pH for molds and yeasts are 2.0 and 2.5, respectively. The maximum pH values for molds and yeasts are 11 and 8.5, respectively.

28.2 MICROBIALLY CAUSED SUGAR PROCESSING PROBLEMS

Microbial problems in sugar processing (excluding microbial diseases of the plant) fall into three basic areas: storage of the harvested crop, processing problems, and quality control problems.

28.2.1 Microorganisms in the Cane (Storage Problems)

Deterioration and loss of sugar in the sugarcane are primarily the result of two major factors, microbial contamination and inversion due to release of plant invertases. By far the more serious problem is microbial deterioration. Bevan and Bond [2] isolated approximately 50 different species of microorganisms from green cane and 17 different organisms from the surface of burned cane. The microbial flora on the surface of the cane not only reflects the cane variety but the geographical location and prior history of the field. Organisms associated with sugarcane will include the gum-producing organisms *Leuconostoc mesenteroides* and *L. dextranicum,* several genera of yeast (*Saccharomyces, Torula,* and *Pichia*); a number of common bacteria, including pseudomonads and bacilli; as well as filamentous fungi such as *Aspergilli, Penicillum,* and *Streptomyces* species.

Irvine [3] discussing the field origins of dextran, noted that *Leuconostoc* enters the cane tissue prior to harvest, where the plant is cracked or damaged. Undamaged standing cane is generally free of internal contamination by this microorganism. Overburning the cane can remove the protective wax surface and produce cracks in the rind where bacteria can enter. Sound, living, whole-stalk cane seldom shows elevated levels of dextran, but burned, killed, or damaged whole-stalk cane will deteriorate rapidly.

Microorganisms can grow rapidly on the surface of burned cane. Within 10 min of burning, xanthomonads, corynebacteria, and bacilli can be found on the cane surface. Significant numbers of fungi and yeast can be found within 24 h of burning. Massive infections are found in the ends of mechanically harvested (chopped) cane within 2 h. The gum-producing organisms leuconostocs, xanthomonads, and Aerobaceteriaceae predominate. Even under favorable harvest and cane storage conditions, significant deterioration occurs rapidly after cutting, as soon as 24 h for whole-stalk cane (Fig. 28.1).

The role of microorganism in the deterioration of juice from frozen cane has received significant attention. Damage to the cane plant following a freeze leads to increased numbers of bacteria, similar to that seen with other forms of crop damage. These organisms, particularly *Leuconostoc,* produce the high levels of the polymer dextran from sucrose, interfering with the crystallization of sugar. Dextran also reduces the yields of alcohol that can be made from molasses produced from freeze-damaged cane, further reducing the value of an already damaged crop [5]. Extensive research in Queensland in the 1960s [6] led to the conclusion that infection control is not practical and the best approach is to reduce the time between harvesting and milling (cut to crush). Even whole stalk cane can show significant loss in overnight storage in mill yards, as much as three times as much sugar is lost per ton of cane after yard storage, due partly to the normal rate of degradation and partly to the extra handling necessitated in yard storage (Fig. 28.2) [7].

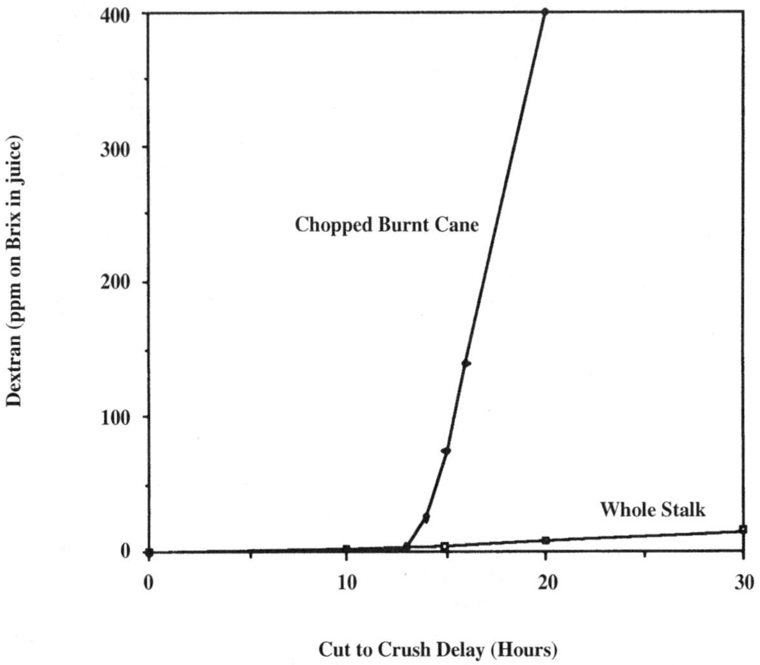

FIGURE 28.1 Rate of deterioration of cane. (From Ref. 4.)

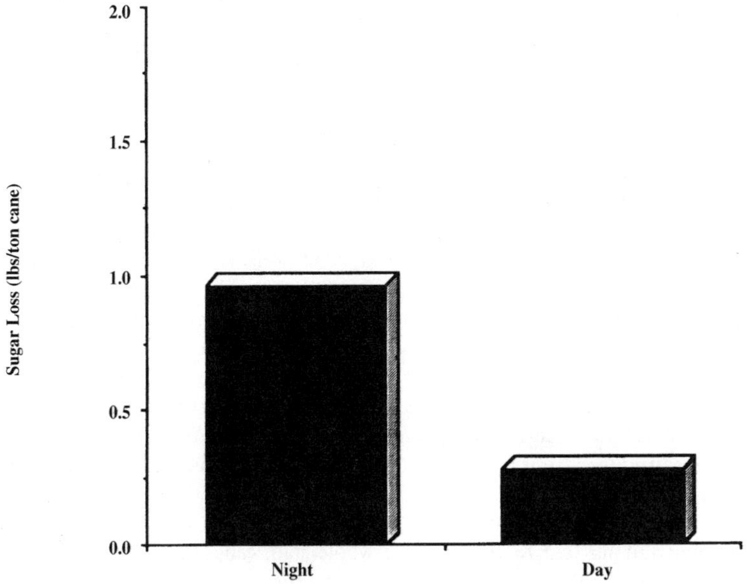

FIGURE 28.2 Sugar losses on wash table as a function of yard storage.

28.2.2 Microorganisms in the Raw Sugar House (Processing Problems)

Sugar factories are environmental microcosms suitable for growth of large numbers of both mesophilic and thermophilic microorganisms. The acid pH values and high sucrose concentrations in sugarcane juice favor growth of the lactic acid bacteria, specifically the streptococci and *Leuconostoc.* These organisms are commonly present in damaged cane and in unprocessed juices. The major contaminant in this class is *Leuconostoc,* which is a common soil microorganism. Processing problems are associated primarily with microbial products, not with the organisms themselves. These products are frequently produced prior to the processing of the raw materials. The microbial loading of sugar process juices reflects the bacterial flora of the source materials. Organisms that do well at low pH generally predominate. A problem specific to the sugar producer is the presence of gum-forming organisms, primarily *Leuconostoc mesenteroides* and *Leuconostoc dextranicum* [8]. Fermentation reactions can occur both inside the harvested plants, prior to sugar purification, and during the production process, producing products that may also have to be considered during sugar manufacture. Some metabolites, such as lactic, acetic, and formic acids, simply add to the pool of non-sucrose-containing materials. Others, such as dextran, can cause serious process disruption. However, in all cases, the presence of microbial metabolites indicates sucrose loss due to enzymatic reactions.

28.2.3 Microorganisms in Sugar Refining (Processing Problems)

In contrast to raw sugar operations, sugar refining differs microbiologically with respect to both the numbers of organisms introduced with the raw material and their viability. Raw sugars have distinct microflora, compromising different species of fungi, bacteria, and yeast; hence they introduce into the sugar refining process a varied microbial potential that can cause multiple points of sucrose losses during refining [9]. Microbiological attack is considered to be a major contributor to sugar loss in the refinery [10]. Sugars from different countries vary greatly in both number and species of microorganisms that they contain, a reflection of the geographic regions where the cane is grown. Microbial surveys of sugar refineries have indicated that spore-forming bacteria predominate [11,12]. Points of primary microbiological concern are those refinery process streams that are stored where conditions are favorable to microbial growth. These include low-density sweetwater streams of both high and low purity, at relatively low temperatures and pH; and affination syrup, which is also maintained at relatively cool temperatures [12].

The principal bacterial species found in the sugar refinery are *Bacillus, Leuconostoc, Streptococcus,* and *Lactobacillus,* and the predominant yeast species are *Torulopsis, Saccharomyces,* and *Hansenula* [13]. Spore-forming bacteria of the genera *Bacillus* and *Clostridium* are capable of surviving the processing conditions because their spores are heat resistant. However, these spores will not germinate until the temperature falls to an acceptable range. Normally, sucrose loss in cane refineries to thermophilic organisms is negligible, because their numbers are low. Nevertheless, they can be a serious problem to end-product users. Thus, standards are placed on acceptable numbers in the final products.

In sweetwaters, large numbers of yeast and lactic acid bacteria may also be present [13]. The predominant organisms here are the mesophilic bacteria [10^7 to 10^8 cfu (colony-forming units)/mL] and meso- and osmophilic yeast (10^6 to 10^7 cfu/mL). The most common species of microorganisms found in raw sugars are *Bacillus subtilis, B. mesentericus, Aerobacter aerogenes,* and species of *Actinomyces, Saccharomyces, Penicillia, Mucor,* and *Aspergillus.* Also commonly found are the thermophilic species *Clostridium nigrift* and *B. stereothermophilus* (the flat sour group) as well as the hydrogen producers *C. thermoputrificum,*

C. thermoaerogenes, C. thermoacidophilus, and *C. thermochainus* [9]. The lactic acid bacteria, particularly the *Leuconostoc,* are omnipresent.

***a. Microbes Introduced in the Raw Sugar* [14].** Microorganisms found in the raw sugar are representative of the raw factories from which the sugar was obtained. These microorganisms are mainly in the form of spores which require many hours to germinate. The bulk of the microorganisms are found in the film of molasses surrounding the crystals. The moisture content of the sugar largely determines how well the organisms grow prior to melting in the refinery. Usually, a safety factor is put into a raw sugar contract, controlling the percent moisture in the raw sugar.

b. Microorganisms in Affination Syrups. The affination syrup is a prime source of contamination in the sugar refinery. It contains an accumulation of organisms from the raw sugar melts. Affination syrups are either boiled to yield remelt sugar or defected and char filtered for soft sugar production. They can be a source of microorganisms that infect the char filters.

Thermophiles are more prevalent in the sugar refinery. Here, operating temperatures are normally high enough to restrict mesophilic growth. The concentration of viable thermophilic spores in affination syrups and refinery sweetwaters is variable and can be very high [12]. There may be as many as several thousand spores per 10 g of syrup. The washed raw sugar will generally have a log less spore population than the affination syrup.

c. Press Filtration. Because of the particulate size removal of conventional plate and frame filtration with diatomaceous earth, proper filtration can cause a dramatic reduction in the numbers of microorganisms in the filtrate. This was originally shown by Cameron and Bigelow [15], who seeded syrup and tested the efficacy of repeated filtration in removing thermophilic spores (Table 28.2).

d. Carbon/Char Decolorization. Another source of microbial contamination can be the base of the char/carbon filter, the blankets, and grids. De Whalley and Scarr [16] found high concentrations of bacteria in the char cloths covering the bottoms of the char filters. They estimated that much of the undetermined losses in sucrose (which may be as much as 0.5% of the raw sugar melted) may be attributed to microbial action.

e. Boiling, Spinning, and Finishing. The temperature in the vacuum pans is too low to destroy microbial spores. After centrifuging, spores accumulate in the syrups, producing high spore counts in No. 3 and No. 4 sugar. It is for this reason that refiner practice is to mix No. 4 syrup with a low-purity sweetwater and press filtering before boiling brown sugar.

TABLE 28.2 Removal of Thermophilic Spores by Filtration and Crystallization

Sample	Number of Spores/mL
Syrup	25,000
First filtrate	2,500
Second filtrate	1,000
Fourth filtrate	1,000
Finished sugar	238 (per gram)

Tilbury et al. [13] surveyed a sugar refinery and found that the most heavily contaminated areas were the dust collectors (Rotoclones) and dirty sweetwater (housewater). Here, mesophilic and osmotolerant yeast were the predominant flora. Sugar loss due to biodeterioration of sweetwater was estimated to be about 0.5% of the refined sugar output. The use of chemical agents to inhibit microbial activity in the United States is discouraged by the Food and Drug Administration. It is recommended that the Rotoclone system be "steamed out" at least weekly and that a high pH be maintained in the sweetwater to minimize microbial growth.

28.2.4 Microorganisms in the Products (Quality Control Problems)

The microbial standards for sugar products are set by the end users. There are no accepted standards for raw sugar since it undergoes further refining. This is also true for non-human-consumption syrups and molasses. The significant groups of microorganisms are classified according to whether they induce undesirable changes in the quality or composition of intermediate and final products, whether they cause sucrose loss during processing, and whether they are of significance because of the deterioration they may cause in the foods or beverages to which they are added. Not all microbial interactions with sugars are deleterious. Some of the volatile and nonvolatile constituents responsible for the flavor and odor of cane sugar products, such as table syrups, molasses, and refinery brown sugar, are due to microbial activity [17].

a. Microorganisms in Sugar. Microorganisms found in raw sugar are a reflection of both local contamination (soil and airborne organisms) and those spore-formers that survive the production process. Because of the high osmotic pressure and low water content, there is very little microbial activity in raw and processed sugar. Organisms that survive in the sugar can cause problems when sugar is used in liquid containing products.

b. Microbial Standards for Sugar. The sugar production industry has not developed specific standards for microbial loading of raw and refined sugar. Rather, the canners' standards (National Canners Association) have been adopted for sugar. The canners' standards are:

- *Flat and sour spores.* In five samples examined, there shall be a maximum of not more than 75 and an average of not more than 50 spores per 10 g.
- *Thermophilic anaerobic spores.* These shall be present in not more than 60% of five samples, and in one sample to the extent of not more than four of the six tubes.
- *Sulfite spoilage bacteria.* These shall be present in not more than 40% of the five samples, and in one sample to the extent of nor more than 5 spores per 10 g. This would be equivalent to two colonies in the six tubes.
- *Carbonated beverage standards* [18]. "Bottlers" granulated sugar shall contain not more than:
  - 200 mesophilic bacteria per 10 g
  - 10 yeast per 10 g
  - 10 mold per 10 g

"Bottlers" liquid sugar shall contain not more than:

- 100 mesophilic bacteria per 10 g (DSE)
- 10 yeast per 10 g (DSE)
- 10 mild per 10 g (DSE)

(DSE = dry sugar equivalent.)

c. Microorganisms in Liquid Products. Determination of microbial contamination of foods requires enumeration of total numbers of organisms present in the sample as well as the types of organisms present. The presence of certain species, the fecal coliform bacteria, are indicative of contamination of the foodstuff with fecal material. Monitoring for fecal bacteria in edible syrups is important in determining its suitability for human consumption. Water testing procedures are carefully prescribed by the American Public Health Association [19]; however, there are no specific procedures for edible syrups; instead, adaptations of water testing methods are used.

Several indicator microorganisms are invariably present in human and animal feces. These are the fecal streptococci, *Clostridium,* and certain species of the anaerobic lactobacilli (*Lactobacillus bifidis*). The presence of the first two species are used as indicators of fecal pollution. The survival time of these organisms are suggestive of the time of contamination. In water, the fecal streptococci do not multiply significantly or survive for long. The lactobacilli are unable to multiply and survive for short periods of time. However, the coliforms may survive for weeks or months, depending on the conditions, and the clostridia, because of their ability to form spores, may survive indefinitely. Detection of any of these organisms is not a perfect index of contamination, but it is an indication of a potential problem. The confirmation of the presence of *Escherichia coli* is probably the best indication of fecal contamination.

The differentiation of *E. coli* from other coliforms is normally made on the ability of this organism to produce indole and degrade methyl red. Given the right media conditions, *E. coli* will produce blue colonies, while all the other coliforms produce red colonies. The presumptive test for the presence of coliforms is not specific but relies on growth and media conditions that select for "coliform-like" organisms. There are a number of other bacteria that will grow in these media, producing false positives, but they are not commonly present in water samples. *Leuconostoc,* which are normally present in significant numbers in these samples, will give false positives in the total coliforms test, leading to rejection of microbially acceptable cane syrups and molasses if the total coliform values are not corrected for the numbers of *Leuconostoc* detected using the same technique [20].

Contamination of molasses (blackstrap or final) with bacteria decreases the ethanol yield from fermentation of cane molasses. Five species of bacteria were isolated from molasses obtained from six Indian distilleries and were identified as *Streptococcus thermophilus, Leuconostoc mesenteroides,* and three species of *Lactobacillus* [5,21].

28.3 MICROBIOLOGICAL METHODS FOR SUGAR PRODUCTS

Microbiological methods for the determination of bacteria in sugar-containing samples can be done either with pour plate technique or, as is now more common, using membrane filtration. The procedural differences are in the selective media used for the various types of microorganisms as well as the incubation times and temperature.

28.3.1 Membrane Filter Technique

Membrane filters of known pore size are used to retain the microorganisms in samples for further culture. Test solutions of sugars are filtered through membranes of known pore size

(0.45 μm). The organisms are retained at the membrane surface and are subsequently detected as colony-forming units by growing on a nutrient medium or pad. Sterile glassware and aseptic technique are used throughout the procedures. The following techniques have been adopted by ICUMSA for the enumeration of mesophilic bacteria, yeast, molds, and thermophilic spore-forming bacteria [22].

a. Mesophile Determination. The majority of procedures use colony counts on nutrient agar at a pH of 6.7 to 7.5 for the enumeration of mesophiles. Incubation temperatures range from 28 to 37°C, but 30°C is probably most common. Plates are read after 48 h. Typically, nutrient agar or plate count media are used. Formulations follow:

- Nutrient agar (1 L)

Beef extract	1 g
Yeast extract	2 g
Peptone	5 g
Sodium chloride	5 g
Agar	15 g
pH	7.4

- Plate count agar (1 L)

Peptone	5 g
Yeast extract	2.5 g
Glucose	1 g
Agar	10 g
pH	7.0

b. Enumeration of Yeast and Molds. Either membrane filter technique or pour plates can be used. Pour plates are favored for high colony counts and/or dark-colored products such as syrups and molasses. The dishes are incubated at 30°C for 72 h. Any of the following media may be used.

- Wort agar (1 L)

Malt extract	15 g
Peptone	0.78 g
Maltose	12.75 g
Dextrin	2.75 g
Glycerol	2.35 g
Potassium hydrogen sulfate	1 g
Ammonium chloride	1 g
Agar	15 g
pH	4.8

- Yeast extract–glucose–chloramphenicol agar (1 L)

Yeast extract	5 g
Glucose	20 g
Chloramphenicol	0.1 g
Agar	15 g
pH	6.6

- Mycophil agar (1 L)

Phytone peptone	10 g
Glucose	10 g
Agar	16 g
pH	4.0

Adjust the pH to 4.0 by adding 15 mL of sterile 10% (v/v) lactic acid per liter of media prior to plating.

c. Enumeration of Thermophilic Spore-Forming Bacteria. Vegetative cells are killed by heating at 100°C for 15 min heat treatment, and the spores are plated on media that will allow the differentiation of acid and non-acid-producing organisms. Flat sours produce a yellow halo around the colonies. The following media may be used for this test.

- Glucose–tryptone media (1 L)

Tryptone	10 g
Glucose	5 g
Bromocresol purple	0.04 g
Agar	12 g
pH	7.0

- Shapton agar (1 L)

Beet extract	3 g
Peptone	5 g
Yeast extract	1 g
Bromocresol purple	0.025 g
Agar	15 g
pH	7.5

d. Estimation of Anaerobic Spore Formers. The membranes are rolled and put into a tube and then covered with a thick layer of nutrient media. The membrane filter is incubated for anaerobes for 72 h at 55°C.

e. Enumeration of Leuconostoc and Other Slime-Formers. Slime-forming bacteria are those that form shiny capsules when grown on sucrose. They include strains of *Bacillus, Streptococcus,* and *Leuconostoc.* ICUMSA methods call for use of either Weman or McClesky–Faville medium. Day and Sarkar [20] have reported on an alternative media for *Leuconostoc* when using membrane filters. Incubation is at 30°C for 48 h.

- Weman medium (1 L)

Raw sugar	40 g
Disodium hydrogen phosphate	2 g
Sodium chloride	0.5 g
Magnesium sulfate	0.1 g
Ferrous sulfate	0.01 g
Precipitated chalk	10 g
Agar	20 g
pH	5.0

- McClesky–Faville medium (1 L)

Tryptone	10 g
Yeast extract	5 g
Raw sugar	100 g
Agar	20 g
pH	6.5

- Day and Sarkar medium (1 L)

Tryptone	10 g
Yeast extract	7 g
Tween 80	1 g
Dipotassium hydrogen phosphate	5 g
Sucrose	50 g
pH	6.7

f. Enumeration of Coliforms. The presence of certain species, the focal coliform bacteria, are indicative of contamination of a foodstuff with fecal material. The monitoring for fecal bacteria in edible syrups is therefore important in determining its suitability for human consumption. Water-testing procedures are carefully prescribed by the American Public Health Association [19]. However, there are no specific procedures for edible syrups; rather, adaptations of water testing methods are used.

There are a number of commercial media available that incorporate the appropriate dyes to differentiate *E. coli* from other coliforms. High numbers of false-positive total coliforms in cane sugar and molasses samples were detected by Day and Sarkar. *Leuconostoc,* which is normally present in significant numbers in these samples gave false positives in the total coliforms test (Table 28.3).

28.3.2 Dextran Effects

Bacterial polysaccharides have been identified as a problem in sugar processing [23] since the studies by Scheiber [24] on the effects of slime production on sugar beet processing more than 100 years ago. He coined the term *dextran* for those polysaccharides that blocked filters and interfered with the crystallization process in the beet sugar industry. Dextrans are polyglucans characterized by a high percentage of one to six linkages. They can vary in size from small (soluble) polymers to extremely large (insoluble) polymers with molecular weights in the millions. Soluble dextrans cause the greatest problems in sugar processing. Dextran-producing bacteria are confined to three genera: *Lactobacillus, Leuconostoc,* and *Streptococcus* [25]. The two most common bacterial species involved in the formation of dextran in sugar mill juices are *Leuconostoc mesenteroides* and *Leuconostoc dextranicum* [8]. Only *Leuconostoc mesenteroides* has been found to contaminate sugar refinery products [26]. Although a wide variety of linkages can be found in dextrans, those dextrans isolated from sugar juices are primarily linear polymers usually containing better than 90% α, 1–6 link-

TABLE 28.3 Microbial Analyses of Louisiana Blackstrap Molasses and Syrups: *Leuconostoc* and Total Coliforms

Sample	Date	*E. coli*	Coliforms	*Leuconostoc*
Molasses	Oct. 30, 1996	0	298	>300
Syrup	Oct. 14, 1996	0	21	40
	Oct. 30, 1996	0	76	102
	Nov. 13, 1996	0	34	88
	Nov. 19, 1996	0	124	132
	Feb. 18, 1997	0	74	86
	Feb. 18, 1997	0	82	100

ages. The dextraotatory property of this polymer affects the polarization of juice, giving inaccurate readings for juice purity and sugar content. Theoretically, for each 333 ppm of dextran in raw sugar there is a polarity increase of 0.15. Analytically, most of the dextran is removed by lead clarification [27], but is not removed by the newer nonlead clarification agents.

Levans (two to six polyfructans) can also be constituents of the polysaccharides isolated from sugarcane juices [28]. They are probably produced by species of *Leuconostoc* [29]. Other contaminating species, such as *Bacillus* and *acetobacter*, are also capable of producing levans. The levan represents a small portion of the total gum produced compared with dextran and is not generally considered to be a problem.

a. Dextran Generation in Cane. The rate of formation of dextran in cane is a function of the ability of microorganisms to infect the cane (i.e., the degree of damage to the stalk) and the other factors that affect microbial growth: specifically, water activity and temperature. There have been many studies on the effects of delay in processing harvested cane. Legendre [30] emphasized the effects of damage on the amount of dextran detected in cane upon processing, and Irvin [31] showed the effects of delayed milling on sugar losses.

b. Dextran Production in the Mill. The production of dextran by *L. mesenteroides* has been a recognized nuisance to the sugar industry for decades [32]. Not only can *Leuconostoc* be introduced into the factory from damaged sugarcane, but during sugar manufacture as well. Where there is poor housekeeping, the walls of the sugar processing equipment can become coated with thick biofilms. These biofilms can be almost pure cultures of *Leuconostoc.* These films continuously shed dextrans into passing process steams. Processing problems associated with dextran include increased juice viscosity [33], poor clarification [34], and crystal elongation [35]. The most damaging effect of dextran formation is seen during the crystal growth process. Sugar crystals grown in the presence of dextran become elongated and are lost through the centrifugal screens. Crystal elongation is particularly noticeable in the low-grade boiling. Low-grade massecuites containing dextran are difficult to centrifuge and show high sugar loss to the final molasses. Magmas made from this sugar are poor footings for commercial sugar. The end result of high dextran concentrations are reductions in factory capacity and sugar recovery.

The economic effects of dextran on the sugar factory can be significant. Depending on the sugar contract, there can be significant penalties assessed against dextra-rich sugar. It is estimated that sugar containing 1000 MAU dextran is unprofitable to produce [36]. The hidden losses due to this polymer are also significant. Day estimated that the sugar losses to molasses were in the range of 2 to 3 points per 1000 ppm dextran in juice [37].

c. Dextran Effects on Sugar Processing. Dextran has a number of effects on sugar processing:

- *Undetermined loss of sugar in juice.* Sucrose loss caused by dextran formation can be calculated from the dextran levels in mixed juice. Direct sucrose loss is approximately 1.9 times the dextran concentration. The rate of dextran formation under normal mill conditions is rapid. It has been reported that a 1-h delay in processing mixed juice can result in a 6.5% loss of its sucrose content [38].
- *Juice clarification.* One of the problems attributed to dextran contamination is slow mud settling rates. Dextran is not removed from the juice during the clarification process.
- *Low-grade boiling.* A comparison of low-grade boilings with and without dextran showed that dextran caused considerable difficulty in boiling the dextran-containing

material. To exhaust the molasses to the same extent, it was necessary to boil at a higher Brix [39].

- *Sugar quality.* High dextran levels in syrup cause crystal elongation. Dextran also reduces crystal growth rates and increases the formation of false grain. The dextran partitions from the syrup into the raw sugar at an estimated concentration of 30% of that in the mother liquor [40].

The effects of dextran occur throughout the sugar production process and are concentration dependent. Even small levels can have economic effects without readily identifiable processing problems (Table 28.4).

d. Dextran in the Sugar Refinery. Dextran that is brought in with the raw sugar, as well as dextran generated by biofilms in processing equipment, can be a problem in sugar refining. Dextran slows the rate of filtration of refined liquors, reducing factor capacity. Dextran in the final refined sugar also causes problems for the producers of sugar-containing products, especially hard candies and cordial liquors [41]. As with raw sugar production dextran passes through the process and a significant portion winds up in the refined sugar.

28.4 MILL SANITATION

The sugar factory and sugar refinery are microbially "dirty" environments. Efforts must be taken to minimize the active microbial loading of the process juices in areas where the temperatures can permit microbial growth and biofilm formation with its ultimate effect on sugar loss. The microorganisms found in sugar mills are not significantly different from those found in other environments and can be controlled in an analogous manner. Exposed surfaces can be kept clean with a combination of high-pressure washing with hot water or steam with the occasional use of biocidal compounds such as chlorinated oxidizing agents Calcium-Hypochlorite (HTH, bleach) to help break up biofilms. This must be done on a routine basis because of the rapid rate of microbial biofilm formation. In areas where surface cleaning is not feasible, judicious use of commercially available food-compatible biocidal agents is required.

28.4.1 Biocides and Biocidal Agents

The term *biocide* denotes chemical agents that have antiseptic, disinfectant, or preservative activity. Factors such as microbial loading, pH, concentration, contact time, temperature, and organic loading affect the biocidal activity and range of these compounds. Different

TABLE 28.4 Effects of Dextran in Juice

Dextran in Juice (ppm / Brix)	Effect
0	None
250	1-point loss in molasses purity
500	2-point loss in molasses purity
750	Production of 250 MAU sugar
1000	Detectable crystal elongation, 5-point loss in molasses purity
1500	Significant operational problems
3000	Severe problems

bacteria have different sensitivities to biocides, the most sensitive being the enteric bacteria. The gram-negative bacteria, especially the pseudomonads, are more resistant. Bacterial spores are the most resistant to these compounds. Chlorine-releasing biocides (such as hypochlorite) and quaternary ammonium compounds are commonly used biocides in sugar processing. The mode of action of the chlorine-releasing compounds is not well understood. They are oxidizing agents and can react with critical proteins and cell components as well as the nucleic acids. Generally, quaternary ammonium compounds disrupt the cell membranes of microorganisms.

The use of biocides or bacteriostats for the prevention of losses due to bacterial action is justified, particularly in areas where it is not possible to routinely wash and clean equipment using hot water or steam. Good housekeeping at the milling station and the use of steam jetting at joints and chain links can solve about half the microbial development problem in the mill, and biochemical agents can be used to control the rest [42]. Cleanliness in the mill is very important, chemicals cannot be relied on to solve all the problem [2].

The permissible combinations of organic agents and dosages that can be used in controlling microorganisms in cane factories in the United States are prescribed by the U.S. Food and Drug Administration [43]. Most commercially available biocides for sugar mill application contain thiocarbamates in one form or another as the active ingredient. The composition of a typical commercial biocide (Midland Laboratories, PCS 6001) is as follows:

Sodium dimethydithiocarbamate	15%
Disodium ethylenebisdithiocarbamate	15%
Inert ingredients	70%

Most commercial biocides have similar compositions.

28.4.2 Preservation of Stored Syrups

Storage of juice during manufacture requires that the temperature be kept above the point for most microbial growth (60°C). Even at high temperatures, juice storage should be avoided. Syrups should be stored at a high enough Brix (above 67 Brix) to minimize water activity and microbial growth. Restriction of water activity is a better method of preservation than restrictive temperature, as high temperatures will select for thermophilic organisms.

28.5 ENZYME CATALYZED SUGAR LOSS

Two classes of enzymes are responsible for most of the enzymatic sugar loss in the sugar production process: the invertases which are hydrolases, and may be of both plant and microbial origin, and the dextransucrases, which are glucosyl transferases and are of microbial origin. In neither case is it possible to ascribe a specific level of sugar loss to a given enzyme activity, but undoubtedly both enzymes are responsible for a significant component of the unclassified losses in sugar manufacture.

28.5.1 Invertase

Sucrose is simultaneously a β-D-fructofuranoside and an α-D-glucoside, and thus is readily hydrolyzed by either a β-D-fructofuranosidase or an α-D-glucosidase. Invertase is the trivial name for β-D-fructrofuranoside fructohydrolase (E.C.3.2.1.26). This enzyme catalyzes the following reaction:

$$\beta\text{-D-fructofuranoside} + H_2O \longleftrightarrow \text{alcohol} + \text{D-fructose}$$

Invertase was first prepared in 1860 from brewer's yeast by alcoholic precipitation [44]. Since then the β-fructofuranosidase activity of yeast has been studied extensively. The enzyme catalyzes the hydrolysis of sugars possessing a terminal unsubstituted β-D-fructofuranosyl residue. These are primarily sucrose, raffinose, and melezitose [45]. Generally, the higher the molecular weight of the fructose portion (afructon), the slower the rate of hydrolysis of the *Saccharomyces* enzyme (Table 28.5).

Invertases occur widely in the plant kingdom and have a general role in the breakdown of sucrose when plants, including sugarcane, have a marked need for hexoses [47], as during the maturation of fruits. Invertase is also a necessary enzyme for the growth of yeast on sucrose. It is used as well in the production of commercial invert syrups. The invertase source can be as simple as killed yeast, as the enzyme is quite stable. The pH optima for invertase ranges from 3.5 to 5.5, depending on the source.

Invert syrups are produced exclusively using immobilized invertase. Probably the first use of immobilized enzymes was the immobilized invertase used in the production of Tates Golden Syrup [48]. The source of commercial invertase is almost invariably a yeast, typically a *Saccharomyces* or *Candida* species. The enzyme is released from microbial cells by autolysis using common salt and an organic solvent. The autolyzed preparation is normally used without further treatment. The quantity of enzyme used in invert syrup preparation is so small that little further refining other than evaporation is necessary for invert syrup production.

Invertase has a specialized use in the manufacture of soft- and liquid-center confectioneries. Such formulations consist mainly of sucrose, but include glucose syrup, invert sugar, and other components. As prepared the filling is stiff and easily coated with chocolate. Invertase is included at a very low dosage and hydrolyzes the sucrose over a period of weeks. This has a number of effects [49]:

1. The center becomes softer or liquefied.
2. The fructose component is hygroscopic, helping prevent loss of moisture during storage.
3. The fructose inhibits crystallization of other components.
4. Inversion of the sucrose increases osmotic pressure sufficiently to inhibit fermentation.

A second group of invertase-type enzymes is produced by some microorganisms. They are characterized by their ability to cleave fructose polymers (inulins), but they also hydrolyze sucrose. There are both yeast and fungal inulinases. Only about 20 yeast species of the

TABLE 28.5 Relative Hydrolysis Rates Yeast Invertase

	Molecular Weight of Fructose Portion	Relative Rate of Hydrolysis
Sucrose	179	100
Raffinose	341	23
Stachyose	503	7

Source: Data from Ref. 46.

144 species capable of using both sucrose and raffinose are capable of hydrolyzing inulin [50]. Common sources of this enzyme include the *Kluyveromyces* species of yeast *K. fragilis* and *K. marxianus* [51], and a wide range of fungi, principally, *Talormyces flavus, T. stipitatus, Chaetomium subspirale, C. sureum, Gelosinospora longespora,* and *Sporotrichum schenkii* [52].

28.5.2 Dextransucrase

Dextransucrase (sucrose 1,6-α-D-glucan 6-α-glycosyltransferase, E.E. 2.4.1.5) is an enzyme secreted by species of *Leuconostoc* and *Streptococcus.* It polymerizes the glucosyl moiety of sucrose to form dextran. The reaction is represented as

$$nC_{12}H_{22}O_{11} \longrightarrow (C_6H_{10}O_5)_n + nC_6H_{12}O_6$$

This enzyme was discovered in 1941 [53]. The hydrolysis of the sucrose molecule provides the energy necessary for condensation of the D-glucosyl units, allowing in vitro dextran synthesis. The enzyme is produced by a wide variety of *Leuconostoc* and streptococci. The end products of the reaction are fructose and α-1–6-linked glucans [54]. Dextransucrase is produced only when these organisms grow on sucrose. The degree of branching of the glucans is both a function of the specific enzyme and of cation concentrations [55]. It is not present if they are grown on glucose or maltose. Dextransucrase is produced as an extracellular enzyme by *Leuconostoc* sp. even when only low levels of sucrose are present. As sucrose concentrations increase, the amount of enzyme that is produced increases. The *Leuconostoc* enzyme has a pH optimum at 5.6 and is active up to 60°C. Because the enzyme is extracellular, *Leuconostoc* biofilms on sugar-processing equipment can continuously shed this enzyme into process streams, where it will remain active until the temperature is high enough to denature the protein.

The enzyme reaction mechanism involves transfer of glucose units from sucrose to a growing glucan chain. The growth of the chain is terminated if an molecule alternative to glucose (an acceptor) is incorporated into the chain. Chain-terminating acceptors include maltose, isomaltose, and nigerose [56]. Shorter, more soluble dextrans are produced in the presence of these acceptors [57].

REFERENCES

1. M. P. Scarr, Ph.D. dissertation, University of London, 1954.
2. D. Bevan and J. Bond, *Proc. Queensl. Soc. Sugar Cane Technol., 38th Conf.,* Watson Ferguson and Co., Brisbane, Queensland, Australia, 1971, pp. 138–143.
3. J. E. Irvine, *Cane Sugar Refin. Res. Proj.,* 1980, pp. 116–120.
4. P. C. Atkins and R. J. McCowage, *Proc. Int. Dextran Workshop,* Sugar Processing Research Institute, New Orleans, LA, 1984, pp. 7–39.
5. Z. O. Lopez, I. E. Moreno, F. A. Fogliati, and H. G. Ayala, *Sugar Azucar,* 83:21–34, 1988.
6. B. G. Wadsworth, *Proc. Queensl. Soc. Sugar Cane Technol., 41st Conf.,* Watson Ferguson and Co., Brisbane, Queensland, Australia, 1974, p. 11.
7. D. F. Day and D. Sarkar, *Sugarcane Research,* Annual Progress Report, Louisiana Agricultural Experiment Station, Baton Rouge, LA, 1988, pp. 178–180.
8. W. L. Owen, *The Microbiology of Sugars, Syrups and Molasses,* Barr-Owen Research Enterprises, Baton Rouge, LA, 1949.
9. G. P. Meade and J. C. P. Chen, *Cane Sugar Handbook,* 10th ed., Wiley, New York, 1977.

10. P. J. Field, Paper 628D, *Proc. Sugar Ind. Technol. Conf.*, Brussels, 1992, p. 130.
11. R. D. Skole, H. Newman, and J. L. Barnwell, *Proc.* 1966 *Tech. Session Cane Sugar Refin. Res. Cane Sugar Refin. Proj.*, New Orleans, LA, 1967, pp. 35–45.
12. G. Snook, B. Tungland, A. Campbell, and J. Kardatzke, 1996 *Proc.* 54*th Annu. Meet. Sugar Ind. Technol.*, La Romana, Dominican Republic, May 7–10, 1996.
13. R. H. Tilbury, C. J. Orbell, C. J. Owen, and M. Hutchinson, *Proc.* 3*rd Int. Biodegradation Symp.*, J. M. Sharpley and A. M. Kaplan, eds., Applied Science Publishers, London, 1976, pp. 533–543.
14. J. P. Chen, *Cane Sugar Handbook*, 11th ed., Wiley, New York, 1985, p. 517.
15. E. J. Cameron and W. D. Bigelow, *Ind. Eng. Chem.*, 23:1330–1333, 1931.
16. H. C. S. de Whalley and M. P. Scarr, *Chem. Ind.*, Aug. 30, 1947, p. 531.
17. M. A. Godshall, *Symp. Chem. Process. Sugar Beet Sugar Cane*, Elsevier, New York, 1987, pp. 236–252.
18. *Quality Specification and Test Procedures for Bottlers*, National Soft Drink Association, June 1975.
19. APHA, *Standard Methods for the Examination of Water and Wastewater*, 12th ed., American Public Health Association, New York, 1967, pp. 567–627.
20. D. F. Day and D. Sarkar, *Sugarcane Research*, Annual Progress Report, Louisiana State University Agricultural Center, Baton Rouge, LA, 1996, pp. 253–255.
21. D. P. Greatal-Singh and D. S. Dahiya, *Proc.* 51*st Annu. Conv. Sugar Technol. Assoc. India*, G33–G39, E. O. M. Kamat, Pune, India, 1988.
22. International Commission for Uniform Methods of Sugar Analysis (ICUMSA), Publications Department, Colney, Norwich, England, GS 2/3-41, 43, 45, 47.
23. R. H. Tilbury, A. H. Walters, and J. J. Elphick, *Biodeterioration of Materials*, Elsevier, Amsterdam, 1968, pp. 717–730.
24. C. Scheiber, *Z. Versuchwesen Dtsch. Zuckerind.*, 24:309–319, 1874.
25. R. L. Sidebotham, Dextrans, in *Advances in Carbohydrate Chemistry and Biochemistry*, Vol. 30, Academic Press, New York, 1974, pp. 371–444.
26. R. A. Kitchen, *Proc. Int. Dextran Workshop*, Sugar Processing Research, Inst., New Orleans, LA, 1984, pp. 53–61.
27. G. A. Bradbury, R. M. Urquhat, J. H. Curtin, and R. J. McCowage, *Sugar J.*, Jan. 1986, pp. 11–13.
28. F. K. E. Imrie and Tilbury, R. H. *Sugar Technol. Rev.*, 1:291–361, 1972.
29. J. F. Robyt and T. F. Walseth, *Carbohydr. Res.*, 68:95–111, 1979.
30. B. Legendre, *J. Austral. Soc. Sugar Cane Technol.*, 5:73–76, 1985.
31. P. C. Irvin, *Proc. Austral. Soc. Sugar Cane Technol.* (*Queensl.*), Townsville Meeting, Watson Ferguson and Co., Brisbane, Queensland, Australia, 1986, pp. 193–199.
32. E. E. Coll, M. A. Clarke, and E. J. Roberts, *Tech. Session Cane Sugar Refin. Res.*, Sugar Processing Research Inst., New Orleans, LA 1978, pp. 92–105.
33. G. L. Geronimos and P. F. Greenfield, *Proc. Queensl. Soc. Sugar Cane Technol.*, 45th Conf., Watson Ferguson and Co., Brisbane, Queensland, Australia, 1978, pp. 119–126.
34. G. P. James and J. M. Cameron, *Proc. Queensl. Soc. Sugar Cane Technol.*, 38th Conf., Watson Ferguson and Co., Brisbane, Queensland, Australia, 1971, pp. 247–250.
35. M. T. Covacevich, G. N. Richards, and G. Stokie, *Proc.* 16*th Congr. Int. Soc. Sugar Cane Technol.*, Sõ Paulo, Brazil, 1977, pp. 2493–2508.
36. B. Beyt, *J. Austral. Soc. Sugar Cane Technol.*, 53, Mar. 1987, p. 14.
37. D. F. Day, *J. Austral. Soc. Sugar Cane Technol.*, 14:53–57, 1994.
38. I. M. Mansour, Y. A. Hamdi, Z. T. Hamid, and H. Toma, *Int. Sugar J.*, 88:1a, 1986.
39. D. Hsu, B. J. Someru, and T. Moritsugu, *Int. Sugar J.*, 91:12a, 1989.
40. D. F. Day, *Sugar J.*, 52(Mar.):9, 1986.

41. G. W. Vane, *Sugar Ind. Technol.,* 40:95–102, 1981.
42. J. P. Chen, *Sugar,* 59(May):65–66, 1964.
43. *Code of Federal Regulations,* Title 21, Section 173.320, Apr. 1988.
44. M. P. E. Berthelot, *C.R. Acad. Sci.,* 50:980–984, 1860.
45. K. Myrback, *Enzymes,* 4:379, 1960.
46. M. Adams, N. K. Richtmyer, and C. S. Hudson, *J. Am. Chem. Soc.,* 65:1369–1380, 1943.
47. T. Ap Rees, in *Plant Biochemistry,* Vol. 11, D. H. Northcote, ed., Butterworth, London, 1974, p. 89.
48. Tate and Lyle, Ltd., and H. C. S. de Whalley, British patent 564270, 1944.
49. C. Bucke, in *Enzymes in Food Processing,* G. Birch, N. Blakebrough, and K. J. Parker, eds., Applied Science Publishers, London, 1981, pp. 51–72.
50. J. A. Barnett, *J. Gen. Microbiol.,* 99:183–190, 1977.
51. J. W. D. Grootwaseink and S. E. Fleming, *Enzyme Microb. Technol.,* 2:45–53, 1980.
52. K. Ishibashi and S. Amao, British patent 1420528, 1976.
53. E. J. Hehre, *Science,* 93:237, 1941.
54. E. J. Hehre, *Adv. Enzymol.,* 11:308, 1951.
55. Y. Tsumuraya, N. Nakamura, and T. Kobayashi, *Agric. Biol. Chem.,* 40:1471–1477, 1976.
56. J. F. Robyt and S. H. Ekland, *Carbohydr. Res.,* 121:279–286, 1983.
57. D. F. Day and D. Kim, U.S. patent 5,229,277, 1993.

CHAPTER 29

Refinery Maintenance Program*

INTRODUCTION

In a sugar refinery, the major cost element is maintenance. Therefore, a refinery should have a well-organized maintenance department with the following functions clearly defined:

1. *Maintenance organization.* This includes an organization chart, job description, optimal worker-to-supervisor ratio, organizational philosophy and attitude, tool/instrument/equipment resources, budget, etc.
2. *Maintenance scheduling and planning.* Statistics should be done on the percent of work orders completed in the planned time period and percentage of frequency that work is completed within estimated time and the backlog time period.
3. *Preventive maintenance.* This should clearly state percent of critical equipment covered by preventive maintenance; percent of preventive maintenance task completed within 1 week of due date; the average time to complete a preventive maintenance task; and personnel responsible for preventive maintenance tasks.
4. *Maintenance training program.* Details of training for supervisor, planner, and craft should be provided. Training interval, format, and material source of instructor should be well thought out. Assessment of the quality and skill level of the maintenance work force should be quantified whenever possible.
5. *Maintenance inventory.* This should cover at least 90% of material requested from the maintenance department. The maximum and minimum level of stored items should be optimized carefully to reduce inventory cost.
6. *Maintenance work order.* In this category, the supervisor should focus on the percent of worker-hours and material charged to a work order; percent of total jobs charged to work orders; percent of work orders completed in 4 weeks and 8 weeks from the date requested; and percent of work orders generated from preventive maintenance inspection.

The tremendous variety of equipment in a refinery from slow-speed, simple, massive, very rugged machinery to highly sophisticated high-speed units and complicated computer-

*By G. Fawcett, updated by Chung Chi Chou.

controlled process equipment presents a daunting challenge to today's maintenance engineer or superintendent. Usually, agricultural and field equipment falls under a separate department from that of mill maintenance, and frequently, packaging maintenance is separate from refinery maintenance. Nevertheless, after deleting these areas, sufficient challenge remains to test the skills of the most talented mechanic or engineer.

29.1 MANAGEMENT PHILOSOPHY

Before organizing a new maintenance function, careful thought should be given to management philosophy and the factors that will affect the utilization of personnel. Although at first it may seem logical to separate production and maintenance as unrelated functions and to further classify maintenance personnel by crafts, the price is always reduced flexibility.

Consideration should be given to assigning as much of the maintenance function as possible to production people. They have a vested interest in keeping their plant in top shape and have a great advantage in being very familiar with the way the equipment functions and sounds under normal conditions. They have the additional advantage of understanding the process, and hence make better troubleshooters than the specialized mechanic who sees the equipment only occasionally.

Most recent installations of integrated, high-speed, paper bag making–filling–sealing–bundling and palletizing lines in the United States and Europe have employed operator–mechanics. The machines and their controls are so complex that a highly specialized mechanic can best operate and maintain them.

When production people are assigned maintenance responsibilities in this fashion, a small maintenance crew with special skills in welding, millwright, electronics, and so on, can supplement the operating personnel when needed. In small plants or for certain specialties, it may be desirable to rely on a reputable contractor. This frequently can be the lowest-cost alternative when contingency arrangements have been made with the contractor in advance and prompt service can be provided.

The advantage of incorporating maintenance as an operating responsibility can be enhanced by a determined program of cross-training, with the ultimate objective of everyone being able to perform every job. Cross-training is a great motivator and can raise the confidence level of the operating crew to remarkable heights and lead to unsurpassed efficiencies.

An attitude of preventive engineering should be encouraged where causes for frequent failures are sought and equipment is redesigned to eliminate the causes. Very often, solutions can be found and implemented by the mechanic or operator on the job.

Also, an attitude of preventive maintenance can be developed where the people on the job become attuned to the process and equipment. With some experience, they will be able to tell when maintenance will be necessary and perform it before it becomes an emergency.

The practical philosophy for a plant maintenance organization will be molded by the existing labor situation and climate whether you are staffing a new plant or reorganizing an old plant. In all cases, maximum flexibility is much to be desired and will contribute to lower costs.

We can classify the types of maintenance work as:

- Emergency repairs involving unscheduled shutdowns or danger to personnel or property
- Routine or normal repairs to restore plant to serviceable condition
- New work involving additions or improvements
- Preventive or predictive maintenance involving regular inspection and repair to prevent breakdowns or malfunction

Depending on the philosophy adopted, these types of jobs can be handled by varying combinations of operating, maintenance, and contractor personnel. The philosophy should also consider the amount of preventive maintenance that is desirable. Preventive maintenance should return substantially more than its cost in the form of reduced overall maintenance cost plus better performance and utilization of equipment. Careful records should be kept to ensure that this is the case. Regular, routine preventive maintenance should be used in cases where breakdown or failure can cause severe damage to the equipment or where downtime is very costly. On the other hand, intuitive or breakdown maintenance may be cost-effective in the case of noncritical equipment, small motors, pumps, gearboxes, and so on [1,2].

The ultimate goal of maintenance programs is to ensure that equipment will perform as designed over an expected specific period of time with minimal maintenance cost. To achieve this goal, both staff and workers need to be well trained in core skills and proper attitude/work ethic. To reduce cost, maintenance needs to be integrated into production. In addition, engineering standards need to be incorporated during development, design, and installation. Contracting and outsourcing of nonroutine activities need to be considered carefully. Causes of plant failures need to be analyzed and eliminated. Feasibility, both economic and engineering aspects, of performing maintenance during shutdown as compared to operating time also needs to be studied. The criteria for measurement of maintenance performance need to be defined and implemented. Most of all, safety considerations must be detailed and incorporated into repair procedures, replacement strategies, and engineering standards in new construction.

29.2 AREA AND CENTRALIZED MAINTENANCE BY ORGANIZATION OR GEOGRAPHY

The management of the maintenance organization can be centralized or decentralized, and the facilities and personnel also can be centralized or decentralized. Frequently, a combination is used. For example, in one plant (Fig. 29.1), area mechanics are assigned to cen-

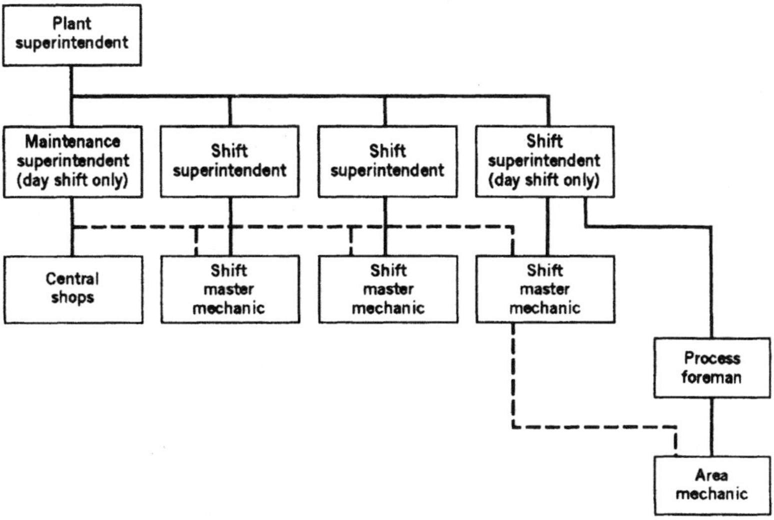

FIGURE 29.1 Combination centralized and area maintenance organization.

trifugals, filter presses, and char house, while the other mechanics are assigned to a central shop and dispatched from there. The area mechanics report to process foremen with line responsibility to the maintenance superintendent, while the shop mechanics report to him directly.

In addition, two or three mechanics and an electrician, plus a master mechanic, are assigned to each shift. The master mechanics report to the shift superintendent (a process person) and have line responsibility to the maintenance superintendent. Shift and maintenance superintendents report to the plant superintendent.

Each shift superintendent and master mechanic has responsibility for preventive maintenance in one-third of the entire plant. They can draw additional resources from the maintenance superintendent as needed.

The size and physical layout of the plant, the distances between shops and process, the age and design of particular parts of the process, as well as the experience and personalities of the personnel and the number of personnel involved are important factors in choosing how to organize or reorganize the maintenance function. If you are organizing a new department, it is always best to staff it very skimpily. Additional people can be added later if necessary, but it is extremely difficult later to reduce the number of people or even to determine if the department is overstaffed. Figure 29.2 shows an example of a craft-oriented maintenance organization for a larger plant. In a small plant, it might be desirable to have no maintenance personnel and rely entirely on operating personnel, and when necessary, outside contractors.

29.3 TYPES OF MAINTENANCE PROGRAMS

Early plant maintenance strategies consisted of running until failure and then fixing the breakdown. Breakdowns always came at an inopportune time, and costs were very high, so something better was sought. *Preventive maintenance* seeks to overcome these problems by scheduling periodic inspection and overhaul regardless of condition of equipment or machinery. Preventive maintenance may require unnecessary overhaul on some units and still

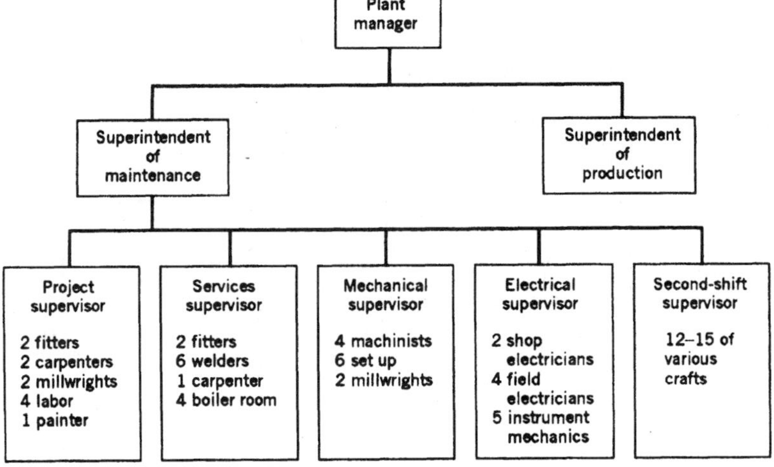

FIGURE 29.2 Centralized craft-oriented maintenance organization. (Adapted from Ref. 3.)

allow breakdowns to occur on others. *Predictive maintenance* monitors characteristics of a machine such as vibration, temperature, and so on, and compares these characteristics against their historical profile and schedules repair based on changes in these profiles. The hope is to avoid breakdowns while extending preventive maintenance intervals.

The objectives and advantages of these maintenance strategies are as follows:

Objectives of Maintenance Program

- To contribute to overall profits
- To ensure a safer plant
- To achieve higher product quality
- To ensure a cleaner plant
- To achieve better scheduling of production, fewer interruptions, and maximize production
- To ensure a well-maintained plant that instills pride and motivates employees
- To develop legacy of skilled and motivated people
- To have spare parts and equipment on hand
- To identify problem equipment, processes, and bottlenecks

Objectives of Preventive Maintenance Program

- To save money (pays for itself) by reducing overall production costs
- To catch problems before breakdown, prevent untimely breakdowns
- To help overall scheduling of maintenance and production
- To maximize equipment life
- To reduce excessive parts inventory
- To provide information for future design and planning
- To ensure a safer plant
- To ensure a cleaner plant

Objectives of Predictive Maintenance Program

- To save money (pays for itself) by reducing overall production costs
- To maximize equipment life between repairs, reduce equipment downtime
- To predict when to repair, what to repair, and show any trend to failure
- To prevent excessive preventive maintenance
- To provide visual report on condition and trend
- To increase production
- To ensure a more effective scheduling of resources and personnel

Tools of Predictive Maintenance

- Computer
- Vibration analysis
- Temperature analysis
- Current analysis
- Resistance analysis
- Pressure analysis
- Oil analysis
- Thermography
- Ultrasonic testing (thickness, cracks)
- Smoke

- Dye
- Speed
- X-ray

Characteristics of a Good Maintenance Program

- Emphasizes good supervision
- Features skilled people
- Provides necessary tools
- Ensures adequate spare parts on hand
- Allows adequate downtime for maintenance
- Provides sufficient training
- Does proper planning
- Features regular lubrication program
- Features regular cleaning program
- Redesigns problem equipment (corrective maintenance)
- Replaces obsolescent equipment
- Keeps spare equipment rebuilt and ready
- Trains operators in emergency response (damage control)
- Does proper repairs (no "band-aids")
- Pays attention to vibrations and changes in sound

29.4 MAINTENANCE SYSTEM

Figure 29.3 shows a flowchart for a maintenance system which can have any or all of the following features:

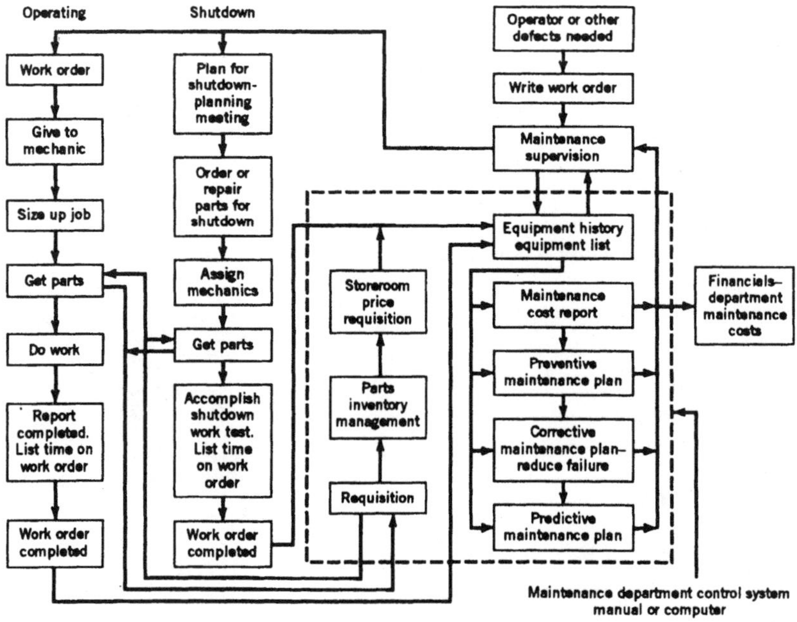

FIGURE 29.3 Maintenance system flowchart.

- Preventive maintenance plan
- Predictive maintenance plan
- Corrective maintenance plan
- Manual or computerized system
- Equipment list
- Work order
- Equipment history
- Parts inventory
- Maintenance cost report
- Sanitation program

29.4.1 Equipment List or Catalog

A system for cataloging all equipment in the plant is necessary for an efficient work order and/or preventive maintenance program. One system divides the plant into functional areas by letter, then major equipment by number, and subequipment by letter and/or number [i.e., F-41-P-2 = dilution tank 41 (major equipment) in area F, pump 1 and pump 2 (subequipment 1 and 2)]. There are many possible systems—the main requirement is to have a system that everyone understands and will use.

29.4.2 Work Order System

A work order system will facilitate planning and control of the maintenance work and personnel and is the keystone of the preventive maintenance system. It is also essential if you want to charge maintenance costs to the various plant departments that use maintenance services.

A simple form can be used showing the date requested, date required for work to be done, a description of the work, and the equipment number (see Fig. 29.4). It is also helpful if the form is numbered serially. The time spent by the mechanic can be entered on the front or back, so labor costs can be entered as equipment history and charged to plant departments. Additional information and planning are required for major jobs and can be entered on a supplemental form, or a separate work order form can be used for major jobs.

As with a preventive maintenance system, a work order system should contribute value to the maintenance function. If it cannot be justified in terms of lower cost, better main-

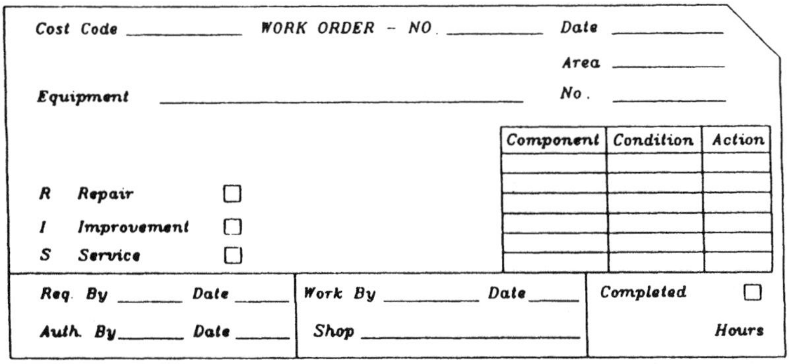
Cost Code ________ WORK ORDER – NO. ________ Date ________

Area ________

Equipment ____________________ No. ________

Component	Condition	Action

R Repair ☐

I Improvement ☐

S Service ☐

Req. By ______ Date ______ | Work By ________ Date ______ | Completed ☐

Auth. By ______ Date ______ | Shop ____________ | Hours

FIGURE 29.4 Sample work order. (From Ref. 3.)

tenance, better utilization of people, higher productivity of the plant, and so on, it should not be used.

29.4.3 Equipment History

The equipment history should give a description of all the repairs made on that piece of equipment and the costs of those repairs. Even minor repairs should be recorded because if they recur frequently, the cost could amount to 20 or 30% of total maintenance costs. Periodic preventive inspections and repairs should also be recorded to help determine if the proper frequency of maintenance is being used on this equipment. These data, when analyzed, provide the basis for the preventive, predictive, and corrective programs. This analysis, considered in conjunction with the cost of low production and of production quality due to shutting down the equipment, indicates how much can be spent to correct a problem permanently.

Traditionally, equipment history cards have been maintained manually with written descriptions of the work done. Keeping histories current required a great deal of tedious work, and therefore they were frequently less than complete. The analysis of this type of history was equally tedious and difficult. C. E. Hutchinson of Westinghouse developed a numerically coded system that greatly simplified the work of keeping the histories and analyzing them. This system provides a common language for identification of the equipment involved, the condition found, and the action taken to correct the problem. This system, shown in Fig. 29.5, facilitates manual or computer entry and analysis. Figure 29.6 shows how a history card can point out a repetitive problem and the need for corrective

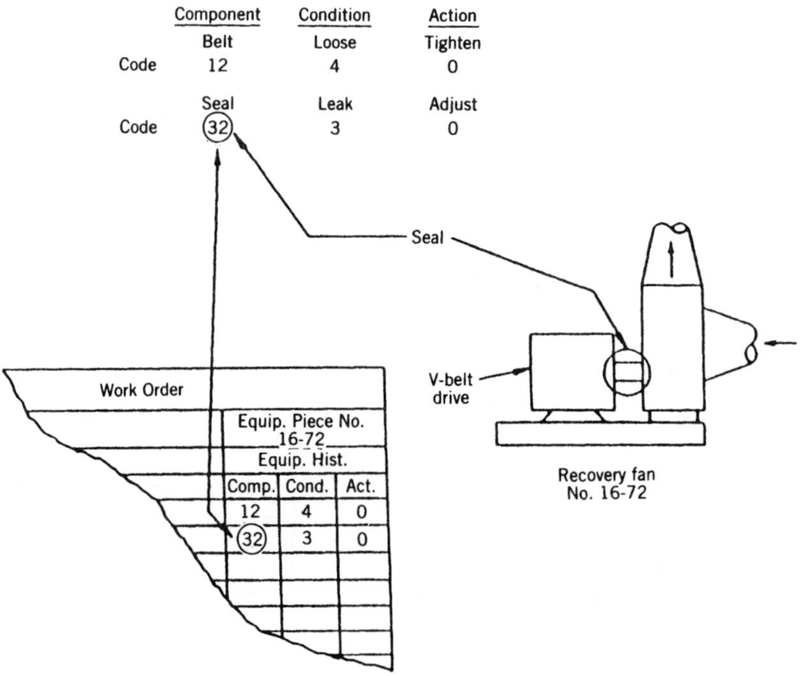

FIGURE 29.5 Repair coding for work order. (From Ref. 3.)

HISTORY CARD RECOVERY FAN NO. 16 - 72												
DATE	TYPE ORDER	COMPONENT										REMARKS
		10	12	14	22	30	32	38	42			
9/26	P		4-0									
11/15	P		4-0									
1/13	P		8-6									
1/19	P		4-0									
2/3	P		4-0									
2/19	P		8-6									
3/1	P						3-0		1-5			Repair broken welds at base

FIGURE 29.6 History-of-repair card. (From Ref. 3.)

maintenance. In this case, component 12 (belt) is frequently loose (condition 4) and requires tightening (action 0).

29.4.4 Parts Inventory System

Downtime (scheduled or breakdown) represents a large part of total maintenance costs. Having necessary spare parts on hand to minimize downtime is essential in controlling maintenance costs. However, large stocks of spare parts are expensive to acquire and maintain, so a good parts system should have as few parts on hand as possible but still cover most needs. Recommendations of manufacturers and experienced mechanics can help in deciding what parts to stock for new installations. Equipment histories can be invaluable in determining what parts to stock for older installations.

Total maintenance costs can be increased by carrying too large a spare parts inventory or by not having a part on hand when needed. The proper balance should be determined from equipment histories and from the maintenance manager's experience and tolerance for risk.

The parts storeroom can be one centralized storeroom or several decentralized units. It can also be separate from general stores or combined. The size, layout, and organization of the plant will influence these choices. It is convenient to have parts located near use points, but there is danger in carrying excess inventory through duplication at different locations. There is also the possibility of requiring additional inventory clerks or of losing control and parts if the stores are unattended.

For an operation of any size, a perpetual inventory system should be kept. Materials should be withdrawn by requisitions that are posted regularly and reordered when minimum stock levels are reached. The system can be manual or computer operated. The costed requisition along with the work order is an essential part of the maintenance cost report and preventive or predictive maintenance systems.

29.4.5 Manual or Computerized Maintenance System

The elements of the system can be manual or computerized. There are many off-the-shelf computerized maintenance management systems (CMMSs) that can handle the entire maintenance system, including work orders, equipment history, parts inventory, and purchasing, planning, and preventive and predictive maintenance scheduling. CMMS programs can interface with accounting systems and central purchasing systems so that cost data can be transferred electronically to purchasing. CMMS systems at different plant locations can be linked so that needed spare parts can be located anywhere in the system. A carefully chosen CMMS system can greatly enhance the cost-effectiveness of the maintenance department and increase the availability and utilization of equipment.

It should be remembered that automating a poor system is certain to yield an automated poor system. An efficient, well-operated manual system is much to be preferred to an automated poor system. However, if you have a good manual system of medium to large size and your people are inclined, intelligent computerization should provide the benefits described above.

29.4.6 Tying It All Together

The organization, systems, tools, and features of a maintenance program outlined above can greatly improve the effectiveness of the maintenance department. What differentiates an excellent program from a mediocre or poor one is the maintenance manager, not the system.

The best preventive maintenance program in the author's memory was operated from a 4 by 6 in. notebook residing in a packaging master mechanic's back pocket. He kept a record of all the work done on a relatively few items of equipment of major importance. His experience had taught him the interval between breakdowns, and he scheduled preventive maintenance accordingly. On less critical equipment, he relied on frequent observation and conversation with operators. Rarely was he caught by an unscheduled outage.

Originally a highly skilled carpenter and a devoted fisherman, he quickly developed welding, metalworking, and mechanical skills when he was tapped to be a packaging mechanic. He built machinery and conveyor systems designed in house, installed machinery purchased, and started up, taught the operators, and maintained all of it. He felt ownership of everything he worked on and stuck with problems doggedly until they got fixed. He was not afraid of any problem or person. As a result, operators and mechanics alike looked up to him as a role model, and he was able to teach his people and develop other leaders. He made the transition easily from his role as mechanic to that of leader and teacher.

To some extent, a strong maintenance system and organization can overcome some of the shortcomings and weaknesses of the maintenance manager; nevertheless, you should seek the most qualified manager you can find. The desired qualities for a maintenance manager are:

- Is mechanically minded
- Is maintenance experienced
- Is process experienced
- Has hands-on style
- Manages by walking around
- Is quick witted
- Is determined
- Is inquisitive
- Feels ownership

- Is a good diagnostician
- Is a good leader
- Has teaching skills
- Is a developer of people

29.5 MEASUREMENT OF MAINTENANCE PERFORMANCE

Since maintenance cost generally is highest among all cost categories, benchmarking should be performed in all components of maintenance cost. A standard procedure should be designed and implemented to benchmark unit cost, reliability, and productivity.

As a basis for benchmarking, many functions have to be recorded and analyzed. The following are some examples for reference: equipment downtime in order of highest to total hours; percentage of critical equipment downtime contributed to production lost; maintenance cost for each piece of equipment; preventive maintenance and predictive maintenance cost for critical equipment; preventive maintenance hours as a percentage of total maintenance hours per equipment; and time report showing total hours spent on emergency, preventive, and regular repair, respectively. To evaluate a planner's efficiency, the following can be used: (1) work order cost estimated vs. actual cost, (2) hours and material planned vs. actual hours and material used, (3) total number of maintenance work orders scheduled as compared to the actual number of work orders completed, and (4) monthly maintenance costs vs. monthly maintenance budget. Other reports, such as (1) comparisons of labor and material costs as percent of total maintenance costs, (2) total costs of outside contractor use, (3) maintenance cost per unit of production, should be produced and analyzed as evaluation tools.

Some organizations evaluate workforce productivity as a measurement of maintenance performance. There are two ways to measure workforce productivity. One method is to establish standard times for most jobs. This approach tends to cause friction between manager and worker. A second method is by activities sampling. This allows management to improve work procedure planning and work conditions, which invariably leads to productivity increases. Work sampling will provide managers with all information necessary to run a cost-effective operation. A typical study consists of recording time from observations of workers, leaders, and foremen's various activities. The activities recorded include direct work (hands-on time), transporting tools or material, obtaining tools or materials, planning and instructing, waiting, traveling, job cleanup, personal time, and late start or early quit. The goal is to minimize nonproductive effort (waiting, traveling and transporting, etc.) to improve performance via improvement of planning, scheduling, purchasing, and storeroom operations. The effectiveness of direct work can be improved by additional training to increase skill levels and by development of standard maintenance procedures. The latter will ensure consistent job performance among workers, provide training for new workers, and expedite the repair function.

29.6 SANITATION PROGRAM

The sanitation program may or may not be part of the maintenance organization. It should be integrated into the management system and should be a prime responsibility of each manager. Regardless of how it fits on the organization chart, it is a critically important function in meeting expectations for a modern food processing plant.

Sanitation must be a way of life for a sugar refinery. It is working and living in a clean environment so as to be able to provide consistently wholesome and high-quality products for its customers. In addition to ethical reasons for employing impeccable sanitation measures, numerous state and federal regulations require these practices. The primary law that applies is the federal Food, Drug, and Cosmetic Act of 1938 and all subsequent amendments. This statute says that food processors and food products may be in violation in two major ways: first, if the products, as manufactured, contain any objectionable extraneous matter, and second, if conditions exist such that the final product could have become contaminated during its preparation by unsanitary conditions in and around the plant, the equipment, or the ingredients used.

A sufficient sanitation environment begins with the staunch backing of top management. It must have adequate funding and properly aligned areas of responsibilities. All sanitation efforts should be carried out under the oversight and direction of a qualified plant sanitarian. To assure full employee participation, top management from each department should serve on a plant sanitation committee, which works in conjunction with the plant sanitarian. This group should be responsible for routine inspections of the plant and regular meetings to evaluate the various sanitation levels found. The committee should assure through follow-up schedules that all identified deficiencies are quickly corrected and a high degree of sanitation excellence is maintained [4,5].

1. *Use of a sanitation program* that contains sufficient steps to prevent product contamination from any cause. The level of participation and cooperation must include every employee.
2. *Good operational methods,* which encompass: (a) proper receipt, transfer, handling, and storage of materials; (b) operational appearance and practices; and (c) delivery procedures. This embodies numerous routine measures, which include, among others, controlling all machinery so that no excessive dust, oil, or spillage of any kind is produced within the food production or storage area, and frequent inspections of filters, screens, or any other equipment that has been installed to prevent product contamination.
3. *Good personnel practices,* which include (a) personal hygiene, (b) personal appearance, and (c) restrictions on the use of tobacco, eating food, and drinking beverages while on the job.
4. *A sound pest control program,* which involves an organized set of procedures for effective eradication and prevention of various infestations that might otherwise be found around a food processing operation.
5. *Good cleaning practices* applied to the cleaning of equipment, the plant, and all storage areas. Equipment must be cleaned so that no contamination of adjacent materials or products may occur. Any messes made during repairs or equipment alterations must be cleaned up immediately. All grease smears and excessive lubricants should be promptly removed.
6. *A continuing preventive maintenance program* must be maintained for all internal equipment, plant structure, and surrounding premises. These guidelines apply:
   a. Provide adequate space for all equipment and working materials. Ensure room for aisles or work spaces between pieces of equipment and wall surfaces.
   b. Provide adequate lighting for all work areas and operations. Lighting fixtures, skylights, or other glass suspended over exposed food surfaces must be of an approved safety type or sufficiently protected to prevent breakage.
   c. Adequate ventilation should be provided. There must not be any condensation on fixtures, ducts, and pipes over equipment or materials that could drop onto food processing areas.
   d. Walls, floors, and ceilings must be easy to clean and maintain. Cracks and crevices should be filled and sealed. Any potential harborages for pests must be eliminated. Rodent-proofing and screening should be employed to prevent entrance of any pests.

e. All equipment must be placed on a preventive maintenance program.
f. Guards, housings, panels, and trim should be removable for cleaning. All product zones, nonproduct zones, and surfaces must be readily accessible for easy cleaning.
g. All new equipment must be of sanitary design according to the latest food-processing standards.
h. Environmental air should be filtered to be free of outside dust and contaminants. No visible dust may be discharged to the outside.
i. Air used in processing must be filtered or cleaned so as to remove particles 50 μm or larger. This air cannot contain any oil or other liquids.
j. Use of lubricants must be controlled, especially on any food-processing equipment. Toxic lubricants are prohibited. Any lubricants used on machinery must be of food grade.
k. Provision for adequate toilet facilities is required, along with associated handwashing and sanitizing stations. Floor drains must be placed where needed.
l. Exterior grounds around the plant must be kept free of litter, weeds, and old or obsolete equipment and materials.
m. Precautions must be taken to exclude harborages for pests from the grounds. Any sources of excessive dirt and other filth must be removed.
n. Roofs and the grounds surrounding the plant must be drained properly so as to prevent seepage of filth, insects, and microbial developments.
o. All plumbing fixtures and service lines must be of adequate size for their intended purposes and meet national or local codes.
p. Potable water for the plant must be provided by an approved supply and should be sampled and tested at least once a year.
q. There must be an adequate sewage disposal system to handle effluent discharged from the plant.

Food processing facilities that adhere to these basic sanitation principles should effectively meet their responsibilities to management, employees, regulatory agencies, and customers, and should provide wholesome, safe, high-quality food products that meet all expectations.

REFERENCES

1. C. H. Becker, *Plant Manager's Handbook,* 1974, Prentice-Hall, Inc. Englewood Cliffs, N.J. pp. 113–122.
2. V. J. Cotz, P.E., *Plant Engineer's Manual and Guide,* 1973, Prentice-Hall, Inc. Englewood Cliffs, N.J. pp. 37–43.
3. L. R. Higgins, P.E., *Maintenance Engineering Handbook,* 1988, McGraw-Hill Book Company, New York pp. 1–9 to 1–50.
4. *AIB Consolidated Standards for Food Safety,* 3rd Edition Revised 1990 American Institute of Baking, 1213 Baker's Way, Manhattan, Kansas, 66502.
5. H. R. Priester, internal papers, Savannah Foods & Industries, Inc., July 1991.

CHAPTER 30

Environmental Quality Assurance*

INTRODUCTION

The raw sugar industry uses large amounts of water, mainly for cane washing and the condensing of vapor. In a sugar refinery, the major uses, in addition to condensers, include char and carbon washings, boiler feedwater, process solution, filters and ion-exchange washings, and noncontact cooling. Particulate emissions include those from boilers, diatomaceous earth and char regeneration kilns, granulated sugar drying, and transport systems. No problems have been attributed to SO_2 and NO_2; natural gas and low-sulfur fuel oils have enabled compliance with limits. However, both have to be monitored. In the United States, the Environmental Protection Agency (EPA) has developed limitation guidelines for both raw sugar factories and cane sugar refineries [1, 2].

The specific water uses of a sugar refinery include affination water, filtercake slurry water, carbon or char column wash water, carbon slurry water, ion-exchange regeneration water, barometric condense water, boiler feed makeup water, truck and car wash water, and floor wash water. The sugar refineries are generally classified as liquid sugar and crystalline sugar facilities. The water balance in these two categories is represented in Figures 30.1 and 30.2, respectively. The wastewater is handled as shown in Figure 30.3.

30.1 MAJOR WASTEWATER ENVIRONMENTAL PARAMETERS

The major parameters for the sugar industry are given in the following subsections.

30.1.1 Biochemical Oxygen Demand

Biochemical oxygen demand (BOD_5) measures the oxygen-consuming capabilities of organic matter (5-day test). During decomposition, organic effluents exert a BOD_5 value that can deplete the oxygen supply. BOD_5 is generally measured and expressed in parts per million or milligrams per liter. The effluents from a sugar refinery can vary from a hundred to

*By Chung Chi Chou.

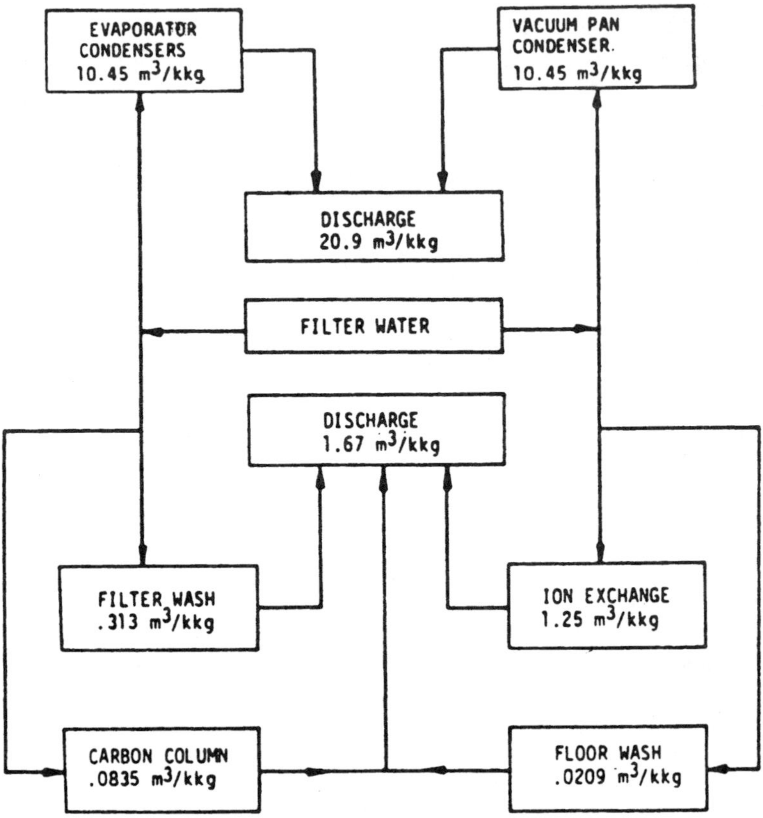

FIGURE 30.1 Water balance in a liquid sugar refinery.

several thousand milligrams per liter. Concentration of individual refinery wastes varies from 10 to about 1000 ppm. Total refinery discharge to surface waters runs from 0 to perhaps 25 ppm average.

BOD_5 is a difficult parameter to measure insofar as refinery discharges are concerned. The precision and accuracy of BOD_5 analysis are bad [3]. EPA's own laboratory manual notes:

Mean (ppm)	Standard Deviation (ppm)	Proportion Mean (%)
175	26	15
2.1	0.7	33

In addition, BOD_5 measurements often do not reflect the biochemical activity characteristics of receiving waters. The method of BOD_5 determination is given in Ref. 4. The BOD_5 value of carbohydrates, according to Bevan and Stevenson [5], is in the ratio BOD_5 1 ppm = carbohydrate 0.833 ppm.

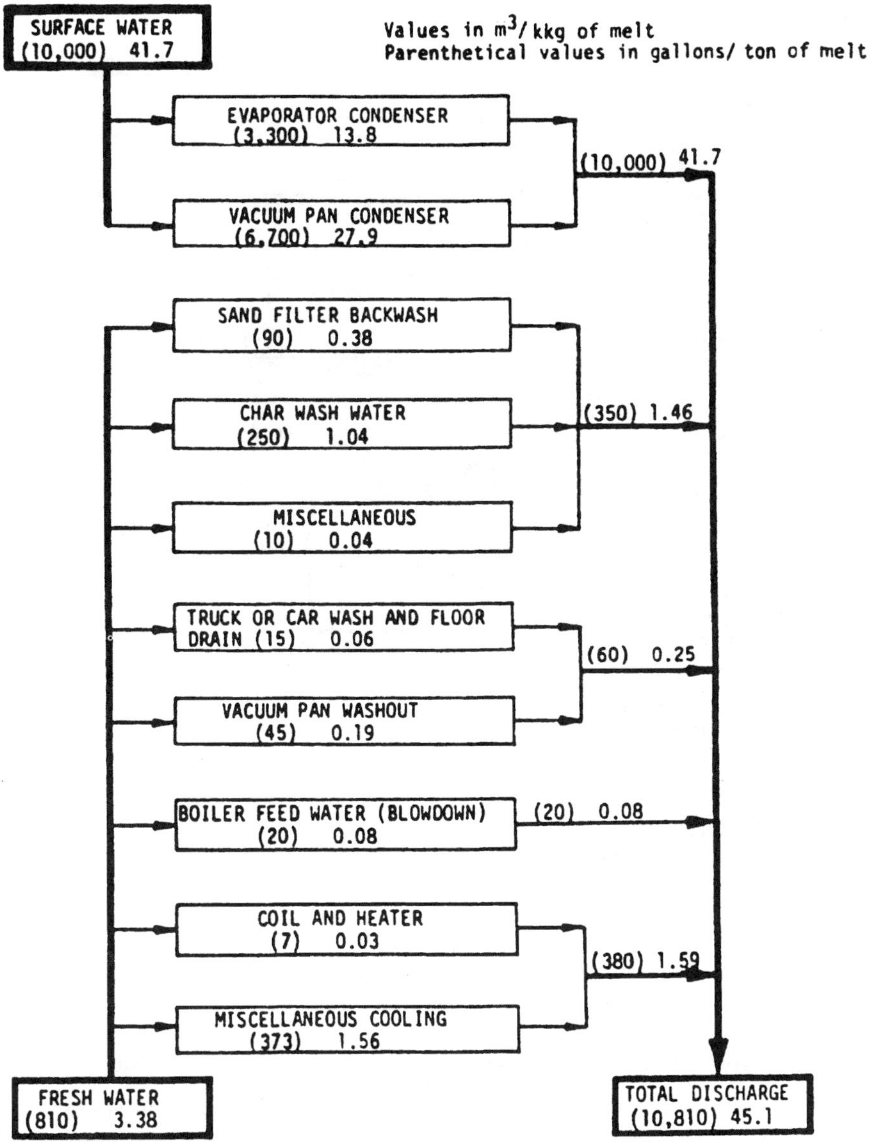

FIGURE 30.2 Water balance for a crystalline sugar refinery.

30.1.2 Discharge Oxygen

Discharge oxygen (DO) is a water quality constituent. When its concentration is appropriate, the living organisms are kept in a state of normal vigor, and the reproduction and population are sustained. Condenser water (90 to 95% of flow) has been heated and deaerated. Char wash, the other major waste, has been heated to 180°F (82°C). DO may not be meaningful for refinery wastes. DO is measured and expressed in parts per million or milligrams per liter. For water bodies of low assimilative capacity, DO is indicative in oxygen depletion. Methods of determination are described in Ref. 6.

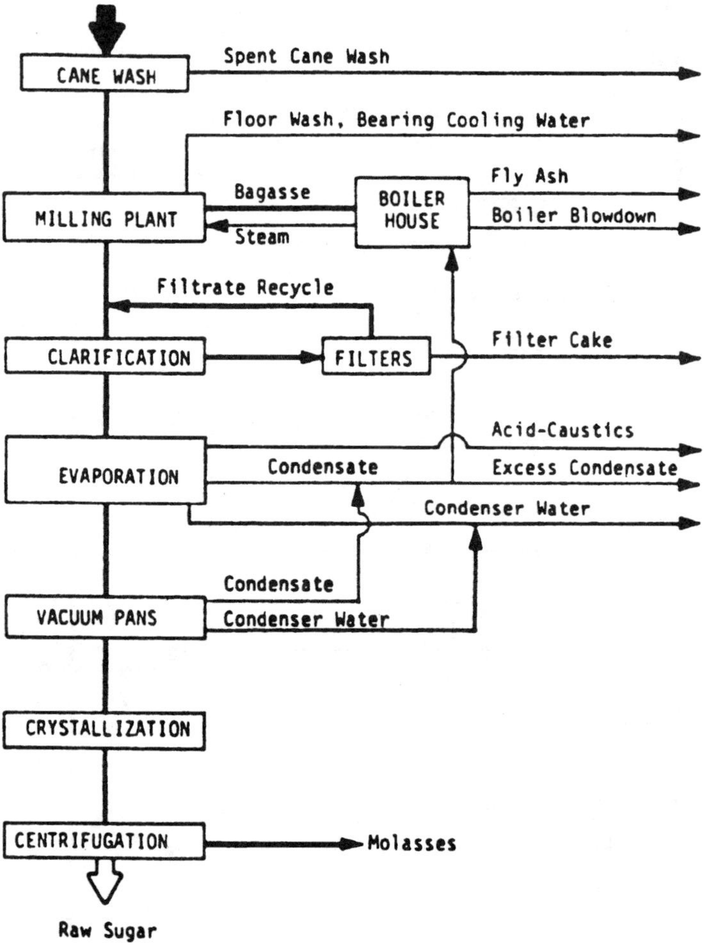

FIGURE 30.3 Simplified wastewater flow diagram for a raw sugar factory.

30.1.3 Total Suspended Solids

Solids in suspension are aesthetically displeasing, but when they settle to form sludge deposits on the stream or lake bed, they are more damaging to the life in water. They cover the bottom of the stream or lake with a blanket of material that destroys the fish-food bottom fauna. Deposits containing organic materials may deplete bottom oxygen supplies and may produce noxious gases such as hydrogen sulfide, carbon dioxide, and methane. In the sugar refinery, most suspended solids are inorganic. Barometric condenser cooling water is essentially free of net suspended solids. Total suspended solids (TSS) is of negligible consequence for most refineries. The TSS in char wash is very low—essentially zero, except for malfunctions. TSS in other process wastes is practically negligible.

30.1.4 pH (Acidity, Alkalinity)

Extreme pH changes can exert stress conditions or kill aquatic life outright. The pH of total process wastes, including condenser water, is practically unchanged in most cases. An exception would occur when $CaCO_3$ is discharged in wastes, in which case a pH rise of perhaps 0.5 might be experienced.

30.1.5 Other Parameters

The chemical oxygen demand (COD) test is widely used to measure the total amount of oxygen required for oxidation of organics. This test can be carried out in a short period of time, unlike the BOD_5 test. But the disadvantage of this test is that it does not distinguish between biologically active and inert organic materials. Under different conditions, COD can be about twice to many times BOD_5, and no definite relationship is yet established. In the presence of chlorides, such as by the intrusion of seawater or in brackish water, the COD test becomes totally out of proportion to the BOD_5 test.

Temperature is a prime regulator of natural process within the water environment. Fish-food organisms are altered severely when temperature approaches or exceeds 90°F (32°C). It is noted that the discharge temperature normally has very little effect on the river's temperature or on aquatic life in most cases.

30.2 SUGAR IN FACTORY OR REFINERY WASTEWATERS

Since sugar and some nonsugars in raw sugars are the sources of BOD_5 in the wastewater, the reduction of sugar and nonsugars in in-plant control is a direct influence to the reduction of pollutants in the effluent. The loss of sugar to the wastewater is a double loss because of the large amount of water used in the condensers of evaporators and pans and in other refining operations; apparently insignificant quantities of sugar distributed through this large volume (millions of gallons per day) may represent heavy money losses. All such wastewaters should be continuously sampled and tested at short intervals—at the most, every hour. An ideal arrangement, practicable in some plants, is to have all wastewaters flow to a central sewer in which a waterwheel type of sampler turned by the stream of water raises and delivers at each revolution a small amount of the sewer water to a receiving tank alongside the wheel. Samples should be controlled every hour, and the receiving tank emptied before the next sample is collected. The laboratory should test this wastewater sample as soon as it is received, reporting any abnormality in the test to the supervisory.

Hourly samples from all evaporator and pan condenser legpipes should also be tested as a cross-check on the main sewer sample.

30.2.1 Sugar in Boiler Feedwater

The feedwater for the steam boilers is derived largely from the steam condensed in the calandria of the first vessel of the multiple effect, the calandrias, and the coils of vacuum pans. Sugar may enter these waters through entrainment with the vapor or through defects that develop in the heating surfaces. Sugar causes the water to foam in the boilers and may lead to accidents. Furthermore, although sugar may not be present in sufficient quantity to endanger the boilers through foaming, it is decomposed by the heat into products that are detrimental to the tubes and shells of the boilers, causing pitting and overheating. Even if sugar is present only in imperceptible amounts, these traces will eventually accumulate on the boiler tubes as a harmful and dangerous carbonaceous deposit. The breakdown of the sugar also forms harmful organic acids.

The addition of either NaOH or Na_2CO_3 to the feedwater to maintain it at pH 8.0 or above is recommended. Frequent testing of boiler feedwater by the α-naphthol test ensures freedom from sugar from unforeseen sources. A pronounced odor develops in the steam if boiler water contains sugar. Under such conditions the contaminated feedwater is turned to the sewer, and the boilers are blown down thoroughly.

Reid and Dunsmore [7] report the development of an apparatus based on the theory that when a sugar solution is heated to 265°C (509°F) for several minutes, there is a change in both pH and conductivity (Fig. 30.4). Since 50 ppm of sugar in water is generally considered as safe for boilers, provided that chemical treatment of boiler feedwater is adequate, this conductivity monitoring device is satisfactory for protecting boilers in the raw sugar factory. As for condensates in a refinery, the sugar solutions may contain insufficient non sugars to trigger a conductivity monitor; a total organic carbon analyzer may be the answer.

30.2.2 α-Naphthol Test for Traces of Sugar in Water

The most widely used test to determine the presence of sugar in condenser waters, boiler feedwaters, factory sewer outflows, and other waters where it may be detrimental or represent a preventable loss is with α-naphthol. It is roughly quantitative if carried out under similar conditions and compared with tests on standard solutions. Add 5 drops of a 20% alcoholic solution of α-naphthol to 2 mL of water (cooled) in a test tube; then by means

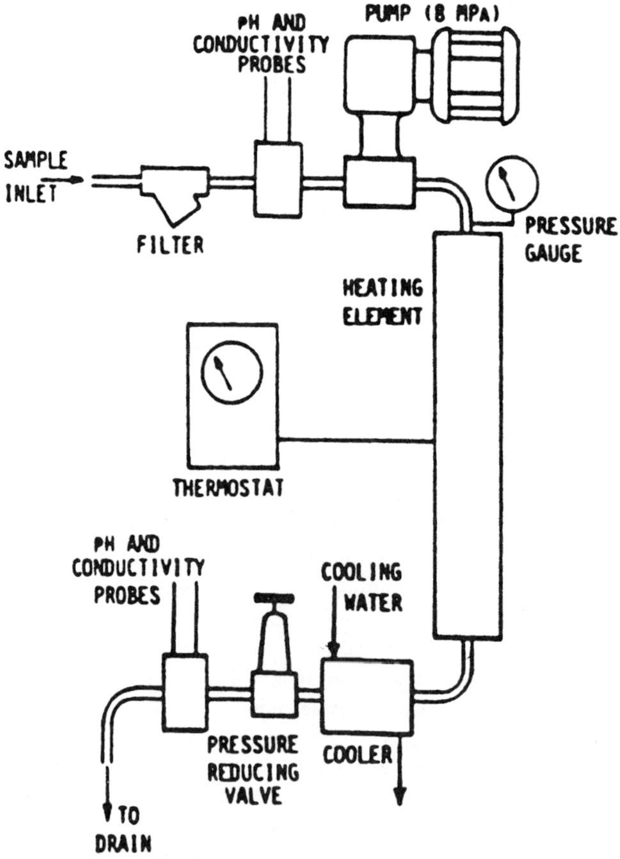

FIGURE 30.4 Apparatus for determining sugar traces.

of a pipette, slowly pour down the side of the tube 5 mL of concentrated sulfuric acid, on which the test water will float. In the presence of sucrose, a violet zone or ring appears at the juncture of the two liquids, the intensity of the color depending on the amount of sucrose present. The acid must be chemically pure, and the α-naphthol should be of good quality. The solution of this reagent darkens on exposure to light and should be freshly prepared at frequent intervals.

The chemist can become familiar with the color variations by making tests on water of the type to be tested, to which has been added known amounts of sugar (e.g., 1 part of sugar to 10,000, 50,000, and 100,000 parts of water). The colors darken on standing.

30.2.3 Quantitative α-Naphthol Test

A quantitative method using the Helige comparator has been devised with a special color disk showing 10, 20, 30, . . . , 100 ppm of sucrose under the conditions of the test. Five drops of α-naphthol and 5 mL of sulfuric acid are added to 2 mL of water (filtered and cooled if necessary), as described earlier, and the mixture is shaken; the color is allowed to develop at room temperature for 3 min and the tube is immersed in cold water for 2 min; then the reading is taken. Any natural color in the water tested is compensated for by having 2 mL of the water mixed with 5 mL of distilled water in the other opening of the comparator. The α-naphthol solution should be checked occasionally against 1:100,000 sugar in distilled water.

The colorimetric method adopted at the British Columbia refinery calls for two solutions of cobaltous chloride and copper sulfate to prepare standards; the sample is compared with standards.

Procedure. Place 5 mL of material to be tested in a test tube or vial of the same diameter as the standards, add 5 drops of a 5% solution of α-naphthol, and while stirring vigorously, add 10 mL of concentrated sulfuric acid. Comparison is made with standards after 2 min standing. The standards for test are given in Chapter 33 of Ref. 8.

30.2.4 Arsenomolybdate Test

A method of developed in Queensland [9] is to mix 5 mL of sample with 1 mL of arsenomolybdate reagent, place the mixture in boiling water for 15 min, then cool it for 10 min. The color developed can be quantified with standards or by spectrophotometer. The preparation of the reagent is given in Chapter 33 of Ref. 8.

Molybdate Test for Trace Sucrose in Water. The molybdate test has been used in the sugar industry for many years to detect trace sugar. Applications include determination of minute quantities of sugar in the condensate for heater, evaporator, and vacuum pan, and in wastewater. The concentrations range from 5 to 500 ppm. The key reagents are 3.85% ammonium molybdate, concentrated hydrochloric acid, and copper sulfate.

To prepare a permanent standard solor solution, a temporary standard has to be made for comparative quantitative analysis as follows: (1) use pure granulated sugar and distilled water to make up a series of sugar solutions with concentrations ranging from 5 to 100 ppm; (2) from each sugar solution, 5 mL is pipetted into each test tube; (3) add 3 mL of molybdate solution and 3 drops of concentrated hydrochloric acid (for hydrolysis of sucrose) to each test tube. Place test tubes in a boiling-water bath for 6 min. The sugar solution in each test tube will develop a color intensity in proportion to its sugar concentration. This series of solutions is used as the temporary standard.

Permanent standards are made to visually match the color intensity of each temporary standard using cupper sulfate in distilled water. A drop of formaldehyde is then put into each of the permanent standard tubes as a preservative. After each tube is tightly sealed and marked, the series of tubes is placed in a rack for comparison with unknown samples for quantitative measurement. These standards will keep 6 months to 1 year without loss of color intensity.

For measurement of trace sugar in water, pipette 5 mL of the sample into a test tube and add 3 mL of molybdate solution and 3 drops of concentrated hydrochloric acid. Place the test tube in a boiling-water bath for 6 min. The intensity of the blue color is an indication of the quantity of sugar present in the sample. Visual comparison of the color intensity of the sample with the permanent standard prepared above would give the result in ppm.

30.2.5 Automatic Detection of Sugar in Wastes

The AutoAnalyzer system (Technicon Controls, Inc., Chauncey, New York) continuously monitors refinery condensates and wastewaters. The reaction involved is that of Hoffman and Wood [10], and the inverted sugars are determined by the reduction of alkaline potassium ferricyanide (yellow) to colorless alkaline potassium ferrocyanide, and the reduction in color (transmitted light) is read out continuously and recorded. The reduction in color is exactly proportional to the amount of sugar present. Detection of 25 ppm is clearly visible by this technique, and the working range is 0 to 400 ppm. Figure 30.5 is a schematic diagram of the instrument used at SuCrest to test 26 different points in the water system.

Revere Sugar Refinery in Boston developed a modified reaction for the AutoAnalyzer that uses 1.0% resorcinol as the reagent instead of the ferri-ferrocyanide reaction [12]. The

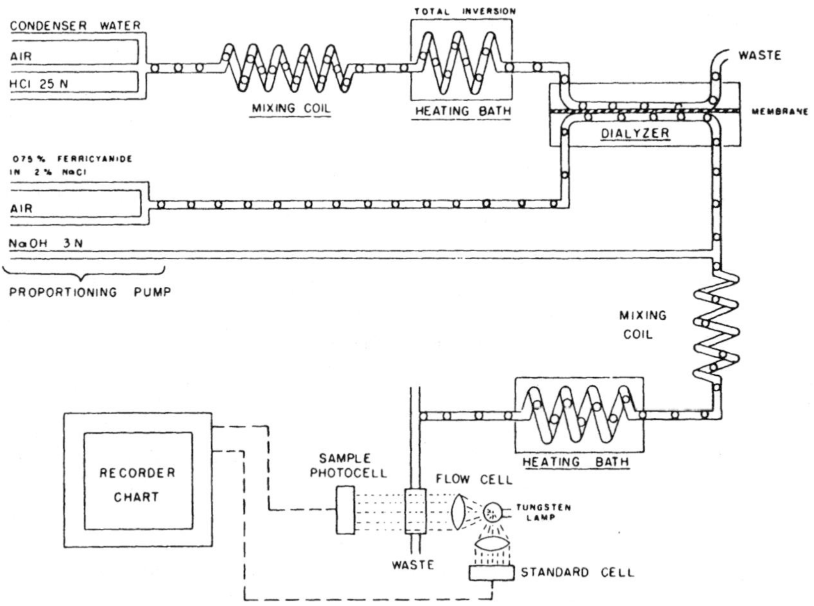

FIGURE 30.5 Testing for sugar in wastewater. (From Ref. 11.)

method operates in an acid instead of alkaline solution and is much better suited for the highly impure water employed for Revere's condensers. Commercial muriatic acid (20°Baumé) is the other reagent used with resorcinol, resulting in low chemical costs; the corrosive action of the strong acid is objectionable, however. The procedure has been applied only to total refinery wastewater, which includes river water, clear wastewater, clarifier mud, filter washings, and activated carbon.

Another detector of sugar is the Enzymax Analyzer of Leeds & Northrop [13]. Lawhead and Sisler have evaluated its performance [14]. The analyzer uses highly purified enzyme catalytic reaction columns. All enzymes (glucose oxidase, invertase, and mutarotase) are immobilized on beads in a column. The reactions produce hydrogen peroxide, which is monitored by electrochemical amperometric detectors. The monitoring of sugar in boiler water or in process water may be carried out as indicated in Figure 30.6. The sensitivity of the analyzer is 0 to 50 ppm for glucose and 0 to 100 ppm for sucrose.

Other continuous sugar detection devices are also in use with sugar refinery effluents. Fowler [15] discusses the thymol sucrose detector, which is basically a single-function recording colorimeter. The principle is that the exothermic reaction of sucrose with thymol in dilute sulfuric acid produces color from light amber to deep red, depending on the increasing concentration of sucrose. The effective measuring range is 5 to 100 ppm sucrose.

When the waste flow contains high suspended solids, an unfiltered optical measurement becomes impossible. Then a Monitor IV autoanalyzer and an ASP II continuous water clarifier would overcome the problem. Details of the two systems are given. Eichhorn [16]

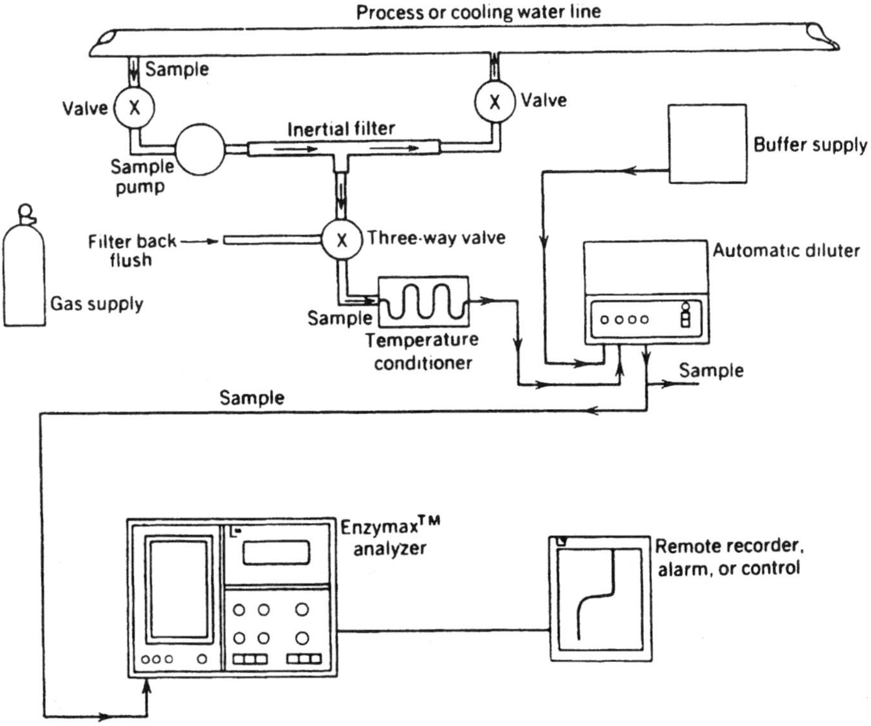

FIGURE 30.6 Sugar monitoring in refinery process or cooling water. (Courtesy of Leeds & Northrop.)

described the Elsdorf control system, in which three sources of waste flow are measured and recorded continuously, before they run into a sampling tank (Fig. 30.7*a*). Then the mixed sample is filtered, cooled,deaerated, polarized, and printed (Fig. 30.7*b*). From the final sugar content, the origin of the sugar losses should be detected.

On-Line Sucrose Detection by YSI Instrument. The sensor technology of the YSI instrument is based on the immobilized enzyme membrane technique developed for the medical field by YSI, Inc. of Yellow Springs, Ohio. The enzyme sensor technology utilizes one or more enzyme-catalyzed reactions to produce hydrogen peroxide which is oxidized electrochemically at the platinum anode. The signal current generated is proportional to the substrate concentration in the sample. To measure the substrate in the sample, you would need to know the approximate range of concentration, the dilution factor, the calibration value, and possible interferring substances. The sample needs to be diluted to the range to match the linear range of the instrument. Color, turbidity, and small particles do not affect the measurement. However, some substance may foul and/or clog the membrane system, resulting in the need to change the sensor frequently and should therefore be avoided. An enzyme consists of protein molecules which are very specific in catalyzing a reaction. In YSI, sensor oxidase enzymes are always involved in producing hydrogen peroxide, which is needed to generate electrochemical current proportional to the concentration of substrate in the sample. The range of calibration curve should approximate that of diluted samples. The following description is an expanded view of the sensor probe.

A substrate injected into the sample chamber is stirred, diluted, and then diffuses through a thin polycarbonate membrane. Once past the membrane, for sucrose analysis, the substrate encounters three types of oxidase enzymes and undergoes the following sequence of reactions:

$$\text{Sucrose} + H_2O \longrightarrow \text{invertase} \longrightarrow \alpha\text{-D-glucose} + \text{fructose}$$

$$\alpha\text{-D-glucose} + O_2 \longrightarrow \text{mutarotase} \longrightarrow \beta\text{-D-glucose}$$

$$\beta\text{-D-glucose} + O_2 \longrightarrow \text{glucose oxidase} \longrightarrow H_2O_2 + \text{D-Glucono-s-lactone}$$

$$H_2O_2 \longrightarrow \text{platinum anode} \longrightarrow 2H^+ + O_2 + 2e^-$$

Hydrogen peroxide diffuses through a thin layer of cellose acetate membrane. The membrane has a molecular weight cutoff point of about 200. It allows H_2O_2 to pass through but prevent other interfering reducing agents from passing across and contributing to sensor current. The electron flow is directly proportional to the H_2O_2 generated, and therefore, to the concentration of sucrose. The YSI instrument has been used successfully in some refineries to detect trace sucrose in refinery water streams.

30.2.6 Monitoring of Vapor for Sugar

The most dangerous source of entrainment loss is the evaporator or vacuum pan condenser. Dale and Lamusse [17] have described the introduction in South Africa of a flame photometer by Marius, used in the beet sugar industry for this specific purpose. It is necessary to predetermine the ratio of sucrose to potassium. Under their conditions, the sucrose/potassium ratio is 85:1 to 100:1 for syrup, 31:1 for A molasses, and 14:1 for B molasses. First, there is the sample probe (Fig. 30.8*A*), then a condensing setup to condense the sample (Fig. 30.8*B*); the liquid sample is sucked into the atomizer through a 0.7 mm-diameter

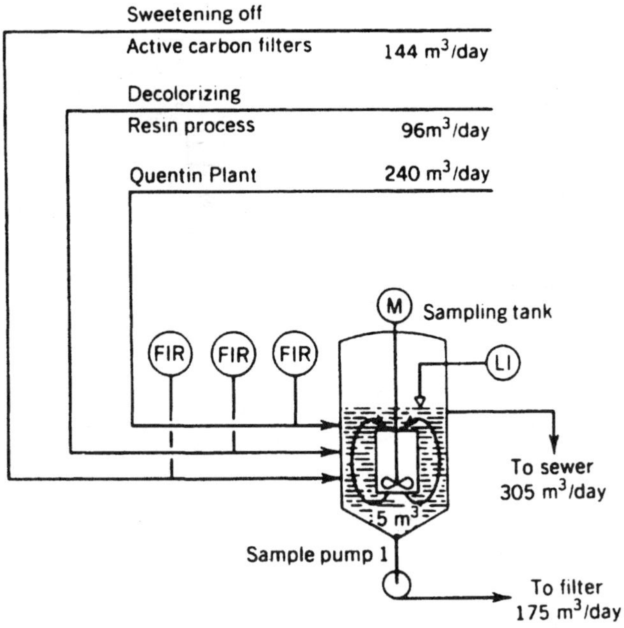

(a)

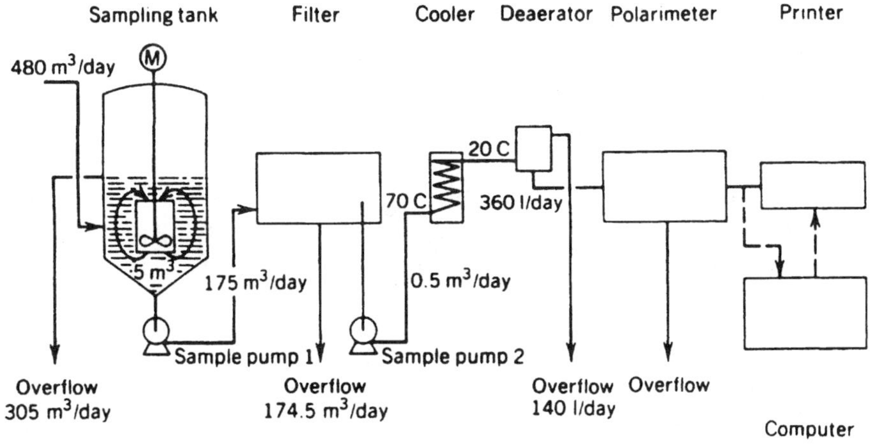

(b)

FIGURE 30.7 (*a*) Wastewater sampling system (Elsdorf); (*b*) continuous recording of sugar losses in wastewater.

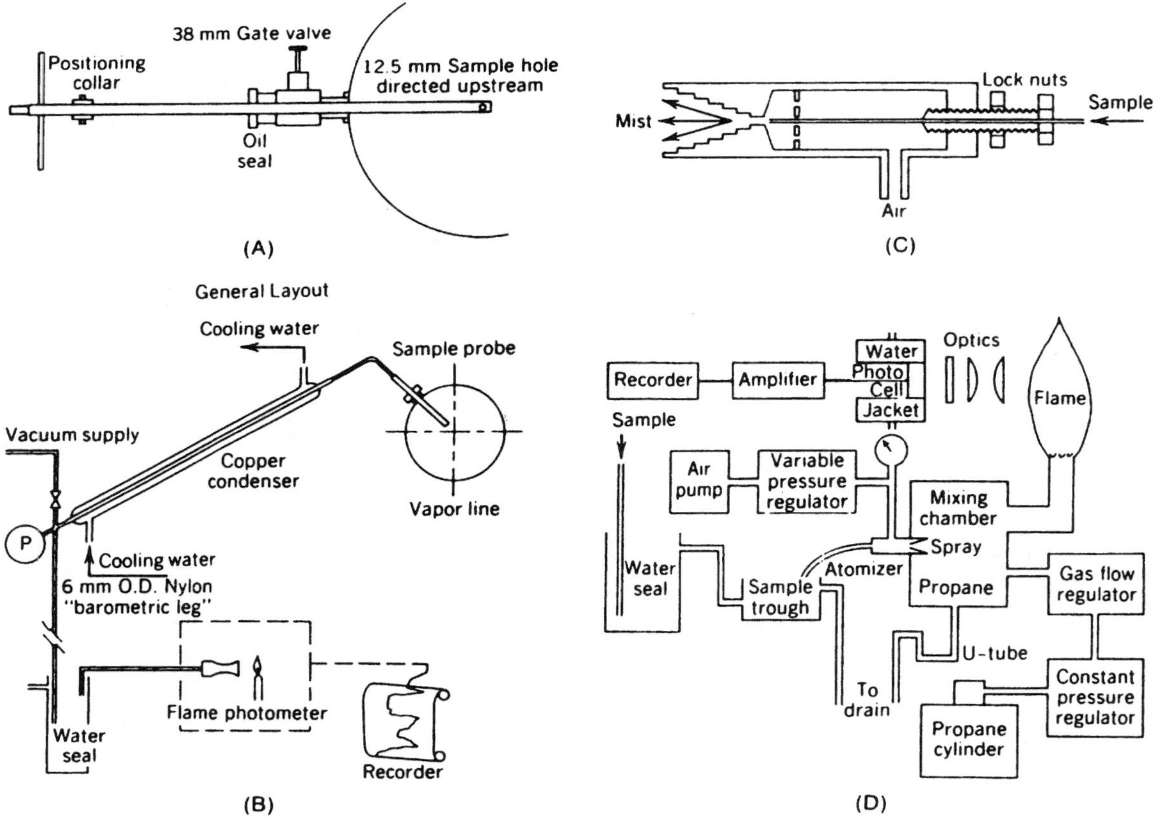

FIGURE 30.8 Monitoring of entrainment by vapor sampling and the use of flame photometer: (*A*) condensing probe; (*B*) condensing of vapor; (*C*) atomizer; (*D*) monitoring the mist with flame photometer.

capillary tube (Fig. 30.8*C*); and in the atomizer, the sample is mixed with compressed air at 1.3 bar absolute pressure, discharging a mist into the mixing chamber, where it is mixed with propane gas and led to the burner (Fig. 30.8*D*) for a flame photometer reading of 768 nm. Examples given indicate how monitoring of sucrose in vapor to the condensers can be used to pinpoint sources of entrainment.

Schaffler [18] points out that the flame photometric method is an indirect procedure and also is extremely difficult to use to measure entrainment in injection condenser waters. Moreover, for a refinery, the liquor purity is high, and thus a potassium flame photometer cannot be used. The continuous sugar-sensitive autoanalyzer developed at the Hulett refinery is therefore recommended. The determination of sucrose is based on its reaction with HCl acid and resorcinol in a 95°C heating bath. Sucrose is inverted to glucose and fructose, but only fructose reacts with resorcinol to form an orange compound, which is measured at 520 nm. Ferric sulfate serves as a sensitizer and also cancels out iron interferences. The sample (20 mL) is treated with basic lead acetate solution (10 drops) and fed into a Technicon AA II autoanalyzer, which has a sampling rate of 30 per hour.

30.3 WASTEWATER TREATMENTS

Reports of Delvaux [19] and Baker [20] discuss in detail both water quality and various treatments designed to help refineries meet local or national standards. The current treatment technology can be categorized into the following groups.

- *Prevention.* Baker discusses the various entrainment separators and shows that decreases in entrainment losses brought about by combined preventive and separation methods can be speculative. The EPA Development Document [2, p. 80] shows that 83.6% overall reduction can be realized from the prevention of entrainment (Fig. 30.9).
- *Impounding.* Sugar refineries are located primarily in municipal areas and, relatively speaking, they have very little available land.
- *Recirculation.* Barometric condenser water in a plant can be recirculated for reuse. Since much land is required for cooling without treatment, this approach is generally prohibited for urban refineries. Where land is available, the water may be spray cooled; otherwise, the water may be recycled through a cooling tower. The blowdown of recirculation varies between 3 and 10% under differing local conditions.

Chemicals and settling aids have been tried and reported by various investigators. The most commonly used material is time. When the pH of water is brought up to 11 to 12, the microbiological activity is significantly slowed or even stopped.

Recirculation of barometric condenser water in cane refineries is most probably unnecessary; in any event the practice is cost-ineffective and uses a very large amount of energy (about 5% of base refinery power load, according to the EPA Refining Development Document [2, p. 140]. Entrainment separation appears to be effective enough to reduce the BOD_5 in condenser water to an unmeasurable level.

30.3.1 Biological Treatment

The biological treatments may be classified as aerobic or anaerobic. One of the treatments is by means of active sludge. Portions of settled sludge are recirculated and mixed with the influent to the aeration section. The anaerobic type functions in the absence of dissolved oxygen, breaking down organic waste to organic acids and alcohols. Finally, the methane

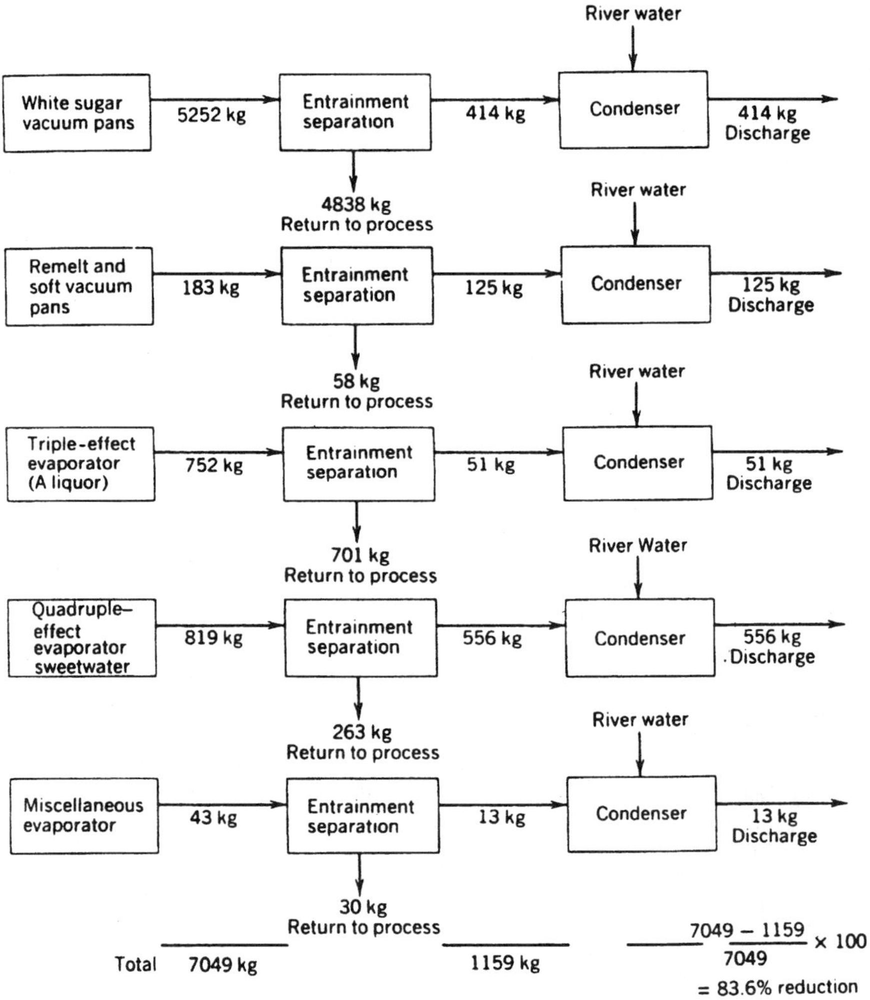

FIGURE 30.9 Entrainment reduction.

bacteria converts the acids and alcohols into carbon dioxide, methane, and hydrogen sulfide. Anaerobic processes are economical because they feature high rates of overall removal of BOD_5 and suspended solids with no power cost (except for pumping) and low land requirements.

In aerobic wastewater treatment, the organic substances are transformed into CO_2 and H_2O in a single metabolic step. But the anaerobic bacteria decompose the substrate only in several steps. Tschersich [21] reports the latest findings showing that anaerobic decomposition proceeds in four stages, producing methane and CO_2 (Fig. 30.10) and describes an anaerobic tank (Fig. 30.11) showing the activated sludge chamber and the gas chamber.

Highly loaded wastewater first passes through a lagoon or a reservoir with a retention time of at least 1 day, so that some carbohydrate compounds may be decomposed by aerobic fermentation before the anaerobic stage.

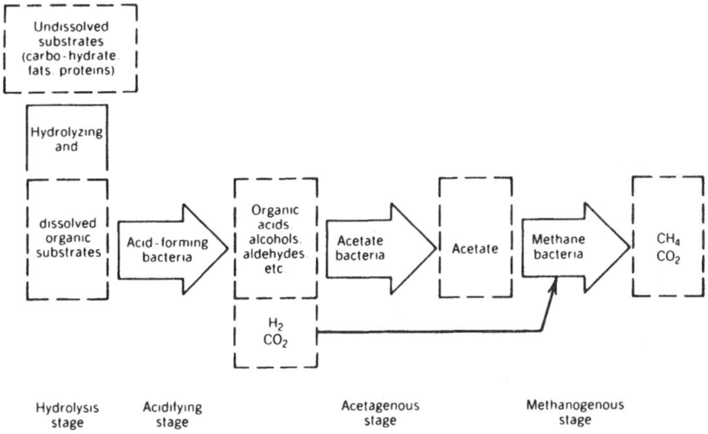

FIGURE 30.10 Anaerobic decomposition of organic substances.

An aerobic system to produce sludge on the activated sludge principle has been operated successfully in Domino's Chalmette Refinery since 1977. A detailed report has been given by Chou et al. [22] The treatment plant is shown in Figure 30.12. The capacity of the combined process is 1,400,000 gal (5299 m^3) for a retention time of 2.8 days. Two storage tanks are used for ammonia (7500 gal; 28.39 m^3) and phosphoric acid nutrients (750 gal; 2.84 m^3). The aerators have the capacity of providing 2 to 4 lb O_2/(hp-h) (1.22 to 2.43 kg/hp · h), propelling the mixed liquor at 30.5 cm/s, equivalent to a complete circulation cycle every 8.7 min. The thickened sludge is discharged without going through an anaerobic stage.

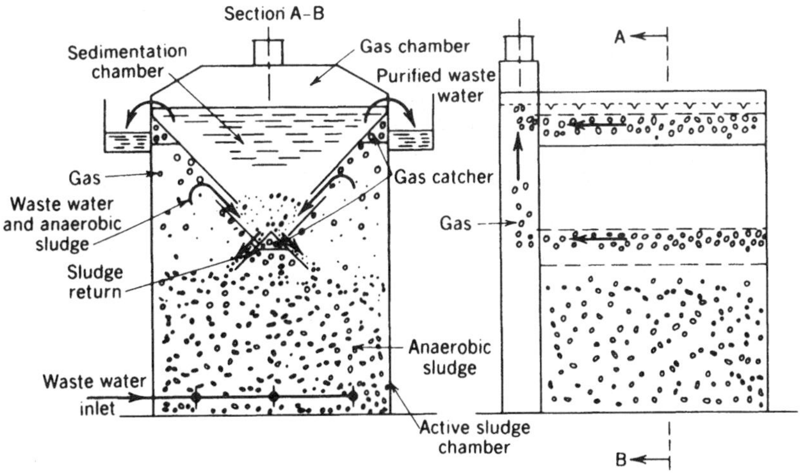

FIGURE 30.11 Anaerobic tank (CSM system).

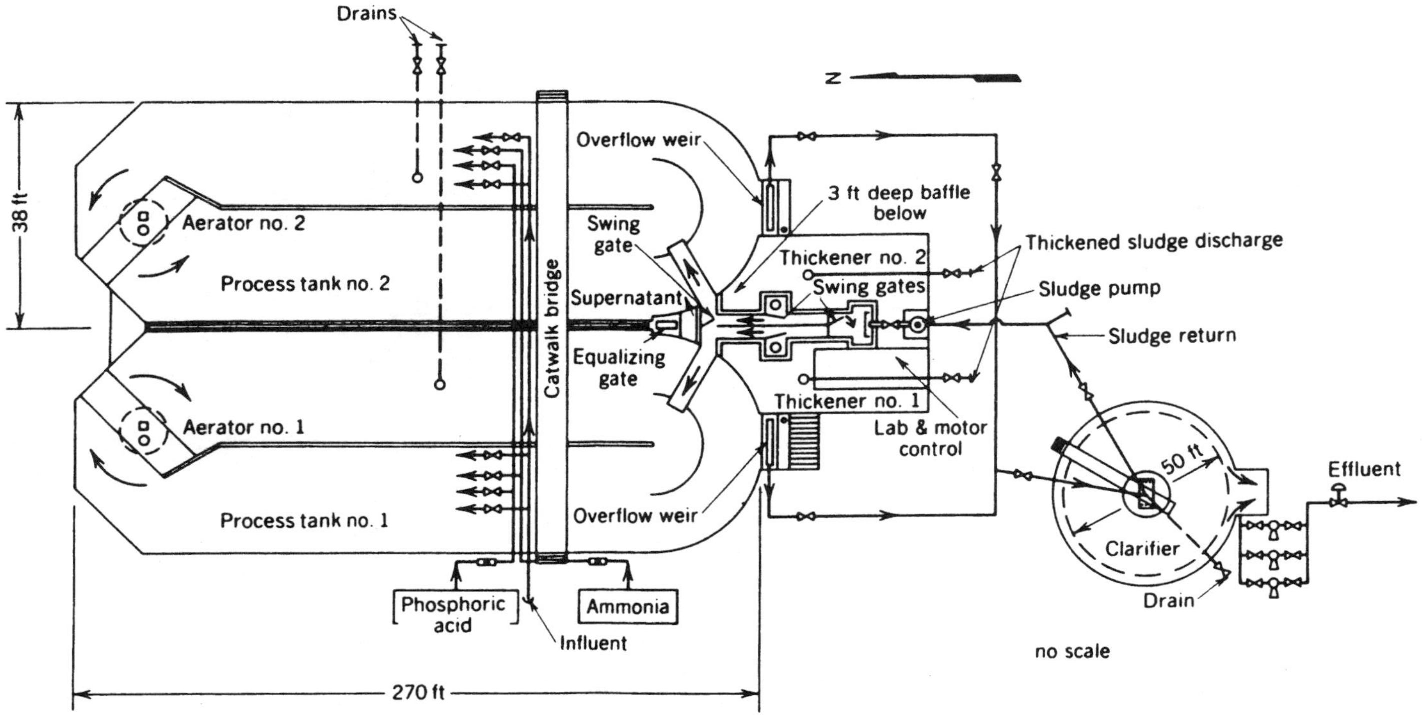

FIGURE 30.12 Aerobic wastewater treatment plant (with activated sludge). (Courtesy of Chalmette, Domino.)

30.3.2 Filtration and Other Treatments

Two types of filtration are in use: sand and diatomaceous earth filtration. There are also microscreening devices, and experiments are carried out that use reverse osmosis, ultrafiltration, and spray irrigation. Where there is ample capacity in a nearby sewage system, arrangements are made to discharge some or all wastes to the sewage system. Some good sources of information are given in the Delvaux report [19].

Development of the up-flow anaerobic sludge blanket (UASB) process for purification of wastewater is reported by Borghans et al. [23]. An UASB reactor is shown in Figure 30.13.

As a roughing filter, a rotating biological contractor (RBC) has been employed satisfactorily by Sadler et al. [24] (Fig. 30.14). This type of contractor is also very effective when used for wastewater treatment [25].

30.3.3 Effluent Limitations

The EPA schedules two stages of achievement: BPCTCA best practical control technology currently available; (generally abbreviated as Bat) by July 1, 1977 and BATEA (best available technology economically achievable; generally abbreviated BAT) by July 1, 1983. Some general limitation figures are given in Table 30.1.

30.4 AIR QUALITY IN SUGAR INDUSTRY

Among the forms of air contaminants that are problematic in the sugar industry are the following:

- *Fly ash.* Fly ash is referred to as particles of ash entrained in combustion gases and carried into the air. The sources of these particles in the sugar industry may be open-air trash burning, solid waste incinerator, and bagasse as fuel in steam boilers.

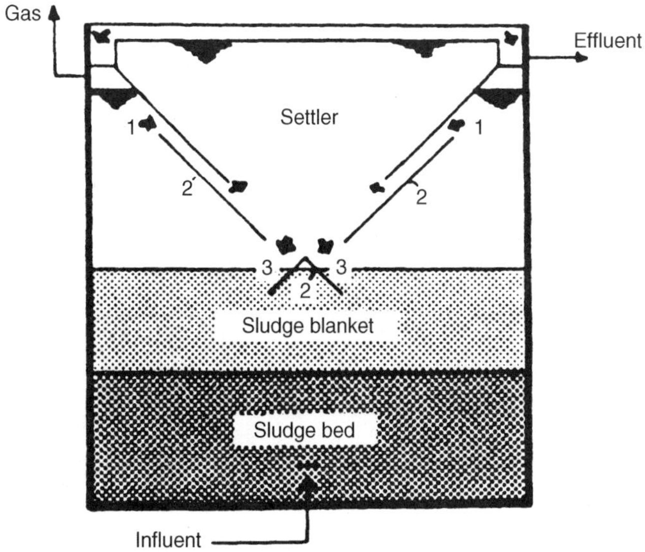

FIGURE 30.13 Schematic drawing of a UASB reactor; 1, sludge–liquid mixture inlet; 2, gas screens; 3, settled sludge return opening.

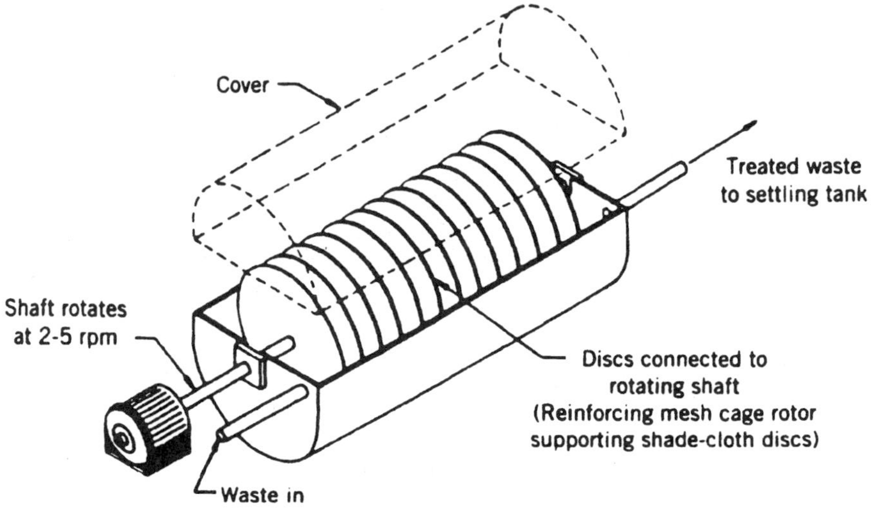

FIGURE 30.14 Rotating biological contractor with shade cloth.

- *Fog*. This is an aerosol of liquid droplets formed near the ground. In the sugar industry, fog may be the result of thermal pollution in connection with spray cooling.
- *Smoke*. This is formed by soot or carbon particles often less than 0.1 μm in diameter as a result of incomplete combustion of carbonaceous fuels. Smoke is the main problem in air pollution in the sugar industry.
- *Particulates*. These originate from granulated sugar drying and handling and from char or carbon regeneration and handling.

30.4.1 Emission Control

There are three types of fuel generally used in raw sugar factories or sugar refineries in the United States: bagasse, natural gas, and heavy oil. In Louisiana, the restrictions on emissions are as follows:

1. The particulate matter from any incinerator should not be in excess of 0.2 grain per standard cubic foot of dry flu gas corrected to 50% excess air or 12% carbon dioxide.
2. The particulate matter from any fuel-burning equipment should not be in excess of 0.6 lb per 10^6 Btu of heat input.

When natural gas is used, it is almost completely burned and does not create air pollution problems.

The next available fuel is bagasse for the raw sugar factories. The emission does not contain toxic materials except the carbon particles resulting from incomplete combustion.

The single-stage mechanical dust collector for bagasse burning boilers, tested in Louisiana, shows collection efficiency ranging from 46 to 97%. The findings of tests carried out in Queensland show the efficiency of dry collectors to be in the range 90 to 92% [26], and in South Africa, more than 92% [27].

The size distribution of particulates from sugar mill boilers in CSR mills in Australia is found to have 35% smaller than 10 μm [28]. The three types of cyclone arrester tested

TABLE 30.1 EPA Effluent Limitations[a]

	BPCTCA		BATEA	
	BOD_5	TSS	BOD_5	TSS
Raw Sugar Factory[a]				
Daily average	1.41	1.41	0.10	0.24
30-day average	0.63	0.47	0.05	0.08
Crystalline Reineries[b]				
Daily maximum	1.02	—	0.18	0.11
30-day averate	0.34	—	0.09	0.035
Liquid Refineries[b,c]				
Daily maximum	0.45	—	0.30	0.09
30-day average	0.15	—	0.15	0.03

[a]All values are expressed in terms of kilograms of pollutant per tonne of field cane.

[b]Those discharging only barometric cooling water.

[c]All values are expressed in terms of Kg of pollutant per tonne of melt.

showed poor efficiencies on dust of 5 μm or less (Fig. 30.15). Two types of wet collectors are being used. One is the wetted-louver arrester, which is based on the design of the CSR evaporator entrainment separator. The dust-laden flue gas follows a zigzag path between stainless-steel corrugated louvers which are washed from a set of oscillating nozzles. In Queensland, a fluidized packing contractor can be operated as a wet collector, the corrosion in the top section with pH of liquid being as low as 3 [29].

In Australia, mechanical cyclone collectors have been used in diameters from 152 to 300 mm or larger. The collection efficiencies are excellent with no blockages, as reported by Ford [30]. These collectors consist of a multiplicity of axial-entry cyclonic tubes; each tube had an inlet ramp, precipitating tube, and clear air tube, as shown in Figure 30.16.

South Africa [31] experienced six years of efficient flue gas scrubbers based on bagasse-fired boilers and four years of scrubbers on dual-fuel (coal/bagasse) boilers, pioneering disasters and proving some good ideas that failed in practice. Three types of scrubbers are presented (Figs. 30.17 to 30.19) that have been successful in operation and have achieved emission rates well below the present permissible maximum. Rauno [32] reports a new sieve plate scrubber which demonstrates that scrubbing efficiencies in excess of 97% are consistently attainable at a pressure drop of less than 75 mm (3 in.) water gauge.

Varieties of fly ash handling plants employed with the modern large suspension bagasse-fired boilers have been used in the Queensland sugar industry [33]. Four of their different systems are shown in Figures 30.20 to 30.23.

Corica et al. [34] Report the latest fly ash handling systems in Australia. They found that anionic flocculants of 20 to 60% hydrolysis are effective for general flocculation. However, in some areas where anionic flocculants are inactive, the addition of aluminum at 1 ppm of Al^{3+} produces efficient flocculation.

In South Africa [35], the use of a multiroller filter to incorporate with small beach-type settlers, surrounding a settling clarifier, is shown in Figure 30.24. The belt path of the multiroller filter, shown in Figure 30.25, consists of two endless stainless steel belts passing around a number of rollers.

In Felixton, South Africa [36], the treatment of smuts water and scrubber water has been improved by the newly designed system to supply recycled water to the scrubbers and ash hoppers and to remove suspended solids (Fig. 30.26). This system can also handle all factory effluence and injection water flow.

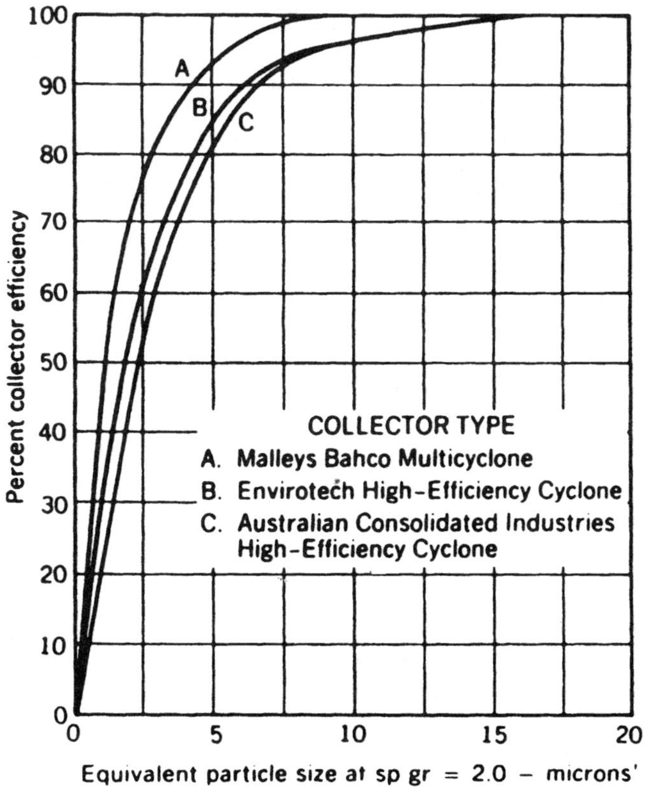

FIGURE 30.15 Efficiences of cyclone-type fly ash collectors.

An effective exhaust scrubbing system for bagasse boiler developed in Taiwan and employed successfully in all sugar mills is reported by Chang and Lee [37]. The system uses a sieve plate (SP) scrubber and reheater to heat exhaust air with waste heat of flue gases to 100°C so that the exhaust out of stack will not be condensed and hazed (Fig. 30.27).

30.5 REGULATORY COMPLIANCE ASSOCIATED WITH CANE SUGAR REFINING OPERATIONS

Cane sugar refineries operating today within United States, its territories, and its possessions must comply with an ever-increasing number of complex federal, state, and local regulations. The impact that agencies such as the Occupational Safety and Health Administration (OSHA), the Environmental Protection Agency (EPA), the Food and Drug Administration (FDA), the U.S. Coast Guard (USCG), and their state and local counterparts can have on the operations of a cane sugar facility can be both dramatic and significant.

30.5.1 Food Plant Sanitation and Food Product Safety

The Federal Food, Drug, and Cosmetic Act is the basic food law of the United States. Its purpose is to assure that all foods are pure and wholesome, safe to eat, produced and held

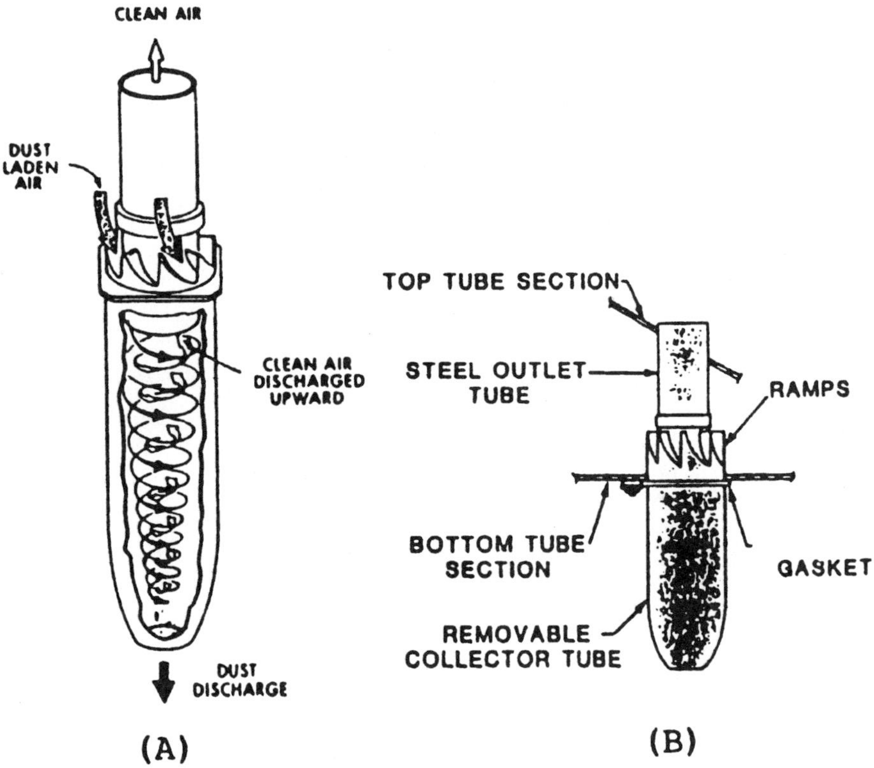

FIGURE 30.16 Mechanical cyclonic collector: (*A*) cutaway of dust collector tube; (*B*) collector tube assembly.

under sanitary conditions, and packaged with truthful and informative labeling. Generally, this law:

- Prohibits adulteration or misbranding of any food, refusal of inspection, and refusal to permit access of documents
- Establishes methods of enforcement
- Defines the meaning of food, food additives, adulterated food, and misbranded food

Regulations. Regulations pertinent to cane sugar operations are given in Title 21 of the *Code of Federal Regulations* (CFR), Parts 1, 7, 70–74, 82, 102, 109, 110, 168, 170, 172–177, 181, 182, 184, 186, and 189. Examples of what these regulations entail and how they affect cane sugar operations on a daily basis can be found in the following two selected subjects.

a. Food Plant Sanitation: 21CFR Part 110—Good Manufacturing Practices

- Personnel
- Building and grounds
- Sanitary facilities
- Equipment and procedures
- Production and process control

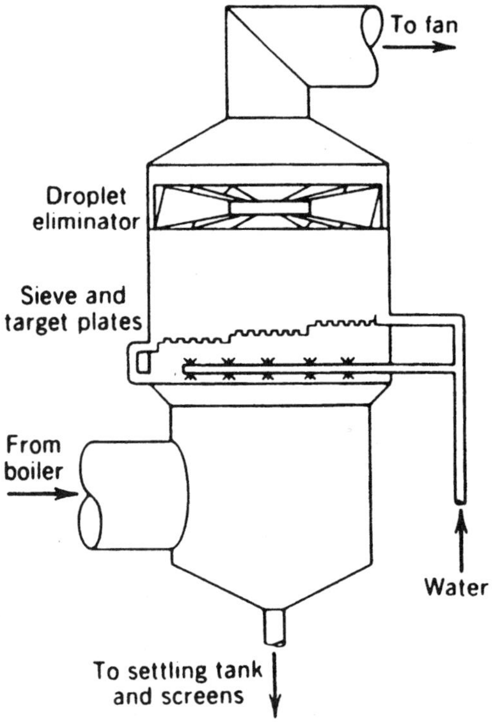

FIGURE 30.17 Peabody scrubber.

b. Food Safety (Additives and Ingredients): 21CFR Part 170—Food Additives. This subject establishes tolerances for use as direct human food ingredients (white granulated sugar is category 41). It also sets forth the ground rules for classification of food additives as "generally recognized as safe" (GRAS).

- 21CFR 172: Direct Addition Food Additives for Human Consumption
  - 21CFR 172.712: Dimethyl dialkyl ammonium chloride for sugar liquor clarification and decolorizing
  - 21CFR 172.816: Methyl flucoside–coconut oil ester for surfactant in molasses
  - 21CFR 172.810: Dioctyl sodium sulfosuccinate for processing aid in unrefined cane sugar production
- 21CFR 173: Secondary Direct Food Additives for Human Consumption
  - 21CFR 173.5: Acrylate–acrylamide resins for clarification of beet or cane sugar juice or liquor
  - 21CFR 173.320: Various chemicals for microorganism control in cane or beet sugar mills
  - 21CFR 174: Indirect Food Additives
- 21CFR 181: Prior Sanctioned Food Ingredients
- 21 CFR 182: Substances Generally Recognized as Safe (GRAS)
- 21CFR 184: Direct Food Substances Affirmed as GRAS (typical are sucrose and invert sugar)

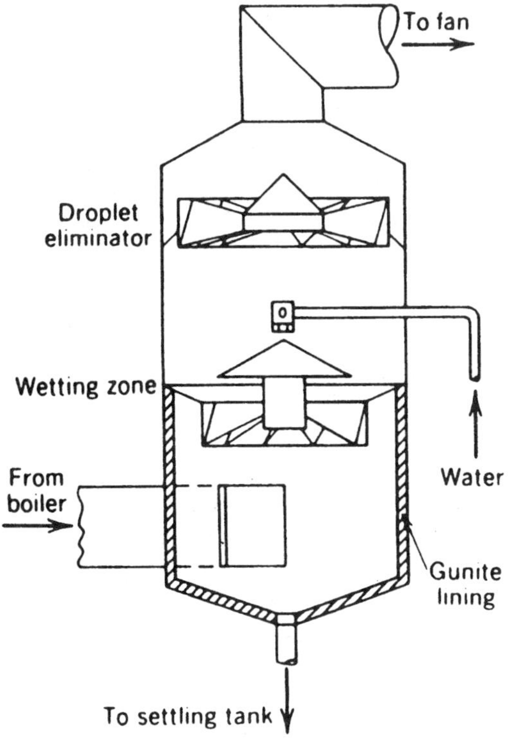

FIGURE 30.18 Brandt–Ducon scrubber.

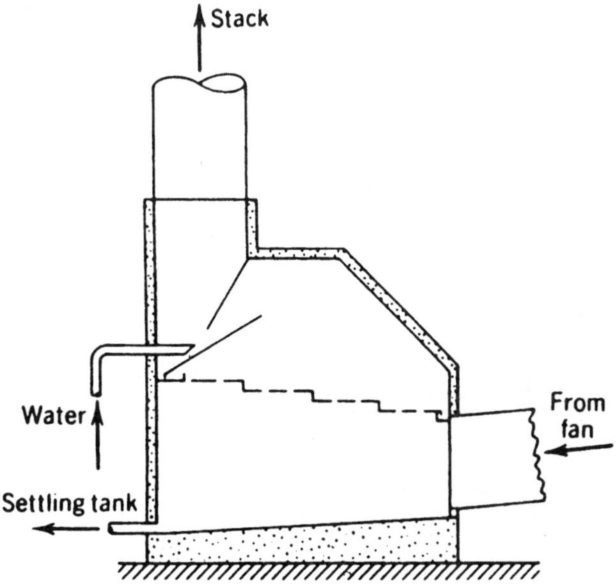

FIGURE 30.19 Tongaat scrubber.

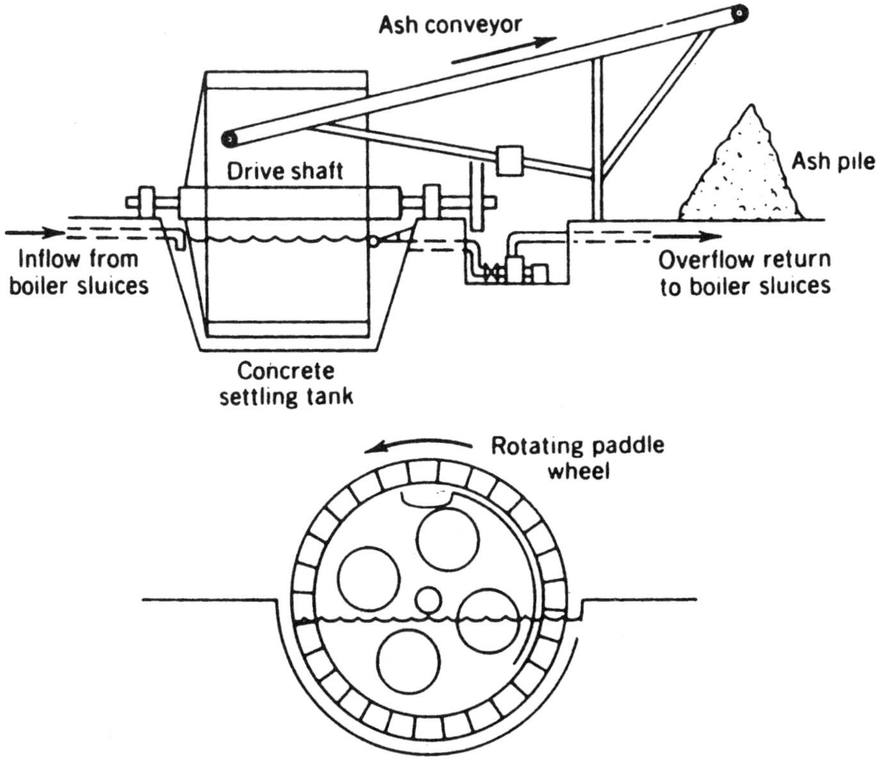

FIGURE 30.20 Mourilyan mill ash removal system.

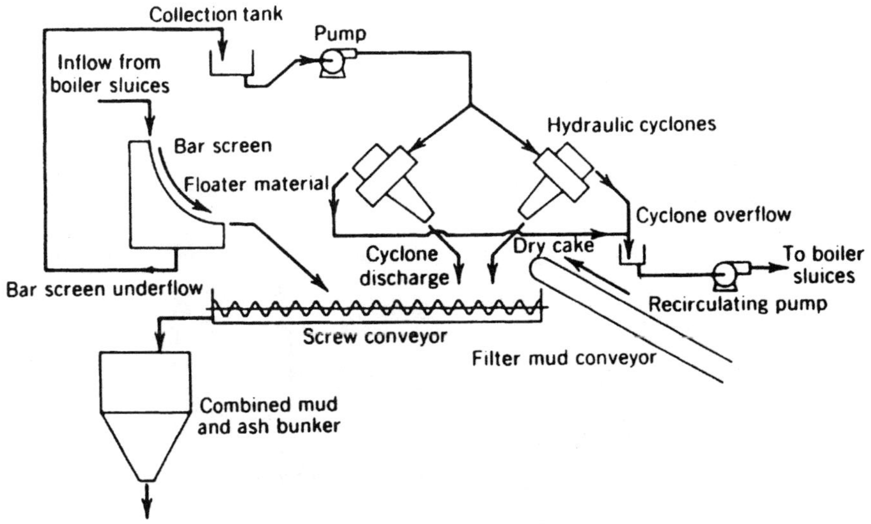

FIGURE 30.21 Hydraulic cyclones in ash handling system.

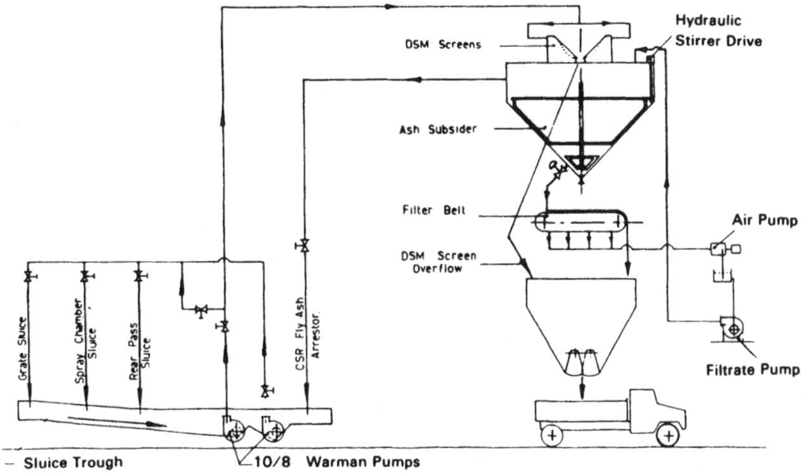

FIGURE 30.22 Race course mill ash handling plant.

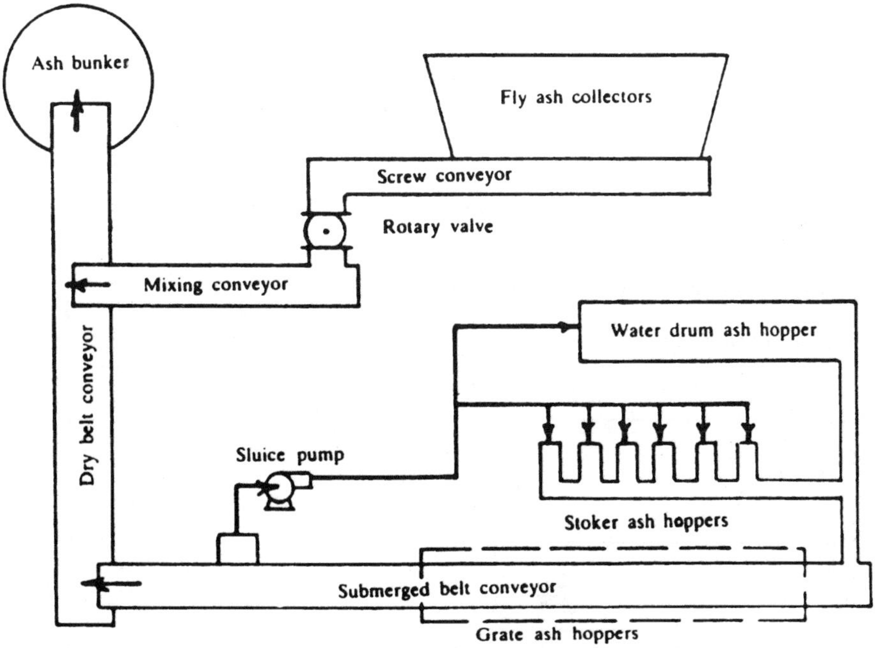

FIGURE 30.23 Dry fly ash removal and coarse-ash sluicing in an ash handling system.

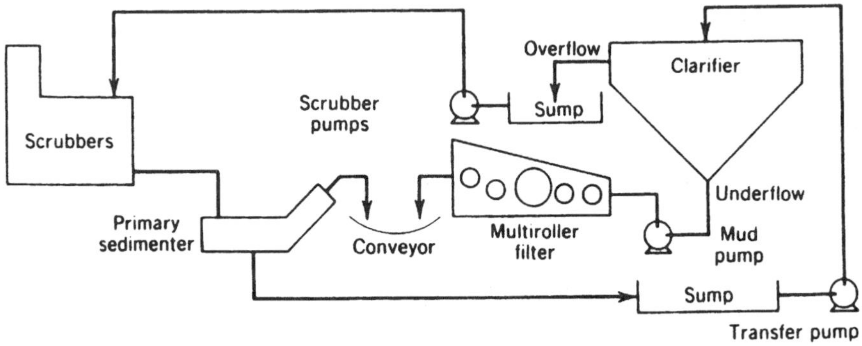

FIGURE 30.24 Schematic of solids separation system for boiler smuts.

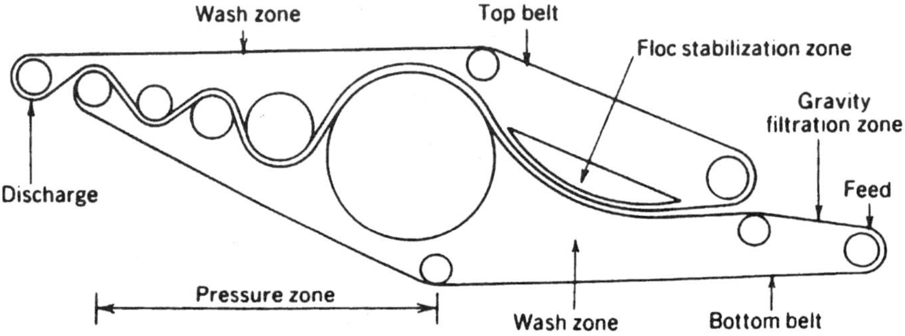

FIGURE 30.25 Belt path through multiroller filter.

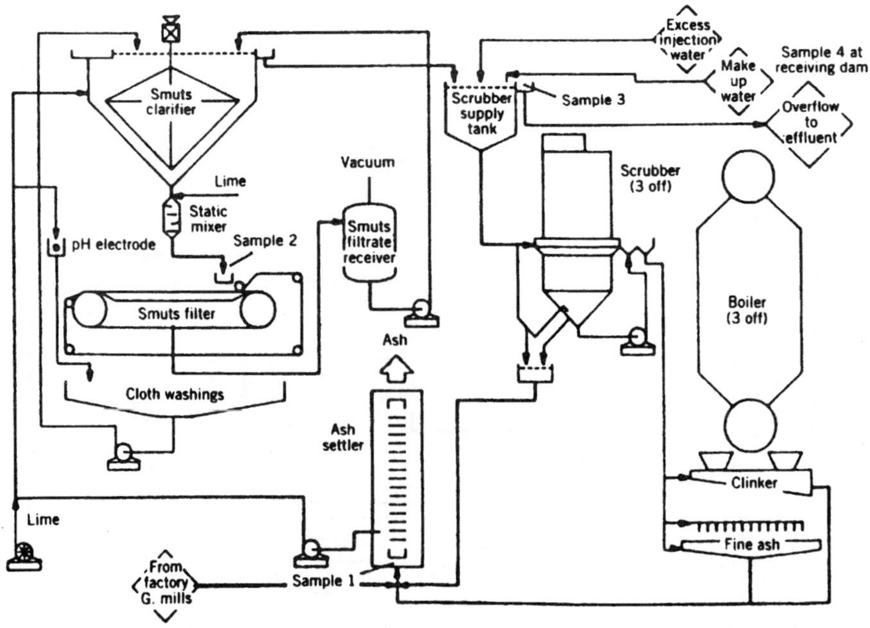

FIGURE 30.26 Layout of Felixtron smuts water treatment and scrubber circuit.

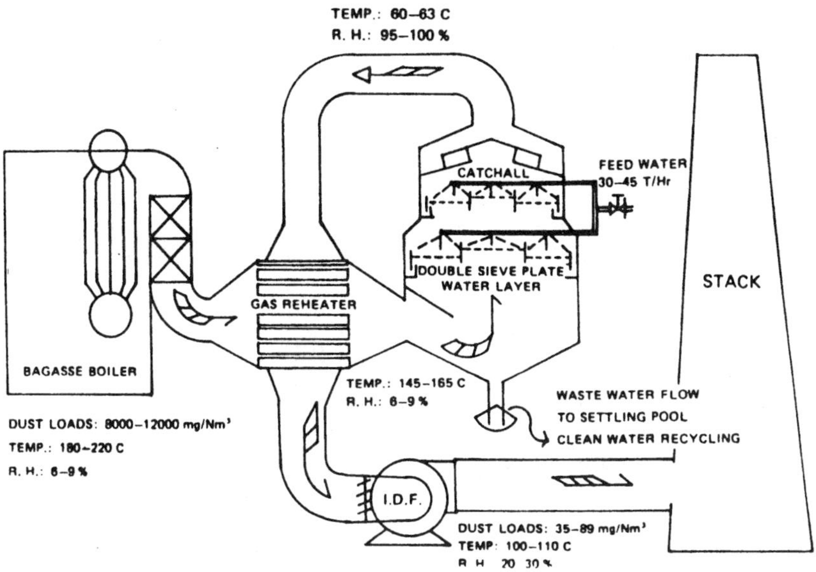

FIGURE 30.27 General arrangement of sieve plate scrubber with reheater. (Courtesy of TSC, Taiwan.)

- 21CFR 186: Direct Food Substances Affirmed as GRAS (typical are dextrans)
- 21CFR 189: Direct Food Substances Affirmed as GRAS (typical is cyclamate)

REFERENCES

1. U.S. EPA, *Cane Sugar Processing,* EPA Development Document, U.S. Environmental Protection Agency, Washington, DC, 1975.
2. U.S. EPA, *Cane Sugar Refining,* EPA Development Document, U.S. Environmental Protection Agency, Washington, DC, 1974.
3. *Proc. Sugar Ind. Technol.,* 1973, pp. 148–155.
4. U.S. EPA, *Methods for Chemical Analysis of Water and Wastes,* Water Quality Office, U.S. Environmental Protection Agency, Washington, DC, 1971.
5. Bevan and Stevenson, *Proc. Queensl. Soc. Sugar Care Technol.,* 1969, p. 320.
6. APHA, *Standard Methods for the Examination of Water and Wastewater,* 13th ed., American Public Health Association, New York, 1971.
7. Reid and Dunsmore, *Proc. S. Afr. Sugar Technol. Assoc.,* 1991, pp. 208–212.
8. J. C. P. Chen and C. C. Chou, *Cane Sugar Handbook,* 12th ed., Wiley, New York, 1993.
9. *Proc. Int. Soc. Sugar Cane Technol.,* 1965, p. 1732.
10. Hoffman and Wood, *Sugar Ind. Technol.,* 9, 1959; *Ann. N.Y. Acad. Sci.,* July 22, 1960.
11. *Sugar J.,* June 1960.
12. *Sugar Ind. Technol.,* 1961, p. 30.
13. Bulletin C2.5112-TP, Leeds & Northrop.
14. Lawhead and Sisler, *Proc. 20th Gen. Meet. Austral. Soc. Sugar Beet Tecnol.,* 1978.
15. Fowler, *SIT,* 1977, pp. 219–231.

16. Eichorn, *Sugar Ind. Technol.,* 1978, pp. 368–385.
17. Dale and Lamusse, *Proc. S. Afr. Sugar Technol. Assoc.,* 1977, pp. 116–118.
18. Schaffler, *Proc. S. Afr. Sugar Technol Assoc.,* 1978, pp. 123–124.
19. Delvaux, *Sugar Technology Review (STR)* 2:95–135, Elsevier, The Netherlands, 1972.
20. Baker, *Sugar Ind. Technol.,* 1973, pp. 139–148.
21. Tschersich, *Int. Sugar J.,* :365–370, 1980.
22. C. C. Chou et al., *Sugar Ind. Technol.,* 1978, pp. 410–472.
23. Borghans et al., *Int. Sugar J.,* 89:143–145, 1978.
24. Sadler et al., *Proc. Austral. Soc. Sugar Cane Technol. (Queensl.),* 1984, pp. 343–349.
25. Stoler, *Time,* Oct. 14, 1985, p. 90.
26. Cullen and Ivin, *Proc. Int. Soc. Sugar Cane Technol.,* 1971, pp. 1552–1559.
27. Camden-Smith, *Proc. S. Afr. Sugar Cane Technol. Assoc.,* 1972, pp. 92–99.
28. Aitken et al., *Int. Sugar J.,* :370–376, 1980.
29. Leung et al., *Proc. Queensl. Soc. Sugar Cane Technol.,* 1972, pp. 169–172.
30. Ford, *Proc. Austral. Soc. Sugar Cane Technol. (Queensl.),* 1989, pp. 133–139.
31. Moor, *Proc. S. Afr. Sugar Technol. Assoc.,* 1977, pp. 133–136.
32. Rauno, *Proc. S. Afr. Sugar Technol. Assoc.,* 1973, pp. 49–50.
33. Sawyer and Cullen, *Proc. Int. Soc. Sugar Cane Technol.,* 1977, pp. 2263–2275.
34. Corica et al., *Proc. Austral. Soc. Sugar Cane Technol. (Queensl.),* 1989, pp. 147–150.
35. Goring, *Proc. S. Afr. Sugar Technol. Assoc.,* 1982, pp. 71–74.
36. *Proc. S. Afr. Sugar Cane Technol. Assoc.,* 1988, pp. 94–98.
37. Chang and Lee, *Taiwan Sugar,* Sept.–Oct. 1991, pp. 9–16.

PART IV

SPECIALTY SUGAR PRODUCTS

CHAPTER 31

Brown or Soft Sugar*

INTRODUCTION

Soft or soft brown sugar is a value-added specialty product that enjoys wide popularity in North America. The name is derived from the texture or mouth feel of the product, which should be smooth and velvety as opposed to the hard, gritty feel of granulated sugar. This texture is the result of a very fine crystalline structure coated in a highly flavored syrup. Soft sugar can be *boiled*, in the traditional manner, by direct recrystallization from dark, low-purity syrups obtained during the refining process, or *blended* by coating very fine granulated sugar crystals with a refiner's syrup. The value of soft sugar to the user is in the unique color and flavor of the product. Soft sugar is used in both the home and food industry to develop a rich molasses-type flavor in cookies, candies, and similar products. It is also used as a coloring agent in a variety of sauces and other ingredients in the food industry.

Three grades of soft sugar are commonly produced today. All are graded somewhat arbitrarily according to color. They can be categorized as light, medium (golden), or dark and conform, generally, to the specifications listed in Table 31.1. Usually, the darker the color, the stronger the flavor of the soft sugar. Products are often custom tailored to suit the requirements of the user. The desirable characteristics of all grades of soft sugars are a spongy texture, free from crystalline appearance. When selecting a soft sugar for use, one must evaluate the smooth flavor and color desired in the finished product.

Soft sugar is sold in 50- to 100-lb bags for the food industry and 1- or 2-lb bags for home and, because of the high water content, require moisture-proof containers, such as multiwalled bags with a laminated polyethylene sheet, to maintain shelf life. Storage conditions must be observed carefully to avoid lumping of the product. Filled bags should be stored in a controlled temperature and humidity environment to avoid drying out. Because of the high viscosity of the syrup coating, the products should be stacked in a manner that avoids compaction.

Other storage factors to be considered include microbiological activity and increases in the color and acidity of the product, which occur with time. Changes in acidity may affect

*By John C. Thompson.

TABLE 31.1 Typical Analysis of Soft Sugar

	Boiled		Blended	
	Medium	Dark	Medium	Dark
% Sucrose	85–93	85–93	90–96	90–96
% Invert	1.5–4.5	1.5–5.0	2–5	2–5
% Ash	1–2	1–2.5	0.3–1	0.3–1
% Nonsugars	2–4.5	2–4.5	1–3	1–3
% Moisture	2–3.5	2–3.5	1–2.5	1–2.5
Color (ICU)	3,000–6,000	7,000–11,000	3,000–6,000	7,000–11,000

producers of preleavened food formulations where the acidity may increase enough to inactivate the leavening agent. In general, users should be advised that soft sugars are not as shelf stable as are granulated products and require special care. In some instances, particularly with seasonal users, emphasis should be placed on good stock rotation practices, and if necessary, shelf-life restrictions should be established and enforced.

31.1 METHODS OF PRODUCTION: BOILED SOFT SUGAR

Besides offering the user a high-quality, premium product, the production of soft sugar offers the refiner several unique opportunities to reduce refining costs, increase production and improve overall net profit. Because soft sugars offer the ability to package nonsucrose and water, soft sugar production has a positive impact on yield. Meikle shows the impact on molasses solids and refined yield for a refinery producing a soft sugar, at various levels of production, with an 11% nonsucrose content and a 4% moisture from a raw sugar containing 1.2% nonsucrose [1]. At levels approaching 13% output, molasses solids fall to zero and yield increases almost three percentage points, from 96.25% to 99.17%. During periods of peak demand, some refiners actually experience a deficit of nonsucrose constituents and must import molasses to overcome any shortfall in soft sugar production.

Soft sugar production also lowers the number of recovery strikes that must be boiled and therefore reduces the amount of sugar recycled back to the process, allowing for higher rates of production or reduced use of processing materials.

Exact methods of production are as varied as the number of refiners, leading to an abundance of regional preferences for both color and flavor of the final product. Methods of production can be grouped into two basic categories, boiled or blended. Boiled sugars can be further subdivided into agglomerated grain and single-crystal styles.

The desired characteristics of the finished product, boiled or blended, is a spongy texture, free from crystalline appearance, a uniform color free from gray-green tones, and a flavor free from harsh, bitter, or salty taste. The refiner also desires low purity and good keeping qualities. These characteristics result from liquors and syrups which are low in total color, free of the greenish casts imparted by polyphenols of iron, relatively low in purity, low in ash, and high in invert sugar.

Ultimately, the color and flavor of the final product will be determined by the nature of the syrups chosen for production and the methods of purification used. The composition of refinery syrups is highly complex and does not permit complete characterization, often changing from one raw sugar cargo to the next even within the same country of origin. Whereas granulated sugar is essentially a single substance, soft sugar and its starting materials can contain many nonsugar components in variable concentrations. Even trace components can have an influence on product quality, particularly sensory factors such as flavor and odor. Although generally described as yellow or brown, color exists as a wide variety of hues from red to green. Some techniques are successful in removing some of the undesirable shades, but the refiner's only real control is in the intensity of the color present in the final product. Almost every decision made in operating a refinery will affect the final characteristics of the soft sugar produced, and with so many variables to consider, it is virtually guaranteed that variations in the final product will exist within a wide margin not normally associated with other refinery products. With so many variables to consider, it is difficult to pinpoint the exact cause of these variations when they arise. It often requires the expertise of knowledgeable people with the experience gained over many years of actual production to ensure a uniformly high quality product.

There are an infinite number of approaches that can be taken for the preparation of the starting materials to produce soft sugar. Each approach depends on the type and quantity of processing equipment available and the volume of soft sugar produced (Table 31.2). To understand why certain approaches are used in certain circumstances or to design for new installations, it is often easiest to start with the finished product and work backward. Ideally, for a boiled soft sugar, massecuite purity should run between 78 and 82. For a 50% yield (on solids) of a soft sugar containing 8% nonsucrose, the resulting runoff syrup will have a purity of 68.

Assuming, for the sake of simplicity, that soft sugar is simply a mixture of pure, colorless sucrose crystals and impure syrup, the soft must contain 75% crystals and 25% runoff syrup:

$$\text{syrup solids in sugar} = \frac{\text{NSS in sugar}}{\text{NSS in syrup}} \times 100\% \tag{31.1}$$

If the soft sugar color required is 4000 ICU, the runoff syrup color must be 16,000, since the sucrose crystal portion contributes no color:

$$\text{color of syrup} = \frac{\text{color of sugar}}{\text{\% syrup solids in sugar}} \times 100\% \tag{31.2}$$

If the syrup content of the sugar is 25% and the crystal content is 75%, the crystal yield of the strike is 75% of 50, or 37.5%. The color of the original massecuite must have been 62.5% of 16,000, or 10,000 ICU. We can now begin to define the characteristics of the starting materials for a 4000 ICU color sugar given that the required final massecuite color is 10,000 at a purity of 80.

Soft strikes can be boiled entirely from a single syrup source or back boiled using some portion of the soft runoff syrup from the previous strike to give the desired final massecuite purity. If the sugar is produced without back boiling, 2 tonnes of syrup solids at 80 purity and 10,000 ICU are required to produce 1 tonne of soft sugar. The most common refinery syrup of this purity is affination syrup, which at 30,000 ICU and a high suspended solids

TABLE 31.2 Characteristics of Starting Materials

	Solids	Sucrose	NSS	Purity	% NSS
Massecuite	100	80	20	80	20
Sugar	50	46	4	92	8
Syrup	50	34	16	68	32

content, requires extensive processing to give the final color and turbidity requirements necessary to produce a 4000 ICU sugar.

On the other hand, a boiling scheme in which 40% of the strike is composed of runoff syrup from the previous strike requires only 1.2 tonnes of 3300 color syrup at 88 purity to produce 1 tonne of soft. Since this material is of a higher starting purity, lower material costs and less equipment are required for purification. A suitable starting material can usually be prepared from a blend of 85% fourth runoff syrup and 15% B molasses, giving a yellow liquor of approximately 88 purity and 12,000 ICU. When one compares the total amount of color that must be removed per tonne of soft sugar produced it is readily apparent that the back-boiling method reduces the processing requirement significantly.

Filtration can consist of direct pressure filtration in leaf-type filters using filter aids such as diatomaceous earth or by batch phosphatation and liming, followed by decantation of the precipitated calcium phosphate and pressure filtration of the supernatant liquor. Most modern installations now use floatation clarifiers to remove suspended solids. These clarifiers, which operate at temperatures between 85 and 90°C, also serve to pasteurize syrups and increase the shelf stability of the final product. Dorn et al. [2] describe a Tate & Lyle clarification system for yellow liquor preparation. Phosphoric acid is added at 350 ppm on solids. The quality of the liquor obtained from clarification of this type is often high enough to be passed directly onto bone char without further pretreatment.

Because of the high ash and low purity of these streams, the anionic flocculants used in soft liquor clarification require much higher charge densities and dosage levels than those used for washed raw liquor clarification. They operate well only over a very limited purity range, leading to frequent clarifier upsets with very slight changes in liquor composition. Recent development in flocculant chemistry has lead to the development of a class of liquid flocculants that can operate over a much greater purity range of soft liquor at much lower dosage levels [3]. Table 31.3 shows relative performance characteristics of this type of flocculant.

Bone char remains the decolorizing agent of choice for the production of soft sugars. Besides its ability to remove polyvalent metal ions such as iron, it also acts as a flavor moderator and is said, in some instances, to add its own unique flavor of its own to the final product. The amount of bone char required will vary with the quantity of material to be processed and the amount of color removal required. A typical refinery producing 8% soft on melt will use about 400 kg of char per tonne of soft produced. Contact time with the liquor is usually twice that for granulated sugar production. Some refiners use a "follow-on" system where filters first complete a fine liquor cycle followed by soft liquor cycle in either a single- or double-pass mode. When multiple passes over char are used, darker colors should always follow lighter-colored materials. Although follow-on systems make better use of existing char capacity, they generate transition materials that must be handled separately. Other refiners designate certain cisterns for soft sugar duty only and settle each cistern with soft sugar liquor, eliminating transition liquors.

TABLE 31.3 Comparative Performance of Clarification Flocculants for Soft Liquor

Parameter	Dry Flocculant	Liquid Flocculant
Purity range	84–89	70–91
% Ash	2–3	0–5
Color/turbidity	40,000 max.	80,000 max.
Dosage (100 % active basis, ppm)	30–150	10–30

Other decolorizing methods have been reported, including the use of nonfunctional ion-exchange resins [4]. In this pilot study, liquor with a color of 50,000 ICU was reduced to 11,000 ICU using Rohm & Haas XAD-16 resin. Taste testing of the resulting liquor showed that it could not be distinguished from liquors decolorized by bone char filtration. It was the author's opinion that such a system could be less expensive than traditional bone char treatment.

Ion-exchange resins have also been used to remove iron. The presence of iron, particularly in the ferric form, is responsible for complexes that contribute to greenness in the final product. These complexes can be decolorized with the addition of phosphoric acid [5] either at the clarification stage or through direct addition to the pan just before the strike is completed.

Soft sugar can be boiled either as single-grain or agglomerated-grain product. Single-grain sugars are boiled in such a way as to give uniform single crystals between 180 and 240 μm. Crystal size distribution less than this gives a muddy or mottled appearance and difficulty in purging. Larger crystals give a gritty appearance and "hard" texture to the finished product. Because of the single-grain nature of the product, there is a limited amount of crystal surface area to carry the syrup. As a result, this type of sugar rarely has a nonsucrose content much greater than 5%. It is possible to increase the overall surface area slightly by creating a second crop of sugar crystals partway through the strike by seeding a second time [6]. This will create a second, smaller distribution of crystals within the original massecuite. Quantity and size of this second crop must be carefully controlled in order not to pack the centrifugals and prevent proper purging. The technique for boiling such a strike is usually divided into three stages: (1) boiling-down charge, (2) graining charge, and (3) final charge. The boiling-down charge is divided into two separate charges, the first having a slightly lower purity than the second. After the first crop of sugar crystals has been established to the desired size, a second charge of higher purity is given and a second crop of crystal is introduced. The first charge can have a purity of 89 and can be followed by a second charge of 94 purity to give an overall charge purity of 91. The final charge is introduced as a diluent to restore fluidity to the masse and reduce the overall purity of the final massecuite.

Agglomerated style soft sugar is made up of very fine sugar crystals between 50 and 75 μm in size which have been allowed to fuse together in the pan to produce stable particles between 500 and 1000 μm. This produces a final product with a much greater amount of surface area which can still be purged efficiently in a batch centrifugal to produce a soft sugar containing 8 to 11% nonsucrose. In addition, agglomerated softs have an almost free-flowing appearance and can readily be packaged in high-speed packaging equipment with relative ease.

The agglomeration takes place immediately after the seed development phase when the pan is allowed to "sit down" with absolutely no agitation for an extended period. After this "sit-down" period, the massecuite is tightened to the desired consistency and liquor feed begins at a slow rate until the desired quantity of liquor solids has been added. Syrup solids from the previous strike are then added until the pan reaches its maximum capacity. The strike is tightened to a high Brix value and dropped. The following procedure is an actual recipe for producing a 1000-ft^3 strike of agglomerated soft sugar. The boiling process is completely automated.

A liquor charge of 35,000 lb of solids at 85 purity and 4000 color is boiled down at 23.5 in. vacuum (150°F) using 10 lb of steam. When Brix reaches 80, steam pressure is reduced to 2 to 3 lb. Charge is grained at 82 Brix using 1500 mL of ball-milled slurry in isopropyl alcohol. Three minutes after seeding, the steam flow is shut off. Four minutes later, the agitator is shut off. Five minutes after the agitator stops, the vacuum pump is shut off to ensure that the massecuite temperature does not fall more than 5°F below the graining temperature. The pan is allowed to sit down in this state for 30 min, at which time the sugar boiler checks the degree of agglomeration by withdrawing a proof and examining it under a microscope. If all the grain is not fully agglomerated, the vacuum pump and steam are turned on and the pan is allowed to tighten up for 15 min. No feed or mechanical agitation is allowed. After this tightening, the pan is allowed to sit down again for 15 min.

After the sit-down period is complete, the vacuum pump is turned on and the steam pressure increased to 3 lb. The pan is allowed to tightened for 20 min then a slow liquor feed begins at a rate of 50,000 lb of solids per hour. The agitator is restarted. After the total addition of 85,000 lb of liquor solids has been completed, syrup feed begins. Syrup purity ranges between 58 and 62 with a color between 10,000 and 16,000. Absolute pressure is ramped down to 4.5 in. absolute (140 to 142°F) over a 1-h period. When the syrup feed is completed, the pan is tightened to 92 to 93 Brix and dropped. Total boiling time is 4 h.

Drop temperature and drop Brix are both critical to the keeping qualities of boiled soft sugar. Drop temperature should be below 135°F. If the massecuite is too hot, sucrose will continue to crystallize from the syrup as the product cools. If this occurs when the sugar crystals are in close contact without agitation, as in a finished package, the result is hardening or lumping. While purging times can be significantly reduced by increasing the drop temperature slightly, this increase in centrifugal throughput can lead to disastrous results in the warehouse.

Water, or more specifically, the water content of the syrup film, plays a major role in most chemical and microbiological processes, which can lead to rapid deterioration of soft sugar in storage. Although the overall moisture of the finished product can be controlled in the centrifugal, it is only the syrup that is being removed at this point. Attempts to reduce water content by increasing the purging time only result in reducing the amount of nonsucrose in the finished product and do not affect the water activity of the syrup film.

Both color and titratable acidity of soft sugar will increase in storage [7]. Color increases can often be dramatic, with as much as a 50% increase within 6 months. Titratable acidity usually stabilizes 120 days after production. Lactic acid formation has been implicated as a factor in both phenomena. Both color and acidity increases can be correlated to the moisture content of the final product, but it is not the only cause, as some high-moisture sugars were shown not to fit the prediction. It is likely that this degradation is more a factor of the water content or water activity of the syrup film itself, so that sugars containing a high syrup content at high Brix show a higher product moisture content but are more stable than sugars with a low syrup content at low Brix and therefore, low overall product moisture.

It is often useful to determine the ratio of water to nonsucrose in the final product. Because both these components exist only in the syrup film, this ratio gives a good indication of the Brix of the syrup film, provided that the purity of the runoff syrup remains relatively constant from strike to strike. This ratio is similar in concept to the safety factor used to determine the stability of raw sugars. Experience has shown that ratios of less than 0.5 result in soft sugars with good storage characteristics, while sugars with ratios in excess of 1.0 are prone to microbiological attack and fermentation.

Soft sugar massecuites are usually purged in batch centrifugals without washing, although the use of continuous machines has also been reported. Batch machines should be fitted with footplates to prevent the possibility of massecuite dripping into the final product and forming hard syrup lumps. Drying/spinning times are varied to reach the desired color level and syrup content. Times ranging between 4 and 6 min are common. Experience has shown that softs which require excessive spinning usually result in a sticky product that is difficult to pack and often gives rise to customer complaints at a later date.

Short conveying distances to the packaging bins are desirable to reduce problems with syrup buildup and lumping. Due to the sticky nature of the product, almost any type of conveying device will give rise to problems. Screw conveyors directly under the centrifugal are often used as a mixing device ensuring that each centrifugal load is homogenized prior to packing. Care must be taken to prevent an excessive syrup buildup on the flights and around hanger bearings. This buildup will eventually fall off and appear as hard dark lumps in the final product. If possible, cooling the sugar before packaging is desirable and helps to avoid lumping and hardening of the product in bag storage.

31.2 METHOD OF PRODUCTION: BLENDED SOFT SUGAR

Soft sugars can also be prepared by blending selected refinery syrups with granulated sugar on a continuous basis. Proponents of blending cite the following advantages of this type of production facility;

1. It automatically meets the changing throughput requirements of the packing station.
2. It reduces load on purification equipment and materials such as bone char and filter presses.
3. It provides more flexible scheduling for both the process and packing departments, based on crew availability as opposed to pan availability.
4. It allows for tighter tolerances in product quality control.

In general, soft sugar blending offers a lower capital-cost alternative to boiling in cases where expanded or new capacity is required. Maintenance costs associated with the blending equipment can also be expected to be less than that associated with vacuum pans, mixers, and batch centrifugals. Blending also offers the potential to make limited runs of special or custom-coated products, either by color or by particle size. Because of the limited surface area involved, blending will not offer the same yield increases as boiling, and unless remelt strikes are used as the substrate, blending offers no reduction in the amount of recovery sugar recirculated back to process.

All that is required to blend soft sugar is a source of granulated sugar, a source of heated syrup, a blender, and associated control equipment. Granulated sugar can be either a special remelt strike that has been dried and scalped, or screened sugar of the appropriate size. When using screened sugar, it is not necessary to remove the through 80-mesh fraction, and in some instances where there is not a strong demand for this product, it can be blended back into the soft sugar instead of being remelted. The average particle size should be the same as for boiled soft sugar, between 180 and 240 μm. Care should be taken in the amount of through 80-mesh fraction incorporated or the product will take on a muddy appearance. Sugar is introduced to the blender using an auger feed with a variable-speed motor.

The percentage of syrup added varies between 15 and 20% by weight of the final product to give the desired color and maintain the moisture level between 2 and 4%. Syrups used for blending should be high in both invert and color and low in ash. If necessary, the invert content can be increased artificially by inversion with phosphoric acid, followed by neutralization with caustic soda. The presence of invert in the final product acts as a humectent to prevent drying and lumping in storage. A high invert content will also prevent any crystallization of sucrose in the syrup film that could lead to bridging and lumping in storage. The Brix of the coating syrup should be between 84 and 86. Because the syrup comprises only 15 to 20% of the final product, the amount of syrup that must be processed is significantly less than that required for boiling soft sugar. To reach the desired final sugar specifications, the color must be higher and the purity lower. For this reason, many blended soft sugars are produced without bone char or other types of decolorizing systems, preferring simple filtration or clarification alone for syrup preparation.

Syrup temperature is also critical for successful blending. Syrup temperature must be maintained between 82 and 85°C [8]. If the temperature is too low, the syrup viscosity lowers blending efficiency. If the temperature is too high, color generation in the syrup becomes significant. A small plate heat exchanger, using hot water as the heating source, is usually sufficient for this function. A portion of the syrup is always recycled back to the storage tank to prevent overheating in the heat exchanger in the event of a process interruption. Syrup is delivered to the blender using a positive-displacement pump with a variable-speed motor.

Most types of continuous blenders are suitable for this application. Typically, a variable-speed double-twin-screw conveyor is used and the syrup is sprayed onto the sugar. As sugar travels down the blender, it is well mingled with the syrup and a uniform product is discharged at the end of the blender directly into the stabilizing hopper of the packaging equipment. Constant care and attention must be given to the blender and any conveying equipment to ensure that the buildup and formation of syrup balls are minimized.

The level of blended sugar inside the packing station hopper is controlled by radio-frequency sensors. If the level in the hopper reaches a high setting, the sugar feed and syrup feed motors are altered automatically to slow production while maintaining the same sugar-to-syrup ratio. Production speeds up automatically when the level in the hopper is low.

31.3 PACKAGING OF SOFT SUGAR

Packing bins should be as steep sided as possible or be emptied mechanically in some fashion. One sugar refinery uses a rotating, flat-bottomed, circular bin fitted with a retractable plough to move sugar from distribution outlets to individual packaging lines. This type of bin is useful when there is very little headroom between the centrifugal and the packing station. Most refiners use some type of flaking device immediately before packing to break up small friable lumps and provide a constant flow to the weigh scales. These can be either rotating screen types or a fixed screen with a rotating beater bar for greater throughput. Again, care must be taken, as any rotating mechanical device will tend to generate a syrup buildup that will fall into the finished product.

Soft sugar can sometimes be handled through auger fill systems, but because the product is so sticky, the auger needs to be very powerful. The required force tends to extrude the product through the auger leading to a buildup of syrup that can break off and enter the bag. Using a vibratory feed system is better, however, weight control for small packages always presents a challenge with this product. In-line check weighing combined with elaborate statistical feedback to the weighing scales is normally used and results in a standard deviation of about 5 g per bag. This results in a typical giveaway per bag between 10 and 15 g.

New multihead electronic computer-controlled weighing scales offer the potential for much higher packaging speeds with better weight control than that of traditional types. Sugar is fed automatically via a dispersion feeder and radial feeders to pool hoppers that stabilize the product before its release to the weighing hoppers. Load cells weigh the sugar in each hopper and transmit the information to a microprocessor that calculates the total weights of many different hopper combinations and selects the one that is equal to or comes close to the required nominal value. The sugar is released from selected hoppers and falls through the discharge chute to the packaging machine. It has been reported [9] that standard deviations can be reduced to 1.5 g, thereby reducing giveaway to 3 to 5 g.

With any system, sugar buildup remains an ongoing problem. In contact with air, the product tends to dry out quickly, causing sticky lumps that require frequent cleaning. Particular attention needs to be focused on the angle of the discharge chutes, which should be a minimum of 60°. All contact surfaces should be coated with nonstick materials such as Teflon.

31.4 PHYSICAL PROPERTIES OF SOFT SUGAR

The value of soft sugar to the user is primarily in the color and flavor of the product. To ensure the consistency of their own products, users require consistency in the products they purchase. A thorough understanding of the methods used to determine the physical attributes of soft sugar, combined with a basic knowledge of the chemical constituents that contribute to its unique flavor and where they originate within the refining process, is essential to maintaining consistent quality of the final product.

Although the color specification for most refined products is measured in solution, this does not always serve to characterize the appearance of soft sugar. As a result, two different approaches to color measurement are often used, the traditional solution color method as well as a reflectance method, which is an attempt to approximate the visual appearance of the product.

Solution color is performed on filtered solution at pH 7 according to the ICUMSA procedure. Because of the ionic nature of much of the colorants in soft sugar, color measurement has a high pH dependency. While a pH of 7 may not be the optimum one, it provides a reference basis for comparative purposes. Failure to adjust the pH to 7 will result in deceptively light color measurements.

Visual appearance is often better judged by using reflectance color methods on samples in the dry state. Visible light can be divided into the primary colors red, yellow, and blue, with red having the longest wavelength and blue the shortest. Objects appear colored because some portion of the visible light reflected from their surface has been adsorbed. For instance, soft sugars appear brown because the blue portion of the spectrum has been adsorbed by the pigments in the syrup film, leaving the yellow and red portions to be reflected. The amount of light adsorbed by the sample can be measured and related to the quantity or the intensity of the color present. This type of measurement is made using a reflectance spectrophotometer. The sample is measured in the dry state. Light from an appropriate source is reflected off the surface of the sample, passed through a monochromatic filter, and the intensity is measured by a suitable detector. Using a red filter ahead of the detector for soft sugar is common, as this correlates well with the reddish-brown or golden brown product which so is pleasing to the eye.

Solution and reflectance measurement correlate poorly, and specifying both a solution color and reflectance color will inevitably lead to difficulties. Solution color is a measure of the total concentration of color within a sample, while reflectance color, since it measures only surface phenomena, is more a measure of the color of the syrup alone. A soft sugar that contains a high percentage of low-colored syrup could have a high solution color but still appear relatively light colored to the eye.

The reflectance measurement, as described, only measures the intensity or saturation of one small component of the visible spectrum. For example, a red filter gives information only to the red–green shade relative to shades of gray. It is possible, by measuring the absorbance at three different wavelengths, to obtain a three-dimensional representation of the color that approximates the shade or hue of color present. Colors are represented by three numbers commonly referred to as *tristimulus values*. They can be plotted on triangular coordinates and identical color shades will always fall within the same region of the spectral graph. For a more complete discussion on this type of measurement system, the reader is directed to an excellent review by Richard Riffer of the California & Hawaiian Sugar Co [10].

Shades that are generally considered aesthetically undesirable are green to green–gray. Greenish hues are believed to result from ferric complexes of catechols, such as chlorogenic acid, that are often present in cane refineries at extremely low concentrations. Greenness can also result from low adsorbent activity. Complexes that exhibit greenness can often be decolorized by adding phosphate, usually in the form of phosphoric acid. Shades of gray are common in soft sugars with a high ash content.

On the other hand, red shades lend a particular visual attractiveness to soft sugars. These red shades are noticeably absent in most raw sugars, which tend to display a more yellow-brown or golden brown appearance. These red constituents appear to be formed in the vacuum pans and sweetwater evaporators as a result of heat processing. It is speculated that the most likely source is the yellow compound chlorogenic acid, which is hydrolyzed to caffeic acid, also yellow, and then oxidized to the red o-quinone. Soft sugars also redden during periods of long storage under warm conditions and the same mechanism is probably involved.

The importance of flavor as a quality parameter should not be overlooked. Although customers rarely comment on flavor characteristics, their expectations are for a product free from harsh, bitter, or salty tastes. Considerable research has been performed in the isolation and identification of constituents that contribute to the flavor of soft sugars. The most important flavorants in soft sugars are formed largely during refining processing and for the most part are not present in raw sugars.

Acetic acid has been identified as the most abundant volatile constituent in brown sugar. Concentrations are found well above the taste and odor threshold. Although not noted as a pleasant flavor by itself, it does contribute to the overall pleasant sensation of the product and is necessary for developing the proper flavor profile. Acetic acid is also responsible for the slightly sharp odor of soft sugar. It arises from bacterial activity in low-purity sweetwaters. Other products arising from microbial activity include lactic acid, ethanol, and diacetyl. Dimethyl sulfide, responsible for the molasses flavor, originates from the cane plant. Caramel flavorants, such as maltol and acetyl formoin, and butterscotch flavorants, such as diacetyl and 2,3-pentanedione, result from sucrose degradation reactions. Table 31.4 is a list of compounds identified in soft sugars, along with their appropriate taste sensation and their possible origin within the refining process [11].

It is relatively easy to acquire the skills necessary to identify the major flavor notes associated with soft sugars. Flavors are typically denoted as acid, bitter, caramel, green cane, licorice, molasses, and burnt or charred. Residual irritation or aftertastes often remain and are described as metallic. Samples for taste testing should be prepared as a fondant to eliminate effects associated with changes in texture and moisture level that also contribute to mouth feel and can play a part in flavor sensations.

The learning process involves achieving, through practice, the ability to identify the taste, composed of several flavors, considered desirable. To accomplish this, a small group of people are required to establish a taste panel. Samples for tasting are prepared synthetically to reproduce first the individual flavors and then the combinations of flavors known to be involved in the production of soft sugars. The tasting process should be slowly broadened to include actual sugar samples considered both pleasant and unpleasant. Samples can be fortified with several artificial flavors to intensify one flavor note or another until the panel is well practiced in identifying individual flavors. The scope can then be broadened to include samples prepared with various refinery syrups to identify the source of both desirable and undesirable characteristics.

TABLE 31.4 Sources of Flavor in Soft Sugar

Compound	Type of Flavor	Source
Acetic acid	Acidic; part of the typical soft sugar flavor	Microbiological, found in sweetwater
Acetyl formoin	Desirable, caramel impact	Carbohydrate degradation
"Charred"	Undesirable if too concentrated	Bone char
Diacetyl	Desirable flavor, sweetness enhancer	Glucose pyrolysis, browning reaction product
Furfural	Caramel, nutty	Carbohydrate degradation
Heptanal, hexanal	Fruity to fatty flavors	Degradation of long-chain fatty acids from sugarcane plant
HMF	May contribute to bitterness or astringency if too concentrated	Acid carbohydrate degradation leads to colorant formation
3-Hexene-1-ol	Green, grassy	Sugarcane plant
Lactic acid	Found in most soft sugars; may enhance acidic notes	Microbiological
Metallic	Undesirable, especially if highly concentrated or in ferrous form	Probably due to iron from equipment, acid corrosion, or rust
Pentyl furan		
3-Methyl butanal, 3-metyl propanal	Found in all sugars; fruity, green flavors; harsh if concentrated	Sterecher degradation of amino acids
Pentyl furan	Can contribute to stale, licorice, or cardboard flavor	Linoleic acid degradation
Phenols and phenolic acids	Various flavors depending upon structure, ranging from spicy-aromatic to smoky	May be metabolic breakdown products; many are colorant precursors

Godshall describes this process in detail [12]. Panelists were trained to identify up to 13 different flavor attributes in a variety of products from several manufacturers. Of these 13, four were found to be significant in describing the flavor of soft sugars. These were described as char, licorice, molasses, and green. Correlation with chemical analysis showed that the best predictors of a favorable soft sugar flavor were low amino nitrogen and chloride content. Total ash, color, and phenolics also had an impact on the desirability of soft sugars when present at high concentrations. However, each of these components is necessary, in the proper proportion, to produce an acceptable brown sugar.

REFERENCES

1. J. L. Meikle, *Sugar Ind. Technol.*, 1982, pp. 140–148.
2. E. L. Dorn et al., *Sugar Ind. Technol.*, 1993, pp. 41–58.
3. T. Craig, *Sugar Ind. Technol.*, 1995, pp. 177–196.
4. W. Rothchild, *Sugar Ind. Technol.*, 1993, pp. 305–324.
5. R. Riffer, *Sugar Processing Research Conference*, 1984, pp. 231–251.

6. C. Baikow, *Manufacture and Refining of Raw Cane Sugar,* Elsevier, Amsterdam, 1967, p. 371.
7. M. A. Godshall et al., *Sugar Processing Research Conference,* 1996, pp. 437–446.
8. G. Belec, *Sugar Ind. Technol.,* 1989, pp. 201–218.
9. J. P. Merle, *Sugar Ind. Technol.,* 1996, pp. 369–373.
10. R. Riffer, *Sugar Processing Research Conference,* 1990, pp. 265–290.
11. M. A. Godshall et al., *Sugar Processing Research Conference,* 1980, pp. 26–47.
12. M. A. Godshall et al., *Sugar Processing Research Conference,* 1984, pp. 22–52.

CHAPTER 32

Areado Soft Sugar Process*

INTRODUCTION

Soft sugars are special sugars consisting of small crystals surrounded by a syrup layer forming a product soft to touch. There are various ways to produce this type of sugar: by boiling small crystals and centrifuging the massecuites in continual machines, by blending small sugar crystals with treated syrup, and by direct crystallization of an oversaturated hot sugar solution provoked by a rapid agitation with or without vacuum. This last process has its origin in old processes to make soft sugars in India (Bura sugar) [1], and in China (Ching Tang and Sha Tang) [2]. Possibly, in the fifteenth and sixteenth centuries Portuguese travelers and navigators took from the East the way of producing this type of sugar, and introduced the process in the West, producing Areado sugar.

In the past, Areado sugar was produced by concentrating a sugar syrup by direct fire heating at high temperature and provoking spontaneous crystallization by manual agitation with wooden paddles in open copper kettles. Due to agitation, water evaporation is enhanced and sugar crystallization starts spontaneously. Heat released from crystallization will provoke further evaporation and crystallization. This process will transform the concentrated hot syrup in a mass of wet soft sugar. The modern process makes use of vacuum pans and vacuum crystallizers, called *areadores,* and is used in Portugal to make a yellow soft sugar [3]. About 10,000 tonnes/yr of yellow Areado sugar are produced in Portugal, representing 3.8% of the total sugar consumed. This process of getting a sugar by spontaneous crystallization from a hot and supersaturated solution was named *transformation,* and the sugar, *transformed sugar* [4]. In different type of the world this process is used to produce transformed sugar with various techniques.

The same type of sugar is produced in Brazil with the local name *Amorfo sugar.* This product is a white soft sugar, 0.25 to 0.35 μm with a color lower than 50 IU, that has enough quality to compete with crystal sugar [5]. The production represents 28% of the Brazilian white sugar market (1993). The Brazilian process does not use vacuum to concentrate the syrup up to 93 to 94 Brix [6], and the agitation process is made in atmospheric

* By Luis San Miguel Bento and Francisco Carlos Bártolo.

agitators called *Batedeiras*. The characteristics of white Amorfo sugar are presented at Table 32.1.

Cassonade sugar is a soft yellow sugar produced in Belgium in a way similar to Areado sugar. The name *Cassonade* came from wood boxes (casses) which the Portuguese transported the soft sugar from Brazil to Europe [7, p. 195]. An average analysis of this sugar produced at Tirlemont refinery, Belgium, is presented in Table 32.1. About 3000 tonnes/yr of this product is sold in Belgium in different packages [8].

A process of producing transformed sugar was developed in England [9]. In this process the hot supersaturated syrup is agitated in a shear device where a rapid crystallization occurs. The nucleated syrup is then transported in a band where sugar crystallizes in a soft open texture. The sugar is then extruded in a particulator.

More recently, a transformed sugar produced from cane juices was presented with the name of Ur-Zucker [10]. The same production process is used to produce a transformed sugar, called Voll-Zucker, from beet juices. The same process of transformation is also used to insert ingredients into agglomerates of tiny crystals of 3 to 30 μm, with a process named *cocrystallization* [11]. This product, produced by Domino Sugar Corporation, presents a surface area 20,000 times greater than normal, 250-μm crystals, and a bulk density of 512 g/L. These features allow distribution of a film of ingredients over the crystals during the transformation process. This cocrystallization process presents important advantages, such as adding an ingredient to soft sugar as flavor, or it can be used to increase dissolving properties of compounds such as fumaric acid.

32.1 THE AREADO SUGAR PROCESS

When sugar is crystallized from a sugar solution, heat is released (i.e., sugar crystallization is an exothermic process). Also, heat of crystallization increases with temperature: At 60°C the heat of crystallization of sucrose is 57.0 J/g, and at 90°C it is 107.6 J/g [9]. If a supersaturated sugar solution is heated at a high temperature and if a quick evaporation of water is provoked by strong agitation or by agitation under vacuum, heat is released to the vapors, and supersaturation increases, initiating a spontaneous crystallization. This crystallization

TABLE 32.1 Characteristics of Transformed Sugars

Parameters	Areado, Portugal	Cassonade, Belgium	Amorfo, Brazil
Type	Yellow	Yellow	White
Polarization	96.6	92.4	99.5
Moisture (%)	2.51	2.77	0.15
Invert sugars (%)	0.12	0.30	0.15
Ash (%)	0.34	1.85	0.10
Color (IU)	1300	4080	20
SO_2 (ppm)	3		10
MA (mm)	0.35		0.30
Coefficient of variation	66		40
pH	7.0	8.5	
Bulk density (g/L)	750–800		700–750[a]

Source: Data from Ref. 5.

[a]From Ref. 6.

release heat that evaporates more water, increasing the supersaturation, and further crystallization occurs with subsequent heat release and water evaporation. This phenomenon continues in a chain reaction until almost all the water is evaporated and the syrup layer is so thick that crystallization almost stops. This process was called as *transformation*. The final product consists in an agglomeration of sugar crystals with a syrup layer that is then dried and cooled. Since further crystallization is stopped, this sugar can be maintained in individual particles without caking. If sugar is stored in humid conditions, water would be absorbed and a subsequent evaporation can cause caking of the sugar. Due to this, sugar must be packed in plastic bags in a dry environment, without humidity variation.

The old Areado sugar process was a true transformation process; that is, Areado sugar was produced from a concentrated sugar solution provoking a rapid crystallization. Amorfo sugar is produced in the same way in Brazil. The modern Areado sugar process comprises three steps. In the first step a sugar solution is crystallized in a vacuum pan by spontaneous crystallization; then this massecuite is heated and discharged into a mixer with vacuum (areadores). In the second step the massecuite is agitated under vacuum, water is further evaporated, the massecuite temperature drops, and further crystallization occur. This crystallization releases heat and the transformation of the remaining syrup in the massecuite starts. In this process the transformation occurs in the mother liquor of the massecuite syrup in the presence of existing crystals. In the third step the sugar is cooled, dried, and sieved before packaging. Performing a heat balance on this process, the theoretical water content of the final sugar produced by this system can be calculated. Results of these calculations are presented at Figure 32.1. In this figure the theoretical moisture of Areado sugar, discharged from the mixer at 60°C and 99% purity, is presented in function of the massecuite Brix and temperature.

32.2 PRODUCTION OF AREADO SUGAR: RAR REFINERY

RAR Sugar Refinery, with a daily capacity of 500 tonnes, receives raw cane sugar and produces white sugar and yellow Areado sugar. The production of yellow Areado sugar represents 2.7% of the total production. The production of yellow Areado sugar consists of three steps:

1. Crystallization to a massecuite of a mixture of liquor and syrup in a vacuum pan
2. Crystallization and drying the massecuite in two vacuum mixers (Areadores) into yellow Areado sugar
3. Final drying, sieving, and packaging of the sugar

The Areado sugar process as practiced at RAR is presented in Figure 32.2.

32.2.1 Boiling

The first step of Areado sugar production is to boil the feed, liquor and syrup, in a 17 m^3 calandria vacuum pan with 120 m^2 of heating surface.

a. Feeding. The strike is started by feeding the mixture of liquor (approximately 10 to 11 m^3) and syrup (4 to 5 m^3). These are the only products and quantities to be fed into the vacuum pan during the boiling. Characteristics of the final mixture are presented in Table 32.2.

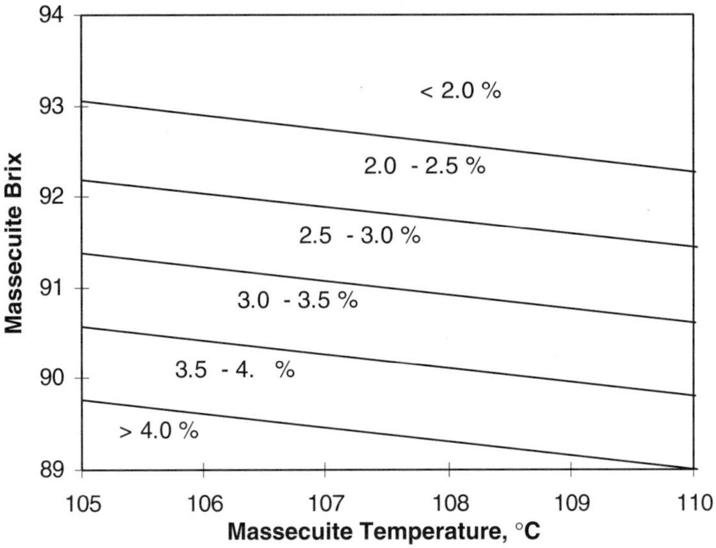

FIGURE 32.1 Theoretical Areado sugar moisture (99° purity, 60°C). (Data from Ref. 3.)

b. Steam On. After feeding the mixture, steam (1 kg/cm^2; 120°C) is fed into the calandria, allowing the temperature of the solution to rise (90 to 95°C). This step is very important to guarantee that the crystals that may come with the solution are all melted.

c. Evaporation and Crystallization. The essential phases are:

1. Start up the vacuum pump. The vacuum rises rapidly, sugar concentration increases, and in about 20 min the labile zone is reached. In this supersaturation zone, crystallization starts by spontaneous nucleation, transforming the solution into a heavy massecuite of aggregated small crystals.
2. Shut down the vacuum pump. The main purpose is to create the right conditions to increase the temperature of the massecuite. It needs to be heated up to 105 to 110°C before being discharged into the two vacuum mixers (Areadores).
3. To maintain good circulation of the massecuite, to increase heat transfer, and also to avoid local overheating, the vacuum is raised within a few seconds, from time to time.
4. Before dropping the massecuite to the mixers, the vacuum is raised so as to thicken the massecuite to 91 to 93 Brix.

d. Steam Off. To avoid sugar degradation and an increase in invert sugars and colorants, the massecuite must be dropped immediately to the vacuum mixers. The standard curves from temperature and Brix of the massecuite and the vacuum during the boiling are represented in Figure 32.3. The total boiling time is calculated from steam on to steam off.

32.2.2 Massecuite Drying

The mixer consists of one cylindrical closed vessel (length = 2.9 m; diameter = 1.5 m; massecuite capacity = 2.5 m^3) with an internal paddle type agitator (paddle angle = 90°;

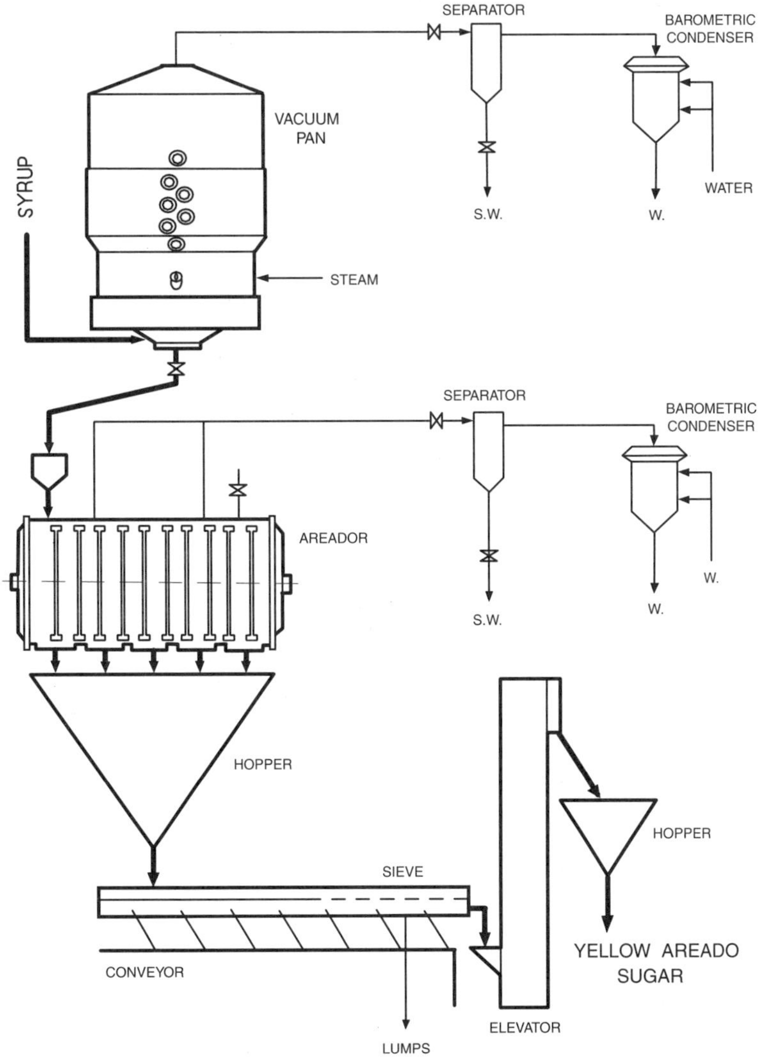

FIGURE 32.2 Areado sugar process flow diagram at RAR.

TABLE 32.2 Characteristics of Feed, Massecuite, and Sugar

Parameter	Feed	Massecuite	Sugar[a]
Temperature (°C)	71–75	105–110	60
Purity (%)	98.5–99.5	98.5–99.8	
Brix (deg)	68.0–72.0	90–93	
Color (IU)	1000–1500		
Ash (%)	0.15–0.50		
Invert (%)	0.12–0.40		
Moisture			2.8–3.4

[a]At Areadores outlet.

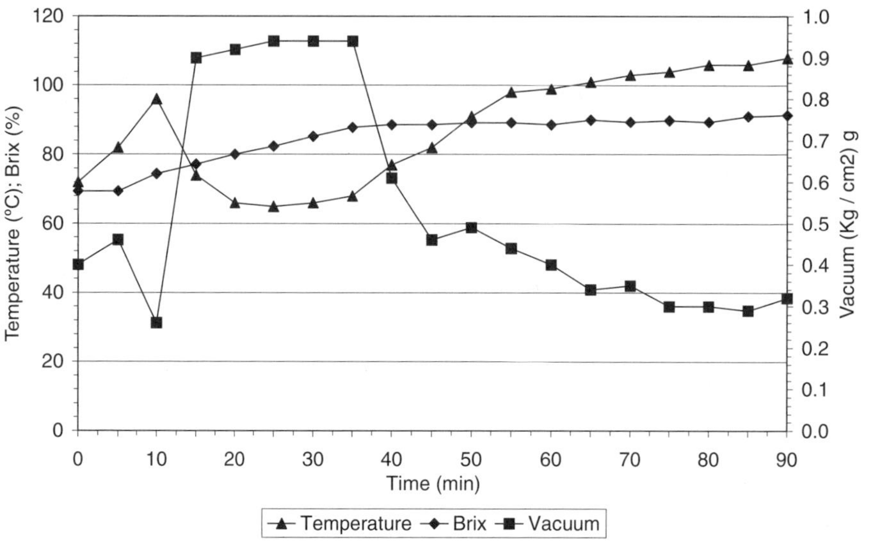

FIGURE 32.3 Boiling data for Areado sugar (RAR).

rotation = 6 rpm; motor = 30 hp). At the top it has a massecuite feeder inlet with hinged cover and a vacuum breaker, and at the bottom, five discharging doors. Other components are sight glasses and vacuum, steam, and purge connections. A multijet condenser is fitted for both mixers. Figures 32.4 presents lateral and front views of RAR Areadores.

Due to the large capacity of the calandria pan compared to the capacity of the two mixers, the discharge of the massecuite is done in two steps. Some care has to be taken with the time that the second half of the massecuite remains in the vacuum pan. Problems such as an increase in invert sugars with the related reduction of the bulk density in the dry sugar may arise. After starting the paddle agitator and feeding the mixers with massecuite up to 10 cm below the axis, the inlet cover is closed and vacuum is raised by opening the vapor valve to the barometric condenser. Crystallization of sucrose present in the syrup starts simultaneously with the water evaporation. The rotating movement created by the internal paddle agitator lifts the massecuite/sugar, increasing the exposed surface area and thus reaching high evaporation rates. The process of sugar drying takes, on average, 15 min. The operation is completed when all massecuite is completely transformed into sugar. This point is done by controlling the amperage of the motor agitator. An amperage increase is observed when all massecuite is transformed.

32.2.3 Conveying, Sieving, and Packaging

The Areado sugar conveying system consists of three grasshopper conveyors. This system enables the hot sugar (60°C) to be naturally cooled and dried before sieving and packing. The sieving process is done in the third part of the conveyor, where all the sugar passes through a screen with a 5-mm aperture. Lumps are removed and dissolved in the melter. An average of 3.5% of lumps is produced for each Areado sugar strike. After sieving, the cooled and dried soft sugar is conveyed through a bucket elevator to the packaging station, where it is bagged in 50-kg bags polypropylene (PP) and low density polyethylene (LDPE) components and packed in 1-kg plastic bags.

(a)

(b)

FIGURE 32.4 RAR Areador mixer: (a) lateral view; (b) front view.

TABLE 32.3 Periods of Time During Boiling and Drying of Areado Sugar

Step	Time (min)
Calandria vacuum pan	10
Steam on / vacuum off, heating up the syrup	
Steam on / vacuum on, evaporation until the labile zone is reached	20
Steam on / vacuum off, heating up the massecuite	50–60
Steam off / vacuum off, discharge the massecuite to Areadores	4
Areador	11
Drying	
Discharging	3
Total time	98–108

Table 32.3 summarizes the various operations of the Areado sugar process done at RAR.

32.3 MAIN FEATURES OF YELLOW AREADO SUGAR

This special yellow sugar is appreciated for its quick-dissolving property and its natural cane sugar flavors and colorants contents. The presence of natural sugarcane compounds can be of nutritional interest, as natural plant colorants have antioxidant properties [12]. These characteristics of yellow Areado sugar explain its principal application in grocery and industrial food production.

The production process of Areado sugar presents advantages both as to capital costs (i.e., simple equipment) and on running costs (i.e., low energy consumption). The overall yield is near 100%, without residual syrup production. For these reasons, the Areado system is an economical process.

REFERENCES

1. L. D. Gupta, K. S. G. Doss, and J. K. P. Agarwal, *Proc. 23rd Conf. Sugar Technol. Assoc. India,* 1954; *Int. Sugar J.,* July 1955, p. 679.
2. T. Yamane, in *Principles of Sugar Technology,* Vol. III, P. Honig, ed., Elsevier, Amsterdam, 1963, Chap. 10.
3. L. S. M. Bento, *Proc. Sugar Ind. Technol. Congr.,* New York, 1983, pp. 222–241.
4. W. M. Nicol, G. W. Vane, M. J. Daniels, *Proc. Sugar Ind. Technol. Cong.,* London, 1978, pp. 396–409.
5. A. Bezerra, *Proc. Sugar Ind. Technol. Congr.,* Toronto, 1993, pp. 299–304.
6. *Sugar Azucar,* May 1977, pp. 55–57.
7. *Histoire Naturelle du Cacao et du Sucre,* Chez Laurent D'Houry, Imprimeur-Libraire, 1719.
8. Raffinerie Tirlemontoise, Tirlemont, Belgium, 1998.
9. W. M. Nicol, G. W. Vane, and M. J. Daniels, *Proc. Sugar Ind. Technol. Congr.,* London, 1978, pp. 396–409.
10. H. Gihering, *Proc. Sugar Process. Res. Inst. Workshop Products Sugarbeet Sugarcane,* Helsinki, 1994, pp. 168–178.
11. A. Ahmed, *Proc. Sugar Process. Res. Inst. Workshop Products Sugarbeet Sugarcane,* Helsinki, 1994, pp. 34–50.
12. J. E. Kinsella, E. Frankel, B. German, and J. Kanner, *Food Technol.,* Apr. 1993, pp. 85–89.

CHAPTER 33

Liquid Sugar Production*

INTRODUCTION

The convenience of handling, storage, and product mix makes liquid sugar very attractive to manufacturers whose products are not affected by the amount of water in liquid sugar products. The advantages of receiving liquid sugar are less labor, no damaged bags, no waste paper, elimination of dust, enhanced cleanliness, correct solids content, and quality control. Liquid sucrose at 67% solids and invert sugar at 77% solids are standard liquid sugars manufactured. Various liquid sugar blends can be made by mixing these two products according to customer needs. This process can be accomplished by the manufacturer or the customer.

There are numerous combinations for the production of liquid sugar. Some of the most common are (1) dissolving low-color refined crystals followed by trap filtration; (2) dissolving refined sugar followed by carbon treatment and filtration; (3) raw sugar purification by clarification, bone char/granular carbon, and/or ion exchange and filtration; and (4) raw sugar, ion exchange, and powdered carbon. The customer's requirement for color, ash, bacteriological standards, percent solids, sucrose, or invert content are all factors in selection of the production method.

When dissolving crystals, the basic selections are whether to dissolve wet or dry sugar, to dissolve to the final density, or to correct the density by evaporation (increase) or dilute (decrease) with water. The choice of color of the sugar to be melted will determine the need for further decolorization.

Heat is generally required for a satisfactory rate of dissolving, and temperatures from 160 to 180°F are used. It is preferable to keep the temperature as low as possible to prevent unnecessary color formation due to elevated temperatures. Sugar crystal melters range from simple manual to automated multistage [1] vessels of the type shown in Figure 33.1. This multistage melter features continuous dissolving and heating in five separate chambers with automatic density control from the final stage. The density can be changed to produce 67 to 77% solids. A wide range of dissolving rates can be controlled utilizing various numbers of stages in the melter.

* By Leon Anhaiser.

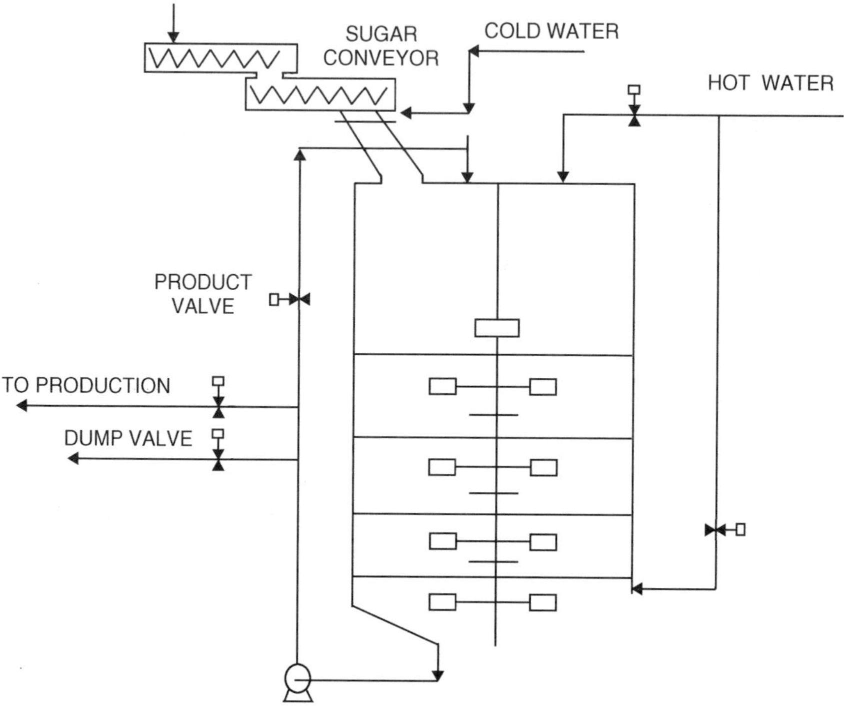

FIGURE 33.1 Multistage melter.

33.1 FILTRATION

Three types of filters used are disposable plug filters, paper cloth filters, and filters that utilize a precoat on the leaves. Disposable plug filters and paper cloth filters are discarded and replaced when they become plugged. Their primary purpose is to remove sediment from the sugar solution. Precoated filter leaves can be used for reducing sediment, microbial organisms, and color. Powdered carbon is a common decolorizing agent added to the melter syrup and/or to the filter leaves after the precoat. Ecosorb [2], a combination of powdered carbon and powdered anion-exchange resin, has been employed on the filter leaves after the filter aid precoat. Sugar syrup is then passed through the layer of Ecosorb to remove the color. This same procedure is effective using high-flow powdered carbon (extremely fine particles removed). The carbon has minimal effect on the differential pressure of the filter leaves. Additional powdered carbon can be added periodically to maintain the necessary level of color removal desired. The filter can be used as a column by recirculating the syrup through the carbon-coated filter leaves until the desired color removal is accomplished.

33.2 INVERSION

Inverting sucrose can be accomplished by batch acid inversion, continuous acid/heat inversion [3], ion exchange, and enzyme. Continuous methods feature a low-acid and/or high-temperature treatment with short retention time. Batch methods can invert from the

50 to 92% level in reasonable time periods. Control is accomplished by polarimetric readings. To stop the inversion, a base is added to neutralize the acid sugar solution. Different invert percentages can be produced by blending with sucrose solutions.

33.3 HEAT EXCHANGER

Since high temperatures increase color in liquid sugar solutions, several methods are used to reduce the temperature prior to storage. The stainless-steel-plate heat exchange is well suited to cool the sugar solution as it is being pumped to storage tanks. Another option is to flash the syrup prior to storage. This reduces the temperature and increases the density at the same time.

33.4 ION-EXCHANGE THEORY

The process of ion exchange is just as it is described—the exchange of ions. As shown in Figure 33.2, ions C in a cationic resin come into contact with a solution containing ions B. The ion-exchange process occurs as some ions C migrate into solution while some B ions assume their place in the ion-exchanger resins. The ion exchanger is shown with negative fixed charges or a cationic exchanger. An anionic exchanger is shown in Figure 33.3. Ions A in an anionic exchanger are exchanged with ions D in solution as the ions move from the initial stage to equilibrium. Notice that same charged ions in the resin (positive in the anion resin) are also in the solution but do not affect the process.

Ion exchanger in not a reaction but a *redistribution* between the ion exchanger and the solution by mass transfer. The process is made easier when the ion exchanger has an open structure to permit easy movement of ions into and out of the exchanger. The ion-exchanger capacity is fixed by the resin, thus determining the balance of the C and B ions in the structure. The total concentration of ions in solution and in the ion exchanger will remain constant. Ion-exchange materials include phenolic, styrene-based, acrylic-based, crystalline zeolite, amorphous alumina-silicates, and heteropolyacids, to name a few.

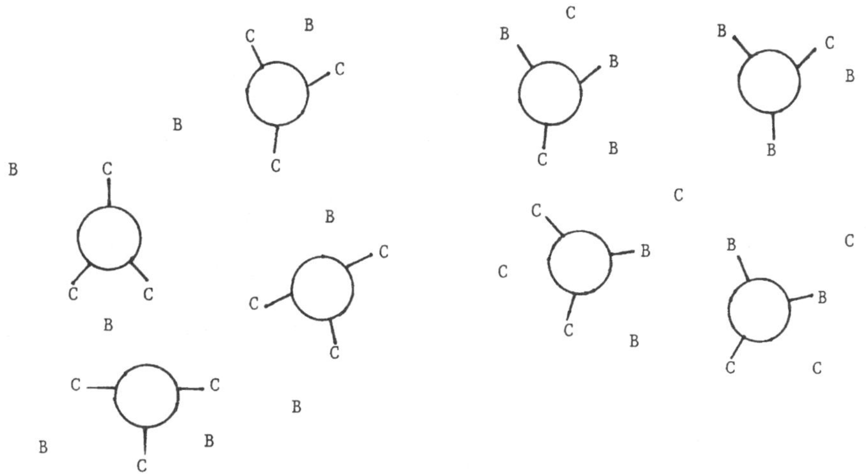

FIGURE 33.2 Ionic exchange.

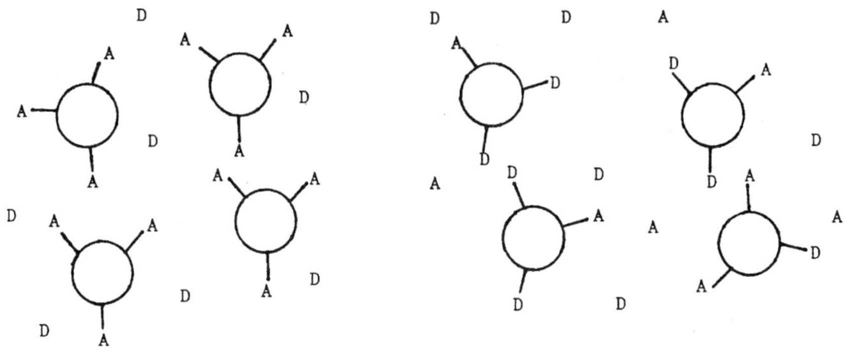

FIGURE 33.3 Anionic exchange.

Macroporous resins have consistent cross-linked macropores which provides ready access to the interior of the resin beads. Their volume changes very little in use, due to swelling conditions. Anionic macroporous acrylic and polystyrenic resins are very suitable for sugar solutions, due to their resistance to fowling and open structure. The color bodies that will be present in the raw sugar solution will be negatively charged and exchanged with negatively charged ions in the resin. The fixed charges in the resins will be positive. Thus, we have an *anionic exchanger.*

Ash impurities in raw sugar solutions can be removed using cationic resins. The fixed charges will be negative in the resins and exchange will be with positive charged ions in the solution. When equilibrium of ions is reached, breakthrough will occur at a very rapid rate. This can be observed directly in the decolorization column samples when viewed in test tubes. Extended cycles after breakthrough show a dramatic color increase. Recovery of the resins to their full potential may require several regeneration cycles to achieve if resins are used in this manner.

33.4.1 Regeneration of Resins

The regeneration process is one of removing the exchanged ions in the resins and replacing them with the original ions. In Figure 33.2 we assume that C ions are hydrogen ions. A solution of hydrochloric acid is passed over the resin to facilitate the exchange of hydrogen ions for the exchanged mineral ions. In a sugar process, a cationic exchanger is used for de-ashing. The result is removal of the mineral ions and a renewed resin ready for the next cycle.

In anionic results, a caustic brine solution can be used to replace the D ions (Fig. 33.3) with hydroxide ions. The type of regenerate will depend on the resin used, the ions being exchanged, and production factors. The percentage of regenerate solution used is critical and should be checked with each resin cycle regeneration. Anionic resins can be used for decolorizing sugar solutions.

33.4.2 Separated and Mixed Resin Columns

Different resin columns are utilized to reduce cost and improve efficiencies. A column can contain one resin only, two resins separated, or two mixed resins. Resins of similar specific gravity can be separated into individual chambers in the upper and lower parts of a column by installing plastic excluders in a separation plate between the top and bottom chambers (Fig. 33.4). These plastic devices have tiny split openings that prevent the resin beads from passing into the next chamber. However, in reality, pressure will force some resins into these excluders and they will have to be removed and emptied manually. This condition can be monitored by observing the differential pressure over the separation plates.

The mixed-bed columns eliminate this problem. However, in the case of a cationic and anionic mixed bed, it is necessary to separate the resins for regeneration. This can be accomplished by utilizing the different specific gravities of the resins. A brine solution is commonly used to float the lighter resins while the heavier resin settles to the bottom of the column.

Mixed-bed columns are designed to allow for regeneration to occur to the cationic and anionic resins by their specific regenerant without affecting the other resin. Special inlet and discharge pipes are designed to minimize the contact of the regenerates with the wrong resin. This must be done to prevent damage to a resin when it comes into contact with an inappropriate regenerate. Mixed-bed columns reduce the cost of materials and space. However, additional headroom above the mixed-bed resins is necessary to allow for the bed expansion during resin separation and mixing.

A liquid such as saturated brine can be used to float the lightweight resins while allowing the heavy resins to settle to the bottom in the separation sequence (Fig. 33.5). This headspace will be used to mix the resins after regeneration using air for agitation. The mixed bed is necessary for the best contact of the resins and the solution. Keeping them separate does not allow for the most efficient use of the resins.

Operational flow direction of the mixed-bed columns is down, to prevent separation of the resins during the exchange cycle. Bottom-to-top flow would suspend the resins and

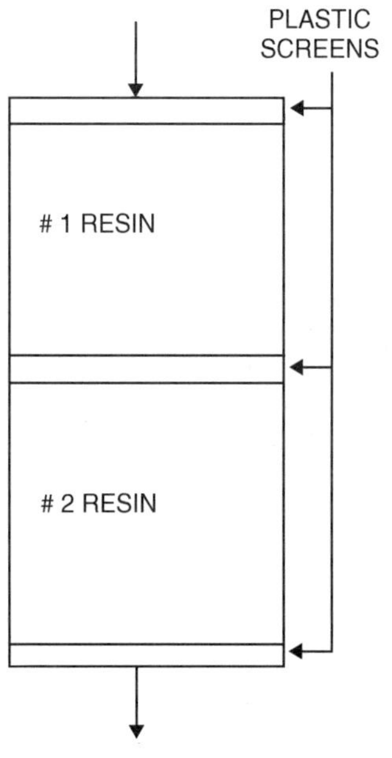

FIGURE 33.4 Separated resin column.

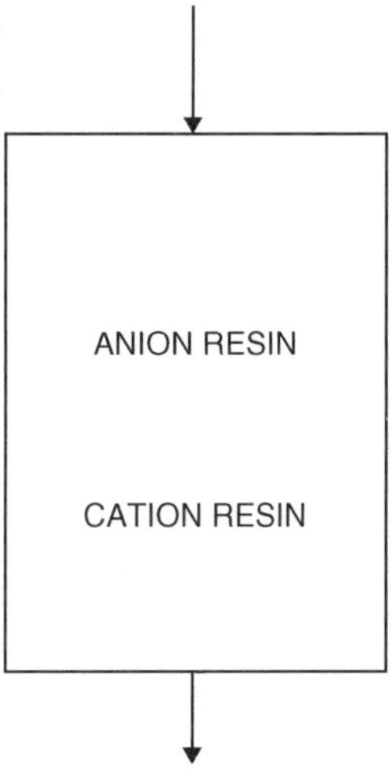

FIGURE 33.5 Mixed-bed column.

could cause separation and loss of resins. Fixed-bed columns can use flow in an up or down arrangement, as the chambers are protected from mixing and no headspace of significance is available.

33.4.3 Process Control Ion Exchange

Modern plant design for an ion-exchange process requires a central computer. An ion-exchange process requires numerous valves and pipes. All activities must be carried out in exactly the correct sequence and concentration. The computer can send signals and keep the process on track if feedback valve performance is part of the design. If this feature is not included, each sequence change must be verified to assure that regeneration, process production, and quality of product will be as expected.

The central computer is best designed with trouble messages and alarms to notify the operator when nonperformance has occurred. All lines and valves should be labeled to enable quick discovery and resolution of any problem. Adjustments should be made directly into the central computer. A sample schedule should be followed to monitor the performance of regeneration and production. These data can be used to evaluate the performance of the resins and enable improvements to be made in the process.

33.5 ION EXCHANGE IN SUGAR PRODUCTION

The use of ion exchange for the production of liquid sugar dates to the 1940s [4]. Ion-exchange resins are used for both color and ash removal (Fig. 33.6). Each ion resin is design specifically to accomplish its task and has its own special regeneration procedure. Columns of resins are normally used in conjunction with other purification methods, such as clarification, bone char, and granular carbon.

A recent refinery [5], utilizing raw sugar, ion exchange, and powdered carbon polishing decolorization proved to meet design criteria for color, ash, microbiological controls, and volume. The simple combination of melter, tanks, columns, and filters resembles a chemical

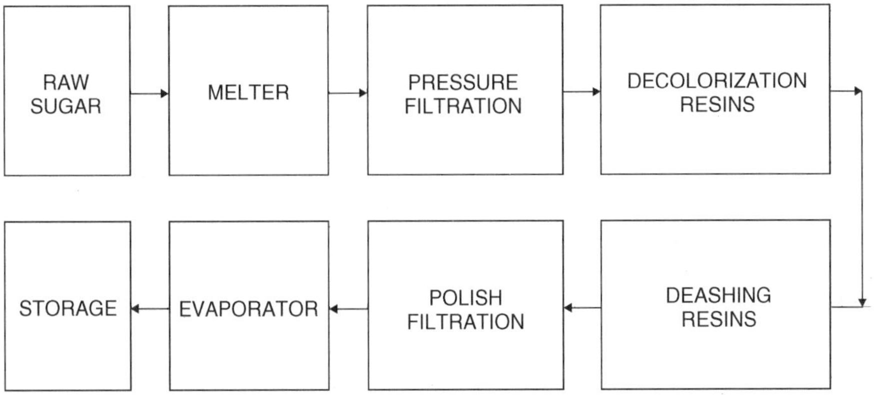

FIGURE 33.6 Flow diagram of an ion-exchange refinery.

plant more than a sugar plant with pans, centrifugals, mixers, granulators, and screens. Plant monitoring and control are directed from a panel that allows for control and monitoring of the entire process from a central location.

Regeneration of resins is performed on a volume or color schedule. Caustic brine solution is pumped onto the decolorizing resins to displace removed colorants and renew the resins for the next cycle. De-ashing resins are treated with brine, acid, and caustic in a controlled fashion, as this is a mixed-bed column and requires specific solutions for the separated resins to be regenerated safely.

33.6 MICROBIOLOGICAL CONTAMINATION

Liquid sugars are protected from microbiological contamination by a variety of methods: tight filtration, heat, pasteurizing, elevated density, and sterilization of equipment and tanks. Sterilization of equipment can be accomplished with chlorine, tamed iodine, hot water, and steam.

Storage tanks are constructed of stainless steel, aluminum, mild steel coated with interior protection, or fiberglass-reinforced plastics and are self-draining. The headspace above the syrup storage in the tanks is kept dry by blowing in heated, filtered, and ultraviolet (UV)-treated air to prevent condensation from forming on the top of the interior of the storage tank and dripping onto the surface of the liquid sugar (Fig. 33.7). In addition, UV lamps are mounted in the ceiling of the storage tanks to control microbial activity in the airspace above the syrup and on the surface of the syrup. These lamps retard the growth of mold on the surface of the syrup and on the walls of the tank.

Periodic cleaning of the tanks is required. They are emptied and cleaned from the interior manually with water hoses or through spray nozzles mounted inside the tank. The size of the tank will determine the number and location of the spray nozzles. Besides hot water, a chemical bactericide is used to ensure that all organisms are eliminated. Many systems operate on a bacteria, yeast, and mold test procedure to determine the time to wash the system or on a 6-month routine schedule. UV lamps have a limited life and are regularly changed on a yearly basis. Some light systems measure the amperes used by the lamps to determine the condition of the UV lights.

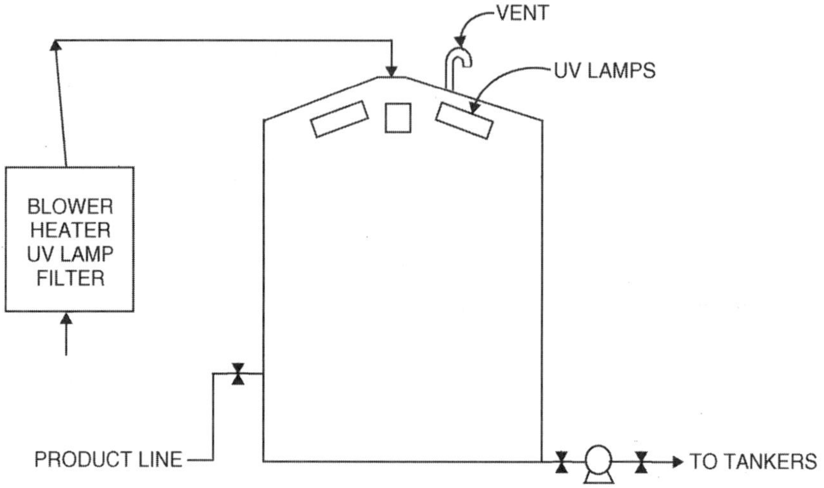

FIGURE 33.7 Typical storage tank.

Liquid storage tanks are equipped with level indicators to monitor the volume of product in the tank. These can be pressure indicators, floats, or sight glasses. A sanitary manway at the base of the tank allows access to the tank for cleaning and repair.

33.7 TRANSFER SYSTEM TO POINTS OF USE FROM STORAGE TANK

The transfer system is composed of a pump, piping, and valves. The pump is constructed of stainless steel, bronze-fitted iron body, or a corrosion-resistant metal or alloy. The pump usually has an internal relief valve or a pressure regulator to prevent damage due to high viscosity or blockage in the pipeline.

The pump may be controlled automatically by a microswitch on a meter, by a pressure switch on the top of an air trap standpipe in the discharge line, or by a pressure tank assembly located at the highest point in the system. The piping should be aluminum or stainless steel with take-apart design. A drain valve must be located at the lowest point of the piping system to enable complete drainage of the pipes. Valves should be stainless steel or nickel-plated $\frac{1}{4}$-turn butterfly design.

Liquid metering devices are usually of the wobble-disk type or the more accurate piston design. The meters are calibrated in either volume or weight measurement, as specified. These devices are automatic in operation and require only setting the desired volume or weight on the register or control device. It is always recommended that the critical endpoint be observed manually.

33.8 STORAGE STABILITY

Sugar syrups are considered to be perishable products and must be stored and handled very carefully. Contamination by color, sediment, or microbiological content will result in recycling of the entire storage system. Proper precautions and an understanding of the weakness of sugar syrups can enable prolonged storage without the loss of quality.

The main source of contamination is from yeast fermentation. Yeast organisms may be introduced into the syrup by carelessness, improper design, or unsanitary conditions.

Once contamination occurs, it is a matter of time until the numbers of organisms will exceed the product specification and a complete cleanout will be required. This involves complete evacuation of the entire sugar system, washing all pipes and tanks before sanitizing with soap and a sanitizing agent such as chlorine or tamed iodine.

It is not possible to predict the length of time that a sugar system will remain free of contamination. Important factors are the bacteriological condition of the syrup produced, pH of the syrup, temperature variations, sanitary features of the storage tank, and operator control. A well-designed and well-controlled sugar system should maintain quality for 6 months to 1 year.

Crystallization of sugar from liquid sugar syrups is not considered a spoilage problem but can cause pump and line stoppages. As long as the syrups are stored between 70 and 100°F, crystallization should not occur. It is when syrups are chilled below 60°F for extended periods of time that crystallization may begin to form. These crystals may grow to very large sizes and cause blockage in strainers or suction lines to pumps.

33.9 TRANSPORTATION

Sterilized shipping containers are a standard requirement. Liquid sugar can be shipped in drums, small containers, or tank truck loads. The most common carriers are 4000 gal tankers. The tankers are prepared for loading by washing with hot water to remove residual sugar from the last load. A specifically designed tank spray washer is lowered down inside the tanker. The spray nozzles are rotated by a small electric motor mounted on the washer. This rotation and high water pressure ensures that all surfaces and the ends of the tanker are well cleaned. Sterilization can be successful using very hot water, steam, or a good bactericide. Cleaning is meant to include the truck pump, hoses, seal caps, and gaskets.

Tankers are normally constructed of stainless steel and both double-conical and rear-discharge tanks are used. The delivery truck can be equipped with its own unloading pump or use a station-located unloading pump. Unloading rates are at the rate of 100 to 130 gal/min. Tankers are loaded through a closed system with a special vent to prevent contamination. In-line meters are used to control the amount of syrup pumped into the tanker. In some systems, in-line blending of syrups is possible and allows for flexibility in preparing loads. Syrup samples are normally taken by the truck driver at a designated sample point of each load of sugar delivered. For bacteriological samples, either heat or chemical sterilization is recommended. These samples are than analyzed by the laboratory to verify the quality of the sugar delivered.

Special care and sample containers are necessary to obtain contamination-free samples. Many factors can cause an incorrect laboratory result: wind, rain, unsanitary sampling points, and failure to take the sample quickly and close the lid without spilling syrup all over the lip of the container and cap. It is wise to remember that bacteria, yeasts, and molds are in the air all around us. They are looking for a home to start their new family. If you give them an opportunity, they will start it in the syrup you have and you will not even know it.

REFERENCES

1. R. W. Chalmers and L. A. Zemanek, Paper 257, *Proc. Sugar Ind. Technol. Conf.*, Vol. 24, 1965.
2. C. C. Chou and G. A. Jasovsky, Paper 630, *Proc. Sugar Ind. Technol. Conf.*, 1992.
3. D. E. Webster, Paper 430, *Proc. Sugar Ind. Technol. Conf.*, 1979.
4. R. Pou and J. Brunet, Paper 483, *Proc. Sugar Ind. Technol. Conf.*, 1982.
5. R. R. Tamaye, Paper 665, *Proc. Sugar Ind. Technol. Conf.*, 1994.

CHAPTER 34

Microcrystalline Sugar*

INTRODUCTION

The process of drying sugar via the transformation of a highly concentrated sugar syrup has been known for many years. Sugar produced in this manner consists of aggregates of microcrystalline sugars with various physical and functional properties that offer considerable commercial opportunities. In addition, with proper process control, the system can tailor-make various products to meet market needs via feed material formulations.

34.1 THEORY

A process operation for tailor-made products requires proper control of the rate of nucleation, crystallization, and thermal balance during the phase-change stage. The rate of nucleation in a perfectly pure sugar solution can be approximated by the general equation of reaction kinetics [1]:

$$\frac{dN}{dt} \times \frac{1}{v} = KnC^n$$

where the term $dN/dt \times 1/v$ represents the number of nuclei formed per unit time and per unit volume, C is the degree of supersaturation expressed in concentration units, and n is the reaction order. The rate of nucleation also depends on the impurities and/or ingredients added, the degree of agitation, and the presence of seed crystals.

The rate of nuclei formation increases very rapidly with increase in the degree of supersaturation or decreasing temperature within certain ranges for each. The rate drops sharply at a lower temperature, due to an increase in viscosity. The rate of crystal growth in terms of weight increase per unit surface, $dM/dt \times 1/S$, is also proportional to absolute supersaturation or with decreasing temperature; it then drops off, again due to the low diffusion rate at a higher viscosity. The peak maximum in the nucleation curve usually takes

*By Chung Chi Chou.

place at a lower temperature than that for the crystal growth curve, as shown in Figure 34.1. It should be noted that at a given degree of supersaturation, both the rate of nucleation and the rate of crystallization increase with increasing temperature.

For a given saturated sugar solution, the faster the rate of nucleation, the greater the number of nuclei formed and the smaller the size of the final sugar crystal. Von Weimarn's equation relates the final crystal diameter d to the degree of supersaturation C as follows:

$$\frac{1}{d} = K \times \frac{c}{C_0}$$

where K is a constant and C_0 is the saturation solubility at a given temperature.

Production of specialty sugars with various functionality is achieved partly via control of the retention time of the process conditions at the peak maximum in both nucleation and crystal growth curves. Another important design and operational variable for the process is the control of heat balance between the heat of crystallization and the latent heat of water evaporation during the transformation of liquid syrup to a solid sugar. The heat of crystallization at the elevated temperature can be estimated via conventional thermodynamic treatment of the heat of "reaction" at different temperature [2]. The latent heat of water evaporation can be derived from the boiling point data of a highly concentrated syrup at various pressures. Further details of the theoretical aspects of the process are described in U.S. Patents 3,194,682, 3, 365,331, and 3,642,535.

34.2 PROCESS DESCRIPTION

The essence of the process has been described in the patents. A patent issued July 13, 1965, to Tippens and Cohen describes the process for making fondant-size crystals of sucrose agglomerates with the flavor of brown sugar. The Miller–Cohen patent, issued January 23, 1968, provides further details of the transformation process. Specific conditions for sugar

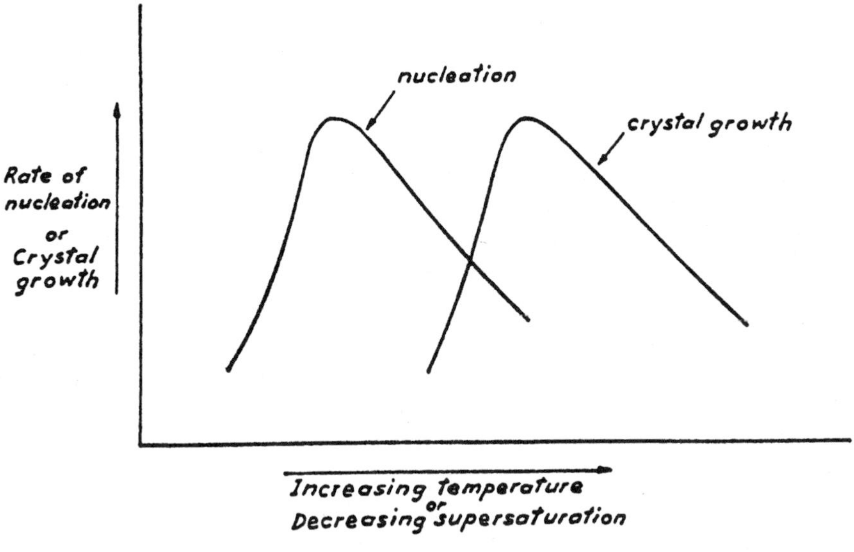

FIGURE 34.1 Rate of nucleation and crystallization as a function of temperature.

crystallization, use of air, retention time, and so on, were discussed. A third patent, issued to Graham, Fonti, and Martinez on February 15, 1972, describes a modified process for the production of a tabletting sugar. The same transformation process is also being used for the production of fondant sugar via control of process parameters and product formulation.

The process consists basically of an evaporator, nucleators/crystallizer, drier, mill, and screen. The evaporator is operated at a temperature in the range 120 to 130°C, with or without vacuum, to produce a concentrated syrup having a solids content of 91 to 97% by weight. Concentrated syrup flows from the evaporator to the vapor–liquid separator, where the vapor is removed by a barometric condenser and the syrup is fed to the nucleator/crystallizer. The retention time of sugar in this unit ranges from 10 to 120 s, depending on the product desired.

Air is introduced into the nucleator/crystallizer to remove vapor released by the crystallization of material therein due to the heat of crystallization. The airflow rate ranges from 9 to 40 ft^3/min per pound of sugar product per minute. If desired, a dual-temperature air system can be used during the nucleation and/or crystallization phases to further maximize the process flexibility and efficiency.

The sugar product in aggregate form, leaving the nucleator/crystallizer with a moisture content ranged from 0.5 to 2.5%, is fed to the drier for further drying if needed. The resulting dried aggregates are then supplied to a cooler, where they are reduced in temperature to about room temperature. The sugar aggregates are then milled and screened to meet product screen specifications. In general, the physical characteristics and functionality of the desired transformed products not only determine the selection of operating conditions, but also affect the design and therefore the capital and operating costs of the process.

34.3 PRODUCT CHARACTERISTICS AND APPLICATIONS

The transformation process produces products with the following improved characteristics: anticaking, antiseparation, compressibility, emulsification, flowability, fondant quality, rate of dissolution, and wettability. Because of this unique functionality, transformation process is used to produce the value-added sugar products discussed below.

34.3.1 Free-Flowing Brown Sugar

Regular soft (brown) sugar, described in Chapter 31, is made of sugar crystals coated with molasses film. These sugars become dry, hard, and lumpy, due to loss of moisture upon storage, particularly after the original package is broken, and the softness of the sugar makes it difficult to automate. The above-described transformation process is used widely to produce brown sugar with free-flowing and anticaking properties, even after a long period of storage. Both granulated and powdered sugar are available commercially. This form of brown sugar adapts very well in many new products of today's convenience-oriented market.

The physical structure of this sugar is unique in that each granule consists of many tiny crystals bound together in a porous spongelike structure. This structure is responsible for many of the functional properties in food application. In general, the product has the following functionalities: free-flowing, convenient carrier for flavors, rapid wet-out rate in cold water, ideal for automatic batching, and color and flavor enhancements. The free-flowing nature of this type of brown sugar makes it most suitable for automatic addition, mixing, and blending in the food industry.

When making free-flowing brown sugar, the additives of choice are those nonsucrose solid commonly found in regular soft sugar. These nonsucrose solids act first as a crystallization inhibitor and then perform their secondary role of providing a unique complementary flavor adjunct to certain classes of icings. A typical product is Brownulated, produced by the Domino Sugar Corporation.

34.3.2 Fondant Sugar

Fondant is a mixture of very fine sugar crystals of size below 44 μm (the grittiness threshold), surrounded by a saturated solution of sweetener components. The composition is largely sucrose, with the remainder being invert sugar and/or corn syrup. The mixture is used to prepare a base for smooth confections and icings.

Fondant has been used in the food, baking, and confectionary industries for a long time. In the beginning, the confection industry used it for candy cream centers. Production was an industry art and was known to but a few master confectioners. The process was laborious, slow, and required the utmost attention. The slightest deviation created an unsavory product. Traditionally, the crystalline sugar was combined with invert sugar and cooked slowly until a fine cream mess was formed. At best, only a few hundred pounds could be made at one time. Improved methods of cooling increased productivity; unfortunately, the manufacturing process was kept as a secret and never became public. As time passed, not only was the confection industry using fondant, but also the food and baking industries. Uniformity and consistency became a problem in mass food production. Therefore, it is essential to have a convenient-to-use icing or fondant in the industry for food application. To meet this demand, several dry fondant sugar products, developed by either the cocrystallization process or co-milling technology, are available in market today. A typical product is Amerfond, marketed by Domino Sugar.

A cream fondant as used in the candy and baking industry is a paste consisting of a suspension of very fine crystals. The size of the crystals is such that when the paste is ground between the teeth, there is no sensation of grit. In general, any particle of 40 μm or less will not produce this gritty feeling. All the specialty sugars produced by the process are made up of crystals that are below 40 μm in size and therefore qualify as fondant sugars. In fact, the majority of the crystals formed are in the range 5 to 10 μm.

These crystals as formed in our process are bound to one another partly by the moisture remaining in the product and partly by nonsucrose if present and form aggregates that are carefully screened and sized to perform their intended function more easily. The combination of small crystals and the airspaces between them as they form their individual aggregates makes these products ideal for direct compaction sugars.

Many specialty products are made from blends of liquid sucrose and specially selected additives. The additive has two distinct roles in the production of these items. The first role of the additives is to inhibit premature crystallization, thus allowing crystallization to take place where we want it. The second role played by these additives is to greatly enhance the functionality of the end products.

In the case of fondant used extensively in the candy and baking industry, the additive used is invert sugar. The finished products contain 5% invert on a dry basis. Invert sugar functions as a humectant that is essential in cream centers and icings. Also in the candy industry, cream centers contain an amount of invert that makes the fondant pliable enough to shape but still firm enough to extrude. In some applications, the user adds invertase to the fondant so that some time after the creams have been coated, more invert will be formed and the creams will become more palatable. In the baking industry, fondant is used in the preparation of high-quality icings. As mentioned, it imparts humectant qualities to the icings, but at the same time, it produces a highly desirable sheen or luster to the finished products.

Bear in mind that the traditional manner for making fondants remains long and complicated and is more an art than a science. Using fondant sugar, the production process becomes the simple addition of water to either one of the products and mixing for 5 min. The resulting fondants are easily reproducible and of excellent quality. Since the invert acts as a humectant, it becomes apparent that these two products are hygroscopic to some degree, and for this reason we do not recommend them to the pharmaceutical trade as a tabletting excipient.

34.3.3 Direct Compacting and Tableting Sugars

The products are widely used in the pharmaceutical industry as a direct compaction sugar. By using the products, the user eliminates the conventional, lengthy process of wet granulation or slugging in the making of tablets. An active ingredient such as a vitamin mix can be added to the sugar with the necessary amount of lubricant, blended for the appropriate number of minutes, and the dry blends fed to the tabletting press. Because of the dextrin additive that is used in the initial liquid sucrose–additive blend, the resulting tablets possess excellent hardness. Of equal importance is the fact that the tablets are very nonhygroscopic, which adds greatly to the stability of the active ingredient. It is not recommended to use this product as a fondant sugar because the dextrin delays the wet-out time of the fondant preparation.

A typical example in this category is Di-Pac (a Domino product). Di-Pac tabletting excipient is the cocrystallization of 97% sucrose and 3% maltodextrins. It is directly compressible into tablets or flakes. It is also readily adaptable to most formulations and meets UPS specifications. The particle size of this product is 3% maximum on U.S. No. 40 and 75% minimum on U.S. No. 100. The bulk density is about 40 lb/ft^3. This direct compacting sugar provides such functionalities as excellent flowability and compressibility, nonhygroscopic, maintaining desired hardness through shelf life, rapid solubility, self-lubricating quality, nonreactive with most medical substances, and providing natural sweetness. It is suggested that this product be used in vitamin and natural food tablets.

34.3.4 Natural Sugar Granules

Molasses, honey, and maple syrup are considered as healthful sweeteners among the general population. These products are enjoying a resurgence in popularity seldom seen in a sweetener. The reasons for renewed interest include the current boom in "natural" and "health" foods. There is also a desire to substitute the natural sweeteners with "great food value" in place of highly refined cane sugar. However, molasses, honey, and maple syrup in original heavy-bodied syrup form have inherent disadvantages in handling and storage. They are very difficult to incorporate with other ingredients in a dry mix, and hence the extensive use of these ingredients in fabricated food products is restricted. For these reasons, Domino sugar Corporation developed natural sugar granules under the trade name Qwik-Flow. These products are made by combining honey, molasses, or maple syrup with an in-process stream of syrup from the cane sugar refinery by the cocrystallization process. The products generally (1) are dry, granulated, free-flowing, and in a noncaking form; (2) are instantly soluble in a liquid system; (3) are composed of 100% natural ingredients; and (4) have multifunctional properties in food applications:

a. Molasses Granules. Molasses granules, a free-flowing dry molasses product, is manufactured by the cocrystallization process whereby edible molasses is incorporated into the sucrose matrix. The product is a noncaking aggregate with a rich molasses flavor and color and the sweetness of refinery syrup. The typical composition is 91% sucrose, 2.5% invert, 3.0% ash, and 0.75% moisture. The bulk density is about 40 lb/ft^3 and the product is shelf stable, free-flowing, noncaking, instantly soluble, and an excellent carrier for homogeneous blending. The suggested applications for this product are in the areas of cake mixes, cookie mixes, cream centers, gravies, sauces, cereals, bread mixes, glazes, candy bars, and seasoning blends.

b. Honey Granules. Honey granules are manufactured by a cocrystallization process whereby a natural, liquid honey was incorporated with selected in-process refinery stream liquor. The product is free-flowing, noncaking, and in granule form. It provides not only sweetness but also a unique honey flavor and taste. It also has the advantages of being an "instantly" soluble property in solution and can be used as a table-top sweetener and as a food ingredient in ready-mix formulation. The typical composition is 92% sucrose, 7% invert sugar, 1.0% ash, and 1.2% moisture. The bulk density and pH of this product is 40 lb/ft^3 and 6.0, respectively. Suggested uses for this product are in the areas of cookie mixes, bread mixes, cake mixes, glazes, cream centers, candy bars, cereals, seasoning blends, variety bread, ice cream, and yogurt topping.

c. Maple Granules. In 1979, Chen et al. invented a method for transforming a pure maple syrup to free-flowing natural granules [3]. Since pure maple syrup is an exotic item and the cost of this material is extremely high, general food uses of this product in its original syrup form is limited. For this reason, a dry maple product with significantly reduced cost was developed by the cocrystallization process by Domino Sugar Corporation. In this process, pure maple syrup was incorporated into a selected refinery sugar stream and a natural flavor was used to enhance flavor of this product. The product is noncaking, free-flowing, with a rich maple flavor and instantly soluble property. The composition of this product is 93% sucrose, 4% invert sugar, 0.5% ash, and 0.5% moisture. Suggested applications are in the areas of table syrup and sweeteners, cake mixes, cookie mixes, bread mixes, fondants and icings, ham and meat products, cereal mixes, seasoning blends, candy bars and tablets, and ice cream toppings.

34.3.5 Amorphous (Transformed) Sugar

A significant quantity of the white sugar consumed in Brazil is amorphous sugar, *amorphous* meaning "without shape." It is really a microcrystalline or fondant sugar and therefore dissolves easily. In amorphous sugar production, 100% yield is achieved with no centrifugal separation of syrup. The very low color liquor is concentrated in an open pan under atmospheric pressure to a Brix of over 92. The syrup is discharged into a beater with arms rotating at 35 to 40 rpm and is transformed into a mass of agglomerated grains with about 2% moisture at 98 to 100°C. It is then dried and cooled by a drum-type drier to a moisture content of 0.1% and about 50 to 55°C. Since there is no separation of mother liquor (syrup), the product from Brazil has a polarization of only about 99.5, ash of 0.08 to 0.15%, and invert 0.05 to 0.25%. In addition to 100% yield, the amorphous sugar production is claimed to be much more energy efficient.

34.3.6 Dry Fondant Sugar by Micropulverization

Dry fondant sugar is manufactured from selected sugar blended with low-dry equivalent (DE) cereal solids. It is then comingled to produce microscopic particles of fondant quality. The composition of this product is 97.5% sucrose, 2.5% low-DE cereal solids, and 0.5% moisture, and the granulation is 99% through U.S. No. 325. The fondant candies can be prepared by the cold process, simply mixing this fondant sugar with liquids. This one-step noncooking eliminates the time-consuming heating method of producing candy cream centers. The product is nongritty and has an excellent extrusion quality. Suggested uses of this product are in the areas of cream centers, mints/wafers, Easter eggs, nut confections, fudges, and frappés.

34.3.7 Ready-to-Use Icing Sugar

In the baking industry, fondant is used as a basis for all high-quality frosting, glazes, and icings. The baker uses fondant sugar to eliminate all perception of grittiness normally associated with confectioner-grade sugar. The grittiness in confectioner's sugar is due to the large particle size, greater than 44 μm. It is commonly accepted that sugar smaller than 44 μm in size is considered as icing sugar. Set&Match manufactured by Domino Sugar Corporation is one of these sugars. This icing sugar is prepared by co-milling a sugar with a low-DE cereal solid. The composition of this product is 89% sugar and 11% low-DE cereal solids. The particle size is 99% through U.S. No. 200 and 98% through U.S. No. 325.

Because of the inclusion of low-DE material, this icing sugar wets out faster, retains and holds more moisture, and will not separate or settle out upon storage as do other icing sugars. This product makes smooth creamy-texture icing and high-quality doughnut glaze with superior adherence and sheen. This suggested application is in the areas of butter cream icing, fruit and nut icing, fudge icing, bakery glazes, pour icings, and doughnut dustings.

REFERENCES

1. A. E. Nielsen, *Kinetics of Precipitation,* Pergamon Press, Oxford, 1964.
2. F. T. Wall, *Chemical Thermodynamics,* 3rd ed., W.H. Freeman, San Francisco, 1974.
3. A. Chen, U.S. patent No. 4,159,210, 1979.

PART V

Chemistry of Sugar Refining

CHAPTER 35

Refining Quality of Raw Sugar*

INTRODUCTION

Sucrose is produced at high levels in certain plants (in particular, sugarcane and sugar beet), and the purpose of the sugar industry is to recover this sucrose for human use, primarily as food. This processing needs to be as efficient as possible, and produce sugar of high quality. For the cane industry, this process usually involves two stages: the production of raw sugar at locations in the cane-growing areas and the refining of this raw sugar, often at a location far away from the growing areas and near major population centers. Raw sugar is a major commodity involving much international trade, and its quality is a critical factor in determining its value and the performance of the refineries.

The purpose of this chapter is to describe the quality characteristics of the product of the raw sugar mill, their impact on refinery performance, and how they are determined by the operation of the raw sugar factory. Many variables are involved, and although general conclusions can be drawn about cause and effect in poor raw sugar quality, this discussion must be descriptive rather than prescriptive. In this chapter we deal with technological rather than commercial factors, and the financial details of raw sugar contracts are dealt with elsewhere in this book. Also not included is discussion of the production of direct-consumption raw sugars or white sugars such as blanco directo. The technology employed for the production of high-quality raw sugars for refining can also be applied to these materials.

The primary objective of a raw sugar factory is to maximize the recovery of raw sugar from cane. The expectations and payments of the farmers are tied to sugar yield. Given the constraints of weather and other crop conditions and the perishability of cane, sugar quality requirements often become of secondary importance, except when a penalty is involved. Raw sugar quality is determined overwhelmingly by the raw sugar producer, and in the last section of this chapter we outline the raw sugar operations related to raw sugar quality. Since knowledge of the magnitude of any penalty often comes much too late for remedial action to be taken, raw sugar mills are often flying blind with regard to raw sugar quality. In the past, the sole aim of sugarcane factories in most cane sugar–producing countries was

* By Stephen J. Clarke.

to produce raw sugar for export, often tied to a specific market. For this type of operation, simple production technology was adequate. Partly as a result of the history and structure of the industry (raw mills separated from refineries) and the remote location of many raw sugar mills, the technology of raw sugar production is minimalized: Use only the minimum technology required to produce an adequate quality of raw sugar at the highest yield. The drive for improved sugar quality cannot be at the expense of significantly reduced recovery or higher production costs. It is very important that raw sugar producers be familiar with refining operations in general, and in particular with the refinery to which the sugar is sent, if this is possible. The improvement in raw sugar quality is a consequence both of refiners setting higher standards and the availability of improved process technology for raw sugar production (e.g., use of polymeric flocculants, good pH control, and high-performance centrifugals).

Many changes have taken place over the last three decades in the trading patterns for sugar, with both the quantities and sources of sugar changing [1,2]. In part this is due to increased beet sugar production, higher energy costs, and the development of alternative sweeteners. Refining companies are now in more competitive situations, with increased demand for high-quality refined sugar leading to a preference on their part for good-quality raw sugar.

Sugar refining is essentially the recrystallization of raw sugar after suitable treatment to remove color, suspended solids, ash, and other undesirable materials. Refining operations must also include the means to recover and recycle sugar from mother liquors which cannot be crystallized to give acceptable white sugar. Higher-quality raw sugars put less demand on the recovery station and therefore reduce refinery operating costs.

Raw cane sugar for refining may be considered in three categories: (1) that refined on site with the raw sugar house being used for recovery, (2) that refined in an operation which is tied to a group of raw mills, and (3) that which may go to a number of refineries which have different processes and requirements. Refiners in the first two categories have the advantage of being able to operate with a relatively limited range of raw sugar and to maintain some control over its quality. Refiners in the last category must have the capability to handle raw sugar of widely varying quality and characteristics. Most of the discussion in this chapter is related to this situation.

Knowledge of the composition of raw sugar is essential in optimizing the refinery operation. Over the last three decades the consequences to the refinery of varying raw sugar quality has received considerable attention. Much detailed study has been applied to this matter, but until recently, relatively little to the relation of these standards to the operation of a raw sugar factory. Generalized recommendations have been made for improving raw sugar quality, but except in a few cases, these have not been followed through. Intensive work is continuing on analyzing the relationships between raw sugar quality and refinery operations, especially for color, polysaccharides, and ash.

There are obviously relationships between the quality of raw sugar entering a refinery and the difficulties encountered by the refinery. The importance of the chemical and physical characteristics of raw sugars (other than the polarization, a measure of sucrose content) has been increasingly recognized. Depending on the type of refinery, some components are more difficult to remove. Different refinery clarification and decolorization stages have different efficiencies for removal of nonsugar components. These factors are dealt with in detail in a later section. Although some quality factors such as dextran and suspended solids are important in all refinery operations; others, such as invert, will be optimum at different levels depending on the refinery product mix. Soft brown sugar production will require a higher level of nonsucrose in the refinery raw material. It is preferable to consider the quality of the raw sugar entering the process rather than the quality of sugar as produced by the mill or received by the refinery. The storage and transportation characteristics of raw sugar are a significant factor, especially for material transported over long distances, often between climatically different areas and/or stored for long periods.

A refinery has an optimum throughput determined by its capacity to handle impurities. Most refineries are forced by economic considerations to operate well above their original design throughputs. Technological changes are in part the reason, but improved raw sugar quality is a significant factor. Under these conditions refiners seek to process raw sugar that allows them to operate at maximum throughput. Even minor differences in raw sugar quality can cause major upsets in performance and throughput. A delightful description of problems in refining damaged raw sugars is to be found in a paper by Somner [3].

If the designed capacity for impurities is exceeded, the consequences will be either reduced refinery throughput or a reduction in white sugar quality, with the latter being unacceptable. With poor sugar there will be less sugar recovered for a given consumption of energy and process materials. If the refinery is limited in throughput, output will suffer. More serious is the loss due to bottlenecking in the recovery system. Lost refinery sales as a consequence swamp the slight raw material price advantage that may be given by the poor raw. Similarly, premiums may be hard to justify.

The No. 14 raw sugar contract, instituted in 1966 and subsequently revised, was the first concerted effort in the U.S. and other markets to relate raw sugar quality to its refinability. This means that the quality parameters have a clear association with refinery operations. Prior to this, some individual refiners had their own standards (including penalties and premiums), and these may still be in place for the sale of domestic raw sugar.

Customer specifications for white sugar quality are becoming more stringent, leading to refiners considering more severe standards for raw sugar. This is particularly the case where much of the refined sugar goes to industrial users. In less developed circumstances the customer may complete the purification, for example by deionization or decolorization.

Decisions made as to raw sugar quality standards (including penalties and premiums) should have a sound technological base. These may include the additional operating costs to the refinery, white sugar quality factors, and environmental concerns related to the process waste generated by the refinery.

Quality standards continue to evolve, and the improved quality of raw sugar as now delivered has become the norm by which its quality is judged. Raw sugar has become more standard, in part due to improved technology in the raw mill, but very poor sugar still arrives at the refinery and can cause major difficulties. Another development is the deliberate application of systems to produce a standard premium-quality raw sugar, with this product being designed to meet specific market requirements.

In this chapter we describe in some detail the rationale behind raw sugar quality standards and the steps that can be taken by the raw mill to meet these standards. Raw sugar producers need feedback on the results obtained for their quality tests. Suppliers should be made aware of the strong and weak points by regular review of their data, as judged against raw sugar quality standards. The quality standards would be more effective if better communications existed between the seller and buyers. Increased emphasis on only one raw sugar quality parameter almost inevitably leads to improvement in the overall quality since increased vigilance is required in the all raw sugar operations.

35.1 SOURCES OF INFORMATION

Although all sugar refining companies have their internal data banks on raw sugar quality, there is much information publicly available, and these have been invaluable sources for the data and discussion presented here. These sources fall into several groups:

1. Reviews and specialized publications that are intermittently produced
   a. Raw cane sugar quality in relation to refining requirements, *Sugar Technology Reviews,* J. A. Watson and W. M. Nicol, Vol. 3, No. 2, 1975. This is a comprehensive review, including the historical development of raw sugar quality standards.
   b. Raw sugar quality and white sugar quality, *Proceedings of the Sugar Processing Research Institute* (SPRI) *Workshops,* edited by M. A. Clarke, 1990. This is a very useful compilation of reports and experiences of a wide range of sugar producers and refiners.
2. Regular proceedings specializing in refinery operations and analytical methods
   a. *Sugar Industry Technologists* (*SIT*). The symposia held in 1979 and 1993, documented in the SIT proceedings, are very informative and give a range of perspectives on this subject.
   b. *Sugar Processing Research Institute* (*SPRI*) *Proceedings.* Much of the recent work on the chemical composition and impact of the various impurities in raw sugar has involved this group, and their reports are a very useful source of information.
3. Proceedings of the congresses of the *International Society of Sugar Cane Technologists* (ISSCT). In some cases papers on raw sugar quality have been grouped together as a symposium (e.g., in 1968).
4. Sugar industry journals in general
   a. *International Sugar Journal*
   b. *Sugar Journal*
   c. *Sugar y Azucar*
   d. *Zuckerindustrie*
   e. Journals and proceedings of national and regional groups of technologists
5. The International Commission for Uniform Methods of Sugar Analysis (ICUMSA) is a comprehensive source of information on the adequacy of analytical procedures. Until 1986 "The refining quality of raw cane sugar" was a subject in its own right, dealing with what these items are and how they should be measured, but not so much with the interpretation of the data since this would vary between refineries. This has since been dispersed among the other subjects.
6. Books related to raw and refinery operations
   a. *Principles of Sugar Technology,* by P. Honig
   b. *Manufacture and Refining of Raw Cane Sugar,* by Baikow
   c. *Chemistry and Processing of Sugarbeet and Sugarcane,* edited by M. A. Clarke and M. A. Godshall
   d. Various editions of the *Cane Sugar Handbook,* by J. C. P. Chen and C. C. Chou

35.2 QUALITY FACTORS AND THEIR IMPACT ON REFINERY OPERATIONS

It is important that both raw sugar producers and refiners appreciate the impact of the quality of their product on the performance of the refinery. To do this adequately, it is necessary to be familiar with the process stages of the refinery and also the quality factors related to them. The following operations involved in refining bulk raw sugar:

- Storage and transportation
- Affination and melting
- Clarification and filtration
- Decolorization
- Crystallization and centrifugation

- Drying, conditioning, and packaging

Other aspects of refining that must be considered are:

- Sampling and analysis
- White sugar quality
- Production of soft brown sugars, etc.
- Recovery operations
- Energy demands
- Environmental factors
- Staffing requirements

The efficiencies of each of the stages are influenced in different ways by the quality of the raw sugar. Each of the impurities in raw sugar (which may be grouped as ash, invert, color, suspended solids, polysaccharides, and other organic compounds) have different impacts on refineries with different process systems (e.g., clarification by carbonation or phosphatation and decolorization with bone char, activated carbon, and/or ion-exchange resin). Distinction should be made between impurities that cause refinery process problems (e.g., polysaccharides and color) and those that yield refined sugar with undesirable components (e.g., heavy metals and pesticides). The latter do not impede refinery operations but can lead to quality problems with the refined sugar.

Depending on the process facilities available, each refiner would like to receive raw sugar with certain quality characteristics. Most of the critical quality parameters are related to the affined or washed sugar since this is the material that enters directly into the refined sugar production process. Raw sugars vary considerably, but the appearance is usually a poor guide to the quality of melt liquor produced from the affined sugar. Each refiner would like to see certain affined sugar characteristics, (e.g., for the Thames refinery, color of <1200 ICU; ash, <0.1%, invert, <0.8%; and starch, <250 ppm) [4]. These goals for sugar quality may be generalized as follows:

- Characteristics consistent enough for uniformity of year-round operation so as to require minimum adjustments to process control and consistent production of refined sugar of the required quality
- Good and predictable filterability
- Polarization sufficiently high to minimize the nonsugar load being processed in the recovery system and to maintain or exceed the designed refinery capacity

For many years there have been such quality factors and the list has grown and will keep on growing as white sugar users have become more critical in their specifications. These generalizations need to be transformed into specific quantifiable technical data, and in their review, Watson and Nicol [5] list 17 factors;

1. Polarization
2. Moisture content
3. Safety factor (or dilution indicator)
4. Total ash
5. Sulfate content
6. Polyvalent anions
7. Color, particularly color that persists into refined sugar
8. Affinability

9. Filterability
10. Starch content
11. Polysaccharide content (other than starch)
12. Insoluble foreign matter
13. Bacterial count
14. Trace elements
15. Insecticides and herbicides
16. Silica
17. Floc-forming constituents

This is a comprehensive list of what the refiner would like to know but is impractical due to the difficulties in analytical procedures, limited manning, and determining acceptable ranges for all the properties mentioned. However, it is a useful guide for improving raw sugar quality, even if the particular item is not part of a contract. Watson and Nicol [5] comment on the list as follows:

1. The effect of the factor must be measurable.
2. It must be possible to translate the effects of the factors into financial terms.
3. The test used to measure the factor must be standard and reproducible.
4. The factor is preferably one that will apply to all refineries and not to one only.
5. The factor must be one over which the raw producer has some control.
6. The cost (and time) of carrying out the tests must not be excessive.
7. The choice of tests should bear in mind the relationships between properties, such as between polarization and ash, filterability and polysaccharides, and so on.

Difficulties arise in that a normal control laboratory may not be able to perform all the tests desired on a raw sugar to assess its refinability completely. The refiner may need to process the sugar immediately upon receipt and before tests can be made. Retroactive tests measure whether the sugar met specifications but are of no value for indicating modification of refinery operation or the necessity of blending sugars. Quality data used to set a value on the sugar need to be reproducible and generally applicable but would not be as comprehensive as those used for planning refinery operations. Refineries are more concerned with obtaining a steady supply of standard-quality sugar that can be refined with minimum difficulty than in purchasing less favorable raws at low prices. If the raw sugar producer just accepts the penalties and makes no effort to improve quality, the exercise has failed.

Refinery-related research has been much involved in developing tests and assessing their reproducibility and application in terms of refinery performance. Improvements in analytical procedures have had a significant impact on these quality factors and in some cases have led to replacement of physical characteristics by analysis of the material primarily responsible (e.g., analysis for individual polysaccharides rather than filterability). Many of these components are at very low levels where useful data require good analysts and equipment and involve methods with considerable inherent variation.

The current standard raw sugar contracts include the factors listed in Table 35.1, the limits indicating values at which penalties are implemented. In part the revision of these raw sugar quality parameters was based on statistical analysis of the data for a wide range of raw sugars. The highest difficulty in meeting the standards has been for polarization and color. The magnitudes of the penalties vary among refining companies. The analytical procedures involved are specified in the contracts and are given in Chapter 37. Methods have varied for measurement of dextran, but the haze test is now the standard procedure. Affination involves removal of the molasses coating on the raw sugar using a sucrose syrup and

TABLE 35.1 Raw Sugar Quality Limits

Factor	Limit(s)
Measured on Whole Raw Sugar	
Polarization	96.00
(values depend on contract)	97.00
Moisture (safety factor)	0.30
Ash (ash % raw sugar vs. standard ash; standard ash = factor × total nonsugar solids)	0.25
Color (ICUMSA 1978; pH 8.5; 420 nm)	6000
Dextran (MAU by haze test)	250
Measured on Affined Raw Sugar	
Grain size (% through U.S. No. 30 or Tyler No. 28 mesh; high number indicates lower quality)	
Color (ICUMSA 1978; pH 8.5; 420 nm)	1500

is parallel to the process in the refinery. This is necessary for grain size determination since the molasses film has to be removed to obtain free-flowing crystals for the sieve analysis. Also, the level of color removal by affination is a good measure of the refining characteristics of the sugar.

Not included in these contracts are considerations of invert level, filterability, starch, and insoluble solids. Future contracts may well include a provision for the starch content of raw sugar. There is overlap between these and the factors already in the contract, and all are important in assessing the raw sugar quality. The temperature of the raw sugar may be a quality factor for sugar delivered straight from processing at the raw mill to the refinery.

The following text deals with each stage of the conversion of raw sugar to refined sugar, outlining the effects of raw sugar quality on each. Many of these quality factors overlap (e.g., a well-purged uniform grain will store well and affine well).

The increased cost of energy has had a significant impact on refinery operations and was the subject of a Sugar Industry Technologists symposium in 1997. Raw sugar quality can significantly impact the energy costs of refining by requiring operation of filters at lower than optimum Brix value by increasing the recycle of material in process and/or by reducing the refining capacity.

35.2.1 Raw Value and Recovery

The first parameter that is looked at in raw sugar is its polarization, and this is the basis for its price. However, in many refineries, the impurities in the affined sugar are as important. These impurities can slow down the melt rate if not removed or transformed, either physically or chemically. Parameters of importance are determined by local circumstances. For example, a refiner receiving raws from a small group of producers whose raw sugar is routinely high in polarization and low in ash can emphasize the development of efficient decolorization technology in the refinery. Refiners taking raw sugars from many sources need to maintain the ability to handle a variety of quality conditions.

The value of raw sugar for tariff and quota purposes is calculated as the *raw value* (RV), which is used to normalize the sugar quality (as 96 polarization sugar) in terms of the quantity of refined (100 polarization) sugar that can be produced from 100 lb of the raw sugar. One formula for the raw value is

$$\text{RV} = (\text{polarization} - 92) \times 1.75 + 93$$

Thus 100 lb of 98.0 polarization raw sugar has a RV of 103.5, meaning that 100 lb of 98 polarization sugar would be expected to yield the same quantity of refined sugar as 103.5 lb of 96 polarization sugar. The assumption is that 107 lb of 96 polarization sugar will produce 100 lb of refined sugar; this is a useful generalization, but actual values can vary quite widely.

The extent of recovery operations is determined primarily by the raw sugar polarization. The impurities in the molasses film will affect primarily the degree of exhaustion achieved in the recovery house. In many refineries this is a secondary consideration and the refinery will not be slowed down to give better molasses exhaustion. Whole raw color is a concern primarily because of the return of high-color remelt from the recovery house.

The actual recovery of refined sugar is determined by many factors, including the refinery efficiency, with the composition and quantity of nonsucrose playing an important role. Most refiners use some relationship between invert and ash and possibly total nonsugars to give a quality value to the sugar. There is a wide range of such relationships, often developed to assess the performance of an individual refinery.

The reducing sugar/ash ratio is commonly used in statistical relationships to predict the purity of final molasses and therefore its quantity. There are several such relationships that have only decent validity under the conditions where they were developed. They do not take into account other factors that limit molasses exhaustibility, notably viscosity increases caused by dextran and other polysaccharides. Part of the reason for the use of invert and ash in these formulas is that they are easily measured and many data are available; the scientific justification is somewhat tenuous. Typical statistical relationships are:

$$\text{expected molasses true purity} = 39.9 - 19.6 \log_{10} (\text{RS/ash}) \qquad \text{South Africa}$$

$$\text{expected molasses true purity} = 42.5 - 12.5 \log_{10} (\text{RS/ash}) \qquad \text{Louisiana}$$

The differences between the equations reflect both variations in molasses quality and processing conditions.

The ratio of invert to ash plus organic nonsugars is also used in formulas for calculation of total sugars in "ideal molasses." The assumption is made that there is no change in the nonsucrose constituents between raw sugars and molasses. If

$$g = \frac{\text{invert}}{\text{ash} + \text{organic nonsugars}}$$

then total sugars in molasses is predicted to be

$$100 \times \frac{(5 + 3g)/3}{3 + g}$$

and total sucrose in molasses is predicted to be

$$100 \times \frac{(5 - g)/3}{3 + g}$$

The ash/total nonsugars ratio is used in contract No. 14 and is intended to predict problems related to scale formation and the level of molasses produced. Although useful, this is an oversimplification, since much of the effect of the nonsugars depends on the levels of individual components. These, especially the ash, vary with cane-producing area, irrigation, raw processing conditions, and other factors and are discussed in a later section. High sulfate levels may cause problems of scaling and loss of efficiency on bone char systems but not excessive molasses production. With varying refinery conditions it is obvious that a nonsugar constituent that is important to one refinery may be of little significance at another.

Various recovery formulas are used in different countries for prediction of refined sugar yield and again involve the levels of reducing sugar and ash. A typical example is

$$\text{yield} = \text{polarization} - (5 \times \text{ash}) - (1 \times \text{invert})$$

Nonsucrose carbohydrates from monosaccharides to polysaccharides can affect the measured polarization of the raw sugar. Invert levels with a glucose/fructose ratio of less than 1.75:1 depress the polarization and with a higher ratio will enhance the polarization. Dextrans are highly dextrorotatory with a specific rotation approximately three times higher than sucrose. Whether this is seen in the measured polarization depends on the clarification method used. Lead salts precipitate most of the dextran but other methods lose most, if not all, of the dextran in solution. Increased polarization values of 0.3° have been measured with sugars containing 100 ppm dextran [6, pp. 223–224].

35.2.2 Affination

The purpose of affination is to remove as completely as possible the impurities in the molasses film surrounding the raw sugar crystal. In good-quality raw sugar this film contains the bulk of the impurities (>80%) and is removed easily. This is the unit operation that should remove by far the greatest proportion of impurities. It becomes uneconomical to remove the impurities at a later stage in process. Any sugar that does not affine well is therefore a problem for the refinery. Refiners need to buy the minimum amount of color and to send the minimum color through the white sugar process.

Quality characteristics that adversely affect the affination are (1) the quality of the grain, (2) the level of impurities in the whole raw, and (3) the presence of insoluble impurities such as bagasse or field soil. The most important is the quality of the raw sugar grain. If there are many fine crystals or damaged grain, they cause blinding in the centrifugals and reduced throughput and impurity removal. Similar problems can arise from the mixing of raws, for example, from several mills shipping out through a central warehouse. Fine insoluble solids have a similar effect in reducing the centrifugal efficiency.

Small lumps and conglomerated crystals do not affine well, and the latter are a major cause of poor affination efficiency. Distinction should be made between materials incorporated into regular crystals and those occluded in the crystal or trapped in conglomerates. Some impurities, such as dextran and some color, are included in the crystal as it grows and are not removed by affination. Affination does not remove all nonsugars to the same extent. Analytical data for raw sugar before and after affination are an important guide to the quality of the raw sugar. The polarization of conglomerated raw sugar of the same mesh size as good grain was significantly lower than for the good grain, and the conglomerated sugar was also higher in polysaccharides. Two raw sugars gave the data shown in Table 35.2. Sugar A meets specifications on both whole and affined color, but sugar B is affined more efficiently as far as color removal is concerned.

Irregular fine grain raws that purge poorly in the affination station often produce white liquors that crystallize slowly and purge poorly in the white sugar centrifugals. Typical examples of this are raw sugars that have been produced from deteriorated cane and contain high levels of polysaccharides. The recirculation of these impurities in the refinery to the affination syrup can further hinder the performance of this operation.

The physical state of the nonsugars in the molasses film also plays a role in the efficiency of affination. Precipitated and insoluble materials such as calcium sulfate or syngenite (mixed salt of calcium and potassium sulfate) which are dried onto the surface of the crystal are more difficult to remove.

One refinery's practice is to attempt to reduce the ash and invert below 0.1% at affination. For each consignment of raw sugar, the optimum spin and wash times are determined and this system is also used to determine whether blending of the raw sugar is desirable. In this context it was observed that both invert and ash in raw sugar decreased but their ratio hardly changed. If ash values in melted sugar rose from 0.15% to 0.45%, there could be at least a 10% reduction in refinery capacity. High ash increased the frequency of boilout of vacuum pans, especially those used for lower-purity strikes.

35.2.3 Clarification and Filtration

These operations are intended primarily to remove suspended solids and some color from the liquor. The two standard processes are carbonation and phosphatation, although some refiners use only lime and filter aid, perhaps with some added carbon. Whether a refinery employs pressure filtration with inert filter aids, phosphoric acid/lime defecation or carbonation, a raw sugar that processes well in one system generally works well in the others. Raw sugar quality factors that affect this operation adversely are principally those related to the filterability of the liquor. There is a considerable body of work on filterability and tests for it, and much of this has been published in the Sugar Industry Technologists proceedings; relevant publications from this source and others are given [7,8].

The bulk separation in phosphatation processes (including the Talofloc system) involves floatation and usually only a polishing filtration step. Filterability is therefore less important to phosphatation refineries and has decreased in importance as more refineries have installed phosphatation systems. Significant correlations have been found for filterability with turbidity and silica content and principally with the levels of polysaccharides in the raw sugar, partially as a consequence of their effect on liquor viscosity. Filterability was a major topic at the ISSCT Congress in 1968. Several filterability tests have been developed, but there are problems with reproducibility and applicability. Even at low concentrations, phosphate, sil-

TABLE 35.2 Affination Analysis (Color Units per Gram)

	Sample A	Sample B
Raw sugar as delivered	4,779	10,360
Coating film	73,900	201,634
(wt %)	(4.5)	(4.1)
Affined sugar	1,460	2,142
Affined color / raw color	0.31	0.21
Film color / affined color	51	94

icate, aluminum, and magnesium have been shown to cause reduction in filtration rates. The South African industry prefers to measure the filtration rate on process carbonated liquor to avoid difficulties with laboratory tests that do not adequately duplicate the refinery conditions. The measurement of turbidity remains problematic, and no standard conditions have yet been established that take into account the slow rate of solubilization of some contributors to turbidity [9].

The effect of polysaccharides on filterability is well established, and the trend has been away from measurement of filterability toward individual polysaccharides that would be the cause(s) of the problem, at least for contract purposes. However, filterability tests remain very useful for comparison of processing systems and raw sugars.

Each class of nonsucrose in raw sugar affects the refining operation differently. It is useful to distinguish between primary and secondary impacts; primary impacts involve a specific chemical feature, such as needle grain or preferential inclusion in the crystal lattice; secondary effects involve the bulk properties, such as increase in viscosity. Aspects of the impact of each are described separately.

Ash, or inorganic components (mostly soluble salts but also including insoluble materials such as silica and calcium sulfate), have an impact on both the recovery operations and the quality of refined sugar produced. Chen has commented on the differences between the various methods for ash determination and the interpretation of these data [10]. Sulfated ash measures all cationic material as the sulfate or oxide and also silica; conductivity ash measures soluble ionic material, but the conditions for determination of conductivity are at high dilution, where some materials may become solubilized (e.g., calcium sulfate or aconitate) when they are not soluble under normal high-Brix process conditions.

The melassigenic effect of the ash components, especially the alkali metal cations, is well known and is the major consequence, in the recovery house, of high ash in whole raw sugar. Ash in fine liquor is the key to the levels of this impurity that can be handled in the white sugar operation. In conventional refinery operations the only step that removes a significant amount of ash after affination is carbonation. Ash removal with phosphatation is minimal, and neither process removes potassium, the major cation in all cane operations. Some de-ashing is achieved with bone char, but a station dedicated to ion-exchange de-ashing has so far proved to be uneconomical, especially when the regeneration requirements are taken into account.

Invert is the major nonsucrose component of most raw sugars, and the level that a refinery can tolerate depends very much on the product mix. Production of soft brown sugars requires the input of invert into the system, or the generation of invert in the refinery; the former is preferred, providing that the invert can be removed efficiently at affination.

The melassigenic impact of invert is ambiguous; it reduces the solubility of sucrose (thermodynamic control) but conversely, it reduces the rate of sucrose crystallization (kinetic control). The latter will be most important when limited recovery house capacity is available. In large recovery houses, such as associated with raw sugar operations, there may be excessive capacity and the tendency to slow the crystallization operation to recover more sugar. However, this can have the adverse effect of increased sucrose degradation and invert production.

Polysaccharides in sugarcane processing have been much studied [11] and fall into two groups, those intrinsic to cane and those produced as a consequence of deterioration and microbial infection. In the older literature these are referred to collectively as *gums*. In the first category are sarkharan, indigenous cane polysaccharide, Robert's glucan, polysaccharide CP, and starch. The impact of polysaccharides on the viscosity of sucrose solutions over a range of concentrations and temperature are shown in Figure 35.1.

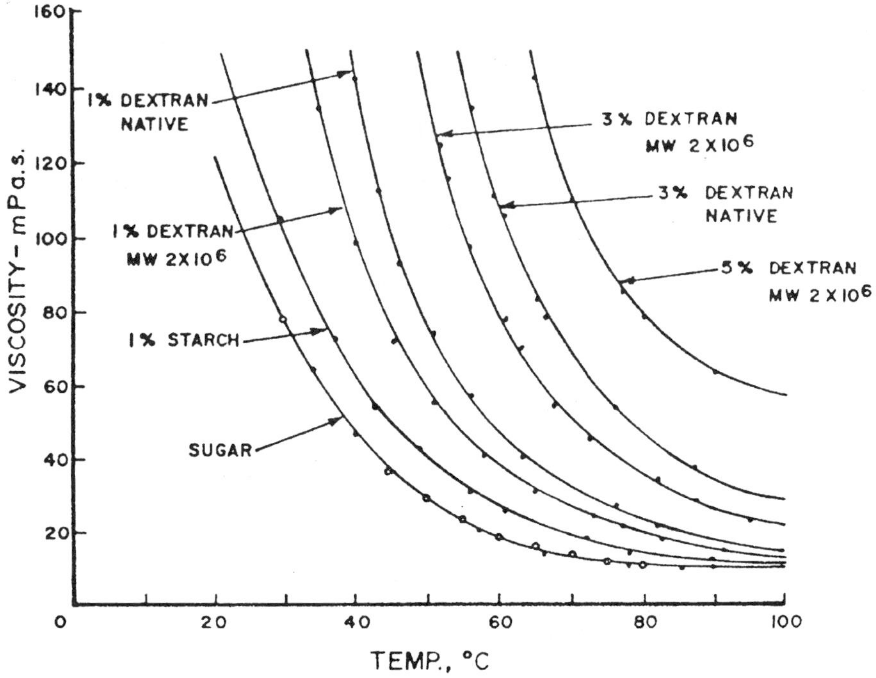

FIGURE 35.1 Nonsugar concentration versus viscosity of sugar solution (65 Brix).

Starch is the most significant of these in causing processing problems [12,13]. It occurs in all cane varieties but usually at levels low enough to be insignificant. Some cane varieties, perhaps influenced by particular growing conditions, have a higher starch level, as found recently in Hawaii and Louisiana. The starch content is highest in immature cane and in the tops and leaves. In the cane most of the starch is in a granular form that becomes gelatinized in process. Starch is included in the growing sugar crystal and is not efficiently removed at affination. Its major consequences in the refinery are in filterability [14] and production of sticky massecuites that do not purge well.

South African studies on the starch levels in raw sugar have shown that amylose is the major starch fraction on the crystal surface, while amylopectin is the major component incorporated into the crystal. Amylose has a greater adverse effect on operations due to its higher intrinsic viscosity. Starch has been implicated in filtration problems, although a new carbonation station seems to have reduced the impact of starch on filtration [15]. Amylose interferes with calcium carbonate crystallization in the carbonation process, and improved control of this crystallization should reduce the adverse impact of the starch.

Dextran, although not present in fresh clean cane, can become a major problem when formed in stale cane and in raw mills (and refineries) operating under unsanitary environments [16]. Sugarcane dextran is of very high molecular weight, and the major effect of the presence of dextran is increased viscosity, leading to slower filtration and crystallization. The effects of dextrans on crystal shape have been much discussed and are complicated by the presence of oligosaccharides also produced by the microorganisms involved in poor sanitary conditions. Several methods for dextran analysis have been developed based on a variety of reactions or properties of the polysaccharide. Quite different results can be obtained on the same sample, but each approach has its value. The haze test is used in raw sugar contracts but the Robert's procedure is claimed to be more useful in predicting refinery problems [17–19]. Detailed studies of the impact of dextran on refinery operations have been reported [20] and are the justification for the penalties imposed on raw sugars with high levels of dextran.

Enzymatic methods for the degradation of both starch and dextran are well established, although there are still some regulatory concerns about the sources of dextranase. These processing aids are rather undesirable as part of routine refinery operations, and their application is much more appropriate in raw sugar operations, where their benefits are greater. These benefits are increased recovery due to reduced massecuite viscosity and improved raw sugar quality.

The scientific method often requires the simplification of material under study, and this leads to attempts to associate specific impurities to specific process problems. In reality, there are often synergistic effects between the various components, and studies identifying these are ongoing: for example, starch and colorants and their impact on filterability and crystal growth [4] and in floc formation [11, p. 387; 21].

35.2.4 Decolorization

The quantity and type of coloring matter in raw sugar are of critical importance to the refiner. Color is a subjective measure of purity and the intensity of the refining activity. If color is removed efficiently, the process required is often assumed to be adequate to remove other impurities, especially organic compounds.

The processing of high (affined)-color raws depends on the decolorization capabilities of the individual refinery. Increased affination sends more sucrose to the recovery operation, with subsequent loss. If liquid sugar or fine liquor sales are a significant component of refinery income, it may be necessary to blend in granulated sugar, a further loss. If inferior final white sugar strikes are boiled, these may be too high to be blended into standard granulated sugar and have to be sold at a discount or recycled back to the processes, another loss.

The colored materials entering the refinery are highly heterogeneous, some being readily removed in the process and others less efficiently. Recent work has been reviewed by Riffer [6, pp. 186–207], who described the characteristics of color in terms of polarity, molecular size, functional groups, and pH sensitivity. Such data are valuable to the work of refiners in removing color efficiently and cheaply. The various types of decolorization system are dealt with elsewhere in the book. Each system has its specific characteristics for removal of different color types. The indicator value (dependence of measured color on pH) has proved to be a useful predictor of refinery performance [22]. In the refinery, dark-colored liquors can result from light raw sugars that decolorize poorly; pH sensitivity was found to be responsible. Recent work has suggested the presence of polysaccharide/color complexes [23].

Color in raw sugar has two basic sources: pigments and color precursors present in the cane being processed and color developed in the process due to carmelization, the Maillard reaction, and so on. Color development problems in process (both raw and refinery) are associated with holding process liquors at high temperatures for extended periods of time. Color prevention is preferable to color removal, but high extraction levels in raw cane mills, especially in diffusers operating at higher temperature, produce more color in the juice. At this stage color is extracted from the cane into the juice rather than being produced by a chemical reaction in the juice [24]. Also, the trend in the industry is toward higher process energy efficiencies (e.g., evaporators operating at higher temperature), another cause of increased color development. If sodium hydrosulfite is used to reduce the raw sugar color (may also be used in the raw sugar vacuum pans to reduce viscosity), the color may return on storage or processing, leading to unanticipated color removal problems.

Other than the color level and type, the raw sugar quality factors that influence the decolorization operation significantly are the levels of invert and ash. Carbonation destroys a small amount of invert during normal operations. Carbonation also removes a substantial amount of color, but higher-than-normal levels of invert may appear as extra color in the carbonation liquor. This may be traced to poor work in the recovery station, but a poorly affinated high invert raw may produce the same result. Phosphatation refineries normally do not detect any significant relationship between the invert level and the color of the purified liquor.

Higher-than-normal ash levels affect the capacity of the char house. For normal raws the ash-absorbing power of the char remains after the decolorizing power is exhausted. If ash is high, the decolorizing capacity is used up rapidly. Higher ash, leading to more color in the char liquor, will require more work on the white sugar centrifugals and therefore reduced throughput and increased recirculation. Not all ash components have the same influence on the decolorization station. Sulfite (sulfur dioxide) and sulfate have a major impact in decreasing the efficiency of char systems. It is recommended that refineries using ion-exchange resins include a test for iron, since this can block the resin and reduce capacity severely.

35.2.5 Crystallization, Including Recovery

The major impact of raw sugar quality on white sugar crystallization performance may be on crystallization rates due to the presence of crystallization-inhibiting materials such as dextran. Slower crystallization rates, reducing capacity, and increased recirculation to the recovery station may occur. As with high ash levels, longer wash on the centrifugals may be necessary.

Recovery stations are also adversely affected by slow crystallization, long pan cycles, poor exhaustion, and problems with purging by the centrifugals. This can lead to slowing down of the entire refinery with recycling of low-grade syrups, which compound the difficulties and lead to increased sugar destruction and reduced yield.

Original impurities in the raw sugar are concentrated in the recovery house. In part, the recovery process acts as a sink for all the reject or recycle material. The products of microbial contamination of the sugar will accumulate in this station. Even if sugar recovery is not a problem, there can be levels of unacceptable materials (e.g., heavy metals) that could render the products, such as brown sugars and edible molasses, unfit for normal use. Also to be considered in this context are the flavor and aroma components of the raw sugar, which determine in part the properties of the refinery brown sugars [25].

35.2.6 White Sugar Quality

Standards for trace elements in white sugar have been in place for many years, and as analytical methods become more sensitive, the industry can expect to be required to monitor the levels of an increasing number of materials. The efficiency of removal of some elements has been investigated, but their presence in raw sugars and the setting of quality parameters has received relatively little attention.

Sulfite (principally from the use of sodium hydrosulfite) in the raw sugar is a potential problem, particularly for brown sugar, since the U.S. Food and Drug Administration now has a sulfite labeling requirement: Food products with more than 10 ppm sulfite must be labeled as such.

There are strict regulations on the presence of biocides, herbicides, insecticides, and pesticides in food products. Refiners are becoming more conscious of the impact of the use of such materials, both in product quality and the pattern for their removal by the process. It is important that sugarcane producers follow the regulations for their use. This could be a particular problem for sugar shipped internationally.

There are a number of raw sugar process additives whose impact on the refinery and sugar quality is poorly understood [e.g., polyacrylate/polymaleate antiscalants, quaternary ammonium compounds for biocides and decolorization, fatty acid esters for surfactant crystallization aids (allowed at 320 ppm in final molasses)]. Use of these materials must follow good practice, and unapproved materials cannot be used.

35.2.7 Environmental Aspects

The major waste materials from a sugar refinery fall into five groups: filter aids, phosphatation scums, carbonate cake, discarded char, and brine from resin decolorization systems. Although none of these materials may be classified as hazardous waste, there is a cost associated with their disposal. Inferior raw sugar will inevitably yield more waste.

35.3 PRODUCTION OF QUALITY RAW SUGAR

As described in Section 35.2, good-quality raw sugar is a prerequisite for efficient refinery operation and production of white sugar that meets all quality specifications. The purpose of this section is to comment on all operations in raw sugar production that are relevant to the production of high-quality raw sugar. It must be stressed that good sugar quality (both raw and refined) is achieved efficiently only with coordination between all stages in the process. It is not adequate to compensate for poor operation in one area with good operation in another. Some compensation is possible, but this is not a desirable approach.

Not all the operations in the raw sugar factory or in the production of cane are under management control. Weather conditions play a large role in the quality of cane as delivered to the mill. It may be necessary to make operational decisions in favor of energy constraints rather than sugar quality. Capacity and equipment limitations are also significant, and industries that have made a commitment to produce good-quality sugar, especially the very high-polarization sugars, have made considerable capital investment to achieve this.

The quality of the laboratory and operational data and their use by the factory management is critical for both good sugar recovery and high sugar quality. Perhaps most important is the necessity for steady factory operation at throughput at or close to capacity, to minimize the time the sugar is in process. The sucrose molecule is more labile under the process conditions of the raw factory than in a refinery, and stoppages can cause reductions in both recovery and quality, often more than appreciated by the factory staff. High temperatures for extended periods of time are a major cause of the unappreciated loss in a raw sugar factory.

Each factory, or a nearby central laboratory, should have the facilities and staff to perform the necessary sugar quality tests. Not all the measurements described in Section 35.2 are required, but certainly essential are data on polarization, moisture, color, ash, grain size, and if an expected problem, dextran. The data should be acted upon promptly or they become merely a matter of historical record. Additional analyses may be necessary to optimize factory performance, as noted in the description of each process stage.

Inefficient factory performance, such as excessive losses in filter cake and high final molasses purity, does not necessarily mean poor sugar quality. However, low sugar recovery is an indicator of lack of attention to the details of factory operation, and this can often lead to poor sugar quality. Any deterioration with loss of sucrose will lead to the formation of undesirable materials with adverse effect on sugar quality, although some are more serious (e.g., dextran more than invert).

In the late 1960s several industry groups began to review the operations of their factories with regard to sugar quality and the steps that could be taken for its improvement. The basic unit operations involved in conversion of cane to raw sugar have little flexibility, but modification of each has made possible very significant improvement in quality, for example the production of a single type of high-polarization raw sugar in South Africa. At the same time there has been an increase in mechanical harvesting, along with higher extraction efficiencies, both of which tend to lower the quality of juice entering the process. Even taking these into account, there has been a general trend toward improved sugar quality, and in this section we describe the options available to raw sugar processors to improve sugar quality. The processing of cane is dealt with in order of operation, including the impact of agronomic and harvesting practices. Emphasis is placed in areas over which the raw sugar producer has some control.

35.3.1 Cane Varieties

Cane varieties are selected on the basis of their agronomic characteristics and sucrose levels. The important agronomic characteristics include yield, vigor of growth, ratooning ability, tolerance of local climatic conditions, and resistance to pests and diseases. Other characteristics, such as brittleness, may be important if the cane is mechanically harvested and/or loaded. Levels of nonsugar in juice (e.g., starch) would be considered of minor importance in the selection of cane varieties.

Varietal differences in chemical composition have been noted in several areas. Legendre [26] measured differences in starch, colorants, and cold tolerance in Louisiana varieties. Color and impurity content of three cane varieties in South Africa have been measured by Lionnet [27] and related to maturity and harvest time, showing that the youngest cane had the lowest color.

Cane processors often have little control over the selection and planting of cane varieties, but they should be aware which varieties are being processed to allow for any different processing character.

35.3.2 Agronomic and Weather Conditions

The ash content and composition of the juice are influenced by the soil type and whether the cane is grown on irrigated or nonirrigated land. Saline and high-ash soils tend to result in canes with higher ash content and possible melassigenic problems.

Sugarcane grown under stress, such as drought, high rainfall, and freezing conditions, will tend to give additional processing problems. Mill management should be prepared for the changes in processing operations that are necessary to cope with these conditions. Of particular concern are conditions that lead to increased polysaccharide levels, either by the cane itself or by microbial infection. Experience in many areas shows that dextran problems are caused predominantly in the field; the current dextran penalty for raw sugar is severe enough that it is uneconomical to process cane with dextran levels above about 10,000 ppm on solids—the value of the sugar is less than the cost of production.

Ripeners have become very important, especially in cane-growing areas where rapid maturation of the cane is desirable. When used properly, they are of great benefit, increasing the sucrose content and reducing the invert content of the cane. Problems due to overlong treatment with Polado have been suspected but not confirmed.

35.3.3 Cane Quality and Payment Systems

Good-quality raw sugar can be made efficiently only from good-quality cane (i.e., cane that is fresh and free of dirt and trash). The importance of paying for cane on the basis of quality cannot be overstated. The mill is paid for sugar on this basis.

The quality parameters should include not only the sucrose content, but the juice purity and the dirt and trash levels in the cane. It is well established that leaves and trash contribute to color and starch in raw sugar and that field soil increases the silica levels. Results from South Africa show little change in color from cane through syrup. Therefore, for steady boiling house operation the overriding determinant of sugar color is the cane. Sugar color is a fairly constant fraction of syrup color, about 6%, for different factories and different seasons. Tops and trash contribute much of the color—2% tops with 4% trash on cane can increase the color by 20%. Some stale and insect-infected cane gives higher color. Stale but clean cane did not show a great increase in color.

If appropriate, the juice from each load of cane should be analyzed for products of deterioration: namely, dextran, ethanol, and titratable acidity. Methods are well established for performing all these tests, and the core sampler system allows each load to be analyzed.

Investigations in Hawaii confirmed that raw sugar produced from burned cane was superior in color, filterability, small crystal content, and polarization. The inference is that trash and excessive soil in unburned cane affects sugar quality adversely. The deterioration rate of burned cane is higher than that for burned cane, and it is even more important that such cane be processed promptly after handling.

35.3.4 Harvesting, Handling, and Storage

Whatever harvesting system is used, it is essential that the cane delivered to the mill be fresh. Billeted cane can deteriorate rapidly, but the high dextran levels initially found in Australian cane cut in this way have been prevented by tight control of harvesting and delivery schedules. The importance of good delivery procedures and recently improved systems are generally well understood but not always followed.

If at all possible the cane should be handled as few times as possible since any damage can cause further deterioration, especially if held for a prolonged time. Similarly, extended storage of cane, especially in large piles with poor aeration, can lead to significant sugar losses and accumulation of deterioration products. Good management of cane harvesting and transportation is essential to maintain good cane quality.

In a number of cane areas using mechanical harvesting for whole-stalk cane it is necessary to wash the cane, especially under wet conditions. Significant losses have been reported in most cases (about 2% of the sucrose in cane), but cane washing is essential to maintain decent factory operation with good-quality clarified juice and to minimize losses in filter cake and maintain the quality of bagasse as fuel. With dirty cane delivered to the mill, cane washing is desirable in maintaining sugar quality.

35.3.5 Extraction

High extraction of sucrose by the mill or diffuser is essential for the profitability of the mill. Recent trends have been toward improved cane preparation and extraction. Both mills and diffusers can achieve high extraction if sufficient imbibition water is used on well-prepared cane. At high extraction levels, the last expressed juice can be of quite low purity (<60) with increased extraction of nonsugars. Good sugar recovery can still be achieved in these circumstances, but high final massecuite viscosities can be a problem, leading to recirculation in the boiling house and color development. Modified boiling schemes become necessary, and this is the basis for the production of high-polarization sugars in South Africa.

Temperatures are higher in diffusers, and there is increased extraction of colored materials at higher temperatures. In the diffuser, color is not formed by high temperature but is extracted; a juice sample from the same cane kept at the same temperature did not show any significant increase in color.

The milling of tops and trash with the cane has been proposed as a source of additional fuel. The advantages of this have to be balanced carefully against the probable reduction in raw sugar quality. Additional clarification and crystallization stages become necessary to produce a high-quality sugar.

Mill sanitation is critical for the production of dextran-free sugar. A poorly sanitized mill (and untreated juice system) can develop and maintain a high infection level by *Leuconostoc mesenteroides,* with a serious impact on factory performance and sugar quality. Chemical treatment with chlorine, quaternary amines, and thiocarbamates is effective but must be combined with steam and hot water treatments to remove the buildup of juice and bagasse around the milling tandem.

35.3.6 Juice Purification

Clarification of the juice is the critical step in conventional raw sugar processing for removal of primarily suspended solids and some soluble material and production of a clear neutralized juice for evaporation and crystallization. Poorly clarified juice is the primary reason for low filterability in raw sugar; juice from deteriorated cane may not clarify well. Both inorganic materials (especially silica) and polysaccharides can pass into the clear juice and into the sugar if there is no further purification stage, as is usual in most raw sugar factories.

Standard procedure is treatment of the raw juice with lime, heating, and settling. Good pH control in lime addition and temperature control on heating for sterilization and flashing to remove air are essential. There are numerous variations on this theme, with excellent results being obtained with saccharate lining. There are differences of opinion on the optimum pH for good clarification, a low pH (6.5 for the clarified juice) giving a lighter-colored but more turbid juice, and higher pH (7.5) giving a darker juice but of higher clarity. Removed in clarification are some inorganic salts (iron, aluminum, etc.), some organic acids, proteins, some polysaccharides, gums, waxes and lipids, mud, and fine bagacillo.

Good clarification may require the addition of phosphate to the raw juice if the level in cane is too low. Rapid settling to give clear juice is enhanced by the use of polymeric flocculants. With the wide variety of such flocculants now available, each mill should evaluate the performance of a range of flocculant types to obtain maximum performance of the clarifier.

Sterilization of the juice by heating is also part of this process, but some thermophilic bacteria can survive, especially if the optimum temperature in the heaters is not maintained. The retention time in the clarifier should be as short as possible consistent with the production of good, clear juice, now possible with short-retention-time single-tray clarifiers. No advantage is gained with long retention times when chemical and microbial loss of sugar and color formation can occur. Steady clarifier operation with the minimum of upset in flow is desirable, and holding juice in the clarifiers during factory stoppages should be avoided.

Filtration of the clear juice is desirable but impractical with current filters. At minimum the juice should be passed through a fine vibrating screen (100/200 mesh stainless steel) to remove bagacillo and other gross suspended solids.

The settled mud solids in the clarifier are filtered to recover juice, rotary vacuum filters being standard equipment. Solids retention on these filters is often poor, due to inadequate mud conditioning, with recycle of fine mud solids to the clear juice. This causes the recycling of hard to flocculate material through the system and increased turbidity in the clear juice. Good filter operation is essential to minimize this recycling and prevent color formation and microbial action in the mud system, the latter occurring especially if the mud temperature is not maintained high enough. Purity differences between mixed, clarified juices and filtrates should be monitored carefully. Lactic acid formation is a useful indicator of poor filter operation. A solution that has proved successful in many cases is a separate filtration system for filtrate.

Floatation clarification of the syrup after evaporation is very effective in removal of suspended solids and also reduces polysaccharides and color. This type of system is especially important for mills with in-house refineries. Good practice is to follow this system with a protecting filter such as a deep-bed filter. A much superior syrup is thus available for crystallization.

35.3.7 Evaporation

Multiple-effect evaporators are standard and must be operated at optimum throughput to minimize the time the juice is held at high temperature. Ample evaporator capacity is essential to produce syrup of high Brix and so minimize the load on the vacuum pans. Both color formation and inversion occur and can be kept to a minimum with good control of juice pH, temperature, flow, level, and so on. A greater increase in invert has been observed in systems with high-volume first effects, but the increases observed would not, in themselves, pose a problem of sugar quality.

35.3.8 Crystallization

All raw sugar mills that produce good-quality raw sugar pay particular attention to operation of the boiling house. All the quality factors described earlier are under the control of the boiling house. A variety of boiling schemes may be devised, but three boiling systems predominate at normal syrup purities. In the simplest system the sugar from the first two strikes (A and B) are used for commercial sugar, with the third or recovery (C) being used as footings for the A and B strikes. This system minimizes remelt and uses less steam than other options. However, this uses the poor-quality grain from the recovery strike (C) in the production of commercial sugar. The C sugar is of poorer filterability and color and is a major determinant of raw sugar quality.

Systems that produce high-quality sugar separate the production of commercial sugar completely from the recovery system, all of the low-grade sugar being remelted. This type of operation requires more pan and centrifugal capacity but produces a consistently high sugar quality and is the basis of the production systems for very high polarization sugar.

With any boiling system the quality of the sugar depends greatly on control of the crystallization equipment, and automated continuous pans are to be preferred. Irregular operation and poor crystal content of strikes will produce uneven and conglomerated grain and excessive recycling of material with increase in color and loss of sucrose.

Good centrifugal operation is essential for both commercial and low-grade massecuites. Properly adjusted sprays with the minimum necessary wash are required for good-quality sugar. Low-grade centrifugals should be operated to minimize recycling of molasses into the boiling house.

35.3.9 Chemical Use

Chemical treatment, in both the field and factory, is becoming more widespread. All regulations for their use should be adhered to strictly and chemical use reduced to the minimum necessary to achieve the desired results.

REFERENCES

1. M. Todd, *Int. Sugar J.*, 99:379–384, 1997.
2. C. C. Chou, *Sugar J.*, June 1990, pp. 14–19.
3. J. E. Somner, *Sugar Ind. Technol.*, 1983, pp. 95–107.
4. M. Donovan, *Proc. Sugar Ind. Technol. Symp.*, 1993.
5. J. A. Watson and W. M. Nicol, *Sugar Technology Reviews*, 1975, 2(3): pp. 69–126.
6. R. Riffer, in *Chemistry and Processing of Sugarbeet and Sugarcane*, M. A. Clarke and M. A. Godshall, eds., Elsevier, Amsterdam, 1988.
7. R. A. Kitchen, in *Chemistry and Processing of Sugarbeet and Sugarcane*, M. A. Clarke and M. A. Godshall, eds., Elsevier, Amsterdam, 1988, pp. 225–228.
8. P. G. Morel du Boil, *Int. Sugar J.*, 99:327–331, 1997.
9. S. J. Clarke, *Proc. Sugar Ind. Technol.*, 1995, pp. 253–262.
10. J. C. P. Chen and C. C. Chou, *Cane Sugar Handbook*, 12th ed., Wiley, New York, 1993, pp. 347, 367–370.
11. M. A. Clarke, E. J. Roberts, and P. J. Garegg, *Proc. Sugar Process. Res. Conf.*, 1996, pp. 368–388.
12. J. A. Devereux and M. A. Clarke, *Proc. Sugar Ind. Technol.*, 1984, pp. 36–59.
13. J. A. Devereux and M. A. Clarke, *Proc. Sugar Process. Res. Conf.*, 1984, pp. 209–230.
14. M. A. Godshall, M. A. Clarke, and C. D. Dooley, *Proc. Sugar Process. Res. Conf.*, 1996, pp. 244–264.
15. E. F. T. Lee and M. Donovan, *Proc. Sugar Process. Res. Conf.*, 1996, pp. 103–120.
16. M. A. Clarke, *Sugar Azucar*, Oct. 1997, pp. 28–40; Nov. 1997, pp. 22–34.
17. R. A. Kitchen, *Proc. Int. Dextran Workshop*, Sugar Processing Research Institute, New Orleans, LA, 1984, pp. 53–61.
18. K. R. Hanson, *Proc. Sugar Ind. Technol.*, 1980, pp. 152–159.
19. M. J. Fowler, *Int. Sugar J.*, 83:74–77, (1981).
20. C. C. Chou and M. Wnukowski, *Proc. Tech. Session Cane Sugar Refin. Res.*, 1980, pp. 1–25.
21. P. G. Morel du Boil, *Int. Sugar J.*, 99:310–314, 1997.
22. N. H. Smith, *Proc. Tech. Session Cane Sugar Refin. Res.*, 1966, pp. 14–29.
23. E. J. Roberts and M. A. Godshall, *Proc. Tech. Session Cane Sugar Refin. Res.*, 1979, pp. 68–80.
24. G. R. E. Lionnet, *Proc. S. Afr. Sugar Technol. Assoc.*, 62:39–41, 1988.
25. M. A. Clarke, M. A. Godshall, R. S. Blanco, and X. M. Miranda, *Int. Sugar J.*, 97:557–561, 1995.
26. B. L. Legendre, *Proc. SPRI Workshop Raw Sugar Quality White Sugar Quality*, M. A. Clarke, ed., 1990, pp. 73–76.
27. G. R. E. Lionnet, *Zuckerindustrie*, 117:39–42, 1992.

CHAPTER 36

Nonsugars and Sugar Refining*

INTRODUCTION

Separation processes are at the very core of the sugar industry. We start with a complex mixture and wish to end up with a highly purified material, and moreover, want to accomplish this task with minimal cost and loss. The strategies that have developed over the past few centuries to attain this goal have succeeded despite limited information about the nature of the impurities being removed. Nevertheless, it is astonishing how much *was* known about the composition of nonsugar constituents in the nineteenth century, from chemists working without the benefit of such powerful analytical tools as chromatography, mass spectroscopy, and atomic absorption. For example, malic acid was first isolated from sugarcane in 1849, succinic in 1888. As is the case today, the incentive for such investigations was the prospect of practical improvement to established procedures. To place the nineteenth-century achievements in perspective, it should be noted that the first true chemical synthesis of sucrose was not accomplished until 1956 [1], although an in vitro synthesis using a phosphorylase from *Pseudomonas saccharophilia* was achieved in 1944 [2].

The nonsugar impurities in raw sugar can be classified in various ways. Since all refineries are controlled on the basis of color removal, we may draw a distinction between colored and uncolored substances, but this alone is not quite satisfactory, since the colorant fraction is highly heterogeneous, as are the uncolored constituents. On the other hand, if we describe the colorants structurally, such as plant phenolics and cinnamic acid derivatives, we must hasten to add that not all such substances are colored, beginning with the parent compounds themselves. For an organic material to be colored, a system of conjugated unsaturation must be present, because the accompanying electron delocalization reduces the energy requirements for absorption from the ultraviolet into the visible zone.

Colored substances in fact account for only a small portion of the nonsugars on a weight basis but occupy our exaggerated focus because they are monitored for process control. Smith estimated that for a 98.5° polarization raw sugar, colorants accounted for about 16 to 20% of the weight of the nonsugars [3]. However, even this relatively low figure includes uncolored high-molecular-weight material that is incidentally complexed or cova-

*By Richard Riffer.

lently bonded to color bodies. Despite low tinctorial power per unit weight, large molecules of this type are of great interest because they display a high tendency to be retained in the crystal and also because of their potential contribution to turbidity.

A case can be made that a proper estimate of color mass should really be weight-averaged, based on the attenuation index of the individual constituents, a task not easily accomplished because of the large number of components in the colorant fraction. In Lambert–Beer colorimetry, the measured absorbance is typically normalized on the weight of colored material, but for sugar samples this information is ordinarily unavailable because few colorants have been characterized as discrete substances. Even if such data were accessible, it would clearly be inappropriate to base a computed absorbancy index on a mixture of variable composition. Consequently, we normalize the absorbance on percent solids in order to put all measurements on the same basis, and the *amount* or mass of color thereby becomes irrelevant. A color measurement performed on a sugar sample is thus conceptually very different from a colorimetric quantitative analysis for a discrete colored substance.

There are still other layers of complexity to be uncovered here: Should we—assuming it were possible—distinguish between substances indigenous to sugarcane and those formed as a consequence of human activity on the plantation or in the mill and refinery? Not only do processing techniques alter substances initially present but also introduce new ones. And how should we classify metabolic by-products of microbiological colonization? This category includes not only soluble polysaccharides and bacterial slimes but also brown sugar flavor components.

The nonsugar substances not considered here are mainly insoluble materials that are macrocontaminants, subject to removal by clarification and filtration operations, such as soil and plant fiber. High-moleculer-weight substances may be soluble or partly so, and these are included, with certain exceptions, such as DNA, carotene, chlorophyll, and xanthophyll. Proteins and polysaccharides are of interest to the degree that they possess appreciable solubility and thus present processing issues.

In this chapter we describe the nonsugar substances as colorants, turbidity-causing matters, nonsucrose carbohydrates, carboxylic acids, nitrogen compounds, and inorganic substances. Flavorants—that is, materials that engage our nasal receptors rather than our taste buds—are of great significance in soft sugar and refiners' syrup manufacture but are minuscule contributors to nonsugar mass. Such substances are beyond the scope of the present chapter and are discussed only briefly (see, however, a review article by Godshall [4]).

36.1 COLORANTS

Decolorization as practiced in the cane sugar industry is almost exclusively a separation rather than a bleaching operation, although some direct consumption sugar processes use sulfite treatment. The bleaching action of sulfur dioxide has been known since the Roman period, although the mechanism of action is very different from that of oxidizing agents such as hypochlorite and peroxide [5].

It is important to recognize that raw sugar is not colored in the same way that iron rust is reddish or coal is black. In these instances, the color is an integral part of the substance, and there is no way that one can separate the red color from rust and still have rust. But sugar is always white—that is, colorless—whether in raw sugar, turbinado, granulated, brown sugar, or molasses. What makes the sugar appear yellow or brown are the colored impurities present. Stated somewhat differently, iron rust and coal absorb significant levels of light in the visible region of the spectrum, whereas sucrose does not.

Another important principle is that colorants tend to be similar in all raw sugars, irrespective of geographical origin. To the degree that differences in the colorant fraction are observed, these are likely to be quantitative rather than qualitative. We are not surprised that a 98° polarization raw sugar exhibits more color than one of 99° polarization, but we should not necessarily expect unique substances to be present. It is possible that one day our knowledge of chemotaxonomy will be so highly developed that a simple laboratory test will differentiate, for example, a Dominican raw sugar from a Taiwanese sugar, but the structures of most present-day purchase contracts offer little incentive for making such distinctions.

The colorants in cane sugar products include not only naturally occurring plant pigments but also a large number of substances produced during processing. For example, colorants in brown sugars are to a large degree artifacts of refining operations, not present in the original raw sugars. Some flavorants also fall into this category, including those of microbiological origin.

Despite the similarities indicated, the colorant fraction is nonetheless highly heterogeneous. Because of differences encountered in raw juice composition from genetic and environmental factors, and almost limitless possibilities for variations in color-forming reactions during sugar manufacture, the probability that two colorant fractions from different sources are exactly alike must be vanishingly small. Furthermore, because no definite or constant composition can be assigned to most colorants, the usual criteria of purity cannot be applied to these substances. However, because we ordinarily see the average properties of a complex mixture, the distinct attributes of individual components tend to be obscured.

Most of the work that has been done toward elucidation of the structures of specific colorants has focused on low-molecular-weight compounds, because these are the most easily isolated and characterized. However, such substances account for only a small portion of the colorant fraction. These relatively simple molecules are nonetheless of great interest, because they are likely to be incorporated into more complex species by further reaction. Moreover, some of these materials display remarkable stability, surviving at trace levels even in granulated sugar.

36.1.1 Phenolics

During a 30-year period beginning about 1950 there was considerable activity directed toward the isolation of sugar colorants, using dialysis, electrophoresis, and chromatographic techniques. Molasses was often used as a concentrated source, but this practice was open to the criticism that molasses colorants are not representative of those in raw sugar. In addition, published articles on the topic did not always distinguish between cane and beet sources, from which the phenolic constituents are very different. Colorants derived from invert or thermally degraded sucrose would be expected to be similar from the two origins, but even here differences in other constituents and in process operations could affect the outcome of the color-forming reactions.

Studies on colorant fractions indicated that they could be categorized into distinct groups: naturally occurring phenolics and their reaction products, melanoidins from reactions between carbohydrates and amino acids, and sugar degradation products. This last division includes distinctly different colorants, depending on such factors as the pH of formation, and contains a subclass, caramelization products, generally considered separately because they result from thermal processes, whereas the other reactions proceed slowly even at room temperature.

A number of naturally occurring pigments in sugarcane were identified in 1970 by Farber and Carpenter [6]:

- Chlorogenic acid
- Caffeic acid
- *p*-Hydroxy cinnamic acid (*p*-coumaric acid)
- 4-Hydroxy-3-methoxy-cinnamic acid (ferulic acid)
- 4-Hydroxy-3,5-dimethoxy-cinnamic acid (sinapic acid)
- 7-Hydroxycoumarin (umbelliferone)
- Kaempferol

These substances are all known phytochemicals, widely distributed in the plant kingdom and by no means unique constituents of sugarcane.

Although the colorants were isolated from sugarcane leaf extracts, many of them were found to survive in refined granulated sugar. Cinnamic acid derivatives, the first five listed above, appear to be especially resistant to removal by known refining techniques. Chlorogenic acid was found by Carpenter's group at trace levels in every sample of granulated examined and was found to turn even the highest-quality sugar faintly yellowish in alkaline solution.

Chlorogenic acid is an important factor in plant metabolism that is widely distributed in both mono-, and dicotyledenous angiosperms. Like other catechols, it forms dark pigments in the presence of ferric iron. Such complexes are strongly colored because they exhibit excited energy levels at separations of about 2 eV, readily available in visible light. The *ferrous* versions of the complexes are uncolored, probably because of the absence of an unpaired 3*d* electron [7].

Using Planck's constant, it can be shown that 2 eV corresponds to a wavelength of 552 nm, visual purple. The ferric–catechol complex actually absorbs red and violet from the two ends of the visible region and hence appears green. In fact, *chlorogenic* means "green forming" from its Greek roots, and despite the name, the molecule contains no chlorine atoms. Because of these spectral characteristics, neither of the two most commonly used wavelengths for sugar color, 420 and 560 nm, is particularly sensitive to iron complex levels. For this reason the visual impact of samples containing significant levels of this type of color can be poorly assessed by standard photometric procedures. Since the eye is not sensitive to a single wavelength, the perceived color intensity of a sugar sample "is related to the integrated light absorption over the visible wavelength range weighted for the photopic retinal response" of the observer [8]. *Photopia* refers to bright light vision mediated by the cones of the retina, as opposed to *scotopia,* mediation by the rods under dim light conditions.

Because of the high stability of the complexes, iron is readily removed in the manufacture of granulated sugar but only with difficulty from brown sugar liquors, which contain relatively high levels of nonsugar chelating groups. Iron plays a significant role in the colorant fraction not only as a constituent of chromophores but also because it strongly catalyzes darkening from hexose degradation products [9].

Carpenter's group subsequently identified additional colorants: coniferin (a cinnamic acid glucoside), coumarin, esculin (a coumarin glucoside), and quercitin and rutin (flavonols) [10]. Once again, these are known substances, widely distributed in the plant kingdom. Such compounds are generally yellow, and more highly colored at higher pH, with the flavones typically being more deeply colored than cinnamic acid derivatives. By 1985, four flavonols and 25 flavones had been identified in sugarcane and in refined products. It is generally believed that about three-fourths of the color in raw sugar is of the phenolic type.

Because of the subtle role played by flavonoids and cinnamic acid derivatives in plant physiology, the development of high-yielding varieties that also display high drought tolerance and specific disease and pesticide resistance—whether by traditional techniques or gene insertion—may have resulted in commercial cane with altered phenolic content. Such substances are antioxidants that may help the plant withstand the direct sunlight that can destroy tissues, and also diminish the impact of insects, slowing their development and making them more vulnerable to predators.

However, in general, artificial selection is a trade-off between the creation of traits desired by humans, such as high sucrose yield, and an unintended but inevitable weakness to drought, disease, and insects. Some high-yielding varieties are also high in starch, which can be troublesome, as shown below. Wild varieties live in a "tough neighborhood" and have been equipped by natural selection to take care of themselves.

Although ease of refining is not necessarily a high priority in the development of new cultivars, the levels of phenolics in raw sugar are known to influence performance on bone charcoal as well as the tendency of color to be retained in the crystal. These observed effects are related not only to the properties of the phenol moiety but also to the size and mobility of the molecule as a whole.

One measure of phenolic levels in the colorant fraction is the *indicator value*, so named because of the color change with pH, defined as the ratio of 420-nm liquor color at pH 9.0 to that at 4.0 [11]. The color change is primarily one of intensity rather than hue, unlike the sharp, brilliant color changes associated with the sulfonaphthaleins. The second derivative of 420-nm color vs. pH displays a single maximum near pH 8.5, which suggests that the pK values for the sensitive species are near this value. In fact, there are also nonphenolic contributors: Caramelization and alkaline degradation products are intermediate in indicator value, with melanoidins at the low end of the scale. Anthocyanins, present in whole cane juice as well as in berries, apples, grapes, red cabbage, and cherries, change color dramatically with shifts in pH, but such substances are sensitive to heat, oxidation, and metal ions and are not believed to be important refinery colorants.

In other work, Smith observed that high-molecular-weight, less pH-sensitive colorants show higher absorption over most of the visible region than that of low-molecular-weight more pH-sensitive colorants [3]. Thus of two liquors with the same 420-nm color, the sample containing less pH-sensitive colorant will be visually darker. In other words, comparison of 420-nm colors is valid only for samples containing similar distributions of colorant indicator value. This stipulation is almost never observed in practice and illustrates one of the pitfalls of using a single wavelength rather than an integrated measurement as an indication of expected ease of refining a raw sugar. Such a measurement, such as at 420 nm, is generally an important quality factor in purchase contracts.

The *hue* of a given colorant sample is an intensive property independent of concentration that can be designated precisely by an absorbance at a specific wavelength. By international convention this is set at 420 nm, although other wavelengths are in use. Since there are no maxima in the visible region for sugar samples, any wavelength is theoretically as good a choice as any other. However, there is also a sensitivity issue, because sugar liquors display strong absorption in the blue region of the spectrum and little in the red. The 420-nm wavelength was chosen because sugar samples have a dominant wavelength in the visual yellow; 560 nm is sometimes preferred because this is approximately the most sensitive wavelength to the human eye. Similarly, because the absorbance is dependent on acidity, it is standard practice to perform the measurement at pH 7.0, although, again, alternative pHs are also used. Consequently, an absorbance reported without a pH and wavelength either indicated or implied is necessarily ambiguous, although a fairly good correlation exists between pH 7.0 color and that at pH 8.5.

The practice of measuring sugar color at 420 and 560 nm can be traced back more than 70 years, to Peters and Phelps [12]. It is sometimes recommended that sugar colors be measured in the ultraviolet region 270 to 280 nm, where most colorant fractions exhibit maxima, and in fact some customers have ultraviolet absorbance specifications. However, in general the argument has prevailed that the visible region is preferable, since the ultimate basis for such a specification is appearance to the human eye.

Since the absolute concentration of the colorant is ordinarily indeterminate, an arbitrary scale based on solids content is used. Note that, as indicated above, the system is valid only when comparing two solutions of identical colorant composition, since the "color" is proportional to the sum of the products of concentration and specific absorptivity of the individual absorbing species. Hence the concept of a standardized color measurement procedure applicable to all sugar samples from all sources is inherently and fundamentally flawed. This is particularly the case when samples of beet and cane origin are compared, and it is clearly incorrect to state that cane and beet sugars of, say, ICUMSA 25, are "identical" in color. What *is* the same is the computed *attenuation index* value. This is distinct from an absorbance and, critically, does not implicitly attribute the diminution of light between source and detector to a single or specific phenomenon.

Deitz extensively studied the linear dependence that exists between the logarithm of the attenuation index and the logarithm of the wavelength for sugar samples [13]. Despite this remarkable uniformity, it is hard to imagine physical grounds for such a phenomenon, although there could be a statistical basis. In any case, the linearity is only approximate, and it is not possible to reliably predict, for example, the 420-nm absorbance from the figure for 560 nm, and moreover, measurements at the two wavelengths differ in precision.

The phenolic group is polar and ionizable (i.e., weakly acidic) and hence can serve as a "handle" for decolorization, whether by bone charcoal, activated carbon, or ion-exchange resin. At the same time, the high reactivity of phenolic compounds allows their participation in complex oxidation and condensation reactions, forming intensely colored polymers. For example, *melanins* are very dark substances that can be formed by reaction of phenols with amino acids. In addition, oxidative enzymes such as polyphenol oxidase in raw juice catalyze darkening reactions that are responsible for a large portion of raw sugar color.

Most colorant molecules, phenolic or not, have weakly acidic properties, and the anionic charge density consequently increases with alkalinity, although the pH zone acceptable for sugar refining is considerably limited. However, considerations of acidity alone cannot fully characterize the ionic charge on such molecules. Williams [14] found that colorants display an almost linear relationship between molecular weight and net charge in the range 300 to 1000 Da, but of course this region encompasses only a small fraction of the total color.

Under conditions of high pH, or in the absence of salts, high-molecular-weight colorants are expanded by mutual repulsion of their charged groups to distended rodlike configurations of increased rotational cross section, similar to the polyethylene glycols. With high ionic strength, however, or at low pH, where the colorants approach their isoelectric points, the molecules contract to a random-coil or globular configuration, so that their molecule sizes are close to those of dextrans of the same molecular weight [15,16] Such transformations clearly influence both kinetic and thermodynamic components of the adsorption process.

More polar colorants should be best adsorbed at polar sites, such as on a bone char hydroxyapatite surface or at resin ion-exchange sites. Less polar species are best removed at carbonaceous sites or on the resin backbone. However, these considerations are thermodynamic ones: The interaction between solute and adsorbent must be stronger than dispersive forces for decolorization to take place. In addition, high-molecular-weight components, particularly near their isoelectric points, have an unfavorable entropy of solution, which serves as a driving force for removal. But kinetic factors are also very important: Diffusion to the adsorption site is rate limiting for higher-molecular-weight substances, particularly at high sugar concentration.

36.1.2 Enzymatic Browning

One way to limit the phenolic contribution to raw sugar color without genetic manipulation of the flavonoid and cinnamic acid precursors is by rapidly inactivating the enzymes in pressed juice with steam. The benefit is obvious: Color that is prevented from being formed in the first place need not be removed. However, this approach has been little exploited to date. Goodacre et al [17] showed that more than half the original color in pressed juice may be derived from the reaction of amino acids with quinones generated by the action of *o*-diphenol: O_2 oxidoreductase enzyme on phenolics.

The fate of chlorogenic acid in the raw juice enzyme environment has been studied extensively and can serve as a model to predict reactions of similar species. The *o*-quinone resulting from enzyme-catalyzed oxidation of the acid is reduced by a secondary *o*-diphenol, and the secondary quinone thus formed polymerizes to dark material. Alternatively, the chlorogenic quinone can react with amino acids or other amino compounds, polymerizing to intensely colored substances [18]. These high-molecular-weight products exhibit reduced pH sensitivity—because phenol functions have disappeared in the processes of polymerization—and enhanced tendency to preferentially boil into the crystal in the manufacture of raw sugar.

Smith [19] found about two dozen colorants in enzyme-inactivated first expressed juice, of which he estimated that above five were the main enzyme substrates, since these substances were either absent or significantly reduced in concentration in normal juice.

o-Quinones, typically red in color, can also be formed in the absence of enzymes, such as in the refinery wherever there is an oxidizing environment: air, elevated temperatures, and especially the presence of traces of iron. Soft sugars darken naturally during transit and storage, particularly during hot weather. The color increase is accompanied by an increase in the red component of the reflectance color, probably from the formation of such quinones. One way to monitor the hue of soft sugar is by plotting the absorbance at three wavelengths, such as visual red, yellow, and green, using triangular coordinates.

The darkening of soft sugar in storage displays approximately zero-order kinetics, indicating a surface reaction with air probably strongly adsorbed on the crystal surface [20]. The color formation is not diffusion-controlled because it displays an appreciable temperature dependence. [Diffusion has a $\sqrt{T}$ dependence, whereas chemical reaction depends on $\exp(-E/RT)$.]

36.1.3 Reflectance Color

In North America brown sugars are often characterized by *reflectance* color, because brightness and luster are considered desirable attributes for such products. Since sugar colorants strongly absorb blue and violet light from the visible spectrum, the reflected light tends to be yellow, yellow-orange, or—at high saturation—brown. However, in this case the "color" depends not only on the absorbing species present but also on grain size, moisture content, and the presence of suspended solids or colloidal substances in inclusions and in the syrup coating the crystals. Nevertheless, at any wavelength, measurement indicates a position intermediate between zero gloss (matte) and a perfect mirror. Although the physical principle does not require the light to be in the visible range, it is hard to imagine a practical significance for other wavelengths, at least for food products.

Red source lamps or red filters are commonly used for such measurements, because red reflectance seems to correlate best with a product that is attractive to the eye—that is, one that is glossy golden brown or reddish brown. Appearance is known to far outweigh flavor characteristics in consumer acceptance and preference. In addition, the red region is relatively sensitive to colloidal impurities, since sugar colorants absorb strongly at blue wavelengths but very weakly at red ones. However, red reflectance measurement does not accurately describe our perception of hue variability, since we do not view the world through a red filter.

Red reflectance will, of course, be highest when green impurities, usually associated with iron contamination, are absent. Through a red filter, red is indistinguishable from white (high reflectance) and green—or more correctly, cyan (blue-green)—appears black (low reflectance). Thus red reflectance detects variability in hue as a linearity or a continuum of grays, an axis for which if x represents redness, $-x$ represents "cyan-ness." Desirable red constituents appear to be formed during refining, probably by invert degradation or formation of *o*-quinones.

Reflectance measurement on sugar samples is of the *diffuse* type. Light from initial interaction with the sample encounters a new interface after traveling a short distance; after repeated interactions, the sample appears uniformly bright in all directions. It is as if the crystals were tiny mirrors, with surfaces inclined statistically in all directions. Our everyday experience with (plane) mirrors is described by a different sort of reflectance, called *specular*. This refers to light reflected from a smooth or polished surface, in which the angles of incidence and reflection are equal, following Huygens' principle. This sort of reflectance is not useful for characterizing sugar samples, as is evidenced by the fact that you can't see your reflection in a cup of sugar.

Outside the United States and Canada it is common to monitor *white* sugar "color" (i.e., gloss) by reflectance measurement. Although the method is attractive because of its speed and simplicity, the correlation with solution color tends to be poor, in part because reflectance is affected by crystal size distribution and the presence of sugar dust. Still, there are commercial systems that can supply a solution color equivalent by reflectance measurement, when the particle size distribution is fairly constant.

The dominant *reflectance* wavelength for white sugars is about 576 nm. An ICUMSA subcommittee found (1970) that a reflectance ratio of 626 nm to 426 nm gave a good correlation with a set of artificially colored standard developed by the Braunschweig Sugar Institute in Germany, and those standards subsequently became the basis for an official analytical method for white sugar reflectance color measurement. The system probably works because the 620-nm wavelength is near the dominant and measures something related to hue, whereas there is very little reflectance at 426 nm, so this wavelength is sensitive to gloss, including the contribution from crystal size distribution.

Since 1970 there has been interest by ICUMSA in evaluating trichromatic or tristimulus measurements for possible correlation with other reflectance techniques. Such systems are appealing because they are designed to be universal and therefore not sensitive to instrument design or idiosyncrasies of human perception. One such system, the L*a*b, described in 1976 by the CIE (Commission Internationale d'Éclairage), permits a common terminology for designation of color in terms of three coefficients.

However, there are limits to any such correspondence because most of the other measurement systems are linear. Although color variation is almost always three-dimensional—brightness, saturation, and hue—not all three are necessarily of practical importance for a given application. Thus the studies conducted so far have shown, not surprisingly, that for sugar samples the correlation between linear and tristimulus systems is to a single trichromatic element—chroma (i.e., brightness or glossiness).

Tristimulus color differences between the colors of sugar samples may be of limited significance, and the technique may be too insensitive to be of practical value. There are theoretical problems as well. The values of the three parameters are not related to the colorant properties in a simple way and in fact designate the color *as observed* under strictly standardized conditions.

Reflectance standards, historically prepared by mixing white sugars with coloring agents, have their limitations, such as in stability and the fact that their color quality may differ from that of sugar samples. Moreover, since most high-quality white sugars are within a narrow color region, only a small portion of the range defined by the standards is often used. Ceramic reference materials may be a superior choice for such work.

36.1.4 Fluorescent Components

Commercial cane sugars always exhibit at least some slight fluorescence, resulting from trace levels of certain colorants and possibly uncolored substances as well. Many sugars display a fluorescence maximum near 440 nm, which is a summation peak from a number of components [21]. Fluorescence, the absorption of light at one wavelength and its emission at a longer one, increases much faster than color in caramelization and invert degradation reactions, but both increase at the same rate in melanoidin-forming reactions. Thus the fluorescent fraction is clearly complex, and the components are not necessarily phenolic. One such substance, DGU, is described in Section 36.1.7.

Fluorescence correlates with structural features also responsible for color: delocalized π-electrons that can be placed in low-lying excited singlet states. The —OH group, which tends to increase the transition probability between the lowest excited singlet state and the ground state, enhances this process. Molecules with a nonbonding pair of valence electrons, such as phenols, often fluoresce because such electrons can be promoted without disruption of bonding. Given a series of aromatic compounds, those that are the most planar, rigid, and sterically uncrowded are the most fluorescent. If it affects the charge of the chromophore, changes in system pH may also influence fluorescence [22].

There can also be inorganic participation. Diamagnetic nonreducible cations can form fluorescent coordinate compounds, but transition metals with unfilled outer *d*-orbitals will quench fluorescence completely.

36.1.5 Caramelization

Maillard reactions can occur even at room temperature, but caramelization normally takes place at elevated temperatures. Because process temperatures rarely exceed 100°C, thermal carbon–carbon σ-cleavage does not occur, and the degradation is a thermolysis but not a true pyrolysis. Traces of impurities strongly catalyze the thermal degradation of sucrose, which is one reason why its reported melting point covers the range 160 to 188°C. A precise melting determination cannot be achieved by observation and requires the use of DTA (differential thermal analysis).

When ice melts, the individual molecules in the water formed are about as close together as they were in the ice (actually slightly closer), but the regular crystalline pattern has been destroyed. The energy required to accomplish this change at constant temperature is the enthalpy of fusion, 80 cal/g. However, in the case of sugar, the forces holding the molecules together in the crystal are so strong that the energy required to disrupt the crystal is nearly as strong as the forces holding some of the individual atoms together in the molecule. Consequently, sugar does not melt "neatly" like ice but instead, tends to degrade. As the melting point is approached, water vapor is formed from hydrogen and oxygen atoms in the molecule, and a graphitelike material of high carbon content is produced. Such systems of conjugated unsaturation and the accompanying extensive electron delocalization have the potential for displaying high color.

Sucrose caramel is a complex mixture of mono-, oligo-, and polysaccharides, together with colored degradation products. The composition depends on reaction time, pH, temperature, and the presence of impurities. Oxygen does not influence the tinctorial power of the caramel formed but can affect its solubility in water or acid [23]. Caramelization produces important flavorants as well as colorants, and these can lend harsh rather than pleasant flavor notes to a brown sugar, refiners' syrup, or commercial caramel color.

Certain of the caramel components may be colloidal, because as the hydroxyl content drops and the carbon content and molecular weight rise, the molecule can less fully interact with water. The reactions involved appear to be a mixture of first-order dehydrations and second-order condensations. Although the reactions are highly complex, it is generally agreed that the first step is the splitting of the glycosidic linkage of sucrose.

Some darkening is inevitable in the thermal operations of sugar refining, particularly in evaporators, but color formation can be minimized by careful process control. Even if the reactions that occur represent only a small sucrose loss, color increases can be significant. The caramels typically are present in small quantities but tend to attach themselves to the crystal surface and can contribute substantially to solution color [24].

Many carbohydrates exhibit patterns of caramelization very similar to those of sugar. For example, thermal degradation of sucrose, starch, lactose, glucose, β-D-glucosides, and even cellulose all yield 1,6-anhydro-β-D-glucopyranose (levoglucosan), which can be further dehydrated to 1,6-anhydro-3,4-dideoxy-delta 3-β-D-pyrosen-2-one (levoglucosenone) [25]. Nucleophilic addition to the double bond in the latter could result in further reaction. Levoglucosan itself is known to pyrolyze to branched-chain species with molecular weights up to 5×10^4 [26], perhaps via the levoglucosenone intermediate. Water elimination, mentioned above in connection with melting, does not require solid-state sugar but only fairly high liquor densities.

Commercial caramels for food color ingredients are produced according to recipes developed over many years, and such manufacture may be considered an art rather than a science. Practices have evolved to generate stable and reproducible color quality, and the mechanisms of darkening and exact composition of the product have been of secondary interest. The manufacturers of caramel color are not eager to publish the secrets of their art, and the complexity of the problem has also contributed to a dearth of published information, although an early article by Salamon and Goldie may be of interest [27].

The starting material for the caramel color of commerce is typically corn syrup rather than sucrose. Such color additives are widely used in the food industry in such products as dark breads and cola beverages and are sometimes used to darken brown sugars produced by blending rather than crystallization. In the United States this practice requires a declaration of caramel as an ingredient on the package label. The sulfite content of commercial caramels may be considerable, although low-sulfite caramels have been available for about 10 years. Manufacturers typically specify the surface charge and isoelectric point of their products, which are important stability parameters for particular food and beverage applications. The color specification may be at 610 nm, according to the Food Chemicals Codex method, suggesting that these products are redder than the substrates formed in the refinery.

36.1.6 Maillard Browning

Such nonenzymatic browning is more than a simple Schiff base condensation between a carbohydrate carbonyl function and an amino acid nitrogen group. It is, in fact, a complex scheme of many reactions, typically in the pH range 4 to 8, that can culminate in formation of large intractable polymers. There is considerable evidence for free-radical participation in accompanying oxidative processes. Sucrose itself, being nonreducing, can engage in such reactions only after inversion. However, low levels of invert are present in all but the highest-purity refinery streams and hence available for darkening.

In a complex medium such as a raw sugar liquor being processed at elevated temperatures, it may not be possible to draw sharp demarcation lines between this form of darkening and other mechanisms of degradation, especially because of the effect of pH. Maillard reactions can also occur in living tissue, between glucose and proteins, and are responsible for several complications of diabetes and of the aging process.

Because Maillard reaction products, sometimes called melanoidins, can be very dark, their contribution to color can be substantial despite the relatively low levels of proteinaceous matter in cane juice, less than 0.5% on solids. The nitrogen content of sugarcane itself amounts to only a few hundredths of 1%, of which about half is present in forms potentially active in color-forming reactions. Beet process streams are higher in nitrogen compounds than those of cane and consequently, more susceptible to such darkening. There is some evidence that some of the reaction products are weakly allergenic, presumably from a protein component.

Amino acids do not react in their cationic forms, only slightly as their zwitterions, and fully only as their anions (i.e., at pH levels above the isoelectric). If the nitrogen compound participating is, in fact, an amino acid, Strecker degradation usually takes place, whereby CO_2 is liberated and an aldehyde or ketone is formed with one carbon atom fewer than in the original amino acid.

The browning can be inhibited by sulfite addition, more likely to be encountered in the United States in the beet industry, but such use here has declined as a result of U.S. Food and Drug Administration (FDA) concerns about residual sulfite in refined sugar products. The FDA estimates that one out of 100 people is sulfite-sensitive. Severely ill asthmatics dependent on corticosteroid medication are especially at risk. Exposure to even low sulfite levels can result in a massive allergic reaction: anaphylactic shock, a life-threatening condition.

The initial step of the reaction is believed to be formation of *N*-(D-glucosyl) amino acids, which are converted to D-fructose amino acids by the Amadori rearrangement [28] (Fig. 36.1). The reactions are usually considered to follow zero-order kinetics, after an initial induction period. The rate is strongly moisture-dependent. At low moisture the reaction is controlled by diffusion of reactants, but at moisture levels greater than those for maximum browning, water dependence becomes negligible.

Numerous studies have been carried out with model compounds, in attempts to elucidate the browning reaction mechanism and rates. Much information has been obtained from these investigations, but the actual process in a multicomponent sugar liquor is far too complex for any comprehensive laboratory simulation. The influence of various amino acids and temperature on the rate of darkening has been studied by Imming et al [29].

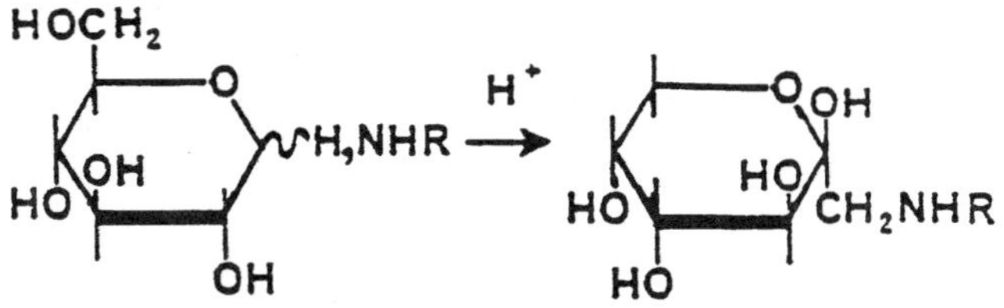

FIGURE 36.1 Initial step in one route of Maillard browning.

Binkley, reporting on melanoidins in cane final molasses, found an empirical formula of $C_{17\text{-}18}H_{26\text{-}27}O_{10}N$, which suggested two six-carbon units bound to a four- to five-carbon amino acid residue [30]. Other researchers have reported similar empirical formulas. Binkley also found an average molecular weight of about 5000, with some material at about 50,000 Da. The carbon/hydrogen ratio suggests a high level of unsaturation, probably resulting from a combination of carbon–carbon double bonds, carbonyls, rings formed by cross-linking, and the inclusion of aromatics. Although the empirical formula corresponding to the elemental analysis does not permit a unique structure to be written, one might predict that a substance with such a formula would be colored.

The familiar condensation of amino acids with carbohydrate carbonyls and complex further reaction forms not only colored polymers but also volatile pyrazines, alcohols, acids, and aldehydes, which are important contributors to the aroma of refiners' syrup, soft sugar, and molasses (see, e.g., Ref. 31). Similar compounds contribute importantly to the flavor and color of many roasted or browned foods, such as cocoa, bread crust, potato chips, brown beans, cooked meats, and coffee. To list just a few examples, the reaction products of carbonyl compounds with amino acids results in the following aromas:

- Aspartic acid
  - At 120° — Bread crust
  - At 180° — Caramel
- Glutamine, asparagine — Butterscotch
- α-Aminobutyric acid — Maple
- Histidine — Toffee

The Maillard reaction is most likely to be encountered at low moisture levels, that is, high Brix. Such browning and flavor development does not occur readily in the microwave oven because the surfaces of the heated foods tend to be moist and the temperature does not ordinarily exceed 100°C.

The Maillard aromas are not necessarily pleasant, particularly some of those formed at higher temperatures, which—not surprisingly—may be described as "burnt sugar." Excessive roasting leads not only to the development of burned or charred flavors and bitterness but also to the loss of desirable flavor components. The olfactory properties are, of course, determined by the sugar as well as the amino acid. Because of their protein source, many amino acids can be found in raw sugar, although in one study performed at C & H Refinery in the United States, aspartic acid and asparagine accounted for about 60% of the total fraction.

A number of important flavorants in sugar products are not nitrogen compounds at all. Maltol is a fructose degradation product. Some phenolic flavor components are astringent and smoky, but there can be complex contributions from furans, acids, ketones, and aldehydes. Certain flavor components are of microbiological origin. High iron levels in soft sugars result in a metallic taste, even though the iron is present in an ionic state.

36.1.7 Invert Degradation Products

When sucrose or fructose is heated under mildly acidic conditions, the cis and trans isomers of 3,4-dideoxyglucosulose-3-ene (DGU) are formed via various routes. DGU, which displays a bright green fluorescence, is the immediate precursor of 5-(hydroxymethyl)-2-furaldehyde (HMF). Although the principal decomposition products of HMF are the colorless levulinic and formic acids, it is also an important color precursor, polymerizing to brown products [32]. Under acidic conditions glucose degrades faster than fructose, but with much less color development.

For sugars sold to beekeepers, it should be noted that HMF levels above 60 ppm may be toxic to honeybees, with a mortality rate proportional to log[HMF] [33]. These authors also studied the rate of HMF formation as a function of pH (Table 36.1). Under alkaline conditions, a completely different set of reactions takes place. In the high-pH region, sugars easily isomerize and condense with amino compounds, as seen above. For sucrose, the first step is dehydration of the fructose moiety. The degradation proceeds via a complex scheme of aldolization, Lobry de Bruyn–van Ekenstein isomerization, β-elimination, α-dicarbonyl cleavage, and benzilic acid rearrangement. However, most of the reaction products are uncolored species. One reason why there is no continuum between high- and low-pH degradation products is that the acid-catalyzed sequence proceeds via reductones, which are unstable at pH 6 and above [34] (Fig. 36.2).

Calcium ion inhibits high-pH color formation, whereas sodium and potassium enhance it. At low pH, sodium and phosphate inhibit darkening, and calcium enhances it; chloride has no impact. Because all of these species are likely to be present in a raw sugar liquor, the predictive value of such information is limited. Although the terms *high* and *low pH* are relative, it should be clear that only pH values fairly close to 7.0–8.5 are appropriate for sugar refining.

Citrate has a striking effect on color formation—complete inhibition at low pH and strong enhancement at high pH [32] (Table 36.2). It should be noted that citrate is the immediate precursor by dehydration of aconitate, the principal organic anion in raw sugar. In some cases these additives probably do not alter the rate of summed sugar degradation reactions so much as redirect them toward more or fewer colorless products.

36.2 NONSUCROSE CARBOHYDRATES

Although glucose and fructose, from sucrose inversion, are probably present in at least trace quantities in all refinery streams and are of keen interest for process control purposes, these substances are beyond the scope of the present chapter, which, instead, focuses primarily on polysaccharides. These are long-chain polymer molecules made up of simple carbohydrate units, such as glucose, in chains either linear or branched. Such diverse materials as starch, cellulose, and dextran are all composed of glucose chains differing only in molecular structure. Polysaccharides contribute to liquor and syrup turbidity and viscosity, with highly branched molecules having a smaller impact. The level of combined polysaccharides in raw sugar is typically 0.1 to 0.2%.

Soluble polysaccharides can give rise to false polarization (pol) contributing to error in the measurement, usually too high. They tend to increase moisture retention, and reduce

TABLE 36.1 HMF Formation as a Function of pH

pH	Relative Rate of HMF Formation	Relative Rate of Inversion[a]
3.30	1.00	1.00
3.15	3.94	1.41
3.04	9.25	1.82
2.88	18.3	2.63
2.70	34[a]	3.98
2.50	57[a]	6.31
2.30	100[a]	10.00

Source: Data from Ref. 33.

[a]Calculated or extrapolated by the author.

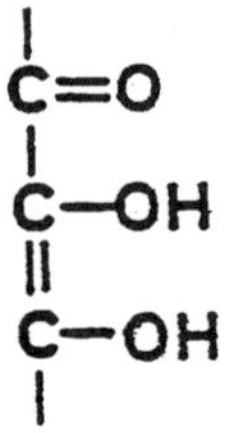

FIGURE 36.2 The reductone structure is an intermediate in low-pH invert degradation.

yield, and can adversely affect crystallization, contributing to false grain, small crystals, conglomerates, and elongated crystals.

Cellulose and hemicellulose, which are components of the cell wall in sugarcane and other plants and provide structural strength to the standing cane plant, are insoluble in water and usually of minor concern in sugar manufacture. They are not considered here. The woody sugarcane fiber that is the residue of milling is, of course, familiar as *bagasse.*

Sucrose itself is a disaccharide (i.e., a two-sugar compound molecule), composed of a glucose unit linked to one of fructose. The molecule can be broken down irreversibly, "inverted," under even mildly acidic conditions. The term *inversion* refers to the property of optical rotation—pol, when measured on a saccharimeter—that accompanies this degradation: the instrument reading gradually changes from positive to negative as the breakdown progresses. However, despite the low pH in the stomach, ingested sucrose is not broken down until it reaches the small intestine, and under catalysis not by acid but by an enzyme.

As every schoolchild is supposed to know, glucose is the primary product of photosynthesis in the leaves of green plants, and in the actively growing portion of the sugarcane its concentration exceeds that of sucrose. Sucrose, formed in a subsequent biosynthetic pathway, provides a means to transfer energy from the leaves to storage in other portions of the plant, such as the stalk of sugarcane or the root of a sugar beet. In mammals, the disaccharide lactose or milk sugar functions in a parallel fashion, to transfer energy from mother to infant. Glucose is the principal energy source in the body and the only sugar

TABLE 36.2 Darkening at High pH

pH	Relative Rate
8.0	1.00
8.5	1.23
9.0	1.59
9.5	2.16
10.0	3.16
10.5	5.18
11.0	10.0
11.5	25.1
12.0	100
12.5	1000

Source: Data from Ref. 35.

normally found in blood. Other simple sugars and complex carbohydrates must be converted to glucose before they can be utilized.

36.2.1 Starch

Sugarcane starch is largely present in the form of spherical granules about 5 μm in diameter, although there is considerable variability in size. Some plant starch granules are as large as 100 μm; that from sago palm is intermediate, about 50 μm. In mature cane stalks it is concentrated at tops and nodes, with considerable quantitative differences among varieties. Unlike sucrose, starch is made in the chloroplasts—the elements of the lead cell that perform photosynthesis; biosynthesis occurs during the day, when there is exposure to light energy.

The linear portion of starch is called *amylose,* and the branched, higher-molecular-weight component is *amylopectin.* The latter structure is approximately spherical with a density decreasing outward (i.e., a Gaussian coil). Amylose is somewhat water soluble and amylopectin more so. The branched structure permits more intimate association with water, and this overrides the opposite effect of molecular weight.

The familiar blue color formed by the action of iodine on "starch" is actually a reaction with amylose; amylopectin provides a red-violet color, the basis for some colorimetric determinations. The chromophore is a nearly linear I_4 unit stabilized within the cavity of the amylose helix. The amylose content of cane starch is about 15% and is preferentially removed during raw sugar manufacture.

As the plant produces starch, it deposits the molecules in successive layers associated through hydrogen bonding into micelle bundles. These highly oriented crystalline areas are responsible for the birefringence of ungelatinized granules. The wetting process is exothermic. As the granule is heated in water, the weaker hydrogen bonds in the amorphous regions are ruptured, destroying the internal structure, and the granule swells with progressive hydration, as endothermic process called *gelatinization.* As more water is imbibed, the amylose molecules are eventually solubilized and leach out into solution, where they may reassociate into aggregates. This congealed paste eventually loses water and shrinks into a mass of unappetizing rubbery consistency, in a process called *retrogradation,* the bane of cooks and pastry chefs. As the entrapped water is expelled, liquid accumulates externally to the gel, in a separation called *syneresis.* Such "weeping" is a major defect in stored gels such as puddings. Freeze–thaw cycles are also known to promote syneresis. Slow recrystallization in amylose regions results in staling in baked goods.

Raw sugar typically contains 200 to 400 ppm, although some poor varieties can contain as much as 2000 ppm. In addition to the strong genetic component, levels in raws are affected by climate, soil, and cane maturity at time of harvest. Levels tend to be lower in raws from irrigated areas and higher in those from wet regions, where burning is likely to be less complete. The use of chemical ripeners is associated with increased starch levels. Variability in starch content is sometimes attributed vaguely to "geography," but it is not clear how the growing plants would know their location, given constant growing conditions.

During milling the starch granules are separated from the plant tissue and become a component of the juice. The granules are gelatinized during heating and liming, and the solubilized starch passes into the clarified juice and is incorporated into the raw sugar.

The threshold for added difficulty in refining is about 250 ppm. Affination is not an effective means of removal, because starch is present throughout the crystal, not concentrated at the molasses film. It retards crystallization, reduces recovery, and impedes filtration. The effect of viscosity is generally negligible, although one characteristic of starch is that its viscosity in a fluid can *increase* with temperature, perhaps contrary to expectation. The deleterious effect is greater on carbonation than on phosphatation. Only a portion is removed in clarification, and the remainder tends to be occluded within the sucrose crystal. α-Amylase enzyme has GRAS status from the FDA and may be used as a processing aid in the United States.

36.2.2 Indigenous Sugarcane Polysaccharide

Indigenous sugarcane polysaccharide (ISP) was first isolated from fresh cane juice in Louisiana in 1964, under conditions that precluded any starch or dextran presence [36]. Although early studies suggested that the structure was composed almost entirely of glucose units, it was later firmly established that the material was in fact an arabinogalactan. As the name suggests, this refers to a polymer containing appreciable levels of the monosaccharides arabinose and galactose. The structure also was found to contain about 8.5% glucuronic acid, exhibiting a strong negative charge down to pH 3, and displaying a negative specific rotation.

ISP, a highly branched cell-wall polysaccharide, has a molecular weight in the range 100,000 to 300,000 and is soluble in cold water. Maximum levels in raw sugar are estimated to be about 2500 ppm, with perhaps 10% surviving in granulated. Associated with this substance in the cell wall is a lower-molecular-weight (below 50,000 Da) glucomannan, a highly branched, very soluble material.

36.2.3 Heterofructans

Heterofructans are high-molecular-weight substances that accumulate in the cane stalk and are of interest primarily because they are reportedly built up of units of fructose and the rare sugar alcohol *galactitol* [37].

36.2.4 Dextran

Unlike the polysaccharides described above, dextran is not found naturally in healthy, sound cane but instead results from bacterial infection by *Leuconostoc mesenteroides.* This organism, of which more than 100 strains exist, is a facultative anaerobe, which means that it is neither killed by oxygen nor does it need oxygen to live. The strain associated with any particular dextran affects the branching architecture, molecular weight, solubility, and specific rotation [38–40].

L. mesenteroides is ubiquitous in soils and can be airborne, but contamination generally requires prior injury to the rind of the cane stalk. Infection readily occurs in damaged sugarcane, such as from frost, storms, burning, and insect or rodent damage; or as a result of delays in cut-to-crush times during poor weather harvesting. In the refinery, dextran is associated with poor housekeeping, such as allowing cold dilute liquors to stand for prolonged periods.

The less well known *L. dextranicum* can form dextran with astonishing rapidity, completely solidifying a 15% solution in less than 48 h. Although such high activity sounds almost diabolical, it should be remembered that this particular organism has evolved to exploit the particular niche of a high-sucrose environment and does so very effectively. For similar reasons, bacteria responsible for dental plaque are found in your mouth, where sugars are regularly present.

Dextrans are glucans (i.e., glucose-containing polysaccharides) in which the main-chain residues are at least 50% α-(1,6)-linked, with α-(1,4) and α-(1,3) linkages at the branch points. The bacteria link up the glucose portions of sucrose molecules into long chains, leaving the fructose as a by-product. Lactic acid is also associated with dextran production.

Most dextrans have high molecular weights, in the range 100,000 to 1,000,000 Da and higher. They are normally soluble in water, insoluble in alcohol, and highly dextrorotatory, with specific rotations of +200° and greater. Since this figure is more than three times that for sucrose, high levels can contribute to a false pol reading. Not only will the refiner be paying for sucrose not received, but the false polarization readings will continue to occur throughout the refinery, particularly in low-purity materials.

Dextran interferes with virtually every refinery operation, decreasing yields and efficiencies. Viscosity increases and filterability decreases. Affination is not particularly effective for removal because dextran tends not to be concentrated in the molasses film. Neither are carbonatation or phosphatation efficacious. Bone char and activated carbon are likewise ineffective at removal, and moreover, the increased viscosity can inhibit diffusion and therefore adversely affect decolorization.

The problems continue in decolorized liquors. There is a pronounced decrease in crystallization rate, particularly in low-grade boilings. Purging of massecuites can be more difficult, as a result of elongated crystals, which pack less densely and hence are less easily washed. Dextran tends to be transferred to the crystal and cannot be effectively separated in the syrup. Low-grade massecuites are particularly difficult to centrifuge, resulting in high final molasses purities. Dextran also has a melassigenic effect, decreasing sucrose recovery. Because of these difficulties in removal, clearly the preferred strategy here is prevention. No bactericides have FDA approval for refinery use.

Dextran persisting in granulated sugar introduces problems in the production of cordial-type beverages, because the alcohol contained in these products decreases its solubility, causing turbidity. High-dextran soft sugar will be sticky and have poor flow characteristics in the refinery conveyance system. High dextran is also known to cause problems in candy-making processes.

Dextran was first identified as a problem in the sugar industry in the late 1950s in Australia. The introduction of chopper harvesters in Queensland in the 1960s resulted in frequent problems, leading to intensive research on the topic. An International Dextran Workshop sponsored by Sugar Processing Research, Inc., was held in New Orleans in 1984. The published proceedings will be of interest to those studying this subject.

Measurement of dextran levels in raw sugar and refinery products is obviously of great interest, but the several procedures used tend not to show close correspondence (Table 36.3), largely because what we term "dextran" is a heterogeneous mixture of substances with similar molecular structure rather than a discrete compound like sucrose. The widely used but time-consuming alcohol haze test is not specific for dextran and is insensitive to material of molecular weight <500,000 (especially <50,000) or levels below 200 ppm. The test is inherently flawed because the result depends on the geometry of the colorimeter detector. Nevertheless, the method has fairly good predictive value for raws likely to present processing problems.

The haze test provides measurement in milli-absorbance units. If conversion to ppm is desired, a calibration curve with standard dextran of molecular weight 100,000 or higher

TABLE 36.3 Dextran Analytical Methods Compared

Method	ppm[a]
Roberts	474
Haze	159
Enzyme / HPLC	174

Source: Data from Ref. 42.

[a]Average of 16 raws.

must be prepared [41], the form of which is exponential rather than linear. Such optical determinations depend on particle size, and if a high polymer such as dextran is "statistically coiled" to resemble a rough sphere or loose tangle of yarn, the diameter is proportional to the square root of the chain length, but the volume occupied by the molecule varies as the chain length to the $\frac{3}{2}$ power.

Gel permeation chromatography and HPLC (high-pressure liquid chromatography) can provide a good indication of molecular weight distribution, but the peaks typically include nondextran components. Enzyme procedures and immunological methods can provide high specificity but are not widely used. The Roberts method, which has AOAC (American Organization of Analytic Chemists) status, isolates the total polysaccharide fraction, then precipitates the dextran with copper sulfate. The method tends to overestimate the dextran level, although its sensitivity is greater than that of the alcohol haze test. Spectroscopic methods, such as NMR (nuclear magnetic resonance) and NIR (near infrared), have also been used.

Dextran should not be confused with *dextrin,* a modified starch used as a thickener and stabilizer in processed foods.

36.2.5 Sarkaran

This polysaccharide, associated with deteriorated cane and assumed to be of microbiological origin, was detected in South Africa mills and determined to be different from dextran or starch [43]. Structural study indicated that it was an α-glucan of molecular weight about 200,000 Da.

36.2.6 Floc Formation

Cloudiness in sweetened beverages is a commonly encountered customer complaint. In the case of acidified soft drinks, the consensus is that negatively charged ISP forms an insoluble adduct with positively charged sugarcane protein. In the case of beet sugar, saponin is the analogous negatively charged species. Other impurities, such as silica, frequently from the water supply, can markedly affect the visual appearance of the floc. Flocs can also result from microbiological contamination from yeasts, molds, or bacteria. In the case of sweetened alcoholic beverages, the alcohol component can result in precipitation of starch, dextran, or other polysaccharides. Charles found, using size exclusion chromatography, that high-molecular-weight dextran (2,000,000 Da) was strongly transferred to the crystal during boiling, and most of this material precipitated at alcohol levels of 30 to 45% [44]. A review on beverage floc was published by Morel du Boil of SMRI, South Africa [45].

36.3 CARBOXYLIC ACIDS

Phenolic acids found in sugarcane have been described in the discussion of colorants above. Some of these substances are uncolored themselves but are precursors to colored species. In raw sugar these substances are present at levels of 0.1 to 15 ppm, and persist only as traces in refined sugar. The carboxylic acid fraction includes a number of components, of which the largest by far is aconitic acid, which is present in raw juice at an average level of 1.5% [46]. Other constituents include citric, malic, oxalic, glycolic, mesaconic, tartaric, succinic, and fumaric acids, all at low levels. Lactic and acetic acids are by-products of microbial infection. The principal significance of these substances with respect to refining is that they tend to depress the pH, requiring lime addition. The calcium salts tend to be insoluble or sparingly so, although calcium acetate is soluble.

36.4 NITROGEN COMPOUNDS

Like all living things, sugarcane contains protein. The small amount of material present in cane juice, about 0.5% on solids, is coagulated and removed during clarification, although low levels persist in raw sugar. Godshall and Roberts estimated such levels to be 20 to 60 ppm [47]. Residuals in refined products are at trace levels, but sufficiently high to be implicated in acid beverage floc (see above). The principal free amino acid in cane juice is aspartic, with lower levels of glutamic, alanine, valine, γ-aminobutyric, threonine, isoleucine, and glycine. In addition, trace levels have been identified of leucine, lysine, serine, arginine, phenylalanine, tyrosine, histidine, proline, methionine, and tryptophan.

36.5 INORGANIC SUBSTANCES

These constituents are sometimes referred to as *ash,* but this term is not indicative of the minerals actually present in sugar samples, because it designates the oxides that result from incineration of the organic matter, including sugar. No such substances are actually present in sugar solutions. For this reason, the U.S. National Committee, an ICUMSA affiliate, in 1993 changed the name of this subject to *inorganic substances,* a term that makes no inference of a particular analytical procedure.

The amount of inorganics present can be measured gravimetrically by weighing the sample before and after combustion, or by a much faster conductance determination, since such substances conduct electricity, whereas sugar does not. Certain nonmineral impurities, such as most color bodies, also conduct, but so slightly that they have little impact on the measurement. The organic conducting molecules are so large compared to inorganic ions that they move relatively sluggishly in the test medium. The empirical correlation between the two methods is only fair, because the composition of the inorganic fraction affects the conductance, whereas the gravimetric procedure is independent of this parameter. The various components have different ionic mobilities, affecting the conductance result. Thus the ICUMSA official method uses a factor of just one significant figure for converting conductance to percent ash. Some ash constituents may be partially lost during combustion, notably chloride and carbonate. Above 600°C, even "nonvolatile" cations such as calcium can be lost.

Specific ion electrodes are available for detection of such species as calcium, magnesium, chloride, nitrate, potassium, and sodium. Individual constituents can also be determined by atomic absorption or ion chromatography. The inorganic fraction includes such heavy metals as lead, cadmium, mercury, and arsenic, which are of concern from a toxicological rather than a refining viewpoint. The levels of such substances in refined sugar are exceedingly low, because crystallization is such a highly effective purification operation, unavailable to almost all other foodstuffs. Even in raw sugar these substances are present below the ppb (parts per billion) level.

Although granulated sugar is among the very purest of the food available to the consumer, one cannot say that the levels of heavy metals are "zero." This is a philosophical issue rather than a technical one: Zero *concentration* cannot be measured in principle, however sensitive the detection instrument. Any analytical procedure for trace contaminants such as heavy metals must necessarily be described in terms of detection limits, a concept that may be baffling to nonchemists. Such limits actually represent a statistical measure of confidence, and the measuring device may in fact be providing numerical results rather than zero. For example, the detection limit may be defined as three times the standard deviation of the method's results on standardized samples.

The inorganic ions of interest in the refinery are those present in raw sugar, the most important being potassium, calcium, magnesium, silica, sulfate, and chloride. At lower concentrations are sodium, phosphate, and iron. Because the concentrations of these substances are higher in the molasses film than in the crystal interior, affination is an effective means of reducing inorganic levels. In a typical operation, only about 10% of the sugar is separated, but about five-sixths of the ash. The low transfer from syrup to crystal in the mill also is of benefit at the point of final crystallization in the refinery.

Bone charcoal will remove a portion of the di- and polyvalent ions (i.e., calcium, magnesium, sulfate, and phosphate), but is ineffectual for the monovalent species potassium, sodium, and chloride. Activated carbon does not remove inorganics, a fact that may be helpful when responding to customer inquiries asking why the industry continues to use animal charcoal. Cation and anion resins are active toward the corresponding charged inorganics, so that decolorizing resins also partially demineralize. This is accomplished more effectively by a mixed-bed system, which removes both anions and cations.

A refinery syrup using no mineral adsorption unit operation (i.e., neither bone charcoal nor resin) will probably produce refined sugars with somewhat elevated levels of inorganics. This does not mean that such sugars are not of high quality, although the products might not be adequate for certain specialized food and pharmaceutical applications. Furthermore, the limits set by some customers will probably be tightening: The ash limit specified for National Formulary sucrose in the 1995 *U.S. Pharmacopeia* was 0.05%, determined gravimetrically, whereas the limit in the monograph harmonized with the European and Japanese pharmacopeias is likely to be 35 μS/cm, corresponding to about 0.015%. The principal problem posed by "high-ash" sugars is the tendency of calcium to cause turbidity in solution or to precipitate other components.

Silica and iron require special comment. A portion of the former is colloidal rather than ionic, which means that customary removal methods dependent on electrostatic attraction, such as bone charcoal and ion-exchange resin, are ineffectual here. Similarly, iron present largely as complexes rather than free ions, is not removed by traditional procedures, including cation-exchange resin. The special refining issues associated with iron have been described above.

Elevated levels of inorganics, especially calcium, sulfate, magnesium, phosphate, and silica, increase the frequency and severity of scale problems on evaporator surfaces. Such deposits typically contain several components and may also incorporate organic substances, such as salts of carboxylic acids and even carbon from burnt sugar.

Associated with the several inorganic constituents is a specific melassigenic coefficient, which is defined as the number of parts of sucrose that will be taken into the molasses at saturation per part of ion present. Potassium and sodium have the greatest effect. This is of particular interest at the raw mill but also affects low-grade boilings and the purity of final refinery molasses.

Decolorization mechanisms dependent at least in part by electrostatic attraction, such as bone charcoal adsorption, are hindered by the presence of high levels of sulfate. Like most colorant molecules, sulfate is negatively charged, and because it is very small relative to color bodies, it can diffuse much faster to cationic calcium sites on the hydroxyapatite surface of the charcoal. Hence it is a highly effective competitor. Monovalent anions, such as chloride, do not fit this category because they are not adsorbed. Addition of a soluble calcium salt, such as chloride, can compensate, but this is not ordinarily cost-effective and in addition risks aggravation of scale problems and possibly, more inversion of sucrose.

36.6 TURBIDITY

It is a common misconception that cloudiness or opalescence in a liquid is invariably associated with particulates. Turbidity is in fact a purely optical phenomenon, describing the scattering of light as it passes through a medium, imparting a haze. The cloudiness may or may not originate in suspended matter. A scattering center is thus actually an optical inhomogeneity, which *may* be due to a colloidal particle or suspended solid in the dispersed phase but can also result, for example, from local thermal density fluctuation. Such scatter is caused by a refractive index gradient in the medium and can be due to, for example, air bubbles when no particulates are present. On the other hand, if the suspended particles have the same refractive index as the bulk medium, there is no gradient and no turbidity.

The fact that there is the potential for a clear solution when particulates are present and a turbid solution when such substances are absent should serve as a caution in evaluating this phenomenon. It may be tempting to visualize turbidity in a measure analogous to that for light absorbance (i.e., color measurement) which of course is the basis for control in all sugar refining. However, the next step in this conceptualization is apt to be an imagined number of particles, each with a characteristic scattering potential. This road can lead one astray because unlike the case for color measurement, light scattering does not obey the Lambert–Beer law. It is possible and in fact commonplace for the turbidity of a sugar sample to *increase* upon dilution with water, thus defying common sense. This rarely happens in the case of color measurement and will occur only if the chromophores change in a fundamental way upon dilution, a deviation from Beer's law. The most common example of this is the pH-sensitive chromate–dichromate equilibrium, which shifts favoring dichromate (450 nm) upon dilution. It is generally assumed that there are no exceptions to the Lambert law. The obvious corollary here is that colorimeters are inappropriate for turbidity measurement, a topic we return to below.

Attempts to define turbidity sometimes run into trouble because a clean demarcation is assumed where none exists. Colloidal suspensions not only scatter light but also absorb a portion, and the very smallest molecules, such as pure water in what appears to be an optically clear solution, both absorb and scatter light to a small degree. The reference work *Standard Methods for the Examination of Water and Wastewater* [48] defines turbidity as the "expression of the optical property that cause light to be scattered and absorbed rather than transmitted in straight lines through the sample." However, the absorption portion of this description would ordinarily be applied to colorimetry rather than turbidimetry.

It is possible to reduce or eliminate turbidity in a liquid sugar by merely raising the Brix level, but we are deluding ourselves if we imagine that such an operation can possibly improve the quality of the product. We are measuring the wrong thing—light scatter rather than particulates. The turbidity will make an unwelcome encore appearance if the customer dilutes the liquid sugar. This is similar to the case for the apparent improvement in color that we sometimes encounter across an evaporator: We haven't "distilled off" colorants; we have merely failed to consider the pH drop associated with the sugar degradation that inevitably accompanies thermal operations. In fact, none of the colorants associated with sugar refining are volatile substances. The colors of sugar solutions are of course known to diminish with lowered pH because the colorants have the properties of indicators. And just as lowering the pH can reduce the absorbancy of a sample but obviously cannot reduce the *amount* of colorant material present, raising the Brix clearly cannot reduce the level of particulates.

Refraction of light is common in our experience, and we are not surprised when a spoon in a glass of water appears to be broken at the water–air interface or when we try to pick up a pebble in a stream and miss our mark. In a turbid solution, by contrast, the interface between media is not necessarily localized at a single surface but can instead be distributed uniformly throughout the sample. A beam of light striking such a system is refracted not once only but many times, in a myriad of such interface interactions that comprise scatter. The scattering centers behave as new light sources and disperse light in all directions.

In a glossy crystalline solid, such as granulated or brown sugar, a similar phenomenon occurs: Light is *reflected* many times from encounters with crystal surfaces, making the sample appear uniformly bright. This is called *diffuse reflectance* and is very different from the *specular reflectance* that is part of our everyday experience with mirrors. The diffuse variety does not reverse images from left to right, and our experience does not lead us to expect to see our reflections in a cup of sugar.

The role of the clarifier in the refinery, then, is ironically *not* really to clarify but rather to remove particulates. This is of course the case whether phosphatation or carbonation is employed. Some decolorization—in fact, a very significant amount—will invariably accompany this removal, because some of the particulates are colored and also because some colored soluble matter will inevitably be incorporated into the mass eliminated. Although we will have later opportunities in the refinery to remove *soluble* material, the clarifier station is our last chance to remove microscopic particulates. Pressure filtration can be a help, but the particles to be removed clearly must be larger than the smallest openings in the filtration medium, generally about 0.5 μm. Clarification has no such limitation. Inadvertent removal of particulates at the top of an adsorbent bed, such as activated carbon or bone char, is undesirable because it will inevitably form a blanket that inhibits free flow.

It must be acknowledged, however, that clarification processes *are* in practice monitored by turbidity measurement. This is because turbidity is easily measured and offers a practical means for assessing the operation, despite the pitfalls. Particle measurement, on the other hand, is fraught with difficulties. Do we measure mass or cross-sectional area? Or perhaps merely count the particles? Does the technique assume that the particles are spherical, which is most assuredly not the case? Should we use weight average or number average? What about differences in shape and refractive index among the species present? And does an instrument even exist that can accomplish the task we have in mind? It is not the case that the monitoring of turbidity to assess clarifier performance is illogical or somehow "wrong," but it is important to recognize what it is we really want to measure, and its physical basis.

36.6.1 Light Scattering

Light is scattered only when its wavelength is greater than the size of a particle in the dispersed phase. What does this mean in practical terms? The visible wavelength range is about 400 (violet) to 800 nm (red), or 0.4 to 0.8 μm. Smaller particles responsible for scatter may be microscopic but are nonetheless about 500 times larger than ordinary small molecules such as sucrose and are in the range of macromolecules. A very significant portion of the suspended solids in raw sugar, perhaps 20 to 30%, lies in this particle-size zone. In the sugar refinery, important members of this class are polysaccharides, especially indigenous sugarcane starch, and dextran resulting from bacterial infection. The particles may be discrete giant molecules or highly dispersed aggregates. Whether they are true colloids is an issue that can be debated, but they are colloid *sized* and like true colloids display a very high surface area per unit weight compared to, for example, filterable material. Thermal degradation of sugar that inadvertently accompanies processing also produces highly colored polymer materials in this size range.

Although such submicron polysaccharide species will tend not to be mechanically caught in ordinary filter media, it must be remembered that the starch and dextran presence is likely to be highly heterogeneous in size, with some portion of sufficiently high molecular weight (or associated with other matter) that it will tend to be insoluble. In general, the *shape* of the macromolecule should be considered as well as its molecular weight. A sphere or random coil molecule is less likely to get past a pressure filter than a rigid rod of the same molecular weight, which may have a dimension of the order of the filter openings. The size and shape of the particle will also influence its optical properties. On the other hand, depending on the focus of one's concern, the shape may be irrelevant: Polysaccharides tend to be gummy substances that can spread out when pressed against a filter cake, occluding many openings and impeding flow.

In addition to starch and dextran, the suspended solids fraction in raw sugar includes pectinlike substances from the plant cell wall, natural gums and waxes, nondextran bacterial slimes, silicates and other soil components, fibrous material from the woody part of the cane, and calcium and magnesium salts of both organic and inorganic anions near their solubility limits. Also present are plant lipids and proteins, DNA fragments, and chlorophyll. We correctly think of crystals as pure or nearly pure substances, and that is of course also the case for raw sugar. However, raw sugar is also a preparation of biological origin and thus apt to display remarkable complexity in its presence of minor or trace components. In the sugar refinery these "minor" impurities are of course a major focus.

If the light wavelength is much smaller than the particle diameter, light is reflected rather than scattered. However, such relatively large particulates will not concern us here, for two reasons. The first is that such materials will tend to separate spontaneously from a sugar liquor without the intervention of a clarifier system, although this might not necessarily occur within a convenient time frame. The second is that such substances are readily removed by ordinary pressure filtration.

For particles of size of the order of the wavelength, scattering is highly complex, and theoretical treatments fail. Particles 50 nm (0.05 μm) in diameter scatter about 25% of 420-nm light, whereas those near 2000 nm (2 μm) scatter about 95%. The particulate fraction in raw sugar is, of course, highly heterogeneous, and much of it would fall into this latter category.

Characteristics of the scattering particle also influence the *color* of light scattered. The intensity of scattered light is inversely proportional to the fourth power of the wavelength, so that about 13 times as much blue light as red light is scattered. Skies on earth are blue because light is scattered ("filtered out") by atmospheric dust, and blue light is scattered the most. An overcast sky is gray because the light is scattered by water droplets of larger size. Red rays are longer and better penetrate the atmosphere. Radio waves are longer still and even more penetrating. They are useful for studying, for example, our galaxy through the interstellar dust. In space the sky is black because there is no dust to scatter visible light.

Infrared photography is useful for aerial surveys because detail ordinarily obscured by scattered light of shorter wavelength is brought out distinctly. Because ordinary sky light contains very little red, black-and-white landscapes photographed using panchromatic film and a red filter show dramatic dark skies in high contrast to white clouds.

At sunset, one side of the Earth is turning away from the sun, and light reaching an observer is at an oblique angle, traveling a longer path through the air. This maximizes absorption and scatter of blue rays, resulting in redness. Air pollution by particulates in smoky industrial districts as well as natural haze increase the amount of absorption and produce redder sunsets. It is a general property of waves—whether of light or sound—that obstacles tend preferentially to filter out short waves and let longer ones pass on undisturbed. Thus the phenomenon that causes red sunsets is essentially the same one that prevents you from hearing the symphony's piccolo player in a large concert hall.

Insoluble material, such as sand in water, interacts with the solvent only marginally, and there is an interface between the two phases. On the other hand, material that is *soluble* in water, such as sugar or salt, interacts intimately with the solvent, forming a single homogeneous phase. Colloids and similar macromolecules are intermediate; they form a uniform dispersed phase, called a "sol," but interact incompletely with the solvent, so that there is an interface between each particle and the solvent. Whether the dispersed phase may properly be called homogeneous—which it of course is, on the macro level at least—is really a philosophical issue rather than a scientific one. Furthermore, in marked contrast to a true solution or pure liquid, in which comparatively few molecules are at a surface (i.e., in contact with air or a container wall), in colloidal suspensions an appreciable fraction of the molecules is located at a solvent interface. The boundary area increases enormously as the particle size decreases. It is such boundaries that result in the local refractive index gradients discussed above. The solvent immediately adjacent to the interface may be highly ordered, with entropy implications for any thermodynamic treatment.

36.6.2 Properties of Macromolecules

In the sugar refinery, the most immediate and dramatic effect of elevated levels of giant molecules is likely to be on viscosity. Syrup viscosities at a given temperature and solids content are relatively insensitive to the sugar composition (i.e., invert vs. sucrose) but highly sensitive to polysaccharide concentration. By way of illustration, a mere 1% aqueous solution of guar gum, a hydrocolloid widely used as a thickener in processed foods, has a viscosity of about 3000 Hz at 25°C, compared to about 300 Hz for 70° Brix sucrose at the same temperature. The resistance to flow in such materials stems largely from their property of binding large numbers of water molecules, *not*—as might be supposed—from intermolecular spaghetti-like entanglements. This characteristic is also responsible for their effectiveness as stabilizers in applications such as commercial icings and glazes, which are intrinsically unstable. Increased viscosities in the refinery not only inhibit liquor flow but can also disrupt the crystallization process. Dextran contamination beyond a low threshold concentration is known to adversely affect virtually every unit operation in the refinery, increasing costs and reducing recovery.

An important property of colloids and colloid-size macromolecules is their resistance to settling, which is a function of particle size (Table 36.4). It can be seen that submicron-sized species do not settle spontaneously within a time frame that is useful for any practical application. Why not? Tiny particles in solution are subject to haphazard buffeting called *Brownian movement* that results from random billiard ball–like collisions. Einstein in 1905 showed that such motion satisfied a statistical law confirming that it resulted from the bombardment of particles by the molecules of the fluid. Occasionally, particles may stick together after colliding, and some clusters may grow large enough to settle or precipitate.

For larger particles of the order of 1 to 8 μm, the sedimentation rate is appreciable, so at some point it is not appropriate to describe such substances as suspended solids. For intermediate particle sizes that settle on the order of minutes, the contribution to turbidity

TABLE 36.4 Sedimentation in Colloidal Systems

Particle Radius (μm)	Sedimentation Rate (cm/s)	Time to Settle Distance of 1 cm
10	0.01	31 s
1	10^{-4}	52 min
0.1	10^{-6}	86 h
0.01	10^{-8}	1 yr
0.001	10^{-10}	100 yr

Source: Based on Ref. 49.

is of course unstable with time. This is most likely to be observed with raw sugars, which must be filtered to obtain a reliable color measurement. In low-pol raw sugar, multiple scattering is likely to occur. That is, light scattered by one particle is rescattered before it leaves the cuvette, resulting in an abnormally high "apparent" turbidity. This can be avoided by diluting the sample before color measurement, not with water but with a purified sugar solution.

Light scattered from particles in motion has a Doppler shift imparted to it, the basis for one means of measuring the particle size of submicron material. The Doppler effect is, of course, the familiar phenomenon that makes the pitch of sirens of approaching ambulances sound higher than that of receding ones and allows astronomers to gauge the distance to galaxies at the far reaches of the universe.

Colloids typically have electrical surface charges. For naturally occurring substances in aqueous solution, the charge is usually *negative,* even for nominally "neutral" species such as starch. If the charge concentration is relatively high, the particles may collide randomly, but they bounce apart again because the like charges repel one another. The magnitude of this charge is called the *zeta potential,* usually expressed in millivolts (mV).

Although colloidal suspensions displaying a high zeta potential will not settle spontaneously, they can be made to settle by human intervention, such as by partially neutralizing the charge or partially desolvating the colloidal molecules at boiling temperatures. In water treatment and certain other applications, a chemical additive such as alum (aluminum sulfate) is used to reduce the zeta potential and promote settling, although overdosing can result in charge reversal and restabilization of the suspension.

We discuss the topic of zeta potential in Chapter 6. However, it should be noted that for some applications it is desirable for the particles *not* to settle, such as in the manufacture of fruit juices, pharmaceuticals, paints, and cosmetics. For such applications, measures may be taken to assure that the zeta potential is maintained at a high value.

In contrast to their potential for dramatic impact on viscosity, macromolecules are typically minor contributors to osmotic pressure, which depends on the concentration of individual particles. A suspension of polysaccharide made up of 100 glucose units per molecule would have only about 1% the osmotic pressure of a glucose solution with the same weight of solute. This difference is rarely put to the test, since one can easily prepare a 50% glucose solution but not a 50% starch solution.

36.6.3 Measurement of Turbidity

When we measure what is commonly called the "color" of a sugar solution, we pass a light beam of unknown intensity and wavelength, usually 420 nm, through a sample of the sugar and then measure the intensity of the light that arrives at the opposite side, where we have placed some sort of photocell detector. Such a colorimeter system treats the space between the light source and the detector as a "black box" and implicitly assumes that any light that doesn't make it to the photocell has been absorbed by impurities in the sugar. However, in sugar work, light scattering is always present. Thus a certain fraction of the light that doesn't arrive at the instrument detector has *not* been absorbed by colorants but instead has been diverted in other directions so that it has missed the photocell target. For this reason, the computed figures should properly be called not the color but rather, the attenuation index. The latter term simply notes that a certain portion of the incident light was detected but does not implicitly attribute any diminution to a single or specific phenomenon.

$$\text{Attenuation} = \text{light lost by absorbance} + \text{light lost by scatter}$$

This is not simply a petty matter of semantics. Color and turbidity are distinct attributes which, although not entirely independent, provide different data for evaluation of the refining process or product quality. The attenuation index by itself is inherently somewhat ambiguous; it's like saying that a man's age in years plus his weight in pounds is 204 (the figure for yours truly). Additional data are needed for such information to be useful, but more important, the source of variability is difficult to identify. It is possible for a man's weight to change drastically within a year, but no matter how hard he tries he can age only one year at a time (assuming he's not in a spaceship approaching the speed of light).

The scattering particles in sugar have a refractive index of about 1.49, that is, about the same as an 80° Brix sugar solution. Thus the clarity of inverted syrups near 77% solids can be exceptionally high. However, upon gradual dilution with water the refractive index gradient between solution and particle increases, and with it the attenuation index. Beyond about 35° Brix, further dilution overpowers the refractive index effect, and the attenuation index decreases, as common sense—in this case, the intuitive aspects of the Beer law—would lead us to expect. But the lesson here is that optical readings will be ambiguous unless the refractive index is kept constant. In practice, this usually means measuring at constant Brix.

Refractive index is an *intensive* property, not an *extensive* one. The index for water depends on the change in the velocity of light as it crosses the air–water interface, and it is of no consequence whether the area of interface is 1 square centimeter or 1 square kilometer. *Snell's law of refraction* (also called *Descartes' law*) states that for light passing from one medium to another, $\sin(I)/\sin(R) = n$, a constant called the *refractive index*, where I is the angle of incidence and R is the angle of refraction.

For the attenuation characteristics of a set of samples to be meaningful, they must be normalized or put on the same basis. By analogy with the Lambert–Beer law used for all colorimetry, we may write

$$a^* = \frac{-\log T}{bc} \qquad \text{(at constant wavelength)}$$

where a, the absorbancy index of colorimetry, will now be called the attenuation *index*, T is the transmittance, b the pathlength in centimeters of the cuvette containing the sample, and c the concentration of absorbing and scattering material. We use an asterisk to distinguish the attenuation index from the absorbancy index. Since absorbance, A, equals $-\log T$, we may make a substitution in the equation and eliminate the logarithm. Note that ordinarily we would have no way to know c, so we will redefine c as the concentration of total solids—sugars plus nonsugars—in g/mL of comparatively small, the approximate value for c is readily available from an refractometric dry substance (RDS) measurement.

Although this transformation might seem suspect in giving us something for nothing (a measurable quantity in place of an indeterminate one), it is mathematically sound. What we have actually done is multiply both sides of the equation by the fraction c/C, where c is the concentration of colorant plus particulates and C is the concentration of sugar:

$$a^* = a\,\frac{c}{C} = \frac{-\log T}{bC}$$

Now everything on the right side can be measured, and the result is proportional to the concentration of colorant and particulate per unit concentration of sugar. Careful readers may notice something inherently fallacious here: The concentration c includes material that does not obey the Lambert–Beer law, even though the equation is in the form of that law. However, we will leave this matter to the philosophers and focus on the pragmatic.

We have not yet addressed the issue of what portion of the attenuation index is due to absorption and what due to scatter (Table 36.5) This is not easily determined but was studied carefully by Rieger and Carpenter [50]. These are, of course, representative samples, and the findings would not necessarily apply to all sugars. However, note that although the light attenuation resulting from scatter is relatively low (about 13%) for *raw* sugar, for *affined* raw the figure approaches two-thirds of the total. What is going on here? We presumably "clean up" the raw sugar at the affination station, yet the effect of particulates increases dramatically.

The molasses film coating the raw sugar crystals is relatively rich in color (and incidentally, ash), so that washing dramatically improves the color. This should not be surprising: in any crystallization, remaining impurities tend to be concentrated in the mother liquor, which in the case of raw sugar is, of course, molasses. This material is always present in higher concentration on the crystal surface than in the crystal interior. (In organic chemistry preparation courses, one is always instructed to "wash the crystals.") However, impurities such as starch and dextran tend to be occluded within the crystal and distributed throughout, so that affination does a relatively poor job of removing these substances. Consequently, their contribution to the attenuation index should increase after washing, and this is what is found. We also know that boiling (i.e., crystallizing) pan feed liquors is not particularly effective at separating polysaccharides, and again the data of Rieger and Carpenter [50] corroborate this.

For our present discussion, what should be of particular concern are the figures for affined raws. However, it is of interest that it is the high-quality white sugars, not the lower-purity softs and raws, for which accurate color measurement is likely to be problematic, and of course it is the former category for which reliable determination is more critical. For such measurement, the refractive index must be kept constant, and if dilution is required, this should be done with extra pure sugar of the same Brix value, not with water.

As has been noted above, if we are presented with an unknown turbid sample, very little can be said about the nature of its particulates or indeed whether or not such substances are even present. However, if we know that the sample is uniform and in addition have some idea of the nature of the material responsible for the turbidity, it is possible to determine not only the molecular weight of the substance but also an estimate of its particle size. The equation that correlates this information was derived by Debye and is beyond the scope of this chapter. However, it may be of interest to know that terms of the equation contain the wavelength of the incident light (to the fourth power), the refractive index of the solvent, Avogadro's number, and a constant related to interaction between particles.

TABLE 36.5 Absorbance Versus Light Scattering

Sugar	Absorbance (436 nm)	Scatter (35° Brix)	% Light Lost by Scatter
Granulated A	0.0077	0.0238	75.5
Granulated B	0.0068	0.0066	49.3
Special quality granulated	0.0331	0.0391	54.2
Tablets	0.0270	0.0182	40.3
Affined raw A	0.2792	0.428	60.5
Affined raw B	0.2154	0.400	65.0
Soft sugar	20.5	2.16	9.53
Raw A	3.53	0.507	12.5
Raw B	4.78	0.782	14.2

Some process monitoring turbidimeters are calibrated in Jackson turbidity units (JTUs). This system is based on the depth of a turbid sample that extinguishes the view of a light source through a tube, or, more correctly, causes the light to diffuse into a uniform glow. Standards are prepared from natural silicates such as kieselguhr. However, the light source for the Jackson system is supposed to be a candle flame, which in addition to being decidedly low-tech, is rich in the yellow-red longer-wavelength end of the visible spectrum, where scattering by small particles is not very effective. In addition, there are serious practical limitations with the Jackson system because of its low sensitivity and the element of subjectivity in determining the extinction point.

36.6.4 Subtractive Strategies

One common practice to determine what is purported to be turbidity is to measure the attenuation index of a sugar liquor before and after filtration, ascribing the difference to suspended solids. The advantages are simplicity and accessibility: namely, the ability to use an ordinary colorimeter. In addition, one obtains a result in ICUMSA color units, although some of us would prefer to call these "pseudo," "quasi," or "apparent" ICUMSA units, because the filtration technique has no ICUMSA status as a *turbidity* determination. However, this technique is based on the assumption that large particles only scatter light and do not absorb it, whereas the smallest only absorb light without scattering it. Both of these suppositions are unfortunately fallacious. It is known that some sugar color is polymeric, and even smaller colorant molecules may be complexed or covalently bonded to polysaccharides or proteinaceous material. This is unsurprising in view of the polar functions and unsaturated groups that are structural features of substances that absorb in the visible region. Phenolics, which are common components of the colorant fraction, readily form stable adducts with proteins. Examples are the tanning of leather, the ease of dying silk and wool, and the difficulty of washing dyes off the skin.

There are other problems as well. At the outset we must remember that turbidity does not obey the Beer–Lambert law, placing in question any procedure based on colorimetry. Colorimeters are designed to measure the colors of true solutions, which one never has in the case of sugar samples, and which is why the preferred term is attenuation index. Or to put it differently, if color is defined as material *in solution* that absorbs in the visible region—which is, after all, the basis of colorimetry—filtration cannot affect the result.

Furthermore, implicit in the method is the notion that filtration eliminates scattering. Although filtration can remove *suspended solids,* it cannot by definition remove colloidal matter, and we know that a very significant fraction of the light-scattering impurities in sugar samples are smaller than the pores in the laboratory filter membrane.

In the filtration procedure, one commonly uses a membrane of pore size a nominal 0.45 μm, since smaller particles are sometimes defined as colloidal. However, it is unrealistic to expect that any such arbitrary cutoff point has a basis in nature. The choice is in fact purely a convenience: Tighter filtrations tend to be impractically slow or to require extreme dilution, whereas looser ones are apt to be ineffectual. For dark samples, even a 0.45-μm filtration may be unacceptably slow, so a 1.2-μm membrane is often used—through which proportionately more suspended matter can, of course, pass, further eroding the rationale behind the technique.

The difficulty of the arbitrary nature of the 0.45-μm filtration (which has ICUMSA status for measuring white sugar *color*) is exacerbated by the wide pore size distribution in commercial membranes and the wide range of hydrophilicity of available membrane materials. The filtration step is known to be an important contributor to poor precision in color measurement. In addition to reproducibility problems introduced by the membrane itself, there is a philosophical issue: It is always possible to filter "a little more" in an effort to approach perfect optical clarity, but for sugar samples the quest never ends, nor can it.

A somewhat different subtractive method is that of the National Soft Drink Association for "bottlers" granulated sugar. In this procedure, the absorbance at 720 nm is doubled and subtracted from the figure at 420 nm, giving a color in RBUs (reference basis units). However, there are no compelling theoretical grounds for the 720-nm turbidity "correction" to the 420-nm color measurement. The factor of 2 is particularly suspect, and the "turbidity" contribution is often exaggerated. Furthermore, the very purest sugars can provide colors that are negative numbers, a situation likely to distress laboratory analysts (and customers), who might choose to deal with this predicament by pretending not to notice the pesky minus sign. The 720-nm measurement was abandoned by ICUMSA in 1970.

Why 720 nm? The rationale was that colorants in white sugars absorb so little light at the red end of the spectrum that any such absorption found can be ascribed to turbidity alone. The key words here are "so little." In fact, some 720-nm absorbance *is* due to color, and the error arising from such a contribution is enlarged by the empirical multiplier used to make the 720-nm factor significant relative to the 420 nm. Another fallacy with this scheme is that while smaller particles scatter shorter (blue) wavelengths relatively intensely while having little effect on longer (red) wavelengths, larger particles do not affect short wavelengths as much as they scatter long wavelengths.

There is, of course, no reason to expect agreement between the ICUMSA and RBU methods. However, the discrepancy might be within the experimental error except in cases of very careful work or for unusual samples. If such a situation is commercially unacceptable, it is best to avoid writing a specification that makes reference to both procedures.

36.6.5 Nephelometry

Because of the complexity of the optical properties of sugar solutions, no general theory has been developed to describe them. However, because of the practical importance of the subject to the industry, various attempts have been made to solve the problem empirically.

In our earlier discussion of the attenuation index, the turbidity was described as the light lost from the transmitted beam. The turbidity may be measured directly by integrating the light scattered over all angles, which can be accomplished with appropriate instrumentation:

$$\text{turbidity} = 2\pi \int_0^{\pi} R \sin \theta \, d\theta$$

where θ is the angle of observation. R, the Rayleigh ratio, is essentially the proportion of scattered to incident light at a particular angle of observation. However, although R is a fundamental parameter describing light scattering by a particular sample, its value will vary among sugar samples of the same type and particularly among different sample types (Table 36.6).

One reason for the sensitivity of scattering angle to sample type has to do with the particle-size distribution of the scattering species. For particles smaller than 0.1, the wavelength of incident light, that is, about 0.04 μm, the pattern is symmetrical (Fig. 36.3). Larger particles, about one-fourth the wavelength of incident light, about 0.1 μm, result in enhanced forward scatter. For particles larger than the incident wavelength, that is, larger than about 0.4 μm, the pattern is complex and displays maxima and minima. The Rayleigh ratio

TABLE 36.6 Typical Rayleigh Ratios, $R\theta$ ($cm^{-1} \times 10^2$)[a]

Sugar	Angle of Observation 10°	30°	60°	90°	135°
Raw	300	50	6	2	1
Affined raw	80	10	2	0.8	0.6
Granulated	1	0.2	0.03	0.02	0.02

Source: Based on data of Ref. 50.

[a]60 Brix at 436 nm.

is also sensitive to the refractive index of the solution. If the shape of the scattering envelope is always the same, measurement is greatly simplified, because the entire envelope can be defined by a measurement at any one point.

The Rayleigh theory of scattering applies to small spherical particles, below about 0.04 μm, where the difference between the refractive index of the particles and the medium is small. These conditions are approximated with highly purified sucrose solutions at high Brix, although the scattering species are not spherical. The extremely complicated Mie–Tyndall treatment can be used for larger spherical particles in the range of 0.02 to about 5 μm, for any refractive index difference. However, as noted above, the heterogeneity of particulates in sugar samples precludes effective theoretical treatment.

Historically, the development of instruments for measuring turbidity has been hampered by the problem that even if the sample were absolutely colorless, the change in transmitted light would probably be so slight that low turbidities could not be measured reliably. On the other hand, if one were to make the measurement at higher concentrations, multiple scattering would interfere with the result.

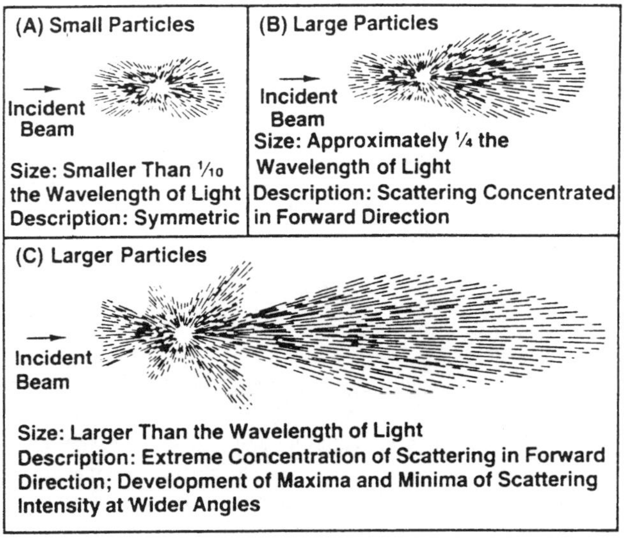

FIGURE 36.3 Angular patterns of scattered intensity from particles of three sizes. (From Ref. 51.)

The solution to this problem was to measure the light scattered at an angle to the incident light beam. A 90° detection angle was chosen because this was considered to be relatively insensitive to variations in particle size of the scattering species. Such instruments are called *nephelometers,* to distinguish them from other turbidimeter designs, and the measured turbidity is reported in nephelometric turbidity units (NTUs). Turbidity expressed in NTUs is preferred by many standard methods, such as the EPA procedure for water.

Although nephelometry is probably the best option for sugar work in terms of simplicity and sensitivity, the turbidity of sugar solutions cannot be fully defined by a single measurement at 90°. Furthermore, since sugar samples scatter much more strongly at forward angles than at right angles, a 90° detector is not the most sensitive configuration. Rieger and Carpenter found the best correlation between turbidity and the Rayleigh ratio when the angle of observation was 20° and pointedly remarked that "the very poor correlation obtained when θ is 90° shows that right-angle scattering measurements are of little value in determining turbidity of commercial sugar liquors" [50]. Of course, this applies to highly purified samples, which approximate Rayleigh scattering and display a scattering envelope that is roughly symmetrical.

Software is available that can be used with a laser light-scattering photometer that can be coupled to any chromatography system. The photometer measures the angular intensity of light scattered by an eluting sample at 15 angles, and these data are used to generate *absolute* molecular weights and sizes at each elution volume. Ordinary right-angle scatter can provide the weight-average molecular weight of small particles from the Rayleigh ratio, and a dissymmetry measurement (scatter intensity at 45° to that at 135°) can provide a molecular dimension when the particle shape is known. A Zimm plot, with the Rayleigh ratio measured at several angles and extrapolated to 0°, can additionally provide the radius of gyration, the molecular shape, and an estimate of polydispersity.

The reliability of nephelometry for white sugar samples of ordinary quality has not been collaboratively tested. At the time of this writing this author is directing an international survey of turbidity measurement of white sugars under the auspices of ICUMSA, which at present does not have an official turbidity method. The results will be published in the *Proceedings of the 22nd Session of ICUMSA,* a meeting held in Berlin in 1998.

The real issue here is whether the heterogeneity of light-scattering species in different sugar samples permits reliable quantified measurement. Solutions of equal suspended solids concentration but different composition in general will not scatter the same amount of light or scatter it in the same way.

Some reference works may state that NTUs and JTUs are interchangeable, but we advise considerable caution on this point. Bear in mind that the Jackson system does not depend on a particular scattering angle, that it is calibrated to a different standard, that it is far more subjective, and that it is much less sensitive to low levels of turbidity. On the other hand, FTUs (formazin turbidity units) *are* equivalent to NTUs, and refer to standard suspensions of formazin that are used almost universally as the primary reference standard to calibrate modern turbidimeters. Formazin is prepared from hexamethylenetetramine and hydrazine sulfate (Fig. 36.4), the latter of which is a systemic poison and suspected carcinogen, and hence should be handled with due care.

36.6.6 Turbidity Standards

Formazin was for many years the only acceptable primary standard for calibrating turbidimeters. If prepared carefully, using ultrafiltered water and properly cleaned glassware, it is accurate to the nearest $\pm 1\%$. Although the reproducibility of the shapes and exact par-

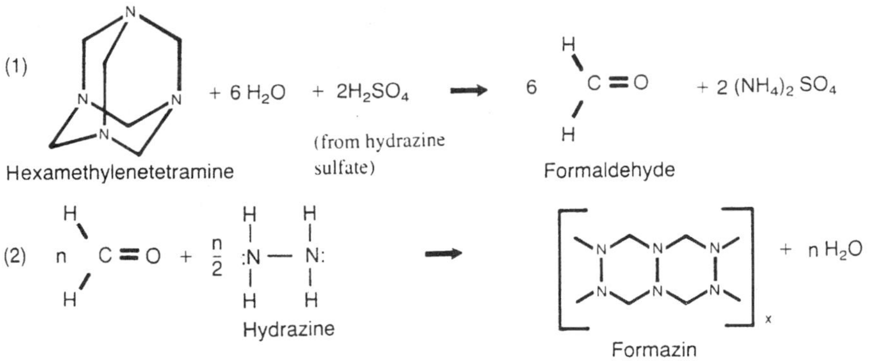

FIGURE 36.4 Preparation of formazin.

ticle-size distribution of the polymer is not easily achievable, the statistical scatter of multiple lots of standards *is* highly reproducible. It is, of course, this summation of individual scattering events that is responsible for a turbidimeter reading.

Formazin displays a wide particle-size distribution, ranging from 0.01 to larger than 10 μm, very much like the range of impurities in raw sugars. The formazin polymer consists of an abundance of particles of disparate shapes, which is inherent to real-world samples, including sugar.

Although one of the starting materials for formazin preparation, hydrazine sulfate, is, as mentioned above, of toxicological concern, only trace levels of this substance remain in solution once the polymer has formed. Nevertheless, suppliers of turbidity standards have addressed this anxiety about exposure to formazin by developing alternative standards. One such system, which has been given EPA approval for use in reporting turbidity, is comprised of styrene divinylbenzene beads of nominal size 0.2 μm with a density matching that of its matrix and hence resistant to settling. They are thus like submicroscopic ion-exchange resin particles. However, the very narrow particle size distribution exhibited by such beads does not approximate real-world systems and in addition makes these standards very wavelength sensitive. In a sense the beads are too nearly perfect, made to such a close tolerance that their application suffers in robustness. As a result, these standards are highly instrument specific and are designed for use only on one specific make and model number of turbidimeter, clearly a serious shortcoming. Their accuracy has been determined to be within ±10% of the value of the standard. Like the formazin 4000 NTU primary standard, they last about one year.

A third standard type, with an indefinite shelf life, consists of metal oxide particles suspended in a rubberlike polymer gel. These reference materials are typically supplied in sealed glass vials. Because the light-scattering properties of such preparations differ from those of formazin, standards of the same nominal NTU valuc will appear to be different to the naked eye but will result in the same instrument response.

REFERENCES

1. R. U. Lemieux and G. Huber, *J. Am. Chem. Soc.,* 78:4117, 1956.
2. W. Z. Hassid et al., *J. Am. Chem. Soc.,* 66:1416, 1944.
3. N. H. Smith, *Proc. Tech. Session Cane Sugar Refin. Res.,* New Orleans, LA, 1966.

4. M. A. Godshall, *Proc. Sugar Ind. Technol.* Durban, South Africa, 1966.
5. R. Riffer, *Proc. Tech. Session Cane Sugar Refin. Res.*, New Orleans, LA, 1980.
6. L. Farber and F. G. Carpenter, *Proc. Tech. Session Cane Sugar Refin. Res.*, Boston, 1970.
7. E. S. Gold, *Inorganic Reactions and Structure,* Holt & Co., New York, 1955, p. 358.
8. K. J. Parker, *Proc. Tech. Session Cane Sugar Refin. Res.*, New Orleans, LA, 1966.
9. R. Riffer, *Am. Chem. Soc. Symp. Chem. Process. Sugar Cane,* New Orleans, LA, 1987.
10. F. G. Carpenter et al., *Proc. Tech. Session Cane Sugar Refin. Res.*, New Orleans, LA, 1972.
11. N. H. Smith, *Proc. Tech. Session Cane Sugar Refin. Res.*, New Orleans, LA, 1964.
12. Peters and Phelps, *Technologic Papers of the Bureau of Standards,* No. 338, 1927.
13. V. R. Deitz, *Bone Char Research Project, Quarterly Report* 25, 1952.
14. J. C. Williams, *C. R. Assem. Gen. Comm. Int. Technol. Sucre.*, 319–327, 1975.
15. Tate & Lyle Ltd., Annual Report, 1979.
16. D. V. Freeland et al., *Int. Sugar J.*, 81:196, 1979.
17. Goodacre et al., *Int. Sugar J.*, 83:11, 1980.
18. D. Gross and J. Coomb, *Int. Sugar J.*, 78:69, 1976.
19. N. H. Smith, *Proc. Tech. Session Cane Sugar Refin. Res.*, New Orleans, LA, 1976.
20. R. Riffer, *Proc. Sugar Process. Res. Conf.*, New Orleans, LA, 1984.
21. J. H. Wall and F. G. Carpenter, *Proc. Tech. Session Cane Sugar Refin. Res.*, Boston, 1978.
22. H. H. Willard et al., *Instrumental Methods of Analysis,* Van Nostrand, New York, 1981, pp. 107–108.
23. M. Palasinski et al., *Starch/Stärke,* 37(9):308, 1985.
24. V. Prey and H. Wesner, *Proc. 15th Gen. Assem. CITS,* 1975.
25. R. Riffer et al., *J. Org. Chem.*, 38:204, 1973.
26. R. D. Guthrie, in *The Carbohydrates: Chemistry and Biochemistry,* W. Pigman and D. Horton, eds., Academic Press, New York, 1972, Vol. IA, p. 431.
27. Salamon and Goldie, *J. Chem. Soc. London,* 19:301, 1900.
28. H. Paulsen and K.-W. Pflughaupt, in *The Carbohydrates: Chemistry and Biochemistry,* W. Pigman and D. Horton, eds., Academic Press, New York, 1980, Vol. IB, p. 899.
29. Imming et al., *Zuckerindustrie,* 119(11), 915–919, 1994.
30. W. W. Binkley, *Zuckerindustrie,* 20(6), 291–295, 1970.
31. J. Pokomy et al., *Nahrung,* 23(9–10):921–917, 1979.
32. F. G. Carpenter and E. J. Roberts, *Proc. Tech. Session Cane Sugar Refin. Res.*, Cherry Hill, NJ, 1974.
33. T. Jachimowicz and G. El Sherbiny, *Apidologie,* 6(2):121–143, 1975.
34. H.-D. Belitz and W. Grosch, *Lehnbuch fur Lebensmittelchemie,* 2nd ed., Springer-Verlag, Berlin, 1987, pp. 213–219.
35. Hangyal et al., *Untersuchung der Kinetik fur Farbstoffbildung im Zuckerhaus,*
36. E. J. Roberts et al., *Proc. Tech. Session Cane Sugar Refin. Res.*, New Orleans, LA, 1964.
37. de Armas et al., *Int. Sugar J.*, 94:147, 1992.
38. R. L. Sidebotham, *Advances in Carbohydrate Chemistry and Biochemistry,* Academic Press, New York, 1974, Vol. 30, pp. 371–444.
39. A. Jeanes et al., *J. Am. Chem. Soc.*, 76:5041–5052, 1954.
40. E. J. Bourne et al., *Carbohydr. Res.*, 33:13–22, 1972.
41. Keniry et al., *Int. Sugar J.*, 71:230–233, 1969.
42. Urquhart et al., *Int. Sugar J.*, 95:1197, 1993.
43. J. Bruijn, *Int. Sugar J.*, 72:195–198, 1970.
44. D. F. Charles, *Int. Sugar J.*, 86:105–109, 1984.
45. P. G. Morel du Boil, *Int. Sugar J.*, 99:310–314, 1997.
46. Roberts and Martin, *Sugar J.*, 22:11–16, 1960.

47. M. A. Godshall and M. A. Roberts, *Proc. Tech. Session Cane Sugar Refin. Res.*, New Orleans, LA, 1976.
48. APHA, AWWA, and WPCF, *Standard Methods for the Examination of Water and Wastewater,* 18th ed., American Public Health Association, American Water Works Association, and Water Pollution Control Federation, Washington, DC, 1992.
49. S. S. Voyutsky, *Colloid Chemistry,* Mir, Moscow, 1978.
50. C. J. Rieger and F. G. Carpenter, *J. Res. Natl. Bur. Stand.,* 63A(3):205–211, 1959.
51. Brumberger et al., *Sci. Technol.,* Nov. 1968, p. 38.

CHAPTER 37

The Analysis of Sugar and Molasses*

INTRODUCTION

The first section of this chapter covers the contract analysis of raw sugar, that is, the procedures generally specified for contract sales of raw sugar in the United States. These methods are from contract form No. 2021-91G of Tate & Lyle North American Sugars, Inc. (Domino Sugar Corporation). Certain procedural suggestions and refinements of technique reflecting the practices of the New York Sugar Trade Laboratory are presented. Also the relationship between the contract polarization procedure and the ICUMSA polarization procedures is discussed.

The second section deals with ICUMSA methods for the analysis of raw and refined sugar. The methods discussed are limited to those that might be encountered most frequently in the classical analysis of sugar: namely, polarization, moisture, ash, color, reducing sugars, and dextran. These methods are discussed in view of the findings presented at the 1998 ICUMSA meeting.

The final section covers briefly the analysis of molasses, specifically those procedures that are of commercial interest in the trading of molasses.

37.1 CONTRACT ANALYSIS OF RAW SUGAR

37.1.1 Polarization

In the late 1980s the Schmidt & Haensch company introduced a polarimeter that operates at a wavelength in the near-infrared region. It thus became possible to analyze raw sugar without the need for clarification, which by the standard ICUMSA procedures had traditionally been accomplished by the use of lead subacetate. A procedure for analyzing sugars by polarization in the near-infrared region (HW-NIR)† was developed at the New York Sugar

---

* By Walter Altenburg.

† The term HW-NIR (high-wavelength polarimetry in the near-infrared region) is used instead of NIR to avoid confusion with NIR spectroscopy methods.

Trade Laboratory (NYSTL). A comparison of over 1500 sugars polarized concurrently by this procedure and the ICUMSA procedure (1979 equivalent dry lead method) was undertaken at the NYSTL. The results of this study were published in the 1991 proceedings of the Sugar Industry Technologists (SIT) [1]. On average the polarization (pol) values for the two procedures were found to be the same within 0.005° polarization. The NYSTL also coordinated a comparative testing program in which a total of 30 sugars, three from each of 10 origins were selected to represent the full range of polarization values normally encountered. These sugars were analyzed in duplicate by both procedures by nine laboratories (eight domestic and one international). Considering the overall average of all the laboratories, the results agreed with the original SIT study within 0.03° polarization. For the three laboratories that had a great deal of experience in running the procedure prior to the comparative testing program, the average results agreed with the SIT study within 0.02° polarization. For those laboratories that had repeatability within the published expectation of 0.10° polarization for the ICUMSA procedure (the same three laboratories), the repeatability and reproducibility of the two procedures were comparable. A summary of the comparative testing program was published in the 1993 proceedings of the U.S. National Committee and the complete details are published in the 1994 proceedings.

The HW-NIR procedure was adopted as part of the Domino Sugar contract in April 1993 and has become the standard procedure for almost all contract purchases of raw sugar in the United States. The adoption of this procedure was based on the empirical data as referenced above. Before the procedure could be accepted by ICUMSA and thereby come to general use internationally, certain theoretical considerations had to be addressed, the most important being the definition of the 100°*Z* point in the near-infrared region. When the HW-NIR procedure was developed, the sugar values of quartz plates in the near-infrared region were determined by application of the Bünnagel equations in accordance with the recommendation of ICUMSA [2]. This was a provisional recommendation of ICUMSA, with the understanding that further study of the optical rotatory dispersion of quartz and sucrose in the near-infrared region was necessary and that the Bünnagel equations may prove to be invalid in this region. In a joint undertaking by the Physikalische-Technische Bundesanstalt, the Schmidt & Haensch company, and the Sugar Institute in Braunschweig, optical rotatory dispersion formulas for quartz and sucrose have been developed that are valid in the near-infrared region and agree with the Bünnagel equations in the region where the latter are applicable. The findings of this study are published in the 1998 proceedings of ICUMSA. By use of these new ICUMSA-approved formulas, the sugar value of quartz plates in the near-infrared region can now be calculated. ICUMSA has also accepted the HW-NIR pol procedure: Recommendation 3 of General Subject 1 of the 1998 ICUMSA proceedings is: "Polarization according to the procedure (P880) described by Altenburg and Chou (1991), modified to give effect to the newly established 100°*Z* point and rotary dispersion of quartz and sucrose at NIR wavelengths should be adopted as an Accepted method." In the report for Subject G-1, it is further written that with the 100°*Z* point now being established at NIR wavelengths, "it is now appropriate for the NIR method to be written up with a view towards collaborative testing of the procedure, and progress towards establishment as a Tentative method."

The polarization values (P589) reported by the NYSTL in the 1991 SIT study were determined by the ICUMSA procedure as published in Ref. 3 with results reported according to the equivalent dry lead method. In this procedure, results are reported in degrees *S* with subtraction of 0.1° to compensate for the "lead error." This was the procedure of the Domino Sugar contract prior to April 1993. The P880 results were determined by the HW-NIR procedure using the extension of the Bünnagel formulas. The standard ICUMSA procedure was changed in 1988, and currently, as written up as Method GS1/2/3-1 (1994) in the

ICUMSA Methods Book [4], reports polarization in degrees Z without any lead correction. Since the effect of changing to degrees Z is to decrease pol results by 0.029% and by not applying the lead correction, the pol value is increased by 0.10°Z, the results by the present ICUMSA lead procedure are approximately 0.071% higher than by the ICUMSA procedure [3] referenced above. The new ICUMSA 100°Z definition in the near infrared has the effect of raising the HW-NIR pol results by 0.063% (a polarimeter calibrated at 882.6 nm using the Bünnagel formulas will show 99.937 for the normal sugar solution). The 1991 SIT study reported the overall average P880–P589 difference for all sugars tested as +0.005. Taking the relationships above into consideration, if we accept the results of the 1991 SIT study, then by a simple algebraic solution, we see that the ICUMSA-accepted HW-NIR procedure of Subject G-1, recommendation 3 (now designated as method GS1/2/3-2) is equivalent to the current ICUMSA lead procedure (GS1/2/3-1) within two decimal places. It should be noted that this numerical equivalency is based on the overall averages presented in the 1991 SIT study. The results for individual sugars and also (as indicated in the SIT study) for different origins of sugar may be different. These differences could result from the different effects of nonsucrose impurities and different levels of nonsugars such as dextran in lead clarified and unclarified solutions. It should be noted that neither polarization procedure professes to be a measure of sucrose content and that adoption of a polarization procedure is by convention. The great preponderance of potential lead-related environment, health, and legal problems make the HW-NIR polarization convention most attractive.

It should also be noted that just as the change of the ICUMSA lead procedure in 1988 did not change the lead-based method then specified in U.S. commercial sugar purchase contracts, the current change in the 100°Z scale in the near-infrared region does not change the method specified in the present commercial sugar purchase contracts, which is the original HW-NIR procedure based on the Bünnagel formulas. If one is using a quartz plate that shows on the certificate sugar values in the near-infrared region calculated according to the current ICUMSA accepted formulas, one must multiply these values by 0.99937 to obtain the sugar values to be used for the contract procedure. Of course, alternatively, one could use the rotation value shown on the certificate for the wavelength at which the plate was calibrated (e.g., 589, 546, or 633 nm) and calculate therefrom the sugar value of the plate in the near infrared by application of the Bünnagel equations. Also, instrument manufacturers will generally provide calculated sugar values of the quartz plate in accordance with customer needs. Therefore, certificates will generally be issued to companies in the United States showing the value to be used to comply with the sugar purchase contract.

The HW-NIR procedure as described in the Domino Contract (2021-91G) is as follows:

Polarization of Raw Sugar (Without Wet Lead Reagent)

1. **Scope and Field of Application.** Polarization is used in many countries as a basis of payment for raw sugar. It is therefore desirable to have a standard method of analysis available to buyers and sellers. It is also desirable to have a method that does not use wet lead reagent. This method details the procedure and apparatus to be used in the determination of polarization by high-wavelength polarimetry in the near-infrared region. The major advantage of this procedure is that it does not require a defecation step, thereby eliminating the use of the wet lead reagent.

2. **Principles of the Method.** This is a physical analysis involving three basic steps: preparation of a "normal" solution of raw sugar in water, clarification of the solution by filtration, and determination of polarization by measurement of the optical rotation of the clarified solution. The "normal" sugar solution is defined by ICUMSA (1970),

Subject 5. The definition of the "International Sugar Scale" is that adopted at the 20th Session of ICUMSA, 1990 (°Z).

3. **Apparatus**

3.1. *Balances.* The balance used for raw sugar polarization is to be of such sensitivity and speed that rapid weighing can be made to ± 0.002 g.

3.2. *Flasks.* Flasks used for raw sugar polarization must conform with the specifications set out by ICUMSA in Subject 2, Appendix 1 (1974) in respect to "flasks used for most precise work." Such flasks have a normal capacity of 100 mL with a tolerance of ± 0.02 mL. When such flasks are not available, a class A flask is used plus a flask correction. The quantity "actual volume of flask -100.00," known as the *flask correction*, should be engraved clearly on the bulb of the flask or recorded separately against the code number of the flask. When corrected flasks are used, the polarization is to be corrected by the flask correction. When corrected flasks are used, the polarization reading is corrected by adding the flask correction. The flask correction may be positive or negative, and the correct sign must be used.

3.3 *Filtration Equipment.* Stemless funnels of corrosion-resistant material are used to filter the solutions prepared for polarization. Funnels are mounted on filtrate receiving glasses or beakers that do not result in excessive splashing, and watch glasses are placed on the funnels to minimize evaporation. The filter paper used is Whatman grade 91, 18.5 cm (Whatman International Ltd., Maidstone, England, catalog number 1091-185), or equivalent. The moisture content of the filter paper is to be in the range of 6 to 8% water determined by drying for 3 h at 100°C. The filter paper is folded into quarters for use (not fluted). The filter aid used is Johns-Manville Celite (filtercel grade) or equivalent.

3.4. *Polarimeter.* The polarimeter used is the Schmidt & Haensch Polartronic NIR Model or equivalent. Calibration of the polarimeter in °*Z* at the near-infrared wavelength (vicinity of 880 nm) is based on provisional recommendations presented at the 20th Session of ICUMSA (1990).

3.5. *Quartz Plates.* The design, material, workmanship, dimensions, and properties of the quartz plates must be of the internationally accepted standard [i.e., of the standard laid down by ICUMSA (1970 and 1974), Subject 6]. The quartz plates used must be either standard plates that have been certified by a recognized authority such as the National Physical Laboratory (Teddington, UK) or the Physikalische-Technische Bundesanstalt (Braunschweig, Germany), or plates that have been compared to a certified plate. If the sugar value of the quartz plate at the polarimeter's operating wavelength is not available from the instrument manufacturer, it can be calculated as follows:

- **(a)** By application of the Bünnagel equation (*Proceedings of the 14th Session of ICUMSA*, 1966, p. 32), the rotation of the standard quartz plate at the near-infrared wavelength is calculated from the rotation of the plate at the wavelength of calibration shown on the certificate of analysis.
- **(b)** The rotation of the normal sucrose solution in degrees *Z* at the near-infrared wavelength is calculated from the rotation of the normal sucrose solution at the Hg wavelength (40.777°) by application of the equation agreed upon at the 14th Session of ICUMSA (1966, p. 17).
- **(c)** The sugar value of the plate in degrees *Z* at the near-infrared wavelength is calculated as (rotation of the quartz plate/rotation of the normal sugar solution) $\times$ 100.

3.6. *Tubes (Cells) and Cover Glasses.* A flow-through tube may be used. Also, unless recommended otherwise by the instrument manufacturer, simple tubes or side-filling tubes may be used. Tubes may be jacketed if needed according to application. Polarimeter tubes and cover glasses must conform to the specifications given by ICUMSA in Subject 2, Appendix 1 (1974), except that the tolerance on lengths of tubes shall either conform with class A or, where it conforms to class B, the actual length (correct within the tolerance specified in class A), must be engraved on the tube. A tube length correction, equal to nominal length divided by actual length, can then be applied as a multiplier to all polarization readings where high precision is required.

3.7. *Constant-Temperature Water Circulator.* A constant-temperature water circulator capable of maintaining a temperature of 20.0 ± 0.2°C is required only if a jacketed tube is deemed necessary. A jacketed tube is required only if polarization readings at 20.0 ± 0.2°C cannot be made by regulating the laboratory temperature.

4. **Reagent.** The reagent is distilled water or demineralized water of equivalent quality.

5. **Procedure**

5.1. *Treatment of Samples.* On arrival in the laboratory and before being opened, packed samples are inspected to see whether (a) the sugars have been packed and sealed so that they could not be affected by changes in the atmospheric humidity between the time of packing and the time of arrival in the laboratory where they are to be analyzed, or (b) the packages have been damaged or tampered with. If as a result of the inspection described above there is reason to believe that a sugar received in the laboratory could differ from the sugar as packed, this must be reported to the person or organization for whom the analysis is to be done, whether or not the analysis is carried out. The humidity of the laboratory in which the samples are unpacked for analysis is to be kept, as far as practical, within the range 65 to 70% RH, and variation must be minimized. The temperature of the room is maintained as close as possible to 20.0°C. Approximately 2 cm of the sample is removed from the top and the sample mixed in its container immediately before weighing.

5.2. *Standardization of the NIR Sugar Polarimeter.* The polarimeter must not be standardized with sucrose solutions. First, the polarimeter is zeroed against air and adjusted in accordance with the manufacturer's instructions to read the proper sugar value of the certified quartz plate at the temperature of observation ($Q_{\text{cert},t}$) [see formula (B) below]. The instrument is then ready for use. However, to ensure the greatest accuracy, the standardization is confirmed and the tube zero is determined both immediately before and after each reading session by the following method:

Immediately before reading the sugar sample, or series of samples, the following procedure is used: The instrument is zeroed against air. The certified quartz plate is then read and its observed value corrected for temperature ($Q_{\text{obs},20}$) is calculated by use of formula (A) below. At least two readings are taken. The plate is handled as little as possible (preferably with gloves) to avoid warming. The instrument zero is then checked and the tube filled with water is read. If the tube reading is not zero, its value, P_w, is recorded and a tube correction will be applied. Alternatively, the instrument may be rezeroed with the tube in place, in which case the effective tube correction is zero for subsequent sample readings (i.e., $P_w = 0$; this is the common practice when using a flow through tube). The combined scale and tube correction to be added to the observed pol is $(Q_{20} - Q_{\text{obs},20}) - P_w$, where Q_{20} is the certified quartz plate value at 20°C.

At the end of the reading session the following is done: if using a flow through tube, the tube is flushed with water and the reading P_w recorded. The tube is then removed, the instrument zeroed against air and the quartz plate is read in the manner described above. The additive correction is calculated as above: $(Q_{20} - Q_{\text{obs},20}) - P_w$. The overall correction to be applied to the polarization is then the average of the value obtained from the initial standardization and the value obtained from the final standardization.

If using a center filling tube to read a series of samples and the initial reading of the water-filled tube is not zero, the value P_w is recorded and the instrument is not rezeroed. This enables one to check the instrument zero (and reset if necessary) between each sample. For the final standardization at the end of the reading session, the instrument is rezeroed if necessary, and the water-filled tube and quartz plate are read as described. The overall correction to be applied is the average of the values obtained from the initial and final standardizations.

The temperature at which the quartz plate readings are taken will be determined and recorded to 0.1°C. If this temperature differs from 20.0°C by more than ±0.5°C, the following temperature corrections apply:

- The plate reading observed when standardizing the instrument is corrected for temperature by the following:

$$Q_{\text{obs},20} = Q_{\text{obs},t} - (T_q - 20) \times 0.000143 \times Q_{20} \tag{A}$$

  where $Q_{\text{obs},20}$ is the observed plate reading corrected for temperature, $Q_{\text{obs},t}$ the observed plate reading at temperature T_q, and Q_{20} the certified value of the quartz plate at 20°C.
- Similarly, the value of a certified quartz plate at temperature T_q may be calculated as follows:

$$Q_{\text{cert},t} = Q_{20} + (T_q - 20) \times 0.000143 \times Q_{20} \tag{B}$$

5.3. *Preparation of the Sample Solution.* 26.000 ± 0.002 g of the sugar is weighed out as rapidly as possible. The sugar is transferred to a 100-mL volumetric flask by washing with distilled water to a volume not exceeding 70 mL; it is then completely dissolved either by hand-swirling or by an automatic shaker. The solution is mixed by gentle swirling and, using the same motion, distilled water is added until the bulb of the flask is full. This solution is allowed to stand for at least 10 min to attain room temperature, preferably in a water bath, temperature controlled at 20°C. Distilled water is then added to about 1 mm below the mark, ensuring that all the neck is washed. Care must be taken that no air bubbles are entrapped and, if necessary, the meniscus defrothed by means of alcohol or ether vapor. The inside of the neck of the flask is dried with a clean roll of filter paper to within a few millimeters of the mark. With the meniscus suitably shaded, the flask is held vertically by the top of the neck, with the calibration mark at eye level, and viewed against a well-lighted background. Distilled water is added dropwise, preferably from a hypodermic needle, until the bottom of the meniscus and the top of the calibration mark just coincide. If fine bagasse or fiber particles are present, the side of the neck is momentarily flicked so that the true position of the meniscus may be sighted. The inside of the neck of the flask is again dried, the flask sealed with a clean, dry stopper, and the flask shaken thoroughly. A clean, dry thermometer is inserted in the flask and the temperature is recorded to 0.1°C. The flask is held by the top of the neck during this measurement.

5.4. *Filtration of the Solutions.* The sugar solution is filtered using a filter paper containing approximately 1 tablespoon (3.5 g) of filtercel grade Celite. Great care must be taken that the filtercel does not get into the filtered solution. This would cause turbidity, and a proper reading would not be obtained. To prevent this, it may be necessary to remove the filter from the funnel when adding the filtercel. Place a watch glass over the filter funnel during filtration to minimize evaporation. Also keep the filter out of drafts and direct sunlight. The first 10 mL of the filtrate is discarded and only sufficient volume collected, usually 50 to 60 mL, as is required for polarization measurement. Sufficient volume of solution is usually obtained within 25 min. If necessary, because of a slow filtration rate, the sample may be allowed to filter up to 45 min. The solution in the filter funnel should not be replenished and none of the filtrate should be returned to the filter. After filtration, the watch glass is placed on the top of the filter glass. If insufficient solution is obtained, an additional sample is prepared.

5.5. *Determination of the Polarization.* A flow-through tube must be used if recommended by the manufacturer. If a different type of tube is used, it is rinsed thoroughly at least twice with the solution to be used and filled in such a way that no air bubbles are entrapped. The caps of the tube are screwed on as tightly, and only as tightly, as necessary to prevent leakage. Tighter screwing may cause the cover glasses to become optically active. The tube should be handled as little as possible, to avoid warming. The tube is placed in the polarimeter or the sample introduced into the flow-through tube. The instrument is allowed to come to equilibrium before reading. It should be ensured that the tube compartment lid is closed when reading. In a similar manner, the quartz plate reading and the water-filled tube reading are determined (see Section 5.2). The observed polarization is corrected by applying the scale correction and tube zero correction, if applicable, as described in Section 5.2.

5.6. *Temperature Corrections.* The standard temperature of ICUMSA polarizations is 20.0°C. Preferably, the water, solution, filtrate, and apparatus used for preparing the solution and measuring the polarization shall be maintained in the temperature range 20.0 ± 0.5°C. Because temperature corrections for polarimetry in the near-infrared region have not been officially established, the temperature of the solution while reading the polarization, t_r, must be kept at 20.0°C ± 0.2°C. This can be accomplished by regulating the room temperature and, if necessary, using a water-jacketed polarimeter tube. If the temperature at which the solution is made to mark is not the same as the temperature at which the polarization is determined, the observed polarization must be corrected for the effect of this temperature difference on the concentration of the solution as made to the mark. This correction should be made by adding to the observed polarization the quantity $(0.027)(t_r - t_m)$, where t_m is the temperature at which the solution is made to the mark in °C and t_r is the temperature of the solution as read in the polarimeter.

5.7. *Results.* The corrections to be applied are those that are necessary to standardize the polarimeter (scale and zero corrections), those that are necessary to allow for irregularities of apparatus (flask and tube corrections), and those that are required to allow for solution preparation and observations at temperatures other than 20.0°C.

The procedure above specifies that the temperature of the pol solution as it is being read must be 20 ± 0.2°C because the temperature corrections in the near infrared had not been officially established at the time of its publication. However, the temperature effects on the rotation of sucrose and quartz in the near-infrared region have now been determined. The temperature coefficients for quartz and sucrose have been published in the 1998 ICUMSA proceedings [5]. The differences between the new formulas for the NIR region and the classical formulas are slight. The classical temperature coefficient for quartz is 0.000144 and the new value for the NIR is 0.000139. As a practical matter this difference is not significant. If the temperature of the quartz plate is kept within ±5°C, the error introduced by use of the classical formula would be less than ±0.0025°Z.

Another way of expressing the quartz temperature correction (Section 5.2 of the procedure) is thus: Subtract from the observed polarization the following: $Q_{obs,t} - Q_{cert,t}$, where $Q_{obs,t}$ is the observed value of the quartz plate at temperature t and $Q_{cert,t}$ is the certified value of the quartz plate at 20°C corrected to temperature t. Substituting for $Q_{cert,t}$, its calculated value, gives the following result to be subtracted from the observed pol value:

$$Q_{obs,t} - Q_{20} - (Q_{20})\ (t_q - 20)\ (0.000139) \qquad \text{(I)}$$

This is the same as the formula given in *ICUMSA Methods Book* [4], Section 9.1 with the classical temperature coefficient of quartz replaced by its value for the NIR.

For sucrose the differences are somewhat greater, the temperature coefficient being about 4% greater in the NIR region. The new formula for the NIR is $P_t/P_{20} = 1.0 - (0.000493)\ (t - 20)$ compared to the classical formula: $P_t/P_{20} = 1.0 - (0.000474)\ (t - 20)$, the difference being: $(0.000019)\ (P_t)\ (t - 20)$. Note that in comparing the formulas above, published in the 1998 ICUMSA proceedings, with the temperature correction formulas of the polarization procedure GS1/2/3-1 of the *ICUMSA Methods Book*, one must keep in mind that the coefficients published in the proceedings are for sucrose alone, and the coefficients of the methods book include the temperature effects of the polarimeter tube. With this in mind we may add 0.000019 to the factors given in Table 1 of the procedure GS1/2/3-1 of the *Methods Book* to obtain corresponding values for use in the NIR. Actually, the ICUMSA HW-NIR polarization procedure, GS1/2/3-2, uses factors which differ from those for the GS1/2/3-1 procedure by 0.000023. The reason for this slight discrepancy is that method GS1/2/3-2 is based on data of Wilson [17] rather than Emmerich because the former are more consistent with the values published in the 1998 ICUMSA proceedings. For example, when using a steel polarimeter tube and Pyrex flasks, we may use the following formula for temperature corrections in the NIR:

$$P_{20} = P_t + (P_t)\ (0.000478)\ (t_r - 20) - (P_t)\ (0.000270)\ (t_m - 20) - (0.004)\ (R)\ (t_r - 20) \qquad \text{(II)}$$

Although the temperature effects on reducing sugars in the NIR have not been studied, the term $-(0.004)\ (R)\ (t_r - 20)$ should be able to be used, as it accounts for only a very small correction. For temperatures ± 5°C for a reducing sugar level of 0.2%, this term would account for a correction of ± 0.004°Z.

If one wishes to express equation (II) in terms of $(t_r - t_m)$, one could use

$$P_{20} = P_t + (P_t)\ (0.000208)\ (t_r - 20) + (P_t)\ (0.000270)\ (t_r - t_m) - (0.004)\ (R)\ (t_r - 20) \qquad \text{(III)}$$

If one makes the assumption that $t_r = t_m$, the equations above would be reduced to

$$P_{20} = P_t + (P_t)\ (0.000208)\ (t_r - 20) - (0.004)\ (R)\ (t_r - 20) \qquad \text{(IV)}$$

Or, for pol values close to 100, and ignoring the reducing sugar term, one could use the approximate formula:

$$P_{20} = P_t + (P_t)\ (0.000208)\ (t_r - 20) + (0.027)\ (t_r - t_m) \qquad \text{(V)}$$

As a practical matter, a measured value for the reducing sugars content is not usually available when analyzing for polarization. An estimated value could be used. However, the reducing sugars temperature correction term is often ignored in practical polarimetry, since it is so small. It should also be noted that the temperature correction formula also has a second-order element [6] that is ignored in the ICUMSA method formulas. Also, the correction for the quartz plate reading given in equation (I) is an additive correction. More exactly, the correction would be expressed as a factor [7]. However, this is not necessary unless the range of the pol measurements is vastly different from the value of the quartz plate.

Modern saccharimeters can be provided with temperature-sensing polarimeter tubes and the instrument can automatically provide temperature corrections for the quartz plate and the pol solutions. The instrument manufacturer's instruction manual will show the type of correction formula available. Some instruments use a formula of the type $P_{20} = P_t + (P_t)\ (K)\ (t_r - 20)$, and with others the following type of formula can be used*:

$$P_{20} = P_t + (P_t)\ (K)\ (t_r - 20) + (P_t)\ (K_2)\ (t_r - 20)^2 + (K_3)\ (t_r - 20)$$

Instruments will come preset with the constants (K values) to be used in the formulas. However, these can be changed by the operator depending on any changes in the basic formulas as described above or in accordance with the assumptions used by the operator. For example, if one were to use a formula of the form $P_{20} = P_t + P_t\ (K)\ (t_r - 20)$ and made the assumption that all operations are carried out at the same temperature ($t_r = t_m$), then (ignoring the reducing sugar term) the constant K would equal 0.000208. If, however, t_m is assumed to be 20°C, as would be the case if the flasks are adjusted to 20°C in a water bath while making to mark, then K would equal 0.000478. Also, if one is measuring t_m and applying the correction $-(P_t)\ (0.000270)\ (t_m - 20)$ [see equation (II)], the constant K would equal 0.000478. With some instruments, there may be a temperature dependence of the instrument itself that will modify the values above. The instrument manufacturers will provide details.

All the foregoing examples of temperature-correction formulas assume the use of a borosilicate glass flask (Pyrex) and a steel polarimeter tube. If a Pyrex polarimeter tube were used with a Pyrex flask, the factor 0.000478 would be replaced by 0.000490 and 0.000208 would be replaced by 0.000220.

In the proper execution of the HW-NIR procedure, all of the normal proper laboratory techniques should, of course, be followed, some of the important points being: Temperature and humidity of the laboratory should be tightly controlled within the required ICUMSA limits and the appropriate temperature corrections applied if necessary. Since a significant volume error can be caused by flasks that do not meet specifications, only calibrated flasks should be used, and if not within ± 0.02, the corrections must be applied. Also, as mentioned in the method, the flasks should be made to mark against a lighted background, and to minimize problems with the meniscus, it is important that the flasks be absolutely clean. Prior to the final making to mark, the contents of the flask should be swirled as water is added to prevent shrinking of the solution after being made to mark.

It is, of course, well known that it is essential to obtain clear solutions for polarization. A solution that is slightly turbid, even if the reading comes to apparent stability, can cause a significant error in the polarization result. What also emphasizes the importance of obtaining absolutely clear solutions for the HW-NIR polarization procedure is the fact that based on preliminary observations at the NYSTL, turbid solutions may react differently with

*The complete temperature correction formula is actually more comprehensive, being third order in terms of $(t - 20)$ and P_t.

different instruments, causing a false high reading on one and a false low reading on another. Great care must therefore be taken to ensure clear solutions. Sometimes a turbid solution may result from a factor beyond the control of the analyst, such as a defective filter paper. However, strict attention to technique is essential. First, it goes without saying that adherence to the method's specifications in regard to the type of filter paper and filtercel is essential. It is also essential to make sure that one discards the full amount of first filtrate specified in the procedure. Also, since even a minute amount of filtercel in the final solution can cause a significant problem with turbidity, as suggested in the method, when filling the filter paper with the filtercel it is advisable that the filter paper be removed from the funnel rather than spooning in the filtercel with the filter paper in place. Although it is not specifically mentioned in the method, after the first portion of the filtrate is collected, the final portion of filtrate should be collected in a clean and dry container. In other words, the funnel is transferred to a new jar rather than the jar just being emptied and replaced beneath the funnel. Also, when the sugar is poured from the flask to the filter apparatus, sometimes some filtercel may become airborne. One should guard against any such particles landing in a jar that will contain the filtrate.

The method states that the Schmidt & Haensch NIR model polarimeter or its equivalent shall be used. At the time of its publication, the Schmidt & Haensch polarimeter was the only available instrument. Polarimeters for use in the NIR are now also available from other manufacturers, including two independent manufacturers in the United States: Rudolph Research Corporation and Rudolph Instrument Co. Also, the Schmidt & Haensch company has recently introduced a quartz wedge instrument that can operate in the NIR region. Quartz wedge instruments are nearly independent of the wavelength of the light source and therefore should not require recalibration due to changes in the light source. When applying temperature corrections, it should be kept in mind that with a quartz wedge instrument, no temperature correction needs to be applied to the quartz plate reading since the temperature effect of quartz plate is compensated by the temperature effect of the quartz wedges of the instrument. Normally with a quartz wedge instrument, one would have to take into account the temperature effect of the quartz wedges when applying the temperature correction formulas to the sugar solution. However, this effect is compensated electronically so the same temperature correction formulas described above for polarimeters can be used without modification for the quartz wedge instrument. It should again be noted that all instruments can be provided with a temperature-sensing polarimeter tube and temperature correction software.

37.1.2 Contract Quality Testing of Raw Sugar

The current methods used in the determination of raw sugar quality for the purpose of setting the final purchase price of most raw sugar sold in the United States are published in the Domino Contract Form 2021-91G Exhibit C, effective January 1, 1991. These methods cover the analysis of moisture, ash, grain size, color (both the raw and centrifugally washed), and dextran. The procedures are described below. When equipment is specified in the contract, it is understood that equivalent equipment may be substituted as availability and circumstances dictate.

Moisture: Factor of Safety

1. Weigh a moisture dish to ± 0.0001 g. A metal dish with a tight-fitting cover should be used.

2. Place approximately 5 g of a representative sample into the dish, cover quickly, and weigh to ±0.0001 g as rapidly as possible.
3. Remove the cover and place the dish and cover in a vacuum oven that has been preheated to 70°C. The sample is heated for 4 h at 70°C at a minimum of 28 in. of vacuum. Air is bled into the vacuum chamber at a rate of 100 mL/min. The oven temperature should be controlled to ±1°C. The air for bleeding is to be predried by passing through a desiccant such as barium oxide or sulfuric acid.
4. After 4 h, reduce the vacuum by eliminating the vacuum source and increasing the airflow through the desiccant. Replace the cover and remove the sample from the oven.
5. Place in a dessicator and cool to room temperature. The sample and dish should be weighed as soon as possible after reaching room temperature.
6. Immediately when cool, weigh to ±0.0001 g.
7. The percent moisture is calculated by the following formula:

$$(\text{weight loss/starting weight}) \times 100 = \%\ \text{moisture}$$

8. The laboratory will report % moisture to the nearest hundredth of a percent for each sample.
9. The factor of safety is calculated in the following way:

$$\frac{\%\ \text{moisture}}{100 - \text{settlement Pol}} = \text{factor of safety}$$

The contract suggests the NAPC vacuum oven (Fisher No. 13-262-1) and 5.5-cm-diameter aluminum weighing dishes (Fisher No. 08-722). It should be noted that some of this model vacuum oven came equipped with an external thermometer well. The proper procedure, however, is to use a thermometer inside the oven in order to obtain the most accurate measurement of temperature. To avoid use of the toxic chemical barium oxide, a drying tower (e.g., Kimble No. 19505-300) filled with indicating Drierite can be used. This is connected to a bubble counter (e.g., Fisher No. 07-368) filled with sulfuric acid. The temperature of the dishes as they cool in the desiccator can be monitored by use of a contact thermometer such as Pacific Transducer Corp. No. 310C.

Sulfated Ash

1. Weigh a preignited and cooled crucible of 50-mL capacity to ±0.0001 g. (A platinum dish may be used in the place of the crucible.)
2. Weigh approximately 4 g of a representative sample to ±0.0001 g in the weighed crucible.
3. Add approximately 1 mL of 1:1 sulfuric acid, in drops, to the sample until completely wetted and heat until the sample is well carbonized. (A 275-W Fisher Infra-Radiator placed about 6 in. above the sample is recommended for this purpose.) Approximately 20 min of heating under the lamp is adequate to prevent splattering when the crucible is placed in the furnace. This procedure must be performed under a laboratory hood.
4. Transfer the carbonized sample to a furnace at 555 ± 30°C and heat until the carbon is burned off. A heating period of about 2 h is sufficient for this ignition.

5. Remove the crucible, and after cooling it to near room temperature, add approximately 0.5 mL of 1:1 sulfuric acid solution, in drops, in such a manner as to wet the material remaining in the crucible.
6. Heat the crucible so that the sulfuric acid solution is evaporated without loss of liquid or solid material through splattering from the crucible. A hot plate is used for this purpose. This procedure must be performed under a laboratory hood.
7. Place the crucible inside the furnace at 555 ± 30°C and heat for 2 h.
8. Remove the crucible from the furnace and cool it to room temperature in a desiccator. Weigh the crucible and contents to ±0.0001 g. The percent ash is calculated in the following way:

$$\frac{\text{weight of ash}}{\text{starting weight}} \times 100 = \%\ \text{ash}$$

9. The laboratory will report % ash for each sample to the nearest hundredth of a percent.

The use of platinum dishes will cut down significantly on the time required for the first sulfation. However, if the cost is prohibitive, Vicor crucibles can be used. Notice should be taken if particles fall from the top of the furnace into the crucible. Some muffle furnaces may have this problem.

Color: ICUMSA Method 4 (Modified)

1. Prepare a 25% solids solution (25 g of sample ± 75 mL of distilled water) of the sugar to be tested.
2. Filter the solution through a 47-mm Millipore filter apparatus using a Whatman GF/C 47-mm glass microfiber filter. Collect the filtrate in a clean dry filter flask. (*Note:* The sample may require changing the filters more than once to collect all the filtrate.)
3. Transfer the filtrate to a clean, dry 150-ml beaker. Adjust the pH of the filtrate to 8.5 ± 0.1 with 0.5 *N* HCl or 0.5 *N* NaOH.
4. Remove entrained air under vacuum or in an ultrasonic cleaner if necessary.
5. Place the solution into one of a previously matched pair of 1-cm absorption cells. (The other cell will contain distilled water and can be used as a zero reference when changing wavelengths.) Determine the absorbance (or -log of the transmittance) at 420 nm and at 720 nm. Record both values.
6. Calculate the color of the solution as follows:

$$\text{color} = \frac{(\text{absorbance}^* \text{ at 420 nm} - 2 \times \text{absorbance at 720 nm}) \times 1000}{0.2764}$$

If the sample is too dark to analyze, further dilution with distilled water and possible pH readjustment will be needed. In this case the calculation would be as follows:

$$\text{color} = \frac{(\text{absorbance}^* \text{ at 420 nm} - 2 \times \text{absorbance at 720 nm}) \times 1000}{\text{specific gravity} \times (\text{Brix}/100) \times \text{cell length}}$$

*(-Log of transmittance) can be substituted if there is no absorbance function on the spectrophotometer.

Note that the value of (Brix × specific gravity)/100 is given as 0.2764 for 25 Brix. This value indicates that true specific gravity is used. True density can be found from *ICUMSA Methods Book* (1990), SPS-4, Table A, column 2 (p). This value, divided by 998.234, yields true specific gravity [8]. This is actually a trivial point. The difference would be only about 1 color unit at most if apparent specific gravity were to be used.

ICUMSA colors are generally read in the range of 20 to 80 % transmittance (0.70 to 0.10 absorbance). Using the 25% solution of raw sugar specified in the method will for many sugars result in a reading that is considerably higher than 0.70 absorbance. The most accurate spectrophotometer readings may not be obtained for very high absorbance solutions. This problem is addressed in step 6 of the procedure, which includes a statement that samples are to be diluted if too dark to read. To quantify this statement, we shall interpret it to mean that the absorbance of the solution should not greatly exceed 0.7 absorbance. Also, since the color value may to some extent be dependent on the dilution, the dilution of samples to a relatively uniform absorbance value is advisable for reproducibility between laboratories. The goal can be to dilute samples to a consistent 0.6 absorbance (this allows for some latitude without going over 0.7). One problem with dilution to constant absorbance instead of constant Brix is that the light-scattering (non-Beer) contribution to the attenuation index becomes ambiguous because the refractive index gradient between sample and suspended solids becomes a variable. However, for most samples, getting the absorbance into the "good precision zone" should take precedence. It should also be kept in mind that if one dilutes below 10 Brix, one loses one significant figure in the Brix value. Therefore, for greatest accuracy with Brix values below 10, one should read refractive index and calculate the Brix corrected to 20°C by use of Tables A and B of SPS-3 in the *ICUMSA Methods Book.*

The dilution of the solution could be accomplished by trial and error. However, to avoid having to rerun samples, the following procedure can be used: An approximate relationship between absorbance of the filtered 25 Brix solution (undiluted, not pH adjusted) and the final color value is first determined. Using several months' data for the absorbance values of the unadjusted solutions (in a 0.5-cm cell) and final color values, a linear regression analysis is run. A formula is thus obtained by which the final color can be roughly estimated from the absorbance reading of the unadjusted solution. Once this formula is obtained, colors can be run as follows.

The unadjusted solution is first read in a 0.5-cm cell and the estimated color is calculated. With this estimated color value and the desired absorbance value of 0.6, and an estimate of the absorbance at 720 (average based on observed data), the formula given in step 6 of the procedure can be solved for the quantity (specific gravity × Brix). Using a calculated table showing the quantities Brix and (specific gravity × Brix), the desired Brix is thus obtained. Solutions are diluted to this Brix before being adjusted to pH and read in the spectrophotometer. The amount of water to be added to the 25 Brix solution to obtain the desired Brix can easily be calculated. If X represents mL of 25 Brix solution to be used, $(100 - X)$ the mL of water to be added (for a total solution of 100 mL), and ρ is the density of the 25 Brix solution, then, by definition, $\text{Brix}/100 = (x)(p)(0.25)/[(x)(\rho) + (100 - x)]$. With the known quantities Brix (calculated desired Brix) and ρ, this equation is solved for x.

The above may seem complicated; however, a computer program can be written that performs the calculations (and the interpolations from the table kept in a file). Using such a program, one can just enter the absorbance of the 25 Brix solution and the output will show the desired Brix and the amounts of 25 Brix solution and water to be mixed.

Further contract quality tests are performed on a sample of centrifugally washed sugar. The following is the procedure for the affination:

Equipment

- *Mixer:* Kitchen-Aid Model K-5, with flat beaters, No. K-5-A-B, manufactured by Kitchen-Aid Inc., 2303 Pipestone Road, Benton Harbor, MI 49022.
- *Centrifugal machine:* Model K with 8-in. basket with draining chamber, manufactured by International Equipment Co., 300 Second Avenue, Needham, MA 02194. If 8-in. basket is unavailable, an 11-in. basket may be substituted. The inside of the basket is to be faced with a metal screen No. 00 mesh 0.020-in. dia., 625 holes/square inch, that can be purchased from Ferguson Perforating & Wire Co., 140 Ernest Street, Providence, RI 02905.

Preparation of the Sample

1. Place 1000 g of well-mixed raw sugar in the mixer. Turn the mixer on to speed 1 (low speed).
2. Gradually add 380 ml of 64.0 Brix granulated sugar syrup at room temperature. (A 64.0 Brix syrup made from high-quality sugar syrup may be substituted for a syrup made from granulated sugar.) The syrup is added slowly from a dispensing burette and must be added at a uniform rate of approximately $4\frac{1}{2}$ min.
3. The raw sugar and syrup continue to mix for an additional 1 min. The total mixing period is $5\frac{1}{2}$ min.
4. Transfer the entire magma at once from the mixer to the laboratory centrifugal machine.
5. Bring the centrifuge up to 3000 rpm in 15 s and spin at 3000 rpm for exactly 2 min. If an 11-in. basket is employed, to maintain the same *g* force, bring the centrifuge up to 2550 rpm in 15 s and spin at 2550 rpm for exactly 2 min.
6. Remove the sugar from the basket and spread it on a clean surface in a thin layer not to exceed $\frac{1}{4}$ in. thick.
7. Immediately after spreading, take representative portions totaling approximately 100 g from all areas in the spread layer and immerse in 75 mL of anhydrous methanol contained in a 250 mL extraction flask. This portion of the sample is to be used for the grain size test.
8. The remaining portion of the spread sample which is to be used for color is mixed periodically (by hand) during drying so that at the end of drying, the sample is well mixed.
9. If the sample is not to be tested immediately, it should be stored in sealed jars.

Although it is not strictly in accordance with the contract specifications, if a centrifuge basket of size other than 8 or 11 in. is used, the rpm value required to obtain the same centrifugal force would be calculated by the following equation, where x is the diameter of the basket:

$$\text{rpm} = \frac{\sqrt{(x/8) \times (\pi \times 8 \times 3000)^2}}{x\pi}$$

Color on the centrifugally washed sample is run in the same manner as for the raw sugar, with the exception that a 50% solids solution is prepared in step 1. Also, a procedure analogous to that described for the raw color can be used to determine the proper dilution for the sample.

The grain size may be run by either of the following two alternative procedures:

Grain Size: Method A

1. The flask containing the sample previously collected for the grain size test is swirled vigorously for 2 min so that the sugar is well mixed with the anhydrous methanol solvent.
2. Drain the solvent from the flask in the manner indicated in Diagram 1. After the solvent has drained, break the vacuum, shake the extraction flask, and place back over the vacuum flask. Repeat two or three times.
3. After draining, return the flask to an upright position, add 50 mL of anhydrous methanol, and repeat the swirling and draining procedure.
4. Repeat the swirling and draining procedure twice, using a 50 mL portion of ethyl ether each time. (*Caution:* Do not use near an open flame.)
5. Spread the drained sugar on absorbent filter paper and allow to air dry. No lumping or caking should occur on drying, soft conglomerates, if any, should be broken by gentle hand pressure. If lumping is observed after drying, discard the sample.

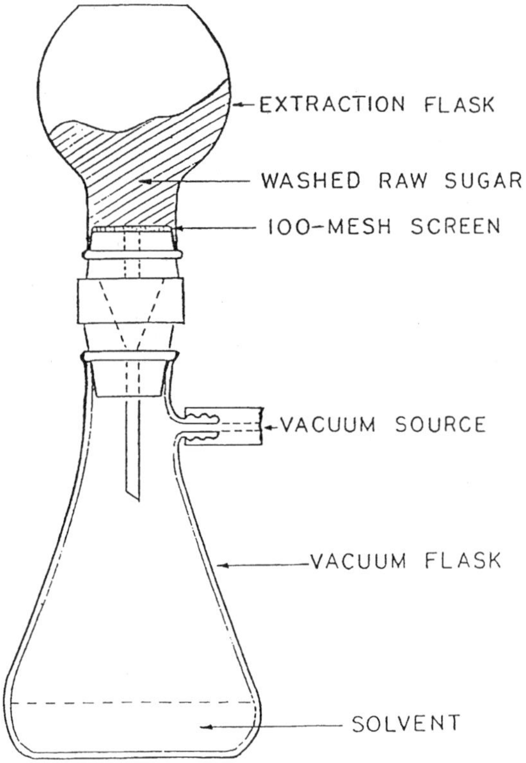

Diagram 1

Begin again, starting with the affination procedure. Whenever a solvent is used for testing, the test should be conducted underneath a laboratory hood and away from any flame or any other heat source.

6. Weigh (to ± 0.1 g) the entire amount of affined raw sugar, which has been washed with solvent and dried.
7. Assemble the screens with a 14-Mesh Tyler as the top screen, followed by the Tyler 20- and 28-mesh screens. A pan (and additional screens if necessary) is added to make up a set of screens that will fit on a mechanical shaker.
8. Place the weighed amount on the 14-mesh Tyler screen.
9. Place the set of screens on a mechanical shaker and run for 5 min.
10. The amount of sample passing through the 28-mesh Tyler screen is determined as follows:

$$\frac{\text{weight through 28 mesh screen}}{\text{starting weight}} \times 100 = \%\text{ through 28-mesh Tyler screen}$$

11. The laboratory will report the % through the 28-mesh screen to the nearest percent.

Grain Size: Method B (Alternative Method)

1. The flask containing the sample previously collected for the grain size test is swirled vigorously for 2 min so that the sugar is well mixed with the anhydrous methanol solvent.
2. Drain the solvent from the flask in the manner indicated in Diagram 1. If a pump is used to provide the vacuum source, it must be rated explosion-proof, such as a Sargent-Welch Dual-Seal No. 1405-W-01 or equivalent. Additionally, the pump must be vented outside the laboratory environment: for example, to a laboratory fume hood. After the solvent has drained, break the vacuum, shake the extraction flask, and place back over the vacuum flask. Repeat two or three times.
3. Repeat the swirling and draining procedure above twice, beginning each methanol wash by returning the flask to its upright position and adding 50 mL of anhydrous methanol.
4. Repeat the swirling and draining procedure for a fourth and final time, again by adding 50 mL of anhydrous methanol. Extreme care must be taken to ensure that the sugar is sufficiently dried by the vacuum. This is accomplished by draining the solvent and subjecting the sugar to the vacuum for a longer period of time than the previous washes. The sequence of breaking the vacuum, shaking the flask, and placing back over the vacuum flask should be repeated a number of times during this last drying step. If the sugar is sufficiently dried, the majority of crystals will not adhere to the sides of the flask or can be dislodged with a minimum of shaking or tapping. The average length of time to complete this fourth anhydrous methanol wash should be in the area of 40 to 50 min.
5. Spread drained sugar on absorbent filter paper and allow to air dry. No lumping or caking should occur on drying. Soft conglomerates, if any, should be broken by gentle hand pressure. If lumping is observed after drying, discard the sample. Begin again, starting with the affination procedure. Whenever a solvent is used for testing, the test should be conducted underneath a laboratory hood and away from any flame or other heat source.
6. Weigh (to ±0.1 g) the entire amount of affined raw sugar that has been washed with solvent and dried.

7. Assemble the screens with a 14-mesh Tyler as the top screen, followed by the Tyler 20- and 28-mesh screens. A pan (and additional screens if necessary) is added to make up a set of screens that will fit on a mechanical shaker.
8. Place the weighed amount on the 14-mesh Tyler screen.
9. Place the set of screens on a mechanical shaker and run for 5 min.
10. The amount of sample passing through the 28-mesh Tyler screen is determined as follows:

$$\frac{\text{weight through 28 mesh screen} \times 100}{\text{starting weight}} = \text{\% through 28-mesh Tyler screen}$$

11. The laboratory will report the % through the 28-mesh screen to the nearest percent.

The initial vacuum source for the grain size test can conveniently be provided by a water aspirator. Also, to save water, the water can be recirculated with a standard well pump, the water reservoir being outside the lab area in order to vent fumes. This will provide sufficient vacuum provided that the water does not become too warm. In any case, the final drying is accomplished by a vacuum pump as described in the procedure. After the liquid has been drawn off, the final drying is most effectively done by placing the flask containing the sugar over an empty 500-mL filtering flask attached to the high-vacuum source.

The screen apertures are specified in accordance with their Tyler designation. Tyler 14-, 20-, and 28-mesh screens are equivalent to U.S. Standard 16, 20, and 30. Care should be taken that the screens used are not damaged. Screens could be damaged by too vigorous cleaning. A screen that is no longer taut in its holder is evidence of possible damage.

The method for the analysis of dextran used in raw sugar purchase contracts since 1981 is the haze procedure developed by Chou and Wnukowski. [9] In a haze procedure, the haze formed by the dextran when alcohol is added to a solution is measured spectrophotometrically. The official ICUMSA procedure is also a haze procedure. The ICUMSA procedure and the contract procedure differ in the way the sample is prepared, the use of deionizing resins, and the amount of time allowed for haze development. By allowing the haze to develop for a full hour in the contract procedure, the timing of reading the sample in the spectrophotometer is not as critical. Also, the ICUMSA procedure expresses dextran in ppm based on a standardization with standard dextran solutions. The contract procedure expresses results in milliabsorbance units (MAU) based directly on the spectrophotometer reading. This avoids the problem of which molecular-weight dextran should be used as a standard. However, with the contract procedure, one must guard against the use of a defective spectrophotometer since variations in spectrophotometers are not compensated for by means of a calibration. The contract dextran procedure is as follows.

Method for the Determination of Dextran in Raw Sugar

Equipment and Reagents

1. Ion-exchange resins: Amberlite IR-120 (H) and any one of the following: Duolite A-368, Duolite A-392, Amberlite IRA-93, or Amberlite IRA-68. These resins normally are supplied wet and should be washed with at least twice their weight in distilled water, drained dry, then washed briefly with acetone for no longer than 2 min, the solvent being removed immediately, as before. The resins are air-dried or oven-dried at low temperature, approximately 30°C, and stored in a closed container.

2. Acid-washed Manville Supercel: Supercel (50 ± 5 g) is added to 1 L of distilled water. Concentrated hydrochloric acid (50 ± 5 mL) is added and the mixture stirred for 5 min. After filtration the Supercel cake is washed with distilled water until the pH of the washings equals that of the distilled water. The Supercel is dried for 6 h at 100°C and stored in a closed container.
3. Trichloroacetic acid J.T. Baker reagent No. 1-0414 (TCA): Trichloroacetic acid (10.0 ± 0.1 g) is dissolved in distilled water and diluted to 100 mL. This reagent will keep for 2 weeks. (*Note:* This reagent attacks protein and should not be allowed to come into contact with skin. Do not pipette TCA by mouth or store it in plastics.)
4. Starch-removing enzyme: Mycolase enzyme, GB Fermentation Industries, Inc., 1 North Broadway, Des Plaines, IL 60016. Or α-Amylase type X-A fungal crude from *Aspergillus oryzae* (Catalog No. A-0273), Sigma Chemical Company, P.O. Box 14508, St. Louis, MO 63178.
5. Alcohol: anhydrous, 200 proof, J.T. Baker reagent No. 9401-1.
6. 25-mL vol. flasks, Corning No. 5660 or equivalent. (At the end of each analysis, the flasks should be washed with acid-cleaning solution, rinsed with distilled water, and dried for future use.)
7. Nessler tubes, Kimble No. 45310A-100 or equivalent. Wash and dry the tubes the same way as described in step 6.
8. Filtering flasks, 1000 mL, Pyrex No. 5340 or equivalent.
9. Burette, 50 mL, Pyrex No. 2137 (right hand) or equivalent.
10. 12.5-mL class A volumetric transfer pipette (custom ordered).
11. Pipette filler, rubber bulb type or equivalent.
12. Millipore funnel No. XX 1004720 and 0.45-μm Millipore filter No. HAWG 047 AO and absorbent pads.
13. Vacuum pump with multiple-outlet connections for filtration (manifold).
14. UV-visible spectrophotometer, two matched 5-cm cells, and two matched 1-cm cells.
15. Jars, wide mouth, 4-oz, flint glass with screw caps.
16. Hot plate stirrer, Corning PC-351 or equivalent, and stirring bars. Hot plate may be used for incubation provided that a water bath is improvised.

Procedure

1. Weigh 23.5 g of whole raw sugar sample into a wide-mouth jar, and add 35 mL of distilled water, insert a magnetic bar, cover, and place on a magnetic stirrer to dissolve.
2. Add 0.05 g of mycolase enzyme or α-amylase to the sample and incubate at 55°C for 1 h in an oven or a water bath, with agitation every 15 min.
3. Following the incubation, add to the sample 5 g of Amberlite IR-120 (H) and 5 g of one of the following: Duolite A-368, Duolite A-392, Amberlite IRA-93, or Amberlite IRA-68 and stir for 30 min.
4. Add 1 g of acid-washed Supercel to the sample, mix, and filter through a Millipore absorbent pad only into a 100-mL Nessler tube placed inside a 1-L filtering flask. Rinse the sample jar with approximately 10 mL of distilled water, allowing the washings to go through the funnel into the Nessler tube. Follow this with two small washings of the funnel and contents, taking care not to exceed 100 mL of total filtrate volume.

5. The sample and washings in the Nessler tube are diluted to the 100-mL mark with distilled water and then 10 mL of TCA is added. The Nessler tube is stoppered and shaken.
6. Filter the above through a 0.45-μm Millipore filter covered with an absorbent pad into a clean Nessler tube inside a 1-L filtering flask, collecting at least 30 mL of filtrate.
7. Pipette 12.5 mL of the filtrate into each of two 25-mL volumetric flasks, designating the first as the control and the second as the sample, respectively. (*Note:* Use a safety pipette filler.) Clean the pipette for the next use by rinsing it with distilled water.
8. To the first flask (the control) add distilled water while swirling the flask to the 25-mL mark. Stopper and shake.
9. To the second flask (the sample) add anhydrous 200 proof alcohol dropwise (from a 50-mL burette) while swirling the flask to the 25-mL mark. Stopper and mix by inverting the flask gently three to five times.
10. Let the sample stand for 60 $\pm$ 2 min from the time of completion of the mixing step.
11. During the waiting period fill two clean 5-cm matched cells with distilled water and the control, respectively. After zeroing the spectrophotometer at 720 nm with the cell containing distilled water, read the absorbance of the control which is designated as *B*.
12. Then save the control by pouring it back into its 25-mL flask for possible future use. Clean the empty cell by rinsing it several times with distilled water and dry it by rinsing it with acetone.
13. At the expiration time of the 60 min period, fill the clean 5-cm cell with the sample. After zeroing the spectrophotometer at 720 nm with the cell containing distilled water, read the absorbance of the sample, which is designated as *A*. Report the results as follows:

$$(A - B) \times 1000$$

When the absorbance of the sample exceeds 0.7 in value both the control and the sample should be reread immediately in 1-cm cells, respectively (after zeroing the spectrophotometer at 720 nm with distilled water in a 1-cm cell). Report the results as follows:

$$(A - B) \times 5000$$

The results represent dextran content expressed in milliabsorbance units (MAU).

Note:

1. To achieve reproducible results, this procedure must be followed precisely.
2. Equivalent equipment and/or reagents may be substituted for those specified in this procedure only after comparability with the designated equipment and/or reagents has been demonstrated. This applies particularly to the alcohol reagent.

An incubator can be conveniently used in step 2 of the procedure. The Duolite A-368 and A-392 resins have been discontinued by the manufacturer. Amberlite IRA-93 has been replaced by IRA-95, and Amberlite IRA-68 has been replaced by IRA-67. The α-Amylase Sigma No. A-0273 is no longer available. This has been replaced by α-Amylase Sigma No. A-6211.

37.2 ICUMSA METHODS OF SUGAR ANALYSIS

37.2.1 Polarization

The ICUMSA procedure for the polarization of raw sugar, Method GS1/2/3-1 (1994), is official. However, the HW-NIR procedure is now also an ICUMSA Accepted procedure. It has been collaboratively tested and written up as Method GS1/2/3-2 (1999). The ICUMSA procedure is essentially the same as the commercial contract HW-NIR procedure except that it is based on the newly defined 100°*Z* scale and employs the current temperature corrections for use at near-infrared wavelengths. Also, as a precaution against obtaining a turbid solution, the ICUMSA procedure specifies that 15 mL of filtrate be discarded rather than 10 mL. When using the ICUMSA HW-NIR procedure, one must calibrate the instrument with a quartz plate, the sugar value of which is determined by the currently accepted rotatory dispersion formulas for quartz and sucrose as presented at the 1998 ICUMSA meeting. The calibration procedure as described in Method GS1/2/3-2 is based on Section SPS-1 of the *ICUMSA Methods Book* (1998). If polarization results have been determined by the raw sugar contract HW-NIR procedure, these may be converted to the ICUMSA values by dividing the contract polarization results by 0.99937.

Polarization of white sugar is carried out by the Braunschweig method (Method GS2/3-1), which has been Official since 1990. For white sugar requiring clarification the Official method is GS1/2/3-1 (the procedure for raw sugar using wet lead). However, these sugars may also be analyzed by the ICUMSA Accepted HW-NIR procedure (Method GS1/2/3-2).

37.2.2 Reducing Sugars

The Knight & Allen procedure for white sugar appears in the 1994 *ICUMSA Methods Book* as Method GS2/3-5 (1994). This has been modified and now appears as Method GS2/3-5 (1997) in the 1998 *ICUMSA Methods Book Supplement.* The method is indicated as suitable for reducing sugar contents up to 0.02%, which is a maximum level in most high-quality white sugars. The procedure also states that it can be used for concentrations up to 0.10% by dilution of the sample with low-invert sucrose. This method has been collaboratively tested under General Subject 2 (white sugar) and the results presented at the 1998 ICUMSA meeting. Since the Codex Alimentarius Commission and the EU allow a level of 0.04% reducing sugars in white sugar, it was decided to use samples in the collaborative study that ranged from 0.0016 to 0.054% in reducing sugar content. The collaborative test showed that the variability was somewhat high, with Horwitz ratios exceeding 2.0 in three of the four samples with less than 0.02% reducing sugar. The results for the higher reducing sugar content (those levels requiring dilution) were even worse. In view of these results, it remains questionable if the method should be used for reducing sugar content over 0.02%. Further collaborative studies of GS2/3-5 will take place in comparison with the Official HPAEC method and an enzymatic method.

The Modified Ofner method for white sugar as it appears in the 1998 *ICUMSA Methods Book Supplement*, Method GS2-6, was also collaboratively tested, with the results presented at the 1998 ICUMSA meeting. Sucrose samples spiked with a known amount of reducing sugars were used. Results were compared with HPAEC (now Official) and enzymatic methods as references. Good agreement was obtained with the reference methods. For the sugars tested, the modified Ofner method gave an overall average of 0.029% reducing sugars compared to the averaged spiked level of 0.027%. The average result for the HPAEC method was 0.031% and for the two enzymatic methods, 0.029% and 0.037%. The method met the requirement of a Horwitz ratio of not more than 2.00 for the two sugars with the highest reducing sugar levels (0.035% and 0.085%), although such results are well above normal levels in white sugar. The average Horwitz ratio for all five sugar samples was 6.70. The average repeatability and reproducibility values were 0.010% and 0.014% reducing sugars, respectively. The Modified Ofner method was confirmed as a Tentative method for determination of reducing sugars in white sugars at the 1998 ICUMSA meeting [10].

Originally, it had been planned to collaboratively test the Berlin Institute method for use with white sugars. However, preliminary tests with the spiked samples showed the Berlin Institute method to give a true recovery only at about 0.025% reducing sugars, with higher values above and lower values below this level. In view of these results, the Berlin Institute method was not included in the collaborative testing program. It would therefore not be considered the method of choice for the determination of reducing sugars in white sugar. It is an EU method for white sugar, and the possibility of replacing it in EU methods by the Modified Ofner method is now being considered.

As a practical consideration, in some applications where high-invert white sugars may be included, it could be an inconvenience to use a method such as the Knight & Allen, which has a very limited range. An analytical laboratory may be faced with having to analyze samples for which the approximate reducing sugar level is unknown.

For reducing sugars in raw sugar, the Lane & Eynon method [Method G/S1/3/7-3 (1997)], the Berlin Institute method, and the Luff Schoorl method were included in a collaborative study the results of which were reported at the 1998 ICUMSA meeting. The Luff Schoorl method met the statistical requirements, with an average Horwitz ratio of 1.41, compared to the limit of 2.0. It also had the best repeatability and reproducibility values of 0.024% and 0.052% reducing sugars, respectively. The Berlin Institute method had a borderline average Horwitz ratio of 2.07 and reproducibility and reproducibility values of 0.031% and 0.080%. The Lane & Eynon method showed poor results, with an average Horwitz ratio of 12.56; reproducibility and repeatability were 0.095% and 0.552%, respectively. On average, the Lane & Eynon gave results 16% higher than the Luff Schoorl method and 26% above the Berlin Institute method.

In view of these results, the Luff Schoorl procedure has been adopted as an Official method for raw sugar, and the Lane & Eynon method downgraded to Accepted. The Luff Schoorl procedure for raw sugar is a modification of the Method GS4/3-9 (1994) and is written up in the Subject 15 report of the 1998 ICUMSA proceedings.

37.2.3 Color

Prior to the 1994 session of ICUMSA, colors for both raw and white sugars had been analyzed by the ICUMSA Method No. 4 at 420 nm (with adjustment of white sugars to pH 7 after 1978) [3]. At present, the official ICUMSA method for color of raw sugar in solution is Method GS1-7 (1994). This is similar to the old ICUMSA Method No. 4 color except that the Brix and cell length to use are specified for different color ranges. Also, Method No. 4 had allowed the use of filter paper with kieselguhr for slow filtering solutions. The official method for white sugar color is Method GS2/3-9 (1994). This is essentially the same as the raw sugar method except that the pH of 7.0 is maintained by dissolving the sugar in the buffer TEA (trihydroxytriethylamine) instead of adjusting the aqueous solution with

NaOH or HCl, and the refractometer Brix reading is adjusted to compensate for the refractive index of the buffer. Although it is true that use of the buffer allows large numbers of samples to be analyzed in a shorter period of time, careful adjustment of pH with the NaOH or HCl and prompt reading of the color thereafter should achieve the same pH result. Some concerns have also been raised by the ICUMSA referee of Subject 7 and others concerning the use of buffers in general and TEA in particular in the measurement of color [11]. To address these concerns, the recommendation of the referee that the scope of Method GS1-7, which is official for raw sugar color, should be extended for tentative application to all sugar samples was approved at the 1998 ICUMSA meeting. Also, while the Method GS2/3-9 had only been intended for use with white sugars, there may have existed some confusion regarding which method to use in borderline cases such as plantation white sugars. Because of the considerable evidence that TEA buffer is unsuitable for color measurement when significant impurities are present, the scope of Method GS2/3-9 now states explicitly that the method is not applicable to samples other than white sugars. Method GS1-7 for raw sugar has also been modified. For colors in the ICUMSA color range of 200 to 500, the use of a 50-mm cell length and 30% solids is permitted as well as the original specification of a 20-mm cell length and 50% solids. This should be of help with poor filtering sugars.

Although the evidence indicates that use of TEA buffer is not satisfactory for raw sugars, the buffer MOPS [3-(*N*-morpholino)propanesulfonic acid], which has a more favorable p*K* value of 7.15 and which is less likely to form certain complexes with the sugar colorants, has been shown to give satisfactory results with raw sugars. A collaborative test of the analysis of raw sugars by the MOPS buffer method [12] has been completed under the auspices of ICUMSA Subject GS-1. This method has been adopted as a Tentative method by ICUMSA.

There has also continued to be interest in the determination of colors without pH adjustment, which had been the ICUMSA practice for white sugars prior to 1978. The main advantage of this procedure is the ability to analyze samples more rapidly. The method of determining color without pH adjustment has been written up as Method GS2/3-10 and was subjected to a collaborative test coordinated by Südzucker AG, Mannheim. Despite objections in principle by some to a procedure without pH adjustment, this method was adopted as Official at the 1998 ICUMSA meeting for white sugars of color not exceeding 60.

37.2.4 Ash

White sugars are analyzed by the conductivity ash procedure, ICUMSA Method GS2/3-17 (1994). Raw sugars can be analyzed either by the conductivity procedure, ICUMSA Method GS1/3/4/7/8-13 (1994), or by the gravimetric sulfated ash procedure, ICUMSA Method, GS1/3/4/7/8-11 (1994). Also, a collaborative test of sulfated ash procedures was completed, which included the ICUMSA Method and the CSR single incineration method. Comparing the overall results, no statistically significant difference was found between these two methods. The results of the collaborative test and details of the CSR procedure can be found in the Subject 7 report of the 1998 ICUMSA proceedings.

In the ICUMSA procedure for conductivity ash, the measured conductivity is multiplied by a factor in order to make the final conductivity ash results correspond approximately to values for sulfated ash. However, because sulfated ash and conductivity ash are not measuring the same entities, each result has its own significance, and one would not expect to be able to exactly predict one result from the other in a manner that would be valid for all sugars. Nonetheless, studies have shown that as a practical matter, on average, the results of conductivity and sulfated ash for raw sugar are quite close. Chou, Mercene, and Altenburg analyzed 100 regular sugar samples by ICUMSA procedure GS1/3/4/7/8-13 and the Domino Contract sulfated ash procedure [13]. The samples analyzed were those received for contract quality testing and were selected so that the proportion of samples from each origin roughly approximated the distribution on the basis of origin of the contract sugar samples normally received by the NYSTL. Each sample was analyzed in duplicate by the NYSTL and Domino Sugar Corporation. Considering the combined data from both labs, the maximum positive conductivity ash—sulfated ash difference was 0.051, and the maximum negative difference was −0.071. These differences between the methods were within the reproducibility of the sulfated ash method (0.052) for all but five samples. The overall average conductivity ash—sulfated ash difference was −0.002, and the average of the absolute values of the differences was 0.020. Furthermore, an analysis of variance (ANOVA) comparing the average conductivity ash results with the average sulfated ash results showed no statistically significant difference between the methods for these data.

37.2.5 Moisture

White sugar is run by Method GS2/1/3-15. Depending on moisture content, raw sugar used to be run by either this method or by Method GS1-9. However, Method GS1-9 has been replaced by Method GS2/1/3-15 and has been deleted from the *ICUMSA Methods Book*.

37.2.6 Dextran

ICUMSA dextran is determined by the procedure GS1-15. This procedure was adopted as Official at the 1994 ICUMSA meeting.

37.3 ANALYSIS OF MOLASSES

ICUMSA now has official methods for the determination of fructose, glucose, and sucrose by gas chromatography (GC) and high-performance ion chromatography (HPIC). However, the methods of choice in the commercial trading of molasses remain the classical procedures. The most common methods used in the United States for the determination of sugars content are the determination of total reducing sugars after hydrolysis (ICUMSA Method GS4/3-7) and the determination of reducing sugars (ICUMSA Method GS4/3-3) by the Lane & Eynon constant-volume procedure. These methods are also written up in the UMTC handbook [14]. A sucrose figure is calculated by multiplying the difference between the total reducing sugars after hydrolysis and the reducing sugars determined before hydrolysis by the factor 0.95. Although this relationship would be accurate for a mixture of pure sucrose and invert sugar, its use with molasses is not expected to give a true sucrose figure but is in common use to satisfy commercial requirements. Also, it should be noted that while GC or HPIC measure the individual components sucrose, glucose, and fructose, the Lane & Eynon total sugars method measures all compounds that before and after acid hydrolysis, reduce Fehling's solution. This may be of commercial interest, since sucrose, glucose, and fructose do not necessarily represent all the fermentable carbon compounds in molasses.

Total reducing sugars after hydrolysis may also be determined by the Luff Schoorl procedure (ICUMSA GS4/3-9). This procedure has a more clearly defined endpoint of the final titration, and therefore may be the method of choice for laboratories that do not routinely carry out the Lane & Eynon procedure. This method may tend to give lower results than the Lane & Eynon procedure [15].

Sucrose may be determined by the double-polarization method of Herzfeld and Clerget, which appears as method GS4/7-1 in the ICUMSA methods book. The result obtained is referred to as "apparent sucrose" because interfering optically active substances are likely to be present. Apparent sucrose may also be obtained by the Jackson & Gillis Method IV, which appears in the UMTC handbook in Section G.1. Both of these methods require the use of lead subacetate and therefore are unavailable to many laboratories which have ceased using lead compounds because of environmental concerns. Much work has been done in investigating alternative clarifying agents, and continued study in this area remains a high-priority concern of ICUMSA in the hope that a totally satisfactory alternative may be found. It should be noted that if the sample is pure enough, such as some high-test (invert) molasses products, the analysis may be able to be performed without any clarifying agent other than filtercel. A limited study of 15 samples showed comparable results for samples analyzed by the Jackson & Gillis Method IV with and without the use of lead subacetate (NYSTL, unpublished results).

There are a number of ways to determine dry substance and apparent dry substance. The most common determinations are Brix by hydrometer (ICUMSA Method GS4-15), refractometer Brix (more properly named refractometric dry substance; ICUMSA Method GS4-13), and dry substance or moisture by vacuum oven drying on sand (ICUMSA Method GS4/7-11). Moisture may also be determined by the Karl Fischer Procedure (ICUMSA Method GS4/7/3-12) and weight per gallon may be determined directly by a method involving weighing a known volume of molasses [16]. Brix by hydrometer is the most common method in the molasses trade. Brix values can be expected to be considerably higher than values obtained by quartz sand drying because of the effects of impurities which have a higher specific gravity than sugars, and the error introduced by diluting the sample. Refractometer dry substance will give a more accurate estimation of true dry substance than Brix and the result is usually intermediate between the Brix result and the quartz sand drying result. When using the ICUMSA Method GS4-13 for refractometric dry substance, note that in the calculation step it is stated that refractometric dry substance (RDS) can be obtained from the appropriate sucrose tables or from the SPS-3 tables of the *ICUMSA Methods Book.* The SPS-3 section includes tables for correction of the RDS based on the % of invert sugar present. To correct for invert, one would apply the correction of Table 2 (page 5) to the refractive index. Apparent RDS would then be obtained from the sucrose Table (Table A, pages 6–7) and corrected for temperature (Table B, page 8). The correction to be applied to the apparent RDS would be calculated from formula No. 3 (page 5) or looked up in Table F (page 15), which is derived from formula No. 3. When reporting results, it should be made clear if an invert correction has been applied.

Unfermentable reducing sugars may be determined by the method of the AOAC, which is also written up in Section K.2 of the UMTC handbook. However, this requires the use of neutral lead acetate solution and therefore is not available to many laboratories.

REFERENCES

1. W. Altenburg and C. C. Chou, *Sugar Ind. Technol.,* 1991, pp. 187–211.
2. *Proc. Int. Comm. Uniform Methods Sugar Anal.,* 1990, p. 203.
3. F. Schneider, *Sugar Analysis ICUMSA Methods,* British Sugar Corporation, Peterborough, England 1979, pp. 125–128.
4. ICUMSA, *Methods Book,* International Commission for Uniform Methods of Sugar Analysis, Colney, Norwich, England, 1994–1998.
5. *Proc. Int. Comm. Uniform Methods Sugar Anal.,* 1998.

6. *Proc Int. Comm. Uniform Methods Sugar Anal.,* 1986, p. 62.
7. Ibid., p. 60.
8. C. A. Browne and F. W. Zerban, *Physical and Chemical Methods of Sugar Analysis,* 3rd ed., Wiley, New York, 1941, p. 49.
9. C. C. Chou and M. Wnukowski, *Cane Sugar Refining Research Project,* 1980, pp. 1–25.
10. *Proc. Int. Comm. Uniform Methods Sugar Anal.,* 1998, Subject 15.
11. Ibid., Subject 7.
12. Ibid., Subject GS-1, Appendix 1.
13. *Sugar Ind. Technol.,* 1996, pp. 139–159.
14. *The Analysis of Molasses,* United Molasses Co., London, 1991, Secs. E.1 and F.1.
15. *Proc. Int. Comm. Uniform Methods Sugar Anal.,* 1994, p. 83.
16. United States Customs Laboratory, No. 502.4, Nov. 1952.
17. *Proc. Int. Comm. Uniform Methods Sugar Anal.,* 1978, p. 154.

APPENDIX

Reference Tables

TABLE 1 Conversions Between U.S. Customary and Metric (SI) Systems

SI to U.S. Customary	U.S. Customary to SI
Length	
1 micron (μm) = 0.001 mm	1 inch (in.) = 25.400 mm
= 0.00004 in.	= 2.540 cm
1 millimeter (mm) = 0.039370 in.	1 foot (ft) = 12 in.
1 centimeter (cm) = 0.3937 in.	= 0.30480 m
1 decimeter (dm) = 3.93701 in.	1 yard (yd) = 3 ft
1 meter (m) = 39.3701 in.	= 0.91440 m
= 3.28084 ft	1 mile (mi) = 1760 yd
= 1.09361 yd	= 5280 ft
1 kilometer (km) = 0.62137 mi	= 1609.34 m
	= 1.609 km
Area	
1 mm^2 = 0.0015500 sq in.	1 sq in. = 645.160 mm^2
1 cm^2 = 0.155 sq in.	= 6.4516 cm^2
1 m^2 = 10.7639 sq ft	1 sq ft = 0.092903 m^2
= 1.196 sq yd	1 sq yd = 9 sq ft
1 hectare (ha) = 2.4710 acres	= 0.8361 m^2
1 km^2 = 0.38610 sq mi	1 acre = 43,560 sq ft
	= 0.40469 ha
	1 sq mi = 640 acres
	= 2.590 km^2
Volume	
1 cm^3 = 1 mL	1 cu in. = 16.3871 mL
= 0.061024 cu in.	1 cu ft = 1728 cu in.
1 m^3 = 35.3147 cu ft	= 7.4805 gal
1 liter (L) = 0.26417 U.S. gal	= 0.028317 m^3
= 0.2200 Imp. gal	= 28.31681 L
= 1.05668 qt	1 U.S. gal. = 231.0 cu in.
= 1 dm^3	= 3.78543 L
	1 Imp. gal = 4.546 L
	1 quart (qt) = 0.94636 L
	1 ounce (oz) = 29.547 mL
Weight	
1 gram (g) = 0.035274 oz (avoird.)	1 ounce (oz) = 28.3495 g
1 kilogram (kg) = 2.20462 lb	1 grain (International) = 0.0648 g
1 quintal = 100 kg	1 pound (lb) (avoird.) = 16 oz
= 2.20462 cwt (U.S.)	= 453.59237 g
= 1.9684 cwt (English)	= 0.453592 kg
1 metric ton (tonne) = 1000 kg	1 Spanish lb (*libra*) = 1.0143 lb (avoird.)
= 2204.62 lb	100 Spanish lb = 46 kg (approx.)
= 1.10231 short ton	1 International lb = 7000.00 grains
= 0.98421 long ton	= 1 lb (avoird.)
	1 cwt (U.S.) = 100 lb
	= 45.36 kg
	1 cwt (English) = 112 lb
	= 50.80 kg
	1 long ton = 2240 lb
	1016.048 kg
	1 short ton = 2000 lb
	= 907.186 kg

TABLE 1 (*Continued*)

SI to U.S. Customary	U.S. Customary to SI
Capacity	
1 m^2 / tonne = 9.765 sq ft / short ton	1 sq ft / short ton = 0.102408 m^2 / tonne
= 10.9366 sq ft / long ton	1 sq ft / long ton = 0.0091436 m^2 / tonne
1 L / tonne = 0.032036 cu ft / short ton	1 cu ft / short ton = 31.215 L / tonne
= 0.03588 cu ft / long ton	1 cu ft / long ton = 27.870 L / tonne
1 m^2 / m^3 = 0.3048 sq ft / cu ft	1 sq ft / cu ft = 3.28084 m^2 / m^3
1 m^2 / hL = 3.048 sq ft / cu ft	= 0.328084 m^2 / hL
Density, Concentration	
1 kg / m^3 = 0.062428 lb / cu ft	1 lb / cu ft = 16.0185 kg / m^3
1 kg / dm^3 = 62.428 lb / cu ft	= 0.01602 kg / dm^3
1 g / ml = 8.3452 lb / gal	1 lb / gal = 0.1198 g / mL
1 ppm = 0.05841 grain / gal	1 grain / gal = 17.12 ppm
= 1 mg / L	
= 1 μg / mL	
Evaporation	
1 kg / m^2 = 0.204816 lb / sq ft	1 lb / sq / ft = 4.88243 kg / m^2
1 kg / m^2 (0 − 100°C) = 0.2428 lb / sq ft (from and at 212°F)	1 lb / (sq ft)(h)(from and at 212°F) = 4.118 kg / (m^2)(h)(from 0 to 100°C)
1 kg / (m^2)(°C) = 0.113786 lb / (sq ft)(°F)	
Heat	
°C = $\frac{5}{9}$(°F − 32)	°F = 1.8°C + 32
1 °C (diff. in temp.) = 1.8°F	1 °F (diff. in temp.) = 0.555556°C
1 kcal = 3.9683 Btu	1 Btu = 252.016 calories (cal)
1 kcal / kg = 1.8 Btu / lb	= 0.252 kcal
1 kcal / (m^2)(h) = 0.368669 Btu / (sq ft)(h)	1 Btu / lb = 0.555556 kcal / kg
1 kcal / (m^2)(h)(°C) = 0.204816 Btu / (sq ft)(h)(°F)	1 Btu / (sq ft)(h) = 2.71246 kcal / (m^2)(h)
1 kcal / (m^2)(h)(°C)(mm)	1 Btu / (sq ft)(h)(°F) = 4.88243 kcal / (m^2)(h)(°C)
= 0.672 Btu / (sq ft)(h)(°F)(ft)	1 Btu / (sq ft)(h)(°F)(ft) = 1.488 kcal / (m^2)(h)(°C)(m)
= 8.0636 Btu / (sq ft)(h)(°F)(in.)	
1 kcal / (m^3)(h) = 0.11237 Btu / (cu ft)(h)	1 Btu / (sq ft)(h)(°F)(in.) = 0.1240 kcal / (m^2)(h)(°C)(m)
	1 Btu / (cu ft)(h) = 8.90 kcal / (m^3)(h)
Pressure	
1 cmHg at 25°C = 13.56 cm water at 20°C	1 in. Hg at 80°F = 1.130 ft water (70°F)
= 13.70 cm water at 50°C	= 1.143 ft water (130°F)
1 mmHg = 0.01933 psi	1 in. Hg = 0.490 psi
1 kg / cm^2 = 14.2233 lb / sq in. (psi)	1 lb / sq in. (psi) = 2.316 ft water (80°F)
1 kg / m^2 = 0.204816 lb / sq ft	1 lb / sq in. = 0.070307 kg / cm^2
1 tonne / dm^2 = 10.2408 short ton / sq ft	1 lb / sq ft = 4.88243 kg / m^2
= 9.1436 long ton / sq ft	1 short ton / sq ft = 0.09765 tonne / dm^2
	1 long ton / sq ft = 0.10937 tonne / dm^2
1 atm = 14.7 psi	1 psi = 0.068 atm
1 bar = 14.5 psi	= 0.069 bar
= 100 kPa	= 6.8948 kPa (kilopascals)
= 0.1 MPa	
= 1.02 kg / cm^2	
1 MPa = Pa × 10^6	
1 kPa = Pa × 10^3	
= 0.1450 psi	
= 1.02 kg / cm^2	

(*continued*)

TABLE 1 (*Continued*)

SI to U.S. Customary	U.S. Customary to SI
Work, Energy	
1 joule (J) = 0.2389 cal	1 ft-lb = 0.138255 kg · m
= 0.7375 ft-lb	1 hp = 550 ft-lb / sec
= 9.4781×10^{-4} Btu	= 76.04 kg · m / s
	= 645.70 W
	1 Btu = 1.0551×10^{3} J
1 kg · m = 7.233 ft-lb	
1 kg · m / s = 0.01315 hp	
1 kW = 1.341 hp	

TABLE 2 Temperature Conversion (Albert Sauveur)[a]

0–49			50–100			110–600			610–1200		
°C		°F	°C		°F	°C		°F	°C		°F
−17.8	**0**	32	5.56	**42**	107.6	28.9	**84**	183.2	177	**350**	662
−17.2	**1**	33.8	6.11	**43**	109.4	29.4	**85**	185.0	182	**360**	680
−16.7	**2**	35.6	6.67	**44**	111.2	30.0	**86**	186.8	188	**370**	698
−16.1	**3**	37.4	7.22	**45**	113.0	30.6	**87**	188.6	193	**380**	716
−15.6	**4**	39.2	7.78	**46**	114.8	31.1	**88**	190.4	199	**390**	734
−15.0	**5**	41.0	8.33	**47**	116.6	31.7	**89**	192.2	204	**400**	752
−14.4	**6**	42.8	8.89	**48**	118.4	32.2	**90**	194.0	210	**410**	770
−13.9	**7**	44.6	9.44	**49**	120.2	32.8	**91**	195.8	216	**420**	788
−13.3	**8**	46.4	10.0	**50**	122.0	33.3	**92**	197.6	221	**430**	806
−12.8	**9**	48.2	10.6	**51**	123.8	33.9	**93**	199.4	227	**440**	824
−12.2	**10**	50.0	11.1	**52**	125.6	34.4	**94**	201.2	232	**450**	842
−11.7	**11**	51.8	11.7	**53**	127.4	35.0	**95**	203.0	238	**460**	860
−11.1	**12**	53.6	12.2	**54**	129.2	35.6	**96**	204.8	243	**470**	878
−10.6	**13**	55.4	12.8	**55**	131.0	36.1	**97**	206.6	249	**480**	896
−10.0	**14**	57.2	13.3	**56**	132.8	36.7	**98**	208.4	254	**490**	914
−9.44	**15**	59.0	13.9	**57**	134.6	37.2	**99**	210.2	260	**500**	932
−8.89	**16**	60.8	14.4	**58**	136.4	37.8	**100**	212.0	266	**510**	950
−8.33	**17**	62.6	15.0	**59**	138.2	43	**110**	230	271	**520**	968
−7.78	**18**	64.4	15.6	**60**	140.0	49	**120**	248	277	**530**	986
−7.22	**19**	66.2	16.1	**61**	141.8	54	**130**	266	282	**540**	1004
−6.67	**20**	68.0	16.7	**62**	143.6	60	**140**	284	288	**550**	1022
−6.11	**21**	69.8	17.2	**63**	145.4	66	**150**	302	293	**560**	1040
−5.56	**22**	71.6	17.8	**64**	147.2	71	**160**	320	299	**570**	1058
−5.00	**23**	73.4	18.3	**65**	149.0	77	**170**	338	304	**580**	1076
−4.44	**24**	75.2	18.9	**66**	150.8	82	**180**	356	310	**590**	1094
−3.89	**25**	77.0	19.4	**67**	152.6	88	**190**	374	316	**600**	1112
−3.33	**26**	78.8	20.0	**68**	154.4	93	**200**	392	321	**610**	1130
−2.78	**27**	80.6	20.6	**69**	156.2	99	**210**	410	327	**620**	1148
−2.22	**28**	82.4	21.1	**70**	158.0	100	**212**	413	332	**630**	1166
−1.67	**29**	84.2	21.7	**71**	159.8	104	**220**	428	338	**640**	1184
−1.11	**30**	86.0	22.2	**72**	161.6	110	**230**	446	343	**650**	1202
−0.56	**31**	87.8	22.8	**73**	163.4	116	**240**	464	349	**660**	1220

TABLE 2 (*Continued*)

9–49			50–100			110–600			610–1200		
°C		°F	°C		°F	°C		°F	°C		°F
0	**32**	89,6	23.3	**74**	165.2	121	**250**	482	354	**670**	1238
0.56	**33**	91.4	23.9	**75**	167.0	127	**260**	500	360	**680**	1256
1.11	**34**	93.2	24.4	**76**	168.8	132	**270**	518	366	**690**	1274
1.67	**35**	95.0	25.0	**77**	170.6	138	**280**	536	371	**700**	1292
2.22	**36**	96.8	25.6	**78**	172.4	143	**290**	554	377	**710**	1310
2.78	**37**	98.6	26.1	**79**	174.2	149	**300**	572	382	**720**	1328
3.33	**38**	100.4	26.7	**80**	176.0	154	**310**	590	388	**730**	1346
3.89	**39**	102.2	27.2	**81**	177.8	160	**320**	608	393	**740**	1364
4.44	**40**	104.0	27.8	**82**	179.6	166	**330**	626	399	**750**	1382
5.00	**41**	105.8	28.3	**83**	181.4	171	**340**	644	404	**760**	1400
410	**770**	1418	454	**850**	1562	504	**940**	1724	571	**1060**	1940
416	**780**	1436	460	**860**	1580	510	**950**	1742	582	**1080**	1976
421	**790**	1454	466	**870**	1598	516	**960**	1760	593	**1100**	2012
427	**800**	1472	471	**880**	1616	521	**970**	1778	604	**1120**	2048
432	**810**	1490	477	**890**	1634	527	**980**	1796	616	**1140**	2084
438	**820**	1508	482	**900**	1652	532	**990**	1814	627	**1160**	2120
443	**830**	1526	488	**910**	1670	538	**1000**	1832	638	**1180**	2156
449	**840**	1544	493	**920**	1688	549	**1020**	1868	649	**1200**	2192
			499	**930**	1706	560	**1040**	1904			

Source: Courtesy of *Chem. Met. Eng.*

[a]The numbers in boldface type refer to the temperature either in degrees Celsius or Fahrenheit which it is desired to convert into the other scale. If converting from degrees Fahrenheit to degrees Celsius the equivalent temperature will be found in the left column, while if converting from degrees Celsius to degrees Fahrenheit, the answer will be found in the column on the right.

$$°C = (F - 32)\left(\frac{5}{9}\right) \qquad °F = (°C \times 1.8) + 32$$

Example:

$$(98.6 - 32)\left(\frac{5}{9}\right) = \frac{66.6}{9} \times 5 = 37°C$$

$$(37 \times 1.8) + 32 = 66.6 + 32 = 98.6°F$$

TABLE 3 Boiling-Point Elevation (or Rise) for Cane products at 760 mmHg Pressure

Total	Elevation in Boiling Point (°C) for Percent Purity of Syrup or Massecuite:							
Solids	**100**	**90**	**80**	**70**	**65**	**50**	**45**	**35**
92	20.5	22	23	24	25	26	27	29
90	19	20	21	22.5	23	24.5	26	27
85	13	14	14.5	15.5	16.5	17.5	18.5	19.5
80	9	10	10.5	11	12	12.5	13.5	14
75	6.5	7	7.5	8	9	9.5	10	10.5
70	5	5.5	6	6.5	7	7.5	8	8
65	4	4	4.5	5	5.5	5.5	6	6.5
60	3	3	3.5	4	4	4.5	5	5
55	2	2.5	3	3	3	3.5	4	4
50	2	2	2	2.5	2.5	3	3	3.5
40	1	1	1.5	1.5	2	2	2	2.5
30	0.5	1	1	1	1	1	1.5	1.5

Source: Courtesy of Hawaiian Sugar Technologists.

TABLE 4 Boiling Point of Water Under Vacuum

Vacuum		Temperature		Vacuum		Temperature	
In. Hg	**mmHg**	**°C**	**°F**	**In. hg**	**mmHg**	**°C**	**°F**
23.62	600	61.5	142.7	25.79	655	52.6	126.7
23.82	605	60.8	141.4	25.98	660	51.6	124.9
24.02	610	60.1	140.2	26.18	665	50.5	122.9
24.21	615	59.3	138.7	26.38	670	49.5	121.1
24.41	620	58.6	137.5	26.58	675	48.3	118.9
24.61	625	57.8	136.1	26.77	680	47.1	116.8
24.80	630	57.0	134.6	26.97	685	45.8	114.5
25.00	635	56.2	133.2	27.17	690	44.5	112.1
25.20	640	55.3	131.5	27.36	695	43.1	109.6
25.39	645	54.5	130.1	27.56	700	41.5	10.67
25.59	650	53.5	128.3				

TABLE 5 Specific Gravity and Brix of Milk of Lime at 27.5°C

Apparent Specific Gravity	Brix	Apparent Specific Gravity	Brix
1.0049	1.3	1.0788	19.2
1.0132	3.4	1.0861	20.8
1.0205	5.2	1.0934	22.4
1.0278	7.0	1.1006	24.0
1.0350	8.8	1.1078	25.6
1.0423	10.6	1.1151	27.2
1.0496	12.4	1.1224	28.8
1.0569	14.1	1.1297	30.3
1.0642	15.8	1.1369	31.8
1.0715	17.5	1.1442	33.3

Source: Data from J. D'Ans and E. Lax, *Taschenbuch für Chemiker und Physiker,* 1949. (Values checked in laboratory by Pieter Honig, according to personal communications, 1960.)

TABLE 6 Solubility of Lime in Sugar Solutions at 25 °C[a]

Sucrose (% of solution)	CaO (% of solution)	Specific Gravity of Solution[b]	Sucrose (% of solution)	CaO (% of solution)	Specific Gravity of Solution[b]
0	0.122	1.001	18	3.54	1.118
2	0.235	1.010	20	4.18	1.133
4	0.43	1.022	22	4.86	1.149
6	0.68	1.034	24	5.57	1.162
8	1.00	1.047	26	6.32	1.174
10	1.39	1.060	28	7.06	1.184
12	1.84	1.074	30	7.83	[c]
14	2.36	1.089	32	8.63	[c]
16	2.94	1.103	34	9.45	[c]

Source: Data compiled by E. E. Coll, 1961.

[a] Sucrose and CaO values from Seidel (Linke), *Solubility of Inorganic Compounds,* 4th ed., Van Nostrand, New York, 1958.

[b] Specific gravity values from data of Table 15, *Cane Sugar Handbook,* 8th ed., McGraw-Hill, New York, 1945.

[c] Forms sucrate.

TABLE 7 Solubility of Pure Sucrose in Water Calculated for Unit Increments of Temperature from the Equation $S = 64.397 + 0.07251t + 0.0020569t^2 - 9.035 \times 10^{-6}\, t^3$

Temp., *t* (°C)	Sucrose by Weight (in air), *S* %	Sucrose per 100 of Water	Temp., *t* (°C)	Sucrose by Weight (in air), *S* %	Sucrose per 100 of Water	Temp., *t* (°C)	Sucrose by Weight (in air), *S* %	Sucrose per 100 of Water
0	64.40	180.9	30	68.18	214.3	60	74.20	287.6
1	64.47	181.5	31	68.35	216.0	61	74.42	291.0
2	64.55	182.1	32	68.53	217.7	62	74.65	294.4
3	64.63	182.7	33	68.70	219.5	63	74.87	297.9
4	64.72	183.4	34	68.88	221.4	64	75.09	301.5
5	64.81	184.2	35	69.07	223.3	65	75.32	305.2
6	64.90	184.9	36	69.25	225.2	66	75.54	308.9
7	65.00	185.7	37	69.44	227.2	67	75.77	312.7
8	65.10	186.6	38	69.63	229.2	68	76.00	316.6
9	65.21	187.5	39	69.82	231.3	69	76.22	320.6
10	65.32	188.4	40	70.01	233.4	70	76.45	324.7
11	65.43	189.3	41	70.20	235.6	71	76.68	328.8
12	65.55	190.3	42	70.40	237.8	72	76.91	333.1
13	65.67	191.3	43	70.60	240.1	73	77.14	337.4
14	65.79	192.3	44	70.80	242.5	74	77.36	341.8
15	65.92	193.4	45	71.00	244.8	75	77.59	346.3
16	66.05	194.5	46	71.20	247.3	76	77.82	350.9
17	66.18	195.7	47	71.41	249.8	77	78.05	355.6
18	66.32	196.9	48	71.62	252.3	78	78.28	360.4
19	66.45	198.1	49	71.83	254.9	79	78.51	365.3
20	66.60	199.4	50	72.04	257.6	80	78.74	370.3
21	66.74	200.7	51	72.25	260.3	81	78.96	375.4
22	66.89	202.0	52	72.46	263.1	82	79.19	380.6
23	67.04	203.4	53	72.67	265.9	83	79.42	385.9
24	67.20	204.8	54	72.89	268.8	84	79.65	391.3
25	67.35	206.3	55	73.10	271.8	85	79.87	396.8
26	67.51	207.8	56	73.32	274.8	86	80.10	402.5
27	67.68	209.4	57	73.54	277.9	87	80.32	408.3
28	67.84	211.0	58	73.76	281.1	88	80.55	414.1
29	68.01	212.6	59	73.98	284.3	89	80.77	420.1

Source: Data from D. F. Charles, *Int. Sugar J.*, Vol LXII:126–131, 1960.

TABLE 8 Solutility of Certain Salts in Water in the Presence of Suucrose[a]

Solution Containing:	5% Sucrose (g)	10% Sucrose (g)	15% Sucrose (g)	20% Sucrose (g)	25% Sucrose (g)
Calcium sulfate	2.095	1.946	1.593	1.539	1.333
Calcium carbonate	0.027	0.036	0.024	0.022	0.008
Calcium oxalate	0.033	0.047	0.012	0.008	0.001
Calcium phosphate	0.029	0.028	0.014	0.018	0.005
Calcium citrate	1.813	1.578	1.505	1.454	1.454
Magnesium carbonate	0.317	0.199	0.194	0.213	0.284

Source: Jacobsthal, *Z. Rübenzuckerind.*, 18: 649; taken from Sidersky's *Traité d'analyse des matrières sucreées,* p. 11.

[a]Note that no temperature is given for this table.

TABLE 9 Most Soluble Mixtures of Sucrose and Invert Sugar

Temperature (°C)	Sucrose (%)	Invert Sugar (%)	Water (%)	Sucrose to 100 g of Water (g)	Invert Sugar to 100 g of Water (g)	Total Sugar to 100 g of Water (g)	Sucrose Alone to 100 g of Water (g)
0	43.7	27.2	29.1	150.2	93.5	243.7	179.2
10	40.9	31.8	27.3	149.8	116.5	266.3	190.5
15	39.1	34.8	26.1	149.8	133.4	283.2	197.9
23.15	36.3	39.9	23.8	152.5	167.6	319.1	208.5
30	33.6	45.4	21.0	160.0	216.2	376.2	213.6
40	31.1	50.7	18.2	170.9	278.6	449.5	238.1
50	27.7	58.0	14.3	193.7	405.6	599.3	260.4

Source: NBS-C440, National Bureau of Standards, Washington, DC, 1942, p. 692. See also Junk, Nelson, and Sherrill, *Food Technology,* 1947, p. 506.

TABLE 10 Viscosity (in Centipoise) of Sucrose Solutions from 0 to 80°C in 5°C Intervals[a]

% Sucrose by Weight in Vacuum	Temperature (°C)								
	0	5	10	15	20	25	30	35	40
20	3.782	3.137	2.642	2.254	1.945	1.695	1.493	1.325	1.184
21	3.977	3.293	2.768	2.357	2.031	1.769	1.555	1.379	1.231
22	4.187	3.460	2.904	2.469	2.124	1.846	1.622	1.436	1.281
23	4.415	3.642	3.050	2.589	2.224	1.931	1.692	1.497	1.333
24	4.661	3.838	3.208	2.719	2.331	2.021	1.769	1.563	1.390
25	4.931	4.051	3.380	2.859	2.447	2.118	1.852	1.634	1.451
26	5.223	4.282	3.565	3.010	2.573	2.223	1.941	1.709	1.516
27	5.542	4.533	3.767	3.175	2.708	2.336	2.037	1.791	1.587
28	5.889	4.807	3.986	3.352	2.855	2.459	2.140	1.880	1.663
29	6.271	5.107	4.225	3.546	3.015	2.592	2.251	1.974	1.744
30	6.692	5.435	4.487	3.757	3.187	2.735	2.373	2.078	1.833
31	7.148	5.794	4.772	3.988	3.376	2.892	2.504	2.188	1.927
32	7.653	6.187	5.084	4.239	3.581	3.062	2.645	2.306	2.029
33	8.214	6.623	5.428	4.515	3.806	3.246	2.799	2.437	2.141
34	8.841	7.106	5.808	4.818	4.052	3.448	2.967	2.578	2.260
35	9.543	7.645	6.230	5.154	4.323	3.670	3.150	2.732	2.390
36	10.31	8.234	6.693	5.524	4.621	3.914	3.353	2.901	2.532
37	11.19	8.904	7.212	5.933	4.950	4.182	3.573	3.083	2.687
38	12.17	9.651	7.791	6.389	5.315	4.476	3.815	3.285	3.856
39	13.27	10.49	8.436	6.895	5.718	4.803	4.082	3.506	3.039
40	14.55	11.44	9.166	7.463	6.167	5.164	4.375	3.747	3.241
41	16.00	12.53	9.992	8.102	6.671	5.565	4.701	4.014	3.461
42	17.67	13.76	10.93	8.821	7.234	6.014	5.063	4.310	3.706
43	19.58	15.17	11.98	9.630	7.867	6.515	5.467	4.639	3.977
44	21.76	16.77	13.18	10.55	8.579	7.077	5.917	5.004	4.277
45	24.29	18.60	14.55	11.59	9.383	7.710	6.421	5.412	4.611

46	27.22	20.72	16.11	12.77	10.30	8.423	6.988	5.869	4.983
47	30.60	23.15	17.91	14.12	11.33	9.231	7.628	6.381	5.400
48	34.56	25.99	19.98	15.67	12.51	10.15	8.350	6.960	5.868
49	39.22	29.30	22.39	17.47	13.87	11.20	9.171	7.613	6.395
50	44.74	33.18	25.21	19.53	15.43	12.40	10.11	8.358	6.991
51	51.29	37.76	28.48	21.94	17.24	13.78	11.18	9.203	7.669
52	59.11	43.18	32.34	24.76	19.34	15.37	12.41	10.17	8.439
53	68.51	49.64	36.91	28.08	21.79	17.23	13.84	11.28	9.321
54	79.92	57.42	42.38	32.00	24.68	19.39	15.49	12.57	10.34
55	93.86	66.82	48.90	36.65	28.07	21.93	17.42	14.06	11.50
56	111.0	78.27	56.79	42.23	32.12	24.92	19.68	15.80	12.86
57	132.3	92.35	66.39	48.96	36.95	28.48	22.35	17.83	14.44
58	159.0	109.5	78.15	57.12	42.78	32.73	25.51	20.22	16.29
59	192.5	131.5	92.70	67.12	49.84	37.83	29.28	23.06	18.46
60	235.7	159.1	110.9	79.49	58.49	44.03	33.82	26.46	21.04
61	291.6	194.2	133.8	94.91	69.16	51.60	39.32	30.53	24.11
62	364.6	239.5	163.0	114.3	82.42	60.92	46.02	35.45	27.80
63	461.6	298.6	200.4	138.9	99.08	72.49	54.27	41.46	32.26
64	591.5	376.5	249.0	170.4	120.1	87.00	64.48	48.84	37.69
65	767.7	480.7	313.1	211.3	147.2	105.4	77.29	57.97	44.36
66	1,013[b]	621.9[b]	398.5[b]	264.9	182.0	128.8	93.45	69.40	52.61
67	1,355[b]	816.1[b]	513.7[b]	336.3	227.8	159.1	114.1	83.82	62.94
68	1,846[b]	1,088[b]	672.1[b]	432.6	288.5	198.7	140.7	102.3	75.97
69	2,561[b]	1,476[b]	892.5[b]	564.0	370.1	251.1	175.6	126.0	92.58
70	3,628[b]	2,038[b]	1,206[b]	746.9	481.6	321.6	221.6	157.0	114.0
71	5,253[b]	2,871[b]	1,658[b]	1,006[b]	636.3	417.8	283.4	198.0	142.0
72	7,792[b]	4,136[b]	2,329[b]	1,379[b]	854.9	551.0	367.6	253.0	178.9
73	11,876[b]	6,103[b]	3,340[b]	1,929[b]	1,170	738.9	484.3	327.9	228.5
74	18,639[b]	9,245[b]	4,906[b]	2,759[b]	1,631	1,009	648.5	431.6	296.0
75	30,207[b,c]	14,428[b]	7,402[b]	4,039[b]	2,328	1,405	884.8	577.4	389.5

(*continued*)

TABLE 10 (*Continued*)

% Sucrose by Weight in Vacuum	Temperature (°C)							
	45	50	55	60	65	70	75	80
20	1.07	0.97	0.88	0.81	0.74	0.68	0.63	0.59
21	1.11	1.00	.91	.84	.77	.71	.65	.61
22	1.15	1.04	.95	.87	.79	.73	.68	.63
23	1.20	1.09	.98	.90	.82	.76	.70	.65
24	1.25	1.13	1.02	.93	.85	.79	.73	.67
25	1.30	1.17	1.06	.97	.89	.82	.75	.70
26	1.36	1.22	1.11	1.01	.92	.85	.78	.72
27	1.42	1.28	1.16	1.05	.96	.88	.81	.75
28	1.49	1.34	1.21	1.10	1.00	.92	.85	.78
29	1.56	1.40	1.26	1.14	1.04	.96	.88	.81
30	1.64	1.47	1.32	1.20	1.09	1.00	.92	.85
31	1.71	1.54	1.38	1.25	1.14	1.04	.96	.88
32	1.80	1.61	1.45	1.31	1.19	1.09	1.00	.92
33	1.89	1.69	1.52	1.37	1.25	1.14	1.04	.96
34	2.00	1.78	1.60	1.44	1.31	1.19	1.09	1.00
35	2.11	1.87	1.67	1.51	1.37	1.25	1.14	1.05
36	2.23	1.98	1.76	1.59	1.44	1.31	1.19	1.10
37	2.36	2.09	1.86	1.67	1.51	1.37	1.25	1.15
38	2.51	2.21	1.97	1.76	1.59	1.44	1.31	1.20
39	2.67	2.35	2.08	1.86	1.67	1.52	1.38	1.26
40	2.84	2.49	2.21	1.97	1.76	1.60	1.45	1.32
41	3.02	2.65	2.34	2.08	1.86	1.68	1.53	1.39
42	3.23	2.82	2.49	2.21	1.97	1.77	1.61	1.46
43	3.45	3.01	2.65	2.35	2.09	1.88	1.69	1.54
44	3.71	3.22	2.83	2.50	2.22	1.99	1.79	1.63
45	3.98	3.46	3.02	2.66	2.36	2.11	1.90	1.71
46	4.29	3.71	3.24	2.85	2.52	2.25	2.01	1.82
47	4.64	4.00	3.48	3.05	2.70	2.40	2.14	1.93
48	5.01	4.32	3.75	3.28	2.89	2.56	2.29	2.05
49	5.45	4.68	4.05	3.53	3.10	2.74	2.44	2.19

50	5.94	5.07	4.38	3.81	3.34	2.94	2.61	2.34
51	6.49	5.52	4.75	4.12	3.60	3.17	2.81	2.50
52	7.11	6.03	5.16	4.47	3.89	3.42	3.02	2.69
53	7.83	6.61	5.64	4.87	4.23	3.70	3.26	2.89
54	8.63	7.27	6.18	5.30	4.60	4.01	3.52	3.12
55	9.57	8.02	6.79	5.81	5.01	4.36	3.82	3.37
56	10.7	8.88	7.50	6.38	5.49	4.76	4.16	3.66
57	11.9	9.88	8.30	7.04	6.03	5.20	4.54	3.98
58	13.4	11.1	9.22	7.80	6.65	5.72	4.97	4.34
59	15.1	12.4	10.3	8.65	7.36	6.30	5.45	4.75
60	17.0	14.0	11.6	9.66	8.17	6.98	6.00	5.20
61	19.4	15.8	13.0	10.9	9.11	7.75	6.64	5.74
62	22.2	17.9	14.8	12.2	10.2	8.63	7.38	6.35
63	25.6	20.5	16.7	13.8	11.5	9.68	8.23	7.05
64	29.7	23.7	19.1	15.7	13.0	10.9	9.21	7.87
65	34.7	27.5	22.0	17.9	14.8	12.4	10.4	8.81
66	40.8	32.1	25.5	20.6	16.9	14.1	11.8	9.93
67	48.4	37.7	29.8	23.9	19.4	16.1	13.4	11.3
68	57.8	44.7	35.1	27.9	22.6	18.4	15.3	12.8
69	69.8	53.3	41.6	32.9	26.3	21.4	17.6	14.7
70	84.9	64.4	49.7	39.0	31.0	25.0	20.4	16.8
71	105	78.4	59.9	46.6	36.7	29.4	23.8	19.5
72	131	96.5	73.0	56.1	43.9	34.9	28.0	22.8
73	165	121	89.7	68.4	52.9	41.7	33.3	26.9
74	209	152	112	84.1	64.6	50.3	39.9	32.0
75	271	193	141	105	79.6	61.4	48.2	38.3

Source: Revised, 1957, by J. F. Swindells, C. F. Snyder, R. C. Hardy, and P. E. Golden, Supplement to NBS-C440, 1942, Table 132, issued July 1958.

[a]These values are based on measurements made at NBS on solutions containing 30, 40, 50, 60, 65, and 70% of sucrose by weight in vacuum, using the value 1.0020 cP for water at 20°C, and are estimated to be accurate to about 0.1%.

[b]Values extrapolated below the temperature range of the measurements.

[c]Incorrect in original table.

[d]*NBS*—S298, 1917.

[e]*Zucker* (several references).

[f]*Ind. Eng. Chem.*, 1930, p. 91.

TABLE 11 Brix, Density, Grams of Sucrose per 100 mL, and Degree Baumé of Sugar Solution at 20°C

Percentage of Sucrose by Weight (Brix) (1)	d_4^{20} (2)	d_{20}^{20} (3)	Grams of Sucrose per 100 mL Weight *in Vacuo* (4)	Degrees Baumé (modulus 45) (5)	Percentage of Sucrose by Weight (Brix) (1)	d_4^{20} (2)	d_{20}^{20} (3)	Grams of Sucrose per 100 mL Weight *in Vacuo* (4)	Degrees Baumé (modulus 45) (5)
0.0	0.998234	1.00000	0.000	0.00	5.5	1.019851	1.02168	5.609	3.07
0.1	0.998622	1.00039	0.100	0.06	5.6	1.020251	1.02208	5.713	3.13
0.2	0.999010	1.00078	0.200	0.11	5.7	1.020651	1.02248	5.818	3.18
0.3	0.999398	1.00117	0.300	0.17	5.8	1.021053	1.02289	5.922	3.24
0.4	0.999786	1.00156	0.400	0.22	5.9	1.021454	1.02329	6.027	3.30
0.5	1.000174	1.00194	0.500	0.28	6.0	1.021855	1.02369	6.131	3.35
0.6	1.000563	1.00233	0.600	0.34	6.1	1.022257	1.02409	6.236	3.41
0.7	1.000952	1.00272	0.701	0.39	6.2	1.022659	1.02450	6.340	3.46
0.8	1.001342	1.00312	0.801	0.45	6.3	1.023061	1.02490	6.445	3.52
0.9	1.001731	1.00351	0.902	0.51	6.4	1.023463	1.02530	6.550	3.57
1.0	1.002120	1.00390	1.002	0.56	6.5	1.023867	1.02571	6.655	3.63
1.1	1.002059	1.00429	1.103	0.62	6.6	1.024270	1.02611	6.760	3.69
1.2	1.002897	1.00468	1.203	0.67	6.7	1.024673	1.02652	6.865	3.74
1.3	1.003286	1.00507	1.304	0.73	6.8	1.025077	1.02692	6.971	3.80
1.4	1.003675	1.00546	1.405	0.79	6.9	1.025481	1.02733	7.076	3.85
1.5	1.004064	1.00585	1.506	0.84	7.0	1.025885	1.02773	7.181	3.91
1.6	1.004453	1.00624	1.607	0.90	7.1	1.026289	1.02814	7.287	3.96
1.7	1.004844	1.00663	1.708	0.95	7.2	1.026694	1.02854	7.392	4.02
1.8	1.005234	1.00702	1.809	1.01	7.3	1.027099	1.02895	7.498	4.08
1.9	1.005624	1.00741	1.911	1.07	7.4	1.027504	1.02936	7.604	4.13
2.0	1.006015	1.00780	2.012	1.12	7.5	1.027910	1.02976	7.709	4.19
2.1	1.006405	1.00819	2.113	1.18	7.6	1.028316	1.03017	7.185	4.24

2.2	1.006796	1.00859	2.215	1.23	7.7	1.028722	1.03058	7.921	4.30
2.3	1.007188	1.00898	2.317	1.29	7.8	1.029128	1.03098	8.027	4.35
2.4	1.007580	1.00937	2.418	1.34	7.9	1.029535	1.03139	8.133	4.41
2.5	1.007972	1.00977	2.520	1.40	8.0	0.029942	1.03180	8.240	4.46
2.6	1.008363	1.01016	2.622	1.46	8.1	1.030349	1.03221	8.346	4.52
2.7	1.008755	1.01055	2.724	1.51	8.2	1.030757	1.03262	8.452	4.58
2.8	1.009148	1.01094	2.826	1.57	8.3	1.031165	1.03303	8.559	4.63
2.9	1.009541	1.01134	2.928	1.62	8.4	1.031573	1.03344	8.665	4.69
3.0	1.009934	1.01173	3.030	1.68	8.5	1.031982	1.03385	8.772	4.74
3.1	1.010327	1.01213	3.132	1.74	8.6	1.032391	1.03426	8.879	4.80
3.2	1.010721	1.01252	3.234	1.79	8.7	1.032800	1.03467	8.985	4.85
3.3	1.011115	1.01292	3.337	1.85	8.8	1.033209	1.03508	9.092	4.91
3.4	1.011510	1.01331	3.439	1.90	8.9	1.033619	1.03549	9.199	4.96
3.5	1.011904	1.01371	3.542	1.96	9.0	1.034029	1.03590	9.306	5.02
3.6	1.012298	1.01410	3.644	2.02	9.1	1.034439	1.03631	9.413	5.07
3.7	1.012694	1.01450	3.747	2.07	9.2	1.034850	1.03672	9.521	5.13
3.8	1.013089	1.01490	3.850	2.13	9.3	1.035260	1.03713	9.628	5.19
3.9	1.013485	1.01529	3.953	2.18	9.4	1.035671	1.03755	9.735	5.24
4.0	1.013881	1.01569	4.056	2.24	9.5	1.036082	1.03796	9.843	5.30
4.1	1.014277	1.01609	4.159	2.29	9.6	1.036494	1.03837	9.950	5.35
4.2	1.014673	1.01649	4.262	2.35	9.7	1.036906	1.03879	10.058	5.41
4.3	1.015070	1.01688	4.365	2.40	9.8	1.037318	1.03920	10.166	5.46
4.4	1.015467	1.01728	4.468	2.46	9.9	1.037730	1.03961	10.274	5.52
4.5	1.015864	1.01768	4.571	2.52	10.0	1.038143	1.04003	10.381	5.57
4.6	1.016261	1.01808	4.675	2.57	10.1	1.038556	1.04044	10.489	5.63
4.7	1.016659	1.01848	4.778	2.63	10.2	1.038970	1.04086	10.597	5.68
4.8	1.017058	1.01888	4.882	2.68	10.3	1.039383	1.04127	10.706	5.74
4.9	1.017456	1.01928	4.986	2.74	10.4	1.039797	1.04169	10.814	5.80
5.0	1.017854	1.01968	5.089	2.79	10.5	1.040212	1.04210	10.922	5.85
5.1	1.018253	1.02008	5.193	2.85	10.6	1.040626	1.04252	11.031	5.91
5.2	1.018652	1.02048	5.297	2.91	10.7	1.041041	1.04293	11.139	5.96
5.3	1.019052	1.02088	5.401	2.96	10.8	1.041456	1.04335	11.248	6.02
5.4	1.019451	1.02128	5.506	3.02	10.9	1.041872	1.04377	11.356	6.07

(*continued*)

TABLE 11 (*Continued*)

Percentage of Sucrose by Weight (Brix) (1)	d_4^{20} (2)	d_{20}^{20} (3)	Grams of Sucrose per 100 mL Weight *in Vacuo* (4)	Degrees Baumé (modulus 45) (5)	Percentage of Sucrose by Weight (Brix) (1)	d_4^{20} (2)	d_{20}^{20} (3)	Grams of Sucrose per 100 mL Weight *in Vacuo* (4)	Degrees Baumé (modulus 45) (5)
11.0	1.042288	1.04418	11.465	6.13	16.5	1.065621	1.06759	17.583	9.17
11.1	1.042704	1.04460	11.574	6.18	16.6	1.066054	1.06802	17.697	9.22
11.2	1.043121	1.04502	11.683	6.24	16.7	1.066487	1.06845	17.810	9.28
11.3	1.043537	1.04544	11.792	6.30	16.8	1.066921	1.06889	17.924	9.33
11.4	1.043954	1.04585	11.901	6.35	16.9	1.067355	1.06933	18.038	9.39
11.5	1.044370	1.04627	12.010	6.41	17.0	1.067789	1.06976	18.152	9.45
11.6	1.044788	1.04669	12.120	6.46	17.1	1.068223	1.07020	18.267	9.50
11.7	1.045206	1.04711	12.229	6.52	17.2	1.068658	1.07063	18.381	9.56
11.8	1.045625	1.04753	12.338	6.57	17.3	1.069093	1.07107	18.495	9.61
11.9	1.046043	1.04795	12.448	6.63	17.4	1.069529	1.07151	18.610	9.67
12.0	1.046462	1.04837	12.558	6.68	17.5	1.069964	1.07194	18.724	9.72
12.1	1.046881	1.04879	12.667	6.74	17.6	1.070400	1.07238	18.839	9.78
12.2	1.047300	1.04921	12.777	6.79	17.7	1.070836	1.07282	18.954	9.83
12.3	1.047720	1.04963	12.887	6.85	17.8	1.071273	1.07325	19.069	9.89
12.4	1.048140	1.05005	12.997	6.90	17.9	1.071710	1.07369	19.184	9.94
12.5	1.048559	1.05047	13.107	6.96	18.0	1.072147	1.07413	19.299	10.00
12.6	1.048980	1.05090	13.217	7.02	18.1	1.072585	1.07457	19.414	10.05
12.7	1.049401	1.05132	13.327	7.07	18.2	1.073023	1.07501	19.529	10.11
12.8	1.049822	1.05174	13.438	7.13	18.3	1.073461	1.07545	19.644	10.16
12.9	1.050243	1.05216	13.548	7.18	18.4	1.073900	1.07589	19.760	10.22
13.0	1.050665	1.05259	13.659	7.24	18.5	1.074338	1.07633	19.875	10.27
13.1	1.051087	1.05301	13.769	7.29	18.6	1.074777	1.07677	19.991	10.33

13.2	1.051510	1.05343	13.880	7.35	18.7	1.075217	1.07721	20.107	10.38
13.3	1.051933	1.05386	13.991	7.40	18.8	1.075657	1.07765	20.022	10.44
13.4	1.052356	1.05428	14.102	7.46	18.9	1.076097	1.07809	20.338	10.49
13.5	1.052778	1.05470	14.213	7.51	19.0	1.076536	1.07853	20.454	10.55
13.6	1.053202	1.05513	14.324	7.57	19.1	1.076987	1.07898	20.570	10.60
13.7	1.053626	1.05556	14.435	7.62	19.2	1.077419	1.07942	20.686	10.66
13.8	1.054050	1.05598	14.546	7.68	19.3	1.077860	1.07986	20.803	10.71
13.9	1.054475	1.05641	14.657	7.73	19.4	1.078302	1.08030	20.919	10.77
14.0	1.054900	1.05683	14.769	7.79	19.5	1.078744	1.08075	21.036	10.82
14.1	1.055325	1.05726	14.880	7.84	19.6	1.079187	1.08119	21.152	10.88
14.2	1.055751	1.05769	14.992	7.90	19.7	1.079629	1.08164	21.269	10.93
14.3	1.056176	1.05811	15.103	7.96	19.8	1.080072	1.08208	21.385	10.99
14.4	1.056602	1.05854	15.215	8.01	19.9	1.080515	1.08252	21.502	11.04
14.5	1.057029	1.05897	15.327	8.07	20.0	1.080959	1.08297	21.619	11.10
14.6	1.057455	1.05940	15.439	8.12	20.1	1.081403	1.08342	21.736	11.15
14.7	1.057882	1.05982	15.551	8.18	20.2	1.081848	1.08386	21.853	11.21
14.8	1.058310	1.06025	15.663	8.23	20.3	1.082292	1.08431	21.971	11.26
14.9	1.058737	1.06068	15.775	8.29	20.4	1.082737	1.08475	22.088	11.32
15.0	1.059165	1.06111	15.887	8.34	20.5	1.083182	1.08520	22.205	11.37
15.1	1.059593	1.06154	16.000	8.40	20.6	1.083628	1.08565	22.323	11.43
15.2	1.060022	1.05197	16.112	8.45	20.7	1.084074	1.08609	22.440	11.48
15.3	1.060451	1.06240	16.225	8.51	20.8	1.084520	1.08654	22.558	11.54
15.4	1.060880	1.06283	16.338	8.56	20.9	1.084967	1.08699	22.676	11.59
15.5	1.061308	1.06326	16.450	8.62	21.0	1.085414	1.08744	22.794	11.65
15.6	1.061738	1.06369	16.563	8.67	21.1	1.085861	1.08789	22.912	11.70
15.7	1.062168	1.06412	16.676	8.73	21.2	1.086309	1.08834	23.030	11.76
15.8	1.062598	1.06455	16.789	8.78	21.3	1.086757	1.08879	23.148	11.81
15.9	1.063029	1.06499	16.902	8.84	21.4	1.087205	1.08923	23.266	11.87
16.0	1.063460	1.06542	17.015	8.89	21.5	1.087652	1.08968	23.385	11.92
16.1	1.063892	1.06585	17.129	8.95	21.6	1.088101	1.09013	23.503	11.98
16.2	1.064324	1.06629	17.242	9.00	21.7	1.088550	1.09058	23.622	12.03
16.3	1.064756	1.06672	17.356	9.06	21.8	1.089000	1.09103	23.740	12.09
16.4	1.065188	1.06715	17.469	9.11	21.9	1.089450	1.09149	23.859	12.14

(*continued*)

TABLE 11 (*Continued*)

Percentage of Sucrose by Weight (Brix) (1)	d_4^{20} (2)	d_{20}^{20} (3)	Grams of Sucrose per 100 mL Weight *in Vacuo* (4)	Degrees Baumé (modulus 45) (5)	Percentage of Sucrose by Weight (Brix) (1)	d_4^{20} (2)	d_{20}^{20} (3)	Grams of Sucrose per 100 mL Weight *in Vacuo* (4)	Degrees Baumé (modulus 45) (5)
22.0	1.089900	1.09194	23.978	12.20	27.5	1.115166	1.11728	30.667	15.20
22.1	1.090351	1.09239	24.097	12.25	27.6	1.115635	1.11775	30.792	15.26
22.2	1.090802	1.09284	24.216	12.31	27.7	1.116104	1.11822	30.916	15.31
22.3	1.091253	1.09329	24.335	12.36	27.8	1.116572	1.11869	31.041	15.37
22.4	1.091704	1.09375	24.454	12.42	27.9	1.117042	1.11916	31.165	15.42
22.5	1.092155	1.09420	24.573	12.47	28.0	1.117512	1.11963	31.290	15.48
22.6	1.092607	1.09465	24.693	12.52	28.1	1.117982	1.12010	31.415	15.53
22.7	1.093060	1.09511	24.812	12.58	28.2	1.118453	1.12058	31.540	15.59
22.8	1.093513	1.09556	24.932	12.63	28.3	1.118923	1.12105	31.666	15.64
22.9	1.093966	1.09602	25.052	12.69	28.4	1.119395	1.12152	31.791	15.69
23.0	1.094420	1.09647	25.172	12.74	28.5	1.119867	1.12199	31.916	15.75
23.1	1.094874	1.09693	25.292	12.80	28.6	1.120339	1.12247	32.042	15.80
23.2	1.095328	1.09738	25.412	12.85	28.7	1.120812	1.12294	32.167	15.86
23.3	1.095782	1.09784	25.532	12.91	28.8	1.121884	1.12341	32.293	15.91
23.4	1.096236	1.09829	25.652	12.96	28.9	1.121757	1.12389	32.419	15.97
23.5	1.096691	1.09875	25.772	13.02	29.0	1.122231	1.12436	32.545	16.02
23.6	1.097147	1.09921	25.893	13.07	29.1	1.122705	1.12484	32.671	16.08
23.7	1.097603	1.09966	26.013	13.13	29.2	1.123179	1.12532	32.797	16.13
23.8	1.098058	1.10012	26.134	13.18	29.3	1.123653	1.12579	32.923	16.18
23.9	1.098514	1.10058	26.255	13.24	29.4	1.124128	1.12627	33.049	16.24
24.0	1.098971	1.10104	26.375	13.29	29.5	1.124603	1.12674	33.176	16.29
24.1	1.099428	1.10149	25.496	13.35	29.6	1.125079	1.12722	33.302	16.35

24.2	1.099886	1.10195	26.617	13.40	29.7	1.125555	1.12770	33.429	16.40
24.3	1.100344	1.10241	26.738	13.46	29.8	1.126030	1.12817	33.556	16.46
24.4	1.100802	1.10287	26.860	13.51	29.9	1.126507	1.12865	33.683	16.51
24.5	1.101259	1.10333	26.981	13.57	30.0	1.126984	1.12913	33.810	16.57
24.6	1.101718	1.10379	27.102	13.62	30.1	1.127461	1.12961	33.937	16.62
24.7	1.102177	1.10425	27.224	13.67	30.2	1.127939	1.13009	34.064	16.67
24.8	1.102637	1.10471	27.345	13.37	30.3	1.128417	1.13057	34.191	16.73
24.9	1.103097	1.10517	27.467	13.78	30.4	1.128896	1.13105	34.318	16.78
25.0	1.103557	1.10564	27.589	13.84	30.5	1.129374	1.13153	34.446	16.84
25.1	1.104017	1.10610	27.710	13.89	30.6	1.129853	1.13201	34.574	16.89
25.2	1.104478	1.10656	27.833	13.95	30.7	1.130332	1.13249	34.701	16.95
25.3	1.104938	1.10702	27.855	14.00	30.8	1.130812	1.13297	34.829	17.00
25.4	1.105400	1.10748	28.077	14.06	30.9	1.131292	1.13345	34.957	17.05
25.5	1.105862	1.10795	28.199	14.11	31.0	1.131773	1.13394	35.085	17.11
25.6	1.106324	1.10841	28.322	14.17	31.1	1.132254	1.13442	35.213	17.16
25.7	1.106786	1.10887	28.444	14.22	31.2	1.132735	1.13490	35.341	17.22
25.8	1.107248	1.10934	28.567	14.28	31.3	1.133216	1.13538	35.470	17.27
25.9	1.107711	1.10980	28.690	14.33	31.4	1.133698	1.13587	35.598	17.33
26.0	1.108175	1.11027	28.813	14.39	31.5	1.134180	1.13635	35.727	17.38
26.1	1.108639	1.11073	28.935	14.44	31.6	1.134663	1.13683	35.855	17.43
26.2	1.109103	1.11120	29.059	14.49	31.7	1.135146	1.13732	35.984	17.49
26.3	1.109568	1.11166	29.182	14.55	31.8	1.135628	1.13780	36.113	17.54
26.4	1.110033	1.11213	29.305	14.60	31.9	1.136112	1.13829	36.242	17.60
26.5	1.110497	1.11260	29.428	14.66	32.0	1.136596	1.13877	36.371	17.65
26.6	1.110963	1.11306	29.552	14.71	32.1	1.137080	1.13926	36.500	17.70
26.7	1.111429	1.11353	29.675	14.77	32.2	1.137565	1.13974	36.630	17.76
26.8	1.111895	1.11400	29.799	14.82	32.3	1.138049	1.14023	35.759	17.81
26.9	1.112361	1.11447	29.923	14.88	32.4	1.138534	1.14072	36.889	17.87
27.0	1.112828	1.11493	30.046	14.93	32.5	1.139020	1.14120	37.018	17.92
27.1	1.113295	1.11540	30.170	14.99	32.6	1.139506	1.14169	37.148	17.98
27.2	1.113863	1.11587	30.294	15.04	32.7	1.139993	1.14218	37.278	18.03
27.3	1.114229	1.11634	30.418	15.09	32.8	1.140479	1.14267	37.408	18.08
27.4	1.114697	1.11681	30.543	15.15	32.9	1.140966	1.14316	37.538	18.14

(continued)

TABLE 11 *(Continued)*

Percentage of Sucrose by Weight (Brix) (1)	d_4^{20} (2)	d_{20}^{20} (3)	Grams of Sucrose per 100 mL Weight *in Vacuo* (4)	Degrees Baumé (modulus 45) (5)	Percentage of Sucrose by Weight (Brix) (1)	d_4^{20} (2)	d_{20}^{20} (3)	Grams of Sucrose per 100 mL Weight *in Vacuo* (4)	Degrees Baumé (modulus 45) (5)
33.0	1.141453	1.14364	37.668	18.19	38.5	1.168800	1.17107	44.999	21.16
33.1	1.141941	1.14413	37.798	18.25	38.6	1.169307	1.17158	45.135	21.21
33.2	1.142429	1.14462	37.929	18.30	38.7	1.169815	1.17209	45.272	21.27
33.3	1.142916	1.14511	38.059	18.36	38.8	1.170322	1.17260	45.408	21.32
33.4	1.143405	1.14560	38.190	18.41	38.9	1.170831	1.17311	45.545	21.38
33.5	1.143894	1.14609	38.320	18.46	39.0	1.171340	1.17362	45.682	21.43
33.6	1.144384	1.14658	38.451	18.52	39.1	1.171849	1.17413	45.819	21.48
33.7	1.144874	1.14708	38.582	18.57	39.2	1.172359	1.17464	45.956	21.54
33.8	1.145363	1.14757	38.713	18.63	39.3	1.172869	1.17515	46.094	21.59
33.9	1.145854	1.14806	38.844	18.68	39.4	1.173379	1.17566	46.231	21.64
34.0	1.146345	1.14855	38.976	18.73	39.5	1.173889	1.17618	46.369	21.70
34.1	1.146836	1.14904	39.107	18.79	39.6	1.174400	1.17669	46.506	21.75
34.2	1.147328	1.14954	39.239	18.84	39.7	1.174911	1.17720	46.644	21.80
34.3	1.147820	1.15003	39.370	18.90	39.8	1.175423	1.17772	46.782	21.86
34.4	1.148313	1.15052	39.502	18.95	39.9	1.175935	1.17823	46.920	21.91
34.5	1.148805	1.15102	39.634	19.00	40.0	1.176447	1.17874	47.058	21.97
34.6	1.149298	1.15151	39.767	19.06	40.1	1.176960	1.17926	47.196	22.02
34.7	1.149792	1.15201	39.898	19.11	40.2	1.177473	1.17977	47.334	22.07
34.8	1.150286	1.15250	40.030	19.17	40.3	1.177987	1.18029	47.472	22.13
34.9	1.150780	1.15300	40.162	19.22	40.4	1.178501	1.18080	47.611	22.18
35.0	1.151275	1.15350	40.295	19.28	40.5	1.179014	1.18132	47.750	22.23
35.1	1.151770	1.15399	40.427	19.33	40.6	1.179527	1.18183	47.889	22.29

35.2	1.152265	1.15449	40.560	19.38	40.7	1.180044	1.18235	48.028	22.34
35.3	1.152760	1.15498	40.692	19.44	40.8	1.180560	1.18287	48.167	22.39
35.4	1.153256	1.15548	40.825	19.49	40.9	1.181076	1.18339	48.306	22.45
35.5	1.153752	1.15598	40.958	19.55	41.0	1.181592	1.18390	48.445	22.50
35.6	1.154249	1.15648	41.091	19.60	41.1	1.182108	1.18442	48.585	22.55
35.7	1.154746	1.15698	41.224	19.65	41.2	1.182625	1.18494	48.724	22.61
35.8	1.155242	1.15747	41.358	19.71	41.3	1.183142	1.18546	48.864	22.66
35.9	1.155740	1.15797	41.491	19.76	41.4	1.183660	1.18598	49.004	22.72
36.0	1.156238	1.15847	41.625	19.81	41.5	1.184178	1.18650	49.143	22.77
36.1	1.156736	1.15897	41.758	19.87	41.6	1.184696	1.18702	49.283	22.82
36.2	1.157235	1.15947	41.892	19.92	41.7	1.185215	1.18754	49.424	22.88
36.3	1.157783	1.15997	42.026	19.98	41.8	1.185734	1.18806	49.564	22.93
36.4	1.158233	1.16047	42.160	20.03	41.9	1.186253	1.18858	49.704	22.98
36.5	1.158733	1.16098	42.294	20.08	42.0	1.186773	1.18910	49.845	23.04
36.6	1.159233	1.16148	42.428	20.14	42.1	1.187293	1.18962	49.985	23.09
36.7	1.159733	1.16198	42.562	20.19	42.2	1.187814	1.19014	50.126	23.14
36.8	1.160233	1.16248	42.697	20.25	42.3	1.188335	1.19062	50.267	23.20
36.9	1.160734	1.16298	42.831	20.30	42.4	1.188856	1.19119	50.408	23.25
37.0	1.161236	1.16349	42.966	20.35	42.5	1.189379	1.19171	50.549	23.30
37.1	1.161738	1.16399	43.100	20.41	42.6	1.189901	1.19224	50.690	23.36
37.2	1.162240	1.16449	43.235	20.36	42.7	1.190423	1.19276	50.831	23.41
37.3	1.162742	1.16500	43.370	20.52	42.8	1.190946	1.19329	50.973	23.46
37.4	1.163245	1.16550	43.505	20.57	42.9	1.191469	1.19381	51.114	23.52
37.5	1.163748	1.16601	43.641	20.62	43.0	1.191993	1.19434	51.256	23.57
37.6	1.164252	1.16651	43.776	20.68	43.1	1.192517	1.19486	51.398	23.62
37.7	1.164756	1.16702	43.911	20.73	43.2	1.193041	1.19539	51.539	23.68
37.8	1.165259	1.16752	44.047	20.78	43.3	1.193565	1.19591	51.681	23.73
37.9	1.165764	1.16803	44.182	20.84	43.4	1.194090	1.19644	51.824	23.78
38.0	1.166269	1.16853	44.318	20.89	43.5	1.194616	1.19697	51.966	23.84
38.1	1.166775	1.16904	44.454	20.94	43.6	1.195141	1.19749	52.108	23.89
38.2	1.167281	1.16955	44.590	21.00	43.7	1.195667	1.19802	52.251	23.94
38.3	1.167786	1.17006	44.726	21.05	43.8	1.196193	1.19855	52.393	24.00
38.4	1.168293	1.17056	44.862	21.11	43.9	1.196720	1.19908	42.536	24.05

(continued)

TABLE 11 (*Continued*)

Percentage of Sucrose by Weight (Brix) (1)	d_4^{20} (2)	d_{20}^{20} (3)	Grams of Sucrose per 100 mL Weight *in Vacuo* (4)	Degrees Baumé (modulus 45) (5)	Percentage of Sucrose by Weight (Brix) (1)	d_4^{20} (2)	d_{20}^{20} (3)	Grams of Sucrose per 100 mL Weight *in Vacuo* (4)	Degrees Baumé (modulus 45) (5)
44.0	1.197247	1.19961	52.679	24.10	49.5	1.226823	1.22927	60.728	27.02
44.1	1.197775	1.20013	52.822	24.16	49.6	1.227371	1.22982	60.878	27.07
44.2	1.198303	1.20066	52.965	24.21	49.7	1.227919	1.23037	61.028	27.12
44.3	1.198832	1.20119	53.108	24.26	49.8	1.228469	1.23092	61.178	27.18
44.4	1.199360	1.20172	53.252	24.32	49.9	1.229018	1.23147	61.328	27.23
44.5	1.199890	1.20226	53.395	24.37	50.0	1.229567	1.23202	61.478	27.28
44.6	1.200420	1.20279	53.539	24.42	50.1	1.230117	1.23257	61.629	27.33
44.7	1.200950	1.20332	53.683	24.48	50.2	1.230668	1.23313	61.780	27.39
44.8	1.201480	1.20385	53.826	24.53	50.3	1.231219	1.23368	61.930	27.44
44.9	1.202010	1.20438	53.970	24.58	50.4	1.231770	1.23423	62.081	27.49
45.0	1.202540	1.20491	54.114	24.63	50.5	1.232322	1.23478	62.232	27.54
45.1	1.203071	1.20545	54.259	24.69	50.6	1.232874	1.23534	62.383	27.60
45.2	1.203603	1.20598	54.403	24.74	50.7	1.233426	1.23489	62.535	27.65
45.3	1.204136	1.20651	54.547	24.79	50.8	1.233979	1.23645	62.686	27.70
45.4	1.204668	1.20705	54.692	24.85	50.9	1.234532	1.23700	62.838	27.75
45.5	1.205200	1.20758	54.837	24.90	51.0	1.235085	1.23756	62.989	27.81
45.6	1.205733	1.20812	54.981	24.95	51.1	1.235639	1.23811	63.141	27.86
45.7	1.206266	1.20865	55.126	25.01	51.2	1.236194	1.23867	63.293	27.91
45.8	1.206801	1.20919	55.272	25.06	51.3	1.236748	1.23922	63.445	27.96
45.9	1.207335	1.20972	55.417	25.11	51.4	1.237303	1.23978	63.597	28.02
46.0	1.207870	1.21026	55.562	25.17	51.5	1.237859	1.24034	63.750	28.07
46.1	1.208405	1.21080	55.708	25.22	51.6	1.238414	1.24089	63.902	28.12

46.2	1.208940	1.21133	55.853	25.27	51.7	1.238970	1.24145	64.055	28.17
46.3	1.209477	1.21187	55.999	25.32	51.8	1.239527	1.24201	64.208	28.23
46.4	1.210013	1.21241	56.145	25.38	51.9	1.240084	1.24257	64.360	28.28
46.5	1.210549	1.21295	56.291	25.43	52.0	1.240641	1.24313	64.513	28.33
46.6	1.211086	1.21349	56.437	25.48	52.1	1.241198	1.24369	64.666	28.38
46.7	1.211623	1.21402	56.583	25.54	52.2	1.241757	1.24425	64.820	28.44
46.8	1.212162	1.21456	56.729	25.59	52.3	1.242315	1.24481	64.973	28.49
46.9	1.212700	1.21510	56.876	25.64	52.4	1.242873	1.24537	65.127	28.54
47.0	1.213238	1.21564	57.022	25.70	52.5	1.243433	1.24593	65.280	28.59
47.1	1.213777	1.21618	57.169	25.75	52.6	1.243992	1.24649	65.433	28.65
47.2	1.214317	1.21673	57.316	25.80	52.7	1.244552	1.24705	65.588	28.70
47.3	1.214856	1.21727	57.463	25.86	52.8	1.245113	1.24761	65.742	28.75
47.4	1.215395	1.21781	57.610	25.91	52.9	1.245673	1.24818	65.896	28.80
47.5	1.215936	1.21835	57.757	25.96	53.0	1.246234	1.24874	66.050	28.86
47.6	1.216476	1.21889	57.904	26.01	53.1	1.246795	1.24930	66.205	28.91
47.7	1.217017	1.21943	58.052	26.07	53.2	1.247358	1.24987	66.359	28.96
47.8	1.217559	1.21998	58.199	26.12	53.3	1.247920	1.25043	66.514	29.01
47.9	1.218101	1.22052	58.347	26.17	53.4	1.248482	1.25099	66.669	29.06
48.0	1.218643	1.22106	58.495	26.23	55.5	1.249046	1.25156	66.824	29.12
48.1	1.219185	1.22161	58.643	26.28	53.6	1.249609	1.25212	66.979	29.17
48.2	1.219729	1.22215	58.791	26.33	53.7	1.250172	1.25269	67.314	29.22
48.3	1.220272	1.22270	58.939	26.38	53.8	1.250737	1.25325	67.290	29.27
48.4	1.220815	1.22324	59.087	26.44	53.9	1.251301	1.25382	67.445	29.32
48.5	1.221360	1.22379	59.236	26.49	54.0	1.251866	1.25439	67.601	29.38
48.6	1.221904	1.22434	59.385	26.54	54.1	1.252431	1.25495	67.757	29.43
48.7	1.222449	1.22488	59.533	26.59	54.2	1.252997	1.25552	67.912	29.48
48.8	1.222995	1.22543	59.682	26.65	54.3	1.253463	1.25609	68.069	29.53
48.9	1.223540	1.22598	59.831	26.70	54.4	1.254129	1.25666	68.225	29.59
49.0	1.224086	1.22652	59.980	26.75	54.5	1.254697	1.25723	68.381	29.64
49.1	1.224632	1.22707	60.129	26.81	54.6	1.255264	1.25780	68.537	29.69
49.2	1.225180	1.22762	60.279	26.86	54.7	1.255831	1.25836	68.694	29.74
49.3	1.225727	1.22817	60.428	26.91	54.8	1.256400	1.25893	68.851	29.80
49.4	1.226274	1.22872	60.578	26.96	54.9	1.256967	1.25950	69.008	29.85

(continued)

TABLE 11 (*Continued*)

Percentage of Sucrose by Weight (Brix) (1)	d_4^{20} (2)	d_{20}^{20} (3)	Grams of Sucrose per 100 mL Weight *in Vacuo* (4)	Degrees Baumé (modulus 145) (5)	Percentage of Sucrose by Weight (Brix) (1)	d_4^{20} (2)	d_{20}^{20} (3)	Grams of Sucrose per 100 mL Weight *in Vacuo* (4)	Degrees Baumé (modulus 145) (5)
55.0	1.257535	1.26007	69.164	29.90	60.5	1.289401	1.29203	78.009	32.74
55.1	1.258104	1.26064	69.322	29.95	60.6	1.289991	1.29262	78.173	32.79
55.2	1.248674	1.26122	69.479	30.00	60.7	1.290581	1.29321	78.338	32.85
55.3	1.259244	1.26179	69.636	30.06	60.8	1.291172	1.29380	78.503	32.90
55.4	1.259815	1.26236	69.794	30.11	60.9	1.291763	1.29439	78.668	32.95
55.5	1.260385	1.26293	69.951	30.16	61.0	1.292354	1.29498	78.833	33.00
55.6	1.260955	1.26350	70.109	30.21	61.1	1.292946	1.29559	78.999	33.05
55.7	1.261527	1.26408	70.267	30.26	61.2	1.293439	1.29618	79.165	33.10
55.8	1.262099	1.26465	70.425	30.32	61.3	1.294131	1.29677	79.330	33.15
55.9	1.262671	1.26522	70.583	30.37	61.4	1.294725	1.29736	79.496	33.20
56.0	1.263243	1.26580	70.742	30.42	61.5	1.295318	1.29796	79.662	33.26
56.1	1.263816	1.26637	70.900	30.47	61.6	1.295911	1.29855	79.828	33.31
56.2	1.263390	1.26695	70.059	30.52	61.7	1.296506	1.29915	79.995	33.36
56.3	1.264963	1.26752	70.217	30.57	61.8	1.297100	1.29975	80.161	33.41
56.4	1.265537	1.26810	70.376	30.63	61.9	1.297696	1.30034	80.328	33.46
56.5	1.266112	1.26868	70.535	30.68	62.0	1.298291	1.30093	80.494	33.51
56.6	1.266686	1.26925	70.694	30.73	62.1	1.298886	1.30153	80.661	33.56
56.7	1.267261	1.26983	70.854	30.78	62.2	1.299483	1.30212	80.828	33.61
56.8	1.267837	1.27041	72.013	30.83	62.3	1.300079	1.30273	80.995	33.67
56.9	1.268413	1.27098	72.173	30.89	62.4	1.300677	1.30334	81.162	33.72
57.0	1.268989	1.27156	72.332	30.94	62.5	1.301274	1.30393	81.329	33.77
57.1	1.269565	1.27214	72.492	30.99	62.6	1.301871	1.30453	81.497	33.82

57.2	1.270143	1.27272	72.652	31.04	62.7	1.302470	1.30513	81.665	33.87
57.3	1.270720	1.27330	72.812	31.09	62.8	1.303068	1.30573	81.833	33.92
57.4	1.271299	1.27388	72.973	31.15	62.9	1.303668	1.30633	82.001	33.97
57.5	1.271877	1.27446	73.133	31.20	63.0	1.304267	1.30694	82.169	34.02
57.6	1.272455	1.27504	73.293	31.25	63.1	1.304867	1.30754	82.337	34.07
57.7	1.273035	1.27562	73.454	31.30	63.2	1.305467	1.30815	82.506	34.12
57.8	1.273614	1.27620	73.615	31.35	63.3	1.306068	1.30875	82.674	34.18
57.9	1.274194	1.27678	73.776	31.40	63.4	1.306669	1.30936	82.843	34.23
58.0	1.274774	1.27736	73.937	31.46	63.5	1.307271	1.30994	83.102	34.28
58.1	1.275354	1.27797	74.098	31.51	63.6	1.307872	1.31055	83.180	34.33
58.2	1.275936	1.27853	74.260	31.56	63.7	1.308475	1.31117	83.350	34.38
58.3	1.276517	1.27911	74.421	31.61	63.8	1.309077	1.31177	83.519	34.43
58.4	1.277098	1.27969	74.583	31.66	63.9	1.309680	1.31237	83.688	34.48
58.5	1.277680	1.28028	74.744	31.71	64.0	1.310282	1.31297	83.858	34.53
58.6	1.278262	1.28086	74.906	31.76	64.1	1.310885	1.31359	84.028	34.58
58.7	1.278844	1.28145	75.068	31.82	64.2	1.311489	1.31418	84.198	34.63
58.8	1.279428	1.28203	75.230	31.87	64.3	1.312093	1.31479	84.367	34.68
58.9	1.280011	1.28262	75.393	31.92	64.4	1.312699	1.31540	84.538	34.74
59.0	1.280595	1.28320	75.555	31.97	64.5	1.313304	1.31600	84.708	34.79
59.1	1.281179	1.28379	75.718	32.02	64.6	1.313909	1.31661	84.879	34.84
59.2	1.281764	1.28437	75.880	32.07	64.7	1.314515	1.31723	85.049	34.89
59.3	1.282349	1.28497	76.043	32.13	64.8	1.315121	1.31784	85.220	34.94
59.4	1.282935	1.28556	76.207	32.18	64.9	1.315728	1.31845	85.391	34.99
59.5	1.283521	1.28614	76.369	32.23	65.0	1.316334	1.31905	85.561	35.04
59.6	1.284107	1.28672	76.533	32.28	65.1	1.316941	1.31966	85.733	35.09
59.7	1.284694	1.28731	76.696	32.33	65.2	1.317549	1.32028	85.904	35.14
59.8	1.285281	1.28789	76.860	32.38	65.3	1.318157	1.32089	86.076	35.19
59.9	1.285869	1.28849	77.024	32.43	65.4	1.318766	1.32150	86.248	35.24
60.0	1.286456	1.28908	77.188	32.49	65.5	1.319374	1.32210	86.419	35.29
60.1	1.287044	1.28966	77.351	32.54	65.6	1.319983	1.32271	86.591	35.34
60.2	1.287633	1.29025	77.515	32.59	65.7	1.320593	1.32332	86.763	35.39
60.3	1.288222	1.29084	77.680	32.64	65.8	1.321203	1.32393	86.935	35.45
60.4	1.288811	1.29143	77.844	32.69	65.9	1.321814	1.32455	87.107	35.50

(continued)

TABLE 11 (*Continued*)

Percentage of Sucrose by Weight (Brix) (1)	d_4^{20} (2)	d_{20}^{20} (3)	Grams of Sucrose per 100 mL Weight *in Vacuo* (4)	Degrees Baumé (modulus 145) (5)	Percentage of Sucrose by Weight (Brix) (1)	d_4^{20} (2)	d_{20}^{20} (3)	Grams of Sucrose per 100 mL Weight *in Vacuo* (4)	Degrees Baumé (modulus 145) (5)
66.0	1.322425	1.32516	87.280	35.55	71.5	1.356612	1.35944	96.998	38.30
66.1	1.323036	1.32577	87.453	35.60	71.6	1.357245	1.36008	97.179	38.35
66.2	1.323648	1.32638	87.626	35.65	71.7	1.357877	1.36072	97.360	38.40
66.3	1.324259	1.32699	87.798	35.70	71.8	1.348511	1.36135	97.541	38.45
66.4	1.324872	1.32759	87.971	35.75	71.9	1.359144	1.36198	97.722	38.50
66.5	1.325484	1.32820	88.142	35.80	72.0	1.359778	1.36261	97.904	38.55
66.6	1.326097	1.32884	88.318	35.85	72.1	1.360413	1.36324	98.085	38.60
66.7	1.326711	1.32945	88.492	35.90	72.2	1.361047	1.36389	98.268	38.65
66.8	1.327325	1.33007	88.666	35.95	72.3	1.361682	1.36452	98.449	38.70
66.9	1.327940	1.33068	88.839	36.00	72.4	1.362317	1.36516	98.632	38.75
67.0	1.328554	1.33129	89.012	36.05	72.5	1.362953	1.36579	98.814	38.80
67.1	1.329170	1.33192	89.187	36.10	72.6	1.363590	1.36643	98.997	38.85
67.2	1.329785	1.33254	89.361	36.15	72.7	1.364226	1.36707	99.179	38.90
67.3	1.330401	1.33315	89.536	36.20	72.8	1.364864	1.36771	99.362	38.95
67.4	1.331017	1.33377	89.711	36.25	72.9	1.365501	1.36836	99.545	39.00
67.5	1.331633	1.33438	89.885	36.30	73.0	1.366139	1.36900	99.728	39.05
67.6	1.332250	1.33500	90.060	36.35	73.1	1.366777	1.36964	99.912	39.10
67.7	1.332868	1.33562	90.235	36.40	73.2	1.367415	1.37028	100.095	39.15
67.8	1.333485	1.33625	90.411	36.45	73.3	1.368054	1.37092	100.278	39.20
67.9	1.334103	1.33686	90.585	36.50	73.4	1.368693	1.37156	100.462	39.25
68.0	1.334722	1.33748	90.761	36.55	73.5	1.369333	1.37220	100.646	39.30
68.1	1.335342	1.33810	90.937	36.61	73.6	1.369973	1.37283	100.827	39.30

68.2	1.334961	1.33872	91.112	36.66	73.7	1.370613	1.37347	101.014	39.39
68.3	1.336581	1.33935	91.288	36.71	73.8	1.371254	1.37411	101.198	39.44
68.4	1.337200	1.33997	91.464	36.76	73.9	1.371894	1.37476	101.383	39.49
68.5	1.337821	1.34059	91.641	36.81	74.0	1.372536	1.37541	101.568	39.54
68.6	1.338441	1.34121	91.817	36.86	74.1	1.373178	1.37605	101.753	39.59
68.7	1.339063	1.34183	91.993	36.91	74.2	1.373810	1.37669	101.937	39.64
68.8	1.339684	1.34245	92.169	36.96	74.3	1.374463	1.37733	102.122	39.69
68.9	1.340306	1.34309	92.347	37.01	74.4	1.375105	1.37798	102.308	39.74
69.0	1.340928	1.34371	92.524	37.06	74.5	1.375749	1.37864	102.493	39.79
69.1	1.341551	1.34433	92.701	37.11	74.6	1.373692	1.37928	102.679	39.84
69.2	1.342174	1.34495	92.878	37.16	74.7	1.377036	1.37993	102.865	39.89
69.3	1.342798	1.34558	93.056	37.21	74.8	1.377680	1.38057	103.050	39.94
69.4	1.343421	1.34621	93.233	37.26	74.9	1.378326	1.38122	103.237	39.99
69.5	1.344046	1.34684	93.411	37.31	75.0	1.378971	1.38187	103.423	40.03
69.6	1.344671	1.34746	93.589	37.36	75.1	1.379617	1.38252	103.609	40.08
69.7	1.345296	1.34809	93.767	34.41	75.2	1.380262	1.38316	103.796	40.13
69.8	1.345922	1.34871	93.945	37.46	75.3	1.380909	1.38381	103.983	40.18
69.9	1.346547	1.34934	94.123	37.51	75.4	1.381555	1.38445	104.170	40.23
70.0	1.347174	1.34997	94.302	37.56	75.5	1.382203	1.38510	104.356	40.28
70.1	1.347801	1.34060	94.481	37.61	75.6	1.382851	1.38575	104.543	40.33
70.2	1.348427	1.35123	94.660	37.66	75.7	1.383499	1.38640	104.731	40.38
70.3	1.349055	1.35186	94.839	37.71	75.8	1.384148	1.38705	104.919	40.43
70.4	1.349682	1.35248	95.017	37.76	75.9	1.384796	1.38770	105.106	40.48
70.5	1.350311	1.35311	95.197	37.81	76.0	1.385446	1.38835	105.294	40.53
70.6	1.350939	1.35375	95.376	37.86	76.1	1.396096	1.38902	105.482	40.57
70.7	1.351568	1.35438	95.556	37.91	76.1	1.386745	1.38967	105.670	40.62
70.8	1.352197	1.35501	95.736	37.96	76.3	1.387396	1.39032	105.859	40.67
70.9	1.352827	1.35564	95.916	38.01	76.4	1.388045	1.39097	106.047	40.72
71.0	1.353456	1.35627	96.096	38.06	76.5	1.388696	1.39162	106.236	40.77
71.1	1.354087	1.35691	96.276	38.11	76.6	1.389347	1.39228	106.424	40.82
71.2	1.354717	1.35754	96.456	38.16	76.7	1.389999	1.39293	106.613	40.87
71.3	1.355349	1.35817	96.636	38.21	76.8	1.390651	1.39358	105.802	40.92
71.4	1.355980	1.35881	96.817	38.26	76.9	1.391303	1.39423	106.991	40.97

(continued)

TABLE 11 (*Continued*)

Percentage of Sucrose by Weight (Brix) (1)	d_4^{20} (2)	d_{20}^{20} (3)	Grams of Sucrose per 100 mL Weight *in Vacuo* (4)	Degrees Baumé (modulus 145) (5)	Percentage of Sucrose by Weight (Brix) (1)	d_4^{20} (2)	d_{20}^{20} (3)	Grams of Sucrose per 100 mL Weight *in Vacuo* (4)	Degrees Baumé (modulus 145) (5)
77.0	1.391956	1.39489	107.181	41.01	82.5	1.428435	1.43148	117.845	43.67
77.1	1.392610	1.39554	107.370	41.06	82.6	1.429109	1.43214	118.044	43.72
77.2	1.393263	1.39619	107.560	41.11	82.7	1.429782	1.43282	118.243	43.77
77.3	1.393917	1.39685	107.750	41.16	82.8	1.430457	1.43350	118.442	43.81
77.4	1.394571	1.39750	107.940	41.21	82.9	1.431131	1.43417	118.641	43.86
77.5	1.395226	1.39816	108.130	41.26	83.0	1.431807	1.43486	118.840	43.91
77.6	1.395881	1.39882	108.320	41.31	83.1	1.432483	1.43553	119.039	43.96
77.7	1.396536	1.39949	108.511	41.36	83.2	1.433158	1.43621	119.239	44.00
77.8	1.397192	1.40014	108.701	41.40	83.3	1.433835	1.43688	119.438	44.05
77.9	1.397848	1.40080	108.892	41.45	83.4	1.434511	1.43756	119.638	44.10
78.0	1.398505	1.40146	109.084	41.50	83.5	1.435188	1.43824	119.838	44.15
78.1	1.399162	1.40211	109.274	41.55	83.6	1.435866	1.43894	120.039	44.19
78.2	1.399819	1.40277	109.466	41.60	83.7	1.436543	1.43961	120.238	44.24
78.3	1.400477	1.40344	109.657	41.65	83.8	1.437222	1.44029	120.439	44.29
78.4	1.401134	1.40409	109.848	41.70	83.9	1.437900	1.44097	120.640	44.34
78.5	1.401793	1.40475	110.041	41.74	84.0	1.438579	1.44165	120.841	44.38
78.6	1.402452	1.40541	110.232	41.79	84.1	1.439259	1.44234	121.042	44.43
78.7	1.403111	1.40607	110.425	41.84	84.2	1.439938	1.44302	121.243	44.48
78.8	1.403771	1.40674	110.617	41.89	84.3	1.440619	1.44370	121.444	44.53
78.9	1.404430	1.40740	110.809	41.94	84.4	1.441229	1.44438	121.646	44.57
79.0	1.405091	1.40806	111.002	41.99	84.5	1.441980	1.44507	121.847	44.62
79.1	1.405752	1.40872	111.195	42.03	84.6	1.442661	1.44575	122.049	44.67

79.2	1.406412	1.40938	111.388	42.08	84.7	1.443342	1.44643	121.251	44.72
79.3	1.407074	1.41005	111.581	42.13	84.8	1.444024	1.44712	121.453	44.76
79.4	1.407735	1.41072	111.775	42.18	84.9	1.444705	1.44780	121.655	44.81
79.5	1.408398	1.41138	111.968	42.23	85.0	1.445388	1.44848	122.858	44.86
79.6	1.409061	1.41204	112.161	42.28	85.1	1.446071	1.44917	123.061	44.91
79.7	1.409723	1.41270	112.354	42.32	85.2	1.446754	1.44985	123.263	44.95
79.8	1.410387	1.41337	112.549	42.37	85.3	1.447438	1.45054	123.466	45.00
79.9	1.411051	1.41404	112.743	42.42	85.4	1.448121	1.45123	123.670	45.05
80.0	1.411715	1.41471	112.938	42.47	85.5	1.448806	1.45191	123.873	45.09
80.1	1.412380	1.41537	113.131	42.52	85.6	1.449491	1.45260	124.076	45.14
80.2	1.413044	1.41603	113.326	42.57	85.7	1.450175	1.45329	124.280	45.19
80.3	1.412709	1.41670	113.521	42.61	85.8	1.450860	1.45397	124.484	45.24
80.4	1.414374	1.41737	113.715	42.66	85.9	1.451545	1.45466	124.688	45.28
80.5	1.415440	1.41804	113.911	42.71	86.0	1.452232	1.45535	124.892	45.33
80.6	1.415706	1.41872	114.106	42.76	86.1	1.452919	1.45604	125.096	45.38
80.7	1.416373	1.41937	114.301	42.81	86.2	1.453605	1.45673	125.301	45.42
80.8	1.417039	1.42004	114.497	42.85	86.3	1.454282	1.45741	125.505	45.47
80.9	1.417707	1.42072	114.692	42.90	86.4	1.454980	1.45810	125.710	45.52
81.0	1.418374	1.42138	114.888	42.95	86.5	1.455668	1.45879	125.915	45.57
81.1	1.419043	1.42205	115.084	43.00	86.6	1.456357	1.45949	126.121	45.61
81.2	1.419711	1.42272	115.280	43.05	86.7	1.457045	1.46018	126.326	45.66
81.3	1.420380	1.42339	115.477	43.10	86.8	1.457735	1.46087	126.531	45.71
81.4	1.421049	1.42406	115.673	43.14	86.9	1.458424	1.46156	126.737	45.75
81.5	1.421719	1.42474	115.870	43.19	87.0	1.459114	1.46225	126.943	45.80
81.6	1.422390	1.42541	116.067	43.24	87.1	1.459805	1.46294	127.149	45.85
81.7	1.423059	1.42608	116.264	43.29	87.2	1.460495	1.46364	127.355	45.89
81.8	1.423730	1.42675	116.461	43.33	87.3	1.461186	1.46433	127.562	45.94
81.9	1.424400	1.42742	116.658	43.38	87.4	1.461877	1.46502	127.768	45.99
82.0	1.425072	1.42810	116.856	43.43	87.5	1.462568	1.46572	127.975	46.03
82.1	1.425744	1.42878	117.053	43.48	87.5	1.463260	1.46641	128.182	46.08
82.2	1.426416	1.42946	117.252	43.53	87.7	1.463953	1.46710	128.389	46.13
82.3	1.427089	1.43013	117.449	43.57	87.8	1.464645	1.46780	128.596	46.17
82.4	1.427761	1.43080	117.647	43.62	87.9	1.465338	1.46849	128.803	46.22

(*continued*)

TABLE 11 (*Continued*)

Percentage of Sucrose by Weight (Brix) (1)	d_4^{20} (2)	d_{20}^{20} (3)	Grams of Sucrose per 100 mL Weight *in Vacuo* (4)	Degrees Baumé (modulus 145) (5)	Percentage of Sucrose by Weight (Brix) (1)	d_4^{20} (2)	d_{20}^{20} (3)	Grams of Sucrose per 100 mL Weight *in Vacuo* (4)	Degrees Baumé (modulus 145) (5)
88.0	1.466032	1.46919	129.011	46.27	91.6	1.491234	1.49447	136.597	47.94
88.1	1.466726	1.46989	129.219	46.31	91.7	1.491941	1.49518	136.811	47.98
88.2	1.467420	1.47058	129.426	46.36	91.8	1.492647	1.49588	137.025	48.03
88.3	1.468115	1.47128	129.635	46.41	91.9	1.493355	1.49659	137.239	48.08
88.4	1.468810	1.47198	129.843	46.45	92.0	1.494063	1.49730	137.454	48.12
88.5	1.469504	1.47267	130.051	46.50	92.1	1.494771	1.49801	137.668	48.17
88.6	1.470200	1.47337	130.260	46.55	92.2	1.495479	1.49872	137.883	48.21
88.7	1.470896	1.47407	130.468	46.59	92.3	1.496188	1.49944	138.098	48.26
88.8	1.471592	1.47477	130.677	46.64	92.4	1.496897	1.50015	138.313	48.30
88.9	1.472289	1.47547	130.886	46.69	92.5	1.497606	1.50086	138.529	48.35
89.0	1.472986	1.47616	131.096	46.73	92.6	1.498316	1.50157	138.744	48.40
89.1	1.473684	1.47686	131.305	46.78	92.7	1.499026	1.50228	138.960	48.44
89.2	1.474381	1.47756	131.515	46.83	92.8	1.499736	1.50299	139.176	48.49
89.3	1.475080	1.47826	131.725	46.87	92.9	1.500447	1.50371	139.392	48.53
89.4	1.475779	1.47897	141.935	46.92	93.0	1.501158	1.50442	139.608	48.58
89.5	1.476477	1.47967	132.145	46.97	93.1	1.501870	1.50513	139.824	48.62
89.6	1.477176	1.48037	132.355	47.01	93.2	1.502582	1.50585	140.041	48.67
89.7	1.477876	1.48107	132.565	47.06	93.3	1.503293	1.50656	140.257	48.72
89.8	1.478575	1.48177	132.776	47.11	93.4	1.504006	1.50728	140.474	48.76
89.9	1.479275	1.48247	132.987	47.15	93.5	1.504719	1.50799	140.691	48.81

90.0	1.479976	1.48317	133.198	47.20	93.6	1.505432	1.50871	140.908	48.84
90.1	1.480677	1.48388	133.409	47.24	93.7	1.506146	1.50942	141.126	48.90
90.2	1.481378	1.48458	133.620	47.29	93.8	1.506859	1.51014	141.343	48.94
90.3	1.482080	1.48529	133.832	47.34	93.9	1.507574	1.51086	141.561	48.99
90.4	1.482782	1.48599	134.043	47.38	94.0	1.508289	1.51157	141.779	49.03
90.5	1.483484	1.48669	134.255	47.43	94.1	1.509004	1.51229	141.997	49.08
90.6	1.484187	1.48740	134.467	47.48	94.2	1.509720	1.51301	142.216	49.12
90.7	1.484890	1.48810	134.680	47.52	94.3	1.510435	1.51372	142.434	49.17
90.8	1.485593	1.48881	134.892	47.57	94.4	1.511151	1.51444	142.653	49.22
90.9	1.486297	1.48951	135.104	47.61	94.5	1.511868	1.51516	142.872	49.26
91.0	1.487002	1.49022	135.317	47.66	94.6	1.512585	1.51588	143.091	49.31
91.1	1.487707	1.49093	135.530	47.71	94.7	1.513302	1.51660	143.310	49.35
91.2	1.488411	1.49164	135.743	47.75	94.8	1.514019	1.51732	143.529	49.40
91.3	1.489117	1.49234	135.956	47.80	94.9	1.514737	1.51804	143.749	49.44
91.4	1.489823	1.49305	136.170	47.84	95.0	1.515455	1.51876	143.968	49.49
91.5	1.490528	1.49376	136.383	47.89					

Source: Columns 1, 3, and 5 are from Table 16 of the 11th edition of the *Cane Sugar Handbook*. Columns 2 and 4 are from Table E, *ICUMSA Methods* (Schneider, 1979).

TABLE 12 Weight per Unit Volume and Weight of Solids per Unit Volume of Sugar Solutions at 20°C

Degree	Apparent Density	Calculated Density		Calculated Solids (Density × Brix / 100)		
Brix	lb / ft³	kg / m³	lb / gal (U.S.)	lb / ft³	kg / m³	lb / gal (U.S.)
0.0	62.253	997.169	8.322	0.000	0.000	0.000
0.2	62.298	997.889	8.328	0.125	1.996	0.017
0.4	62.350	998.722	8.335	0.249	3.995	0.033
0.6	62.395	999.443	8.341	0.374	5.997	0.050
0.8	62.447	1000.276	8.348	0.500	8.002	0.067
1.0	62.492	1000.997	8.354	0.625	10.010	0.084
1.2	62.545	1001.846	8.361	0.751	12.022	0.100
1.4	62.590	1002.567	8.367	0.876	14.036	0.177
1.6	62.642	1003.400	8.374	1.002	16.054	0.134
1.8	62.687	1004.120	8.380	1.128	18.074	0.151
2.0	62.739	1004.953	8.387	1.255	20.099	0.168
2.2	62.784	1005.674	8.393	1.381	22.125	0.185
2.4	62.836	1006.507	8.400	1.508	24.156	0.202
2.6	62.881	1007.228	8.406	1.635	26.188	0.219
2.8	62.934	1008.077	8.413	1.762	28.226	0.236
3.0	62.978	1008.782	8.419	1.889	30.263	0.253
3.2	63.031	1009.631	8.426	2.017	32.308	0.270
3.4	63.076	1010.351	8.432	2.145	34.352	0.287
3.6	63.128	1011.184	8.439	2.273	36.403	0.304
3.8	63.180	1012.017	8.446	2.401	38.457	0.321
4.0	63.225	1012.738	8.452	2.529	40.510	0.338
4.2	63.278	1013.587	8.459	2.658	42.571	0.355
4.4	63.323	1014.308	8.465	2.786	44.630	0.372
4.6	63.375	1015.141	8.472	2.915	46.696	0.390
4.8	63.427	1015.974	8.479	3.044	48.767	0.407
5.0	63.472	1016.694	8.485	3.174	50.835	0.424
5.2	63.525	1017.543	8.492	3.303	52.912	0.442
5.4	63.577	1018.376	8.499	3.433	54.992	0.459
5.6	63.622	1019.097	8.505	3.563	57.069	0.476
5.8	63.674	1019.930	8.512	3.693	59.156	0.494
6.0	63.727	1020.779	8.519	3.824	61.247	0.511
6.2	63.779	1021.612	8.526	3.954	63.340	0.529
6.4	63.834	1022.493	8.533	4.085	65.440	0.546
6.6	63.876	1023.166	8.539	4.216	67.529	0.564
6.8	63.929	1024.015	8.546	4.347	69.633	0.581
7.0	63.973	1024.720	8.552	4.478	71.730	0.599
7.2	64.026	1025.568	8.559	4.610	73.841	0.616
7.4	64.078	1026.401	8.566	4.742	75.954	0.634
7.6	64.130	1027.234	8.573	4.874	78.070	0.652
7.8	64.183	1028.083	8.580	5.006	80.190	0.669
8.0	64.228	1028.804	8.586	5.138	82.304	0.687
8.2	64.280	1029.637	8.593	5.271	84.430	0.705
8.4	64.332	1030.470	8.600	5.404	86.559	0.722
8.6	64.385	1031.319	8.607	5.537	88.693	0.740
8.8	64.437	1032.152	8.614	5.670	90.829	0.758
9.0	64.482	1032.873	8.620	5.803	92.959	0.776
9.2	64.534	1033.706	8.627	5.937	95.101	0.794
9.4	64.587	1034.555	8.634	6.071	97.248	0.812

TABLE 12 *(Continued)*

Degree Brix	Apparent Density	Calculated Density		Calculated Solids (Density × Brix / 100)		
	lb / ft³	kg / m³	lb / gal (U.S.)	lb / ft³	kg / m³	lb / gal (U.S.)
9.6	64.639	1035.388	8.641	6.205	99.397	0.830
9.8	64.692	1036.236	8.648	6.340	101.551	0.848
10.0	64.744	1037.069	8.655	6.474	103.707	0.866
10.2	64.796	1037.902	8.662	6.609	105.866	0.884
10.4	64.849	1038.751	8.669	6.744	108.030	0.902
10.6	64.894	1039.472	8.675	6.879	110.184	0.920
10.8	64.946	1040.305	8.682	7.014	112.353	0.938
11.0	65.006	1041.266	8.690	7.151	114.539	0.956
11.2	65.051	1041.987	8.696	7.286	116.703	0.974
11.4	65.103	1042.820	8.703	7.422	118.881	0.992
11.6	65.155	1043.653	8.710	7.558	121.064	1.010
11.8	65.208	1044.502	8.717	7.695	123.251	1.029
12.0	65.260	1045.335	8.724	7.831	125.440	1.047
12.2	65.312	1046.168	8.731	7.968	127.632	1.065
12.4	65.365	1047.017	8.738	8.105	129.830	1.084
12.6	65.417	1047.850	8.745	8.243	132.029	1.102
12.8	65.470	1048.698	8.752	8.380	134.233	1.120
13.0	65.522	1049.531	8.759	8.518	136.439	1.139
13.2	65.574	1050.364	8.766	8.656	138.648	1.157
13.4	65.627	1051.213	8.773	8.794	140.863	1.176
13.6	65.686	1052.158	8.781	8.933	143.094	1.194
13.8	65.739	1053.007	8.788	9.072	145.315	1.213
14.0	65.791	1053.840	8.795	9.211	147.538	1.231
14.2	65.844	1054.689	8.802	9.350	149.766	1.250
14.4	65.896	1055.522	8.809	9.489	151.995	1.268
14.6	65.948	1056.355	8.816	9.628	154.228	1.287
14.8	66.001	1057.204	8.823	9.768	156.466	1.306
15.0	66.053	1058.037	8.830	9.908	158.706	1.325
15.2	66.105	1058.870	8.837	10.048	160.948	1.343
15.4	66.165	1059.831	8.845	10.189	163.214	1.362
15.6	66.218	1060.680	8.852	10.330	165.466	1.381
15.8	66.270	1061.513	8.859	10.471	167.719	1.400
16.0	66.322	1062.346	8.866	10.612	169.975	1.419
16.2	66.375	1063.195	8.873	10.753	172.238	1.437
16.4	66.434	1064.140	8.881	10.895	174.519	1.456
16.6	66.487	1064.989	8.888	11.037	176.788	1.475
16.8	66.539	1065.822	8.895	11.179	179.058	1.494
17.0	66.592	1066.671	8.902	11.321	181.334	1.513
17.2	66.644	1067.504	8.909	11.463	183.611	1.532
17.4	66.704	1068.465	8.917	11.606	185.913	1.552
17.6	66.756	1069.298	8.924	11.749	188.196	1.571
17.8	66.809	1070.147	8.931	11.892	190.486	1.590
18.0	66.868	1071.092	8.939	12.036	192.796	1.609
18.2	66.921	1071.941	8.946	12.180	195.093	1.628
18.4	66.973	1072.774	8.953	12.323	197.390	1.647
18.6	67.033	1073.735	8.961	12.468	199.715	1.667
18.8	67.085	1074.568	8.968	12.612	202.019	1.686
19.0	67.138	1075.416	8.975	12.756	204.329	1.705
19.2	67.198	1076.378	8.983	12.902	206.664	1.725

(continued)

TABLE 12 (*Continued*)

Degree Brix	Apparent Density	Calculated Density		Calculated Solids (Density × Brix / 100)		
	lb / ft³	kg / m³	lb / gal (U.S.)	lb / ft³	kg / m³	lb / gal (U.S.)
19.4	67.250	1077.211	8.990	13.047	208.979	1.744
19.6	67.302	1078.043	8.997	13.191	211.297	1.763
19.8	67.362	1079.005	9.005	13.338	213.643	1.783
20.0	67.414	1079.837	9.012	13.483	215.967	1.802
20.2	67.474	1080.799	9.020	13.630	218.321	1.822
20.4	67.527	1081.647	9.027	13.776	220.656	1.842
20.6	67.579	1082.480	9.034	13.921	222.991	1.861
20.8	67.639	1083.442	9.042	14.069	225.356	1.881
21.0	67.691	1084.274	9.049	14.215	227.698	1.900
21.2	67.751	1085.236	9.057	14.363	230.070	1.920
21.4	67.803	1086.068	9.064	14.510	232.419	1.940
21.6	67.863	1087.030	9.072	14.658	234.798	1.960
21.8	67.916	1087.878	9.079	14.806	237.158	1.979
22.0	67.975	1088.824	9.087	14.955	239.541	1.999
22.2	68.028	1089.673	9.094	15.102	241.907	2.019
22.4	68.088	1090.634	9.102	15.252	244.302	2.039
22.6	68.140	1091.467	9.109	15.400	246.671	2.059
22.8	68.200	1092.428	9.117	15.550	249.073	2.079
23.0	68.260	1093.389	9.125	15.700	251.479	2.099
23.2	68.312	1094.222	9.132	15.848	253.859	2.119
23.4	68.372	1095.183	9.140	15.999	256.273	2.139
23.6	68.424	1096.016	9.147	16.148	258.660	2.159
23.8	68.484	1096.977	9.155	16.299	261.080	2.179
24.0	68.544	1097.938	0.163	16.451	263.505	2.199
24.2	68.596	1098.771	9.170	16.600	265.903	2.219
24.4	68.656	1099.732	9.178	16.752	268.335	2.239
24.6	68.709	1100.581	9.185	16.902	270.743	2.260
24.8	68.768	1101.526	9.193	17.054	273.178	2.280
25.0	68.828	1102.487	9.201	17.207	275.622	2.300
25.2	68.881	1103.336	9.208	17.358	278.041	2.320
25.4	68.940	1104.281	9.216	17.511	280.487	2.341
25.6	69.000	1105.242	9.224	17.664	282.942	2.361
25.8	69.060	1106.203	9.232	17.817	285.400	2.382
26.0	69.113	1107.052	9.239	17.969	287.834	2.402
26.2	69.172	1107.997	9.247	18.123	290.295	2.423
26.4	69.232	1108.958	9.255	18.277	292.765	2.443
26.6	69.292	1109.919	9.263	18.432	295.239	2.464
26.8	69.344	1110.752	9.270	18.584	297.682	2.484
27.0	69.404	1111.713	9.278	18.739	300.163	2.505
27.2	69.464	1112.674	9.286	18.894	302.647	2.526
27.4	69.524	1113.635	9.294	19.050	305.136	2.547
27.6	69.584	1114.597	9.302	19.205	307.629	2.567
27.8	69.636	1115.429	9.309	19.359	310.089	2.588
28.0	69.696	1116.391	9.317	19.515	312.589	2.609
28.2	69.756	1117.352	9.325	19.671	315.093	2.630
28.4	69.816	1118.313	9.333	19.828	317.601	2.651
28.6	69.876	1119.274	9.341	19.985	320.112	2.672
28.8	69.935	1120.219	9.349	20.141	322.623	2.692
29.0	69.995	1121.180	9.357	20.299	325.142	2.714

TABLE 12 *(Continued)*

Degree Brix	Apparent Density	Calculated Density		Calculated Solids (Density × Brix / 100)		
	lb / ft³	kg / m³	lb / gal (U.S.)	lb / ft³	kg / m³	lb / gal (U.S.)
29.2	70.055	1122.141	9.365	20.456	327.665	2.735
29.4	70.107	1122.974	9.372	20.611	330.154	2.755
29.6	70.167	1123.935	9.380	20.769	332.685	2.776
29.8	70.227	1124.896	9.388	20.928	335.219	2.798
30.0	70.287	1125.857	9.396	21.086	337.757	2.819
30.2	70.347	1126.818	9.404	21.245	340.299	2.840
30.4	70.407	1127.779	9.412	21.404	342.845	2.861
30.6	70.466	1128.724	9.420	21.563	345.390	2.882
30.8	70.526	1129.685	9.428	21.722	347.943	2.904
31.0	70.586	1130.647	9.436	21.882	350.500	2.925
31.2	70.646	1131.608	9.444	22.042	353.062	2.947
31.4	70.706	1132.569	9.452	22.202	355.627	2.968
31.6	70.766	1133.530	9.460	22.362	358.195	2.989
31.8	70.826	1134.491	9.468	22.523	360.768	3.011
32.0	70.893	1135.564	9.477	22.686	363.381	3.033
32.2	70.953	1136.525	9.485	22.847	365.961	3.054
32.4	71.013	1137.486	9.493	23.008	368.546	3.076
32.6	71.072	1138.431	9.501	23.169	371.129	3.097
32.8	71.132	1139.392	9.509	23.331	373.721	3.119
33.0	71.192	1140.353	9.517	23.493	376.317	3.141
33.2	71.252	1141.315	9.525	23.656	378.916	3.162
33.4	71.312	1142.276	9.533	23.818	381.520	3.184
33.6	71.372	1143.237	9.541	23.981	384.128	3.206
33.8	71.439	1144.310	9.550	24.146	386.777	3.228
34.0	71.499	1145.271	9.558	24.310	389.392	3.250
34.2	71.559	1146.232	9.566	24.473	392.011	3.272
34.4	71.618	1147.177	9.574	24.637	394.629	3.293
34.6	71.686	1148.266	9.583	24.803	397.300	3.316
34.8	71.746	1149.227	9.591	24.968	399.931	3.338
35.0	71.806	1150.189	9.599	25.132	402.566	3.360
35.2	71.865	1151.134	9.607	25.296	405.199	3.382
35.4	71.932	1152.207	9.616	25.464	407.881	3.404
35.6	71.993	1153.184	9.624	25.630	410.533	3.426
35.8	72.052	1154.129	9.632	25.795	413.178	3.448
36.0	72.112	1155.090	9.640	25.960	415.832	3.470
36.2	72.180	1156.179	9.649	26.129	418.537	3.493
36.4	72.239	1157.124	9.657	26.295	421.193	3.515
36.6	72.299	1158.085	9.665	26.461	423.859	3.537
36.8	72.367	1159.175	9.674	26.631	426.576	3.560
37.0	72.426	1160.120	9.682	26.798	429.244	3.582
37.2	72.494	1161.209	9.691	26.968	431.970	3.605
37.4	72.554	1162.170	9.699	27.135	434.652	3.627
37.6	72.613	1163.115	9.707	27.302	437.331	3.650
37.8	72.681	1164.204	9.716	27.473	440.069	3.673
38.0	72.741	1165.165	9.724	27.642	442.763	3.695
38.2	72.808	1166.239	9.733	27.813	445.503	3.718
38.4	72.868	1167.200	9.741	27.981	448.205	3.741
38.6	72.928	1168.161	9.749	28.150	450.910	3.763
38.8	72.995	1169.234	9.758	28.322	453.663	3.786

(continued)

TABLE 12 *(Continued)*

Degree Brix	Apparent Density lb/ft³	Calculated Density kg/m³	Calculated Density lb/gal (U.S.)	Calculated Solids (Density × Brix/100) lb/ft³	Calculated Solids (Density × Brix/100) kg/m³	Calculated Solids (Density × Brix/100) lb/gal (U.S.)
39.0	73.055	1170.195	9.766	28.491	456.376	3.809
39.2	73.122	1171.268	9.775	28.664	459.137	3.832
39.4	73.182	1172.229	9.783	28.834	461.858	3.855
39.6	73.249	1173.302	9.792	29.007	464.628	3.878
39.8	73.317	1174.392	9.801	29.180	467.408	3.901
40.0	73.376	1175.337	9.809	29.350	470.135	3.924
40.2	73.444	1176.426	9.818	29.524	472.923	3.947
40.4	73.504	1177.387	9.826	29.696	475.664	3.970
40.6	73.571	1178.460	9.835	29.870	478.455	3.993
40.8	73.631	1179.421	9.843	30.041	481.204	4.016
41.0	73.698	1180.495	9.852	30.216	484.003	4.039
41.2	73.765	1181.568	9.861	30.391	486.806	4.063
41.4	73.825	1182.529	9.869	30.564	489.567	4.086
41.6	73.893	1183.618	9.878	30.739	492.385	4.109
41.8	73.960	1184.691	9.887	30.915	495.201	4.133
42.0	74.020	1185.652	9.895	31.088	497.974	4.156
42.2	74.087	1186.726	9.904	31.265	500.798	4.179
42.4	74.154	1187.799	9.913	31.441	503.627	4.203
42.6	74.214	1188.760	9.921	31.615	506.412	4.226
42.8	74.282	1189.849	9.930	31.793	509.255	4.250
43.0	74.349	1190.922	9.939	31.970	512.097	4.274
43.2	74.416	1191.995	9.948	32.148	514.942	4.298
43.4	74.476	1192.957	9.956	32.323	517.743	4.321
43.6	74.543	1194.030	9.965	32.501	520.597	4.345
43.8	74.611	1195.119	9.974	32.680	523.462	4.369
44.0	74.678	1196.192	9.983	32.858	526.325	4.393
44.2	74.745	1197.265	9.992	33.037	529.191	4.416
44.4	74.805	1198.226	10.000	33.213	532.013	4.440
44.6	74.873	1199.316	10.009	33.393	534.895	4.464
44.8	74.940	1200.389	10.018	33.573	537.774	4.488
45.0	75.007	1201.462	10.027	33.753	540.658	4.512
45.2	75.074	1202.535	10.036	33.933	543.546	4.536
45.4	75.142	1203.625	10.045	34.114	546.446	4.560
45.6	75.209	1204.698	10.054	34.295	549.342	4.585
45.8	75.276	1205.771	10.063	34.476	552.243	4.609
46.0	75.336	1206.732	10.071	34.655	555.097	4.633
46.2	75.404	1207.821	10.080	34.837	558.013	4.657
46.4	75.471	1208.894	10.089	35.019	560.927	4.681
46.6	75.538	1209.968	10.098	35.201	563.845	4.706
46.8	75.606	1211.057	10.107	35.384	566.775	4.730
47.0	75.673	1212.130	10.116	35.566	569.701	4.755
47.2	75.740	1213.203	10.125	35.749	572.632	4.779
47.4	75.808	1214.293	10.134	35.933	575.575	4.804
47.6	75.875	1215.366	10.143	36.117	578.514	4.828
47.8	75.942	1216.439	10.152	36.300	581.458	4.853
48.0	76.017	1217.640	10.162	36.488	584.467	4.878
48.2	76.077	1218.601	10.170	36.669	587.366	4.902
48.4	76.144	1219.675	10.179	36.854	590.323	4.927
48.6	76.219	1220.876	10.189	37.042	593.346	4.952

TABLE 12 (*Continued*)

Degree Brix	Apparent Density	Calculated Density		Calculated Solids (Density × Brix / 100)		
	lb / ft³	kg / m³	lb / gal (U.S.)	lb / ft³	kg / m³	lb / gal (U.S.)
48.8	76.286	1221.949	10.198	37.228	596.311	4.977
49.0	76.354	1223.038	10.207	37.413	599.289	5.001
49.2	76.421	1224.112	10.216	37.599	602.263	5.026
49.4	76.488	1225.185	10.225	37.785	605.241	5.051
49.6	76.556	1226.274	10.234	37.972	608.232	5.076
49.8	76.623	1227.347	10.243	38.158	611.219	5.101
50.0	76.690	1228.420	10.252	38.345	614.210	5.126
50.2	76.765	1229.622	10.262	38.536	617.270	5.152
50.4	76.832	1230.695	10.271	38.723	620.270	5.177
50.6	76.900	1231.784	10.280	38.911	623.283	5.202
50.8	76.975	1232.986	10.290	39.103	626.357	5.227
51.0	77.042	1234.059	10.299	39.291	629.370	5.252
51.2	77.109	1235.132	10.308	39.480	632.388	5.278
51.4	77.177	1236.221	10.317	39.669	635.418	5.303
51.6	77.244	1237.294	10.326	39.858	638.444	5.328
51.8	77.319	1238.496	10.336	40.051	641.541	5.354
52.0	77.386	1239.569	10.345	40.241	644.576	5.379
52.2	77.453	1240.642	10.354	40.430	647.615	5.405
52.4	77.528	1241.844	10.364	40.625	650.726	5.431
52.6	77.595	1242.917	10.373	40.815	653.774	5.456
52.8	77.663	1244.006	10.382	41.006	656.835	5.482
53.0	77.738	1245.207	10.392	41.201	659.960	5.508
53.2	77.805	1246.280	10.401	41.392	663.021	5.533
53.4	77.872	1247.354	10.410	41.584	666.087	5.559
53.6	77.947	1248.555	10.420	41.780	669.226	5.585
53.8	78.014	1249.628	10.429	41.972	672.300	5.611
54.0	78.089	1250.830	10.439	42.168	675.448	5.637
54.2	78.156	1251.903	10.448	42.361	678.531	5.663
54.4	78.231	1253.104	10.458	42.558	681.689	5.689
54.6	78.299	1254.193	10.467	42.751	684.790	5.715
54.8	78.366	1255.267	10.476	42.945	687.886	5.741
55.0	78.441	1256.468	10.486	43.143	691.057	5.767
55.2	78.508	1257.541	10.495	43.336	694.163	5.793
55.4	78.583	1258.742	10.505	43.535	697.343	5.820
55.6	78.658	1259.944	10.515	43.734	700.529	5.846
55.8	78.725	1261.017	10.524	43.929	703.648	5.872
56.0	78.800	1262.218	10.534	44.128	706.842	5.899
56.2	78.867	1263.292	10.543	44.323	709.970	5.925
56.4	78.942	1264.493	10.553	44.523	713.174	5.952
56.6	79.009	1265.566	10.562	44.719	716.310	5.978
56.8	79.084	1266.768	10.572	44.920	719.524	6.005
57.0	79.151	1267.841	10.581	45.116	722.669	6.031
57.2	79.226	1269.042	10.591	45.317	725.892	6.058
57.4	79.301	1270.243	10.601	45.519	729.120	6.085
57.6	79.368	1271.317	10.610	45.716	732.278	6.111
57.8	79.443	1272.518	10.620	45.918	735.515	6.138
58.0	79.518	1273.719	10.630	46.120	738.757	6.165
58.2	79.593	1274.921	10.640	46.323	742.004	6.193
58.4	79.660	1275.994	10.649	46.521	745.180	6.219

(*continued*)

TABLE 12 (*Continued*)

Degree Brix	Apparent Density lb/ft³	Calculated Density kg/m³	Calculated Density lb/gal (U.S.)	Calculated Solids (Density × Brix/100) lb/ft³	Calculated Solids kg/m³	Calculated Solids lb/gal (U.S.)
58.6	79.735	1277.195	10.659	46.725	748.436	6.246
58.8	79.810	1278.397	10.669	46.928	751.697	6.273
59.0	79.877	1279.470	10.678	47.127	754.887	6.300
59.2	79.952	1280.671	10.688	47.332	758.157	6.327
59.4	80.027	1281.872	10.698	47.536	761.432	6.355
59.6	80.101	1283.058	10.708	47.740	764.702	6.382
59.8	80.176	1284.259	10.718	47.945	767.987	6.409
60.0	80.244	1285.348	10.727	48.146	771.209	6.436
60.2	80.318	1286.534	10.737	48.351	774.493	6.464
60.4	80.393	1287.735	10.747	48.557	777.792	6.491
60.6	80.468	1288.936	10.757	48.764	781.095	6.519
60.8	80.543	1290.138	10.767	48.970	784.404	6.546
61.0	80.618	1291.339	10.777	49.177	787.717	6.574
61.2	80.685	1292.412	10.786	49.379	790.956	6.601
61.4	80.760	1293.614	10.796	49.587	794.279	6.629
61.6	80.834	1294.799	10.806	49.794	797.596	6.656
61.8	80.909	1296.000	10.816	50.002	800.928	6.684
62.0	80.984	1297.202	10.826	50.210	804.265	6.712
62.2	81.059	1298.403	10.836	50.419	807.607	6.740
62.4	81.134	1299.604	10.846	50.628	810.953	6.768
62.6	81.209	1300.806	10.856	50.837	814.304	6.796
62.8	81.283	1301.991	10.866	51.046	817.650	6.824
63.0	81.358	1303.192	10.876	51.256	821.011	6.852
63.2	81.443	1304.554	10.887	51.472	824.478	6.881
63.4	81.508	1305.595	10.896	51.676	827.747	6.908
63.6	81.583	1306.796	10.906	51.887	831.123	6.936
63.8	81.657	1307.982	10.916	52.097	834.492	6.964
64.0	81.732	1309.183	10.926	52.308	837.877	6.993
64.2	81.807	1310.385	10.936	52.520	841.267	7.021
64.4	81.882	1311.586	10.946	52.732	844.661	7.049
64.6	81.957	1312.787	10.956	52.944	848.061	7.078
64.8	82.039	1314.101	10.967	53.161	851.537	7.107
65.0	82.114	1315.302	10.977	53.374	854.946	7.135
65.2	82.188	1316.487	10.987	53.587	858.350	7.163
65.4	82.263	1317.689	10.997	53.800	861.768	7.192
65.6	82.338	1318.890	11.007	54.014	865.192	7.221
65.8	82.413	1320.091	11.017	54.228	868.620	7.249
66.0	82.488	1321.293	11.027	54.442	827.053	7.278
66.2	82.570	1322.606	11.038	54.661	875.565	7.307
66.4	82.645	1323.808	11.048	54.876	879.008	7.336
66.6	82.720	1325.009	11.058	55.092	882.456	7.365
66.8	82.794	1326.194	11.068	55.306	885.898	7.393
67.0	82.877	1327.524	11.079	55.528	889.441	7.423
67.2	82.951	1328.709	11.089	55.743	892.893	7.452
67.4	83.026	1329.910	11.099	55.960	896.360	7.481
67.6	83.109	1331.240	11.110	56.182	899.918	7.510
67.8	83.183	1332.425	11.120	56.398	903.384	7.539
68.0	83.258	1333.627	11.130	56.615	906.866	7.568
68.2	83.333	1334.828	11.140	56.833	910.353	7.597

TABLE 12 (*Continued*)

Degree Brix	Apparent Density lb / ft³	Calculated Density kg / m³	Calculated Density lb / gal (U.S.)	Calculated Solids (Density × Brix / 100) lb / ft³	Calculated Solids (Density × Brix / 100) kg / m³	Calculated Solids (Density × Brix / 100) lb / gal (U.S.)
68.4	83.415	1336.141	11.151	57.056	913.921	7.627
68.6	83.490	1337.343	11.161	57.274	917.417	7.656
68.8	83.572	1338.656	11.172	57.498	920.996	7.686
69.0	83.647	1339.858	11.182	57.716	924.502	7.716
69.2	83.722	1341.059	11.192	57.936	928.013	7.745
69.4	83.804	1342.372	11.203	58.160	931.606	7.775
69.6	83.879	1343.574	11.213	58.380	935.127	7.804
69.8	83.961	1344.887	11.224	58.605	938.731	7.834
70.0	84.036	1346.089	11.234	58.825	942.262	7.864
70.2	84.118	1347.402	11.245	59.051	945.876	7.894
70.4	84.193	1348.603	11.255	59.272	949.417	7.923
70.6	84.268	1349.805	11.265	59.493	952.962	7.953
70.8	84.350	1351.118	11.276	59.720	956.592	7.983
71.0	84.425	1352.320	11.286	59.942	960.147	8.013
71.2	84.507	1353.633	11.297	60.169	963.787	8.043
71.4	84.590	1354.963	11.308	60.397	967.443	8.074
71.6	84.664	1356.148	11.318	60.619	971.002	8.104
71.8	84.747	1357.477	11.329	60.848	974.669	8.134
72.0	84.822	1358.679	11.339	61.072	978.249	8.164
72.2	84.904	1359.992	11.350	61.301	981.914	8.195
72.4	84.979	1361.194	11.360	61.525	985.504	8.225
72.6	85.061	1362.507	11.371	61.754	989.180	8.255
72.8	85.143	1363.821	11.382	61.984	992.861	8.286
73.0	85.218	1365.022	11.392	62.209	996.466	8.316
73.2	85.300	1366.335	11.403	62.440	1000.158	8.347
73.4	85.383	1367.665	11.414	62.671	1003.866	8.378
73.6	85.457	1368.850	11.424	62.896	1007.474	8.408
73.8	85.540	1370.180	11.435	63.129	1011.193	8.439
74.0	85.622	1371.493	11.446	63.360	1014.905	8.470
74.2	85.697	1372.695	11.456	63.587	1018.539	8.500
74.4	85.799	1374.008	11.467	63.820	1022.262	8.531
74.6	85.861	1375.321	11.478	64.052	1025.990	8.563
74.8	85.944	1376.651	11.489	64.286	1029.735	8.594
75.0	86.018	1377.836	11.499	64.514	1033.377	8.624
75.2	86.101	1379.166	11.510	64.748	1037.133	8.656
75.4	86.183	1380.479	11.521	64.982	1040.881	8.687
75.6	86.265	1381.793	11.532	65.216	1044.635	8.718
75.8	86.348	1383.122	11.543	65.454	1048.407	8.750
76.0	86.430	1384.436	11.554	65.687	1052.171	8.781
76.2	86.505	1385.637	11.564	65.917	1055.855	8.812
76.4	86.587	1386.951	11.575	66.152	1059.630	8.843
76.6	86.669	1388.264	11.586	66.388	1063.410	8.875
76.8	86.752	1389.594	11.597	66.626	1067.208	8.907
77.0	86.834	1390.907	11.608	66.862	1070.998	8.938
77.2	86.916	1392.220	11.619	67.099	1074.794	8.970
77.4	86.998	1393.534	11.630	67.336	1078.595	9.002
77.6	87.081	1394.863	11.641	67.575	1082.414	9.033
77.8	87.163	1396.117	11.652	67.813	1086.226	9.065
78.0	87.238	1397.378	11.662	68.046	1089.955	9.096

(*continued*)

TABLE 12 (*Continued*)

Degree Brix	Apparent Density	Calculated Density		Calculated Solids (Density × Brix / 100)		
	lb / ft³	kg / m³	lb / gal (U.S.)	lb / ft³	kg / m³	lb / gal (U.S.)
78.2	87.320	1398.692	11.673	68.284	1093.777	9.128
78.4	87.402	1400.005	11.684	68.523	1097.604	9.160
78.6	87.485	1401.335	11.695	68.763	1101.449	9.192
78.8	87.567	1402.648	11.706	69.003	1105.287	9.224
79.0	87.647	1403.930	11.717	69.241	1109.104	9.256
79.2	87.732	1405.291	11.728	69.484	1112.991	9.289
79.4	87.821	1406.717	11.740	69.730	1116.933	9.322
79.6	87.904	1408.046	11.751	69.972	1120.805	9.354
79.8	87.986	1409.360	11.762	70.213	1124.669	9.386
80.0	88.068	1410.673	11.773	70.454	1128.539	9.418
80.2	88.150	1411.987	11.784	70.696	1132.413	9.451
80.4	88.233	1413.316	11.795	70.939	1136.306	9.483
80.6	88.315	1414.630	11.806	71.182	1140.192	9.516
80.8	88.397	1415.943	11.817	71.425	1144.082	9.548
81.0	88.480	1417.273	11.828	71.669	1147.991	9.581
81.2	88.562	1418.586	11.839	71.912	1151.892	9.613
81.4	88.652	1420.028	11.851	72.163	1155.903	9.647
81.6	88.732	1421.309	11.862	72.405	1159.788	9.679
81.8	88.816	1422.655	11.873	72.651	1163.732	9.712
82.0	88.898	1423.968	11.884	72.896	1167.654	9.745
82.2	88.981	1425.298	11.895	73.142	1171.595	9.778
82.4	89.071	1426.739	11.907	73.395	1175.633	9.811
82.6	89.153	1428.053	11.918	73.640	1179.572	9.844
82.8	89.235	1429.366	11.929	73.887	1183.515	9.877
83.0	89.317	1430.680	11.940	74.133	1187.464	9.910
83.2	89.407	1432.121	11.952	74.387	1191.525	9.944
83.4	89.489	1433.435	11.963	74.634	1195.485	9.977
83.6	89.572	1434.764	11.974	74.882	1199.463	10.010
83.8	89.662	1436.206	11.986	75.137	1203.541	10.044
84.0	89.744	1437.519	11.997	75.385	1207.516	10.078
84.2	89.826	1438.833	12.008	75.633	1211.497	10.111
84.4	89.916	1440.274	12.020	75.889	1215.592	10.145
84.6	89.998	1441.588	12.031	76.138	1219.583	10.178
84.8	90.080	1442.901	12.042	76.388	1223.580	10.212
85.0	90.170	1444.343	12.054	76.645	1227.692	10.246
85.2	90.252	1445.657	12.065	76.895	1231.699	10.279
85.4	90.342	1447.098	12.077	77.152	1235.822	10.314
85.6	90.425	1448.428	12.088	77.404	1239.854	10.347
85.8	90.507	1449.741	12.099	77.655	1243.878	10.381
86.0	90.597	1451.183	12.111	77.913	1248.017	10.416
86.2	90.679	1452.496	12.122	78.165	1252.052	10.449
86.4	90.769	1453.938	12.134	78.424	1256.202	10.484
86.6	90.851	1455.251	12.145	78.677	1260.248	10.518
86.8	90.941	1456.693	12.157	78.937	1264.409	10.552
87.0	91.023	1458.006	12.168	79.190	1268.466	10.586
87.2	91.113	1459.448	12.180	79.451	1272.639	10.621
87.4	91.195	1460.762	12.191	79.704	1276.706	10.655
87.6	91.285	1462.203	12.203	79.966	1280.890	10.690
87.8	91.375	1463.645	12.215	80.227	1285.080	10.725
88.0	91.457	1464.958	12.226	80.482	1289.163	10.759

TABLE 12 (*Continued*)

Degree Brix	Apparent Density lb/ft³	Calculated Density kg/m³	Calculated Density lb/gal (U.S.)	Calculated Solids (Density × Brix/100) lb/ft³	Calculated Solids (Density × Brix/100) kg/m³	Calculated Solids (Density × Brix/100) lb/gal (U.S.)
88.2	91.547	1466.400	12.238	80.744	1293.365	10.794
88.4	91.639	1467.874	12.250	81.009	1297.600	10.829
88.6	91.719	1469.155	12.261	81.263	1301.671	10.863
88.8	91.808	1470.581	12.273	81.526	1305.876	10.898
89.0	91.891	1471.910	12.284	81.723	1310.000	10.933
89.2	91.980	1473.336	12.296	82.046	1314.215	10.968
89.4	92.063	1474.665	12.307	82.304	1318.351	11.002
89.6	92.153	1476.107	12.319	82.569	1322.592	11.038
89.8	92.242	1477.532	12.331	82.833	1326.824	11.073
90.0	92.325	1478.862	12.342	83.093	1330.976	11.108
90.2	92.414	1480.287	12.354	83.357	1335.219	11.143
90.4	92.504	1481.729	12.366	83.624	1339.483	11.179
90.6	92.594	1483.171	12.378	83.890	1343.753	11.214
90.8	92.676	1484.484	12.389	84.150	1347.912	11.249
91.0	92.766	1485.926	12.401	84.417	1352.192	11.285
91.2	92.856	1487.367	12.413	84.685	1356.479	11.321
91.4	92.945	1488.793	12.425	84.952	1360.757	11.356
91.6	93.035	1490.235	12.437	85.220	1365.055	11.392
91.8	93.118	1491.564	12.448	85.482	1369.256	11.427
92.0	93.207	1492.990	12.460	85.750	1373.551	11.463
92.2	93.297	1494.431	12.472	86.020	1377.866	11.499
92.4	93.387	1495.873	12.484	86.290	1382.187	11.535
92.6	93.477	1497.315	12.496	86.560	1386.513	11.571
92.8	93.559	1498.628	12.507	86.823	1390.727	11.607
93.0	93.649	1500.070	12.519	87.094	1395.065	11.643
93.2	93.738	1501.495	12.531	87.364	1399.394	11.679
93.4	93.828	1502.937	12.543	87.635	1403.743	11.715
93.6	93.918	1504.379	12.555	87.907	1408.098	11.751
93.8	94.008	1505.820	12.567	88.180	1412.459	11.788
94.0	94.097	1507.246	12.579	88.451	1416.811	11.824
94.2	94.187	1508.687	12.591	88.724	1421.183	11.861
94.4	94.277	1510.129	12.603	88.997	1425.562	11.897
94.6	94.367	1511.571	12.615	89.271	1429.946	11.934
94.8	94.457	1513.012	12.627	89.545	1434.336	11.970
95.0	94.546	1514.438	12.639	89.819	1438.716	12.007

TABLE 13 Density, Apparent Density, and Weight per Unit Volume of Water at Temperatures 0 to 100°C

Temperature (°C)	Apparent Density (kg/L)	Calculated Density kg/m³	lb/ft³	lb/gal (U.S.)
0	0.99881	998.81	62.355	8.3357
1	0.99887	998.87	62.359	8.3362
2	0.99891	998.91	62.362	8.3365
3	0.99893	998.93	62.363	8.3367
4	0.99894	998.94	62.364	8.3368
5	0.99893	998.93	62.363	8.3367
6	0.99891	998.91	62.362	8.3365
7	0.99887	998.87	62.359	8.3362
8	0.99882	998.82	62.356	8.3358
9	0.99875	998.75	62.352	8.3352
10	0.99867	998.67	62.347	8.3345
11	0.99857	998.57	62.340	8.3337
12	0.99846	998.46	62.334	8.3328
13	0.99834	998.34	62.326	8.3318
14	0.99821	998.21	62.318	8.3307
15	0.99807	998.07	62.309	8.3295
16	0.99791	997.91	62.299	8.3282
17	0.99774	997.74	62.289	8.3268
18	0.99756	997.56	62.277	8.3253
19	0.99737	997.37	62.266	8.3237
20	0.99717	997.17	62.253	8.3220
21	0.99696	996.96	62.240	8.3203
22	0.99674	996.74	62.226	8.3184
23	0.99650	996.50	62.211	8.3164
24	0.99626	996.26	62.196	8.3144
25	0.99601	996.01	62.181	8.3123
26	0.99575	995.75	62.164	8.3102
27	0.99548	995.48	62.148	8,3079
28	0.99520	995.20	62.130	8.3056
29	0.99491	994.91	62.112	8.3032
30	0.99461	994.61	62.093	8.3007
31	0.99430	994.30	62.074	8.2981
32	0.99399	993.99	62.055	8.2955
33	0.99367	993.67	62.035	8.2928
34	0.99334	993.34	62.014	8.2901
35	0.99300	993.00	61.993	8.2872
36	0.99265	992.65	61.971	8.2843
37	0.99229	992.29	61.948	8.2813
38	0.99193	991.93	61.926	8.2783
39	0.99156	991.56	61.903	8.2752
40	0.99118	991.18	61.879	8.2720
41	0.99080	990.80	61.855	8.2689
42	0.99041	990.41	61.831	8.2656
43	0.99001	990.01	61.806	8.2623
44	0.98960	989.60	61.780	8.2589
45	0.98918	989.18	61.754	8.2553

TABLE 13 (*Continued*)

Temperature (°C)	Apparent Density (kg/L)	Calculated Density		
		kg/m³	lb/ft³	lb/gal (U.S.)
46	0.98876	988.76	61.728	8.2518
47	0.98833	988.33	61.701	8.2483
48	0.98790	987.90	61.674	8.2447
49	0.98746	987.46	61.647	8.2410
50	0.98701	987.01	61.619	8.2372
51	0.98656	986.56	61.591	8.2335
52	0.98610	986.10	61.562	8.2296
53	0.98563	985.63	61.533	8.2257
54	0.98515	985.15	61.503	8.2217
55	0.98467	984.67	61.473	8.2177
56	0.98418	984.18	61.442	8.2136
57	0.98369	983.69	61.412	8.2095
58	0.98319	983.19	61.380	8.2054
59	0.98269	982.69	61.349	8.2012
60	0.98218	982.18	61.317	8.1969
61	0.98166	981.66	61.285	8.1926
62	0.98114	981.14	61.252	8.1882
63	0.98061	980.61	61.219	8.1838
64	0.98007	980.07	61.186	8.1793
65	0.97953	979.53	61.152	8.1748
66	0.97899	978.99	61.118	8.1703
67	0.97844	978.44	61.084	8.1657
68	0.97788	977.88	61.049	8.1610
69	0.97732	977.32	61.014	8.1564
70	0.97675	976.75	60.978	8.1516
71	0.97617	976.17	60.942	8.1468
72	0.97559	975.59	60.906	8.1419
73	0.97501	975.01	60.870	8.1371
74	0.97442	974.42	60.833	8.1322
75	0.97383	973.83	60.796	8.1272
76	0.97323	973.23	60.759	8.1222
77	0.97262	972.62	60.720	8.1171
78	0.97200	972.00	60.682	8.1120
79	0.97138	971.38	60.643	8.1068
80	0.97076	970.76	60.604	8.1016
81	0.97013	970.13	60.565	8.0964
82	0.96950	969.50	60.526	8.0911
83	0.96887	968.87	60.486	8.0858
84	0.96823	968.23	60.446	8.0805
85	0.96758	967.58	60.406	8.0751
86	0.96693	966.93	60.365	8.0697
87	0.96627	966.27	60.324	8.0641
88	0.96561	965.61	60.283	8.0586
89	0.96494	964.94	60.241	8.0530
90	0.96427	964.27	60.199	8.0475

(*continued*)

TABLE 13 (*Continued*)

Temperature (°C)	Apparent Density (kg/L)	Calculated Density		
		kg/m³	lb/ft³	lb/gal (U.S.)
91	0.96360	963.60	60.157	8.0419
92	0.96292	962.92	60.115	8.0362
93	0.96223	962.23	60.072	8.0304
94	0.96154	961.54	60.029	8.0247
95	0.96085	960.85	59.986	8.0189
96	0.96015	960.15	59.942	8.0131
97	0.95944	959.44	59.898	8.0071
98	0.95873	958.73	59.853	8.0012
99	0.95802	958.02	59.809	7.9953
100	0.95731	957.31	59.765	7.9894

TABLE 14 International Refractive Index Scale of ICUMSA (1974) for Pure Sucrose Solutions at 20°C and 589 nm[a]

Sucrose g/100 g Brix	0.0	0.1	0.2	0.3	0.4	0.5	0.6	0.7	0.8	0.9
0	1.332986	1.333129	1.333272	1.333415	1.333558	1.333702	1.333845	1.333989	1.334132	1.334276
1	1.334420	1.334564	1.334708	1.334852	1.334996	1.335141	1.335285	1.335430	1.335574	1.335719
2	1.335864	1.336009	1.336154	1.336300	1.336445	1.336590	1.336736	1.336882	1.337028	1.337174
3	1.337320	1.337466	1.337612	1.337758	1.337905	1.338051	1.338198	1.338345	1.338492	1.338639
4	1.338786	1.338933	1.339081	1.339228	1.339376	1.339524	1.339671	1.339819	1.339967	1.340116
5	1.340264	1.340412	1.340561	1.340709	1.340858	1.341007	1.341156	1.341305	1.341454	1.341604
6	1.341753	1.341903	1.342052	1.342202	1.342352	1.342502	1.342652	1.342802	1.342952	1.343103
7	1.343253	1.343404	1.343555	1.343706	1.343857	1.344008	1.344159	1.344311	1.344462	1.344614
8	1.344765	1.344917	1.345069	1.345221	1.345373	1.345526	1.345678	1.345831	1.345983	1.346136
9	1.346289	1.346442	1.346595	1.346748	1.346902	1.347055	1.347209	1.347362	1.347516	1.347670
10	1.347824	1.347978	1.348133	1.348287	1.348442	1.348596	1.348751	1.348906	1.349061	1.349216
11	1.349371	1.349527	1.349682	1.349838	1.349993	1.350149	1.350305	1.350461	1.350617	1.350774
12	1.350930	1.351087	1.351243	1.351400	1.351557	1.351714	1.351871	1.352029	1.352186	1.352343
13	1.352501	1.352659	1.352817	1.352975	1.353133	1.353291	1.353449	1.353608	1.353767	1.353925
14	1.354084	1.354243	1.354402	1.354561	1.354721	1.354880	1.355040	1.355199	1.355359	1.355519
15	1.355679	1.355840	1.356000	1.356160	1.356321	1.356482	1.356642	1.356803	1.356964	1.357126
16	1.357287	1.357448	1.357610	1.357772	1.357933	1.358095	1.358257	1.358420	1.358582	1.358744
17	1.358907	1.359070	1.359232	1.359395	1.359558	1.359722	1.359885	1.360048	1.360212	1.360376
18	1.360539	1.360703	1.360867	1.361032	1.361196	1.361360	1.361525	1.361690	1.361854	1.362019
19	1.362185	1.362350	1.362515	1.362681	1.362846	1.363012	1.363178	1.363344	1.363510	1.363676
20	1.363842	1.364009	1.364176	1.364342	1.364509	1.364676	1.364843	1.365011	1.365178	1.365346
21	1.365513	1.365681	1.365849	1.366017	1.366185	1.366354	1.366522	1.366691	1.366859	1.367028
22	1.367197	1.367366	1.367535	1.367705	1.367874	1.368044	1.368214	1.368384	1.368554	1.368724
23	1.368894	1.369064	1.369235	1.369406	1.369576	1.369747	1.369918	1.370090	1.370261	1.370342
24	1.370604	1.370776	1.370948	1.371120	1.371292	1.371464	1.371637	1.371809	1.371982	1.372155
25	1.372328	1.372501	1.372674	1.372847	1.373021	1.373194	1.373368	1.373542	1.373716	1.373890
26	1.374065	1.374239	1.374414	1.374588	1.374763	1.374938	1.375113	1.375288	1.375464	1.375639
27	1.375815	1.375991	1.376167	1.376343	1.376519	1.376695	1.376872	1.377049	1.377225	1.377402

(*continued*)

TABLE 14 (*Continued*)

Sucrose g/100 g Brix	0.0	0.1	0.2	0.3	0.4	0.5	0.6	0.7	0.8	0.9
28	1.377579	1.377756	1.377934	1.378111	1.378289	1.378467	1.378644	1.378822	1.379001	1.379179
29	1.379357	1.379536	1.379715	1.379893	1.380072	1.380251	1.380431	1.380610	1.380790	1.380969
30	1.381149	1.381329	1.381509	1.381689	1.381870	1.382050	1.382231	1.382412	1.382593	1.382774
31	1.382955	1.383137	1.383318	1.383500	1.383682	1.383863	1.384046	1.384228	1.384410	1.384593
32	1.384775	1.384958	1.385141	1.385324	1.385507	1.385691	1.385874	1.386058	1.386242	1.386426
33	1.386610	1.386794	1.386978	1.387163	1.387348	1.387532	1.387717	1.387902	1.388088	1.388273
34	1.388459	1.388644	1.388830	1.389016	1.389202	1.389388	1.389575	1.389761	1.389948	1.390135
35	1.390322	1.390509	1.390696	1.390884	1.391071	1.391259	1.391447	1.391635	1.391823	1.392011
36	1.392200	1.392388	1.392577	1.392766	1.392955	1.393144	1.393334	1.393523	1.393713	1.393903
37	1.394092	1.394283	1.394473	1.394663	1.394854	1.395044	1.395235	1.395426	1.395617	1.395809
38	1.396000	1.396192	1.396383	1.396575	1.396767	1.396959	1.397152	1.397344	1.397537	1.397730
39	1.397922	1.398116	1.398309	1.398502	1.398696	1.398889	1.399083	1.399277	1.399471	1.399666
40	1.399860	1.400055	1.400249	1.400444	1.400639	1.400834	1.401030	1.401225	1.401421	1.401617
41	1.401813	1.402009	1.402205	1.402401	1.402598	1.402795	1.402992	1.403189	1.403386	1.403583
42	1.403781	1.403978	1.404176	1.404374	1.404572	1.404770	1.404969	1.405167	1.405366	1.405565
43	1.405764	1.405963	1.406163	1.406362	1.406562	1.406762	1.406961	1.407162	1.407362	1.407562
44	1.407763	1.407964	1.408165	1.408366	1.408567	1.408768	1.408970	1.409171	1.409373	1.409575
45	1.409777	1.409980	1.410182	1.410385	1.410588	1.410790	1.410994	1.411197	1.411400	1.411604
46	1.411807	1.412011	1.412215	1.412420	1.412624	1.412828	1.413033	1.413238	1.413443	1.413648
47	1.413853	1.414059	1.414265	1.414470	1.414676	1.414882	1.415089	1.415295	1.415502	1.415708
48	1.415915	1.416122	1.416330	1.416537	1.416744	1.416952	1.417160	1.417368	1.417576	1.417785
49	1.417993	1.418202	1.418411	1.418620	1.418829	1.419038	1.419247	1.419457	1.419667	1.419877
50	1.420087	1.420297	1.420508	1.420718	1.420929	1.421140	1.421351	1.421562	1.421774	1.421985
51	1.422197	1.422409	1.422621	1.422833	1.423046	1.423258	1.423471	1.423684	1.423897	1.424110
52	1.424323	1.424537	1.424750	1.424964	1.425178	1.425393	1.425607	1.425821	1.426036	1.426251
53	1.426466	1.426681	1.426896	1.427112	1.427328	1.427543	1.427759	1.427975	1.428192	1.428408
54	1.428625	1.428842	1.429059	1.429276	1.429493	1.429711	1.429928	1.430146	1.430364	1.430582
55	1.430800	1.431019	1.431238	1.431456	1.431675	1.431894	1.432114	1.432333	1.432553	1.432773

56	1.432993	1.433213	1.433433	1.433653	1.433874	1.434095	1.434316	1.434537	1.434758	1.434980
57	1.435201	1.435423	1.435645	1.435867	1.436089	1.436312	1.436534	1.436757	1.436980	1.437203
58	1.437427	1.437650	1.437874	1.438098	1.438322	1.438546	1.438770	1.438994	1.439219	1.439444
59	1.439669	1.439894	1.440119	1.440345	1.440571	1.440796	1.441022	1.441248	1.441475	1.441701
60	1.441928	1.442155	1.442382	1.442609	1.442836	1.443064	1.443292	1.443519	1.443747	1.443976
61	1.444204	1.444432	1.444661	1.444890	1.445119	1.445348	1.445578	1.445807	1.446037	1.446267
62	1.446497	1.446727	1.446957	1.447188	1.447419	1.447650	1.447881	1.448112	1.448343	1.448575
63	1.448807	1.449039	1.449271	1.449503	1.449736	1.449968	1.450201	1.450434	1.450667	1.450900
64	1.451134	1.451367	1.451601	1.451835	1.452069	1.452304	1.452538	1.452773	1.453008	1.453243
65	1.453478	1.453713	1.453949	1.454184	1.454420	1.454656	1.454893	1.455129	1.455365	1.455602
66	1.455839	1.456076	1.456313	1.456551	1.456788	1.457026	1.457264	1.457502	1.457740	1.457979
67	1.458217	1.458456	1.458695	1.458934	1.459174	1.459413	1.459653	1.459893	1.460133	1.460373
68	1.460613	1.460854	1.461094	1.461335	1.461576	1.461817	1.462059	1.462300	1.462542	1.462784
69	1.463026	1.463268	1.463511	1.463753	1.463996	1.464239	1.464482	1.464725	1.464969	1.465212
70	1.465456	1.465700	1.465944	1.466188	1.466433	1.466678	1.466922	1.467167	1.467413	1.467658
71	1.467903	1.468149	1.468395	1.468641	1.468887	1.469134	1.469380	1.469627	1.469874	1.470121
72	1.470368	1.470616	1.470863	1.471111	1.471359	1.471607	1.471855	1.472104	1.472352	1.472601
73	1.472850	1.473099	1.473349	1.473598	1.473848	1.474098	1.474348	1.474598	1.474848	1.475099
74	1.475349	1.475600	1.475851	1.476103	1.476354	1.476606	1.476857	1.477109	1.477361	1.477614
75	1.477866	1.478119	1.478371	1.478624	1.478878	1.479131	1.479384	1.479638	1.479892	1.480146
76	1.480400	1.480654	1.480909	1.481163	1.481418	1.481673	1.481929	1.482184	1.482440	1.482695
77	1.482951	1.483207	1.483463	1.483720	1.483976	1.484233	1.484490	1.484747	1.485005	1.485262
78	1.485520	1.485777	1.486035	1.486294	1.486552	1.486810	1.487069	1.487328	1.487587	1.487846
79	1.488105	1.488365	1.488625	1.488884	1.489144	1.489405	1.489665	1.489926	1.490186	1.490447
80	1.490708	1.490970	1.491231	1.491493	1.491754	1.492016	1.492278	1.492541	1.492803	1.493066
81	1.493328	1.493591	1.493855	1.494118	1.494381	1.494645	1.404909	1.495173	1.495437	1.495701
82	1.495966	1.496230	1.496495	1.496760	1.497026	1.497291	1.497556	1.497822	1.498088	1.498354
83	1.498620	1.498887	1.499153	1.499420	1.499687	1.499954	1.500221	1.500488	1.400756	1.501024
84	1.501292	1.501560	1.501828	1.502096	1.502365	1.502634	1.502903	1.503172	1.503441	1.503711
85	1.503980									

Source: The data are those of Rosenbruch and Emmerich, as reported in Schneider, *ICUMSA Methods,* 1979, pp. 234–236.

[a]The table gives values of refractive index against air, with solution concentrations corrected to vacuum.

TABLE 15 Temperature Corrections for Refractometric Sucrose (Dry Substance) Measurements at 589 nm[a]

Temperature (°C)	0	5	10	15	20	25	30	35	40	45	50	55	60	65	70	75	80	85
							Subtract from the Measured Value											
15	0.29	0.31	0.32	0.33	0.34	0.35	0.36	0.37	0.37	0.38	0.38	0.38	0.38	0.38	0.38	0.38	0.37	0.37
16	0.24	0.25	0.26	0.27	0.28	0.28	0.29	0.30	0.30	0.30	0.31	0.31	0.31	0.31	0.31	0.30	0.30	0.30
17	0.18	0.19	0.20	0.20	0.21	0.21	0.22	0.22	0.23	0.23	0.23	0.23	0.23	0.23	0.23	0.23	0.23	0.22
18	0.12	0.13	0.13	0.14	0.14	0.14	0.15	0.15	0.15	0.15	0.15	0.15	0.15	0.15	0.15	0.15	0.15	0.15
19	0.06	0.06	0.07	0.07	0.07	0.07	0.07	0.08	0.08	0.08	0.08	0.08	0.08	0.08	0.08	0.08	0.08	0.07
							Add to the Measured Value											
21	0.06	0.07	0.07	0.07	0.07	0.07	0.08	0.08	0.08	0.08	0.08	0.08	0.08	0.08	0.08	0.08	0.08	0.07
22	0.13	0.14	0.14	0.14	0.15	0.15	0.15	0.15	0.16	0.16	0.16	0.16	0.16	0.16	0.15	0.15	0.15	0.15
23	0.20	0.21	0.21	0.22	0.22	0.23	0.23	0.23	0.23	0.24	0.24	0.24	0.24	0.23	0.23	0.23	0.23	0.22
24	0.27	0.28	0.29	0.29	0.30	0.30	0.31	0.31	0.31	0.32	0.32	0.32	0.32	0.31	0.31	0.31	0.30	0.30
25	0.34	0.35	0.36	0.37	0.38	0.38	0.39	0.39	0.40	0.40	0.40	0.40	0.40	0.39	0.39	0.39	0.38	0.37
26	0.42	0.43	0.44	0.45	0.46	0.46	0.47	0.47	0.48	0.48	0.48	0.48	0.48	0.47	0.47	0.46	0.46	0.45
27	0.50	0.51	0.52	0.53	0.54	0.55	0.55	0.56	0.56	0.56	0.56	0.56	0.56	0.55	0.55	0.54	0.53	0.52
28	0.58	0.59	0.60	0.61	0.62	0.63	0.64	0.64	0.64	0.65	0.65	0.64	0.64	0.64	0.63	0.62	0.61	0.60
29	0.66	0.67	0.68	0.69	0.70	0.71	0.72	0.73	0.73	0.73	0.73	0.73	0.72	0.72	0.71	0.70	0.69	0.68
30	0.74	0.75	0.77	0.78	0.79	0.80	0.81	0.81	0.81	0.82	0.81	0.81	0.81	0.80	0.79	0.78	0.77	0.75
31	0.83	0.84	0.85	0.87	0.88	0.89	0.89	0.90	0.90	0.90	0.90	0.90	0.89	0.88	0.87	0.86	0.84	0.83
32	0.91	0.93	0.94	0.95	0.96	0.97	0.98	0.99	0.99	0.99	0.99	0.98	0.97	0.96	0.95	0.94	0.92	0.90
33	1.00	1.02	1.03	1.04	1.05	1.06	1.07	1.08	1.08	1.08	1.07	1.07	1.06	1.05	1.03	1.02	1.00	0.98
34	1.10	1.11	1.12	1.13	1.15	1.15	1.16	1.17	1.17	1.17	1.16	1.15	1.14	1.13	1.12	1.10	1.08	1.06
35	1.19	1.20	1.22	1.23	1.24	1.25	1.25	1.26	1.26	1.25	1.25	1.24	1.23	1.21	1.20	1.18	1.16	1.13
36	1.29	1.30	1.31	1.32	1.33	1.34	1.35	1.35	1.35	1.35	1.34	1.33	1.32	1.30	1.28	1.26	1.24	1.21
37	1.38	1.40	1.41	1.42	1.43	1.44	1.44	1.44	1.44	1.44	1.43	1.42	1.40	1.38	1.36	1.34	1.32	1.29
38	1.48	1.50	1.51	1.52	1.53	1.53	1.54	1.54	1.53	1.53	1.52	1.51	1.49	1.47	1.45	1.42	1.39	1.36
39	1.49	1.60	1.61	1.62	1.62	1.63	1.63	1.63	1.63	1.62	1.61	1.60	1.58	1.56	1.53	1.50	1.47	1.44
40	1.69	1.70	1.71	1.72	1.72	1.73	1.73	1.73	1.72	1.71	1.70	1.69	1.67	1.64	1.62	1.59	1.55	1.52

Source: Calculated by Rosenbruch, as reported in Schneider, *ICUMSA Methods,* 1979, p. 237.

[a]Reference temperature; 20°C.

TABLE 16 Refractive Indices of Invert Sugar Solutions

Invert Sugar by Weight in Air (%)	Temperature (°C)			
	15	20	25	30
0	1.33339	1.33299	1.33250	1.33194
5	1.34066	1.34020	1.33969	1.33908
10	1.34821	1.34768	1.34714	1.34648
15	1.35603	1.35544	1.35487	1.35416
20	1.36413	1.36348	1.36287	1.36213
25	1.37252	1.37183	1.37118	1.37038
30	1.38122	1.38048	1.37978	1.37895
35	1.39022	1.38944	1.38869	1.38782
40	1.39955	1.39873	1.39793	1.39702
45	1.40920	1.40835	1.40750	1.40655
50	1.41919	1.41830	1.41740	1.41642
55	1.42952	1.42860	1.42766	1.42664
60	1.44020	1.43926	1.43827	1.43722
65	1.45125	1.45027	1.44925	1.44817
70	1.46266	1.46166	1.46060	1.45950
75	1.47446	1.47343	1.47235	1.47121
80	1.48664	1.48559	1.48449	1.48333
85	1.49922	1.49815	1.49703	1.49584

Source: As reported to ICUMSA, 1958, Subj. 7, p. 26 by C. F. Snyder.

TABLE 17 International Refractive Index Scale of ICUMSA (1990) for Pure Invert Sugar Solutions at 20°C and 589 nm[a]

Invert Sugar (g/100 g)	0.0	0.1	0.2	0.3	0.4	0.5	0.6	0.7	0.8	0.9
0	1.332988	1.333129	1.333271	1.333413	1.333555	1.333698	1.333840	1.333982	1.334125	1.334267
1	1.334410	1.334553	1.334696	1.334839	1.334982	1.335125	1.335268	1.335412	1.335555	1.335699
2	1.335843	1.335986	1.336130	1.336275	1.336419	1.336563	1.336707	1.336852	1.336996	1.337141
3	1.337286	1.337431	1.337576	1.337721	1.337866	1.338011	1.338157	1.338302	1.338448	1.338594
4	1.338740	1.338886	1.339032	1.339178	1.339324	1.339470	1.339617	1.339764	1.339910	1.340057
5	1.340204	1.340351	1.340498	1.340645	1.340793	1.340940	1.341088	1.341235	1.341383	1.341531
6	1.341679	1.341827	1.341975	1.342124	1.342272	1.342421	1.342569	1.342718	1.342867	1.343016
7	1.343165	1.343314	1.343463	1.343613	1.343762	1.343912	1.344061	1.344211	1.344361	1.344511
8	1.344661	1.344811	1.344962	1.345112	1.345263	1.345414	1.345564	1.345715	1.345866	1.346017
9	1.346169	1.346320	1.346471	1.346623	1.346774	1.346926	1.347078	1.347230	1.347382	1.347534
10	1.347687	1.347839	1.347992	1.348144	1.348297	1.348450	1.348603	1.348756	1.348909	1.349062
11	1.349216	1.349369	1.349523	1.349677	1.349831	1.349985	1.350139	1.350293	1.350447	1.350602
12	1.350756	1.350911	1.351065	1.351220	1.351375	1.351530	1.351685	1.351841	1.351996	1.352152
13	1.352307	1.352463	1.352619	1.352775	1.352931	1.353087	1.353243	1.353400	1.353556	1.353713
14	1.353870	1.354027	1.354184	1.354341	1.354498	1.354655	1.354813	1.354970	1.355128	1.355286
15	1.355444	1.355602	1.355760	1.355918	1.356076	1.356235	1.356393	1.356552	1.356711	1.356870
16	1.357029	1.357188	1.357347	1.357506	1.357666	1.357825	1.357985	1.358145	1.358305	1.358465
17	1.358625	1.358785	1.358946	1.369106	1.359267	1.359428	1.359588	1.359749	1.359911	1.360072
18	1.350233	1.360395	1.360556	1.360718	1.360880	1.361041	1.361203	1.361366	1.361528	1.361690
19	1.361853	1.362015	1.362178	1.362341	1.362504	1.362667	1.362830	1.362993	1.363157	1.363320
20	1.363484	1.363648	1.363812	1.363976	1.364140	1.364304	1.364468	1.364633	1.364797	1.364962
21	1.365127	1.365292	1.365457	1.365622	1.365788	1.365953	1.366119	1.366284	1.366450	1.366616
22	1.366782	1.366948	1.367114	1.367281	1.367447	1.367614	1.367781	1.367948	1.368115	1.368282
23	1.368449	1.368616	1.368784	1.368951	1.369119	1.369287	1.369455	1.369623	1.369791	1.369959
24	1.370128	1.370296	1.370465	1.370634	1.370803	1.370972	1.371141	1.371310	1.371480	1.371649
25	1.371819	1.371989	1.372159	1.372329	1.372499	1.372669	1.372839	1.373010	1.373181	1.373351
26	1.373522	1.373693	1.373864	1.374036	1.374207	1.374379	1.374550	1.374722	1.374894	1.375066
27	1.375238	1.375410	1.375583	1.375755	1.375928	1.376101	1.376273	1.376446	1.376620	1.376793
28	1.376966	1.377140	1.377313	1.377487	1.377661	1.377835	1.378009	1.378183	1.378358	1.378532

29	1.378707	1.378882	1.379056	1.379231	1.379407	1.379582	1.379757	1.379933	1.380108	1.380284
30	1.380460	1.380636	1.380812	1.380989	1.381165	1.381341	1.381518	1.381695	1.381872	1.382049
31	1.382226	1.382403	1.382581	1.382758	1.382936	1.383114	1.383292	1.383470	1.383648	1.383826
32	1.384005	1.384183	1.384362	1.384541	1.384720	1.384899	1.385078	1.385258	1.385437	1.385617
33	1.385797	1.385976	1.386157	1.386337	1.386517	1.386697	1.386878	1.387059	1.387239	1.387420
34	1.387601	1.387782	1.387964	1.388145	1.388327	1.388509	1.388690	1.388872	1.389054	1.389237
35	1.389419	1.389602	1.389784	1.389967	1.390150	1.390333	1.390516	1.390699	1.390883	1.391066
36	1.391250	1.391434	1.391618	1.391802	1.391986	1.392171	1.392355	1.392540	1.392724	1.392909
37	1.393094	1.393279	1.393465	1.393650	1.393836	1.394021	1.394207	1.394393	1.394579	1.394765
38	1.394952	1.395138	1.395325	1.395512	1.395699	1.395886	1.396073	1.396260	1.396448	1.396635
39	1.396823	1.397011	1.397199	1.397387	1.397575	1.397763	1.397952	1.398141	1.398329	1.398518
40	1.398707	1.398897	1.399086	1.399275	1.399465	1.399655	1.399845	1.400035	1.400225	1.400415
41	1.400605	1.400796	1.400987	1.401177	1.401368	1.401560	1.401751	1.401942	1.402134	1.402325
42	1.402517	1.402709	1.402901	1.403093	1.403286	1.403478	1.403671	1.403864	1.404056	1.404249
43	1.404443	1.404636	1.404829	1.405023	1.405217	1.405411	1.405605	1.405799	1.405993	1.406187
44	1.406382	1.406577	1.406771	1.406966	1.407161	1.407357	1.407552	1.407748	1.407943	1.408139
45	1.408335	1.408531	1.408727	1.408924	1.409120	1.309317	1.409514	1.409711	1.409908	1.410105
46	1.410302	1.410500	1.410697	1.410895	1.411093	1.411291	1.411489	1.411687	1.411886	1.412084
47	1.412283	1.412482	1.412681	1.412880	1.413080	1.413279	1.413479	1.413678	1.413878	1.414078
48	1.414278	1.414479	1.414679	1.414880	1.415080	1.415281	1.415482	1.415683	1.415885	1.416086
49	1.416288	1.416489	1.416691	1.416893	1.417095	1.417298	1.417500	1.417702	1.417905	1.418108
50	1.418311	1.418514	1.418717	1.418921	1.419124	1.419328	1.419532	1.419736	1.419940	1.420144
51	1.420349	1.420553	1.420758	1.420963	1.421168	1.421373	1.421578	1.421783	1.421989	1.422195
52	1.422400	1.422606	1.422813	1.423019	1.423225	1.423432	1.423638	1.423845	1.424052	1.424259
53	1.424467	1.424674	1.424881	1.425089	1.425297	1.425505	1.425713	1.425921	1.426130	1.426338
54	1.426547	1.426756	1.426965	1.427174	1.427383	1.427593	1.427802	1.428012	1.428222	1.428432
55	1.428642	1.428852	1.429062	1.429273	1.429484	1.429694	1.429905	1.430117	1.430328	1.430539
56	1.430751	1.430962	1.431174	1.431386	1.431598	1.431811	1.432023	1.432236	1.432448	1.432661
57	1.432874	1.433087	1.433301	1.433514	1.433728	1.433941	1.434155	1.434369	1.434583	1.434798
58	1.435012	1.435227	1.435441	1.435656	1.435871	1.436086	1.436302	1.436517	1.436733	1.436948
59	1.437164	1.537380	1.437596	1.437813	1.438029	1.438246	1.438463	1.438679	1.438896	1.439114
60	1.439331	1.439548	1.439766	1.439984	1.440202	1.440420	1.440638	1.440856	1.441075	1.441293

(continued)

TABLE 17 (*Continued*)

Invert Sugar (g/100 g)	0.0	0.1	0.2	0.3	0.4	0.5	0.6	0.7	0.8	0.9
61	1.441512	1.441731	1.441950	1.422169	1.442388	1.442608	1.442827	1.443047	1.443267	1.443487
62	1.443707	1.443928	1.444148	1.444369	1.444589	1.444810	1.445031	1.445252	1.445474	1.445695
63	1.445917	1.446139	1.446360	1.446583	1.446805	1.447027	1.447249	1.447472	1.447695	1.447918
64	1.448141	1.448364	1.448587	1.448811	1.449034	1.449258	1.449482	1.449706	1.449930	1.450155
65	1.450379	1.450604	1.450828	1.451053	1.451278	1.451503	1.451729	1.451954	1.452180	1.452406
66	1.452631	1.452858	1.453084	1.453310	1.453536	1.453763	1.453990	1.454217	1.454444	1.454671
67	1.454898	1.455126	1.455353	1.455581	1.455809	1.456037	1.456265	1.456493	1.456722	1.456950
68	1.457179	1.457408	1.457637	1.457866	1.458095	1.458325	1.458554	1.458784	1.459014	1.459244
69	1.459474	1.459704	1.459934	1.460165	1.460396	1.460626	1.460857	1.461088	1.461320	1.461551
70	1.461783	1.46201	1.46225	1.46248	1.46271	1.46294	1.46317	1.46314	1.46364	1.46387
71	1.46411	1.46434	1.46457	1.46480	1.46504	1.46527	1.46551	1.46574	1.46597	1.46621
72	1.46644	1.46668	1.46691	1.46715	1.46738	1.46762	1.46785	1.46809	1.46832	1.46856
73	1.46879	1.46903	1.46926	1.46950	1.46974	1.46997	1.47021	1.47045	1.47068	1.47092
74	1.47116	1.47139	1.47163	1.47187	1.48211	1.47234	1.46258	1.47282	1.47306	1.47330
75	1.47353	1.47377	1.47401	1.47425	1.47449	1.47473	1.47497	1.47251	1.47545	1.47568
76	1.47592	1.47616	1.47640	1.47664	1.47688	1.47712	1.47737	1.47761	1.47785	1.47809
77	1.47833	1.47857	1.47881	1.47905	1.47929	1.47954	1.47978	1.48002	1.48026	1.48050
78	1.48075	1.48099	1.48123	1.48147	1.48172	1,48196	1.48220	1.48245	1.48269	1.48293
79	1.48318	1.48342	1.48366	1.48391	1.48415	1.48440	1.48464	1.48488	1.48513	1.48537
80	1.48562	1.4859	1.4861	1.4864	1.4866	1.4868	1.4871	1.4873	1.4876	1.4878
81	1.4881	1.4883	1.4886	1.4888	1.4891	1.4893	1.4896	1.4898	1.4900	1.4903
82	1.4905	1.4908	1.4910	1.4913	1.4915	1.4918	1.4920	1.4923	1.4925	1.4928
83	1.4930	1.4933	1.4935	1.4938	1.4940	1.4943	1.4945	1.4948	1.4950	1.4953
84	1.4955	1.4958	1.4960	1.4963	1.4965	1.4968	1.4970	1.4973	1.4975	1.4978
85	1.4980									

[a]Values of refractive index against air with invert sugar mass fraction.

About the Editor/Author

Dr. Chung Chi Chou, the editor, and contributor of many chapters in this handbook on cane sugar refining, was born in Kaoshiung, Taiwan, in 1936. He graduated from Chengkung University in 1959 with a Bachelor of Science degree in chemical engineering, following which he worked as a chemical engineer at the Chisan Sugar Factory of the Taiwan Sugar Corporation for about five years. Subsequently, he moved to the United States, where he received a doctorate degree in physical chemistry from Baylor University in 1968. His doctoral thesis was on the theory of adsorption onto solid surfaces. He then began a long career at the then American Sugar Company (predecessor of the Amstar Corporation, now Domino Sugar, a part of the Tate & Lyle Group). He moved rapidly through the ranks, starting out as a research scientist in the R&D division, then Technical Manager, and finally, Technical Director with responsibilities in process technology and analytical methods development and governmental regulatory compliance with respect to food product safety. After 30 years with Domino, he retired, and recently was named the Managing Director of Sugar Processing Research Institute in New Orleans, Louisiana.

Dr. Chou was twice the recipient of the George & Eleanore Meade Award, in 1972 and 1978, as well as of the prestigious Crystal Award in Sugar Technology of the Sugar Industry Technologists in 1989. He was the Chairman of the New York Sugar Trade Laboratory and Chairman of the U.S. National Committee on Sugar Analysis. He has served as a member of the executive committee and board of directors of various sugar organizations, including the Sugar Processing Research Institute and Sugar Industry Technologists. He has some 40 publications in professional publications, and is, in addition, the co-editor, along with James C. P. Chen, of the 12th edition of the *Cane Sugar Handbook.*

Since 1992, he has been an advisory professor of South China University of Technology in the People's Republic of China, since 1986 a lecturer at the Cane Refiners Institute at Nicholls State University, and Adjunct Professor at Louisiana State University since 1993.

As the editor of the handbook, Dr. Chou has provided a reference tool which will be valuable to the experienced technologist as well as an invaluable aid in training those who are new to the industry.

Jack Lay
President, Refined Sugars, Inc.
Yonkers, New York

(Formerly, President/CEO of Domino Sugar Corporation)

Index